AF574536

Cardiac Electrophysiology and Arrhythmias

CURRENT TOPICS IN CARDIOLOGY

Series Editor

Shahbudin H. Rahimtoola, MB, FRCP

George C. Griffith Professor of Cardiology
and Professor of Medicine
Chief, Division of Cardiology
University of Southern California
Los Angeles, California

Acute Myocardial Infarction

Bernard J. Gersh, MB, ChB, DPhil, FRCP
Shahbudin H. Rahimtoola, MB, FRCP

Cardiac Electrophysiology and Arrhythmias

Charles Fisch, MD
Borys Surawicz, MD

Cardiac Electrophysiology and Arrhythmias

Edited by
Charles Fisch, MD
Director, Cardiovascular Division
Krannert Institute of Cardiology
Indianapolis, Indiana

Borys Surawicz, MD
Krannert Institute of Cardiology
Indianapolis, Indiana

Elsevier
New York • Amsterdam • London • Tokyo

Elsevier Science Publishing Co., Inc.
655 Avenue of the Americas
New York, New York 10010

Sole distributors outside the United States and Canada:

Elsevier Science Publishers B.V.
P.O. Box 211, 1000 AE Amsterdam, The Netherlands

This book is printed on acid-free paper.

Library of Congress Cataloging-in-Publication Data

Cardiac electrophysiology and arrhythmias / edited by Charles Fisch, Borys Surawicz.
p. cm.—(Current topics in cardiology)
Includes bibliographical references and index.
ISBN 0-444-01601-5 (hard cover : alk. paper)
1. Arrhythmia. 2. Heart conduction system. 3. Electrophysiology.
4. Heart—physiopathology. I. Fisch, Charles. II. Surawicz, Borys, 1917– . III. Series.
[DNLM: 1. Arrhythmia—physiopathology. 2. Arrhythmia—therapy.
3. Electrophysiology. WG 330 C26921]
RC685.A65C2852 1991
616.1'28—dc20
DNLM/DLC
for Library of Congress 91-13152
CIP

Current printing (last digit):
10 9 8 7 6 5 4 3 2 1

Manufactured in the United States of America

Contents

Foreword

The fund of knowledge about the physiology and pathophysiology of cardiac arrhythmias has progressed at a rapid pace over a short period of time. This has been matched by significant progress in diagnostic and therapeutic modalities.

One goal of our CURRENT TOPICS IN CARDIOLOGY series is to present comprehensive, in-depth updates of information in selected areas of cardiology. By providing their own contributions along with those of other experts in the field, Drs. Charles Fisch and Borys Surawicz have done a superb job of editing this excellent update on cardiac electrophysiology and arrhythmias. Consistent with the other volumes in the series, it is hoped that this work will prove valuable to practicing physicians and specialists, investigators, and academicians.

Shahbudin H. Rahimtoola, MB, FRCP

Preface

Cardiac Electrophysiology and Arrhythmias is a comprehensive update of present knowledge in the broad field of investigative and practical electrocardiology. It includes 33 chapters organized in six sections. The first two sections deal with basic electrophysiology and mechanisms underlying cardiac arrhythmias. The subsequent four clinical sections include: clinical arrhythmias; diagnostic procedures; pharmacologic treatment; and nonpharmacologic treatment.

This volume offers the reader a spectrum of the progress in electrocardiology from cell membrane research to practical therapy. Each chapter describes new developments that are of potential interest to the medical student, physician in training, clinical investigator, and physician caring for patients with heart disease.

We are indebted to the contributors for providing their material to us; their promptness and cooperation made it possible to publish a timely text. We acknowledge the professionalism of the Elsevier Science Publishing Company, Inc., and the editorial assistance of Mrs. Erna Martin.

Charles Fisch, MD
Borys Surawicz, MD

Contributors

J. A. Abildskov, MD
Cardiovascular Research and Training Institute, University of Utah, Building 500, Salt Lake City, Utah 84112

Justus Anumonwo, PhD
Department of Pharmacology, SUNY Health Science Center at Syracuse, 766 Irving Avenue, Syracuse, New York 13210

Evgeny P. Anyukhovsky, PhD
Department of Pharmacology, Columbia University, College of Physicians and Surgeons, 630 West 168 Street, P&S 7-517, New York, New York 10032

Gust H. Bardy, MD
University of Washington Medical Center, Mailstop RG-22, Seattle, Washington 98195

A. Bayes de Luna, MD
Hospital de la Santa Creu i Sant Pau, Department of Cardiology, Universidad Autonoma de Barcelona, Avda. San Antonio Maria Claret 167, Barcelona 08025, Spain

Edward J. Berbari, PhD
University of Oklahoma Health Sciences Center, Department of Medicine, Division of Cardiology and Department of Veterans Affairs Medical Center, Oklahoma City, Oklahoma 73190

Alan D. Bernstein, EngScD
Newark Beth Israel Medical Center, 201 Lyons Avenue, Newark, New Jersey 07112

Mohamed Boutjdir, PhD
SUNY Health Science Center at Brooklyn, 450 Clarkson Avenue, P.O. Box 1199, Brooklyn, New York 11203

Pedro Brugada, MD
Cardiovascular Center, Postgraduate School of Cardiology, OLV Hospital, Moorselbaan 164, B9300 Aalst, Belgium

Agustin Castellanos, MD
Division of Cardiology, D39, Department of Medicine, University of Miami School of Medicine, P.O. Box 016960, Miami, Florida 33101

James L. Cox, MD
Suite 3108 Queeny Tower, Barnes Hospital Plaza, St. Louis, Missouri 63110

James W. Cox, Jr., MD
Professor of Medicine, Cardiology Section, Louisiana State University School of Medicine, P.O. Box 33932, Shreveport, Louisiana 71130

John P. DiMarco, MD, PhD
Cardiology Division, Box 158, University of Virginia Medical Center, Charlottesville, Virginia 22908

Anne H. Dougherty, MD
University of Texas Medical School, P.O. Box 20708, Houston, Texas 77225

Nabil E. El-Sherif, MD
SUNY Health Science Center at Brooklyn, 450 Clarkson Avenue, P.O. Box 1199, Brooklyn, New York 11203

T. Bruce Ferguson, Jr., MD
Suite 3108 Queeny Tower, Barnes Hospital Plaza, St. Louis, Missouri 63110

M. Fiol, MD
Hospital de la Santa Creu i Sant Pau, Department of Cardiology, Universidad Autonoma de Barcelona, Avda. San Antonio Maria Claret 167, Barcelona 08025, Spain

Charles Fisch, MD
Krannert Institute of Cardiology, 1001 West Tenth Street, Indianapolis, Indiana 46202

Guy Fontaine, MD
Hospital Jean Rostand, 39 rue Jean Le Galleu, 94200 Ivry, France

R. Frank
Hospital Jean Rostand, 39 rue Jean Le Galleu, 94200 Ivry, France

Leonard S. Gettes, MD
Division of Cardiology, CB# 7075, Burnett-Womack Building, University of North Carolina at Chapel Hill, Chapel Hill, North Carolina 27599-7045

William S. Gough, PhD
SUNY Health Science Center at Brooklyn, 450 Clarkson Avenue, P.O. Box 1199, Brooklyn, New York 11203

Thomas B. Graboys, MD
Lown Cardiovascular Group, 21 Longwood Avenue, Brookline, Massachusetts 12146

Larry S. Green, MD
Cardiovascular Research and Training Institute, University of Utah, Building 500, Salt Lake City, Utah 84112

Y. Grosgogeat
Hospital Jean Rostand, 39 rue Jean Le Galleu, 94200 Ivry, France

J. Guindo, MD
Hospital de la Santa Creu i Sant Pau, Department of Cardiology, Universidad Autonoma de Barcelona, Avda. San Antonio Maria Claret 167, Barcelona 08025, Spain

John J. Hayes, MD
Marshfield Clinic, 1000 North Oak Avenue, Marshfield, WI 54449

Philip E. Hill, MD
Cardiology Division, Box 158, University of Virginia Medical Center, Charlottesville, Virginia 22908

Bruce G. Hook, MD
Hospital of the University of Pennsylvania, 3400 Spruce Street, Philadelphia, Pennsylvania 19104

Raymond E. Ideker, MD, PhD
Duke University Medical Center, P.O. Box 3140, Durham, North Carolina 27710

Alberto Interian, Jr., MD
Division of Cardiology, Department of Medicine, University of Miami School of Medicine, Miami, Florida 33101

T. Iwa, Jr.
Hospital Jean Rostand, 39 rue Jean Le Galleu, 94200 Ivry, France

Jose Jalife, MD
Department of Pharmacology, SUNY Health Science Center at Syracuse, 766 Irving Avenue, Syracuse, New York 13210

Eric E. Johnson, MD
Duke University Medical Center, P.O. Box 3140, Durham, North Carolina 27710

Mark E. Josephson, MD
Hospital of the University of Pennsylvania, 3400 Spruce Street, Philadelphia, Pennsylvania 19104

George J. Klein, MD
Arrhythmia Service, Cardiac Investigation Unit, University Hospital, P.O. Box 5339, Postal Station A, London, Ontario, Canada N6A 5A5

Chien-Sun Kuo, MD
Department of Medicine, Division of Cardiology, University of Kentucky, Lexington, Kentucky 40536

Ralph Lazzara, MD
University of Oklahoma Health Sciences Center, Department of Medicine, Division of Cardiology and Department of Veterans Affairs Medical Center, Oklahoma City, Oklahoma

Michael D. Lesh, MD
Department of Medicine, University of California, San Francisco, California 94143

Paul L. McHenry, MD
Krannert Institute of Cardiology, 1001 West Tenth Street, Indianapolis, Indiana 46202

Frits L. Meijler, MD, FACC
Interuniversity Cardiology Institute of the Netherlands, P.O. Box 19258, 3501 DG, The Netherlands

E. Neil Moore, DVM, PhD
School of Veterinary Medicine, University of Pennsylvania, Philadelphia, Pennsylvania 19104

Robert J. Myerburg, MD
Division of Cardiology, Department of Medicine, University of Miami School of Medicine, Miami, Florida 33101

Gerald V. Naccarelli, MD
University of Texas Medical School, P.O. Box 20708, Houston, Texas 77225

Andrea Natale, MD
Arrhythmia Service, Cardiac Investigation Unit, University Hospital, P.O. Box 5339, Postal Station A, London, Ontario, Canada N6A 5A5

Victor Parsonnet, MD
Newark Beth Israel Medical Center, 201 Lyons Avenue, Newark, New Jersey 07112

Eugene Patterson, PhD
University of Oklahoma Health Sciences Center, Department of Medicine, Division of Cardiology and Department of Veterans Affairs Medical Center, Oklahoma City, Oklahoma 73190

Milton L. Pressler, MD
Krannert Institute of Cardiology, 1001 West Tenth Street, Indianapolis, Indiana 46202

David P. Rardon, MD
Krannert Institute of Cardiology, 1001 West Tenth Street, Indianapolis, Indiana 46202

Shmuel Ravid, MD
Lown Cardiovascular Group, 21 Longwood Avenue, Brookline, Massachusetts 12146

Pratap C. Reddy, MD
Professor of Medicine, Cardiology Section, Louisiana State University School of Medicine, P.O. Box 33932, Shreveport, Louisiana 71130

Mark Restivo, PhD
SUNY Health Science Center at Brooklyn, 450 Clarkson Avenue, P.O. Box 1199, Brooklyn, New York 11203

Michael R. Rosen, MD
Department of Pharmacology, Columbia University, College of Physicians and Surgeons, 630 West 168 Street, P&S 7-517, New York, New York 10032

J. M. Dominguez de Rozas, MD
Hospital de la Santa Creu i Sant Pau, Department of Cardiology, Universidad Autonoma de Barcelona, Avda. San Antonio Maria Claret 167, Barcelona 08025, Spain

Sanjeev Saksena, MD, FACC
Director, Arrhythmia & Pacemaker Service, Eastern Heart Institute, 350 Boulevard, Passaic, New Jersey 07055

Benjamin J. Scherlag, PhD
University of Oklahoma Health Sciences Center, Department of Medicine, Division of Cardiology and Department of Veterans Affairs Medical Center, Oklahoma City, Oklahoma 73190

Michael B. Simson, MD
Hospital of the University of Pennsylvania, 3400 Spruce Street, Philadelphia, Pennsylvania 19104

Joep Smeets, MD
Cardiovascular Center, Postgraduate School of Cardiology, OLV Hospital, Moorselbaan 164, B9300 Aalst, Belgium

Joseph Spear, PhD
Department of Medicine, University of California, San Francisco, California 94143

Marshall S. Stanton, MD
Mayo Clinic, 200 First Street SW, Rochester, Minnesota 55905

Borys Surawicz, MD
Krannert Institute of Cardiology, 1001 West Tenth Street, Indianapolis, Indiana 46202

Hans de Swart, MD
Cardiovascular Center, Postgraduate School of Cardiology, OLV Hospital, Moorselbaan 164, B9300 Aalst, Belgium

J. Tonet
Hospital Jean Rostand, 39 rue Jean Le Galleu, 94200 Ivry, France

Andres Varro, MD, PhD
Children's Hospital Research Foundation, Cincinnati, Ohio 45229-2899

Albert L. Waldo, MD
Division of Cardiology, University Hospitals of Cleveland, 2074 Abington Road, Cleveland, Ohio 44106

Hein J. J. Wellens, MD
University Hospital Maastricht, University of Limburg, P.O. Box 1918, 6201 BX Maastricht, The Netherlands

Fred H. M. Wittkampf, MD
Interuniversity Cardiology Institute of the Netherlands, P.O. Box 19258, 3501 DG, The Netherlands

Raymond Yee, MD
Arrhythmia Service, Cardiac Investigation Unit, University Hospital, P.O. Box 5339, Postal Station A, London, Ontario, Canada N6A 5A5

Liagat Zaman, MD
Division of Cardiology, Department of Medicine, University of Miami School of Medicine, Miami, Florida 33101

PART I

Mechanisms

Chapter **1**

Cardiac Resting and Action Potentials: Current Concepts

David P. Rardon, MD, and Milton L. Pressler, MD

The normal rhythm of the heart originates from cyclical changes in the membrane potential of cardiac cells. Analogous to all other excitable tissues, the membrane potential (V_m) stems from the unequal distribution of ions across a selectively permeable cell membrane. However, cardiac cells alone have intrinsic rhythmicity because they possess specialized pacemaking ion channels. The activity of pacemaker channels (see below) produces a gradual change in V_m and thereby instability in membrane permeability when V_m depolarizes to a critical value ("threshold"). At threshold, the resting selectivity for potassium (K^+) of the cardiac cell membrane is changed to selectivity for sodium (Na^+), and the upstroke of the action potential results. A chorus of other permeability changes follows this event, and the type, magnitude, timing, and regulators of the resulting ionic currents are responsible for the contour, duration, and refractory period of the cardiac action potential. Other well-known electrical phenomena such as response of the heart to adrenaline, autonomic regulation of the heart rate, and effects of paired stimuli derive from the behavior of specific ion channels and their response to various stimuli. Pathophysiologic alterations in these ionic currents or in their flow from cell-to-cell may produce cardiac arrhythmias secondary to enhanced automaticity, reentry, or abnormal automatic mechanisms.

For almost 30 years, a quantitative description of the exact magnitude and sequence of ionic currents responsible for inscribing the cardiac action potential (AP) has defied theoreticians and experimentalists alike. Initial attempts to modify the kinetics of the Na^+ and K^+ currents measured in nerve axons were successful in reproducing the shape but not the exact behavior of the AP.[1] Later models[2,3] incorporated calcium (Ca^{2+}) currents and additional K^+ currents not observed in nervous tissues. Additional characteristics of the cardiac AP were simulated, but other deficiencies remained: (a) the kinetics of several K^+ currents (especially I_{K1}) were incorrect because accumulation and depletion of K^+ from narrow extracellular spaces were not accounted for[4]; (b) pumps (for example, Na^+/K^+-ATPase) and an exchanger (for example, Na/Ca exchange) were missing[5] that generate changes in transmembrane potential (that is, "electrogenic" processes); and (c) the flow of current through the pacemaker channel was incorrectly described.[6] Recently, some of these problems have been remedied[7–9] thanks to the explosion of knowledge from single-cell voltage clamp and patch clamp studies.[10] From measurements of single-channel events in native and artificial membranes and by cloning the channel proteins, our experimental knowledge of channel properties is quickly outstripping our ability to assimilate and integrate the components into a model of cardiac electrical activity. Consequently, this chapter will not attempt to describe all of the new information on ionic events responsible for the AP, but rather give an overview of new information that is thought to play a major role in generation of the cardiac AP.

THE RESTING POTENTIAL

In excitable cells, including cardiac cells, the resting membrane potential is largely determined by the concentration gradient for K^+ ions across the cell membrane (see Fig. 1.1A). The cell membrane is permeable to K^+ ions, but relatively impermeable to other ions such as Na^+, Ca^{2+}, and chloride (Cl^-). Outward diffusion of K^+ ions causes the intracellular space to be negatively charged with respect to the extracellular space. Electroneutrality cannot be maintained by the

655 Avenue of the Americas, New York, NY 10010
Current Topics in Cardiology

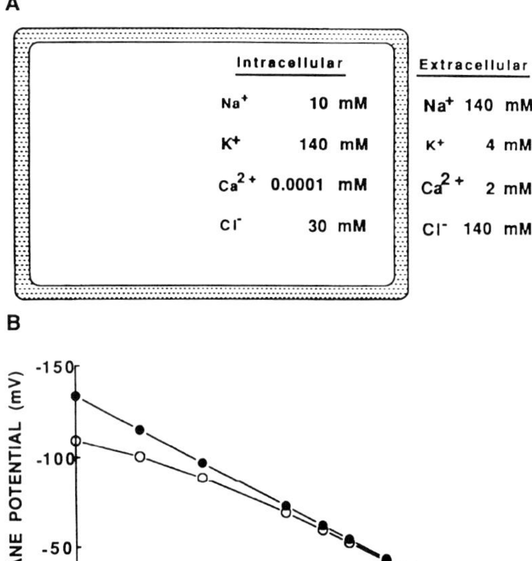

FIGURE 1.1A. Approximate concentrations of Na^+, K^+, Ca^{2+}, and Cl^- in the extracellular and intracellular fluid compartments. **B.** Effects of $[K^+]_o$ on resting membrane potential. •—• was determined from the Nernst equation in which resting membrane potential only depends on $[K^+]_o/[K^+]_i$. ○—○ was calculated from the Goldman equation in which resting membrane potential varies with the gradients of both K^+ and Na^+.

outward movement of anions because these are largely associated with cellular proteins.[11] The resultant negative charge opposes the further diffusion of K^+ out of the cell. The intracellular potential at which the net passive flux of K^+ ions is zero is called the equilibrium potential for K^+ ions, E_K, and its value is given by the Nernst equation,[12]

$$E_K = (RT/F) \ln [K^+]_o/[K^+]_i$$

where R is the gas constant, T is absolute temperature, F is the Faraday constant, and $[K^+]_o$ and $[K^+]_i$ are the extracellular and intracellular concentrations of K^+ ions, respectively. At 36°C, E_K will change 61.4 mV per 10-fold change in either $[K^+]_o$ or $[K^+]_i$. The membrane potentials of resting Purkinje fibers and atrial and ventricular myocardial fibers are well approximated by the Nernst equation when $[K^+]_o$ is higher than 10 mM.[13] At lower $[K^+]_o$, the inward movement of Na^+ influences the resting membrane potential to a greater amount.

The depolarizing influence of Na^+ is described by the Goldman[14] and Hodgkin and Katz[15] "constant field" equations for the resting potential, V_r, of a cell permeable to both K^+ and Na^+,

$$V_r = \frac{RT}{F} \ln \frac{[K^+]_o + P_{Na}/P_K[Na^+]_o}{[K^+]_i + P_{Na}/P_K[Na^+]_i}$$

where P_{Na}/P_K is the ratio of the Na^+ to the K^+ permeability coefficient of the cell membrane. Figure 1.1B illustrates V_r when calculated for a membrane permeable only to K^+ compared to a membrane permeable to both Na^+ and K^+.

The resting potential may play an important role in the genesis of arrhythmias and conduction disturbances under pathologic conditions. For example, a localized increase in $[K^+]_o$ occurs during myocardial ischemia, which depolarizes cells in the ischemic zone,[16] that may contribute to ischemic-induced arrhythmias or conduction abnormalities via secondary changes in Na, K, and pacemaker channels.

THE NORMAL CARDIAC ACTION POTENTIAL

In most excitable cells, the AP can be divided into three phases: upstroke, plateau, and repolarization. Different ionic currents predominate during each phase. The initial rapid depolarization in the cell membrane is mediated primarily by the inward movement of Na^+ through voltage-sensitive ion channels. The plateau is more complicated and develops from two simultaneous events: influx of Ca^{2+} through voltage-sensitive Ca^{2+} channels and closure of K^+ channels responsible for the high K^+ permeability at the resting potential. Ca^{2+} entering the cell during this phase of the cardiac action potential initiates excitation–contraction coupling[17] and modulates multiple Ca^{2+}-activated biochemical processes. Lastly, repolarization of the action potential occurs when voltage-sensitive K^+ channels open and outward movement of K^+ overcomes the residual influx of Na^+ and Ca^{2+}.

The normal APs from sinus node, atrial myocardium, AV node, Purkinje fibers, and ventricular muscle are shown in Figure 1.2. The general tenets outlined above apply to all these various cardiac cells even though their APs differ significantly in configuration, upstroke velocity, and the presence or absence of spontaneous diastolic depolarization. Each type of AP resembles the others in that all arise from sequential changes in Na^+ and/or Ca^{2+} influx coupled to K^+ efflux which then repeats with each heartbeat.

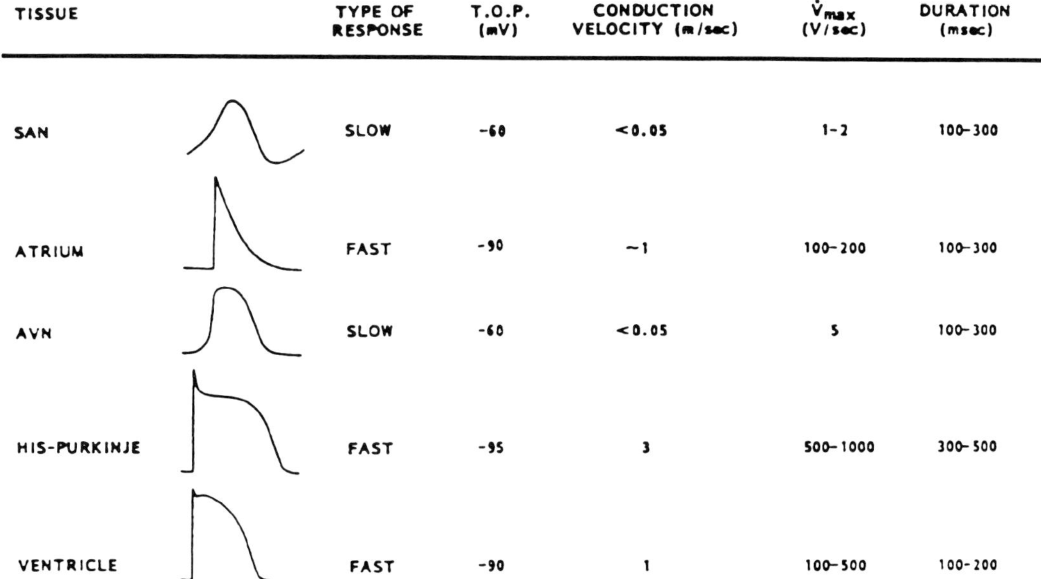

TISSUE	TYPE OF RESPONSE	T.O.P. (mV)	CONDUCTION VELOCITY (m/sec)	$\dot{V}_{max}$ (V/sec)	DURATION (msec)
SAN	SLOW	-60	<0.05	1-2	100-300
ATRIUM	FAST	-90	~1	100-200	100-300
AVN	SLOW	-60	<0.05	5	100-300
HIS-PURKINJE	FAST	-95	3	500-1000	300-500
VENTRICLE	FAST	-90	1	100-500	100-200

FIGURE 1.2 Action potentials and characteristics from sinus (SAN), atrium, atrioventricular node (AVN), His-Purkinje, and ventricular myocardial cells.

To begin discussion of the ionic currents responsible for the normal cardiac action potential, we shall use a computer model.[5] Figure 1.3 illustrates the normal cardiac Purkinje fiber AP together with the voltage-gated, exchanger, background, and pump currents responsible for its generation. The upstroke of the action potential (phase 0) is produced by a very large inward Na^+ current, i_{Na} (not shown in Fig. 1.3), which also contributes to the plateau a small steady current, the so-called window current or late Na^+ current. The early rapid repolarization of the action potential (phase 1) is secondary to a transient outward K^+ current, i_{to}. During the plateau of the AP (phase 2), other ionic currents are of importance. The inward Ca^{2+} current, i_{Ca}, is activated and the subsequent rise in $[Ca^{2+}]_i$ engenders an inward Na/Ca exchange current, i_{NaCa}. The repolarization of the action potential is primarily due to an efflux of K^+ (see repolarization currents below). Spontaneous diastolic depolarization (phase 4) occurs from the opening of pacemaker channels (i_f). Changes in the inward rectifier current, i_{K1}, influence the rate of diastolic depolarization.

The ionic currents responsible for the generation of sinus node, atrial, AV node, and ventricular cells differ to varying degrees from those outlined for cardiac Purkinje cells (see reviews 9,18). Major differences in ion channels in non–Purkinje cardiac cells include the following: (a) Sinus and AV nodal cells lack or have no significant inward Na^+ or inward rectifying K^+ channels. Therefore, the upstroke of these APs is generated entirely by i_{Ca}. (b) In sinus and AV nodal cells, spontaneous diastolic depolarization arises from i_{Ca} and, to a lesser extent, the pacemaker current, i_f. (c) Normal atrial and ventricular myocardial cells have no spontaneous diastolic depolarization.

CARDIAC SODIUM CHANNELS

The fast upstroke of atrial muscle, ventricular muscle, and Purkinje fibers is secondary to the rapid influx of Na^+. i_{Na} is the primary determinant of the rate of rise of phase 0 and the amplitude of the AP. For a number of circumstances, conduction velocity (Θ) of the cardiac impulse changes in proportion to the rate of rise ($\dot{V}_{max}$) of the action potential ($\Theta / \sqrt{\dot{V}_{max}}$).[19] Therefore, interventions that reduce the magnitude of i_{Na}, such as depolarization of the resting membrane potential (that is, elevation of $[K^+]_o$) or Na^+ channel blockers such as tetrodotoxin (TTX) or lidocaine, will decrease AP amplitude, reduce $\dot{V}_{max}$, and decrease the velocity of the propagated impulse.

i_{Na} kinetics were first determined by Hodgkin and Huxley[20] in squid axons. Even today, many generalizations from the classic Hodgkin–Huxley model[20] of squid axon Na^+ channel also apply

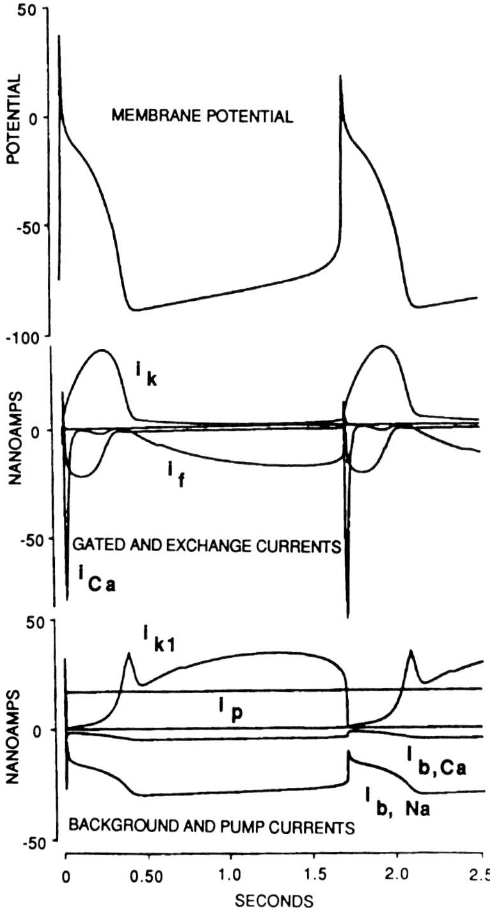

FIGURE 1.3 Rhythmic activity computed using the DiFrancesco–Noble model of cardiac Purkinje fibers. **Top.** Variation in membrane potential. **Middle.** Variations in gated and exchange currents. The delayed K^+ current, i_K, activates during the plateau and its decay contributes a small amount of current during the pacemaker depolarization. The nonspecific cation channel, i_f, is activated as the cell repolarizes and contributes significantly to the pacemaker current. The action potential upstroke is generated by a very large inward Na^+ current, which also contributes a small steady current "late Na^+ current" to the plateau (because of its large size of ~10,000 nA, i_{Na} is not depicted here). During the upstroke of the action potential, the Ca^+ current, i_{Ca}, is activated and the subsequent rise in $[Ca^+]$ activates the inward Na/Ca exchange current, i_{NaCa}. **Bottom.** Background and pump currents. The Na^+ pump contributes a steady level of outward current, i_p, during the whole cycle. The inward-rectifier K^+ current, i_{K1}, carries a significant current during the pacemaker depolarization that largely balances the current carried by i_f. Finally the background Na^+ and Ca^{2+} fluxes, $i_{b,Ca}$ and $i_{b,Na}$, are depicted. *Reprinted from Noble, with permission.*

to the cardiac Na^+ channel. The Hodgkin–Huxley model postulated that Na^+ channels exist in either closed, open, or inactivated states. Transitions to another state of the channel were assumed to occur independent of the existing state of the channel. Controlling activation and inactivation of the channel were activation (m) and inactivation (h) gates, respectively. The m gates are voltage-dependent and open very rapidly upon depolarization. The inactivation gates also activate upon depolarization, but the kinetics of closing the channel are slower. The magnitude and duration of the i_{Na} are mainly dependent on the proportion of channels that are inactivated.

Recent data suggest that the kinetic properties of the cardiac Na^+ channel are more complicated than predicted by the Hodgkin–Huxley model.[21–23] If only a single open or closed state of the Na^+ existed, the distribution of times spent in a state would be described by a single exponential equation.[24] In contrast, inactivation of the cardiac Na^+ channel is best described by at least two exponentials, suggesting the presence of two inactivated states of the channel.[21–23,25] Single-channel studies[26,27] support two phases of Na^+ channel inactivation. The fast phase allows channels to open briefly (1 to 2 ms) without reopening. Slow components of the i_{Na} occur either as isolated brief openings or as bursts of openings that are several hundred milliseconds in duration.[28–30] The slow component of i_{Na} may contribute to the slowly inactivating current that flows during the AP plateau. This may account for many previous reports that the Na^+ channel blocking agent, TTX, abbreviates the action potential duration (APD) of Purkinje and ventricular myocardial cells.[31–33]

Other studies have shown that β-adrenergic stimulation modulates the cardiac Na^+ channel. Schubert et al.[34] demonstrated that isoproterenol strongly inhibits cardiac Na^+ channels when the membrane potential is depolarized. The effects of β-agonists were mediated via a G-protein. Though the concentration of isoproterenol needed was high (~1 μM), the inhibitory effects on i_{Na} may be a potentially important mechanism for slowing conduction (and development of reentry) in partially depolarized myocardium. Thus, the increased risk of ventricular arrhythmias in patients with myocardial infarction and elevated catecholamine levels[35,36] might be partly explained by β-adrenergic-induced Na^+ current depression.

Recordings of single Na^+ channels have provided new information on the mechanisms by which specific receptor ligands modify channel function. Lidocaine has been shown to cause use-dependent block of Na^+ channels primarily by blocking a nonconducting state of the channel.[37,38] Kolhardt and Fichtner[39] found that amiodorone and propafenone caused use-dependent block of neonatal rat Na^+ channels without decreasing the mean open time. The result of Na^+ channel inhibition with any of these pharmacological agents is reduction in action potential amplitude, $\dot{V}_{max}$, and conduction velocity.

CARDIAC CALCIUM CHANNELS

Different types of voltage-sensitive Ca^{2+} channels (classified as L, T, and N) are known to exist. The L-type channel is activated at membrane potentials more positive than −10 mV and conducts a long-lasting current of conductance of ~25 picosiemens (pS).[40,41] The T-type channel is activated at more negative V_m (that is, positive to −70 mV) and is characterized by relatively transient currents of small conductance (~9 pS). The N-type channel activates at intermediate V_m (positive to −30 mV) and conducts a comparatively brief current of intermediate size (~15 pS). In cardiac muscle, both L- and T-type, but not N-type, channels have been observed.[42–44] Evidence also supports the existence of subtypes or isoforms of the different varieties of voltage-sensitive Ca^{2+} channels. The L-type Ca^{2+} channel present in the heart differs in certain pharmacological and electrophysiological properties from the L-type Ca^{2+} channel present in skeletal muscle.[45,46]

The L-type Ca^+ channel is selectively effected by a group of compounds referred to as Ca^{2+} effectors. These drugs belong to three distinct chemical classes: (a) the 1,4-dihydropyridine derivatives like nitrendipine, nimodipine, and Bay K 8644; (b) the phenylalkylamines such as verapamil, D-600, and D-888; and (c) the benzothiazepines, typified by diltiazem. All of the drugs used clinically were developed to inhibit the Ca^{2+} channel but activating compounds (for example, Bay K 8644) also have been identified. The binding of dihydropyridines to Ca^{2+} channels is strongly voltage dependent. These inhibitors have higher affinity for depolarized than polarized channels.[47–49] This suggests dihydropyridines bind better to the inactivated form of the channel than to the resting form.

The Ca^{2+} channel effectors that inhibit the L-type channel will obviously alter the AP characteristics of cardiac tissues. These agents inhibit SA and AV nodal AP amplitude and rate of firing because nodal APs primarily arise from Ca^{2+} influx. In addition, they lower and shorten the plateau (phase 2) of atrial, ventricular, and Purkinje fiber APs.

β-Adrenergic receptor agonists such as isoproterenol increase i_{Ca}, which in turn heightens the plateau and shortens the APD of cardiac Purkinje fibers (for reviews see refs. 50,51). The effects of β-agonists on i_{Ca} are believed to be mediated by cyclic AMP-mediated phosphorylation of the L-type Ca^{2+} channel. Microinjection of the purified catalytic subunit of cyclic AMP-dependent protein kinase into isolated myocytes[52,53] mimicks the effects of elevation of intracellular cyclic AMP and can be inhibited by phosphoprotein phosphatase 1 or 2A. Phosphorylation of the L-type channels increases the probability that the channel is in the open state.[50,51] Cyclic AMP-dependent phosphorylation of L-type channels is mediated by G-proteins. The G_s-protein has been shown to activate directly the channel independent of activation of protein kinases.[54]

CHANNELS INVOLVED IN REPOLARIZATION OF THE ACTION POTENTIAL

The plateau (phase 2) and rapid repolarization (phase 3) of the ventricular and His-Purkinje APs include contributions from Na^+, Ca^{2+}, and K^+ channels, the Na/K pump, and Na/Ca exchange.[9] Table 1.1 lists nine of the major ionic events. Transmembrane movement of Na^+ is partly responsible for maintenance of the AP plateau. Cardiac tissue is distinct from nerve in that not all of the TTX-sensitive Na^+ channels inactivate during prolonged depolarization. Thus, Na^+ current ("late Na^+ current" or "window current") continues to enter the cell at plateau potentials and contributes to the plateau's long duration.[31,55] APD is also lengthened by Na^+ entry via the Na/Ca exchanger.[56] Na/Ca exchange (stoichiometry, 3 Na^+ : 1 Ca^{2+}) is probably the main cellular mechanism for extrusion of Ca^{2+} gained during the AP.[57] Persistence of a net inward driving force for Na^+ during most of the action potential and the electrogenic stoichiometry of the exchange leads to a net i_{Na} for most of the plateau.[9,58,59] Na/Ca exchange seems to be responsible for the late phase of repolarization observed in atrial APs and rat ventricular APs.[9,57] In the opposite direction, Na/K pump activity also generates a current that affects the transmembrane potential, that is, it produces a net outward Na^+ current. Under normal conditions, the activity of the Na/K pump changes little during the AP because $[Na^+]_i$ (the principal regulatory factor once $[K^+]_o > 2$ mM) is only trivially altered by excitation[9,60,61]. However, Na/K pump activity may be quite important to the shortening of APD when Na^+_i rises by 5 to 10 mM in ischemic or digitalis-intoxicated cells.[9,61]

Voltage-gated calcium currents formerly re-

TABLE 1.1 Plateau and Repolarizing Currents in the Ventricle

Current Type	Direction	Effect on AP	Activators	Blockers
Late Na^+ current (window current)	Inward Na^+	Lengthens AP, raises plateau	Depolarize > −60 mV	TTX ≈ 30 μM
Na/K ATPase (pump current)	Net outward Na^+	Shortens AP	$Na^+{}_i$ (K_m ~ 10 mM); $K^+{}_o$ (K_m ~ 1 mM)	Cardiac glycosides
i_{NaCa} (Na/Ca exchange)	Net inward Na^+	Lengthens AP	?Phospholipids; ?$Ca^{2+}{}_i$	La^{3+}; amiloride and derivatives; bepridil
L-type i_{Ca} (slow inward current	Inward Ca^{2+}	Lengthens AP, raises plateau	β-Agonists BAY K 8644	Dihydropyridines; verapamil; diltiazem
I_A or I_{to} (transient outward)	Outward K^+	Phase 1; notch	Depolarize > −30 mV	4-AP, quinidine
i_K or i_X (delayed rectifier)	Outward K^+	Shortens AP; HR effects on AP	Depolarize > −45 mV β-agonists	TEA, Cs^+, Ba^{2+}
i_{K1} (inward rectifier)	Small outward K^+	Shortens AP; sets V_r	Hyperpolarize negative to E_K	TEA, Cs^+, Ba^{2+}
$i_{K(Ca)}$ (Ca activated)	Outward K^+	Shortens AP	$Ca^{2+}{}_i$ (K_m ~ 1 μM); Depolarize > −50 mV	TEA, charybdotoxin, Ba^{2+}
$i_{K(ATP)}$ (ATP sensitive)	Outward K^+	Shortens AP in ischemia	Diazoxide, Nicorandil, Pinacidil	ATP_i ≈ 500 μM, glibenclamide, tolbutamide

Abbreviations: AP = action potential; TTX = tetrodotoxin; $Na^+{}_i$ = intracellular [Na^+]; $K^+{}_o$ = extracellular [K^+]; $Ca^{2+}{}_i$ = intracellular [Ca^{2+}]; K_m = half-maximal activation; 4-AP = 4-aminopyridine; HR = heart rate; TEA = tetraethylammonium; V_r = resting potential; E_K = equilibrium potential for K^+. Only selected activators and blockers are shown. For references, see citations in text.

ferred to as "slow inward current" or (i_{si}), are crucial for excitation–contraction coupling, providing an influx of Ca^{2+} that triggers the release of Ca^{2+} from the sarcoplasmic reticulum (SR).[62] As described above, i_{Ca} in the heart is generated by two different ion channels, now commonly referred to as T-type and L-type channels.[43,62] i_{Ca} is responsible for part of the inward current that maintains the plateau of the AP at depolarized potentials. i_{Ca} may contribute less current to the later stages of the AP plateau than previously thought because measurements in single cells[63,64] have shown that i_{Ca} peaks within 2 to 3 ms and is largely inactivated after 20 to 30 ms. However, i_{Ca} affects the repolarization of the AP in another aspect: The rise of [Ca^{2+}] in the cytosol activates a specific K^+ channel whose opening is part of phase 3 repolarization of the AP.[65] Thus, the activity of Ca^{2+} channels is self-limiting in two ways: Ca^{2+} channels not only inactivate as a consequence of depolarization and Ca^{2+} movement through the channel[62] but also trigger the opening of another channel that restores the transmembrane potential toward its resting value. i_{Ca} arising from L-type Ca^{2+} channels is thought to be the most clinically significant Ca^{2+} current because β-adrenergic agonists like isoproterenol and Ca^{2+} channel blockers (verapamil, nifedipine, diltiazem), respectively, activate and block this channel.[62]

K^+ channel activity during phases 2 and 3 of the ventricular and Purkinje fiber APs is one of the most complicated subjects in cellular electrophysiology and remains incompletely understood to this day. Mammalian cardiac ventricular and Purkinje myocytes express at least seven or eight different K^+ channels,[66] several of which are unique to cardiovascular tissues. Table 1.1 lists five of the K^+ currents that seem to be major contributors to the repolarization of the AP. All of these channels provide outwardly directed current which acts to restore the transmembrane potential toward its resting value. They differ in electrical properties of activation, inactivation, rectification, and sensitivity to external K^+.[66] With regard to their primary activation mechanism, K^+ channels can be classified into three basic categories: (a) voltage-activated; (b) ion-activated; and (c) ligand-activated. The transient outward current, i_{to},[67,68] is a time-dependent outward K^+ current activated by depolarization positive to approximately − 20mV.[69–71] Characteristic of this channel is its slow inactivation, usually described by two exponential components,[68] and its inhibition by 4-aminopyridine (4-AP).[72] Conjointly, activation of i_{to}, inactivation of i_{Na}, and current flow into recesses of the cell membrane ("clefts") account for phase 1 repolarization from the AP peak to the plateau. Activation of i_{to} is believed to inscribe the "notch" in the AP observed at the junction of phases 1 and 2 and to lower the amplitude of the plateau.[70,72] The presence of i_{to} in ventricular epicardium but not endocardium is thought to be responsible for the distinct spike-and-dome appearance of epicardial action potentials, a clear-cut difference in AP contour from the endocardium.[73] The significance of the difference in expression of i_{to} in endocardial and epicardial myocytes is unknown but i_{to} may play a role in the heterogeneity of APD from endocardium to epicardium during myocardial ischemia/infarction.

The delayed rectifier, i_K,[74–76] and the inward rectifier, i_{K1},[77–79] are two additional K^+ channels primarily activated by changes in voltage. Both i_K and i_{K1} are outward currents but i_K is activated by depolarization and is delayed in onset whereas i_{K1} changes very rapidly with potential, is activated by hyperpolarization, and is sensitive to changes in external K^+.[66] During the depolarization of the AP upstroke, i_{K1} channels close and markedly lower the resting K^+ conductance (permeability) of the cell membrane.[77] The closure of i_{K1} channels affects the development of the plateau because it reduces the inward current required to maintain the transmembrane potential at a depolarized level. The delayed opening of i_K channels during the plateau provides much of the outward current that initiates and sustains phase 3 repolarization.[9] Furthermore, the hyperpolarization occurring during phase 3 repolarization activates i_{K1} which in turn amplifies and accelerates the return of the membrane potential to its resting value.

The activity of two other K^+ channels, Ca activated i_{KCa} and ATP sensitive $i_{K(ATP)}$, is principally controlled by the intracellular concentrations of Ca^{2+} and ATP, respectively. i_{KCa} channels[65,80] are opened by the rise in intracellular Ca^{2+} from i_{Ca} and SR Ca^{2+} release during EC coupling. Several subtypes of Ca^{2+}-activated K^+ channels are known,[81] differing in single-channel conductance and sensitivity to blockers. i_{KCa} channels seem to be part of the cell's feedback mechanism to close Ca^{2+} channels that open during the initial portion of the AP plateau. Outward repolarizing current flowing through i_{KCa} channels shortens the duration of the AP and helps to end the depolarization that initiated the opening of Ca^{2+} channels. An illustration of the effects of Ca^{2+} on APD is shown in Figure 1.4. Opening of i_{KCa} channels is an important part of the mechanism for APD shortening under pathological circumstances of Ca^{2+} overload (for example, cell injury). ATP-sensitive K^+ channels[82,83] are inactivated by normal millimolar concentrations of intracellular ATP (ATP_i). The opening of these channels whenever ATP_i falls below 0.5 mM probably contributes to the marked shortening of the APD during myocardial ischemia. Experiments using specific activators

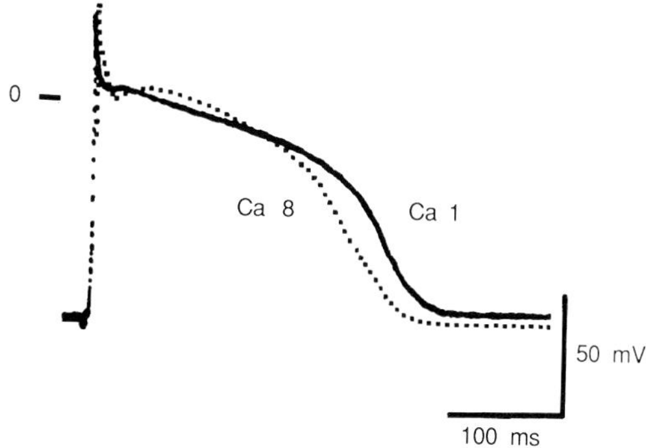

FIGURE 1.4 Effects of Ca^{2+} on the AP of a dog Purkinje fiber. Solid lines represent AP in $[Ca^{2+}]_o$ of 1 mM; dotted lines represent record in $[Ca^{2+}]_o$ of 8 mM. Increasing $[Ca^{2+}]_o$ raised the plateau potential, shortened APD, and slightly hyperpolarized the maximum diastolic potential. Changes in $[Ca^{2+}]_o$ have been shown to elevate $[Ca^{2+}]_i$, and opening of Ca^{2+}-activated K^+ channels may explain the decrease in APD.

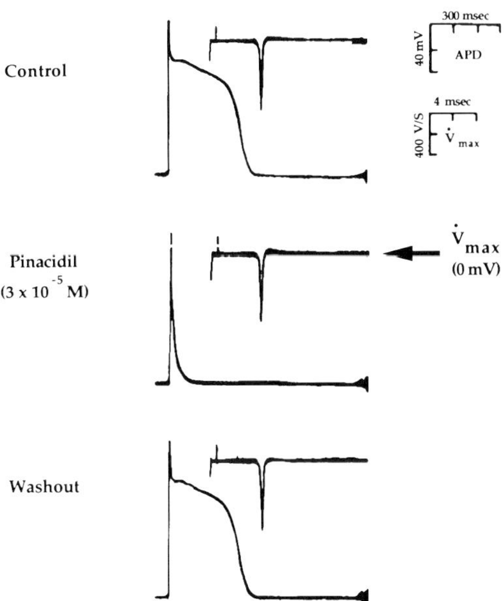

FIGURE 1.5 Effects of pinacidil, an activator of ATP-sensitive K^+ channels, on the AP of a dog Purkinje fiber. Superfusion with pinacidil reversibly shortened the APD by 83% with no change in resting potential or maximum upstroke velocity ($\dot{V}_{max}$). *Reprinted from Steinberg et al.,[84] with permission.*

of $i_{K(ATP)}$ channels (diazoxide, pinacidil)[84] profoundly alter APD without affecting the AP upstroke (see Fig. 1.5). Recent evidence[85] suggests that blockers of $i_{K(ATP)}$ channels may have efficacy for arrhythmias arising during ischemia by reducing K^+ loss or perhaps by maintaining APD in the ischemic region.

PACEMAKER CURRENTS

Identifying which currents are responsible for the spontaneous diastolic depolarization underlying the pacemaking function of cardiac tissue has been a source of contentious debate for over 20 years.[86] The main argument was whether the gradual depolarization during diastole stemmed from the time-dependent inactivation of an outward current[87] or the time-dependent activation of an inward current.[6] Present thinking is that both mechanisms participate in diastolic depolarization within the sinus node but rate control is mainly conferred by the inward current component, termed i_f.[9] i_f is a nonspecific cation channel carrying both Na^+ and K^+ that is activated (opened) by hyperpolarization.[88] Its magnitude and kinetics are sensitive to cholinergic and adrenergic neurotransmitters (for example, acetylcholine, norepinephrine, epinephrine).[89] i_f is present in both nodal and Purkinje cells but has a much larger magnitude (and consequently greater influence on diastolic depolarization) at the more negative diastolic potentials in the Purkinje fiber.[9] Slow reduction in the amount of outward K^+ current through i_K, the delayed rectifier channel, is also a major contributor to diastolic depolarization in the sinus node.[9,89] In Purkinje fibers, i_K plays very little role in diastolic depolarization but outward current through i_{K1} partially offsets the inward current via i_f and thereby influences the rate of diastolic depolarization.[9] Changes in i_{K1} may explain the experimental observation that the rate of diastolic depolarization depends on $[K^+]_o$. Pacemaker activity of the Purkinje fiber is depressed by increasing $[K^+]_o$ and this may come about via $[K^+]_o$-induced increases in i_{K1}. Regardless, all of these currents are by themselves inadequate to depolarize the cell to the point of excitation. The final "surge" in inward current just prior to the upstroke of the AP probably arises from i_{Ca} in nodal cells and i_{Na} in Purkinje cells. A brief review of the various components of the pacemaker potential can be found in Noble.[9]

SUMMARY

The electrical characteristics of all excitable tissues emanate from the inherent properties of the cell membrane, its molecular constituents, and the ionic microenvironment across the membrane. Cardiac cells express several distinct membrane proteins that form ion channels with unique properties. For example, cardiac cells alone possess intrinsic rhythmicity due to the presence of pacemaker channels. At least two Na^+, two Ca^{2+}, seven or eight K^+, and one nonspecific cation channel are known to exist in cardiac tissue. The activity of these channels under constantly changing conditions of potential, extracellular and intracellular ions, and messenger substances produces the familiar electrical impulse, the AP. Not all cells of the heart express the same types of channels and thus the properties and contour of the AP vary with location. This chapter provides an overview of current concepts of cardiac cellular potentials. These concepts continue to evolve and the next 5 to 10 years will witness further refinements in our understanding.

ACKNOWLEDGMENTS

The authors want to express their gratitude to Dr. Charles Fisch for his advice and encouragement. This work was supported in part by grants 881031 and 881032 from the American Heart Association, Dallas, TX, and Indiana Affiliate Inc., by the Veterans Administration, Washington, DC, and by the Herman C. Krannert Fund.

REFERENCES

1. Noble D: A modification of the Hodgkin–Huxley equations applicable to Purkinje fibre action and pace-maker potentials. *J Physiol (Lond)* 1962;160:317–352.
2. McAllister RE, Noble D, Tsien RW: Reconstruction of the electrical activity of cardiac Purkinje fibres. *J Physiol (Lond)* 1975;251:1–59.
3. Beeler GW, Reuter H: Reconstruction of the action potential of ventricular myocardial fibres. *J Physiol (Lond)* 1977;268:177–210.
4. Cohen I, Kline R: K^+ fluctuations in the extracellular spaces of cardiac muscle. Evidence from the voltage clamp and extracellular K^+-selective microelectrodes. *Circ Res* 1982;50:1–16.
5. DiFrancesco D, Noble D: A model of cardiac electrical activity incorporating ionic pumps and concentration changes. *Phil Trans R Soc B* 1985;307:353–398.
6. DiFrancesco D: A new interpretation of the pacemaker current in calf Purkinje fibres. *J Physiol (Lond)* 1981;314:359–376.
7. Hilgemann DW, Noble D: Excitation-contraction coupling and extracellular calcium transients in rabbit atrium; Reconstruction of basic cellular mechanisms. *Proc R Soc B* 1987;230:163–205.
8. Noble D, DiFrancesco D, Denyer JC: Ionic mechanisms in normal and abnormal cardiac pacemaker activity, in Jacklet JW (ed): *Cellular and Neuronal Oscillators*. New York, Marcel Dekker, 1989, pp 59–85.
9. Noble D: Ionic mechanisms in normal cardiac activity, in Zipes DP, Jalife J (eds): *Cardiac Electrophysiology. From Cell to Bedside*. Philadelphia, WB Saunders, 1990, pp 163–171.
10. Hamill O, Marty A, Neher E, et al.: Improved patch-

clamp techniques for high-resolution current recording from cells and cell-free membrane patches. *Pflügers Arch* 1981;391:85–100.
11. Adrian RH: Potassium chloride movement and the membrane potential of frog muscle. *J Physiol (Lond)* 1960;151:154–185.
12. Hodgkin AL: Ionic movements and electrical activity in giant nerve fibres. *Proc R Soc Lond B* 1957;148:1–37.
13. Gadsby DC, Cranefield PF: Two levels of resting potential in cardiac Purkinje fibers. *J Gen Physiol* 1977;70:725–746.
14. Goldman DE: Potential, impedence and rectification in membranes. *J Gen Physiol* 1943;27:37–60.
15. Hodgkin AL, Katz B: The effect of sodium ions on the electrical activity of the giant axon of the squid. *J Physiol (Lond)* 1949;108:37–77.
16. Kléber AG: Extracellular potassium accumulation in acute myocardial ischemia. *J Mol Cell Cardiol* 1984;16:389–394.
17. Fabiato A: Time and calcium dependence of activation and inactivation of calcium-induced release of calcium from the sarcoplasmic reticulum of a skinned canine cardiac Purkinje cell. *J Gen Physiol* 1985;85:247–289.
18. Noble D, Noble SJ: A model of S.A. node electrical activity using a modification of the DiFrancesco-Noble (1984) equations. *Proc R Soc Lond B* 1984;222:295–304.
19. Walton MK, Fozzard HA: The conducted action potential. Models and comparison to experiments. *Biophys J* 1983;44:9–26.
20. Hodgkin AL, Huxley AF: A quantitative description of membrane current and its application to conduction and excitation in nerve. *J Physiol (Lond)* 1952;117:500–544.
21. Benndorf K, Nilius B: Inactivation of sodium channels in isolated myocardial mouse cells. *Eur Biophys J* 1987;15:117–127.
22. Brown AM, Lee KS, Powell T: Sodium current in single rat heart muscle cells. *J Physiol (Lond)* 1981;318:479–500.
23. Follmer CH, Ten Eick RE, Yeh JZ: Sodium current kinetics in cat atrial myocytes. *J Physiol (Lond)* 1987;384:169–197.
24. Horn R, Vandenberg CA: Statistical properties of single sodium channels. *J Gen Physiol* 1984;84:505–534.
25. Makielski JC, Sheets MF, Hanck DA, et al.: Sodium current in voltage clamped internally perfused canine cardiac Purkinje cells. *Biophys J* 1987;52:1–8.
26. Kirsch GE, Brown AM: Kinetic properties of single sodium channels in rat heart and rat brain. *J Gen Physiol* 1989;93:85–99.
27. Kunze DL, Lacerda AE, Wilson DL, et al.: Cardiac Na currents and the inactivating, reopening and waiting properties of single cardiac Na channels. *J Gen Physiol* 1985;86:691–719.
28. Patlak JB, Ortiz M: Slow currents through single sodium channels of the adult rat heart. *J Gen Physiol* 1985;86:89–104.
29. Grant AO, Starmer CF: Mechanisms of closure of cardiac sodium channels in rabbit ventricular myocytes: Single-channel analysis. *Circ Res* 1987;60:897–913.
30. Kiyosue T, Arita M: Late sodium current and its contribution to action potential configuration in guinea pig ventricular myocytes. *Circ Res* 1989;64:389–397.
31. Coraboeuf E, Deroabaix E, Coulombe A: Effect of tetrodotoxin on action potentials of the conduction system in dog heart. *Am J Physiol* 1979;236:H561–H567.
32. Carmeliet E: Voltage dependent block by tetrodotoxin of the sodium channel in rabbit cardiac Purkinje fibers. *Biophys J* 1983;51:389–401.
33. Wasserstrom JA, Salata JJ: Basis for tetrodotoxin and lidocaine effects on action potentials in dog ventricular myocytes. *Am J Physiol* 1988; 254:H1157–H1166.
34. Schubert B, Van Dangen AMJ, Kirsch GE, Brown AM: β-Adrenergic inhibition of cardiac sodium channels by dual G-protein pathways. *Science* 1989;245:516–519.
35. Jewitt DE, Mercer CJ, Reid D, et al.: Free noradrenaline and adrenaline secretions in relation to the development of cardiac arrhythmias and heart failure in patients with acute myocardial infarction. *Lancet* 1969;1:635–641.
36. Videback J, Christensen NJ, Sterndorff B: Serial determination of plasma catecholamines in myocardial infarction. *Circulation* 1972;46:846–855.
37. Grant AO, Dietz MA, Gilliam FR III, Starmer CF: Blockade of cardiac sodium channels by lidocaine: Single channel analysis. *Circ Res* 1989;65:723–760.
38. Zilberter YI, Khodorov BI, Motin LG, Papin AA: Slow inactivation of single sodium channels depends on the pathway of fast inactivation, in Sellin LC, Libelius R, Thesleff S (eds): *Neuromuscular Junction.* Elsevier Science Publishing Co., Inc., New York, 1989, pp 43–50.
39. Kohlhardt M, Fichtner H: Block of single cardiac Na^+ channels by antiarrhythmic drugs: The effect of amiodarone, propafenone and diprafenone. *J Membrane Biol* 1988;102:105–119.
40. Nowycky MC, Fox AP, Tsien RW: Three types of neuronal calcium channels with different calcium agonist sensitivity. *Nature* 1985;316:440–443.
41. Fox AP, Nowycky MC, Tsien RW: Single channel recordings of three types of calcium channels in chick sensory neurones. *J Physiol (Lond)* 1987;394:173–200.
42. Bean BP: Two kinds of calcium channels in canine atrial cells. Difference in kinetics, selectivity and pharmacology. *J Gen Physiol* 1985;86:1–30.
43. Nilius B, Hess P, Lansman JB, et al.: A novel type of cardiac calcium channel in ventricular cells. *Nature* 1985;316:443–446.
44. Mitra R, Morad M: Two types of calcium channels in guinea pig ventricular myocytes. *Proc Natl Acad Sci USA* 1986;93:5340–5344.
45. Glossman H, Ferry DR, Goll A, Rombusch M: Molecular pharmacology of the calcium channel: evidence for subtypes, multiple drug receptor sites, channel subunits, and the development of a radiolodinated 1,4-dihydropyridine calcium channel label, [125_I] 10 dipine. *J Cardiovasc Pharmacol* 1984;6:S608–S621.
46. Rosenberg RL, Hess P, Reeves JP, et al.: Calcium channels in planar lipid bilayers: New insights into the mechanisms of permeation and gating. *Science* 1986;231:1564–1566.
47. Bean BP: Nitrendipine block of cardiac calcium channels: high affinity binding to the inactivated state. *Proc Natl Acad Sci USA* 1984;81:6388–6392.
48. Sanguinetti MC, Kass RS: Voltage dependent block of calcium current in the calf cardiac Purkinje fiber by dihydropyridine calcium channel antagonists. *Circ Res* 1984;55:336–348.

49. Sanguinetti MC, Krafte DS, Kass RS: Voltage-dependent modulation of Ca channel current in heart cells by Bay K8644. *J Gen Physiol* 1986;88:369–392.
50. Reuter H: Calcium channel modulation by neurotransmitters, enzymes and drugs. *Nature* 1983;301:569–574.
51. Tsien RW, Bean BP, Hess P, et al.: Mechanisms of calcium channel modulation by beta-adrenergic agents and dihydropyridine calcium agonists. *J Mol Cell Cardiol* 1986;18:691–710.
52. Osterrieder W, Brum G, Hescheler J, et al.: Injection of subunits of cyclic AMP-dependent protein kinase into cardiac myocytes modulates Ca^{2+} current. *Nature* 1982;298:576–578.
53. Brum G, Osterrieder W, Trautwein W: β-Adrenergic increase in the calcium conductance of cardiac myocytes studied with the patch clamp. *Pflügers Arch* 1984;401:111–118.
54. Yantani A, Codina J, Imoto Y, et al.: A G protein directly regulates mammalian cardiac calcium channels. *Science* 1987;238:1288–1292.
55. Attwell D, Cohen I, Eisner D, Ohba M, Ojeda C: The steady state TTX-sensitive ("window") sodium current in cardiac Purkinje fibres. *Pflügers Arch* 1979;379:137–142.
56. Egan TM, Noble D, Noble SJ, et al.: Sodium–calcium exchange during the action potential in guinea-pig ventricular cells. *J Physiol (Lond)* 1989;411:639–661.
57. Powell T, Noble D: Calcium movements during each heart beat. *Mol Cell Biochem* 1989;89:103–108.
58. Kimura J, Noma A, Irisawa H: Na-Ca exchange current in mammalian heart cells. *Nature* 1989;319:596–597.
59. Mechmann S, Pott L: Identification of Na-Ca exchange current in single cardiac myocytes. *Nature* 1986;319:597–599.
60. Gadsby DC: Activation of electrogenic Na^+/K^+ exchange by extracellular K^+ in canine cardiac Purkinje fibers. *Proc Natl Acad Sci USA* 1980;77:4035–4039.
61. Gadsby DC: The Na/K pump of cardiac myocytes, in Zipes DP, Jalife J (eds): *Cardiac Electrophysiology. From Cell to Bedside.* Philadelphia, WB Saunders, 1990, pp 35–51.
62. Hess P: Elementary properties of cardiac calcium channels: A brief review. *Can J Physiol Pharmacol* 1988;66:1218–1223.
63. Isenberg G, Klöckner U: Calcium currents of isolated bovine ventricular myocytes are fast and of large amplitude. *Pflügers Arch* 1982;395:30–41.
64. Mitchell MR, Powell T, Terrar DA, Twist VW: Characteristics of the second inward current in cells isolated from rat ventricular muscle. *Proc R Soc Lond B* 1983;219:447–469.
65. Callewaert G, Vereecke J, Carmeliet E: Existence of a calcium-dependent potassium channel in the membrane of cow cardiac Purkinje cells. *Pflügers Arch* 1986;406:424–426.
66. Carmeliet E: K^+ channels in cardiac cells: Mechanisms of activation, inactivation, rectification and K^+_e sensitivity. *Pflügers Arch* 1989;414(suppl 1):S88–S92.
67. Dudel J, Peper K, Rüdel R, Trautwein W: The dynamic chloride component of membrane current in Purkinje fibers. *Pflügers Arch* 1967;295:197–212.
68. Fozzard HA, Hiraoka M: The positive dynamic current and its inactivation properties in cardiac Purkinje fibres. *J Physiol (Lond)* 1973;234:569–586.
69. Kenyon JL, Gibbons WR: Effects of low-chloride solutions on action potentials of sheep cardiac Purkinje fibers. *J Gen Physiol* 1977;70:635–660.
70. Coraboeuf E, Carmeliet E: Existence of two transient outward currents in sheep cardiac Purkinje fibers. *Pflügers Arch* 1982;392:352–359.
71. Josephson IR, Sanchez-Chapula J, Brown AM: Early outward current in rat single ventricular cells. *Circ Res* 1984;54:157–162.
72. Kenyon JL, Gibbons WR: 4-aminopyridine and the early outward current of sheep cardiac Purkinje fibers. *J Gen Physiol* 1979;73:139–157.
73. Litovsky SH, Antzelevitch C: Transient outward current prominent in canine ventricular epicardium but not endocardium. *Circ Res* 1988;62:116–126.
74. McAllister RE, Noble D: The time and voltage dependence of the slow outward current in cardiac Purkinje fibres. *J Physiol (Lond)* 1966;186:632–662.
75. Clapham DE, DeFelice LJ: Voltage-activated K channels in embryonic chick heart. *Biophys J* 1984;45:40–42.
76. Gintant GA, Datyner NB, Cohen IS: Gating of delayed rectification in acutely isolated canine cardiac Purkinje myocytes: Evidence for a single voltage-gated conductance. *Biophys J* 1985;48:1059–1064.
77. Hutter OF, Noble D: Rectifying properties of heart muscle. *Nature* 1960;188:495.
78. Carmeliet E: Induction and removal of inward-going rectification in sheep cardiac Purkinje fibres. *J Physiol (Lond)* 1982;327:285–308.
79. Kurachi Y: Voltage-dependent activation of the inward-rectifier potassium channel in the ventricular membrane of guinea-pig heart. *J Physiol (Lond)* 1985;366:365–385.
80. Siegelbaum SA, Tsien RW, Kass RS: Role of intracellular calcium in the transient outward current of calf Purkinje fibres. *Nature* 1977;269:611–613.
81. Latorre R, Oberhauser A, Labarca P, Alvarez O: Varieties of calcium-activated potassium channels. *Annu Rev Physiol* 1989;51:385–399.
82. Noma A: ATP-regulated K^+ channels in cardiac muscle. *Nature* 1983;305:147–148.
83. Escande D: The pharmacology of ATP-sensitive K^+ channels in the heart. *Pflügers Arch* 1989;414 (suppl 1):S93–S98.
84. Steinberg MI, Ertel P, Smallwood JK, et al.: The relation between vascular relaxant and cardiac electrophysiological effects of pinacidil. *J Cardiovasc Pharm* 1988;12(suppl 2):S30–S40.
85. Kantor PF, Coetzee WA, Carmeliet E, et al.: Reduction of ischemic K^+ loss and arrhythmias in rat hearts. Effect of glibenclamide, a sulfonylurea. *Circ Res* 1990;66:478–485.
86. Weidmann S: Microelectrode recordings: Reflections on errors of interpretation, in Zipes DP, Jalife J (eds): *Cardiac Electrophysiology and Arrhythmias.* Orlando, FL, Grune & Stratton, 1985, pp 61–64.
87. Vassalle M: Analysis of cardiac pacemaker potential using a "voltage clamp" technique. *Am J Physiol* 1966;210:1335–1341.
88. DiFrancesco D: A study of the ionic nature of the pace-maker current in calf Purkinje fibres. *J Physiol (Lond)* 1981;314:377–393.
89. DiFrancesco D: Current i_f and the neuronal modulation of heart rate, in Zipes DP, Jalife J (eds): *Cardiac Electrophysiology. From Cell to Bedside.* Philadelphia, WB Saunders, 1990, pp 28–35.

Chapter **2**

The Electrophysiology of Acute Ischemia

Leonard S. Gettes, MD

Acute myocardial ischemia causes a series of metabolic, ionic, neurohumoral, and structural effects that result in changes in the electrophysiology of the heart. These changes are manifested in single cells, membrane patches, isolated fibers, and intact hearts as alterations in the functional characteristics of ionic channels, changes in the ionic currents responsible for the action potential (AP), and changes in the AP. The changes are also expressed as alterations in active and passive membrane properties, in conduction, refractoriness, and automaticity, and development of spontaneous ectopic impulses and rhythms. Several recent reviews[1-3] have detailed the electrophysiologic changes induced by acute ischemia in great depth and have provided extensive and up-to-date bibliographies. Rather than repeat the material contained in these reviews, I have chosen to address the following three questions, which I believe are fundamental to appreciating the pathophysiology of ischemia-related arrhythmias. (a) What are the electrophysiologic alterations that cause the changes in conduction, refractoriness, and automaticity induced by acute myocardial ischemia? (b) What is (are) the factor(s) responsible for the premature ectopic impulses that are ubiquitous in the setting of acute ischemia and often initiate sustained life-threatening ventricular arrhythmias? and (c) How do the hearts that develop ventricular fibrillation differ from those that do not?

WHAT ARE THE ELECTROPHYSIOLOGIC ALTERATIONS THAT CAUSE THE CHANGES IN CONDUCTION, REFRACTORINESS, AND AUTOMATICITY INDUCED BY ACUTE MYOCARDIAL ISCHEMIA?

Conduction

Conduction of the cardiac impulse depends on a variety of active and passive membrane properties. These include the resting and threshold membrane potentials, the strength of the sodium inward current, the $\dot{V}_{max}$ of the AP upstroke, and the magnitude of the intracellular and extracellular resistances (Fig. 2.1). Each of these parameters is determined by one or more additional factors, many of which are influenced by the metabolic, ionic, and neurohumoral events precipitated by acute ischemia. Of particular importance is the role of the resting membrane potential (RMP), the relationships between the RMP and the sodium (Na^+) current (i_{Na}), the relation between i_{Na} and the maximum rate of rise of the action potential upstroke ($\dot{V}_{max}$), and the relation between $\dot{V}_{max}$, extracellular and intracellular resistances (r_o and r_i), and conduction: that is, resting V_m versus i_{Na}, i_{Na} versus $\dot{V}_{max}$, and $\dot{V}_{max}$, $r_o + r_i$ versus θ.

655 Avenue of the Americas, New York, NY 10010
Current Topics in Cardiology

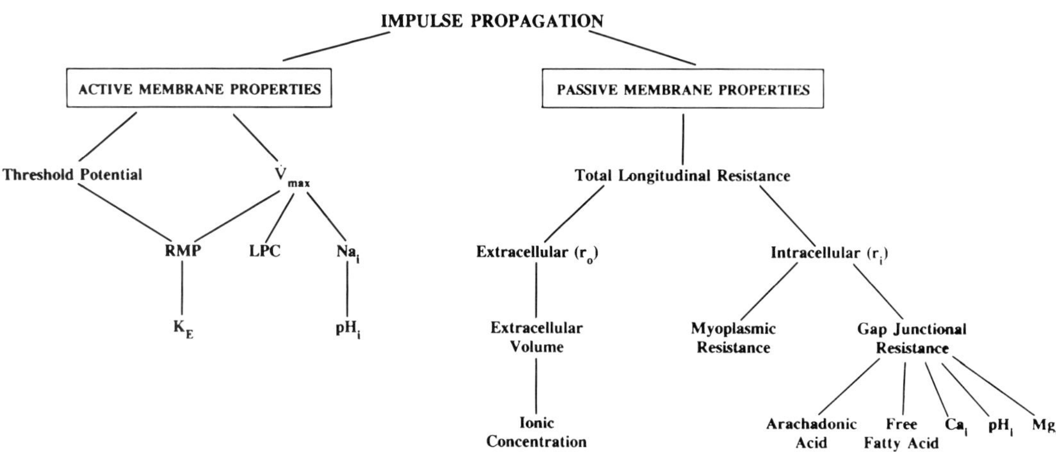

FIGURE 2.1 Line drawing depicting factors that determine the speed of impulse propagation. RMP = resting membrane potential, LPC = lysophosphotydelcholine, K_E = extracellular potassium concentration. See discussion in text.

The RMP is of primary importance because it determines the availability of the Na^+ channels and their opening, closing, and reactivation kinetics. The RMP is also one of the factors that determines the driving force for the inward i_{Na} and therefore $\dot{V}_{max}$ of the AP upstroke. The RMP in normal cells is between -80 and -90 mV. This closely approximates the potassium (K^+) equilibrium potential as determined by the Nernst equation ($E_K = -61.5 \log_{10} [K]_i/[K]_o$). When the myocardium is made acutely ischemic by coronary occlusion, high-energy phosphates are hydrolyzed, intracellular and extracellular pH fall, and extracellular K^+ ($[K^+]_o$) rises (Fig. 2.2). This rise is triphasic with an initial phase beginning within seconds of the acute interruption of coronary flow and lasting approximately 10 min.[4] During this time, the RMP falls slightly in excess of that predicted by the changes in K^+ equilibrium potential (E_K) calculated from the measured change in $[K^+]_o$ and the anticipated change in intracellular K^+ ($[K^+]_i$).[5] This additional fall in RMP is most likely explained by the associated fall in extracellular pH and by the fall in P_{O_2}, both of which have been shown to cause small but definite changes in RMP.[6,7] The initial effect of the decreased RMP results in voltage difference between the RMP and the threshold potential. This effect will speed conduction provided it is not counteracted by a fall in $\dot{V}_{max}$ or by changes in r_o and r_i. It is important to note that $\dot{V}_{max}$ does not decrease until the RMP has decreased from about -85 mV to approximately -75 mV (Fig. 2.3). As predicted by the Nernst equation, this 10 mV fall in RMP requires a 50% rise in $[K^+]_o$, that is, from a normal control value of approximately 4.0 to 6.0 mM. Within this K^+ range, supernormal conduction has been observed in most species[6,8–10] and this mechanism probably explains the speeding of conduction that occurs immediately following the onset of acute ischemia.[11] This period of accelerated conduction is of brief duration and is followed by a period of conduction slowing that is often associated with activation block. This conduction slowing is due, in large part, to a decrease in the i_{Na} and as a result, a decrease in $\dot{V}_{max}$ of the AP upstroke. The magnitude of i_{Na} depends upon the conductance of the available Na^+ channels (g_{Na}) and on the electrical driving force for the Na^+ ion, which is defined as the difference between its equilibrium potential (E_{Na}) and the RMP (E). Thus, the i_{Na} is related to $g_{Na} \cdot (E_{Na} - E)$. The E_{Na} as determined by the Nernst equation is approximately $+40$ mV. Thus, during diastole, when the V_m is -85 mV, the driving force is large. However, no current flows during diastole because g_{Na} is very low. g_{Na} begins to rise when the fibers depolarize to the threshold potential and a propagated AP is generated. The strength of i_{Na} determines the magnitude of $\dot{V}_{max}$ of the AP upstroke. The relationship between i_{Na} and $\dot{V}_{max}$ approaches linearity as conditions tend to become more physiologic.[12] It has been shown that changes in temperature affect this relationship, but the role of other factors, such as changes in intracellular or extracellular ions or the neurotransmitters is uncertain. It is known that i_{Na} and $\dot{V}_{max}$ decrease as the RMP becomes less negative in accordance with the well-defined inactivation curve of the

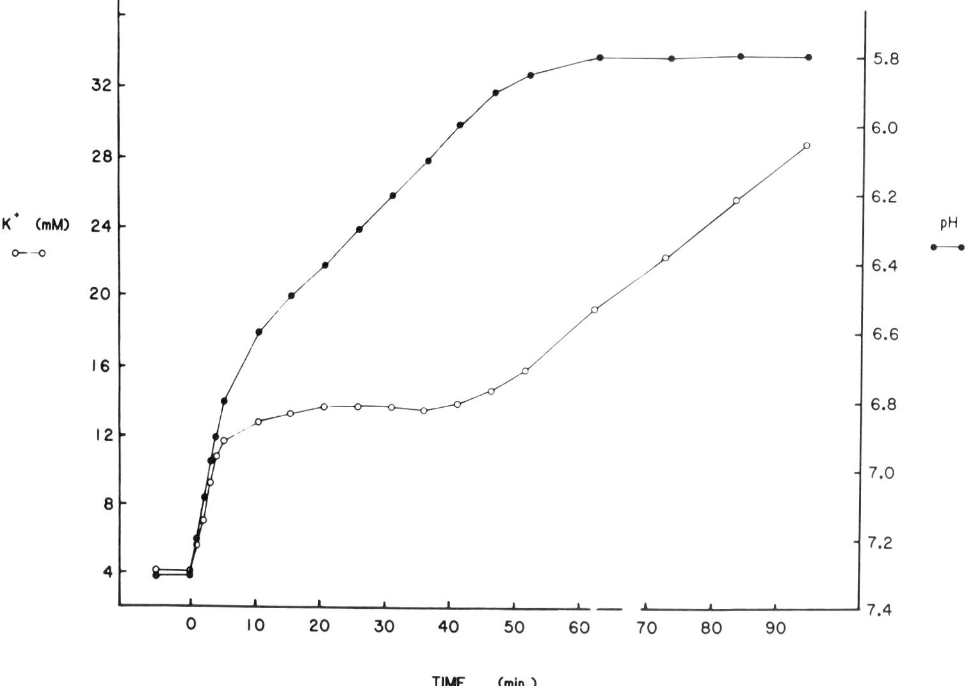

FIGURE 2.2 Time course of changes in midmyocardial $[K^+]_o$ (open symbols) and pH (solid symbols) within the center of the ischemic zone following ligation of the left anterior descending coronary artery in the anesthetized open-chested pig. *Reprinted from Gettes et al.,*[1] *with permission.*

i_{Na}[12–14] (Fig. 2.3). However, the effects of changes in intracellular pH and $[Ca^{2+}]_i$ on these relationships have not been reported.

A decrease in $\dot{V}_{max}$ will cause a predictable decrease in conduction velocity provided there is no change in the difference between RMP and threshold potential and in the values of extracellular and intracellular longitudinal resistances (r_o and r_i). The factors that influence only $\dot{V}_{max}$ such as reduction in $[Na^+]_o$, class I antiarrhythmic drugs, and additional increase in $[K^+]_o$ cause a decrease in conduction velocity, which is expressed by the relationship $\Theta^2 \sim \dot{V}_{max}$.[15] When r_o and r_i also change, then the relationship becomes $\Theta^2 \sim \dot{V}_{max}/r_o + r_i$. This is an oversimplified statement of the cable equation, which defines the movement of an electrical impulse along a unidimensional cable of infinite length.[16] The slowing of conduction that occurs during the early phase of ischemia is more pronounced than anticipated by the rise in $[K^+]_o$[4,6,17] (Fig. 2.4). Three factors contribute to the K^+-independent component of the initial conduction slowing: (a) a greater change in the RMP than caused solely by the changes in $[K^+]_o$ and $[K^+]_i$; (b) a greater change in $\dot{V}_{max}$ than that caused solely by the change in RMP; and (c) a greater change in conduction than that caused solely by the change in $\dot{V}_{max}$.

As indicated above, the changes in P_{O_2} and pH that characterize acute ischemia cause slight changes in RMP. As a result the ischemia induced changes in RMP and $\dot{V}_{max}$ are more marked than anticipated solely by the changes in K^+. In addition, the change in pH induces a shift in the RMP–$\dot{V}_{max}$ relationship, which results in $\dot{V}_{max}$ being depressed more than one would anticipate solely by the change in RMP.[6] An increase in $[Na^+]_i$ will occur both from inhibition of the Na/K pump and/or from $[Na^+]_i$–hydrogen exchange.[18] The driving force for the i_{Na} would decline and by this mechanism cause a decrease in $\dot{V}_{max}$, independent of any change in RMP. Although it is known that an increase in $[Na^+]_i$ does occur in acute ischemia,[19–21] it most likely does not occur rapidly enough to contribute to the early changes in $\dot{V}_{max}$ and conduction.[5,22]

A possible factor contributing to the early K^+-independent effects on RMP and $\dot{V}_{max}$ is the accumulation of lysophosphoglycerides, particularly lysophosphotydylcholene in the sarcolemmal membrane.[23,24] These products of fatty acid

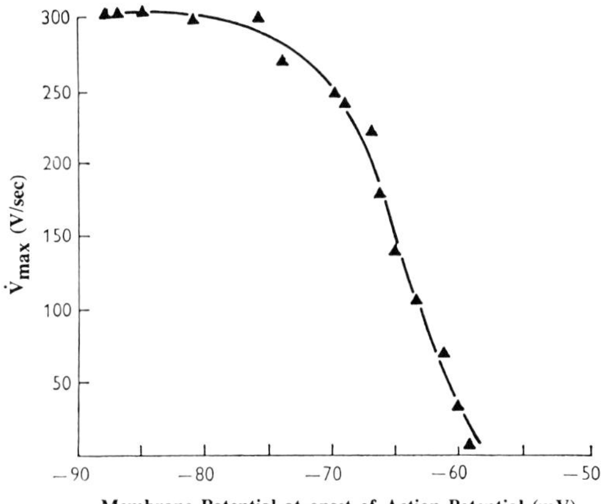

FIGURE 2.3 Effect of changes in V_m at the onset of the AP induced by increasing $[K^+]_o$ on the maximum rate of rise of the action potential upstroke ($\dot{V}_{max}$) in isolated superfused guinea pig papillary muscle. *Reprinted from Gettes et al.,*[14] with permission.

metabolism have been demonstrated to cause changes in RMP and $\dot{V}_{max}$ which may be independent of any changes in K^+ and are more marked at acidic pHs. The time course of this accumulation appears appropriate to play this role during the early stage of acute ischemia.

The fact that the changes in conduction are more marked than anticipated by the change in $\dot{V}_{max}$ can be attributed to changes in the passive membrane properties, particularly extracellular resistance. It has been shown that extracellular resistance (r_o) rises immediately following interruption of coronary flow,[25] a phenomenon that is attributed to the collapse of the microvasculature with the resultant diminution of the extracellular compartment. This effect will slow the extracellular component of the local circuit required for a propagated excitation wave. It is also clear that resistance at the gap junction increases as a consequence of acute ischemia and produces cellular uncoupling.[25–27] However, these studies have shown that r_i does not rise for approximately 10 min after the onset of either true or simulated ischemia (Fig. 2.5), a time course that is too slow to contribute to the $\dot{V}_{max}$ independent component of the conduction slowing that occurs during the early stage of acute ischemia.

Two additional characteristics of the initial conduction slowing need to be stressed. The first is that they are inhomogeneous and the second is that they are stimulation frequency or rate dependent. The inhomogeneous nature of the changes in conduction relate to and are caused by the inhomogeneous nature of the ionic changes. These inhomogeneities in K^+ and pH are present not only across the lateral margin of the ischemic zone[4,28] but also at the endocardial surface[29] between the various layers of the myocardium, that is, subendocardium, midmyocardium, and subepicardium,[4] and between closely spaced sites within the same layer[28,30,31] (Fig. 2.6). The most likely mechanism for the inhomogeneities existing at the lateral and subendocardial border areas is the diffusion of K^+ and CO_2 out of the ischemic zone and the diffusion of O_2 into the ischemic zone.[30] However, this mechanism is an unlikely cause of inhomogeneities existing in the center. Here, the inhomogeneous rate at which high-energy phosphates are hydrolyzed is the more likely explanation.[32] The result of these inhomogeneities is to alter the sequence of activation and to permit the creation

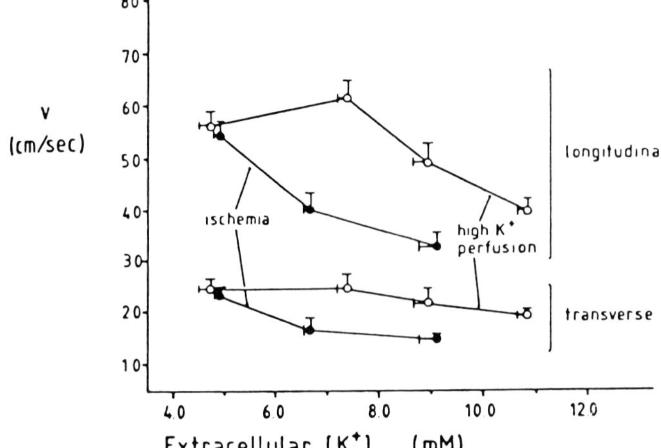

FIGURE 2.4 Changes in longitudinal and transverse intraventricular conduction occurring at various levels of increased myocardial $[K^+]_o$ induced by ischemia (closed symbols) or perfusion with high K^+ solution (open symbols) in the isolated perfused porcine heart. *Reprinted from Kleber et al.,*[17] *with permission.*

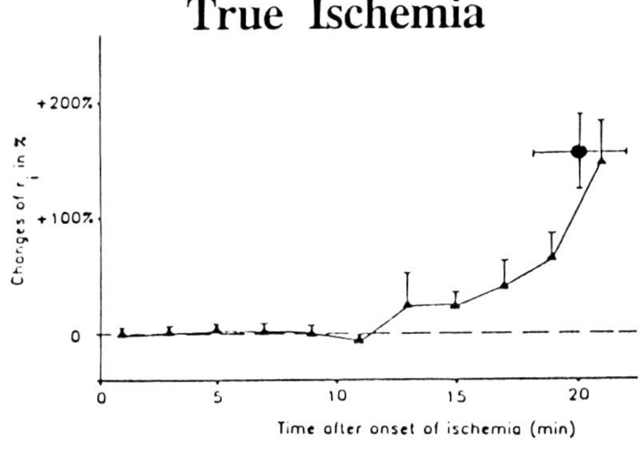

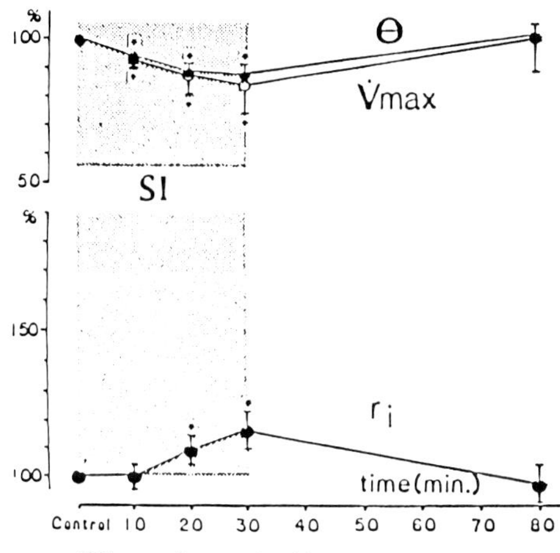

FIGURE 2.5 The upper panel shows the changes in internal longitudinal resistance (r_i) following cessation of arterial perfusion in the isolated rabbit papillary muscle. *Reprinted from Kleber et al.,*[25] *with permission.* The lower panel shows changes in r_i as well as changes in conduction velocity (Θ) and the maximum rate of rise of the AP upstroke ($\dot{V}_{max}$) following exposure to superfusing solutions modified to simulate ischemia (K = 9.0 mM; pH = 6.5; PO_2 < 20 mm Hg; glucose = 0) in the isolated superfused guinea pig papillary muscle. *Reprinted from Hiramatsu et al.,*[26] *with permission.* Note that the onset of cellular uncoupling, that is, the rise in r_i, begins approximately 10 min after the onset of either true or simulated ischemia.

of islands of activation block which can then lead to the development of reentry circuits.[30] It is also possible that the generation of electrical currents—the injury currents—which result from the inhomogeneous changes in AP characteristics can influence the resting level of RMP and thereby affect the $\dot{V}_{max}$ and the conduction velocity.[33]

The rate-dependent conduction changes reflect the rate-dependent changes of several active and passive membrane properties induced by acute ischemia. These include the fall of $\dot{V}_{max}$ and the rise of r_i.[26] There is still dispute as to whether the rise in $[K^+]_o$, that is, the dominant factor responsible for the initial slowing of conduction, is itself rate dependent. In the in situ pig and Langendorff-perfused guinea pig heart, the ischemia-induced rise in K^+ is not rate dependent within stimulation frequencies of approximately 1 to 3 per second.[4,5] However, in more isolated preparations such as the perfused septum[34] and perfused papillary muscle,[27] rate dependence of the K^+ rise within this stimulation frequency range has been noted. It is possible that some of the rate-dependent changes in $\dot{V}_{max}$ and r_i may be due to increases in $[Na^+]_i$ and $[Ca^{2+}]_i$ which have been demonstrated even in nonischemic preparations.[35,36] The importance of the rate dependency is that it will cause greater conduction slowing at more rapid rates and facilitate an earlier likely occurrence of reentry.[37]

Many of the ischemia-induced changes in active and passive membrane properties that affect impulse conduction are influenced by drugs or interventions that influence $[Ca^{2+}]_i$. In general, agents that tend to lessen the rise in $[Ca^{2+}]_i$ such as pretreatment with verapamil tend to lessen and retard the changes in $[K^+]_o$ and pH, the changes in $\dot{V}_{max}$ and r_i, and the change in conduction. In addition, these drugs prevent ventricular fibrillation.[26,27,28,38,39] Agents that tend to exaggerate the rise in $[Ca^{2+}]_i$ such as pretreatment with Bay K 8644 tend to worsen and speed these ionic and electrophysiologic changes.[38,40] It is not yet clear as to whether these effects reflect a primary role of changes in $[Ca^{2+}]_i$ on the electrical phenomena themselves, or whether the rate of utilization of high energy phosphates and the subsequent metabolic and ionic changes that underlie the conduction changes are influenced by the changes in $[Ca^{2+}]_i$. In this scenario, the influence of drugs such as verapamil and Bay K on conduction would be indirect secondary to a primary effect on the rate of utilization of the high-energy phosphates.

Following the initial period of conduction slowing, there is a period during which resting potential, $\dot{V}_{max}$, and conduction improve spontaneously.[4,41,42] This period corresponds to the plateau or second phase of the rise in $[K^+]_o$ and may be due to the effect of locally released myocardial catecholamines[43,44] which will stimulate the Na/K pump[45] leading to a reduction of $[Na^+]_i$ and a reduction in $[K^+]_o$. These ionic changes undoubtedly contribute to the spontaneous improvement in RMP, $\dot{V}_{max}$ of the AP upstroke, and conduction which occur during this period. The

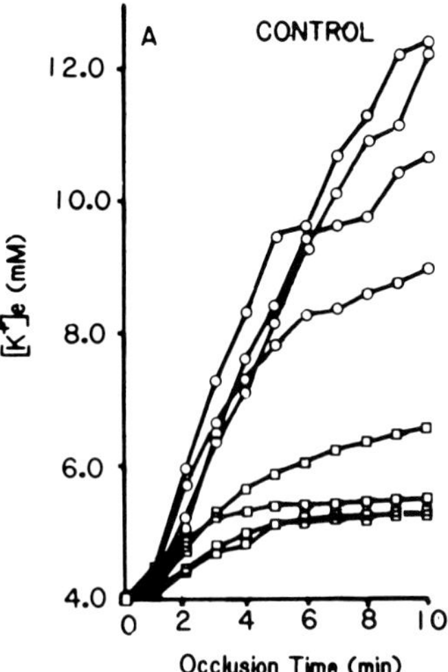

FIGURE 2.6 Changes in midmyocardial $[K^+]_o$ following ligation of the left anterior descending coronary artery as recorded simultaneously by eight K^+-sensitive electrodes located in the midmyocardium of the anesthetized open-chested pig. The upper four tracings were recorded from electrodes located in the center of the ischemic zone. The lower four tracings were recorded from electrodes located at the lateral margin, that is, within 5 mm of the cyanotic border. *Reprinted from Fleet et al.,*[28] *with permission.*

period of transient improvement lasts 10 to 15 min, giving way to a period of progressive conduction slowing, culminating in excitability and activation block in the center of the ischemic zone. This period of activation slowing occurs as $[K^+]_o$ again rises, albeit at a rate which is slower that during the initial phase.[4] This component of the rise in $[K^+]_o$ is associated with a decrease in intracellular and extracellular pH to approximately 6.0. At the same time $[Na^+]_i$ and $[Ca^{2+}]_i$ begin to increase.[19–21,46,47] These metabolic and ionic abnormalities result in further decrease in RMP and $\dot{V}_{max}$ and a progressive increase in r_i, indicating cell-to-cell uncoupling. Recent information suggests that $[Mg^{2+}]_i$ may also rise[48,49] and contribute to the cellular uncoupling which in turn contributes to the slowing of impulse conduction. This phase culminates with irreversible changes and cell death.

Refractory Period Changes

The refractory period may be defined as the time required for a cell to regain its excitability following a preceding depolarization. This time interval includes the time required for the membrane to be repolarized to that transmembrane potential level at which the i_{Na} is capable of being reactivated which in normal ventricular myocardium is approximately −60 mV, and the time at that V_m required for the reactivation to occur.[14] Thus, refractoriness has two components, a voltage-dependent component (−60 mV) and a time-dependent component (the time at that voltage required for reactivation to occur). This time-dependent component is also voltage dependent,[14,50] and as the RMP is changed as by ischemia or by an increase in $[K^+]_o$, the time constant for reactivation also becomes prolonged[14,50] (Fig. 2.7). The components of acute ischemia, particularly hypoxia and the increase in $[K^+]_o$, shorten the action potential plateau by increasing the conductance of the i_{K^+} (ref. in 1) and thus lessen the time required for the membrane to return to its reactivation voltage level. At the same time, the change in RMP, augmented by hypoxia and acidosis, lengthens the time required for reactivation to occur at that voltage level. As a result of these changes, the refractory period is first shortened[51,52] and then lengthened[41,53] by acute ischemia. As the metabolic and ionic changes associated with acute ischemia are inhomogeneous, so are the changes in refractoriness across and within the ischemic zone. These inhomogeneities contribute to the

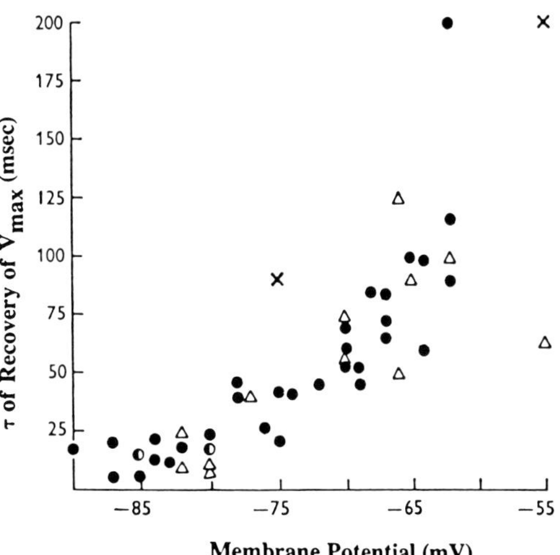

FIGURE 2.7 Effect of change in RMP induced by increasing $[K^+]_o$ on the time constant (τ) of the recovery of the maximum rate of rise of the AP upstroke ($\dot{V}_{max}$) in isolated superfused trabeculae or papillary muscles of guinea pigs (closed circles), sheep (open triangles), calf (open circles), and pigs (crosses). In all species, the reactivation time constant becomes prolonged as the RMP becomes less negative. *Reprinted from Gettes et al.,*[14] *with permission.*

long recognized increase in the dispersion of the recovery of excitability that occurs in the setting of acute ischemia.[54–56] It has been recently suggested that cellular uncoupling might also increase the dispersion of the recovery of excitability by altering the electrotonic effects contributing to APD.[57] The combination of the inhomogeneities in conduction and the inhomogeneities in refractoriness provide the substrate capable of initiating and supporting reentry.

Automaticity

Automaticity is the term applied to the spontaneous depolarization that occurs during electrical diastole or phase IV of the action potential in pacemaker and potentially pacemaker cells. This type of spontaneous activity occurs characteristically at membrane potential levels less than −60 mV and is exemplified by the pacemaker cells of the SA node.[58] Spontaneous diastolic depolarization also occurs in the −80 mV range in cells in the His-Purkinje system. This type of spontaneous activity is accentuated by a reduction in $[K^+]_o$.[59] In addition, abnormal spontaneous activity may occur as a result of early and delayed afterdepolarizations (EADs and DADs).[60] If these afterdepolarizations are of subthreshold strength, they may be undetected. However, if they are of threshold strength they may then induce a single propagated response or trigger a burst of repetitive activity referred to as "triggered automaticity." This type of spontaneous activity can be induced by agents known to cause an increase in $[Ca^{2+}]_i$ such as digitalis[61] and Bay K 8644.[62] It is believed that oscillations in $[Ca^{2+}]_i$ are the underlying mechanism[63] and that the transient inward current responsible for the spontaneous activity is carried predominantly by Na^+ ions.[64]

The various components of acute ischemia are each capable of altering automaticity. The increase in $[K^+]_o$ suppresses almost all types of abnormal automaticity, although it does not appear to influence the normal automaticity present in pacemaker cells. Almost all of the other components of acute ischemia tend to enhance abnormal automaticity. The increase in β-adrenergic stimulation occurring about 10 min after coronary occlusion is capable of enhancing automaticity due to spontaneous diastolic depolarization and to afterdepolarizations. However, this may be counteracted by an increase in α-adrenergic agonists. Myocardial fiber stretch that may occur as a result of the dyskinesis associated with acute ischemia has been shown experimentally to cause spontaneous activity in ventricular myocardium,[65] and an increase in $[Ca^{2+}]_i$ as is known to occur in the later stages of acute ischemia is a principal cause of afterdepolarizations.

Electrical currents of injury are generated during electrical diastole by the inhomogeneous changes in RMP.[33,66,67] These currents flow across the various margins and within the center of the ischemic zone and are capable of inducing and enhancing spontaneous diastolic depolarization[68] (Fig. 2.8). These currents of injury may also bring subthreshold afterdepolarizations to threshold and either cause a single propagated response or trigger a burst of spontaneous activity. In the experimental preparations simulated ischemia has been shown to be capable of causing early afterdepolarizations and triggered activity in Purkinje fibers.[69] Thus, acute ischemia can enhance automaticity by a variety of mechanisms.

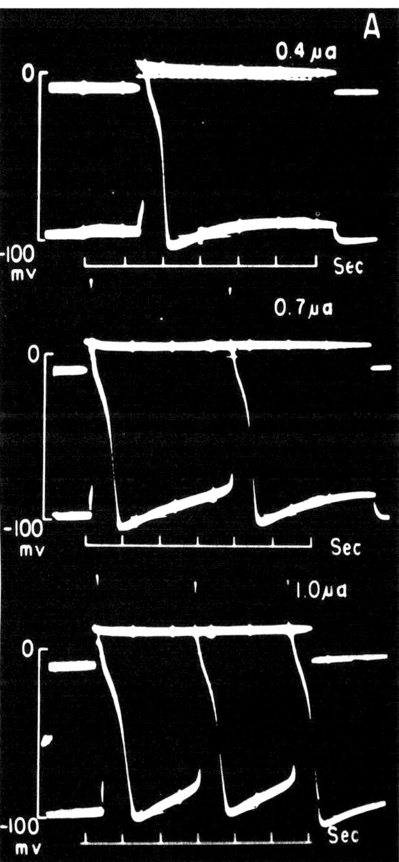

FIGURE 2.8 Increase in spontaneous diastolic depolarization induced by the application of direct current pulses of increasing strength in sheep Purkinje fibers. *Reprinted from Trautwein and Kassebaum,[68] with permission.*

WHAT ARE THE CAUSES OF THE UBIQUITOUS EXTRASYSTOLES?

As indicated above, the metabolic, ionic, neurohumoral, and contractile changes associated with acute coronary occlusion result in changes in conduction, refractoriness, and spontaneous activity which can create a substrate for reentry and of alter automaticity. These changes explain the virtually universal appearance of ventricular premature complexes in the setting of acute no-flow ischemia. It has been estimated that approximately two-thirds of the premature complexes that occur in the setting of acute myocardial ischemia are due to spontaneous reentry, and that the remaining one-third are due to a non-reentrant mechanism, presumably enhanced automaticity.[(70)] However, it is important to recognize that the differentiation of enhanced automaticity from reentry, particularly microreentry or reflection,[(2,71)] may be difficult in the absence of the direct recording of cellular electrical events.

Spontaneous reentry has been demonstrated to occur at virtually all regions of the ischemic zone, including the lateral border,[(33,72,73)] the subepicardium,[(74)] the subendocardium,[(75)] and intramurally[(70,74)] in the center of the ischemic zone. On the other hand, the location at which spontaneous activity occurs has not been defined. The future use of sophisticated laser and optical techniques and voltage-sensitive dyes may provide the ability to determine the mechanism and site of spontaneous activity with less trauma and greater accuracy than that associated with the use of multiple electrodes.

WHAT SEPARATES THOSE HEARTS THAT DEVELOP VENTRICULAR FIBRILLATION FROM THOSE THAT DO NOT?

Thus far, this review has attempted to show that the metabolic, ionic, and neurohumoral events that characterize acute ischemia are inhomogeneous and induce a series of inhomogeneous electrophysiologic changes capable of creating the substrate that will produce and support reentry as well as changes capable of producing spontaneous activity. The ventricular premature beats that are virtually ubiquitous in the setting of acute ischemia both in experimental animals and in man can be attributed to these changes. Frequently, these premature beats will progress to sustain ventricular tachyarrhythmias and to ventricular fibrillation. The occurrence of these arrhythmias has been shown to have a three-phase time course that was originally described by Harris[(76)] and subsequently expanded upon by Kaplinsky et al.[(77)] The first state of arrhythmia induction described by Harris occurred within the first 30 min of the ischemic period and was followed by a quiescent period of several hours. During the early phase, that is, stage one, the arrhythmias have been attributed primarily to reentry. Kaplinsky and coworkers further subdivided this phase into three subphases, each lasting approximately 10 min. The first 10 min of the ischemic period was highly arrhythmogenic. As detailed above, K^+ rises rapidly and inhomogeneously during this period, conduction slows dramatically and inhomogeneously and to a degree that is greater than explained purely by the K^+ rise. The refractory period first shortens and then lengthens in an inhomogeneous fashion, increasing the dispersion of the recovery of excitability, and injury currents are generated. The second 10-min phase which is relatively arrhythmia-free corresponds to the plateau phase of the K^+ rise. During this period, the inhomogeneity of the K^+ changes lessens. There is an increase in myocardial and circulating catecholamines that partially normalize many of the electrophysiologic changes occurring in the prior 10-min interval. The changes in refractoriness are lessened and conduction improves. The third 10-min phase is again arrhythmogenic. The onset of this phase corresponds to a fall in intracellular and extracellular pH to approximately 6.0. At this point, K^+ begins to rise again, albeit at a slower rate and more homogeneously than during the first phase, and cellular uncoupling commences, a process that further slows conduction. The cells become irreversibly damaged during this phase, which probably terminates when the uncoupling process is complete. In spite of the substantial increase in our understanding of the biochemistry and electrophysiology of the ischemic process, it has not been possible to determine which hearts will, or will not, develop sustained ventricular tachycardia or ventricular fibrillation within each of these phases. Nor has it been possible to predict which human patients will die suddenly during the acute stage of an ischemic event or infarction. One possible factor is the size of the ischemic zone. It has been long appreciated that small hearts do not support ventricular fibrillation as opposed to larger hearts.[(78)] This principle implies that the smaller the mass of tissue involved, the less likely the heart is to fibrillate. However, it is not clear as to whether a small acutely ischemic zone does in fact offer protection from ventricular tachycardia or fibrillation or whether it merely identifies a subgroup in which these arrhythmias are less common. Other factors that may play a role include the speed, magnitude, and inhomogeneity of the metabolic, ionic, and neurohumoral changes, and the location of the spontaneously occurring premature beats relative to inhomogeneities. In our laboratory, we have been unable to correlate the ventricular tachyarrhythmias, including ventricular fibrillation, which occur following acute occlusion of the left anterior descending coronary artery in the open-chested pig, to the magnitude or rate at which myocardial $[K^+]_o$ or pH change. Nor have we been able to link the presence of these arrhythmias to the magnitude of the inhomogeneities in the change or by inference, to the inhomogeneities in the changes in conduction and refractoriness, or to the magnitude of the injury currents that occur as a result of the inhomogeneous changes. No link has been found between the onset of arrhythmias and the size or characteristics of the border zone, although it has been clearly shown that most cases of sustained ventricular tachyarrhythmia or ventricular fibrillation in the experimental animals are initiated by premature beats arising in the normal myocardium adjacent to the lateral border of the ischemic zone.[(79,80)] It has been demonstrated by computer simulation that only slight changes in the dispersion of refractoriness are required to allow a premature beat to reenter to cause fibrillation.[(81,82)] In other models, fibrillation has been induced by changing pacemaker characteristics in some cells and coupling characteristics in others.[(83,84)] What is not known is whether the premature beats that initiate ventricular fibrillation differ in any significant way from those that do not, or whether the hearts that develop ventricular fibrillation differ in any significant or fundamental way from those that do not. In the absence of acute myocardial ischemia or a prior myocardial insult in which the substrate for reentry is present, the "safety factor" which prevents ventricular premature beats from causing sustained life-threatening arrhythmias is large, and one has to go to rather extraordinary means to induce ventricular fibrillation. One only has to recognize the low incidence of ventricular fibrillation induced in the electrophysiologic laboratory in spite of introduction of multiple premature beats and burst pacing at multiple sites. Acute myocardial ischemia results in an ionically and electrophysiologically heterogeneous substrate that is capable of spawning premature beats from multiple ischemic and peri-ischemic regions and that is capable of initiating and sustaining reentry. As a result, the "safety factor" disappears, and ventricular fibrillation can occur spontaneously and/or be easily induced by applied stimuli. Whether or not one of the ubiquitous premature beats that occur with great frequency in the setting of acute myocardial ischemia will, in fact, reenter to cause spontaneous ventricular fibrillation and sudden death may be a random event that requires the coexistence of several critical factors. For instance, it may be necessary that (a) the premature beat arise in a region that encourages propagation in a pathway containing an area of unidirectional block, (b) the relationship between conduction velocity within this pathway and the duration of the refractory period in the area of block be such that reentry can occur, and (c) the location of the reentry circuit be such that the spreading impulses involve the entire myocardium. In a sense, one may think of the development of ventricular fibrillation during acute ischemia as analogous to the development of paroxysmal supraventricular tachycardia in patients with dual AV junctional pathways localized either within or outside the AV node. In both, the substrate capable of supporting reentry exists, and, in both, premature beats that do not promote reentry occur frequently. For reentry to occur, the initiating beat must occur within the defined time span and subtle changes in the relationship between the timing of the premature beats and the arrhythmogenic substrate must occur to allow one of these premature beats to reenter and cause either supraventricular tachycardia in the patient with dual AV junctional pathways or ventricular fibrillation in the patient with acute ischemia.

The electrophysiologic changes induced by acute ischemia are now quite well understood and explicable in terms of the extracellular and intracellular metabolic, ionic, neurohumoral, and structural changes. What remains elusive is the ability to identify the specific factors that determine whether or not ventricular fibrillation will occur. Such identification might then allow the development of strategies capable of preventing the sudden cardiac death from occurring in the setting of acute myocardial ischemia.

ACKNOWLEDGMENTS

Some of the studies referred to in this chapter were supported by National Heart, Lung, and Blood Institutes grants P01 27430 and R37 HL 38885.

REFERENCES

1. Gettes LS, Cascio WE: Effect of acute ischemia on cardiac electrophysiology, in Fozzard H: *The Heart and Cardiovascular System. Scientific Foundations* (2d ed.). New York, Raven Press, (in press).
2. Janse MJ, Wit AL: Electrophysiological mechanisms of ventricular arrhythmias resulting from myocardial ischemia and infusion. *Physiol Rev* 1989; 69:1049–1169.

3. Kleber AG, Janse MJ: Impulse propagation in myocardial ischemia, in Zipes DP, Jalife J: *Cardiac Electrophysiology: From Cell to Bedside*, Philadelphia, WB Saunders, 1990, pp 156–161.
4. Hill JL, Gettes LS: Effect of acute coronary artery occlusion on local myocardial extracellular K^+ activity in swine. *Circulation* 1980;61:768–778.
5. Kleber AG: Resting membrane potential, extracellular potassium activity and intracellular sodium activity during acute global ischemia in isolated perfused guinea pig hearts. *Circ Res* 1983;52:442–450.
6. Kagiyama Y, Hill JL, Gettes LS: Interaction of acidosis and increased extracellular potassium on action potential characteristics and conduction in guinea pig ventricular muscle. *Circ Res* 1982;51:614–623.
7. Ruiz-Ceretti E, Ragault P, Leblanc N, Ponce Zumino AZ: Effects of hypoxia and altered K-0 on the membrane potential of rabbit ventricle. *J Mol Cell Cardiol* 1983;15:845–854.
8. Gettes LS, Buchanan JW Jr, Saito T, Kagiyama Y, Oshita S, Fujino T: Studies concerned with slow conduction, in Zipes DP, Jalife J (eds): *Cardiac Electrophysiology and Arrhythmias*. Orlando, FL, Grune & Stratton, 1985, pp 81–87.
9. Spear JF, Mopore EN: Supernormal excitability and conduction in the His-Purkinje system of the dog. *Circ Res* 1974;35:782–792.
10. Dominguez G, Fozzard HA: Influence of extracellular K^+ concentration on cable properties and excitability of sheep cardiac Purkinje fibers. *Circ Res* 1970;26:565–574.
11. Arnsdorf MF, Schreiner E, Gambetta M, Friedlander I, Childers RW: Electrophysiological changes in the canine atrium and ventricle during progressive hyperkalemia: Electrocardiographical correlates and the in vivo validation of in vitro predictions. *Cardiovasc Res* 1977;11:409–418.
12. Sheets MF, Hanck DA, Fozzard HA: Nonlinear relation between Vmax and INa in canine cardiac Purkinje cells. *Circ Res* 1988;63:386–398.
13. Weidmann S: The effect of the cardiac membrane potential on the rapid availability of the sodium-carrying system. *J Physiol (Lond)* 1955;127:213–224.
14. Gettes LS, Reuter H: Slow recovery from inactivation of inward currents in mammalian myocardial fibres. *J Physiol (Lond)* 1974;240:703–724.
15. Buchanan JW Jr, Saito T, Gettes LS: The effects of antiarrhythmic drugs, stimulation frequency, and potassium-induced resting membrane potential changes on conduction velocity and dV/dt max in guinea pig myocardium. *Circ Res* 1985;56:696–703.
16. Pressler ML: Passive electrical properties of cardiac tissue, in Zipes DP, Jalife J (eds): *Cardiac Electrophysiology: From Cell to Bedside*. Philadelphia, WB Saunders, 1990, pp 108–122.
17. Kleber AG, Janse MJ, Wilms-Schopman FJG, Wilde AAM, Coronel R: Changes in conduction velocity during acute ischemia in ventricular myocardium of the isolated porcine heart. *Circulation* 1986;73:189–198.
18. Deitmer JW, Ellis D: The intracellular sodium activity of cardiac Purkinje fibers during inhibition and reaction of the Na-K pump. *J Physiol (Lond)* 1978;284:241–259.
19. Neubauer S, Balschi JA, Springer CS, Smith TW, Ingwall JS: Intracellular Na^+ accumulation in hypoxic vs. ischemic rat heart: Evidence for Na^+-H^+ exchange. *Circulation* 1987;76(suppl IV):IV-56(abstract).
20. Pike MM, Kitakase M, Marban E: Increase in intracellular free sodium concentration during ischemia revealed by 23Na NMR in perfused ferret hearts. *Circulation* 1988;78:II-151–II-151(abstract).
21. Tani M, Neely JR: Role in intracellular Na^+ in Ca^{2+} overload and depressed recovery of ventricular function of reperfused ischemic rat hearts: Possible involvement of H^+-Na and Na^+-Ca^{2+} exchange. *Circ Res* 1989;65:1045–1056.
22. Wilde AAM, Kleber AG: The combined effects of hypoxia, high K^+, and acidosis on the intracellular sodium activity and resting potential in guinea pig papillary muscle. *Circ Res* 1986;58:249–256.
23. Corr PB, Gross R, Sobel BE: Amphipathic metabolites and membrane dysfunction in ischemic myocardium. *Circ Res* 1984;55:135–154.
24. Corr PB, Dobmeyer DJ: Amphipathic lipid metabolites and arrhythmogenesis: A perspective, in Rosen MR, Palti Y (eds): *Lethal Arrhythmias Resulting From Myocardial Ischemia and Infarction*. Boston/Dordrecht/Lancaster, Kluwer Academic Publishers, 1989, pp 91–104.
25. Kleber AG, Riegger CB, Janse MJ: Electrical uncoupling and increase of extracellular resistance after induction of ischemia in isolated, arterially perfused rabbit papillary muscle. *Circ Res* 1987; 61:271–279.
26. Hiramatsu Y, Buchanan JW Jr, Knisley SB, Koch GG, Kropp S, Gettes LS: Influence of rate-dependent cellular uncoupling on conduction change during simulated ischemia in guinea pig papillary muscles: Effect of verapamil. *Circ Res* 1989;65:95–102.
27. Cascio WE, Yan G-X, Kleber AG: Passive electrical properties, mechanical activity and extracellular potassium in arterially perfused and ischemic rabbit ventricular muscle: Effects of calcium entry blockade or hypocalcemia. *Circ Res* 1990;66:1461–1473.
28. Fleet WF, Johnson TA, Graebner CA, Engle CL, Gettes LS: Effects of verapamil on ischemia-induced changes in extracellular K^+, pH, and local activation in the pig. *Circulation* 1986;73:837–846.
29. Wilensky RL, Tranum-Jensen J, Coronel R, Wilde AAM, Fiolet JWT, Janse MJ: The subendocardial border zone during acute ischemia of the rabbit heart: An electrophysiologic, metabolic, and morphologic correlative study. *Circulation* 1986;74: 1137–1146.
30. Coronel R, Fiolet JWT, Wilms-Schopman FJG, et al.: Distribution of extracellular potassium and its relation to electrophysiologic changes during acute myocardial ischemia in the isolated perfused porcine heart. *Circulation* 1988;77:1125–1138.
31. Johnson TA, Engle CL, Jenkins MG, Gettes LS: Magnitude of potassium inhomogeneity during acute ischemia in the pig. *Circulation* 1988;78 (suppl II):II-638(abstract).
32. Reimer KA, Jennings RB: Myocardial ischemia, hypoxia and infarction, in Fozzard HA, Haber E, Jennings RB, Katz AM, Morgan, HE (eds): *The Heart and Cardiovascular System. Scientific Foundations*. New York, Raven Press, 1986, pp 1133–1201.

33. Janse MJ, Kleber AG: Electrophysiological changes and ventricular arrhythmias in the early phase of regional myocardial ischemia. *Circ Res* 1981; 49:1069–1081.
34. Weiss J, Shine KI: Extracellular K^+ accumulation during myocardial ischemia in isolated rabbit heart. *Am J Physiol* 1982;242:H619–H628.
35. Cohen CJ, Fozzard HA, Sheu SS: Increase in intracellular sodium ion activity during stimulation in mammalian cardiac muscle. *Circ Res* 1982;50:651–662.
36. Lado MG, Sheu SS, Fozzard HA: Changes in intracellular Ca^{2+} activity with stimulation in sheep cardiac Purkinje strands. *Am J Physiol* 1982;243:133–137.
37. Hope RR, Williams DO, El-Sherif N, Lazzara R, Scherlag BJ: The efficacy of antiarrhythmic agents during acute myocardial ischemia and the role of heart rate. *Circulation* 1974;50:507–514.
38. Kabell G: Modulation of conduction slowing in ischemic rabbit myocardium by calcium-channel activation and blockade. *Circulation* 1988;77:1385–1394.
39. Kaumann AJ, Aramendia P: Prevention of ventricular fibrillation induced by coronary ligation. *J Pharmacol Exp Ther* 1968;164:326–332.
40. Maruyama T, Sun W, Buchanan J, Gettes L: Resting membrane potential-independent depression of $\dot{V}max$ by acute ischemia: Role of internal calcium. *Circulation* 1990;82(suppl III):III-745.
41. Downar E, Janse MJ, Durrer D: The effect of acute coronary artery occlusion on supepicardial transmembrane potentials in the intact porcine heart. *Circulation* 1977;56:217–224.
42. Morena H, Janse MJ, Fiolet JWT, Krieger WJG, Crijns H, Durrer D: Comparison of the effects of regional ischemia, hypoxia, hyperkalemia, and acidosis on intracellular and extracellular potentials and metabolism in the isolated porcine heart. *Circ Res* 1980;46:634–646.
43. Schomig A, Dart AM, Dietz R, Mayer E, Kubler W: Release of endogenous catecholamines in the ischemic myocardium of the rat. Part A: Locally mediated release. *Circ Res* 1984;55:689–701.
44. Wilde AA, Peters RJG, Janse MJ: Catecholamine release and potassium accumulation in the isolated globally ischemic rabbit heart. *J Mol Cell Cardiol* 1988;20:887–896.
45. Ellingsen O, Vengen OA, Kjeldsen SE, Eide I, Ilebekk A: Myocardial potassium uptake and catecholamine release during cardiac sympathetic nerve stimulation. *Cardiovasc Res* 1987;21:892–901.
46. Steenbergen C, Murphy E, Levy L, London RE: Elevation in cytosolic free calcium concentration early in myocardial ischemia in perfused rat heart. *Circ Res* 1987;60:700–707.
47. Marban E, Kitakaze M, Kusuoka H, Porterfield JK, Yue DT, Chacko VP: Intracellular free calcium concentration measured with 19F NMR spectroscopy in intact ferret hearts. *Proc Natl Acad Sci USA* 1987;84:6005–6009.
48. Murphy E, Steenbergen C, Levy L, Raju B, London R: Cytosolic free magnesium levels during myocardial ischemia. *J Mol Cell Cardiol* 1988;20(suppl III):52(abstract).
49. Borchgrevink PC, Bergan AS, Bakoy OE, Jynge P: Magnesium and reperfusion of ischemic rat heart as assessed by 31P-NMR. *Am J Physiol* 1989;256:H195–H204.
50. Kodama I, Wilde A, Janse MJ, Durrer D, Yamada K: Combined effects of hypoxia, hyperkalemia and acidosis on membrane action potential and excitability of guinea-pig ventricular muscle. *J Mol Cell Cardiol* 1984;16:247–259.
51. Brooks CM, Gilbert JL, Greenspan ME, Lange G, Mazzella HM: Excitability and electrical response of ischemic heart muscle. *Am J Physiol* 1960;198:1143–1147.
52. Tsuchida T: Experimental studies on the excitability of ventricular musculature in infarcted region. *Jpn Heart J* 1965;65:152–164.
53. Elharrar V, Zipes DP: Cardiac electrophysiologic alterations during myocardial ischemia. *Am J Physiol* 1977;233:H329–H345.
54. Han J, Moe GK: Nonuniform recovery of excitability of ventricular muscle. *Circ Res* 1964;14:44–60.
55. Avitall BS, Naimi AH, Brilla AH, Levine HJ: A computerized system for measuring dispersion of repolarization in the intact heart. *J Appl Physiol* 1974;37:456–458.
56. Kuo CS, Reddy CP, Munakata K, Surawicz B: Arrhythmias dependent predominantly on dispersion of repolarization, in Zipes DP, Jalife J (eds): *Cardiac Electrophysiology and Arrhythmias.* Orlando, FL, Grune & Stratton, 1985, pp 277–286.
57. Lesh MD, Pring M, Spear JF: Cellular uncoupling can unmask dispersion of action potential duration in ventricular myocardium: A computer modeling study. *Circ Res* 1989;65:1426–1440.
58. Cranefield PF: The conduction of the cardiac impulse. The slow response and cardiac arrhythmias. Mount Kisco, NY, Futura, 1975.
59. Gettes LS, Surawicz B, Shiue JD: Effect of high K^+, low K^+ and quinidine on QRS duration and ventricular action potential. *Am J Physiol* 1963;203:1135–1140.
60. Wit AL, Rosen MR: Afterdepolarizations and triggered activity, in Fozzard HA, Haber E, Jennings RB, Katz AM, Morgan HE (eds): *The Heart and Cardiovascular System. Scientific Foundations.* New York, Raven Press, 1986, pp 1449–1490.
61. Kass RS, Tsien RW, Weingart R: Ionic basis of transient inward current induced by stropthanidin in cardiac Purkinje fibres. *J Physiol (Lond)* 1978;281:209–226.
62. January CT, Riddle JM, Salata JJ: A model for early afterdepolarizations: induction with the Ca^{2+} channel agonist Bay K 8644. *Circ Res* 1988;62:563–571.
63. Kass RS, Tsien RW: Fluctuations in membrane current driven by intracellular calcium in cardiac Purkinje fibers. *Biophys J* 1982;38:259–262.
64. DiFrancesco D, Ferroni A, Mazzanti M, Tromba C: Properties of the hyperpolarizing-activated current (i_f) in cells isolated from the rabbit sino-atrial node. *J Physiol (Lond)* 1986;377:61–88.
65. Kaufmann R, Theophile U: Automatie-fordernde Dehnungseffekte an Purkinje-Faden, Papillarmuskeln und Vorhoftrabekeln von Rhesus-Affen. *Pflügers Arch ges Physiol* 1967;297:174–189.
66. Kleber AG, Janse MJ, van Capelle FJL, Durrer D: Mechanism and time course of S-T and T-Q segment changes during acute regional myocardial ischemia in the pig heart determined by extracellular and intracellular recordings. *Circ Res* 1978;42:603–613.

67. Janse MJ, van Capelle FJL, Morsink H, et al.: Flow of "injury" current and patterns of excitation during early ventricular arrhythmias in acute regional myocardial ischemia in isolated porcine and canine hearts. Evidence for two different arrhythmogenic mechanisms. *Circ Res* 1980;47:151–165.
68. Trautwein W, Kassebaum DG: On the mechanism of spontaneous impulse generation in the pacemaker of the heart. *J Gen Physiol* 1961;45:317–330.
69. Coraboeuf E, Deroubaix E, Hoerter J: Control of ionic permeabilities in normal and ischemic heart. *Circ Res* 1976;38:I92–I98.
70. Pogwizd SM, Corr PB: Reentrant and nonreentrant mechanisms contribute to arrhythmogenesis during early myocardial ischemia: results using three-dimensional mapping. *Circ Res* 1987;61:352–371.
71. Rozanski GJ, Jalife J, Moe GK: Reflected reentry in nonhomogeneous ventricular muscle as a mechanism of cardiac arrhythmias. *Circulation* 1984;69:163–173.
72. Harris AS, Rojas AG: The initiation of ventricular fibrillation due to coronary occlusion. *Exp Med Surg* 1943;1:105–122.
73. Ideker RE, Klein GJ, Harrison L, et al.: The transition to ventricular fibrillation induced by reperfusion after acute ischemia in the dog: A period of organized epicardial activation. *Circulation* 1981; 63:1371–1379.
74. Kaplinsky E, Ogawa S, Kmetzo J, Balke CW, Dreifus LS: Intramyocardial activation in early ventricular arrhythmias following coronary artery ligation. *J Electrocardiol* 1980;13:1–6.
75. Kaplinsky E, Ogawa S, Balke W, Dreifus LS: Role of endocardial activation in malignant ventricular arrhythmias associated with acute ischemia. *J Electrocardiol* 1979;12:299–306.
76. Harris AS: Delayed development of ventricular ectopic rhythms following experimental coronary occlusion. *Circulation* 1950;1:1318–1325.
77. Kaplinsky E, Ogawa S, Balke CW, Dreifus LS: Two periods of early ventricular arrhythmia in the canine acute myocardial infarction model. *Circulation* 1979;60:397–403.
78. Garry WE: The nature of fibrillary contraction of the heart—its relation to tissue mass and form. *Am J Physiol* 1914;33:397–414.
79. Burgess MJ, Williams D, Ershler P: Influence of test site on ventricular fibrillation threshold. *Am Heart J* 1977;94:55–61.
80. Kuo CS, Munakata K, Reddy CP, Surawicz B: Characteristics and possible mechanism of ventricular arrhythmia dependent on the dispersion of action potential durations. *Circulation* 1983;67:1356–1367.
81. Moe GK, Rheinboldt WC, Abildskov JA: A computer model of atrial fibrillation. *Am Heart J* 1964;67:200–220.
82. Allessie MA, Bonke FIM, Schopman FJG: Circus movement in rabbit atrial muscle as a mechanism of tachycardia. III. The "leading circle" concept: a new model of circus movement in cardiac tissue without the involvement of an anatomical obstacle. *Circ Res* 1977;41:9–18.
83. Janse MJ, van Capelle FJL: Electrotonic interactions across an inexcitable region as a cause of ectopic activity in acute regional myocardial ischemia. A study in intact porcine and canine hearts and computer models. *Circ Res* 1982;50:527–537.
84. van Capelle FJL, Durrer D: Computer simulation of arrhythmias in a network of coupled excitable elements. *Circ Res* 1980;47:454–466.

Chapter **3**

The Importance of Cell Coupling and Anisotropy to Conduction in Cardiac Tissue

Michael D. Lesh, MD, Joseph F. Spear, PhD, and E. Neil Moore, DVM, PhD

A variety of clinical arrhythmias, from atrial flutter to ventricular tachycardia, are caused by reentry. It has long been known that slow conduction is a requirement for the occurrence of reentry. In addition, reentry must be initiated by a beat, usually premature, which encounters unidirectional block. There has been substantial effort in cardiac electrophysiology directed toward understanding how slow conduction and unidirectional block occur in normal and diseased cardiac tissue. Traditionally, much of this effort was aimed toward understanding how modification of the active and passive membrane ionic properties, and specifically alterations in the inward sodium current (i_{Na}), led to slow conduction. Increasingly, attention is being focused on the role of cell-to-cell coupling in the transmission of electrical impulses from one cell to the next and on the importance of abnormal cell coupling to arrhythmogenesis. In this chapter, we will discuss the significance of the directional differences in cell-to-cell coupling, known as anisotropy, that occur in cardiac tissue. In addition, we will explore how abnormalities in cell-to-cell coupling that result in nonuniform anisotropy might be involved in cardiac arrhythmias. We will describe work with a computer model of nonuniformly anisotropic myocardium which is being used to understand the mechanism of reentry in myocardium with disrupted cell coupling. Finally, we will consider how pharmacologic modification of anisotropy could be important as a mechanism of antiarrhythmic (or proarrhythmic) drug action.

ANISOTROPIC CONDUCTION

Definition of Anisotropy

Any property of a physical system that differs depending on the direction in which it is measured is termed "anisotropy." In the heart, anisotropy generally refers to the differences in conduction velocity that are related to the direction of propagation. Heart muscle consists of overlapping layers of elongated fibers. In any given layer, the velocity of a conducted impulse is two to four times as fast when conduction proceeds along the myocardial fiber axis than when it is forced to travel transversely to the long axis of the fibers. Thus, in normal myocardium, the geometric arrangement of cell bundles leads to anisotropic conduction.

A distinction is typically made between uniform and nonuniform anisotropy.[1] Conduction in uniformly anisotropic myocardium proceeds smoothly in all directions, albeit faster in a longitudinal rather than a transverse direction to the fiber axis. In the case of nonuniform anisotropy, there is asynchronous or discontinuous activation of adjacent cells or groups of cells that gives rise to an irregular pattern of activation. Implicit in these definitions is the size scale on which measurements of activation time and conduction velocity are made. When using extracellular electrodes with intraelectrode spacing on the order of millimeters or centimeters, conduction in normal tissue appears uniformly anisotropic. However, at a very fine size scale with electrode spacing of several hundred microns,[2] even appar-

655 Avenue of the Americas, New York, NY 10010
Current Topics in Cardiology

ently uniform anisotropy reveals its microscopic nonuniform and discontinuous nature. Discontinuous conduction occurs because of the nonuniform spatial distribution of cell-to-cell connections. Microscale inhomogeneities lead to microscopic nonuniformities in conduction, even in normal tissue. Inhomogeneities of cellular structure on a larger size scale occur in such conditions as myocardial infarction.[3–5] Since reentrant tachycardia is rare in normal tissue, an important and unanswered question is this: On what size scale do inhomogeneities of cellular coupling distort the electrophysiology sufficiently for arrhythmogenesis?

Although anisotropy of conduction velocity is generally attributed to directional differences in intracellular resistivity, there is also evidence of directional differences in electrical capacitance[6] and evidence that the anisotropic resistivity of the extracellular space may be oriented in a way that is orthogonal to that of the intracellular compartment.[7] Other factors such as local inhomogeneities of extracellular potassium ion concentration ($[K^+]_o$) also can occur. How multiple electrophysiologic variables that occur in a spatially heterogeneous fashion combine to produce arrhythmia is incompletely understood.

Local Circuit Theory

The electrical properties of cardiac cells are far from uniform. The cytoplasm has very low resistance to current flow. The cell membrane, if considered in isolation from ionic and gap junction channels, has an extremely high resistance to direct current flow because its phospholipid bilayer acts as an effective insulator. For cardiac impulse propagation to occur, current must flow from cell to cell and this occurs at specialized organelles known as gap junctions or nexuses, which have a much lower resistance to current flow than nonjunctional cell membrane.

To understand adequately the role of cell coupling in propagation of the cardiac impulse, it is important to consider the local circuit theory of conduction. This theory was originally developed to describe electrical propagation in nerve fibers,[8] but it is applicable to conduction in cardiac muscle as well. After an initial provoking stimulus such as externally applied current or the automatic depolarization of a pacemaker cell, the cardiac action potential (AP) conducts along a given cell because i_{Na} responsible for the AP upstroke generates sufficient intracellular current to depolarize resting membrane downstream. The depolarized membrane causes a further inrush of i_{Na}. To complete this local circuit, the current must cross the membrane and return to the original active site via the extracellular space. Eventually, this regenerative response encounters the end of a cell. For conduction to continue, current must flow from one cell to the next. Gap junctions provide a low resistance path for cell-to-cell current flow and thus provide the pathway by which the local circuit can be completed in cardiac tissue.

Gap Junctions

The structure and property of cardiac gap junctions has been reviewed by DeMello.[9] Gap junctions, which occur at intercalated disks, consist of numerous connexons, each of which contains six identical protein subunits surrounding a central hydrophilic hemichannel. The juxtaposition of two connexons between adjacent cells completes the channel pore. The connexons in the nexal membrane of two contiguous cells are separated by a very small (2 to 4 nm) intracellular gap which permits the diffusion of ions and small molecules. It is the diffusion of ions, particularly K^+, that provides for electrical current flow. In addition, the exchange of metabolites and possibly other chemical signaling agents between cells could occur at gap junctions.

Gap junctions in mammalian myocardium are generally considered to be ohmic. That is, they have a linear voltage current relationship and do not modulate their conductivity in response to transmembrane voltage. However, recent experiments using voltage-clamped pairs of isolated myocytes[10,11] have shown that overall junctional resistance results from the gating open and closed of individual channels with response times on the order of milliseconds. Net junctional conductance is the product of channel open times and single-channel conductivity. Regulation of gap junctional conductance may therefore occur via a change in the percentage of time individual channels are opened or via alteration in the conductance of an open channel. A number of substances and conditions regulate junctional conductance and therefore affect anisotropic conduction. Cytoplasmic acidosis such as occurs in ischemia results in an increase in coupling resistance. Perhaps the most important modulator of junctional conductance is intracellular calcium ($[Ca^{2+}]_i$).[12–14] Other alterations, such as drugs or ischemia, may act ultimately through $[Ca^{2+}]_i$ to modulate junctional resistance.

Other uncouplers include cardiac glycosides,[12] metabolic inhibitors such as 2,4-dinitrophenol, and strontium ion. Aliphatic alcohols, such as octanol and heptanol, inhibit cellular coupling probably via their incorporation into the phospholipid bilayer, resulting in a decrease in gap junctional pore size.[12,15] By selectively increasing junctional resistance, heptanol and octanol have proven useful tools to electrophysiologists evaluating the contribution of cell-to-cell coupling to conduction. For example, Joyner[16] showed that octanol preferentially slows propagation at the Purkinje muscle junction because of a restricted pathway for current flow at that junction.

Relationship Between Cellular Coupling and Anisotropic Conduction

The spatial distribution of gap junctions within layers of the myocardium and the organization of fiber layers to form the three-dimensional structure of the heart gives rise to anisotropic conduction. Ventricular and atrial myocardial cells are longer than they are wide. Cellular coupling via gap junctions occurs where the cells are joined either end-to-end or side-to-side. However, because the myoplasm has a lower resistivity than the gap junction, and because more gap junctional resistors are encountered per unit distance transverse to the myocardial fiber axis, the effect of intracellular resistivity is less for a given distance along myocardial fibers than for the same distance transverse to the fibers. If all other factors remain the same, the higher the resistance in a given direction, the lower the conduction velocity. Therefore, increased resistivity transverse to fiber orientation causes conduction to proceed slower than when an impulse propagates along fiber orientation.

Because of this relationship, anything that affects gap junctional resistivity will affect conduction transverse to fiber orientation to a greater degree than conduction longitudinal to fiber axis. Balke[17] and coworkers evaluated the effect of heptanol on anisotropic conduction in isolated perfused strips of normal canine epicardium. In this preparation, longitudinal conduction velocity is approximately 0.55 meters/s whereas transverse conduction velocity is normally 0.23 meters/s. By treating these preparations with relatively low doses of heptanol (1 to 1.5 mmol), conduction in the longitudinal direction fell by 20% but conduction in the transverse direction fell by almost 50%. This indicates that while gap junctional resistivity is a determinate of conduction in both directions, gap junction resistivity is a more important contributor to conduction in the transverse direction.

If increased $[Ca^{2+}]_i$ causes uncoupling, then interventions that decrease$[Ca^{2+}]_i$ might be expected to result in a normalization of anisotropic conduction. Kadish[18] showed that procainamide results in a decreased ratio of longitudinal-to-transverse conduction velocity, that is, a relative normalization of anisotropic conduction. A putative mechanism involves $[Ca^{2+}]_i$. By blocking i_{Na}, procainamide results in a decrease in $[Ca^{2+}]_i$, thus leading to a decrease in gap junction resistivity. Alterations in anisotropy may be an important and relatively unexplored mechanism of antiarrhythmic drug action.

INFLUENCE OF ANISOTROPY ON REFRACTORINESS AND UNIDIRECTIONAL BLOCK

Although slow conduction is a requirement for the perpetuation of sustained reentry, it alone is not sufficient for the initiation of that reentry. In addition to slow conduction, unidirectional block and recovery of excitability distal to the site of block must occur. Cellular coupling and anisotropy have been proposed as important factors in conduction block. Spach[19] showed that the anisotropic properties of cardiac muscle may cause block of a premature impulse because of a reduced safety factor for conduction longitudinal to fiber orientation, despite faster conduction in that direction. That is, conduction that is transverse to fiber orientation, though slow, was found to be more resistant to disturbances in membrane properties than faster longitudinal conduction. Such longitudinal block was found to occur despite no increase in the measured refractory periods at the site of block. Recently, Dillon[20] proposed that a decreased safety factor longitudinal to fiber axis was involved in the initiation of sustained ventricular tachycardia in a canine infarct model. In their study of the characteristics of conduction in the epicardial border zone overlying a transmural infarction, the combination of conduction block of a premature stimulus in the longitudinal direction and very slow conduction that is transverse to the fiber axis was implicated in establishing sustained reentry.

In addition to reduced safety factor secondary to anisotropy as a mechanism of block, the anisotropic tissue structure and cellular coupling can electrotonically modulate action potential duration (APD) and thus dynamically affect local dispersion of refractoriness. Myocardial refractoriness and APD are dependent on the pattern and timing of activation. In the sucrose gap experiments of Jalife and Moe,[21] when there was high coupling resistance in the sucrose gap, propagation between the proximal and distal segments failed and electrotonic influences shortened the AP of the proximal segment. As coupling resistance was lowered to a critical value, propagation occurred from the proximal to the distal region. With propagation established, the AP of the proximal segment was markedly prolonged as a result of electrotonic influences from the distal segment.

Veenstra and DeHaan[11] provided an experimental example of electrotonic coupling of repolarization in a preparation that consisted of cultured chick embryo ventricular cells of different ages. When studied in isolation, aggregates of 7-day-old embryo myocytes have longer intrinsic APDs than those of 4-day-old embryos. When the aggregates were brought into contact in a perfusion bath, new gap junctions were formed over a period of several hours, and resulted in improved coupling. The recorded APDs of the two aggregates became nearly identical at a value intermediate between that of the 4-day-old and that of the 7-day-old embryos.

Electrotonic influence of adjacent tissue on a given region is also responsible for the observation that APD is longer at the site of origin of a conducted wavefront than at some distance from the site of origin.[22] Furthermore, at a site of collision of two approaching wavefronts, APD is reduced. Burgess et al.[23] examined the relationship between anisotropy and nonuniform activation and repolarization in the canine pulmonary conus. By varying the pacing site and therefore the direction of activation with respect to fiber orientation in the conus, they varied the uniformity of activation at a given test site. They found that the degree of nonuniformity of activation correlated positively with directionally dependent differences in refractory period at that test site. These experiments may provide some explanation for pacing site-dependent induction of sustained tachycardia that is observed in the clinical electrophysiology laboratory. Doherty[24] found that of 38 patients with clinical ventricular arrhythmias that could not be induced using programmed stimulation from the right ventricular apex, 53% had ventricular arrhythmias induced when programmed stimulation was performed from the right ventricular outflow tract. It is intriguing to speculate that such pacing site-dependent induction of arrhythmias may result from the interaction between wavefront direction and nonuniform anisotropy in the substrate for reentry.

Lesh et al.[25] examined the ability of cellular coupling and anisotropy to mask or unmask intrinsic differences in APD. They used a computer model of conduction as a two-dimensional grid of cells. The model employed the Beeler–Reuter[26] equations for transmembrane ionic current flow. In addition, each cell was coupled to its neighbor with an independent resistor representing gap junction conductance. Using this model of ventricular myocardium, Lesh found that anisotropy can modulate the way APD and refractory periods are manifested. In regions where tight coupling occurred, the APDs of neighboring cells were uniform. However, when modeled cells were progressively uncoupled, intrinsic differences in APD between adjacent cells became manifest, because local current flow between regions that tended to repolarize early and those that remained depolarized longer was impeded. This study also suggested that a further reduction in the degree of cell coupling could exacerbate tissue inhomogeneity to the point at which conduction block would occur in areas of prolonged APD.

One can imagine a statistical phenomenon in which there is normally dispersion of cellular ionic properties, but these differences are masked by effective cell-to-cell coupling. Pathological uncoupling, such as that which might result from an infarction, would tend to allow for dispersion of recovery to become manifest and would make unidirectional block of a premature beat statistically more likely to occur.

The importance of electrotonic modulation of APD in cardiac arrhythmogenesis in vivo has not been completely studied because of the technical difficulties of simultaneously recording from a large number of cells. However, since unidirectional block between adjacent regions of tissue is a requirement for the establishment of reentry, attenuated cell coupling may facilitate such block.

ANISOTROPY AND CARDIAC ARRHYTHMIAS

While the basic electrophysiologic properties of anisotropic slow conduction have been examined in a number of preparations, less is known about the specific role of anisotropy in clinical or experimental arrhythmias. However, in a number of instances there is considerable evidence that anisotropic conduction is important. While in the setting of acute myocardial ischemia, abnormal conduction in the ischemic zone appears due to depolarization of the membrane and a decrease in the magnitude of the rapid i_{Na}, electrical uncoupling of ischemic cells also occurs. Conduction in the ischemic zone is nonuniform and discontinuous, which produces the characteristic fractionated electrograms recorded from this area.

In addition to abnormal membrane ionic properties, anisotropy appears to contribute to ventricular arrhythmias in the acute ischemic setting in two ways. Initially, poor coupling between the depolarized ischemic region and the adjacent nonischemic region allows for the maintenance of spatial differences in resting membrane potential (RMP). This difference produces electrotonic current flow which may bring some cells to threshold, resulting in focal excitation. Further uncoupling of the ischemic cells may result in dissociated slow conduction, which would then support sustained reentrant tachycardia.

Once myocardial ischemia leads to subacute and chronic infarction, cardiac muscle is replaced by an ingrowth of collagenous septa. In the setting of chronic infarction weeks or months following the acute ischemic episode, the membrane ionic properties of the surviving cells return to normal and have a normal RMP and AP upstroke.[27,28] Despite the fact that surviving cells are normal, conduction through the chronic infarct region is highly nonuniform, with regions of very slow conduction. Spear,[29] using a canine model in which left anterior descending coronary artery (LAD) occlusion produced a ventricular infarction, found that space constant, a measure of cell-to-cell coupling resistance, was markedly reduced in the infarct region. Therefore, slow conduction in the setting of chronic infarction resulted from poor cell-to-cell coupling rather than any abnormality in the membrane ionic properties of the cells themselves. Dillon[20] showed that slow conduction in the epicardial border zone of healed canine infarctions is adequate to support sustained reentrant tachycardia. In the epicardial border zone, slow conduction in a direction transverse to fiber orientation resulting from compromised cell coupling provided pivot lines around which the tachycardia circulated. More recently, Saltman[30] has shown that both fixed and functional lines of transverse uncoupling can occur to the epicardial border zone and both types of block can provide a circuit adequate for sustained reentry. Others have also shown the importance of nonuniform anisotropy on the induction and perpetuation of ventricular tachycardia. Cardinal[31,32] showed in an experimental canine preparation that propagation transverse to fiber orientation accounted for the slow conduction required for reentry and that the line of block around which the reentry loop circulated was oriented parallel to the fiber axis. When figure-of-eight reentry was observed in a canine infarct model by El-Sherif,[33] the narrow common return path was also generally oriented along the fiber axis.

The importance of the canine infarct model is that anatomical features of the epicardial border zone are similar to the subendocardial origin of ventricular tachycardia in patients with coronary artery disease.[5,34,35] Substantial evidence favors reentry as the mechanism for sustained unimorphic ventricular tachycardia that occurs in patients in the setting of healed myocardial infarction.[36] The anatomical substrate for human ventricular tachycardia is a microreentry circuit occupying approximately 1 to 4 cm^2 of subendocardium. In this region, a markedly heterogeneous admixture of scar interdigitating with relatively normal cells is seen.[5,37] For a reentry circuit to occupy such a relatively circumscribed volume of myocardium, it is likely that the slow conduction observed in nonuniformly anisotropic tissue plays a crucial role in establishing the electrophysiologic requirements for such reentry. More work is currently under way on the role of anisotropic conduction in human ventricular tachycardia. The availability of mapping systems using multiple simultaneous electrodes should make investigation possible in patients undergoing surgical resection to cure their ventricular tachycardia.

COMPUTER MODEL OF ANISOTROPIC VENTRICULAR MYOCARDIUM

Electrical propagation in ventricular muscle results from current flow between cells on a microscopic size scale. Disruption of this microscale spread of current in cardiac tissue might be expected to cause cardiac arrhythmias. However, we are limited in a detailed understanding of the mechanism of such arrhythmias because of the difficulty in measuring microscale voltage and current using physical electrodes in biologic preparations. This limitation has motivated the development of computer models of electrical propagation. We have developed such a model to test the hypothesis that abnormal cell-to-cell electrical coupling plays a primary role in the initiation and perpetuation of reentrant tachycardia in the setting of chronic, healed myocardial infarctions. We have implemented a finite element, electrical equivalent circuit model of ventricular muscle (Figure 3.1). We simulate myocardium as a network of resistively coupled current sources, each representing a small portion of cell membrane. The nonlinear time and voltage dependence of transmembrane current that produces each AP is represented by a system of differential equations.[26] The cells are connected with gap junctions which are represented by independent resistors in the model. In this way, we can simulate the discrete nature of cell-to-cell electrical interaction and account for both uniform and nonuniform anisotropic coupling. By separately implementing the membrane ionic properties

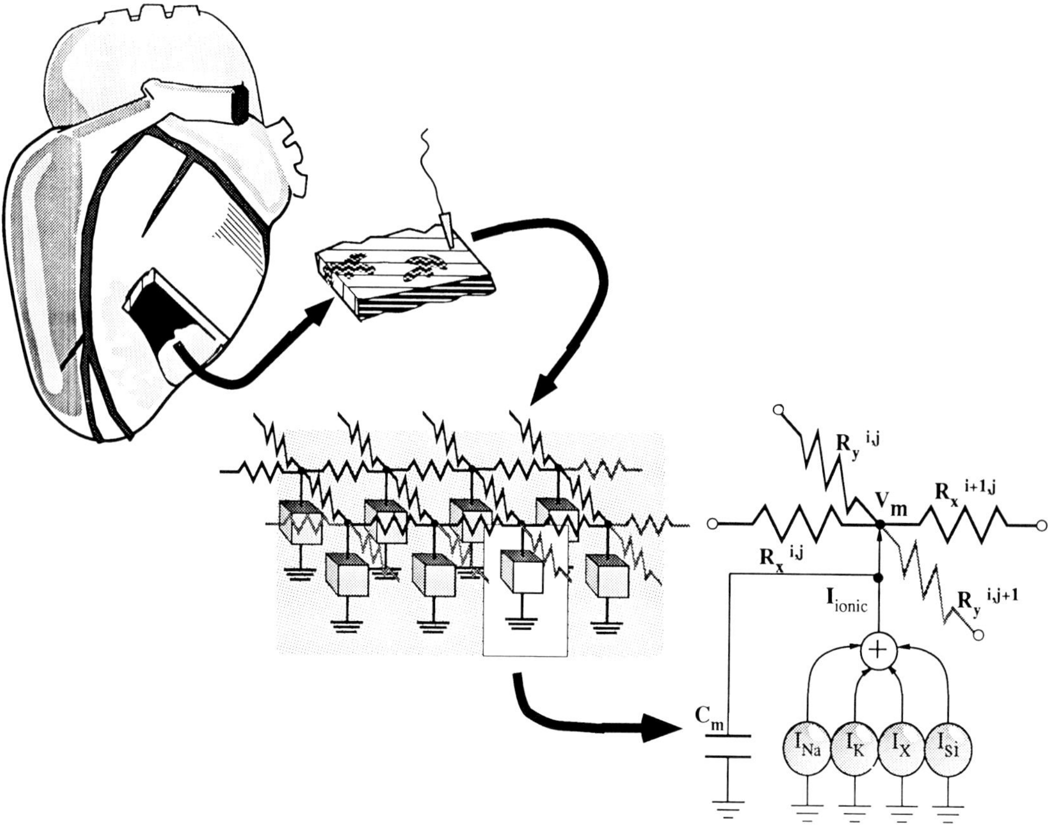

FIGURE 3.1. Electrical equivalent circuit model of a nonuniformly anisotropic sheet of ventricular myocardium. A two-dimensional sheet representing a thin layer of subendocardium is simulated. Membrane patches obeying Beeler–Reuter kinetics are coupled with axial resistors. A representative element is shown at the right. R_x and R_y = longitudinal and transverse resistivity, V_m = transmembrane voltage, C_m = membrane capacitance, I_{ionic} = total transmembrane ionic current, I_{Na} = sodium current, I_K = inward rectified current, I_x = delayed rectifier current, I_{si} = slow inward current.

and the passive resistivity of gap junctions, alterations in either individually or in both together can be studied with regard to their importance in the initiation of reentrant tachycardia. Pacing is simulated by injection of current pulses at appropriate times and locations during the simulation. All experiments were carried out on a Cray Y-MP supercomputer.

To simplify both the implementation of the model and the interpretation of the results, a two-dimensional sheet of muscle is illustrated here. Figure 3.2 shows the effect of changing the ratio of longitudinal to transverse resistivity on the resulting pattern of conduction. In each experiment, pacing is performed from the midpoint of one edge of a two-dimensional sheet of ventricular muscle. In experiment A, uniform isotropy is simulated. That is, the resistivity values in the longitudinal and transverse directions are equal. In this case, semicircular isochrones result, indicating equal conduction velocity in both the longitudinal and transverse directions. In experiments B through D, uniform anisotropy is simulated. The ratio of longitudinal to transverse resistivity was 1 : 2, 1 : 3, and 1 : 4, respectively. We note that as the transverse resistivity is increased, the isochrones become more elongated. This indicates progressive slowing of conduction in the transverse direction as resistivity in this direction is increased. Thus, using the computer model, we have demonstrated that changes in cellular coupling alone can account for anisotropic transverse slow conduction.

We can also use the model to understand conditions required for sustained reentry. Figure 3.3 shows a typical experimental arrangement of nonuniform anisotropy. A 20 × 6 mm sheet of anisotropic ventricular myocardium employing 12,000 Beeler–Reuter elements had overall longitudinal resistivity of 2.5 Kohm·cm and trans-

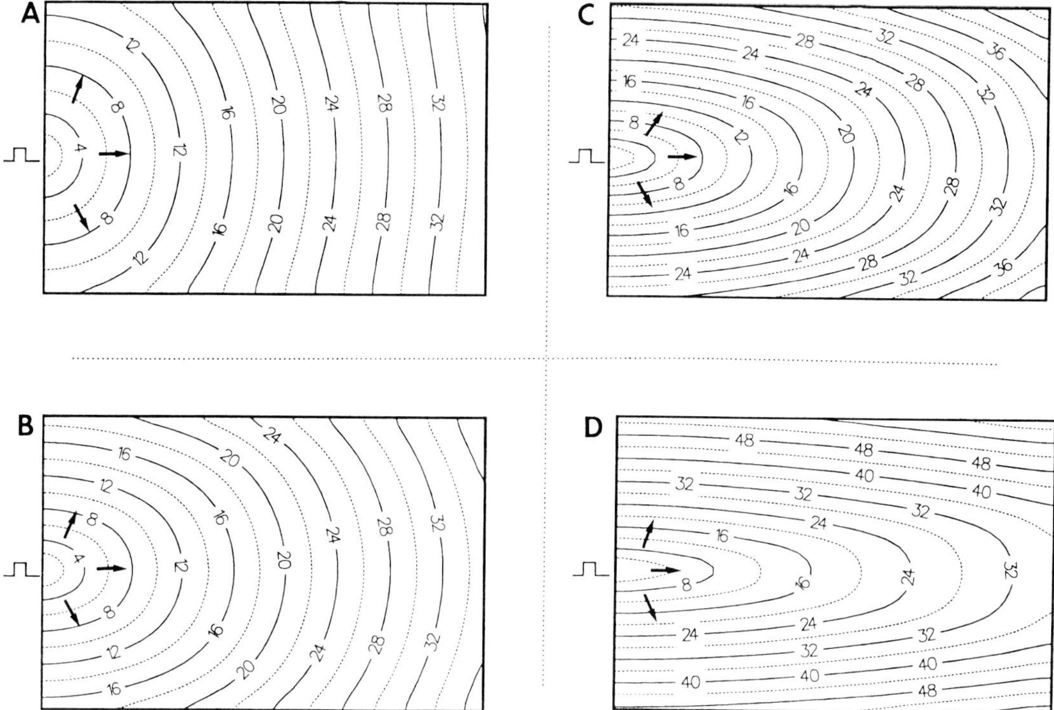

FIGURE 3.2. Four experiments in a simulated sheet of ventricular myocardium. **A.** $R_L = R_T = 2.5$ Kohm·cm. **B.** $R_L{:}R_T = 1{:}2$. **C.** $R_L{:}R_T = 1{:}3$. **D.** $R_L{:}R_T = 1{:}4$. Note that as transverse resistivity is increased, conduction progressively slows in the transverse direction. Membrane ionic properties were uniform throughout in all experiments.

verse resistivity of 10 Kohm·cm. In addition to the overall anisotropy, nonuniform uncoupling was specified as parallel lines of functional transverse uncoupling. As described by Dillon,[20] these were not lines of fixed block (that is, zones of infinite resistivity) but rather lines of moderately impaired transverse coupling. The lines of transverse uncoupling defined an isthmus in the center of the sheet. At one end of the isthmus, a region of relative dispersion of APD was specified by allowing cells in this region to have an increased magnitude of g_{si}, the magnitude of the fully activated slow inward, or calcium (Ca^{2+}), current in the Beeler–Reuter membrane model. Pacing was performed along one edge with a basic drive of 1,000 ms and a single premature beat introduced at a coupling interval of 310 ms. The results are shown in Figure 3.4. The basic drive beat (S_1) proceeds smoothly across the sheet. However, the premature beat (S_2) encounters the region of longer APs and blocks. Conduction proceeds outside the lines of transverse uncoupling (functional block). At the distal end of the isthmus, conduction proceeds beyond the ends of the lines of functional block and then continues in a retrograde fashion through the isthmus. Because the lines of function block are of sufficient length, the region at the mouth of the isthmus at which S_2 blocked has now recovered excitability and reentrant tachycardia is initiated

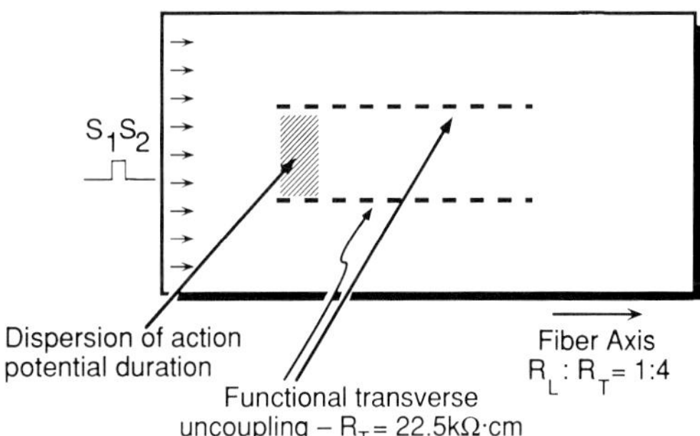

FIGURE 3.3. Arrangement of computer model for determining conditions required for sustained reentry. A 20 × 6 mm sheet of nonuniformly anisotropic ventricular myocardium is simulated. Pacing (S_1S_2) is performed from one edge of the sheet. Lines of functional transverse uncoupling define a conducting isthmus. A region of local increase in APD is specified (hatched) by increasing by 20% g_{si} in the Beeler–Reuter model.

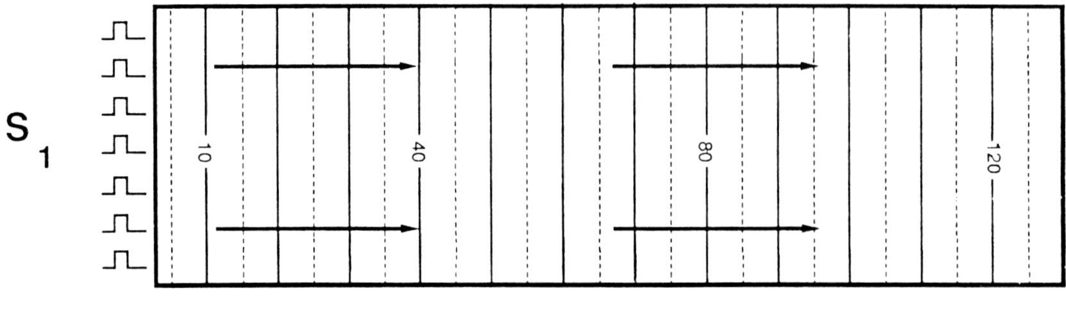

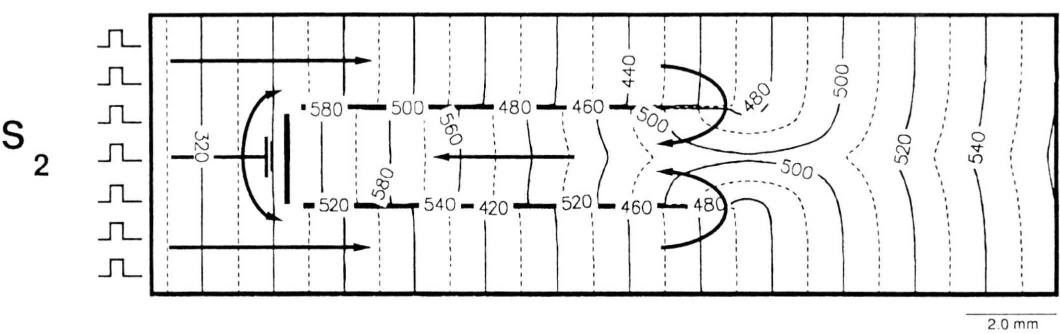

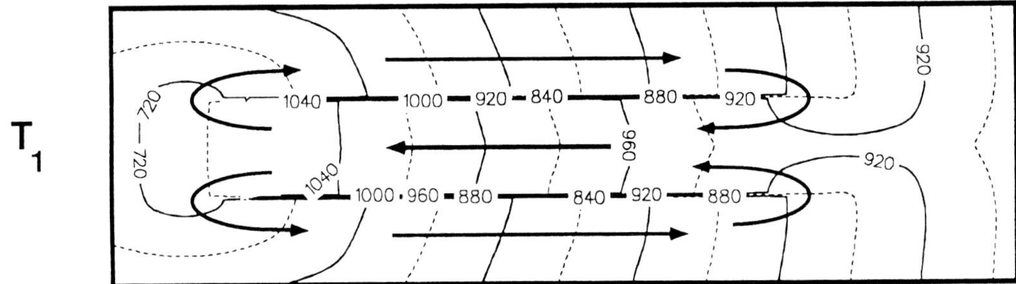

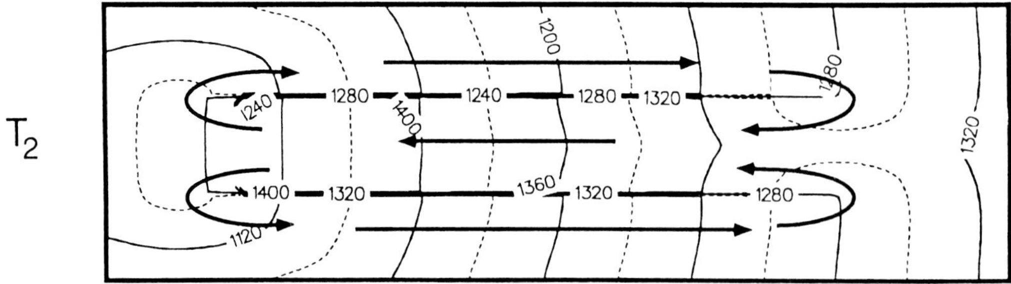

FIGURE 3.4. For the experimental arrangement in Figure 3.3, isochronal contour maps for the last beats of a basic drive (S_1) (CL = 1,000 ms), and a premature beat (S_2) at a coupling interval of 310 ms. The premature beat blocks at the mouth of the isthmus. The isthmus is protected by lines of functional block and tachycardia is induced (T_1 and T_2).

(T_1). The propagating impulse is able to recirculate on the outside of the isthmus in the antegrade direction and back through the center of the isthmus in the retrograde direction (T_2). Thus, in a simple model of nonuniform anisotropy, sustained reentrant tachycardia is initiated using programmed stimulation in a fashion quite analogous to that observed in biologic preparations. The conditions required for reentry were lines of functional transverse uncoupling and a region of abnormally long APD. In other experiments, we shortened the length of the resistive barriers. When the barriers are of insufficient length, there is inadequate time for recovery of excitability at the site of block at the mouth of the isthmus, unidirectional block becomes bidirectional block, and reentry cannot occur. Thus, computer modeling is a useful tool in understanding the mechanism of reentrant tachycardia in the setting of nonuniform anisotropic myocardium.

IMPORTANCE OF ANISOTROPY IN THERAPY OF ARRHYTHMIAS

We have discussed the concept that abnormalities in cell-to-cell electrical continuity, that is, nonuniform anisotropy, exist in a number of pathological states and the likelihood that anisotropy plays a prominent role in clinical arrhythmias, including ventricular tachycardia in the setting of healed infarction. If this is the case, then these results may have important therapeutic implications. Antiarrhythmic drugs have generally been classified by their ability to alter active membrane ionic properties such as the i_{Na} or repolarizing currents. Less is known about the direct and indirect effects of antiarrhythmic agents on cellular coupling and anisotropy, although such effects may be as important if not more important to their mechanism of the antiarrhythmic action. For example, we can speculate that if anisotropy plays a role in unidirectional block and the initiation of sustained reentry, any drug that preferentially depresses conduction in the longitudinal direction may be able to convert unidirectional block occurring longitudinally to bidirectional block, eliminating the substrate for sustained reentry.

Another potential approach to antiarrhythmic therapy based on our knowledge of anisotropy is to alter gap junctional conductance and completely eliminate regions of slow conduction resulting from tenuous coupling. One can speculate that conduction pathways that contain very high gap junctional resistance would be very sensitive to agents that further modify junctional resistance. A pharmacologic agent could then chemically "ablate" this slowly conducting region and eliminate these tenuously coupled areas from the conduction path. Spear et al[38] showed this to be possible using heptanol in infarcted canine epicardium. Pharmacologically enhancing junctional conductance in areas where it is depressed would represent, in theory, an alternative method.

In addition to the possible antiarrhythmic role of pharmacologic agents that affect cellular coupling, we should be aware of the possibility that drug-induced changes in coupling could be proarrhythmic. Since slow conduction is important for reentry, a drug that causes further conduction slowing as a result of cellular uncoupling could make reentry more, rather than less, likely.

In the future, we might use the knowledge that has been gained regarding the role of anisotropy in cardiac conduction disturbances and arrhythmias to design new antiarrhythmic drugs which may include agents whose primary mode of action is to alter cellular coupling. In this way, we could change the arrhythmogenic substrate sufficiently to prevent or abolish cardiac arrhythmias.

ACKNOWLEDGMENT

We wish to thank Carol Inaba for her careful preparation of this manuscript.

REFERENCES

1. Spach MS, Dolber PC, Heidlage JF: Influence of the passive anisotropic properties on directional differences in propagation following modification of the sodium conductance in human atrial muscle. A model of reentry based on anisotropic discontinuous propagation. *Circ Res* 1988;62:811–832
2. Spach MS, Dolber PC, Heidlage JF: Resolution of discontinuous versus continuous propagation: Microscopic mapping of the derivatives of extracellular potential waveforms, in Zipes D, Jalife J (eds): *Cardiac Electrophysiology From Cell to Bedside*. Philadelphia, WB Saunders, 1990, pp 139–148.
3. Ursell PC, Gardner PI, Albala A, Fenoglio JJ, Wit AL: Structural and electrophysiological changes in the epicardial border zone of chronic myocardial infarcts during infarct healing. *Circ Res* 1985;56:436–451.
4. Michelson EL, Spear JF, Moore EN: Electrophysiologic and anatomic correlates of sustained ventricular tachyarrhythmias in a model of chronic myocardial infarction. *Am J Cardiol* 1980;45:583–590.
5. Fenoglio JJ Jr, Pham TD, Harken AH, et al.: Recurrent sustained ventricular tachycardia: structure and ultrastructure of subendocardial regions where tachycardia originates. *Circulation* 1983;68:518–533.
6. Spach MS, Dolber PC, Heidlage JF, Kootsey JM, Johnson EA: Propagating depolarization in anisotropic human and canine muscle: Apparent differences in membrane capacitance. *Circ Res* 1987; 60:206–219.

7. Sepulveda NG, Walker CF, Heath RG: Finite element analysis of current pathways with implanted electrodes. *J Biomed Eng* 1983;5:41–48.
8. Hodgkins AL, Rushton WAH: The electrical constants of crustacean nerve fibre. *Proc Soc Lond Ser B* 1946;33:444–479.
9. DeMello WC: Cell-to-cell communication in heart and other tissues. *Prog Biophys Mol Biol* 1982; 39:147–182.
10. Weingart R, Maurer P: Action potential transfer in cell pairs isolated from adult rat and guinea pig ventricles. *Circ Res* 1988;63:72–80.
11. Veenstra RD, DeHaan RL: Cardiac gap junction channel activity in embryonic chick ventricle cells. *Am J Physiol 254 (Heart Circ Physiol)* 1988;23:H170–H180.
12. Deleze J, Herve JC: Effect of several uncouplers of cell-to-cell communication on gap junction morphology in mammalian heart. *J Membr Biol* 1983; 74:203–215.
13. Weingart R: The actions of ouabain on intercellular coupling and conduction velocity in mammalian ventricular muscle. *J Physiol (Lond)* 1977;264:341–365.
14. Noma A, Tsuboig N: Dependence of junctional conductance on proton, calcium and magnesium ions in paired cardiac cells of guinea pig. *J Physiol (Lond)* 1987;382:193–211.
15. Bernadini G, Peracchia C, Peracchia LL: Reversible effects of heptanol on gap junction structure and cell-to-cell electrical coupling. *Eur J Cell Biol* 1984;34:307–312.
16. Joyner RW, Overhold ED: Effects of octanol on canine subendocardial Purkinje-ventricular transmission. *Am J Physiol* 1985;249:H1228–H1231.
17. Balke CW, Lesh MD, Spear JF, Kadish A, Levin JH, Moore EN: Effects of cellular uncoupling on conduction in anisotropic canine ventricular myocardium. *Circ Res* 1988;63:879–892.
18. Kadish AH, Spear JF, Levine JH, Moore EN: The effects of procainamide on conduction in anisotropic canine ventricular myocardium. *Circulation* 1986;74:616–625.
19. Spach MS, Miller WT III, Dolber PC, et al.: The functional role of structural complexities in the propagation of depolarization in the atrium of the dog: Cardiac conduction disturbances due to discontinuities of effective axial resistivity. *Circ Res* 1982;50:175–191.
20. Dillon SM, Allessie MA, Ursell PC, Wit AL: Influences of anisotropic tissue structure on reentrant circuits in the epicardial border zone of subacute canine infarcts. *Circ Res* 1988;63:182–206.
21. Jalife J, Moe GK: Excitation, conduction, and reflections of impulses in isolated bovine and canine cardiac Purkinje fibers. *Circ Res* 1981;49:233–247.
22. Abildskov JA: Effects of activation sequence on the local recovery of ventricular excitability in the dog. *Circ Res* 1976;38:240–243.
23. Burgess JM, Steinhaus BM, Spitzer KW, Ershler PR: Non-uniform epicardial activation and repolarization properties of in vivo canine pulmonary conus. *Circ Res* 1988;62:233–246.
24. Doherty JU, Kienzle MG, Waxman HL, Buston AE, Marchlinski FE, Josephson ME: Programmed ventricular stimulation at a second right ventricular site: An analysis of 100 patients, with special reference to sensitivity, specificity and characteristics of patients with induced ventricular tachycardia. *Am J Cardiol* 1983;52:1184–1189.
25. Lesh MD, Pring M, Spear JF: Cellular uncoupling can unmask dispersion of action potential duration in ventricular myocardium. A computer modeling study. *Circ Res* 1989;65:1426–1440.
26. Beeler GW, Reuter H: Reconstruction of the action potential of ventricular myocardial fibers. *J Physiol (Lond)* 1977;268:177–210.
27. Gardner PI, Ursell PC, Fenoglio JJ, Wit AL: Electrophysiologic and anatomic basis for fractionated electrograms recorded from healed myocardial infarcts. *Circulation* 1985;72:596–611.
28. Wit AL, Dillon S, Ursell PC: Influence of anisotropic tissue structure on reentrant ventricular tachycardia, in Brugada P, Wellen HJJ (eds): *Cardiac Arrhythmias: Where To Go From Here?* New York, Futura Publishing, 1987, pp 27–50.
29. Spear JF, Michelson EL, Moore EN: Reduced space constant in slowly conducting regions of chronically infarcted canine myocardium. *Circ Res* 1983;176–185.
30. Saltman AE, Coromilas J, Ursell PC, Wit AL: Both functional and anatomical reentry cause ventricular tachycardia in health myocardial infarcts. *Circulation* 1990;82:III-453(abstract).
31. Cardinal R, Savard P, Carson DL, et al.: Mapping of ventricular tachycardia induced by programmed stimulation in canine preparation of myocardial infarction. *Circulation* 1984;70:136–148.
32. Cardinal R, Vermeulen M, Shenasa M, et al.: Anisotropic conduction and functional dissociation of ischemic tissue during reentrant ventricular tachycardia in canine myocardial infarction. *Circulation* 1988;77:1162.
33. El-Sherif N, Smith A, Evans K: Canine ventricular arrhythmias in the late myocardial infarction period: Epicardial mapping of reentrant circuits. *Circ Res* 1981;49:255–265.
34. Josephson ME, Horowitz LN, Farshidi A, et al.: Recurrent sustained ventricular tachycardia. 1. Mechanisms. *Circulation* 1978;57:431–440.
35. Josephson ME, Horowitz LN, Fashidi A, et al.: Recurrent sustained ventricular tachycardia. 2. Endocardial mapping. *Circulation* 1978;57:440–447.
36. Josephson ME, Buxton AE, Marchlinski FE, et al.: Sustained ventricular tachycardia in coronary artery disease—evidence for reentrant mechanism, in Zipes DP, Jalife J (eds): *Cardiac Electrophysiology and Arrhythmias.* Orlando, FL, Grune and Stratton 1985, pp 409–418.
37. Richards DA, Blake GJ, Spear JF, et al.: Electrophysiologic substrate for ventricular tachycardia: correlation of properties in vivo and in vitro. *Circulation* 1984;69:369–381.
38. Spear JF, Balke W, Lesh MD, Kadish AH, Levine JL, Moore EN: Effect of cellular uncoupling by heptanol on conduction in infarcted myocardium. *Circ Res* 1990;66:202–217.

Chapter **4**

Initiation and Modulation of Pacemaker Activity in Cardiac Cells

Justus Anumonwo, PhD, and José Jalife, MD

INTRODUCTION

The rhythmically contracting heart has aroused curiosity since the times of Greek philosophers (around 600 to 200 B.C.; see ref. 1 for an extensive review). However, despite attempts made at understanding the underlying mechanisms, the process remains essentially mysterious. The human heart beats on the average once every second, and under normal conditions each beat is initiated spontaneously in the so-called pacemaker region of the sinoatrial node (SAN), as a single impulse that propagates to the rest of the heart to initiate contraction.

The SAN was recognized as far back as the early 1900s as the normal site of impulse initiation,[2] but technical difficulties limited any adequate electrophysiological study of the initiation process until relatively recently. Thus, for a long time, the mechanisms responsible for initiating each impulse remained unclear. Spontaneous diastolic depolarization, the hallmark of pacemaker activity, was first recorded at about the beginning of the 1950s.[3,4] However the recordings were done using the simple microelectrode technique, a procedure that is inadequate for investigating ionic mechanisms. The first insight into these mechanisms came after the development of the voltage-clamp technique,[5] which provided the tool to begin the dissection of the transmembrane ionic currents in pacemaking cells. The recently developed single-cell[6] and single-channel[7] recording techniques have proved invaluable to physiologists in obtaining the first clear insight into cardiac pacemaker mechanisms.

Another equally baffling issue on pacemaker activity is the process by which the several thousand spontaneously active cells of the SAN coordinate or *synchronize* their electrical activity to generate the single impulse that is propagated during each cardiac cycle. Although mechanical and electrical schemes have been proposed to explain it, evidence from several studies tends to support the idea that electrical coupling underlies synchronization between the pacemaker cells of the SAN. This chapter will discuss the initiation, nervous control, and dynamic modulation of pacemaker activity in cardiac cells, with a focus on the primary pacemaker, the SAN.

INITIATION OF PACEMAKER ACTIVITY IN THE HEART

Electrical impulses underlying normal cardiac contractions are automatic, that is, the initiating impulse is generated spontaneously by a group of cells within the SAN. However, impulses can also be generated spontaneously from other areas of the heart, such as in extranodal areas of the atrium, in the atrioventricular node (AVN), or in the Purkinje fibers. Furthermore, rhythmic cardiac excitation can also result from impulses that are repetitively induced by other mechanisms.[8] Thus, there are *normal* and *abnormal* mechanisms of impulse generation in the heart. The various ways by which impulses are generated by these mechanisms are discussed in the following section.

Excitation by Normal Impulse Generation

The normal heart rhythm is initiated by automatic impulses originating from the SAN. Classically, it is thought that electrical impulses gen-

655 Avenue of the Americas, New York, NY 10010

erated from other (subsidiary) pacemakers are overdrive suppressed under normal conditions by the dominant SAN activity. However, since automaticity is a natural attribute of such subsidiary cells, pacemaker activity in them can therefore be considered "normal."

Automatic Activity

The sinus node was first identified by Keith and Flack[2] as a small oval- or rectangular-shaped piece of tissue in the right atrium. It is located in an area bounded by the superior and inferior vena cavae on two sides and by the crista terminalis and the interatrial septum on the other sides. The prevailing theory that normal pacemaker activity in the heart originates from one focus, which is in a small group of cells within the SAN, was developed on the basis of very early studies[1] which reported that the SAN is the site of the first electrical negativity during the cardiac cycle. More extensive research has since confirmed those original findings. For example, intracellular microelectrode studies on the SAN[9] reported that SAN cells have the following properties: (a) a relatively low maximum diastolic potential in the range of -50 to -60 millivolts, (b) an ability to produce rhythmic action potentials (APs) as a result of slow depolarization of membrane potential during diastole, (c) action potential duration (APD) of 100 to 300 ms (depending on the species), and (d) spontaneous cycle lengths that vary over a broad range of values.

In another extensive study, using successive microelectrode impalements, Bleeker et al.[10] systematically mapped the SAN and reported that in the central nodal region, there are areas of dominant and of latent pacemakers. These workers used earliest activation of upstroke, rate of diastolic depolarization, slow-phase zero depolarization, and a smooth transition from diastolic to systolic depolarization as criteria for making a distinction between dominant and latent pacemakers. Bleeker and his colleagues then concluded that a group of about 5,000 cells constitute the dominant pacemaker region. The cells occupy a region of about 0.3 mm^2 within the central nodal area, and are of one type, the so-called typical nodal cells (or P-cells).[11] However, the issue of dominant and latent pacemakers should be addressed with some caution for there is evidence[12] that in the dog atrium, areas spanning a region three to four times the size of the SAN, which are located around the superior vena cava and the right atrial junction, are all capable of impulse initiation. These areas, in a hierarchical order, are each capable of controlling excitation of the heart in a given range of heart rates.

With the relatively recent development of the single-cell isolation procedure, the electrophysiological data previously obtained from multicellular preparations, such as the spontaneous APs and the values of maximum diastolic potential and of spontaneous cycle length, have essentially been confirmed in isolated SAN cells from a number of animal species.[13–15,39] However, these data remain controversial given the heterogeneity in morphology and electrophysiology of the SAN,[10,12] as well as the limitations of the single-cell isolation procedure. Moreover, with most cell isolation techniques, it is sometimes difficult to tell the exact origin of an isolated cell.

In the AVN there are cells that are spontaneously active. Diastolic depolarization was recorded in AVN whole-tissue preparations by Hoffman and Cranfield.[16] Single-cell electrophysiological data have subsequently demonstrated spontaneous electrical activity in some AVN cells. Although the AVN pacemaker cells are usually suggested to be located in the N and NH regions,[17] there are conflicting data on this issue (for a review, see ref. 18. In one study,[19] pacemaker activity, with the usually attendant overdrive suppression, was observed in all the regions of the AVN, and it was suggested that the overdrive suppression could conceal spontaneous activity in cells proximal to the atrium, that is, in the AN cells. It is difficult at the present time to make any definite statement about the region of the AVN that has the pacemaker cells. Systematic optical or microelectrode mapping experiments that are followed by enzymatic cell isolation and patch clamp recording will probably be necessary.

Classically, it is often thought that following extirpation of, or damage to, the SAN, control of the heart rate is taken over by the AVN, a concept based on the nature of the developed rhythm, as well as the electrocardiographic patterns, which are indicative of impulse initiation from such a locus. This suggestion might be an oversimplification or may even be flawed given the observation of Boineau et al.[12] as well as the report that in the dog, pacemaker activity is present in areas of the inferior right atrium.[20,21] These subsidiary atrial pacemakers have been shown to take over rhythm generation in animals following extirpation of the SAN tissue.[22]

Atrioventricular valves (AVVs) have automatic activity.[23–27] These studies demonstrated that within the AVVs, there are cells that are normally spontaneously active. Pacemaker activity in the tissues from the tricuspid and mitral valves was studied systematically[22] and it was shown that, although qualitatively similar to that present in the SAN, automatic activity in the AVV differs quantitatively in that it has a higher dV/dt, and a more negative maximum diastolic potential (MDP). In a recent study,[27] myocytes were isolated from the rabbit tricuspid valve and some of those earlier observations were confirmed. For example, spontaneous electrical activity was present in some of the isolated cells, and the spontaneously active cells were shown to have an MDP of −82 millivolts (compared to −65 millivolts for the SAN) and a maximum dV/dt value of 6.3 V/s.

The presence of myocytes with automatic activity in the AVVs has always been a curious observation.[26] The origin of these cells is not clear, but has been suggested to be atrial[28] or AV nodal.[27,29] Although automatic activity is an inherent property of these cells, the physiological role of such automaticity is usually considered only in connection with the generation of abnormal rhythms in the heart, such as by the generation of ectopic impulses.

Purkinje fibers show diastolic depolarization in normal Tyrode solution.[3,16,30] However, the incidence of spontaneous activity is greatly increased in the presence of low external potassium concentrations ($[K^+]_o$), or in the presence of norepinephrine.[13] In a recent study, Callewaert et al.[31] isolated cardiac Purkinje cells from several mammalian species, including dog, sheep, and cow. Spontaneous activity was shown to be present in the rabbit Purkinje cells exposed to normal Tyrode solution; however, it could only be induced in some sheep cells after lowering $[K^+]_o$. In the spontaneously active cells, MDP was between −70 and −85 mV and the overshoot was 30 mV. The upstroke of the action potentials was fast, with a maximum rate of depolarization of between 150 and 750 V/s.

Purkinje cell pacemaker activity, similar to other subsidiary pacemakers, is normally suppressed and only becomes functional in the event of a damage to atrioventricular conduction. The mechanism for the suppression of Purkinje pacemaker activity has been suggested to be due to an accumulation of K^+.[1] We will now consider rhythmic excitation of the heart by mechanisms involving abnormal impulse generation.

Excitation by Abnormal Impulse Generation

Rhythmicity in the heart can result from impulses that are generated repetitively from mechanisms that do not involve cells that are "normally" automatic. Such mechanisms include reentrant activity,[32] triggered activity,[8] and depolarization-induced automaticity. Since such mechanisms are usually associated with arrhythmogenesis and not with the normal rhythmic excitation of the heart, the automaticity is considered abnormal. The rhythms generated by abnormal automaticity last for variable lengths of time and can self-terminate, or can be stopped by a single pulse. These criteria are somewhat arbitrary and have low specificity, particularly with regard to termination of the rhythm by a single appropriately timed stimulus.[33] Nevertheless, an important difference between normal automaticity and that present in depolarized cells is the demonstration of the phenomenon of overdrive suppression in the former, but not in the latter.[8] These abnormal mechanisms for the repetitive excitation of the heart will be discussed only briefly in this section because only their role in the repetitive excitation of the heart is being considered. A wide variety of excellent literature exists on these individual topics (for reviews, see refs. 8, 32–34).

Reentrant Activity

Reentrant activity (or reentry) was first demonstrated by Mines[35] and later characterized by Schmitt and Erlanger[36] in the turtle ventricle. Under certain conditions in the heart, an impulse is capable of reexciting a region through which it had previously passed. This reexcitation may occur more than once, resulting in a repetitive excitation. The classical prerequisites for reexcitation are a region of unidirectional block and an unusually slow conduction. The reexcitation may occur through a loop of fibers, such as in circus movement reentry,[32] or by reflection within a fiber.[37] However, theoretical studies based on wave propagation in other types of excitable media have provided a new interpretation for reentry. These studies suggest that rotors and "spiral" waves may be the basis for reentrant rhythm disturbances.[110,111] Recent experimental work using microelectrodes and optical mapping in a two-dimensional cardiac muscle preparation[112] has demonstrated the applicability of these theoretical concepts.

Reentrant activity does not depend on the presence of diastolic depolarization and can be terminated by an appropriately timed stimulus. Termination of the activity may result from a stimulus-induced prolongation of refractory period of the circuit of reentry, or from a collision between the rotating process and a new wave front created by the stimulus. The role of reentry in generating arrhythmias is widely known.[32] In fact, it is thought that it is the mechanism responsible for most clinically observed ventricular tachyarrythmias.

Triggered Activity

The heart can be excited repetitively by a mechanism that does not depend on the normal mechanisms for spontaneous activity or on reentrant activity. Such a mechanism is triggered activity in which excitation depends on depolarizing afterpotentials, such as early or delayed afterdepolarizations,[8] which lead to runs of several APs not unlike those generated spontaneously. For example, the APs are preceded by slow depolarizations to the threshold of the following AP. As originally described by Cranefield,[8] triggered activity depends on a depolarizing input from a preceding AP, which serves as the "trigger" for the initiation of the nondriven impulses. These impulses have been implicated in some clinically observed bursts of tachycardia and extrasystoles.

Depolarization-Induced Automatic Activity

Automatic activity can be induced by depolarizing cardiac Purkinje[38] and ventricular preparations.[34] The depolarization can be achieved by an application of a constant current pulse or by exposure of the preparation to barium ions (Ba^{2+}). It has been suggested that depolarization-induced automaticity is a function of both diastolic potential and the reversal potential of the pacemaker current.[34] Thus, a reduction in membrane potential (V_m) without a concomitant shift in the reversal potential of the pacemaker current will lead to automaticity. In a relatively recent study, Delmar and Jalife[40] used the sucrose gap technique to investigate the mechanism of Ba^{2+}-induced automaticity in well-polarized cat papillary muscle. They showed that the ionic mechanism of the automatic activity was different from that responsible for the activity in depolarized preparations, and was due to a Ba^{2+}-induced time-dependent blockade of the inward rectifier current, i_{K1}. This has been confirmed in single-cell experiments by Imoto et al.[41] Repetitive excitation of the heart by depolarization-induced automatic activity may be the result of injury current across the myocardium, and is considered arrhythmogenic.[34]

IONIC CURRENTS IN DIASTOLIC DEPOLARIZATION

Complete understanding of the ionic bases of normal pacemaker activity in the SAN has been hampered by technical limitations which probably have been responsible for conflicting electrophysiological data reported from several laboratories. First, there is the technical problem in the identification and isolation of "true" pacemaker cells, given the heterogeneity of the sinus node.[10] Second, with the technical improvements in electrophysiological experimentation, such as that which came with the introduction of the voltage- and patch-clamping, it has become clear that appropriate separation of membrane currents, so as to study their individual roles in pacemaking, is sometimes a formidable task. The conflicting electrophysiological data reported by various authors might indeed be attributable to limitations inherent to these clamp procedures.

These problems notwithstanding, important insights into pacemaker mechanisms have been achieved within the last decade. For example, among the several ionic currents that could be involved in pacemaker activity in the mammalian heart, it appears that three strong contenders have emerged: the decay of the delayed rectifier current (i_K), the calcium current (i_{Ca}), and the hyperpolarization-activated current (i_F). These currents, as well as the others such as the i_{Na} and the transient outward and background currents, which have been shown to be present in cardiac pacemaker cells, have been extensively reviewed recently by Giles et al.[13,39] Moreover, there is increasing evidence[42] that apart from these three current systems, the Na/K pump and the Na/Ca exchanger currents might contribute to the background current which, in association with the decay of i_K, will play a role in diastolic depolarization. Although these background currents might be small in amplitude, given that SAN pacemaker cells have a capacitance of about 60 pF, and that the rate of change of voltage during phase 4 depolarization is about 0.05 V/s, an inward current of about 3 pA will suffice to generate the pacemaker potential.[13] In the following section, the three main voltage- and time-dependent currents, as well as the background currents, will be discussed.

Delayed Rectifier Current

Underlying the repolarization phase of the cardiac action potential is i_K.[30,43] During an action potential, i_K becomes activated positive to −40 mV and subsequently decays with repolarization. In the presence of a background current, this decay in i_K leads to diastolic depolarization in pacemaking cells.[38,44] Although those studies were carried out in whole-tissue preparations, subsequent single-cell data have confirmed their findings in isolated amphibian and mammalian pacemaking cells.[39,43,45] The kinetics of the decay of the current in the single-cell preparations show a single or a multiexponential process in mammalian,[43,45] but a single exponential time course in amphibian, pacemaker cells.[39]

Despite these extensive studies on i_K, very few single-channel data are available on the current, a situation which is attributable to the small value of the unitary conductance.[46] In the study of Shibasaki,[46] single-channel conductance and kinetic properties of i_K were measured in rabbit SAN cells. A channel density of 0.7/μm^2 and a unitary conductance of 11.1 pS at a $[K^+]_o$ of 150 mM were reported. In a physiological solution, that is, $[K^+]_o$ = 5.4 mM, the unitary conductance was then determined to be equal to 1.6 pS. The role of i_K in pacemaker activity thus seems reasonably well established by electrophysiological data. However, it must be stated that some questions remain to be answered. For example, Noma and Irisawa[47] observed in their tissue strip experiments that, after a quiescent period, the preparations resumed pacemaker activity, suggesting that the decay of i_K after an AP might not be sufficient to generate diastolic depolarization. Furthermore, Yanagihara et al.[48] reported that pacemaker activity was preserved in the presence of Ba^{2+} (that is, after eliminating i_K). These authors also reported that the current underwent little or no change in the pacemaker potential range. Finally, there is also the issue of the number of exponentials describing the decay of the current, which has varied depending on the investigator. Clearly, considerable electrophysiological data are needed to resolve these issues.

Calcium Current

Early studies on multicellular preparations of Purkinje fibers suggested that pacemaker activity results from a decay of K^+ conductance (g_K) leading to diastolic depolarization.[38] This hypothesis was however challenged in the mid-1970s by the experimental data presented by Noma and Irisawa,[47] working with strips of pacemaking tissue from the rabbit SAN. These workers observed that pacemaker activity resumed in the tissue strips that were previously quiescent following sectioning, and therefore attributed diastolic depolarization to a process other than the decay of g_K. The investigators demonstrated the presence of a transient inward current in the preparation and then suggested that given the high input impedance of pacemaker cells, slight changes in membrane current will generate oscillations in V_m. Following that study, several investigations have observed a tetrodotoxin (TTX)-insensitive, depolarization-induced, transient inward current in cardiac pacemaker whole-tissue and single-cell preparations[13,44,45] that is probably carried by Ca^{2+}.

In cardiac cells, the i_{Ca} flows through two types of channels, T- and L-types, which mediate, respectively, the transient and the long-lasting currents.[49,50] The T-type Ca^{2+} channels carry a rapidly inactivating i_{Ca}, which is activated by weak depolarizations. The channel is relatively insensitive to dihydropyridines or cadmium but shows sensitivity to nickel.[50] The L-type channel, on the other hand, is activated by strong depolarizations and is slowly inactivating, thus allowing sustained depolarization which is partly responsible for the plateau phase of the cardiac AP. The L-type current has been well studied: It corresponds to the conventional i_{Ca}, sensitive to dihydropyridines. Because it is activated by strong depolarization, the L-type current is usually separated from the T-type current under voltage-clamp conditions by holding the membrane potential at two levels, -80 mV and -40 mV, while applying depolarizing pulses to elicit i_{Ca}.[49,50] The difference current under these conditions is the T-type current. Hagiwara et al.[50] showed that when depolarizing to -40 mV (holding potential = -80 mV), the T-current had an amplitude of about 2.1 pA/pF. With 100 mM $BaCl_2$ in the patch pipette, the conductances of the T- and L-type channels were reported as 8.5 pS and 16.0 pS, respectively.

Hyperpolarization-Activated Current

The i_F is a nonspecific, Na^+- and K^+-dependent inward current present in several types of cells, including SAN pacemaker cells (for a review, see ref. 51). The i_F is usually referred to as the pacemaker current,[52] and is also said to mediate the modulation of pacemaker activity by catecholamines. The subscript, F, stands for funny, because of the funny properties of the current.[53] In fact, it is indeed funny that this very controversial current is referred to as the pacemaker current because the current is not present in certain cardiac pacemaker cells.[13] Moreover, in several computer simulations as well as in experiments,[27,42] blockade of the channel did not result in the arrest of pacemaker activity. It is generally thought that since the MDP value in SAN pacemaker cells is fairly depolarized (-65 mV) and thus outside the activation voltage of the current, the current might not contribute to diastolic phase 4 depolarization. In fact, the i_F is generally regarded as the main current responsible for diastolic depolarization in Purkinje cells, considering their well-polarized (about -85 mV) MDP value.[54] However, it is now almost certain that some amount of the current is activated during diastolic depolarization in SAN pacemaker cells.[14] Although this amount might be small, as previously discussed, such small currents may indeed be enough to generate a significant depolarization. Perhaps the main issue against the current is that its time course of activation is too slow, even for the equally slow process of diastolic depolarization.

The i_F was originally termed i_{K2} and was described as an outward i_K.[30] The reinterpretation was brought about by the numerous inconsistencies found with the properties of the current. For example, the supposed i_K disappeared in Na^+-free solutions. Also, the value of the reversal potential varied from one investigator to another. One major factor responsible for these unusual properties was the effect of accumulation/depletion in the cardiac preparations that were used for these initial experiments. In a very systematic study, these issues were addressed by DiFrancesco,[55] who demonstrated that the i_{K2} was indeed an inward current, and it was then shown to be the same current as i_F.

i_F has subsequently been found and extensively characterized in several cardiac whole-tissue and single-cell preparations.[14,26,27,51,56] The current is nonspecific (Na^+- and K^+-dependent) and has a reversal potential of -20 mV in the SAN.[57] The current starts activating at about -40 to -50 mV, and is strongly depressed by a low concentration (1 mM) of cesium ions. There is very limited single-channel data on the i_F. DiFrancesco[57] (1986) recently reported the first single-channel recordings of i_F from SAN cells of the rabbit and he showed that the channel conductance is small, which would explain the prior inability to record it. It was shown that with 70 mM KCl and 70 mM NaCl in the external solution, single-channel conductance was 1 pS, a value rather small for a nonselective channel.

Na/K Pump and Na/Ca Exchanger Currents

It is certain that there are background inward and outward currents in cardiac pacemaker cells which are important in diastolic depolarization in these cells.[13,42,48] In the Yanagihara et al.[48] Hodgkin–Huxley-type mathematical model of the SAN, it was reported that when the background current (inward or outward) was eliminated, pacemaker activity stopped. Further support for the inward component in the pacemaker cell is the observation that during repolarization, a process mediated by the potassium current, i_K, the maximum diastolic potential does not reach the reversal potential of potassium, which is about -90 mV. This observation suggests[13] that there must be an inward current in these cells that tends to cause a slight depolarization.

Although the origin of the background current is not clear, Irisawa and Hagiwara[42] describe two possible mechanisms. The Na/K pump, with a coupling ratio of 3 : 2, will tend to cause an outward current. In the guinea pig ventricular cell, the pump current was calculated as about 160 pA at 0 mV, and with an $[Na^+]_i$ of 34 to 41 mM. The second mechanism, the Na/Ca exchanger current, is an inward current throughout the range of diastolic depolarization, and will therefore be expected to participate in the membrane depolarization. We shall now consider how the three main ionic currents (i_K, i_{Ca}, and i_F), as well as the background exchanger current, have been proposed to act in concert to generate diastolic depolarization in pacemaker cells.[42]

Interplay of Ionic Currents in Diastolic Depolarization

There is still some controversy as to the exact role of the ionic currents in the process of diastolic depolarization of cardiac pacemaking cells. However, a picture that is emerging from various experiments and computer simulations is that there is an interplay between these ionic currents, and that each current is activated at some point during diastolic (phase 4) depolarization, depending on when depolarization courses through the voltage ranges at which the current is normally activated.[13,27,42,55] Given that there are quantitative differences in the electrophysiological properties of the various cardiac pacemaking cells, it is relevant at this point to consider the manner in which the different ionic currents might play different roles in the pacemaker activity of these various cells. The current views as to these roles in diastolic depolarization in the primary pacemaker cell (the SAN cell)[42] and in a subsidiary pacemaker cell (the tricuspid valve cell)[27] will be the focus.

Figure 4.1A shows diastolic depolarization of a SAN cell. From the onset of phase 4, that is, at the MDP (approximately -65 mV), three current systems will operate in concert to begin the process of diastolic depolarization. These currents include i_K, i_F, and the exchanger (Na/Ca) background current. Midway through phase 4, the $i_{Ca(T)}$ is activated, and is subsequently followed by the activation of $i_{Ca(L)}$ just before the threshold for upstroke is reached. The later activation of calcium (T- and L-) currents depends on when phase 4 depolarization leads the membrane into the voltage ranges at which they are activated, which are, respectively, -55 and -30 mV. In Figure 4.1B, phase 4 depolarization in the tricuspid valve is shown. In a typical tricuspid valve cell, the MDP is much more negative (-82

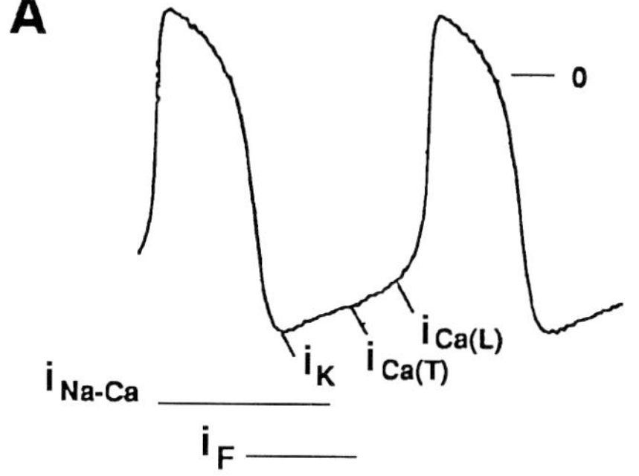

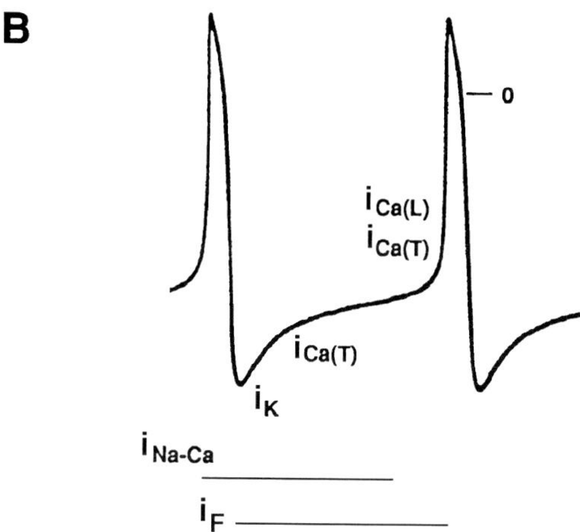

FIGURE 4.1 Ionic currents involved in diastolic depolarization of pacemaker cells in the heart. **A.** APs from the sinus node. Diastolic depolarization is emphasized, including the roles of the different ionic currents in the depolarization process. *Modified from Irisawa and Hagiwara.*[42] **B.** The same process for the tricuspid valve. This process may apply to other subsidiary pacemakers, for example, the Purkinje cell, in which these ionic currents are demonstrable.

mV) than that in the SAN, and phase 4 is characterized by a biphasic time course, suggesting a complex interplay between several ionic currents.[25–27] Anumonwo et al.[27] showed the presence as well as a time-dependent inactivation of the inward rectifier current, i_{K1}, within the pacemaker potential range of tricuspid valve cells. On the basis of this, they suggested that the inactivation process, in conjunction with other current mechanisms, may be responsible for diastolic depolarization in the valve cell. In the valve cell, similar to the situation in the SAN cell, i_K, i_F, and the exchanger (Na/Ca) background current are expected to play similar roles in initiating phase 4 depolarization. However, given the relatively more negative MDP values, quantitative differences are expected in the relative amounts of the currents activated. For example, an activation of a very significant amount of i_F is expected in the valve cell. Perhaps the activation of a relatively large amount of the i_F and/or the inactivation of the i_{K1} may explain the rapid depolarization (phase D_1)[58] seen at the beginning of diastolic depolarization in these cells. Furthermore, because the take-off potential in the valve is about −56 mV, and given that the activation voltage range of $i_{Ca(T)}$ is −55 to −60 mV, a limited amount of $i_{Ca(T)}$ will be activated just before threshold, with full activation of both T- and L-type currents occurring at upstroke.[27]

The MDP value in the Purkinje cell is also quite negative (−75 to −85 mV). Therefore, it has been suggested[54] that the i_F is the primary inward current causing diastolic depolarization in these cells. Recently, DiFrancesco and Noble[59] carried out a mathematical reconstruction of pacemaker activity of the Purkinje cell and estimated the relative contributions of the i_K and i_F. x and y gating variables were assigned, respectively, to the decay of the i_K and the activation of the i_F (see Figure 1 in ref. 54). The results showed that there was a strong activation of the i_F at the range of voltages of pacemaker depolarization. In a more recent study,[60] $i_{Ca(T)}$ and $i_{Ca(L)}$ were characterized in the Purkinje cell. In well-polarized Purkinje cells, it is expected that the take-off potential will be similar to that in the valve cell. Therefore, in Purkinje fibers these channels should play a role similar to that in the valve, that is, essentially the T-type channel being partially activated during phase 4, with full activation of both the T- and L-types occurring at the upstroke.

Nervous Control of Ionic Currents Underlying Pacemaker Activity

Autonomic modulation of pacemaker activity depends primarily on the control of membrane permeability in diastole. Specifically, this modulation is achieved by the interplay of cholinergic and β-adrenergic control of the ionic currents involved in diastolic depolarization. Postsynaptic vagal fibers release acetylcholine (ACh), causing a muscarinic receptor-mediated increase in K^+ permeability, as well as a decrease in Ca^{2+} permeability in the SAN cells. The resultant effect is, usually, negative chronotropy on pacemaker activity.

On the other hand, sympathetic stimulation results in the release of norepinephrine which combines with the β-adrenergic receptors to cause an increase in membrane permeability to Ca^{2+} and K^{+}. Sympathetic stimulation has a positive chronotropic effect on pacemaker activity. There are several excellent reviews (see ref. 61) on the effect of neurotransmitters on cardiac cell membrane currents. The following section will focus on the effects of adrenergic and cholinergic stimulations on the main pacemaker current systems in the SAN.

Adrenergic Control

Voltage-clamp studies in SAN tissue(53,62) and single-cell preparations(50) have shown that adrenergic stimulation-induced increases in i_{Ca} underlie the observed positive chronotropy. Noma et al.(62) found that the stimulation increased the limiting i_{Ca} conductance without affecting the kinetics of the current. Hagiwara et al.(50) determined the effect of the β-adrenergic agonist isoprenaline on the two types of i_{Ca} and showed that, at both the macroscopic and single-channel current levels, the effect of the agonist was on $i_{Ca(L)}$, and not $i_{Ca(T)}$.

Noma et al.(62) in their SAN preparations reported that, similar to its effect on the i_{Ca}, adrenergic stimulation also caused an increase in i_K by increasing the limiting conductance without modifying the current kinetics. In another study, Brown and DiFrancesco(52) used a Ca^{2+} channel blocker to show that the increase in i_K is possibly linked to an increase in i_{Ca}: In the presence of D600, there was no detectable effect on i_K decay tails. How would adrenergic stimulation have a positive chronotropic effect on the SAN if it causes an increase in an inward (Ca^{2+}) as well as an outward (K^{+}) current? It is possible that the increase in i_K could oppose the larger overshoot caused by the adrenaline-induced increase in i_{Ca}.(63)

Several studies suggest that the i_F is increased following adrenergic stimulation.(14,53,57,62) The increase in i_F is also associated with catecholamine-induced positive shift in the activation curve of the current without a change in the fully activated I-V curve.(14,57) DiFrancesco(57) determined the effect of adrenaline on i_F single-channel currents in the rabbit SAN cells and showed that the adrenaline-induced modulation of the current was mediated by an increased probability of channel opening with no change in conductance. These results were taken to indicate that the effect of adrenaline is on current kinetics, rather than on current amplitude. It is possible that the main role of i_F in pacemaking might be to serve as the means by which pacemaker rate is modulated by catecholamines. It is known that some SAN cells are more sensitive to catecholamines than others,(10) and that in the presence of catecholamines one observes a shift in pacemaker site to subsidiary pacemaker cells. Thus, it is possible that in the presence of a catecholamines, the increase in i_F in the subsidiary pacemaker cells will cause them to become the dominant pacemakers.

The mechanism by which adrenaline mediates its effects on these ionic currents is reasonably well known.(61) The binding of the β-adrenergic agonists to the β-receptor activates adenylate cyclase and results in an increase in cyclic AMP. The cyclic AMP activates a protein kinase which then phosphorylates the ion channels.

Cholinergic Control

Pacemaker cells in the SAN respond to vagal stimulation, or to exogenously applied ACh by membrane hyperpolarization,(63–65) as well as by an inhibition of pacemaker rate.(65) The initial studies(61) on the slowing effect of vagal stimulation on pacemaker rate showed that this was due to the muscarinic receptor-mediated decrease and increase in, respectively, the g_{Ca} and the g_K. Recent studies, however,(65,66) have shown that the suppression of i_F might be equally involved in the process of slowing the pacemaker. How are these different ionic conductances affected by ACh?

An indication that the i_{Ca} is sensitive to ACh came from the observation that in nodal cells, the final phase of the upstroke was suppressed by ACh.(67) Subsequently, voltage-clamp data(61,68,69) have shown that the slow i_{Ca} is inhibited in the presence of ACh. This inhibition seems to be the result of a reduction in the maximum conductance, while current kinetics were unaffected.(68) A possible mechanism for this inhibition is that ACh interferes with the enzyme adenylate cyclase in the β-adrenergic cascade. Given the implication of i_{Ca} in phase 4 depolarization, an inhibition of the current would be expected to slow down the depolarization, and therefore the frequency of pacemaker activity.

ACh increases membrane conductance to K^{+} by activating a specific channel, an ACh-activated K^{+} channel, $i_{K(ACh)}$,(70–72) which has gating kinetics much faster than that of the inward rectifier, i_{K1}. The kinetic properties of $i_{K(ACh)}$ were extensively studied by Sakmann et al.(72) In that study, a type of K^{+} channel rarely present in ventricular cells was found in SAN and AVN cells. The channel had a low probability of opening, which was greatly increased in the presence of ACh. The channel had a very brief (1 ms) mean open time. It was also shown from single-channel and from whole-cell recording that $i_{K(ACh)}$ undergoes a strong rectification. For example, in the presence of 20 mM K^{+} in the patch pipette, the single-channel chord conductance was markedly less in the outward than in the inward direction (5 pS compared to 25 pS). Recent experimental studies on the mechanisms of $i_{K(ACh)}$ have shown that although activation of the i_K involves GTP-dependent proteins,[73] no intracellular mechanisms, such as cyclic GMP, are involved.[62,74,75]

Originally, it was thought that the chronotropic effect on the heart following vagal stimulation was due to ACh-induced changes in the i_K and i_{Ca}. However, relatively recent studies[65,66,76] have shown that the inhibition of pacemaker activity by ACh can also be linked to an effect of ACh on the hyperpolarization-activated current, i_F. Low concentrations of ACh (0.03 to 1 μm) inhibited the current, shifting current activation to more negative voltages.[65] Furthermore, in the presence of ACh, pacemaker activity was inhibited. The involvement of the muscarinic receptor was indicated by the observation that the ACh-induced inhibition was antagonized by atropine. In the presence of pertussis toxin, this ACh-induced inhibition of current was abolished, suggesting the involvement of a G-protein.[65,66]

Unlike $i_{K(ACh)}$, ACh-induced modulation of the i_F channel does involve a second messenger.[76] In this regard, it was shown that in pacemaker cells preloaded with cyclic AMP or a phosphodiesterase inhibitor, isobutylmethylxanthine, ACh-induced current inhibition was abolished. The workers concluded that cyclic AMP was the second messenger involved in the ACh-induced inhibition process. They also concluded that this inhibition occurred via a decrease in cyclic AMP, resulting from an inhibition of a high adenylate cyclase activity normally present at rest.

DYNAMIC MODULATION OF PACEMAKER ACTIVITY

The individual pacemaking cells of the SAN, in isolation, have slightly differing natural frequencies, but together, coordinate their rhythms to produce a singular impulse that is propagated to the rest of the heart during each cardiac cycle. Furthermore, this coordinated discharge in the SAN responds to a reflex autonomic input from the vagus nerve following each systolic pressure wave. The mechanisms responsible for the coordinated discharge of SAN cells as well as the reflexly mediated control by the vagus have been studied by several laboratories.[77–81] Two major hypotheses—a mechanical and an electrical hypotheses—have emerged as a result of those studies. One scheme in the mechanical hypothesis assumes that a dominant cell depolarizes first and then its contraction will stretch the membrane and accelerate the depolarization of the neighboring pacemaker cells, causing them to advance their subsequent discharge.[77] Another scheme[82] suggests that synchronization results from the entrainment of the SAN cells to the rhythmic pulsation of the sinus node artery. However, several considerations have led to the suggestion that such mechanisms are unlikely. For example, the electromechanical latency required for impulse transmission between cells is relatively long,[83] and is much longer than the time needed for transmission of an impulse between pacemaker cells in the SAN. Furthermore, it is unlikely that the poor contractile machinery of pacemaker cells[10] is adequate for such a function. Moreover, the presence of such a "leading" pacemaker cell remains to be shown.

Several lines of evidence indicate that the underlying mechanism for synchronization between pacemaker cells is electrical. One scheme in the electrical hypothesis[84] suggests that dynamic changes in the electric field in the SAN, in the absence of any physical contact between cells, can result in a coordinated firing of pacemaker cells. This scheme proposes that following a depolarization of the leading pacemaker cell, there is a hyperpolarization of the surrounding extracellular space, which then accelerates the depolarization of the neighboring cells to threshold. However, it is unlikely that this is the case[83] since successful transmission of the excitation will require the leading pacemaker cell to depolarize in less than a millisecond. On the other hand, support for the idea that cell-to-cell communication and pacemaker coordination result in part from resistive type of electrical coupling is overwhelming. Some of the important work supporting such a scheme is discussed in the following section.

Electrotonic Interactions Underlie Synchronization in the Sinus Node

Results from several electrophysiological studies[85–89] have suggested that phase-dependent, electrotonically mediated interactions through gap junctions[90] are the basis of synchronization of electrical activity in various kinds of cardiac cells. Certain conditions are required if synchronization is to be mediated by an electrical process: for example, there must be evidence for these low-resistance pathways. Reports from early electron microscopy work[11] had indicated that gap junctions were rare or totally absent in the SAN, and led to the proposal of the mechanical hypothesis.[77] More recent studies, however, have shown that gap junctions are indeed present in the mammalian sinus node tissue.[10,91] Although those studies showed that the distribution of gap junction in the SAN was sparse and also variable compared to the distribution in other cardiac tissues, there was one gap junction between any two cells. Because of this observation, it was questioned whether such small nexal contacts could allow for synchronization. However, it has been estimated[92] that synchrony between two pacemaker cells can be maintained in the presence of just a single nexal contact. As shown by the studies discussed, it is now generally accepted that the mechanism underlying synchronization is electrical in nature.

Phase-Response Curves

On the basis of oscillator theory,[93] it has been established that interactions between two or more electrically coupled pacemaker cells must be phase-dependent for synchronization to occur. Such interactions can be readily studied by the use of perturbation techniques, in which external input is applied to a spontaneously active pacemaker cell to determine its dynamic behavior. An example of this is presented in Figure 4.2, which shows the response of a simulated pacemaker cell to a brief depolarizing current pulse scanning the spontaneous cycle. Panels A and B are membrane potential recordings; the spontaneous cycle length was 610 ms. In panel A, a brief (50 ms) subthreshold depolarizing current pulse was applied at a relatively early time or phase (Φ) in the cell cycle and resulted in a transient prolongation or positive phase shift (ΔΦ) of the cell cycle. In panel B, when the same pulse was applied relatively late in the cycle, it resulted in cycle abbreviation or negative ΔΦ. If such a brief current pulse is used to scan the entire cell cycle, and then the various values of phase shifts are plotted against their corresponding phases, the resulting plot is known as a phase-response curve (Figure 4.2C). This plot is important in that it predicts the entrainment properties of the cell in the presence of the current pulse. For example, the limits of the curve determine the maximum abbreviation or prolongation of the cycle length that is possible with such a current pulse.

Phase Resetting and Entrainment in Cardiac Pacemakers

Several experimental studies[78,85,94] and computer simulations[87,95] have indeed shown that electrotonic interactions do occur between cardiac oscillators. These studies demonstrated that pacemaker cells behave like coupled oscillators, and that electrotonic current inputs among two or more such cells affect their respective periodicities in a phase-dependent manner, resulting in mutual entrainment and coordinated behavior.

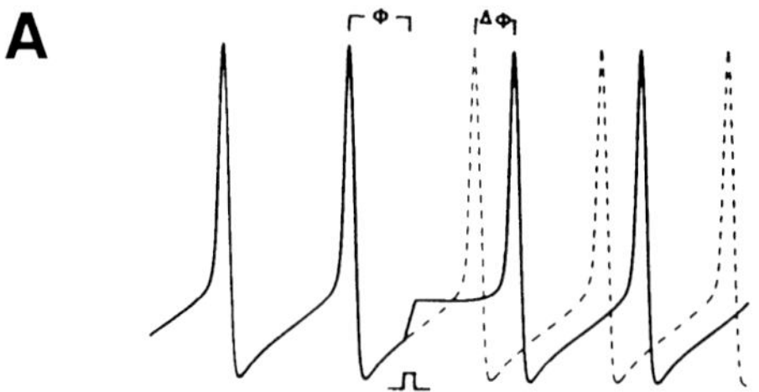

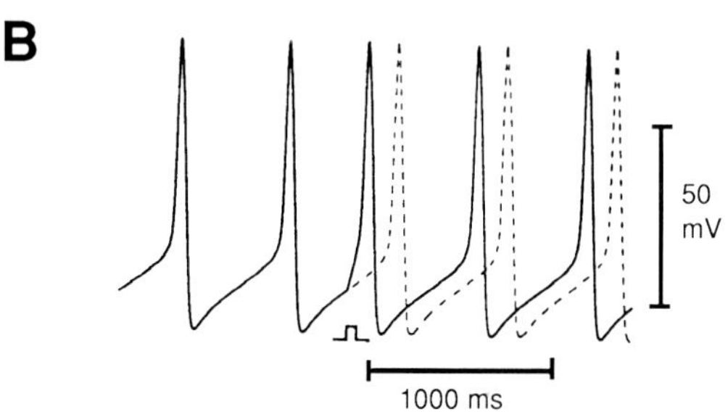

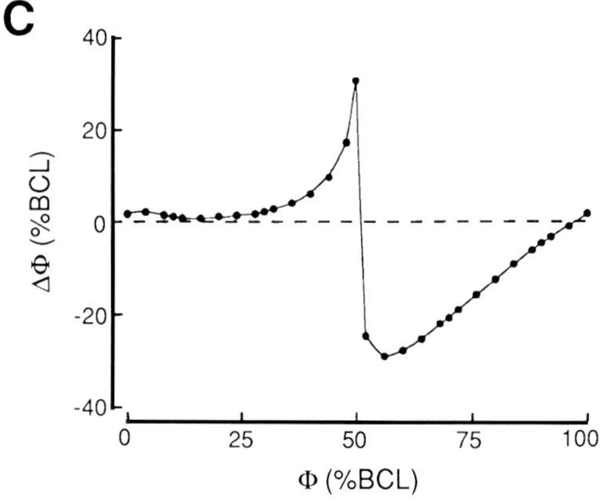

FIGURE 4.2 The phasic effects of a brief depolarizing current pulse on the oscillations of a simulated pacemaker cell. Simulations were carried out using equations from Yanagihara et al. **A** and **B.** APs from the simulated cell with a spontaneous cycle length of 610 ms (cell was under the influence of a bias hyperpolarizing current). Brief (50 ms, −0.2 μA/cm^2) subthreshold depolarizing current pulse was applied at a given time or phase (Φ) in the cell cycle and resulted in a transient prolongation (**A**) or abbreviation (**B**), which are denoted as phase shift (ΔΦ) of the cell cycle. The dotted lines represent the control oscillations of the cell. If such a brief current pulse is used to scan the entire cell cycle, and then the various values of phase shifts are plotted against their corresponding phases (both expressed as percent of the pacemaker cycle length), the resulting plot is known as a phase-response curve (Fig. 4.2C). *Reprinted from Jalife and Michaels,*[108] *with permission.*

In relation to ventricular arrhythmias also, there is evidence for the phenomenon of phase resetting in the electrical interactions occurring between pacemaker sites. Such an activity has been shown to result in some abnormalities in the cardiac activation pattern. For example, electrotonic modulation of ventricular parasystolic activity has been discussed in connection with the development of closely coupled premature beats.[96–98]

In a sucrose gap study of modulated parasystole, for example, Jalife and Moe[96] investigated the effect of local circuit currents across the region of block (the gap) on the pacemaker activity of a Purkinje fiber. The experiments showed that subthreshold depolarizations, spontaneous or evoked APs, had a biphasic effect on pacemaker cycle length that was predictable from the phase-response curve. This study was further extended by Moe et al.[97] in their model of modulated parasystole. Moe et al.[97] developed the model on the basis of the phase-response curve, in which a parasystolic focus was allowed to interact with an intervening sinus beat. The model predicted the ectopic activity and entrainment patterns expected from the interactions. In fact, the model revealed several features of protected pacemakers, some of which were initially thought to be inconsistent with parasystole. Results of those initial studies have been extensively applied in models of ventricular parasystole[99] and have enabled less confounding interpretation and analysis of parasystole.

Tuganowski et al.[100] and later Jalife[85] demonstrated that coordinated activity in the SAN is electrical in nature. Using the sucrose gap technique on rabbit sinus node tissue,[83,100] it was shown that electrotonic interaction between two pacemaker sites separated by the gap resulted in a mutual entrainment of the two sites. Figure 4.3 shows the effect of electrical coupling between two pacemaker sites in one of the sucrose gap studies.[85] Two sites, one fast (F) and the other slow (S), were coupled across the sucrose gap through a shunt, whose resistance could be varied at will. In all panels, top traces were recorded from F and bottom traces from S pacemaker sites. In the top panel, APs were recorded while Tyrode solution was being superfused on all segments. The recordings show that there is synchronization between the two sites. Panel B was recorded following a 30-min superfusion of the central segment with sucrose (shunt resistance was also set at infinity) and shows a loss of the synchronized rhythm. Panel C was recorded a few seconds later, after the shunt resistor was set at 0 Ω. This allowed interaction through the shunt and mutual entrainment,

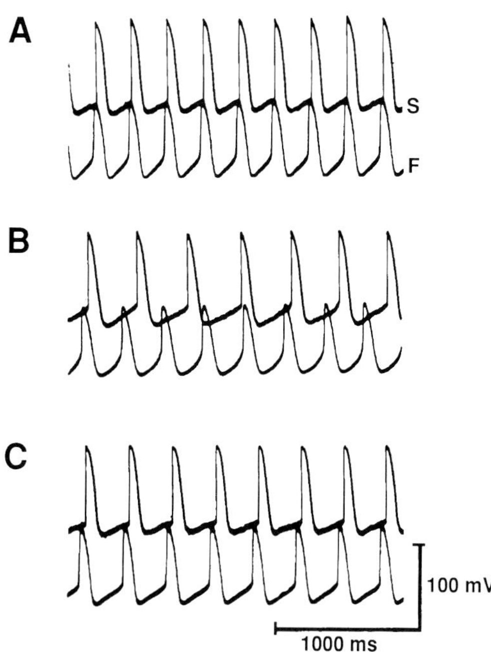

FIGURE 4.3 Sucrose gap experiment demonstrating the effect of electrical coupling on mutual entrainment between two pacemaker sites in the SAN tissue. Two sites, one fast (F) and the other slow (S), were coupled through a shunt, variable resistor. **A.** Potentials recording was carried out while Tyrode was being superfused in all the segments and shows synchronization between the F and the S. **B.** After 30-min superfusion of the central segment with sucrose (with shunt resistance set at infinity), a loss of the synchronized rhythm can be seen. **C.** Recorded a few seconds later, the shunt resistor was set at 0 Ω, demonstrating the electrical nature of the synchronization mechanism. *Reprinted from Jalife[85] with permission.*

which demonstrates the electrical nature of the synchronization mechanism. In the experiment, various patterns, for example, 1 : 1, 2 : 1, 3 : 2, and so forth, were obtained by varying the value of the shunt resistance.

Michaels et al.[101] used computer simulations of the electrical activity of the SAN to show that coordination and conduction in this tissue is mediated by phase-dependent interactions of the cells. They showed that electrical coupling in a two-dimensional array of simulated pacemaker cells resulted in mutual entrainment to a common frequency. Strikingly, following the synchronization of activity in the array, there emerged a dominant pacemaker region. These results led Michaels et al.[101] to propose that synchronization of pacemaker activity in the SAN is by a "democratic" process involving the mutual interaction among the pacemaking cells of the SAN.

Some insight into phase resetting and entrainment properties of cardiac pacemakers has also been obtained from studies in which pacemaking cell aggregates from the chick ventricle have been subjected to periodic forcing. For example, Guevara et al.[86,102] showed that single pulses were capable of resetting, in a phase-dependent manner, the pacemaker activity in the aggregates. Furthermore, they showed that repetitive application of the pulse over a broad range of stimulus frequencies resulted in the entrainment of activity with phase-locked patterns such as 1 : 1, 2 : 1, and more complex patterns. All of these studies have provided very strong evidence for the electrical nature of synchronization of electrical activity in the SAN. Also, in general, these studies were carried out in multicellular pacemaking preparations, and the results were used to infer the presence of such a mechanism at the single-cell level. However, a very recent work from our laboratory[94] provided direct evidence that such interactions do indeed occur at the single-cell level.

Anumonwo et al.[94] studied the phase-resetting and entrainment properties of single pacemaker cells. These workers used computer simulations in a model of the rabbit sinus node cell, as well as isolated rabbit sinus nodal cells. Spontaneous electrical activity in the cell model was reconstructed using Hodgkin and Huxley–type equations describing time- and voltage-dependent membrane currents. Single subthreshold current pulses (depolarizing or hyperpolarizing) were used to scan the spontaneous cycle of the cells. The pulses perturbed the subsequent discharge and produced temporary phasic changes in pacemaker period which enabled the construction of phase-response curves. On the basis of these results, the entrainment characteristics of the cells were then studied. Application of repetitive pulses caused phasic changes in spontaneous cycle of the cell and resulted in stable 1 : 1 entrainment at a range of basic cycle length around the spontaneous cycle length, or a 2 : 1 pattern at basic cycle length values about half the spontaneous cycle length. Wenckebach-like patterns, for example, 5 : 4, 4 : 3, and 3 : 2, were observed between the two entrainment zones. The demonstration of phase resetting in single cells provides the basis for phenomena such as mutual entrainment between electrically coupled pacemaker cells, apparent intranodal conduction, and reflex vagal control of heart rate. Therefore, there is enough evidence from computer simulations, as well as from morphological and electrophysiological studies, supporting the hypothesis that synchronization of their electrical activity is by a process that involves electrotonic, phase-dependent interactions among pacemaker cells of the SAN.

Phasic Vagal Control of Pacemaker Activity

The heart rate is dynamically regulated by a feedback mechanism involving the reflex discharge of the vagus nerve. It is known that during systole, pressure on the carotid sinus, due to pulse waves, results in the discharge of the vagus.[103,104] In the early 1930s, Brown and Eccles,[105] while investigating the effect of vagal stimulation on the heart, showed that brief vagal trains affected the frequency of the heart rhythm, an effect that was shown to depend on the intensity and the time of arrival of the train. These observations have subsequently been extensively investigated in several experimental studies,[79,106–108] as well as in computer simulations.[87,109] These studies showed that vagal discharges applied repetitively to the SAN are capable of entraining the already entrained pacemaker cells. In these studies the vagal input was mimicked with either hyperpolarizing pulses, or with simulated, brief ACh pulses. In all cases, the phase-response curve plotted under these conditions could be used to determine the entrainment properties of the SAN during repetitive application of the vagal input. Additionally, using the vagal trains at different interpulse intervals, these studies showed that during entrainment, the SAN could be forced to beat at frequencies faster or slower than its intrinsic rate.

The vagal discharges considered so far were applied repetitively to the SAN, without consideration to the phase of application. However, in vivo, the situation is different; vagal input to the SAN occurs as brief discharges with a fixed phasic relation to the cardiac cycle. This notwithstanding, the rhythm resulting from such an interaction can also be predicted from the phasic mechanisms underlying the phase-response curve. In fact, some vagus–SAN interactions seen clinically have been reproduced remarkably well using pulses applied at fixed coupling intervals (FCIs) to the SAN pacemaker cycle.[87] In Figure 4.4 is shown the entrainment of simulated SAN pacemaker activity by simulated brief ACh pulses (50 ms, 1×10^{-6} M) using such a protocol. The control cycle in panel A was 318 ms. At an FCI of 140 ms (panel B), a stable entrained pacemaker cycle of 445 ms was obtained. Increasing the FCI to 260 ms (panel C) resulted in an entrained cycle of 570 ms. A further increase in the FCI to 270 ms (panel D) led to the development of an arrhythmic pat-

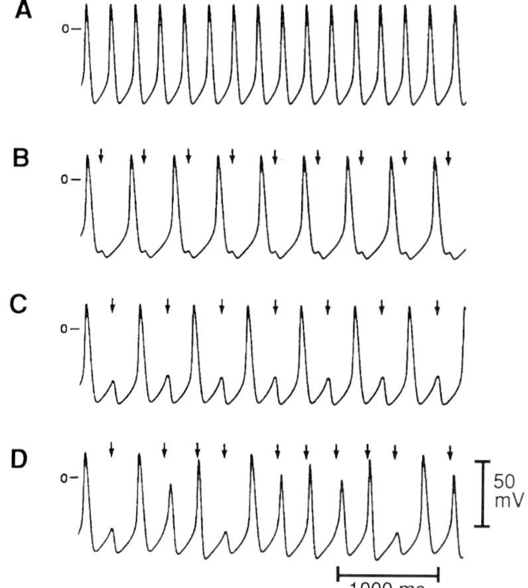

FIGURE 4.4 Experiment demonstrating the effect of vagal pulses (applied at fixed coupling intervals, FCIs) on a simulated pacemaker cell. Control pacemaker cycle length was 318 ms (**A**). ACh pulses (1×10^{-6}, 50 ms) were applied at an FCI = 140 ms (**B**), 260 ms (**C**), or 270 ms (**C**). See text for more details. *Reprinted from Michaels et al.,*[87] *with permission.*

tern. The arrhythmic pattern shown in panel D is similar to the clinically described "sick sinus" syndrome. As previously suggested,[87] it is possible that in vivo, the small amplitudes of some of the SAN discharges such as those in panel C might prevent such impulses from activating the atria, resulting in sinoatrial block.

CONCLUSIONS

Considering how vital the process of pacemaker activity is to life, it is unfortunate that the underlying mechanisms for its initiation and modulation are incompletely understood. However, with the relatively recent development of very sophisticated tools, cardiac electrophysiologists are now beginning to have clues to the mechanisms. For example, a number of membrane ionic currents have been implicated in the process. However, given the inherent complexity of the pacemaking process itself, as well as the limitations of the tools used to investigate it, several controversies remain to be resolved.

Data from several experimental studies and computer simulations suggest that the dynamic modulation between pacemaker cells, as well as the reflexly mediated dynamic control of pacemaker activity by the autonomic nervous system, is electrical in nature, involving a process of phase-dependent electrotonic interaction between these cells.

ACKNOWLEDGMENTS

This work was supported by grants HL-29439, HL-40923, and HL-40991 from the National Heart, Lung, and Blood Institutes and a fellowship to JA from the American Heart Association, New York State Affiliate.

REFERENCES

1. Brooks CMC, Lu HH: *The Sinoatrial Pacemaker of the Heart.* Springfield, IL, Charles C. Thomas, 1972.
2. Keith A, Flack M: The form and nature of the muscular connections between the primary divisions of the vertebrate heart. *J Anat Physiol* 1907;41:172–189.
3. Coraboeuf E, Weidmann S: Potentiel de repos et potentiel d'action du muscle cardiaque mesure a l'aide d'electrodes internes. *CR Soc Bio* 1949;143:1329–1331.
4. Draper MH, Weidmann S: Cardiac resting and action potentials recorded with an intracellular electrode. *J Physiol (Lond)* 1951;115:74–94.
5. Cole KS: Some physical aspects of bioelectric phenomena. *Proc Natl Acad Sci (USA)* 1949;35:558–566.
6. Hamill OP, Marty A, Neher E, Sakmann B, Sigworth FJ: Improved patch clamp techniques for high resolution current recording from cells and cell-free membrane patches. *Pflügers Arch* 1981;391:85–100.
7. Neher E, Sakmann B: Single channel currents recorded from membrane of denervated frog muscle fibers. *Nature (Lond)* 1976;260:779–802.
8. Cranefield P: Action potentials, afterpotentials, and arrhythmias. *Circ Res* 1977;41:415–423.
9. West TC: Ultramicroelectrode recording from the cardiac pacemaker. *J Pharm Exp Ther* 1955; 115:283–290.
10. Bleeker WK, MacKaay AJC, Masson-Pevet M, Bouman LN, Becker AE: Functional and morphological organization of the rabbit sinus node. *Circ Res* 1980;46:11–22.
11. James TN, Sherf L, Fine G, Morales AR: Comparative ultrastructure of the sinus node in man and dog. *Circulation* 1966;24:139–163.
12. Boineau JP, Schuessler RB, Henkel DB, Miller CB, Brockus CB, Wylds AC: Widespread distribution and rate differentiation of atrial pacemaker complex. *Am J Physiol* 1980;239:H406–H415.
13. Giles W, van Ginneken A, Shibata EF: Ionic currents underlying cardiac pacemaker activity: A summary of voltage-clamp data from single cells, in Nathan RD (ed): *Cardiac Muscle: The Regulation of Excitation and Contraction.* New York, Academic Press, 1986, pp 1–27.
14. DiFrancesco D, Ferroni A, Mazzanti M, Tromba C: Properties of the hyperpolarizing-activated current (I_f) in cells isolated from the rabbit sino-atrial node. *J Physiol (Lond)* 1986;377:61–88.
15. Nathan R: Two electrophysiologically distinct types of cultured pacemaker cells from rabbit sinoatrial node. *Am J Physiol* 1986;259(19):H325–H329.

16. Hoffman B, Cranfield P: *Electrophysiology of the Heart*. New York, McGraw-Hill, 1960.
17. Paes de Carvalho A, De Almeida DF: Spread of activity through the atrioventricular node. *Circ Res* 1960;8:801–809.
18. Meijler F, Janse M: Morphology and electrophysiology of the mammalian atrioventricular node. *Physiol Rev* 1988;68(2):608–647.
19. Tse W: Evidence of presence of automatic fibers in the canine atrioventricular node. *Am J Physiol* 1973;225:716–723.
20. Jones SB, Euler DE, Randall WC, Brynjolfsson G, Hardie EL: Atrial ectopic foci in the canine heart: hierarchy of pacemaker automaticity. *Am J Physiol* 1980;238(7):H788–H793.
21. Rozanski G, Lipsius S, Randall W: Functional characteristics of sinoatrial and subsidiary pacemaker activity in the right atrium. *Circulation* 1983; 67(6):1378–1387.
22. Randall W, Rinkema LE, Jones SB, Moran JF, Brynjolfsson G: Functional characterization of atrial pacemaker activity. *Am J Physiol* 1982;242(11): H989–H106.
23. Wit AL, Fenoglio JJ, Wagner BM, Bassett AL: Electrophysiological properties of cardiac muscle in the anterior mitral valve leaflet and adjacent atrium in the dog. Possible implications for the genesis of atrial dysrhythmias. *Circ Res* 1973;32:731–745.
24. Makarychev V, Kosharskaya I, Ul'Yaniskii L: Automatic activity of pacemaker cells of the atrioventricular valves in the rabbit heart. *Biull Eksp Biol Med* 1976;81:646–649.
25. Rozanski G, Jalife J: Automatic activity in atrioventricular valve leaflets of rabbit heart. *Am J Physiol* 1986;250:H397–H406.
26. Anumonwo JMB: Ionic currents underlying pacemaker activity in the rabbit tricuspid valve (Dissertation). Syracuse, NY: SUNY Health Science Center at Syracuse, 1988.
27. Anumonwo JMB, Delmar M, Jalife J: Electrophysiology of single heart cells from the rabbit tricuspid valve. *J Physiol (Lond)* 1990;425:145–167.
28. Gross L, Kugel M: Topographic anatomy and histology of the valves in the human heart. *Am J Pathol* 1931;7:445–474.
29. Anderson RH, Janse MJ, Van Cappelle FJL, Billette J, Becker AE, Durrer D: A combined morphological and electrophysiological study of the atrioventricular node of the rabbit heart. *Circ Res* 1974;35:909–922.
30. Noble D, Tsien WR: The kinetic and rectifier properties of the slow potassium current in cardiac Purkinje fibers. *J Physiol (Lond)* 1968;195:185–214.
31. Callewaert J, Verecke F, Verdonck, Carmeliet E: Isolated cardiac Purkinje cells, in Noble D, Powel T (eds): *Electrophysiology of Single Cardiac Cells.* New York, Academic Press, 1987, pp 187–222.
32. Moe G: Evidence for reentry as a mechanism of cardiac arrhythmias. *Rev Physiol Biochem Pharmacol* 1975;72:55–81.
33. Jalife J, Antzelevitch C: Phase resetting and annihilation of pacemaker activity in cardiac tissue. *Science* 1979;206:695–697.
34. Katzung B, Morgenstern J: Effects of extracellular potassium on ventricular automaticity and evidence for a pacemaker current in mammalian ventricular myocardium. *Circ Res* 1977;40(1):105–111.
35. Mines GR: On dynamic equilibrium in the heart. *J Physiol (Lond)* 1913;46:349–382.
36. Schmitt FO, Erlanger J: Directional differences in the conduction of the impulse through heart muscle and their possible relationship to extrasystolic and fibrillatory contractions. *Am J Physiol* 1929;87:326–347.
37. Antzelevitch C, Jalife J, Moe G: Characteristics of reflection as a mechanism of reentrant arrhythmias and its relationship to parasystole. *Circulation* 1980;61:182–191.
38. Hauswirth O, Noble D, Tsien RW: The mechanism of oscillatory activity at low membrane potentials in cardiac Purkinje fibers. *J Physiol (Lond)* 1969;200:255–265.
39. Shibata G: Ionic currents which generate the spontaneous diastolic depolarizations in individual cardiac pacemaker cells. *Proc Natl Acad Sci (USA)* 1985;82:7796–7800.
40. Delmar M, Jalife J: Low Ba-induced pacemaker current in well polarized cat papillary muscle. *Am J Physiol* 1987;252:H258–H268.
41. Imoto Y, Ehara T, Matsuura H: Voltage- and time-dependent block of I_{K1} underlying Ba^{2+}-induced ventricular automaticity. *Am J Physiol* 1987;252:H325–H333.
42. Irisawa H, Hagiwara N: Pacemaker mechanism of mammalian sinoatrial node cells, in Mazgalev T, Dreifus L, Michelson E (eds): *Electrophysiology of Sinoatrial and Atrioventricular Nodes.* New York, Allan R. Liss, 1988, pp 33–52.
43. Matsuura H, Ehara T, Imoto Y: An analysis of the delayed outward current in single ventricular cells of the guinea pig. *Pflügers Arch* 1987;410:596–603.
44. Brown HF, Kimura J, Noble S: The relative contributions of various time-dependent membrane currents to pacemaker activity in the sinoatrial node, in Bowman L, Jongsma H (eds): *Cardiac Rate and Rhythm.* Boston, Martinus Nijhof Publishing, 1982, pp 53–68.
45. Nakayama T, Kurachi Y, Noma A, Irisawa H: Action potential and membrane currents of single pacemaker cells of the rabbit heart. *Pflügers Arch* 1984;402:248–257.
46. Shibasaki T: Conductance and kinetics of delayed rectifier potassium channels in nodal cells of the rabbit heart. *J Physiol (Lond)* 1987;387:227–250.
47. Noma A, Irisawa H: Membrane currents in the rabbit sinoatrial node cell studied by the double microelectrode. *Pflügers Arch* 1976;364:251–258.
48. Yanagihara K, Noma A, Irisawa H: Reconstruction of sino-atrial node pacemaker potential based on the voltage clamp experiments. *Jap J Physiol* 1980;30:841–857.
49. Bean B: Two kinds of calcium channels in canine atrial cells: differences in kinetics, selectivity, and pharmacology. *J Gen Physiol* 1985;86:1–30.
50. Hagiwara N, Irisawa H, Kameyama M: Contribution of two types of calcium currents to the pacemaker potentials of rabbit sino-atrial node cells. *J Physiol (Lond)* 1988;395:85–100.
51. Di Francesco D: The cardiac hyperpolarizing activated current, I_f: Origins and developments. *Prog Biophys Molec Biol* 1982;46:163–183.
52. Brown H, Di Francesco D: Voltage clamp investigations of membrane currents underlying pace-

maker activity in rabbit sino-atrial node. *J Physiol* 1980;308:331–351.
53. Brown H, Di Francesco D, Noble S: Cardiac pacemaker oscillations and its modulation by autonomic transmitters. *J Exp Biol* 1979;81:175–204.
54. Noble D: Ionic basis of rhythmic activity in the heart, in Zipes D, Jalife J (eds): *Cardiac Electrophysiology and Arrhythmias.* Orlando, FL, Grune & Stratton, 1985, pp 3–11.
55. Di Francesco D: A new interpretation of the pacemaker current i_{K2} in Purkinje fibres. *J Physiol* 1981;314:359–376.
56. Callewaert G, Carmeliet E, Vereeke J: Single cardiac Purkinje cells: General electrophysiology and voltage clamp analysis of the pacemaker current. *J Physiol (Lond)* 1984;349:643–661.
57. Di Francesco D: Characterization of single pacemaker channels in cardiac sino-atrial cells. *Nature (Lond)* 1986;324:470–473.
58. Rozanski GJ: Electrophysiological properties of automatic fibers in rabbit atrioventricular valves. *Am J Physiol* 1987;253:H720–H727.
59. Di Francesco D, Noble D: Implications of the reinterpretation of i_{K2} for the modelling of electrical activity of pacemaker tissues in the heart, in Bouman LN, Jongsma HJ (eds): *Cardiac Rate and Rhythm: Development in Cardiovascular Medicine.* The Hague, Martinus Nijhoff, 1982, pp 93–128.
60. Gea-Ny S, Boyden P: Multiple types of Ca^{2+} currents in single canine Purkinje cells. *Circ Res* 1989; 65:1735–1750.
61. Trautwein W, Osterrieder W: Mechanisms of beta-adrenergic modulation and cholinergic control of Ca and K currents in the heart, in Nathan RD (ed): *Cardiac Muscle: The Regulation of Excitation and Contraction.* Florida, Academic Press, 1986, pp 87–128.
62. Noma A, Trautwein W: Relaxation of the ACh-induced potassium current in the rabbit sinoatrial node cell. *Pflügers Arch* 1978;377:193–200.
63. Brown HF: Electrophysiology of the sinoatrial node. *Physiol Rev* 1982;62:505–530.
64. Carmeliet E, Mubagwa K: Changes by acetylcholine of membrane currents in rabbit cardiac Purkinje fibres. *J Physiol (Lond)* 1986;371:201–217.
65. Di Francesco D, Tromba C: Muscarinic control of the hyperpolarization-activated current (i_F) induced by acetylcholine in rabbit sinoatrial node myocytes. *J Physiol (Lond)* 1988a;405:493–510.
66. Di Francesco D, Tromba C: Acetylcholine inhibits the activation of the cardiac pacemaker current, i_F. *Pflügers Arch* 1987;410:139–142.
67. Paes de Carvalho AP, Hoffman BF, de Carvalho MP: Two components of the cardiac action potential. I. Voltage-time course and the effect of acetylcholine on atrial and nodal cells of the rabbit heart. *J Gen Physiol* 1969;54:607–635.
68. Ikemoto Y, Goto M: Nature of the negative inotropic effect of acetylcholine on the myocardium. An elucidation of the bullfrog atrium. *Proc Jap Acad* 1975;51:501–505.
69. Giles WR, Noble SJ: Changes in membrane currents in bull frog atrium produced by acetylcholine. *J Physiol (Lond)* 1976;261:103–123.
70. Noma A, Trautwein W: Relaxation of the ACh-induced potassium current in the rabbit sinoatrial node cell. *Pflügers Arch* 1978;377:193–200.
71. Osterrieder W, Noma A, Trautwein W: On the kinetics of the potassium channel activated by acetylcholine in the S-A node of the rabbit heart. *Pflügers Arch* 1980;386:101–109.
72. Sakmann B, Noma A, Trautwein W: Acetylcholine activation of single muscarinic K^+ channels in isolated pacemaker cells of the mammalian heart. *Nature* 1983;303:250–253.
73. Breitwieser GE, Szabo G: Uncoupling of cardiac muscarinic and beta-adrenergic receptors from ion channels by a guanine nucleotide analogue. *Nature* 1985;280:235–236.
74. Soejima M, Noma A: Mode of regulation of ACh-sensitive K-channel by the muscarinic receptor in rabbit atrial cells. *Pflügers Arch* 1984;400:424–431.
75. Pfaffinger P, Martin J, Hunter D, Nathanson N, Hille B: G.T.P.-binding proteins couple cardiac muscarinic receptors to a K^+ channel. *Nature (Lond)* 1985;317:536–538.
76. Di Francesco D, Tromba C: Muscarinic control of the hyperpolarization-activated current (i_F) in rabbit sino-atrial node myocytes. *J Physiol (Lond)* 1951;115:74–94.
77. Pollack GH: Cardiac pacemaking: An obligatory role of catecholamines? *Science* 1977;196:731–738.
78. Jalife J, Hamilton AJ, Lamanna VR, Moe GK: Effects of current flow on pacemaker activity of the isolated kitten sinoatrial node. *Am J Physiol* 1980;238:H307–H316.
79. Jalife J, Slenter VAJ, Salata JJ: Dynamic vagal control of pacemaker activity in the mammalian sinoatrial node. *Circ Res* 1983;52:642–656.
80. Delmar M, Jalife J, Michaels D: Effects of changes in excitability and intercellular coupling on synchronization in the rabbit sinoatrial node. *J Physiol (Lond)* 1986;370:127–150.
81. Michaels DC, Matyas EP, Jalife J: Dynamic interactions and mutual synchronization of sinoatrial node pacemaker cells. *Circ Res* 1986;58:706–720.
82. James TN: Cardiac innervation: Anatomic and pharmacologic relations. *Bull NY Acad Med* 1967; 43:1041–1086.
83. Sperelakis N: Propagation mechanisms in heart. *Annu Rev Physiol* 1979;41:441–457.
84. Mann JE, Sperelakis N: Further development of a model for electrical transmission between myocardial cells not connected by low resistance pathways. *J Electrocardiol* 1979;12:23–33.
85. Jalife J: Mutual entrainment and electrical coupling as mechanisms for synchronous firing of rabbit sinoatrial pace-maker cells. *J Physiol (Lond)* 1984;356:221–243.
86. Guevara MR, Shrier A, Glass L: Phase resetting of spontaneously beating embryonic ventricular heart cell aggregates. *Am J Physiol* 1986;251:H1298–H1305.
87. Michaels DC, Matyas EP, Jalife J: A mathematical model of the effects of acetylcholine pulses on sinoatrial pacemaker activity. *Circ Res* 1984;55:89–101.
88. Veenstra RD, De Haan RL: Electrotonic interactions between aggregates of chick embryo cardiac pacemaker cells. *Am J Physiol* 1986;250:H453–H463.
89. Guevara MR, Shrier A: Phase resetting in a model of cardiac Purkinje fiber. *Biophys J* 1987;52:165–175.
90. De Mello WC: Intercellular communication and

junctional permeability, in Bittar E (ed): *Membrane Structure and Function*. New York, John Wiley and Sons, 1980, pp 128–170.

91. Masson-Pevet M, Bleeker WK, MacKaay AJC, Gross D, Bowman LN: Ultrastructural and functional aspects of rabbit sinoatrial node, in Bonke FIM (ed): *The Sinus Node. Structure, Function and Clinical Relevance*. The Hague, Martinus Nijohof, 1979, pp 195–211.
92. De Haan RL: In vitro models of entrainment of cardiac cells, in Bouman L, Jongsma H (eds): *Cardiac Rate and Rhythm*. Boston, Martinus Nijhof Publishing, 1982, pp 323–361.
93. Winfree AT: *The Geometry of Biological Time*. New York, Springer-Verlag, 1980.
94. Anumonwo JMB, Delmar M, Michaels DC, Jalife J: Phase resetting and entrainment properties of single sinus nodal cells. *Circ Res* 1991;68:1138–1153.
95. Michaels DC, Chialvo DR, Matyas EP, Jalife J: Chaotic activity in a mathematical model of the vagally driven sinoatrial node. *Circ Res* 1989;65:1350–1360.
96. Jalife J, Moe G: A biologic model of parasystole. *Am J Cardiol* 1979;43:761–772.
97. Moe GK, Jalife J, Mueller WJ, Moe B: A mathematical model of parasystole and its implication to clinical arrhythmias. *Circulation* 1977;56(6):968–979.
98. Jalife J, Moe GK: Effect of electrotonic potentials on pacemaker activity of canine Purkinje fibers in relation to parasystole. *Circ Res* 1976;39:801–808.
99. Courtemanche M, Glass L, Rosengarten MD, Goldberger AL: Beyond pure parasystole: Promises and problems in modelling complex cardiac arrhythmias. *Am J Physiol* 1989;257:H693–H706.
100. Tuganoski W, Kopeck P, Tarnowski W: Mechanisms of intercellular synchronization in the rabbit sinus node. *Experentia* 1981;37:485–487.
101. Michaels DC, Matyas EP, Jalife J: Mechanisms of sinoatrial pacemaker synchronization: A new hypothesis. *Circ Res* 1987;61:704–714.
102. Guevara MR, Shrier A, Glass L: Phase-locked rhythms in periodically stimulated heart cell aggregates. *Am J Physiol* 1988;254:H1–H10.
103. Green JH: Cardiac vagal efferent activity in the cat. *J Physiol (Lond)* 1959;149:47P–49P.
104. Jewett DK: Activity of single efferent fibres in the cervical vagus nerve of the dog, with special reference to special cardioinhibitory fibers. *J Physiol (Lond)* 1964;175:321–357.
105. Brown GL, Eccles JC: The action of a single vagal volley on the rhythm of the heartbeat. *J Physiol* 1934;82:211–241.
106. Levy MN, Martin PJ, Iano T: Paradoxical effect of the vagus nerve on heart rate in dogs. *Circ Res* 1969;25:303–314.
107. Suga H, Oshima M: Periodic variation of the heart rate caused by repetitive electric stimulation of cardiac vagus nerve. *Nippon Seirigaku Zasshi* 1969;31:33–34.
108. Jalife J, Michaels D: Phase dependent interactions of cardiac pacemakers as mechanisms of control and synchronization in the heart, in Zipes D, Jalife J (eds): *Cardiac Electrophysiology and Arrhythmias*. Orlando, FL, Grune & Stratton, 1985, pp 109–119.
109. Greco EC, Clark JW Jr: Changes in membrane currents in bull frog atrium produced by acetylcholine. *J Physiol (Lond)* 1976;23:192–199.
110. Winfree A: Electrical instability in cardiac muscle: Phase singularities and rotors. *J Theor Biol* 1989;138:353–405.
111. Frazier DW, Wolf PD, Wharton JM, Tabg ASL, Smith WM, Ideker RE: Stimulus-induced critical point mechanism for electrical initiation of reentry in normal canine myocardium. *J Clin Invest* 1983;83:1039–1052.
112. Davidenko JM, Kent PF, Chialvo DR, Michaels DC, Jalife J: Sustained vortex-like waves in normal isolated ventricular muscle. *Proc Natl Acad Sci USA* 1990;8785–8789.

Chapter **5**

Automaticity

Borys Surawicz, MD

Automaticity is an inherent property of cardiac muscle. Automatic action is achieved by means of spontaneous diastolic depolarization which is controlled by ionic transmembrane pacemaker currents and modulated by the autonomic nervous system and electrotonic interaction with the neighboring muscle fibers. The network of specialized cardiac fibers capable of generating diastolic depolarization is headed in the mammalian hearts by the sinoatrial node (SAN), followed by automatic fibers in the atria, the atrioventricular junction, and the His bundle–Purkinje fiber network system.

In the more highly developed hearts, such as those of birds, reptiles, or mammals, the pacemaker activity initiates a propagated impulse that precedes contraction, a process known as electromechanical coupling. In contrast, in a more primitive molluscan heart, the electrical pacemaker activity follows the mechanical stretch resulting in a mechanoelectrical coupling. In some invertebrate hearts that lack localized pacemakers, any myocardial cell can act as a pacemaker, and the pacemaker sites change frequently. In the crustacean heart, the cardiac ganglion is the pacemaker whereas in fish, reptiles, and amphibians, the pacemaker is located within sinus venosus. The SAN in the mammalian heart contains specialized P cells found mainly in the central region of the node. Numerous nerve endings are located near these P cells. An almost universal characteristic of pacemaker cells is their stretch sensitivity and the stretch-induced acceleration of activity.[1] Pacemaker cells are also sensitive to temperature, pressure, and various tactile, osmotic, and chemical stimuli. These properties enable them to serve as receptors for sensing time, pressure, volume, and chemical signals in the circulatory system.[2]

HIERARCHY OF PACEMAKERS

The SAN is the dominant pacemaker in the mammalian heart. It is the fastest pacemaker with the steepest slope of diastolic depolarization. Human SAN has multiple sites of impulse generation and shifts of the pacemaker site in the SAN occurs frequently after rapid atrial pacing or carotid sinus massage.[3] The slower subsidiary pacemakers are located within the internodal atrial tracts, the atrioventricular (AV) junction (consisting of the AV node and the His bundle), and the network of Purkinje fibers. This network extends from the bifurcation of the His bundle to the terminal twigs which penetrate the myocardium to various depths, depending on the species. The pacemaker fibers in the SAN and in certain parts of the AVN are small and poorly coupled to each other; they repolarize to membrane diastolic potential of only about −40 to −55 mV, and conduct the impulse at a much slower rate than the nonautomatic fibers. The low level of Maximal diastolic potential (MDP) is attributed to a high ratio of sodium (Na^+) to potassium (K^+) permeability (p_{Na}/p_K). In contrast, the subsidiary pacemaker fibers in the atria and in the Purkinje fibers are larger than those of the neighboring myocardium, and their MDP is either the same or more negative than the resting membrane potential (RMP) of the nonautomatic fibers. As a result, the Purkinje fibers conduct the impulse three to four times as rapidly as the ventricular myocardium.

The intrinsic rate of pacemaker firing is determined by their anatomic location. The automatic activity becomes progressively slower along the path from the SAN to the distal Purkinje fibers (Fig. 5.1). The purpose of such hierarchy is to create an efficient system of escape mech-

Published 1991 by Elsevier Science Publishing Co., Inc.
655 Avenue of the Americas, New York, NY 10010
Current Topics in Cardiology

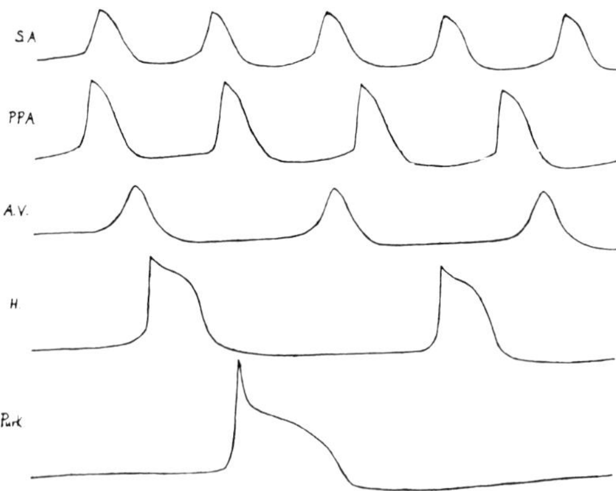

FIGURE 5.1 Diagrammatic representation of the hierarchy of the five types of pacemaker fibers: SA = sinoatrial node; PPA = potential pacemaker in the atria; A.V. = AV junction (lower part of AV node); H = His bundle; Purk = Purkinje fiber. Note the decreasing rate of discharge as one proceeds from the SAN caudally. *Modified from Hoffman and Cranefield,* Electrophysiology of the Heart, *New York, McGraw-Hill, 1960.*

anisms that spring into action when separated anatomically or functionally from the more rapidly firing, proximally located automatic fibers. Accordingly, the failure of the SAN will result in subsidiary pacemakers. Should the latter fail, the pacemaker will shift into the AV junction, and an impulse in both bundle branches will activate one of the escape pacemakers in the Purkinje network. It would appear that the abundance of the Purkinje fibers should protect the heart from standstill, even after very extensive damage of the ventricular myocardium. Unfortunately the safeguard is not always adequate because simultaneous failure of all escape pacemakers can occur resulting in cardiac arrest. This is one of the mechanisms of sudden cardiac death.

MECHANISM OF DIASTOLIC DEPOLARIZATION

Slow diastolic depolarization, which brings the membrane potential to the threshold potential for rapid depolarization, is caused by one or more cardiac membrane currents, and the mechanisms differ among different types of automatic fibers.

Purkinje Fibers

The earliest studies of the pacemaker current were attempted in the Purkinje fibers. The underlying assumption is that when the fiber is quiescent at the level of RMP the inward current carried by Na^+ (i_{Na}) and the outward current carried by K^+ (i_K) are in equilibrium, that is:

$$i_{Na} = i_K$$

i_{Na} and i_K can be described as follows:

$$i_{Na} = (E - E_{Na}) \times g_{Na}$$

$$i_K = (E - E_K) \times g_K$$

where E = resting membrane potential, E_{Na} and E_K are the equilibrium potentials* for the respective ions, and g_{Na} and g_K are conductances for the respective ions. Since $(E - E_{Na})$ is large and $(E - E_K)$ is very small, it is obvious that the equilibrium is caused by a high ratio of g_{Na}/g_K.

Pacemaker activity will occur when i_{Na} exceeds i_K, which may be caused by an increase in g_{Na} or a decrease in g_K. To distinguish between the two putative mechanisms of diastolic depolarization, that is, the effect of increasing i_{Na} versus that of decreasing i_K, small rectangular current pulses were led into the fiber during diastolic depolarization and a steady polarizing current was superimposed so that changes in membrane input resistance could be followed during diastole at the same membrane potential.[4] This procedure, which was later modified by the use of voltage clamp,[5] revealed a time- and voltage-dependent decrease in membrane conductance during diastole. These findings were more consistent with a decrease in g_K than with an increase in g_{Na}.[6] Thus, it was assumed that the pacemaker current was an outward i_K which gradually decayed in the presence of a stable background inward i_{Na}. Subsequent analysis of the voltage-clamp studies in Purkinje fibers identified a separate time- and voltage-dependent membrane "pacemaker current"[7] which was activated at the level of RMP and deactivated when the potential became more negative. This was believed to be a nearly pure i_K because its reversal potential was close to E_K.

For a number of years, the pacemaker current, designated as i_{K2}, was regarded as the best-characterized membrane current in the heart, and the paper that described this current[7] became a citation classic.[8] However, more recently the existence of i_{K2} was put in doubt, and the mechanism of pacemaker current in the Pur-

* The equilibrium potential of an ion is the potential at which the influx of the ion across the membrane is balanced by the efflux of the ion so that the net ionic current is zero. E_{Na} is approximately +55 mV and E_k is approximately −85 mV, at normal extracellular concentrations of these ions.

kinje fibers underwent reinterpretation. The error was attributed to the neglected phenomenon of accumulation/depletion in the extracellular clefts in the Purkinje fibers.

Deep and tortuous clefts separating the Purkinje fibers can be viewed as stagnant pools where the concentrations of ions differ from those in the bulk of the extracellular fluid. This is particularly important for the extracellular K^+ concentration $[K^+]_o$ because it is low. It has been suspected that the $[K^+]$ changes occurring in the clefts during depolarization and repolarization could distort the time course of ionic currents flowing in the vicinity of E_K. Indeed, it has been shown that when brief positive current pulses, aimed at depleting the clefts of K^+, were applied before hyperpolarizing the membrane, the reversal potential of i_{K2} became more negative.

In the meantime, two groups of investigators [9,10] discovered in the SAN a current that was activated upon hyperpolarization with a time constant of 2 to 4 s, and they named it a hyperpolarization-activated current, alias i_F (see later). Di Francesco[11] postulated that i_F was the real pacemaker current in the Purkinje fibers whereas i_{K2} was an artifact caused by an overlapping i_F and K^+ depletion current. Indeed when i_K was blocked by barium (Ba^{2+}), i_F became manifest in the Purkinje fibers. The current was decreased by lowering $[Na^+]_o$, and in contrast to i_{K2} did not reverse at E_K even at high $[K^+]$.[11]

The earlier findings of declining conductance during diastolic depolarization could be explained by the overlapping voltage-dependent decline of the i_K caused by K^+ depletion. The previously described reversal potential attributed to i_{K2}[7] turned out to be an artifact simulated by the net zero current at the crossing of the increasing inward i_F and the decaying "depletion" of the inward rectifying K current. Application of Ba^{2+} abolished this spurious reversal potential even after large hyperpolarization.[11]

The discovery of a depolarizing current activated by hyperpolarization revealed an exceptional current because all heretofore known membrane currents had in common activation upon depolarization. Hence, the authors gave the current the name i_F, where F stands for funny. The i_F was first discovered in the frog sinus venosus,[12] later in the rabbit SAN,[9,10] rabbit AVN,[13] and in isolated Purkinje fibers from dog, sheep, and cow hearts.[14] The demonstration of i_F activated within the range of -70 to -120 mV in the isolated Purkinje fibers was important because these fibers are devoid of intercellular clefts. The reversal potential of i_F is about -20 mV, which is midway between E_{Na} and E_K. This and other evidences suggested that the current was carried by both Na^+ and K^+ ions but, in the pacemaker range, predominantly by Na ions.[15] Similar current systems have been recorded in noncardiac tissues, for example, photoreceptors in salamander. The biological function of this current is assumed to represent a safety mechanism consisting of a negative feedback which limits the level of hyperpolarization. For instance, in the retina such current, by limiting the hyperpolarization produced by light stimulation, restores the level of transmitter release that takes place in the dark.[11] Today, it is believed that the i_F is the most important current responsible for diastolic depolarization in Purkinje fibers (Fig. 5.2). However, other pacemakers appear to have other mechanisms of diastolic depolarization.

Sinoatrial Node

The information about the membrane currents in these fibers was obtained later than it was in Purkinje fibers because the small size of the pacemaker cells in the SAN made it difficult to apply the voltage clamp. The successful voltage clamp was first accomplished in small pieces of rabbit SAN,[16] and later in isolated rabbit SAN cells.[17] Under normal conditions the MDP of the SAN is not sufficiently negative to activate i_F, and this current plays no role in the mechanism of pacemaker activity.

The mechanism of diastolic depolarization in SAN appears to be more complex than in Purkinje fibers, and requires two major components, namely a decay of an outward i_K and an inward current. In one study, pacemaker activity was attributed to the decaying time-dependent i_K activated within the range of -60 to $+10$mV.[18] In another study, this current could be divided into two components.[19] However, cardiac cells contain several ligand-gated K channels, such as the ATP-sensitive channel, the acetylcholine (ACh)-activated channel, the Na^+-activated channel, and perhaps other channels. It is not known to what extent they may contribute to the pacemaker activity.[20]

Another major component of the mechanism of diastolic depolarization is a depolarizing inward current, which is most likely a calcium current (i_{si}). This current probably plays a greater role during the last third of diastolic depolarization when membrane potential is closer to the activation threshold of i_{si}.[21] Figure 5.3 shows the arrangement of currents used to reconstruct diastolic depolarization in the SAN in the model of Noble and Di Francesco.[15]

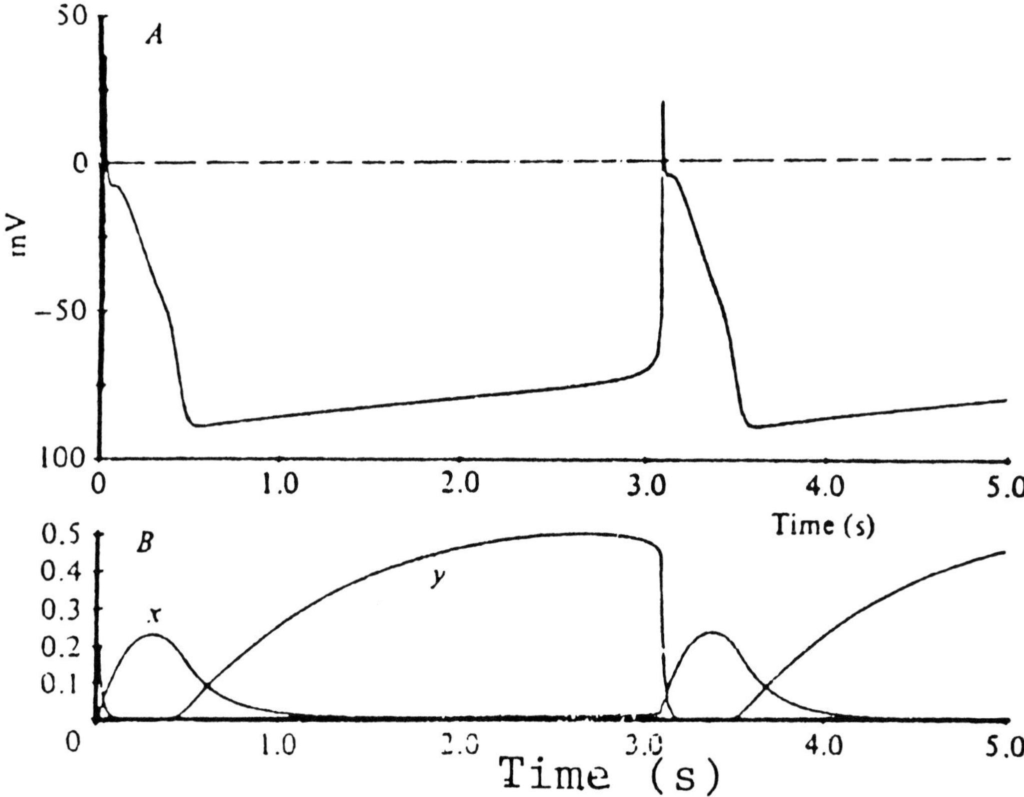

FIGURE 5.2 Reconstruction of the ionic mechanism responsible for the pacemaker activity in the Purkinje fiber (**A**). **B.** i_F (controlled by the gating *y* process) plays a major role in generating the diastolic depolarization whereas the role of decaying i_K (controlled by the *x* gating process) is small. *Reprinted from Figure 1 in Noble D: Ionic bases of rhythmic activity in the heart, in* Cardiac Electrophysiology and Arrhythmias, *New York, Grune & Stratton, 1985.*

Subsidiary Atrial Pacemaker

Subsidiary atrial pacemakers develop spontaneous diastolic depolarization after removal or suppression of SAN. In the dog, these pacemakers have been studied at the junction of the right atrium with the inferior vena cava and along the posterior internodal pathway extending from the SAN to the coronary sinus, and in the cat at the eustachian ridge of the right atrium. The underlying mechanism is primarily a time- and voltage-dependent i_F current activated by hyperpolarization with participation of increasing i_{si} and declining i_K.[22]

Atrioventricular Junction

In the dog automatic activity can arise in three different parts of the AV junction, namely the approaches to AV node (perinodal fibers), AV node, and His bundle.[23] In the rabbit, the pacemaker was always located in the lower portion of the AV node, and the pacemaker rate increased when the AV node was isolated from both atrial tissue and the His bundle. This suggests that normally these fibers are electrotonically suppressed by the connecting myocardium.[24]

Automaticity in Atrioventricular Valves

The AV valves contain sparsely arranged cardiac fibers that are in direct continuity with atrial myocardium. The MDP of these fibers is about −75 to −80 mV. They have pacemaker properties and may play a role in cardiac arrhythmias.[25,26]

Autonomic Nervous Control of Pacemaker Activity

β-Adrenergic stimulation and adrenaline increase the slope of diastolic depolarization and enhance automaticity,[27–32] whereas vagal stimulation and ACh decrease the rate of diastolic depolarization and cause hyperpolarization.[33]

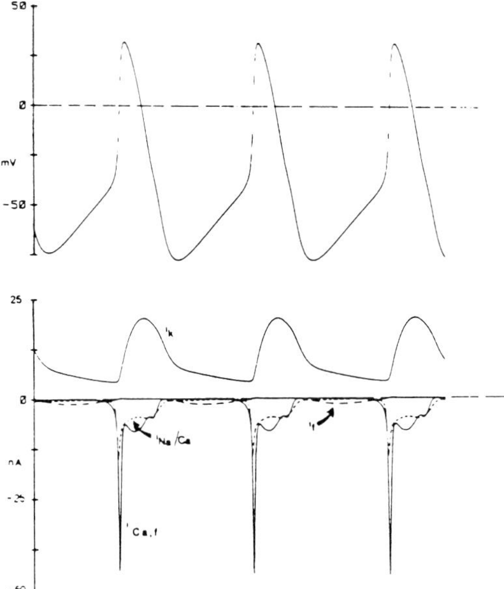

FIGURE 5.3 Reconstruction of the ionic mechanism responsible for the pacemaker activity in the SAN (**A**). **B.** The decay of i_K plays a major role in generating the diastolic depolarization with a small contribution of i_F. In this diagram the calcium current ($i_{Ca'f}$) plays a role only in generating the upstroke of the AP. *Reprinted from Figure 2 in Noble D: Ionic bases of rhythmic activity in the heart, in* Cardiac Electrophysiology and Arrhythmias. *New York, Grune & Stratton, 1985.*

However, all adrenergic agonists produce a biphasic response, that is, a decreased rate at low and an increased rate at high concentrations.[28,29] The slowing at low concentrations is an α-adrenergic effect because it was abolished by phentolamine.[28] Under certain conditions vagal stimulation may paradoxically increase the rate of pacemaker discharge. This occurs in the presence of atropine,[34] and also in the absence of atropine, stimulation of the vagus may lead to postvagal tachycardia.[35–37] This postvagal tachycardia has been attributed to a reflex sympathetic stimulation caused by hypotension, but a more likely mechanism is the excitation of parasympathetic fibers that leads to liberation of catecholamines from extramural sources.[35,37]

In terms of membrane currents, the effect of adrenaline on SAN is attributed to an increased i_{si} because it can be abolished by blocking this current.[21] In Purkinje fibers, adrenaline also enhances i_F. The effect of ACh is attributed to an increased g_K,[38] but in some species ACh also decreases i_{si}.[36] ACh has no effect on i_F in the absence of β-adrenergic stimulation, but after β-adrenergic stimulation ACh returns i_F to control level. These actions prevent excessive slowing of pacemaker activity in the presence of high vagal tone and prevent excessive increase in rate when both parasympathetic and sympathetic tone are high.[39]

For the discussion of the phenomena of overdrive suppression and acceleration the reader is referred elsewhere.[40]

Other Influences

Automaticity is depressed by high $[K_o^+]$ and enhanced by low $[K_o^+]$.[41] Sensitivity of different pacemakers to K^+ varies. Pacemaker activity of the SAN persists at high $[K^+]$ (20 mM) when atrial activity is suppressed. Pacemaker activity is sensitive to hypoxia and metabolic inhibitors but is resistant to cold. When excised and immersed in Tyrode's solution at 0 to 3°C, the SAN survives for several days and retains normal sensitivity to neurotransmitters.[42] The slope of diastolic depolarization and the rate of firing increase progressively from 32°C to 43°C with a Q value of 3.2.[43] Histamine accelerates ventricular[44] and SAN automaticity[45] by stimulating the H_2 receptors. Alkaline milieu enhances and acid milieu suppresses automaticity.[46] Volatile anesthetics attenuate sympatomimetic action.[47] The specific bradycardia-inducing agents alindil and falipamil slow the rate without involvement of β-adrenergic and cholinergic receptors.[48,49]

Modulation of Pacemaker Activity

Brief depolarizing or hyperpolarizing current pulses cause transient shifts in the slope of diastolic depolarization. The magnitude and direction of the phase shift varies as a function of polarity, current strength, and, most importantly, the phase at which the modifying current pulse is applied.[50–52] These changes reflect voltage-dependent effect on the underlying pacemaker current. For example, brief subthreshold depolarization applied during the early course of diastolic depolarization may temporarily reactivate the decaying outward current, which would lead to hyperpolarization and consequently a delay in approaching the threshold. In contrast, brief hyperpolarization applied during the same period may accelerate the decay of the outward current, increase the depolarizing effect of the background inward current, and consequently accelerate the approach to threshold. Current pulses applied during the late phase of diastolic depolarization have opposite effects, presumably by the action on the i_{si}. Therefore, the corrected course of diastolic depolarization depends on the phase of diastolic depolarization. It has been shown that cardiac pacemakers, like other oscillatory systems, display a phase-dependent sensitivity to the input of various correction factors. Therefore, each perturbation, such as electrotonic depolarizations and neurotransmitters,[51,52] can produce a phase-response curve that typically consists of a period of acceleration followed by a period of retardation. The characteristics of the phase-response curve permit a quantitative description of any given correction factor that modulates the rate and rhythm of pacemaker activity.[50]

ABNORMAL AUTOMATICITY

Abnormal automaticity can arise in any depolarized Purkinje, atrial, or ventricular fiber. Depolarization can be achieved by application of depolarizing current or various experimental manipulations that affect either the slow diastolic depolarization or the regenerative upstrokes.[53–57] In animal models, the application of depolarizing current pulses drives the membrane potential to a range where the inward i_{Ca} and outward i_K are nearly equal, so that the inward current partially counterbalances the hyperpolarizing effect of the background i_K.[58]

Purkinje Fibers

In the Purkinje fibers, rhythmic automatic depolarizations are generated in the voltage range of -65 to -10 mV, presumably by the decreasing influence of the outward i_K, which is normally responsible for the repolarization to the resting potential. The inward current responsible for the depolarization may be an incompletely inactivated i_{Na} or calcium current, i_{si}. The abnormal automaticity can be elicited in quiescent Purkinje fibers by various interventions which decrease g_K. These include, among others: low $[K^+]$, hypoxic acidic Tyrode's solution, loading with tetraethylammonium chloride (TEAC), application of Ba^{2+} which decreases the outward current[59,60] and increases the slow inward current,[61] anoxia, Cl-free solutions, solutions of very low ionic strength, and aconitine. They are

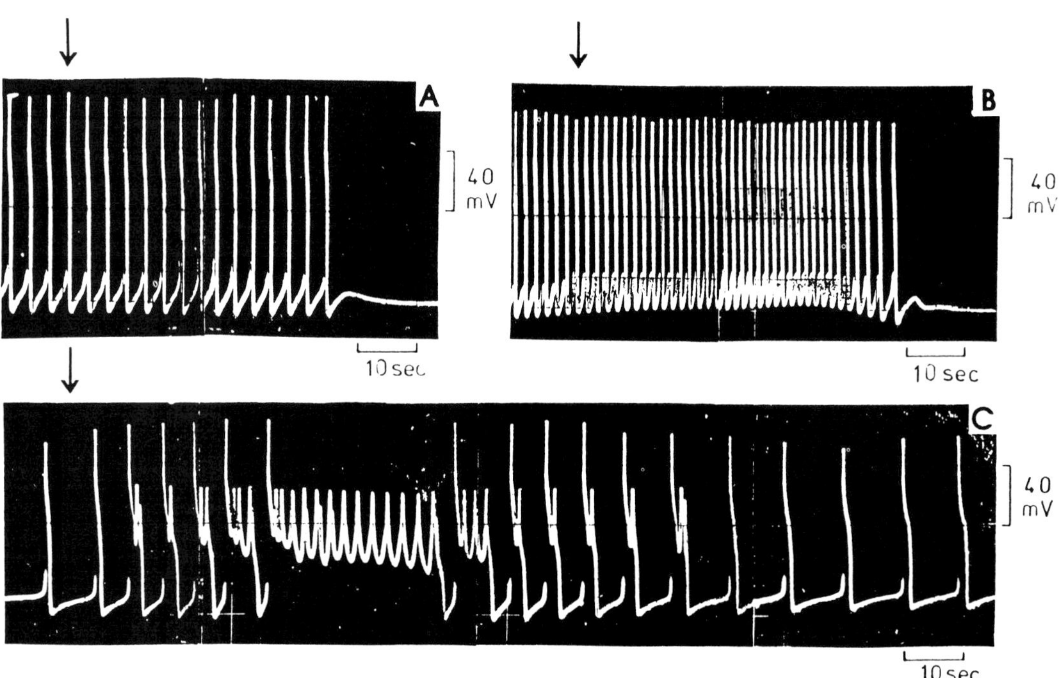

FIGURE 5.4 Action potentials from canine Purkinje fibers. The effect of changes from low K^+ (0.54 mM) to high K^+ (10.8 mM) solution (**A**), from low K^+ to control (5.4 mM K^+) without tetraethylammonium chloride (TEAC) (**B**), and from low K^+ to control solution in a TEAC-loaded fiber (**C**). In each panel the arrows mark the time of solution change and the recording is continuous. Note that the change to high K^+ (**A**) causes an earlier arrest of pacemaker activity than change to control solution (**B**), and that arrest does not occur in the TEAC-loaded fiber. Further note two types of automatic activity occurring from two different levels of diastolic membrane potential in **C**. *Reprinted from Ito and Surawicz,[63] with permission.*

also observed in excised fibers that have not recovered from the initial disturbance of excision.[62] In depolarized fibers, the activity can be enhanced by epinephrine presumably due to an increase in i_{si}.

In TEAC-loaded fibers, abnormal automaticity was facilitated by low K_o and suppressed by high K_o.[63] By adding a critical amount of K^+, one can stop the abnormal automatic activity without repolarizing to the MDP so that the fiber remains depolarized at a membrane potential of about −30 to −40 mV, where the outward and the inward current appear to be in an approximate balance. Under these conditions, a very small depolarization will initiate abnormal automaticity and a very small hyperpolarization will rapidly increase the rate of repolarization and bring the fiber to normal RMP (Fig. 5.4). Thus, we deal with two levels of resting potential which means two levels of membrane potential at which the net membrane current is zero. Automaticity can originate at each of these two levels.[64] The current–voltage relation in the region between two levels of resting potential corresponds to the negative chord conductance, which means that in this region membrane resistance increases as the membrane potential becomes less negative. It is assumed that the experimental interventions that decrease g_K abolish the negative chord conductance, and thereby facilitate the shift from the more negative to the less negative membrane potential.[65]

Ventricular Myocardial Fibers

The mechanism of abnormal automaticity is intimately related to decreased g_K. From his study of embryonic heart cells, Sperelakis[66] concluded that "all nonpacemaker myocardial cells in culture are capable of being converted into pacemaker cells by a decrease in g_K, and all pacemaker cells can be converted into nonpacemaker cells by an increase in g_K." Some of the experimental procedures used to transform the myocardial nonpacemaker fibers into pacemaker fibers include: (a) a decrease in $[K^+]_o$ alone or together with a decreased $[Ca^{2+}]$; (b) application of blockers of various i_K, for example, barium, cesium and 4-amino pyridine; (c) application of aconitine; (d) stretch; (e) application of agents that prolong inactivation of i_{Na}, for example, anthopleurin A[67,68]; and (f) application of electric currents.

Detailed description of these procedures has been reviewed elsewhere.[69,70] Although by definition, abnormal automaticity originates in depolarized myocardium, the level of depolarization varies in different studies. In some preparations, application of electric currents induced automatic activity at levels more negative than −60 mV, that is, the levels at which the rapid inward i_{Na} could be responsible for the maintenance of depolarizations.[55] However, in the majority of studies, the threshold potential for automatic depolarizations was less negative, and the depolarizations were dependent on slow inward current. This type of automaticity was suppressed by verapamil, manganese, and increased $[K_o^+]$[56] but not appreciably affected by the Na^+ channel blocker lidocaine.[71]

Similar to Purkinje fibers, the conditions that cause abnormal automaticity by blocking i_K cause an increase in the net inward current in the region of negative chord conductance (Fig. 5.5). Similar effect can be achieved by prolonging the activation of i_{Na}[67] (Fig. 5.6B), a procedure that also induces abnormal automaticity.[68]

Atrial Fibers

Abnormal automaticity in the atrial fibers has been induced by the same procedures as in the

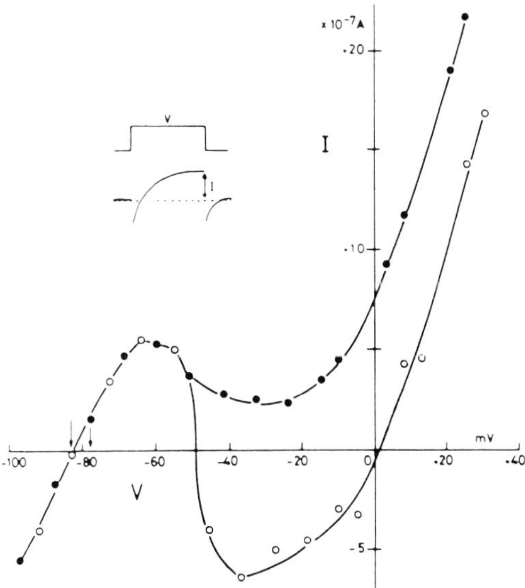

FIGURE 5.5 Effect of anthopleurin-A on the current–voltage relation of the outward current recorded at the end of 1 s clamp from the holding potential of −83 mV to various levels of membrane potential in dog ventricular muscle fiber. The current was measured from current zero as shown in the inset. Note that anthopleurin (open circles) decreases the outward current compared to control (solid circles) within a wide range of membrane potentials. *Reprinted from Hashimoto et al.,*[67] *with permission.*

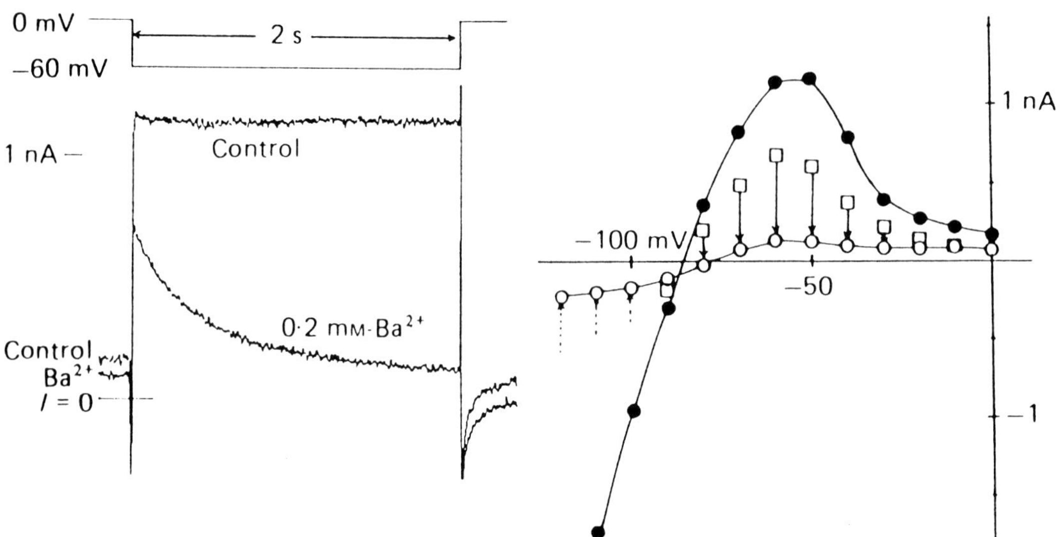

FIGURE 5.6 Effect of Ba^{2+} on the membrane current during repolarization from the holding potential of 0 mV to −60 mV (**A**), and on the current–voltage relation (**B**) in a guinea pig myocyte. Note that in **A**, Ba^{2+} decreases the positive membrane current at the onset of clamp and causes a decaying outward current. In **B**, Ba^{2+} decreases the outward current during a wide range of membrane potentials, both at the end of 2 s (solid versus open dots) and at the onset of clamp (points of arrows arising from the control values marked by squares). *Reprinted from Hirano and Hiraoka,*[60] *with permission.*

ventricular fibers.[69] Recently, abnormal automaticity has been studied in depolarized human atrial tissues[58,72,73] where it was promoted by chamber dilation.[73]

AUTOMATIC RHYTHMS IN ANIMAL MODELS

As discussed earlier, automatic rhythms can develop from a wide range of MDP depolarizations arising from the level of about −70 to −90 mV and generated by rapid i_{Na}, and depolarizations from the MDP less negative than −60 mV generated by slow inward current, that is, predominantly i_{Ca}. Within the intermediate range of MDP, that is, between −60 and −70 mV, the upstroke of action potential (AP) may be generated by an incompletely inactivated rapid i_{Na}, a partially activated i_{Ca}, or both.

Another way of subdividing automatic rhythms is based on differences in the mode of onset of automatic depolarization that can arise from smooth diastolic depolarizations, early afterdepolarizations, or late diastolic depolarizations. In the latter case, AP can be generated by a single late diastolic depolarization reaching the threshold of activation, or by two or more successive diastolic depolarizations of gradually increasing amplitude.

The characteristics of automatic activity depend also on the mode of initiation, which can be either spontaneous or triggered by single or repetitive stimuli. Thus, a complete description of an automatic rhythm should include all three characteristics, that is, the level of MDP, the preceding course of diastolic depolarization, and the mode of initiation. Such information can be derived indirectly by studying tissues isolated from the sites of assumed automatic foci in models of experimental arrhythmia.

Normal automaticity in Purkinje fibers has been studied in escape rhythms occurring in animals with complete AV block. Abnormal automaticity has been investigated in various cardiac tissues but most often in Purkinje fibers from infarcted myocardium of dogs and cats after coronary ligation.[74–76] The normal, that is, slow, escape rhythms form an uninterrupted series of depolarizations that respond to overdrive first by suppression and then by subsequent gradual resumption of automatic activity.[77] The abnormal, that is, fast, automaticity which can also occur spontaneously in dogs with complete AV block[78] is either intermittent or continuous, and is usually initiated (triggered) by an impulse with different morphology. Abnormal automaticity tends to undergo moderate deceleration before ceasing, and when no longer present can be brought back by fast driving. The abnormal rhythms are suppressed by fast driving, but the suppression is often preceded by transient acceleration. They can be progressively accelerated by repeated short periods of driving during recovery from prolonged overdrive, and by sympathetic stimulation or norepinephrine.[77] Rhythms of this type are suppressed by verapamil.[79] Studies of triggered activity in dog Purkinje fibers from 1-day-old myocardial infarction[74] have shown that this activity arose from delayed afterdepolarizations, and that it was initiated either by stimulated impulses or by the background slow normal Purkinje automaticity. With increasing rate of stimulation, the rate of triggered automatic pacemaker increased, and the coupling interval of the first depolarization decreased. Varying degrees of entrance and exit block around sites of triggered activity were commonly present, and some rhythms assumed the pattern of parasystole.[74] Also, patterns of modulated parasystole have been observed in the same type of preparation.[76] The amplitude of delayed afterdepolarizations underlying the automatic rhythms was increased by epinephrine and Ca^{2+} which facilitated the triggered activity, and decreased by verapamil which attenuated or suppressed the triggered activity.[74] Consistent with these observations in vitro is the finding that sympathetic stimulation accelerates ventricular tachycardia in dogs with 1-day-old myocardial infarction.[80] Triggered activity arising from delayed diastolic afterdepolarizations was also recorded in Purkinje fibers from chronic infarcts in cats.[75] In this preparation, the amplitude of delayed afterdepolarizations was augmented by increased $[Ca^{2+}]_o$, isoproterenol, and phenylephrine in the presence of propranolol.[75]

The delayed afterdepolarizations underlying triggered automaticity in Purkinje fibers from the infarcted myocardium are similar to delayed afterdepolarizations induced by digitalis but in the setting of myocardial ischemia they are probably caused by Ca^{2+} overload, or accumulation of lysophosphoglycerides.[81]

The response to overdrive stimulation has been considered helpful in the differential diagnosis of the mechanism of automaticity because normal automaticity is consistently suppressed by overdrive, whereas abnormal Purkinje fiber automaticity arising from MDP of -60 to -70 mV is only slightly suppressed by overdrive, and abnormal automaticity arising from MDP less negative than -60 mV is not suppressed at all.[82] However, in the atrial fibers from coronary sinus abnormal triggered automaticity was suppressed by overdrive,[83] and as mentioned earlier abnormal automaticity in the Purkinje fibers was also partially suppressed by overdrive.[77] Thus, the absence of suppression mitigates against normal automaticity, but the presence of suppression is less helpful in the differential diagnosis.

Single premature stimuli were found to reset nontriggered automatic rhythms originating at all levels of MDP; however, in triggered automatic rhythm induced by Ba^{2+}, single premature stimuli shortened the first return cycle or produced a short arrest.[84]

AUTOMATIC AND TRIGGERED RHYTHMS IN HUMANS

The mechanism of an ectopic rhythm cannot be deduced from surface ECG. Intracardiac recordings can sometimes pinpoint the mechanism, but even this procedure leaves the question of mechanism open in most cases.

In the clinical setting automatic rhythms can be suspected when they: (a) originate in the conducting system; (b) are parasystolic; (c) respond in a predictable manner to an electrophysiologic or a pharmacologic intervention known to affect automaticity. Triggered rhythms are suspected when the arrhythmia simulates the behavior of triggered rhythms elicited in the laboratories of cellular electrophysiology.

Origin in the Specialized Conducting System

Slow ectopic escape rhythms are nearly certainly automatic but all parts of the conducting system are also capable of generating accelerated automatic rhythms. The mere recognition that an arrhythmia originates in the conducting system provides no assurance that automaticity is the underlying mechanism. The suspicion of an automatic rhythm is heightened when a slow escape rhythm coexists with an accelerated rhythm from the same focus, that is, accelerated AV junctional rhythm.[85] Sinus rhythm is usually automatic but sinus tachycardia may be caused by reentry, particularly in symptomatic patients.[86] Most ectopic atrial arrhythmias appear to be automatic, but atrial reentry is known to occur.[87] In the cases of AV junctional or fascicular tachycardia, the probability of automaticity is increased when tachycardia is precipitated by

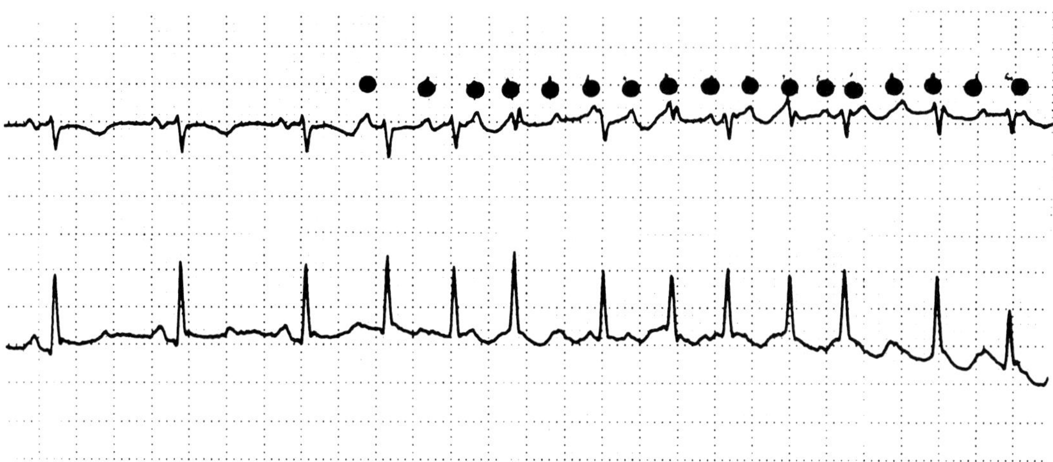

FIGURE 5.7 An example of presumably automatic atrial tachycardia with P-wave morphology similar to that of sinus rhythm. Note the gradual increase in rate of the atrial ectopic rhythm ("warming up" effect) and the nearly constant cycle after the third ectopic P wave. Two synchronous leads from an ambulatory recording.

factors known to increase automaticity in vitro, for example, digitalis toxicity, hypokalemia, or sympathetic stimulation. Also the appearance of incessant tachycardia in childhood or adolescence favors the mechanism of automaticity.[88]

Parasystolic Rhythms

Parasystole is usually attributed to automaticity,[89–91] but the proof of a parasystolic rhythm is difficult to establish unless the parasystolic activity is slow, regular, and well-"protected" from modulating influences of background activity. The diagnostic certainty lessens with increasing irregularity, decreasing entrance block ("protection"), increasing exit block, and presence of intermittence. Rapid parasystolic activity has been postulated in patients with uni- and multifocal AV junctional or ventricular tachycardia with or without exit block in a number of case reports.[89–99] However, even the most laborious calculations and the most ingenious deductions cannot pro-

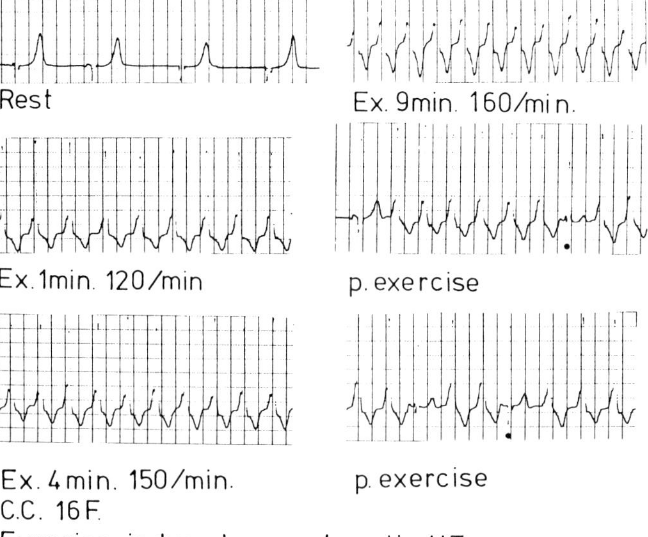

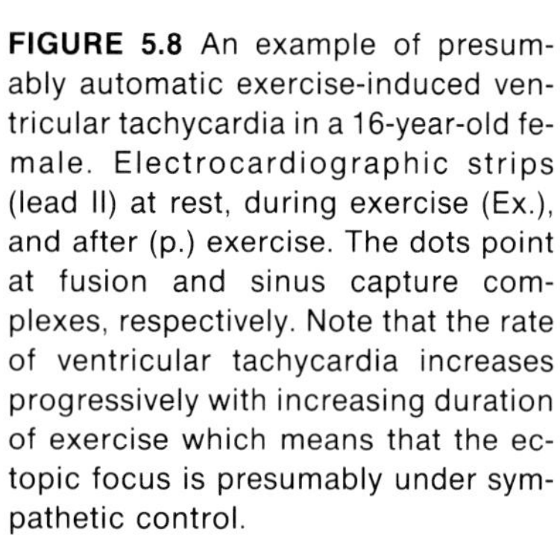

FIGURE 5.8 An example of presumably automatic exercise-induced ventricular tachycardia in a 16-year-old female. Electrocardiographic strips (lead II) at rest, during exercise (Ex.), and after (p.) exercise. The dots point at fusion and sinus capture complexes, respectively. Note that the rate of ventricular tachycardia increases progressively with increasing duration of exercise which means that the ectopic focus is presumably under sympathetic control.

vide a definite proof of either parasystole or automaticity in such cases.

Clinical Characteristics

The following characteristics suggest automaticity in patients with ectopic atrial tachycardias[100]: (a) initiation by premature atrial depolarization which exhibits no delay in AV nodal conduction; (b) atrial cycle during tachycardia is not dependent on AV nodal conduction; (c) progressive "warming up" at the onset; (d) premature stimulation during tachycardia resets the atrial cycle; and (e) single premature stimuli and overdrive fail to suppress the tachycardia.

Automatic supraventricular tachycardias tend to be incessant,[101] and the intermittent sinus rhythm is either absent or rare. P-wave configuration of the initiating premature complex is identical to P waves of succeeding complexes. After the initial progressive shortening of cycle ("warm up" phenomenon) P–P intervals do not vary by more than 50 ms unless an exit block is present (Fig. 5.7). The rate of tachycardia varies from day to day and from hour to hour,[101] being influenced by changes in autonomic tone. Characteristically, though not invariably, overdrive pacing captures the atrium but upon cessation of pacing tachycardia immediately resumes without an intervening sinus rhythm.[101]

Automatic mechanism has been suspected in patients with accelerated ventricular rhythm or ventricular tachycardia occurring within 24 h after onset of acute myocardial infarction because in this setting ventricular tachycardia could not be induced by premature stimulation.[102] Similarly, automatic mechanism has been postulated in a group of patients with exercise-provocable ventricular tachycardia of long duration characterized by left bundle branch block pattern morphology. These tachycardias are induced by isoproterenol but not by programmed electrical stimulation[103] and they respond to treatment with β-blockers.[103] An example of such tachycardia is shown (Fig. 5.8).

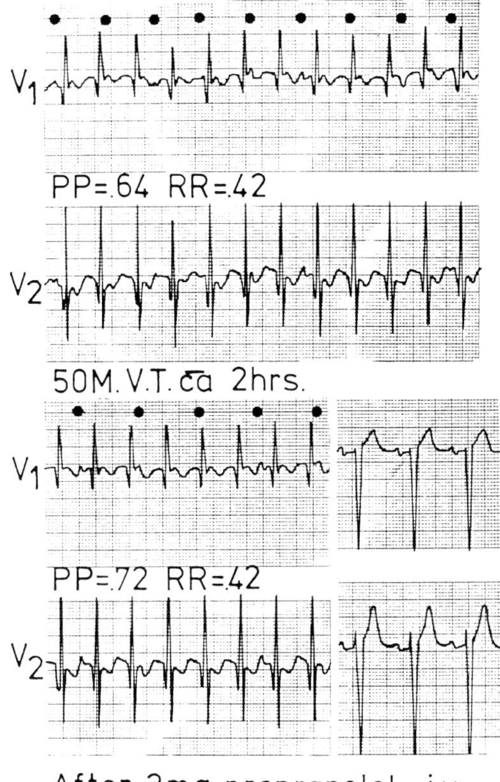

FIGURE 5.9 Effect of propranolol on ventricular tachycardia documented by AV dissociation (black dots indicate P waves). The upper two strips show leads V1 and V2 during tachycardia of approximately 2 h duration resistant to lidocaine and procainamide in a 50-year-old male. The two lower strips on the left show slowing of sinus rhythm during administration of 3 mg of propranolol intravenously. Shortly after completion of propranolol infusion, ventricular tachycardia stopped as indicated by the strips showing sinus rhythm on the lower right side. Automatic mechanism of tachycardia is suspected because it responded to treatment with propranolol, even though tachycardia ceased abruptly without prior slowing of ventricular rate.

Pharmacologic Observations

β-Blocking therapy can suppress spontaneous ventricular arrhythmia (Fig. 5.9) and prevent inducibility of ventricular tachycardia[104] in a number of patients. However, such observations do not prove the automaticity dependence on increased sympathetic stimulation because various direct and indirect actions of β-blocking can affect arrhythmias precipitated by other mechanisms. The same applies to facilitation of ventricular tachycardia induction by isoproterenol,[105,106] and to the observations that certain ventricular arrhythmias can be suppressed by phenylephrine,[107,108] Valsalva maneuver,[109,110] or carotid sinus stimulation.[111]

Does Triggered Activity Have a Role in the Genesis of Cardiac Arrhythmias?

This question was answered by Rosen and Reder[112] with a qualified "yes" because of the similarity of certain digitalis-induced arrhyth-

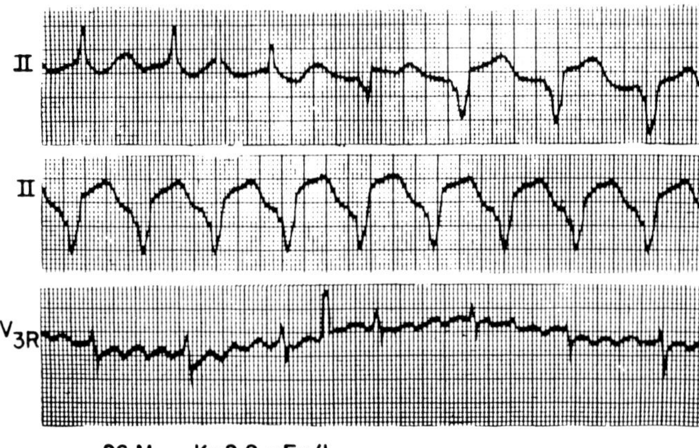

FIGURE 5.10 Example of possible triggered automaticity in a 26-year-old male with hypokalemia and presumed digitalis toxicity. The two upper strips show a continuous record during atrial fibrillation and accelerated AV junctional rhythm which is followed by a fusion complex and subsequent ventricular tachycardia at a slightly more rapid rate. The lowest strip shows resumption of accelerated AV junctional rhythm. It may be assumed that ventricular tachycardia is triggered by the AV junctional rhythm which presumably initiated delayed afterdepolarizations in the setting of digitalis toxicity.

mias, in particular the accelerated AV junctional escape rhythms[85,113] to the cellular electrophysiologic characteristics of triggered activity associated with delayed afterdepolarizations. Triggered automaticity may be suspected when one type of fast ectopic rhythm precipitates another type of ectopic rhythm without evidence of overdrive (Fig. 5.10). Triggered activity was also suspected in a subset of patients with ventricular tachycardia originating in the right ventricular outflow tract (left bundle branch block morphology with QRS axis in the frontal plane directed inferiorly) in whom there was inverse relation of the initiating coupling interval to the first tachycardia interval.[106]

Another possible evidence in favor of triggered activity is the suppression by Ca^{2+} channel-blocking drugs. Table 5.1 lists arrhythmias reported to be suppressible by verapamil. Numerically, the largest group is probably represented by patients with exercise-induced ventricular tachycardias.[114,115] However, the exercise-induced ventricular tachycardias tend to respond also to β-blockers, adenosine, and vagal stimulation. Other categories include multiform accelerated ventricular ectopic rhythm in acute myocardial infarction,[116,117] sustained ventricular tachycardia with right bundle branch morphology and left axis in the frontal plane in young patients, and possible torsade de pointes in patients with acute myocardial infarction and those with normal QT interval and R on T phenomenon.

Triggered activity has been also implicated in patients with ventricular tachycardia suppressed by adenosine but not by verapamil.[118] The suggested mechanism of action in such cases is decreased production of cyclic AMP.[118]

TABLE 5.1 Ventricular Tachycardias Suppressed by Verapamil

1. Exercise-induced (rate 130/220/min)*
2. Sustained VT with RBBB and LAD morphology (usually young age)
3. Multiform variety of accelerated ventricular ectopic rhythm in acute myocardial infarction (ventricular rate 50–125/min)
4. Torsade de pointes in acute myocardial infarction (?)
5. Torsade de pointes with normal QT and R on T (?)

* Tend to respond also to β-blockers, adenosine, and vagal stimulation.

VT = ventricular tachycardia; RBBB = right bundle branch block; LAD = left axis deviation.

REFERENCES

1. Brooks C, Lu H, Lange G, Mangi R, Shaw R, Geoly K: Effect of localized stretch of the sinoatrial node region of the dog heart. *Am J Physiol* 1966;211:1197–1202.
2. Irisawa H: Comparative physiology of the cardiac pacemaker mechanism. *Physiol Rev* 1978;58:461–498.
3. Gomes JA, Winters SL: The origins of the sinus node pacemaker complex in man: Demonstration of dominant and subsidiary foci. *J Am Coll Cardiol* 1987;9:45–52.
4. Trautwein W, Kassebaum DG: On the mechanism of spontaneous impulse generation in the pacemaker of the heart. *J Gen Physiol* 1961;45:317–330.
5. Vassalle M: Analysis of cardiac pacemaker potential using a "voltage lamp" technique. *Am J Physiol* 1966;210:1335–1341.
6. Weidmann S: Allgemeine Elektrophysiologie bei Stillstand und Wiedereinsetzen der Herztatigkeit. *Verh Deutsch Ges Krelslaufforschg* 1964;30:1–10.
7. Noble D, Tsien RW: The kinetics and rectifier properties of the slow potassium current in cardiac Purkinje fibres. *J Physiol* 1968;195:185–214.

8. Noble D: The cardiac "pacemaker" current. *Current Contents* 1989, 30 July, p 21.
9. Yanagihara K, Irisawa H: Inward current activated during hyperpolarization in the rabbit sinoatrial node cell. *Pflügers Arch* 1980;385:11–19.
10. Brown H, DiFrancesco D: Voltage-clamp investigations of membrane currents underlying pacemaker activity in rabbit sino-atrial node. *J Physiol* 1980;308:331–351.
11. DiFrancesco D: The cardiac hyperpolarizing-activated current, i_f origins and developments. *Prog Biophys Mol Biol* 1985;46:163–183.
12. Brown HF, Giles W, Noble SJ: Membrane currents underlying activity in frog sinus venosus. *J Physiol* 1977;271:783–816.
13. Noma A, Irisawa H, Kokobun S, Kotake H, Nishimura M, Watanabe Y: Slow current systems in the AV node of the rabbit heart. *Nature* 1980;285:228–229.
14. Callewaert G, Carmeliet E, Vereecke J: Single cardiac Purkinje cells: General electrophysiology and voltage-clamp analysis of the pace-maker current. *J Physiol* 1984;349:643–661.
15. Noble D: The surprising heart: A review of recent progress in cardiac electrophysiology. *J Physiol* 1984;353:1–50.
16. Noma A, Irisawa H: Membrane currents in the rabbit sinoatrial node cell as studied by the double mictroelectrode method. *Pflügers Arch* 1976;366:45–52.
17. DiFrancesco D, Ferroni A, Mazzanti M, Tromba C: Properties of the hyperpolarizing-activated current (i_f) in cells isolated from the rabbit sino-atrial node. *J Physiol* 1986;377:61–88.
18. Yanagihara K, Irisawa H: Potassium current during the pacemaker depolarization in rabbit sinoatrial node cell. *Pflügers Arch* 1980;388:255–260.
19. DiFrancesco D, Noma A, Trautwein W: Kinetics and magnitude of the time dependent K current in the rabbit sinoatrial node: Effect of external potassium. *Pflügers Arch* 1979;381:271–279.
20. Irisawa H: Membrane currents in cardiac pacemaker tissues, *Experientia* 1987;43:1131–1135.
21. Noma A, Kotake H, Irisawa H: Slow inward current and its role mediating the chronotropic effect of epinephrine in the rabbit sinoatrial node. *Pflügers Arch* 1980;388:1–9.
22. Rubenstein DS, Lipsius SL: Mechanisms of automaticity in subsidiary pacemakers from cat right atrium. *Circ Res* 1989;64:648–657.
23. Tse WW: Adrenergic potentiation on spontaneous activity of canine paranodal fibers. *Am J Physiol* 1984;16:H415–H421.
24. Kirchhof CF, Bonke FI, Allessie MA: Evidence for the presence of electrotonic depression of pacemakers in the rabbit atrioventricular node. The effects of uncoupling from the surrounding myocardium. *Basic Res Cardiol* 1988;83:190–201.
25. Rozanski GY, Lipsius SL: Electrophysiology of functional subsidiary pacemakers in canine right atrium. *Am J Physiol* 1985;249:H594–H603.
26. Rozanski G: Electrophysiological properties of automatic fibers in rabbit atrioventricular valves. *Am J Physiol* 1987;253:H720–H72.
27. Carmeliet E: Electrophysiological effects of catecholamines in the heart, in Riemersma and Oliver (eds): *Catecholamines in the Nonischaemic and Ischaemic Myocardium.* Elsevier/North Holland Biomed Press, 1982, pp 77–88.
28. Hoffman BF: Neural influences on cardiac excitability and rhythm, in Randall WC (ed): *Neural Regulation of the Heart.* New York, Oxford University, 1977, pp 289–312.
29. Rosen MR, Gelband H, Hoffman BF: Effects of phentolamine on electrophysiologic properties of isolated canine Purkinje fibers. *J Pharmacol Exp Ther* 1971;179:586–593.
30. Rosen MR, Hordof AY, Ilvento JP, Danilo P Jr: Effects of adrenergic amines on electrophysiological properties and automaticity of neonatal and adult canine Purkinje fibers. *Circ Res* 1977;40:390–400.
31. Mary-Rabine L, Hordof AY, Danilo P Jr, Malm JR, Rosen MR: Mechanisms for impulse initiation in isolated human atrial fibers. *Circ Res* 1980;47:267–277.
32. Tse W: Effect of epinephrine on automaticity of the canine atrioventricular node. *Am J Physiol* 1975; 229:34–37.
33. Bailey JC, Greenspan K, Elizari MV, Anderson GJ, Fisch C: Effects of acetylcholine on automaticity and conduction in the proximal portion of the His-Purkinje specialized conduction system of the dog. *Circ Res* 1972;30:210–216.
34. Urthaler F, Neely BH, Hageman GR, Smith LR: Differential effects of sympathetic activity on AV junctional automaticity and AV conduction. *Basic Res Cardiol* 1986;81:497–507.
35. Vassalle M: The acceleratory action of the vagus on the sinus node, in Bonke FIM (ed): *The Sinus Node: Structure, Function and Clinical Relevance.* The Hague, Martinus Nijhoff, 1978, pp 279–289.
36. Vassalle M: Cardiac automaticity in normal and some abnormal conditions, in Surawicz B, Reddy CP, Prystowsky EN (eds): *Tachycardias.* Boston/The Hague, Martinus Nijoff, 1984, pp 55–56.
37. Prystowsky EN, Zipes DP: Postvagal tachycardia. *Am J Cardiol* 1985;55:995–999.
38. Dudel J, Trautwein W: Der Mechanismus der automatischen rhythmischen Impulsbildung der Herzmuskelfaser. *Pflügers Arch* 1958;267:553–563.
39. Chang F, Gao J, Tromba C, Cohen I, DiFrancesco D: Acetylcholine reverses effects of beta agonists on pacemaker current in canine cardiac Purkinje fibers but has no direct action. A difference between primary and secondary pacemakers. *Circ Res* 1990; 66:633–636.
40. Valenzuela F, Vassalle M: Interaction between overdrive excitation and overdrive suppression in canine Purkinje fibers. *Cardiovas Res* 1983;18:608–619.
41. Vassalle M, Greenspan K, Jomain S, Hoffman BF: Effects of potassium on automaticity and conduction of canine hearts. *Am J Physiol* 1964;207:334–340.
42. Nishi K, Yoshikawa Y, Takenaka F, Akaike N: Electrical activity of sinoatrial node cells of the rabbit surviving a long exposure to cold Tyrode's solution. *Circ Res* 1977;41:242–247.
43. Yamagishi S, Sano T: Effect of temperature on pacemaker activity of rabbit sinus node. *Am J Physiol* 1967;212:829–834.
44. Levi R, Zavecz JH: Acceleration of idioventricular rhythms by histamine in guinea pig heart. *Circ Res* 1979;44:847–855.
45. Sanchez-Chapula J, Elizalde A: Characterization of the effects of histamine on the transmembrane electrical activity of guinea-pig and rabbit SA- and AV-

node cells. *Naunyn Schmiedebergs Arch Pharmacol* 1987;336:218–223.
46. Bogaert PP, Vereecke LS, Carmeliet EE: The effect of raised pH on pacemaker activity and ionic currents in cardiac Purkinje fibers. *Pflügers Arch* 1978;375:45–52.
47. Stowe DF, Dufic Z, Bosnjak ZJ, Kalbfleisch JH, Kampine JP: Volatile anesthetics attenuate sympathomimetic actions on the guinea pig SA node. *Anesthesiology* 1988;68:887–894.
48. Snyders DJ, Van Bogaert PP: Alinidine modifies the pacemaker current in sheep Purkinje fibers. *Pflügers Arch* 1987;410:83–91.
49. Kobinger W, Lillie C: Specific bradycardic agents—A novel pharmacological class? *Eur Heart J* 1987;8 (suppl L):7–15.
50. Jalife J, Hamilton AY, Lamanna VR, Moe GK: Effects of current flow on pacemaker activity of the isolated kitten sinoatrial node. *Am J Physiol* 1980;238:H307–H316.
51. Jalife J, Moe GK: Phasic effects of vagal stimulation on pacemaker activity of the isolated sinus node of the young cat. *Circ Res* 1979;45:595–607.
52. Jalife J, Slenter AJ, Salata JJ, Michaels DC: Dynamic vagal control of pacemaker activity in the mammalian sinoatrial node. *Circ Res* 1983;52:642–656.
53. Carmeliet E, Vereecke J: Electrogenesis of the action potential and automaticity, in *The Handbook of Physiology, The Cardiovascular System,* vol I. Bethesda MD, Am Physiol Soc, 1979, pp 269–334.
54. Imanishi S: Calcium-sensitive discharges in canine Purkinje fibers. *Jpn J Physiol* 1971;21:443–463.
55. Katzung BG: Effects of extracellular calcium and sodium on depolarization-induced automaticity in guinea pig papillary muscles. *Circ Res* 1975;37:118–127.
56. Imanishi S, Surawicz B: Automatic activity in depolarized guinea pig ventricular myocardium: Characteristics and mechanisms. *Circ Res* 1976;39:751–759.
57. Hiraoka M, Okamoto Y: Two types of abnormal automaticity in canine ventricular muscle fibers. *Arch Pharmacol* 1981;317:339–344.
58. Escande D, Coraboeuf E, Planch'e C, Lacour-Gayer F: Effects of potassium conductance inhibitors on spontaneous diastolic depolarization and abnormal automaticity in human atrial fibers. *Basic Res Cardiol* 1986;81:244–257.
59. Imoto Y, Ehara T, Matsuura H: Voltage- and time-dependent block of i_{K1} underlying Ba^{2+}-induced ventricular automaticity. *Am J Physiol* 1987; 252:H325–H333.
60. Hirano Y, Hiraoka M: Barium-induced automatic activity in isolated ventricular myocytes from guinea-pig hearts. *J Physiol* 1988;395:455–472.
61. Shibata J: The effects of barium on the action potential and the membrane current of sheep heart Purkinje fibers. *J Pharmacol Exp Ther* 1973;183:418–426.
62. Hauswirth O, Noble D, Tsien RW: The mechanism of oscillatory activity at low membrane potentials in cardiac Purkinje fibres. *J Physiol* 1969;200:255–265.
63. Ito S, Surawicz B: Transient, "paradoxical" effects of increasing extracellular K^+ concentration on transmembrane potential in canine cardiac Purkinje fibers. *Circ Res* 1977;41:799–807.
64. Wiggins JR, Cranefield PF: Two levels of resting potential in canine cardiac Purkinje fibers exposed to sodium-free solutions. *Circ Res* 1976;39:466–474.
65. Gadsby DC, Cranefield PF: Two levels of resting potential in cardiac Purkinje fibers. *J Gen Physiol* 1977;70:725–746.
66. Sperelakis N: Electrical properties of embryonic heart cells, in De Mello WC (ed): *Electrical Phenomena Of the Heart.* New York and London, Academic Press, 1972, pp 1–61.
67. Hashimoto K, Ochi R, Inui J, Miura Y: The ionic mechanism of prolongation of action potential duration of cardiac ventricular muscle by anthopleurin-A, and its relationship to the inotropic effect. *J Pharmacol Exper Therap* 1980;215:479–485.
68. El-Sherif N, Zeiler RH, Craelius W, Gough WB, Henkin R: QTU prolongation and polymorphic ventricular tachyarrhythmias due to bradycardia-dependent early afterdepolarizations. *Circ Res* 1988; 63:286–305.
69. Surawicz B: Depolarization-induced automaticity in atrial and ventricular myocardial fibers, in Zipes DP, Bailey JC, Elharrar V (eds): *The Slow Inward Current and Cardiac Arrhythmias.* The Hague/Boston/London, Martinus Nijhoff Publishers, 1980, pp 375–396.
70. Surawicz B: Electrophysiologic substrate of torsade de pointes: Dispersion of repolarization or early afterdepolarization? *J Am Coll Cardiol* 1989;14:172–184.
71. Imanishi S, McAllister RG Jr, Surawicz B: The effects of verapamil and lidocaine on the automatic depolarization in guinea-pig ventricular myocardium. *J Pharmacol Exper Therap* 1978;207:294–303.
72. Kimura T, Imanishi S, Arita M, Hadama T, Shirabe J: Two different mechanisms of automaticity in diseased human atrial fibers. *Jpn J Physiol* 1988; 38:851–867.
73. Escande D, Coraboeuf E, Planch'e C: Abnormal pacemaking is modulated by sarcoplasmic reticulum in partially-depolarized myocardial from dilated right atria in humans. *J Mol Cell Cardiol* 1987; 19:231–241.
74. El-Sherif N, Gough WB, Zeiler RH, Mehra R: Triggered ventricular rhythms in 1 day-old myocardial infarction in the dog. *Circ Res* 1983;52:566–579.
75. Kimura S, Bassett AL, Kohya T, Kozlovskis PL, Myerburg RJ: Automaticity, triggered activity, and responses to adrenergic stimulation in cat subendocardial Purkinje fibers after healing of myocardial infarction. *Circulation* 1987;75:651–660.
76. Rosenthal JE: Contribution of depolarized foci with variable conduction impairment to arrhythmogenesis in 1 day old infarcted canine cardiac tissue: An in vitro study. *J Am Coll Cardiol* 1986;8:648–656.
77. Vassalle M, Knob RE, Cummins M, Lara GA, Castro C, Stuckey JH: An analysis of fast idioventricular rhythm in the dog. *Circ Res* 1977;41:218–226.
78. Hope RR, Scherlag BJ, El-Sherif N, Lazzara R: Hierarchy of ventricular pacemakers. *Circ Res* 1976;39:883–888.
79. Ilvento JP, Provet J, Danilo P, Rosen MR: Fast and slow idioventricular rhythms in the canine heart: A study of their mechanism using antiarrhythmic

drugs and electrophysiologic testing. *Am J Cardiol* 1982;49:1909–1916.
80. Martins JB: Autonomic control of ventricular tachycardia: Sympathetic neural influence on spontaneous tachycardia 24 hours after coronary occlusion. *Circulation* 1985;72:933–942.
81. Pogwizd SM, Onufer JR, Kramer JB, Corr PB: Induction of delayed afterdepolarization and triggered activity in canine Purkinje fibers by lysophosphoglycerides. *Circ Res* 1986;59:416–426.
82. Dangman KH, Hoffman BF: Studies on overdrive stimulation of canine cardiac Purkinje fibers: Maximal diastolic potential as a determinant of the response. *J Am Coll Cardiol* 1983;2:1183–1190.
83. Johnson N, Danilo P, Wit AL, Rosen MR: Characteristics of initiation and termination of catecholamine-induced triggered activity in atrial fibers of the coronary sinus. *Circulation* 1986;74:1168–1179.
84. Dangman KH, Hoffman BF: The effects of single premature stimuli on automatic and triggered rhythms in isolated canine Purkinje fibers. *Circulation* 1985;71:813–822.
85. Knoebel SB, Fisch C: Accelerated junctional escape. A clinical and electrocardiographic study. *Circulation* 1974;50:151–158.
86. Gomes JA, Hariman RF, Kang PS, et al: Sustained symptomatic sinus node reentrant tachycardia: Incidence, clinical significance, electrophysiologic observations and the effect of antiarrhythmic agents. *J Am Coll Cardiol* 1985;5:45–57.
87. Haines DE, DiMarco JP: Sustained intraatrial reentrant tachycardia: Clinical, electrocardiographic and electrophysiologic characteristics and long-term follow-up. *J Am Coll Cardiol* 1990;15:1345–1354.
88. Gillette PC: The mechanisms of supraventricular tachycardia in children. *Circulation* 1976;54:133–139.
89. Scherf D, Choi KH, Bahadori A, Orphanos RP: Parasystole. *Am J Cardiol* 1963;12:527–538.
90. Surawicz B, MacDonald MG: Ventricular ectopic beats with fixed and variable coupling. *Am J Cardiol* 1964;13:198–208.
91. Schamroth L, Surawicz B: Concealed interpolated A-V junctional extrasystoles and A-V junctional parasystole. *Am J Cardiol* 1971;27:703–707.
92. El-Sherif N, Samet P: Multiform ventricular ectopic rhythm. *Circulation* 1975;51:492–504.
93. Ogawa S, Dreifus LS, Watanabe Y: Rapid ventricular parasystole with possible multilevel exit block. Evidence for multiple parasystolic activity. *J Electrocardiol* 1981;14:309–314.
94. Robles de Medina E: Multifocal ventricular parasystolic tachycardia. *Br Heart J* 1976;38:1047–1052.
95. Robles de Medina EO, Delmar M, Sicouri S, Jalife J: Modulated parasystole as a mechanism of ventricular ectopic activity leading to ventricular fibrillation. *Am J Cardiol* 1989;63:1326–1332.
96. Kuo CS, Surawicz B: Coexistence of ventricular parasystole and ventricular couplets: Mechanism and clinical significance. *Am J Cardiol* 1979;44:435–440.
97. Fisch C: Parasystole, in Fisch C (ed): *Electrocardiography of Arrhythmias.* Philadelphia, London, Lea & Febiger, 1990, pp 169–196.
98. Oreto G, Arrigo F, Coglitore S, Giannetto M, Consolo F: Parasystolic ventricular tachycardia with variable exit block. *J Electrocardiol* 1982;15:411–416.
99. Lightfoot PR: His bundle parasystole: A form of A-V junctional parasystole. *Chest* 1974;66:695–701.
100. Goldreyer BN, Gallagher JJ, Damato AN: The electrophysiologic demonstration of atrial ectopic tachycardia in man. *Am Heart J* 1973;85:205–215.
101. Reddy CP: Supraventricular/ectopic tachycardia due to mechanism other than reentry, in Surawicz B, Reddy CP, Prystowsky EN (eds): *Tachycardias.* Boston, The Hague, Marinus Nijhoff, 1984, pp 173–183.
102. Wellens HJJ, Lie KI, Durrer D: Further observations on ventricular tachycardia as studied by electrical stimulation of the heart. *Circulation* 1974;39:647–653.
103. Palileo EV, Ashley WW, Swiryn S, Bauernfeind RA, Strasberg B, Petropoulos AT, Rosen KM: Exercise provocable right ventricular outflow tract tachycardia. *Am Heart J* 1982;104:185–192.
104. Huikuri HV, Cox M, Interian A, Kessler KM, Glicksman F, Castellanos A, Myerburg RJ: Efficacy of intravenous propranolol for suppresion of inducibility of ventricular tachyarrhythmias with different electrophysiologic characteristics in coronary artery disease. *Am J Cardiol* 1989;64:1305–1309.
105. Reddy CP, Gettes LS: Use of isoproterenol as an aid to electric induction of chronic recurrent ventricular tachycardia. *Am J Cardiol* 1979;44:705–713.
106. Ritchie AH, Kerr CR, Qi A, Yeung-Lai-Wah J: Nonsustained ventricular tachycardia arising from the right ventricular outflow tract. *Am J Cardiol* 1989;64:594–598.
107. Waxman MB, Downar E, Berman ND, Felderhof CH: Phenylephrine (Neo-synephrine) terminated ventricular tachycardia. *Circulation* 1974;50:655–664.
108. Weiss T, Lattin GM, Engelman K: Vagally mediated suppression of premature ventricular contractions in man. *Am Heart J* 1975;89:700–707.
109. Waxman MB, Wald RW, Finley JP, Bonet JF, Downar E, Sharma AD: Valsalva termination of ventricular tachycardia. *Circulation* 1980;62:843–851.
110. Fu LT, Nozaki M, Iinuma H, Watanabe T, Koyama S: Contribution of vagal tone to initiation and termination of ventricular tachycardia: Report of three cases. *Clin Cardiol* 1980;3:137–142.
111. Waxman MB, Wald RW: Termination of ventricular tachycardia by an increase in cardiac vagal drive. *Circulation* 1977;56:385–391.
112. Rosen MR, Reder RF: Does triggered activity have a role in the genesis of cardiac arrhythmias? *Ann Intern Med* 1981;94:794–801.
113. Rosen MR, Fisch C, Hoffman BF, Danilo P, Lovelace DE, Knoebel SB: Can accelerated atrioventricular junctional escape rhythms be explained by delayed afterdepolarizations? *Am J Cardiol* 1980;45:1272–1284.
114. Wu D, Kou HC, Hung JS: Exercise-triggered paroxysmal ventricular tachycardia. *Ann Intern Med* 1981;95:410–414.
115. Woelfel A, Foster JR, McAllister RG, Simpson RJ,

Gettes LS: Efficacy of verapamil in exercise-induced ventricular tachycardia. *Am J Cardiol* 1985;56:292–297.

116. Sclarovsky S, Strasberg B, Fuchs J, Lewin RF, Arditi A, Klainman E, Kracoff OH, Agmon J: Multiform accelerated idioventricular rhythm in acute myocardial infarction: Electrocardiographic characteristics and response to verapamil. *Am J Cardiol* 1983;52:43–47.

117. Grenadier E, Alpan G, Maor N, Keidar S, Binenboim C, Margulies T, Palant A: Polymorphous ventricular tachycardia in acute myocardial infarction. *Am J Cardiol* 1984;53:1280–1283.

118. Lerman BB, Belardinelli L, West GA, Berne RM, DiMarco JP: Adenosine-sensitive ventricular tachycardia: Evidence suggesting cyclic AMP-mediated triggered activity. *Circulation* 1986;74:270–280.

Chapter **6**

Arrhythmias Triggered by Afterdepolarizations

Michael R. Rosen, MD, and Evgeny P. Anyukhovsky, PhD

Afterdepolarizations are oscillations in membrane potential (V_m) that are induced by a preceding action potential (AP). Unlike automaticity, they cannot occur de novo. Although oscillatory activity was recognized in the early twentieth century, the first experiments that identified it as contributing to specific arrhythmias were performed independently by Segers and by Bozler in the 1940s.[1,2] Little note was paid to these studies in subsequent years as investigators concentrated on reentry and automaticity as the two principal mechanisms of arrhythmogenesis. Interest rekindled in the 1970s, and subsequently a great deal has been learned about the mechanisms responsible for oscillatory activity as well as its contributions to arrhythmias. Arrhythmias induced by afterdepolarizations are referred to as "triggered," reflecting the fact that they are not truly spontaneous but rather are induced by a preceding electrical event. Moreover, the characteristics of afterdepolarizations and resultant triggered arrhythmias are such that the previous history of the rhythmic activity in a preparation or an intact heart determines not only whether triggering of the arrhythmia will occur, but the cycle length characteristics of the arrhythmia. Extensive reviews on the subject have been provided in recent years,[3–5] and our purpose in this chapter will be to highlight areas that are most readily applicable to the genesis of cardiac arrhythmias.

DELAYED AFTERDEPOLARIZATIONS

Delayed afterdepolarizations occur after completion of phase 3 repolarization (Fig. 6.1). In the early 1970s, investigators in three different laboratories reported nearly simultaneously on the occurrence of these oscillations as a result of exposure of Purkinje fibers and of atrial tissues to digitalis.[6–11] The oscillations were variously termed low-amplitude potentials,[6,7] transient depolarizations,[8–10] and enhanced diastolic depolarization.[11] In retrospect, earlier studies had described the same phenomenon, but had interpreted it as being a variant on phase 4 depolarization and automaticity (see, for example, refs. 12 and 13). Confusion over terminology had largely ended by the end of the 1970s, at which time the terms now in general use, delayed afterdepolarizations[14,15] or oscillatory afterpotentials,[16] had come into use.

Digitalis-Induced Delayed Afterdepolarizations

Most investigators have used either ouabain or *N*-acetylstrophanthidin to induce delayed afterdepolarizations. On exposure of a Purkinje fiber to digitalis, the events that occur initially are depolarization of the resting membrane potential (RMP) and acceleration of repolarization (sometimes preceded by prolongation of repolarization). The loss of V_m is attributed to poisoning of the sodium/potassium (Na/K) pump by digitalis. The resultant loss of K^+ from the cell's interior results in a decreased intracellular to extracellular K^+ concentration ($[K^+]_i/[K^+]_o$) ratio and, as predicted by the Nernst equation, the V_m depolarizes. The accumulation of K^+ in the clefts outside the cells also contributes to the acceleration of repolarization, as elevating $[K^+]_o$ increases the K^+ conductance (g_K) of the membrane.

These effects of toxic concentrations of digitalis on the transmembrane potential have been known for some time. The abnormal impulse initiation that occurred as toxicity progressed had been thought to result from abnormal automaticity. However, studies using voltage-clamp techniques indicated that the normal pacemaker po-

655 Avenue of the Americas, New York, NY 10010
Current Topics in Cardiology

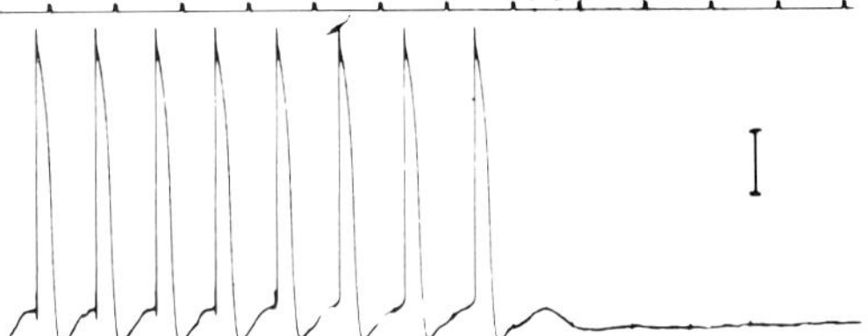

FIGURE 6.1 Delayed afterdepolarizations induced by superfusing a Purkinje fiber with ouabain. Each driven AP is followed by an afterdepolarization, which is seen most clearly when drive is stopped. Time marks = 1 s; vertical calibration = 20 mV.

the afterdepolarizations attain threshold and initiate a triggered AP or sequence of APs.

The pattern of the relationship between drive cycle length, afterdepolarization amplitude, and CI suggests that any rhythm induced by delayed afterdepolarizations in the intact heart might be expected to show greater frequency as heart rate increases, as well as a faster rate as preexisting heart rate increases. Although this is true, there is an important caveat, based on the fact that sequences of afterdepolarizations occur

tential was depressed by digitalis and that the afterdepolarizations resulted from a completely different mechanism.[17,18] This mechanism has been described as follows[18]: as Na/K ATPase is poisoned, the cell accumulates Na^+ intracellularly while losing K^+. The increase in $[Na^+]_i$ causes a shift in the Na/calcium (Ca) exchange mechanism such that it extrudes increasing amounts of Na^+, while accumulating Ca^{2+}. More free Ca^{2+} becomes available as a result of its release from stores in the sarcoplasmic reticulum and mitochondria. As the free intracellular Ca^{2+} concentration is elevated, it increases the monovalent cation conductance of the membrane. It has been suggested that the result is a transient inward current (i_{ti}) carried by Na^+, which then induces the afterdepolarization.[3] An alternative hypothesis suggests that it is the Na/Ca exchanger that generates the current responsible for the afterdepolarization. Specifically, Shimoni and Giles[19] presented electrophysiological evidence based on simultaneous recordings of exchanger current and i_{ti} showing that these two currents can be separated and suggesting that they are controlled by different mechanisms. Although these results suggest the Na/Ca exchanger may not always be directly involved in generating the i_{ti} currents, this area of investigation is proceding to determine whether one or both mechanisms is operative.

The Phenomenology of Digitalis-Induced Delayed Afterdepolarizations

Digitalis-induced delayed afterdepolarizations can occur singly (Fig. 6.1) or in sequences of two, three, or more.[6–10] They have a specific cycle length dependence, as demonstrated in Figure 6.2. Note that as drive cycle length decreases, the amplitude of the afterdepolarizations increases and their coupling interval (CI) to the preceding AP shortens. When amplitude is sufficiently large,

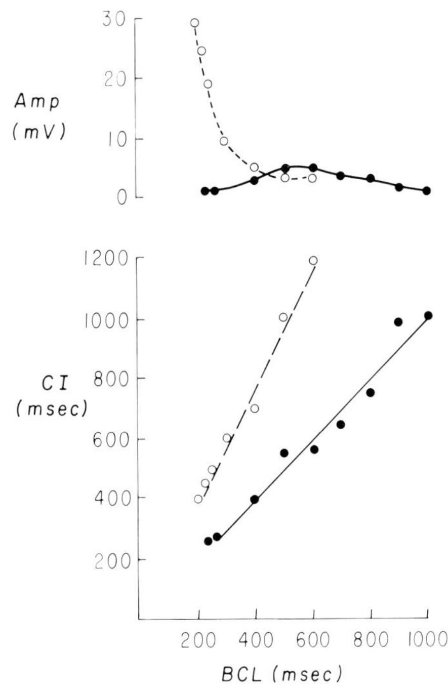

FIGURE 6.2 Relationship of the amplitude and coupling interval (CI) of delayed afterdepolarizations to the AP that induced them. In this figure, records were made from a preparation that generated two delayed afterdepolarizations on cessation of drive. For the first (filled circles), the afterdepolarization amplitude increases to a maximum at a basic cycle length (BCL) of approximately 550 ms, and then decreases in amplitude at shorter cycle lengths. For the second (open circles) amplitude continues to increase at shorter cycle lengths. The second afterdepolarization generated triggered APs at the three shortest cycle lengths. Note that the CI for both afterdepolarizations shortens in linear fashion as drive cycle length is decreased. *Revised from M. Rosen, P. Danilo: Digitalis induced delayed afterdepolarizations, in Zipes D, Bailey J, Elharrar V (eds):* The Slow Inward Current and Cardiac Arrhythmias. *Boston, Nijhoff, 1980, p 421.*

and either the first or the second in the sequence may attain threshold. As shown in Figure 6.2, the expected linear relationship between CI for an afterdepolarization and its resultant triggered beat may be decidedly nonlinear if there is a shift from one to another delayed afterdepolarization as the triggering event. Hence, whereas it is true that a premature AP whose CI shortens as preceding drive cycle length shortens is characteristic of digitalis-induced triggered activity, it also is true that triggered rhythms may show no consistent coupling relationship to the basic drive if there is a shift from one to another afterdepolarization.

As digitalis toxicity increases, there is a tendency to see not just single afterdepolarizations attaining threshold and inducing APs but sequences of APs. When the afterdepolarizations induce single APs, the likelihood is that the pattern will be a strict bigeminy at any cycle length at which triggering occurs. With increasing toxicity there may be trigeminy, quadrigeminy, and so forth, through bursts of rapid rhythms hundreds of beats in duration.

Once a triggered rhythm commences it may stop abruptly, or following a period during which its rate increases or decreases.[20] Attempts to overdrive pace may be met by an increase in the rate of the triggered activity to approximate that of the overdrive. At some critical cycle length, it usually is possible to terminate the rhythm, and when this cycle length is identified, it tends to be reproducible. The pacing-induced termination of digitalis-induced triggered rhythms often is preceded by sequences of "afterbeats" which may range from 1 or 2 up to 25 beats in duration.[20] These are characterized by somewhat unstable cycle lengths, and may themselves represent triggered beats arising from other foci or reentry.

One important variable that may influence the occurrence and amplitude of delayed afterdepolarizations is the RMP. It has been demonstrated that maximal afterdepolarization amplitude is reached when V_m is in the -70 to -75 mV range.[21] When the membrane is moved to higher or lower potentials, the amplitude of afterdepolarizations tends to diminish and they cease to fire. Alternatively, if a preparation is at a low or high V_m and afterdepolarizations are small, any event that moves V_m into the -70 to -75 mV range will increase their amplitude and may induce them to initiate APs.[21]

Pharmacological Effects on Delayed Afterdepolarizations

A number of drugs have been demonstrated to suppress or to amplify digitalis-induced delayed afterdepolarizations. Drugs that depress the rapid i_{Na} that contributes to phase 0 of the AP (for example, lidocaine,[22] quinidine, and procainamide[23]) suppress the afterdepolarizations, possibly by reducing the level of $[Na^+]_i$ and, secondarily, $[Ca^{2+}]_i$. Alternatively, they may suppress i_{ti} directly. Ca^{2+} channel blockers like verapamil suppress delayed afterdepolarizations,[24] presumably by reducing intracellular Ca^{2+}.

The desirability of a drug that suppresses delayed afterdepolarizations without affecting other arrhythmogenic mechanisms has long been recognized. Such a drug would not only be therapeutic but its action would diagnose triggered activity to the exclusion of other mechanisms. However, the only drug that thus far has shown such properties is the antitumor drug, doxorubicin,[25,26] whose toxicity is such that it cannot be considered for use as an antiarrhythmic agent.

Digitalis-Induced Triggered Rhythms in Intact Animals

Studies in the intact canine heart have attempted to distinguish between triggered activity and other mechanisms for arrhythmogenesis. It is clear from studies of dogs in complete heart block that administration of digitalis suppresses normal idioventricular pacemakers prior to onset of triggered rhythms.[27,28] The arrhythmias that occur tend to be inducible at fast rather than at slow drive rates, or following premature stimulation at short rather than long CIs. The CIs for the ectopic beats bear a linear relationship to the preceding basic drive or sinus rhythm, such that as the preceding rate increases, so does the rate of the arrhythmia. Doxorubicin[26] as well as the Ca^{2+} overload blocker flunarazine[29] tend to suppress these rhythms while leaving rhythms known to be automatic unaffected. All this information is consistent with the cellular electrophysiologic studies of digitalis-induced delayed afterdepolarizations described above.

Catecholamine-Induced Delayed Afterdepolarizations and Triggered Activity

α-Adrenergic catecholamines have been shown to induce delayed afterdepolarizations in ventricular specialized conducting fibers in the presence of high $[Ca^{2+}]_o$.[30] In contrast, β-adrenergic catecholamines can induce such afterdepolarizations in atrial and ventricular tissues, even when the $[Ca^{2+}]$ is normal.[3] This probably reflects a better efficiency of the β-adrenergic system in elevating $[Ca^{2+}]_i$, which appears to be the common denominator for the occurrence of different types of delayed afterdepolarizations.[3]

Perhaps the largest body of work on catecholamine-induced delayed afterdepolarizations has considered the effects of β-adrenergic amines on the coronary sinus.[31,32] Isolated preparations of coronary sinus placed in the tissue bath are either quiescent or generate very slow automatic rhythms, generally at depolarized membrane potentials. In the presence of β-adrenergic agonists, these fibers tend to hyperpolarize. Pacing then induces delayed afterdepolarizations and triggered activity. In contrast to digitalis-induced delayed afterdepolarizations, those induced by catecholamine tend to be single rather than occurring in sequences. Their amplitude increases and their coupling to the preceding AP shortens as drive rate is increased. When they reach threshold, these afterdepolarizations, like those described for digitalis, can induce bigeminal, trigeminal, and so forth, premature APs, as well as tachyarrhythmias. For all of these the CI for the first triggered AP decreases as drive rate increases.[32]

Unlike digitalis-induced delayed afterdepolarizations and triggered rhythms, those resulting from catecholamine effects in the coronary sinus arise from the first delayed afterdepolarization in a sequence nearly exclusively (indeed, it is rare to see sequences of delayed afterdepolarizations here). Hence, the complicating factor occurring with digitalis, where triggered beats may arise from the first or second afterdepolarization and thereby obscure the ability to detect a linear relationship between drive and triggered CIs does not occur. Also, unlike digitalis effects in the Purkinje system, the catecholamine-treated coronary sinus preparations show a consistent CI relationship only for the first beat.[32] Thereafter, the rhythm tends to slow down (or increase) in rate, establishing a steady state that will remain rather constant despite a wide range of initiating cycle lengths. Termination of the rhythm may follow a period of increasing or decreasing rate or may be abrupt. Attempts to terminate the rhythm with pacing will—like the digitalis situation—result in varying sequences of "afterbeats" having unstable cycle lengths.

V_m is an important determinant of triggered rhythms in the coronary sinus. These tend to increase in amplitude as V_m depolarizes.[31,32] Hence, driving at a constant cycle length may induce delayed afterdepolarizations having varying amplitude as V_m changes.

In studies of arrhythmias triggered from the coronary sinus in the intact animal, experiments have been performed by infusing catecholamine into the great cardiac vein of dogs and pacing the atrium.[33] It is clear that such infusions are associated with ectopic P waves having the same characteristics as P waves stimulated by pacing the coronary sinus region. Moreover, they bear a CI relationship to the preceding paced rhythm such that the first ectopic beat shows a shorter CI as drive rate increases. This is in contrast to ectopic atrial impulses, which showed the characteristic overdrive suppression of automatic rhythms. Because of limitations in the model, sustained tachycardias have not been inducible. Nonetheless, the phenomenology of the single and multiple premature beats elicited has been entirely consistent with triggered activity demonstrated in isolated coronary sinus, as described above.

Delayed Afterdepolarizations and Triggered Activity in Ischemia and Infarction

Harris[34] showed that ventricular arrhythmias after coronary artery occlusion occurred in two distinct phases separated by a period of regular sinus rhythm. The first, or early, phase of arrhythmias begins within minutes after the occlusion and lasts about 30 min, often degenerating into ventricular fibrillation. The arrhythmias of the second, or delayed, phase usually begin 4 to 8 h after occlusion, reach a plateau 3 to 4 h later, and last for 2 to 4 days.[35] Harris et al.[36] speculated that different electrophysiologic mechanisms might be responsible for these two phases of ischemia-related arrhythmias, a speculation supported by later evidence.

There are many experimental proofs that the major mechanism for early arrhythmias is circus movement reentry within the ischemic myocardium.[37–39] However, the possibility that nonreentrant mechanisms (in particular delayed afterdepolarizations and triggered activity) may also cause ventricular arrhythmias during the early postocclusion period cannot be ruled out. Pogwizd et al.[40] showed that lysophosphoglycerides that accumulate in ischemic myocardium during acute coronary occlusion,[41,42] combined with acidosis and elevated $[K^+]$ can induce delayed afterdepolarizations and triggered activity in isolated Purkinje fibers. Prolonged mild hypoxia and acidosis induce $[Ca^{2+}]_i$ overload and also produce delayed afterdepolarizations and triggered activity.[43] When Purkinje fibers are subjected to a superfusate mimicking ischemic conditions (hypoxia, elevated lactate, acidosis, zero substrate) followed by reperfusion with a normal solution, delayed afterdepolarizations are observed.[44] These results have been interpreted as suggesting that delayed afterdepolarizations and triggered activity may cause premature ventricular depolarizations during the early stage of myocardial infarction. However, this possibility remains speculative because there are no direct proofs that delayed afterdepolarizations occur in the intact heart in the early postocclusion period.

One day after coronary artery occlusion in dogs, ventricular muscle cells die, whereas Purkinje fibers survive on the endocardial surface of transmural infarcts.[45,46] These subendocardial Purkinje fibers exhibit abnormal impulse generation and are the source of arrhythmias.[47–49] The focal origin of ventricular arrhythmias was demonstrated in mapping experiments of the endocardial surface of the 24 h infarction zone.[50] The question posed by El-Sherif et al.[50] was whether abnormal automaticity or triggered activity was responsible for abnormal impulse generation in infarcted subendocardial Purkinje fibers. The authors showed that delayed afterdepolarizations and triggered activity could be induced by electrically stimulating subendocardial Purkinje fibers in quiescent endocardial preparations from 1-day-old canine infarcts. This activity was completely suppressed by verapamil. On the basis of these results they suggested triggered activity to be the primary mechanism of ventricular arrhythmias in the late stage of myocardial infarction. However, this suggestion was not uniformly confirmed in experiments on heart in situ. Verapamil (0.3 mg/kg) either had practically no effect on ventricular arrhythmias in the 24 h-old canine infarct,[51] or these arrhythmias could be suppressed in approximately half of the animals by a combination of propranolol and verapamil in high doses, which induced dramatic decreases in arterial pressure.[52]

Attempts to distinguish between abnormal automaticity and triggered activity were also made by using drugs that suppress one or the other mechanism.[53] Therapeutic concentrations of lidocaine, which suppress delayed afterdepolarizations and normal automaticity but have little effect on abnormal automaticity, had practically no effect both on sustained rhythmic activity of isolated subendocardial preparations excised from the infarct and on ventricular tachycardia 24 h after coronary occlusion in conscious dogs. Movicizine, which suppresses abnormal automaticity and delayed afterdepolarizations but not normal automaticity, depressed sustained rhythmic activity of the isolated preparations as well as ventricular tachycardia in intact dogs. Doxorubicin, which specifically suppresses delayed afterdepolarizations and triggered activity, had no effect on delayed ventricular arrhythmias.[26] These results are consistent with the proposal that the predominant rhythm in this stage of myocardial infarction is caused by abnormal automaticity (although triggered activity was seen in a subset of isolated tissues, especially 48 to 72 h postinfarction).

Recently, Anyukhovsky et al.[54] have investigated the influence of changes in ionic milieu on sustained rhythmic activity in specimens isolated from infarcted subendocardium 24 h after coronary artery occlusion. Changes in $[Na^+]_o$ and/or $[Ca^{2+}]_o$ of the superfusate allowed discrimination between two types of responses suggestive of two different mechanisms for sustained rhythmic activity. In one group of specimens, the rate of activity increased as $[Ca^{2+}]_o$ increased and/or $[Na^+]_o$ decreased. The activity decreased when $[Ca^{2+}]_o$ was lowered or $[Na^+]_o$ increased. These results are consistent with activity triggered by delayed afterdepolarizations. In a second group of specimens, the opposite responses to changes in $[Na^+]_o$ and $[Ca^{2+}]_o$ were observed. This rhythm appeared not be triggered by delayed afterdepolarizations and was attributed to abnormal automaticity.[55] This study suggests that 24 h after coronary artery ligation subendocardial Purkinje fibers with both abnormal automaticity and triggered activity can be presented in the infarcted area.

Given that both abnormal automaticity and delayed afterdepolarizations-induced triggered activity may occur 24 h after coronary ligation, what are the conditions that predispose to one or another mechanism? Ferrier[21] has shown that the optimal V_m range for development of delayed afterdepolarizations is −75 to −70 mV (see above), whereas abnormal automaticity usually emerges at more positive potentials.[53,55] Anyukhovsky et al.[54] have also registered more negative maximal diastolic potential (−68 ± 0.8 mV) in fibers in which abnormal activity was identified as triggered than in the fibers with abnormal automaticity (−62 ± 1.0 mV). Other important modifying factors are temperature (automaticity predominates in the range of 39°C, delayed afterdepolarizations at 36°C) and infarct size (smaller infarcts usually show more negative V_m, and tend to be less automatic and show more triggering).

In sum, the results presented above suggest that delayed afterdepolarizations and triggered activity can induce a subset of ventricular premature depolarizations and tachyarrhythemias during the period of delayed arrhythmias after coronary artery occlusion.

EARLY AFTERDEPOLARIZATIONS

Early afterdepolarizations are oscillations in membrane potential that occur during phase 2 or 3 of repolarization; that is, they precede full repolarization (Fig. 6.3). These have not been studied as intensively as delayed afterdepolarizations, but they are eliciting increased interest because certain characteristics of their cycle length dependence have suggested they might be responsible for a subset of pleomorphic tachycardias including torsade de pointes.

Early afterdepolarizations may be induced by any event that prolongs repolarization and/or by events that increase inward currents carried by Ca^{2+} during the plateau or decrease repolarizing currents carried by K^+.[3] Recent studies by January and associates have made use of Bay K 8644, which consistently and reproducibly induces early afterdepolarizations during phase 2 and early phase 3.[56,57] The basis for this action is an increase in transsarcolemmal i_{Ca}, carried via L-type Ca^{2+} channels, which are opened by Bay K 8644. In contrast, cesium (Cs^{2+}) blocks the delayed rectifying i_K, prolongs AP duration (APD), and induces early afterdepolarizations which tend to arise late during phase 3.[58] This action is potentiated by lowering $[K^+]$ and/or by slowing the stimulation rate of isolated tissues, both actions which decrease the g_K of the membrane. Certain drugs which decrease g_K and prolong repolarization also, typically, induce early afterdepolarizations, including quinidine and *N*-acetyl procainamide.[3]

As is the case for delayed afterdepolarizations, early afterdepolarizations have a typical cycle length dependence with respect to preceding basic drive periods[58] (Figs. 6.3, 6.4). However, the relationship is opposite that for delayed afterdepolarizations: That is, with the early afterdepolarizations, the afterpotentials increase in amplitude and tend more readily to initiate triggered action potentials as drive rate *slows*. The relationship between drive cycle length and the initiation of ectopic impulses by early afterdepolarizations has been studied most completely with Cs^{2+},[58,59] although the effects of quinidine,[60] and Bay K 8644[56,57] have been reported, as well.

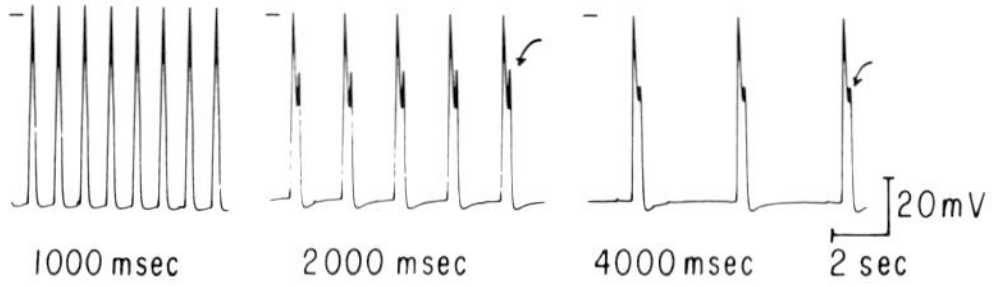

FIGURE 6.3 Cs^{2+}-induced early afterdepolarizations occurring at the termination of the AP plateau (arrows). Note that no afterdepolarizations are present at BCL = 1,000 ms, large afterdepolarizations are present at 2,000 ms, and they decrease in size at 4,000 ms. *Revised from Damiano and Rosen,*[58] *with permission.*

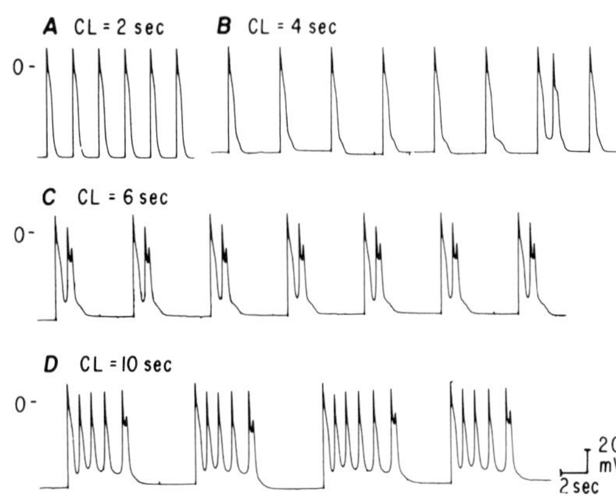

FIGURE 6.4 Cs^{2+}-induced early afterdepolarizations and triggered activity. **A.** There are no afterdepolarizations at cycle length = 2 s. **B.** Afterdepolarizations appear as a shoulder on phase 3, one of which induces an AP. **C.** A bigeminal pattern of triggered APs is seen as cycle length is increased to 6 s. **D.** At a still longer cycle length there are bursts of triggered activity. *Reprinted from Damiano and Rosen,*[58] *with permission of the American Heart Association.*

As drive rate is reduced, early afterdepolarizations become larger in amplitude and the level of V_m at which they are initiated tends to become more positive. At a critical cycle length they will reach threshold, and the pattern of coupling of the APs induced is such that bigeminy, trigeminy, and so forth, are the patterns seen. As drive rate slows, the rate of the tachyarrhythmias induced increases, such that an overall pattern of bradycardia-induced ectopy and tachyarrhythmias emerges.

The exceptionally slow drive rates required to induce the arrhythmias (for example, often less than 20 beats/min) obviously suggest that in a heart with a normally functioning sinus mechanism, the rate could never be low enough to favor the occurrence of such a triggered rhythm. However, if a parasystolic focus is protected by entry block which is firing at a low rate, or if conduction block to a site in the heart is sufficient that only 3 : 1 or 4 : 1 or less conduction to that site is occurring at normal sinus rates, then the rate characteristics required for early afterdepolarizations to attain threshold would be met. Similarly, if there is a normal sinus mechanism with occasional long sinus pauses, the same conditions are met. Finally, even at normal sinus rates, if there is abnormal prolongation of repolarization (as in the presence of quinidine and low $[K^+]_o$) early afterdepolarizations can occur and elicit triggered beats.

Once sustained triggering occurs, it may or may not cease spontaneously, usually associated with slowing of rate of the triggered rhythm and/or hyperpolarization of the membrane. Attempts to overdrive pace such triggered rhythms may or may not meet with success, depending on the V_m at which the arrhythmia is occurring.[58] At relatively high V_ms, where the APs are dependent on the rapid i_{Na} for their upstrokes, there tends to be overdrive suppression based on Na/K pump stimulation, but at low V_ms, where pump stimulation does not occur, the rhythms may not be overdriven. In manifesting this characteristic, arrhythmias triggered by early afterdepolarizations are similar to automatic rhythms. Indeed, it has been argued that the only difference between early afterdepolarization-induced and automatic rhythms is in their initiation. The former depend on interruption of repolarization by an oscillation to induce them at a given V_m; the latter depend on spontaneous depolarization of the membrane to a specific V_m to induce them.

Prevention and termination of early afterdepolarization-induced rhythms also can be brought about by pharmacologic means.[3] Any agent that accelerates repolarization, such as lidocaine, will tend to terminate them. Similarly, Ca^{2+} channel blockers like verapamil have been effective. Agents like β-adrenergic catecholamines that both increase inward i_{Ca} and accelerate repolarization may have inconsistent effects, sometimes enhancing and sometimes suppressing. α-Adrenergic catecholamines, which prolong repolarization, tend to enhance Cs^{2+}-induced early afterdepolarizations.[61]

Cs^{2+}-induced triggered activity has been studied in intact animals.[59] These animals develop a long Q-T interval and pleomorphic tachycardias that sometimes resemble torsade de pointes. Similar results have been seen with *N*-acetylprocainamide. It must be stressed that no simple model relating prolonged repolarization to early afterdepolarizations and torsade de pointes has been developed to date. Indeed, studies with quinidine and with ketanserin—both of which prolong repolarization—have stressed the need for an additional pathogenic factor.[62,63] In these models, the application of aconitine to two sites on the epicardium, thereby inducing two independent oscillators, was required if torsade was to occur.

Clinical Implications

Although arrhythmias triggered by early and delayed afterdepolarizations have been studied extensively in isolated tissues and intact animals, they have been less actively studied in patients. The types of studies done to date have depended on an understanding that for rhythms triggered by delayed afterdepolarizations the first triggered beat (and in some instances the subsequent triggered rhythm) should show a shortening of its CI as the preceding paced or dominant cardiac rhythm increases in rate.[64,65] Such phenomenology has been tested using statistical methods for evaluating electrocardiograms, and with electrophysiologic testing. To date it appears that some accelerated AV junctional escape rhythms, some ventricular tachycardias occurring in the setting of atherosclerotic heart disease, and some catecholamine-sensitive tachycardias in patients with otherwise healthy hearts may be explained by triggered activity secondary to delayed afterdepolarizations.[3]

Two lines of evidence have been used to link early afterdepolarizations to cardiac arrhythmias in patients. The first is their role in generating pleomorphic ventricular tachycardias in the setting of delayed repolarization in animal experiments. The similarity of this phenomenon to the occurrence of pleomorphic tachycardias including torsade de pointes in patients with abnormally long Q-T intervals has been noted.[59] Moreover, in a subset of patient studies, monophasic AP recordings have been made that demonstrate oscillations (which might, however, be artifactual) during repolarization.

It is obvious from the above that triggered activity is suspected to cause only a minority of the cardiac arrhythmias that one might see in a general clinical population. However, given the lethality of at least a subset of the suspected rhythms as well as the need to understand mechanisms for different types of arrhythmias if we are to arrive at improved modalities for therapy and prevention, it is important that our understanding of the clinical importance of these arrhythmias continue to be refined.

ACKNOWLEDGMENT

The collaboration between Drs. Rosen and Anyukhovsky is supported by the US-USSR Exchange Program on Sudden Cardiac Death (Program Area 5). Certain of the studies referred to were supported by USPHS-NHLBI grant HL-28223.

REFERENCES

1. Bozler E: The initiation of impulses in cardiac muscle. *Am J Physiol* 1943;138:273–282.
2. Segers M: Le battement auto-entretenu du coeur. *Arch Int Pharmacodyn* 1947;75:144–156.
3. Wit AL, Rosen MR: Afterdepolarizations and triggered activity, in Fozzard H, Haber E, Jennings R, Katz A, Morgan H (eds): *The Heart and Cardiovascular System*. New York, Raven Press, 1986, pp 1449–1491.
4. January CT, Fozzard HA: Delayed afterdepolarizations in heart muscle: Mechanisms and relevance. *Pharmacol Rev* 1988;40:219–227.
5. Cranefield PF, Aronson RW: *Cardiac Arrhythmias: The Role of Triggered Activity and Other Mechanisms*. Mt. Kisco, NY, Futura Publishing, 1988.
6. Rosen MR, Gelband HB, Merker C, Hoffman BF: Mechanics of digitalis toxicity. Effects of ouabain on phase 4 of canine Purkinje fiber transmembrane potentials. *Circulation* 1973;47:681–689.
7. Rosen MR, Gelband HB, Hoffman BF: Correlation between effects of ouabain on the canine electrocardiogram and transmembrane potentials of isolated Purkinje fibers. *Circulation* 1973;47:65–72.
8. Ferrier GR, Saunders JH, Mendez C: A cellular mechanism for the generation of ventricular arrhythmias by acetylstrophanthidin. *Circ Res* 1973;32:600–609.
9. Saunders JH, Ferrier GR, Moe GK: Conduction block associated with transient depolarizations induced by acetylstrophanthidin in isolated canine Purkinje fibers. *Circ Res* 1973;32:610–617.
10. Hashimoto K, Moe GK: Transient depolarizations induced by acetylstrophanthidin in specialized tissue of dog atrium and ventricle. *Circ Res* 1973;32:618–624.
11. Davis LD: Effects of changes in cycle length on diastolic depolarization produced by ouabain in canine Purkinje fibers. *Circ Res* 1973;32:206–214.
12. Muller P: Ouabain effects on cardiac contraction, action potential and cellular potassium. *Circ Res* 1965;17:46–56.
13. Vassalle M, Karis J, Hoffman BF: Toxic effects of ouabain on Purkinje fibers and ventricular muscle fibers. *Am J Physiol* 1962;203:433–439.
14. Cranefield PF: *The Conduction of the Cardiac Impulse: The Slow Response and Cardiac Arrhythmias*. Mt. Kisco, NY, Futura Publishing, 1975.
15. Cranefield PF: Action potentials, afterpotentials and arrhythmias. *Circ Res* 1977;41:415–423.
16. Ferrier GR: Digitalis arrhythmias: Role of oscillatory afterpotentials. *Prog Cardiovasc Dis* 1977;19:459–474.
17. Aronson RS, Gelles JM: The effect of ouabain, dinitrophenol and lithium on the pacemaker current in sheep cardiac Purkinje fibers. *Circ Res* 1977; 40:517–524.
18. Tsien RW, Kass RS, Weingart R: Cellular and subcellular mechanisms of cardiac pacemaker oscillations. *J Exp Biol* 1979;81:205–215.
19. Shimoni Y, Giles W: Separation of Na-Ca exchange and transient inward currents in heart cells. *Am J Physiol* 1987;253:H1330–H1333.
20. Moak JP, Rosen MR: Induction and termination of triggered activity by pacing in isolated canine Purkinje fibers. *Circulation* 1984;69:149–162.
21. Ferrier GR: Effects of transmembrane potential on oscillatory afterpotentials induced by acetylstrophanthidin in canine ventricular tissues. *J Pharmacol Exp Ther* 1980;215:332–341.
22. Rosen MR, Danilo P: Effects of tetrodotoxin, lidocaine, verapamil and AHR-2666 on ouabain-induced delayed afterdepolarizations in canine Purkinje fibers. *Circ Res* 1980;46:117–124.
23. Hewitt K, Gessman L, Rosen MR: Effects of procainamide, quinidine and ethmozin on delayed afterdepolarizations. *Eur J Pharmacol* 1983;96:21–28.
24. Rosen MR, Ilvento JP, Gelband H, Merker C: Effect of verapamil on electrophysiologic properties of canine cardiac Purkinje fibers. *J Pharmacol Exp Ther* 1974;189:414–422.
25. Binah O, Cohen IS, Rosen MR: The effects of adriamycin on normal and ouabain-toxic canine Purkinje and ventricular muscle fibers. *Circ Res* 1983;53:655–662.
26. LeMarec H, Spinelli W, Rosen MR: Effects of doxorubicin on ventricular tachycardia. *Circulation* 1986;74:881–889.
27. Wittenberg SM, Streuli F, Klocke FJ: Acceleration of ventricular pacemakers by transient increases in heart rate in dogs during ouabain administration. *Circ Res* 1970;26:705–716.
28. Gorgels APM, Beekman HDM, Brugada P, Dassen WRM, Richards DAB, Wellens HJJ: Extrastimulus related shortening of the first postpacing interval in digitalis induced ventricular tachycardia. *J Am Coll Cardiol* 1983;1:840–857.
29. Vos M, Gorgels A, Leunissen J, Wellens H: Flunarizine allows differentiation between mechanisms of arrhythmias in the intact heart. *Circulation* 1990; 81:343–349.
30. Kimura S, Cameron JS, Kozlovskis PL, Bassett AL, Myerburg RJ: Delayed afterdepolarizations and triggered activity induced in feline Purkinje fibers by α-adrenergic stimulation in the presence of elevated calcium levels. *Circulation* 1984;70:1074–1082.
31. Wit AL, Cranefield PF: Triggered and automatic activity in the canine coronary sinus. *Circ Res* 1977;41:435–445.
32. Johnson N, Danilo P, Wit A, Rosen M: Response to pacing of triggered activity occurring in catecholamine-treated canine coronary sinus. *Circulation* 1986;74:1168–1179.
33. Malfatto G, Rosen TS, Rosen MR: The response to overdrive pacing of triggered atrial and ventricular arrhythmias in the canine heart. *Circulation* 1988;77:1139–1148.
34. Harris AS: Delayed development of ventricular ectopic rhythm following experimental coronary occlusion. *Circulation* 1950;1:1318–1328.
35. Rosenshtraukh LV, Urthaler F, Anyukhovsky EP, Beloshapko GG, Hageman GR, James TN: Serial production of controlled periods of temporary heart block used to unmask and assess latent ventricular automaticity during experimental acute myocardial ischemia. *J Am Coll Cardiol* 1986;8:95A–103A.
36. Harris AS, Bisteni A, Russel RA, Brigham JC, Firestone JE: Excitatory factors in ventricular tachycardia resulting from myocardial ischemia: Potassium a major excitant. *Science* 1954;119:200–203.
37. Janse MJ, Van Capelle FJL, Morsink H, Kleber AG, Wilms-Schopman FJG, Cardinal R, Nauman D'Alnoncourt C, Durrer D: Flow of "injury" current and patterns of excitation during early ventricular arrhythmias in acute regional myocardial ischemia in

isolated porcine and canine hearts. Evidence for 2 different arrhythmogenic mechanisms. *Circ Res* 1980;47:151–165.

38. Janse MJ, Kleber AG: Electrophysiological changes and ventricular arrhythmias in the early phase regional myocardial ischemia. *Circ Res* 1981;49:1069–1081.
39. Pogwizd SM, Corr PB: Reentrant and nonreentrant mechanisms contribute to arrhythmogenesis during early myocardial ischemia: Results using three dimensional mapping. *Circ Res* 1987;61:352–371.
40. Pogwizd SM, Onufer JR, Kramer JB, Sobel BE, Corr PB: Induction of delayed depolarizations and triggered activity in canine Purkinje fibers by lysophosphoglycerides. *Circ Res* 1986;59:416–425.
41. Sobel BE, Corr PB, Robinson AK, Goldstein RA, Witkowski FX, Klein MS: Accumulation of lysophosphoglycerides with arrhythmogenic properties in ischemic myocardium. *J Clin Invest* 1978;62:546–553.
42. Shaikh NA, Downar E: Time course of changes in porcine myocardial phospholipid levels during ischemia: A reassessment of the lysophospholipid hypothesis. *Circ Res* 1981;49:316–325.
43. Adamantidis MM, Caron JF, Dupuis BA: Triggered activity induced by combined mild hypoxia and acidosis in guinea pig Purkinje fibers. *J Mol Cell Cardiol* 1986;18:1287–1299.
44. Ferrier GR, Moffat MR, Lukas A: Possible mechanisms of ventricular arrhythmias elicited by ischemia followed by reperfusion: Studies on isolated canine ventricular tissues. *Circ Res* 1985;56:184–194.
45. Fenoglio JJ Jr, Karagueuzian HS, Friedman PL, Albala A, Wit AL: Time course of infarct growth toward the endocardium after coronary occlusion. *Am J Physiol* 1979;236:H353–H370.
46. Friedman PL, Stewart JR, Fenoglio JJ, Wit AL: Survival of subendocardial Purkinje fibers after extensive myocardial infarction in dogs: In vitro and in vivo correlation. *Circ Res* 1973;33:597–611.
47. Friedman PL, Stewart JR, Wit AL: Spontaneous and induced cardiac arrhythmias in subendocardial Purkinje fibers surviving extensive myocardial infarction in dogs. *Circ Res* 1973;33:612–626.
48. Lazzara RN, El-Sherif N, Scherlag BJ: Electrophysiological properties of canine Purkinje cells in one-day-old myocardial infarction. *Circ Res* 1973; 33:722–734.
49. Horowitz LN, Spear JF, Moore EN: Subendocardial origin of ventricular arrhythmias in 24-hour-old experimental myocardial infarction. *Circulation* 1976;53:56–63.
50. El-Sherif N, Gough WB, Zeiler RH, Mehra R: Triggered ventricular rhythm in 1-day-old myocardial infarction in the dog. *Circ Res* 1983;52:566–579.
51. Hashimoto K, Shibuya T, Satoh H, Imai S: Quantitative analysis of the antiarrhythmic effect of drugs on ventricular arrhythmias by the determination of minimum effective plasma concentration. *Jpn Circ J* 1983;47:92–97.
52. El-Sherif N, Gough WB, Zeiler RH, Mehra R: Ventricular rhythms in one-day-old canine infarction are due to triggered activity. *Circulation* 1982;66 (suppl 2):357(abstract).
53. LeMarec H, Dangman KH, Danilo P Jr, Rosen MR: An evaluation of automaticity and triggered activity in the canine heart one to four days after myocardial infarction. *Circulation* 1985;71:1224–1236.
54. Anyukhovsky EP, Urthaler F, Beloshapko GG: Influence of ionic modification on electrical activity of Purkinje fibers obtained from dogs with 1-day-old myocardial infarction. *J Mol Cell Cardiol* 1991; (in press).
55. Katzung BG: Effects of extracellular calcium and sodium on depolarization-induced automaticity in guinea pig papillary muscle. *Circ Res* 1975;37:118–127.
56. January CT, Riddle MJ: Early afterdepolarizations: Mechanism of induction and block; A role for L-type Ca^{++} current. *Circ Res* 1989;64:977–990.
57. January CT, Riddle MJ, Salata JJ: A model for early afterdepolarizations: Induction with the Ca^{2+} channel agonist Bay K 8644. *Circ Res* 1988;62:563–571.
58. Damiano BP, Rosen MR: Effects of pacing on triggered activity induced by early afterdepolarizations. *Circulation* 1984;69:1013–1025.
59. Brachmann J, Scherlag BJ, Rosenshtraukh LV, Lazzara R: Bradycardia-dependent triggered activity: Relevance to drug-induced multiform ventricular tachycardia. *Circulation* 1983;68:846–856.
60. Roden DM, Hoffman BF: Action potential prolongation and induction of abnormal automaticity by low quinidine concentrations in canine Purkinje fibers. Relationship to potassium and cycle length. *Circ Res* 1985;56:857–867.
61. Ben David J, Zipes D: Alpha adrenoceptor subtype antagonist modulates cesium-induced early afterdepolarizations and ventricular tachyarrhythmias in dogs. *Circulation* 1988;78(suppl 2):157 (abstract).
62. Leichter D, Danilo P Jr, Boyden P, Rosen TS, Rosen MR: A canine model of torsades de pointes. *PACE* 1988;11:2235–2245.
63. Zaza A, Malfatto G, Rosen MR: Electrophysiologic effects of ketanserin on canine Purkinje fibers. *J Pharmacol Exp Ther* 1989;250:397–405.
64. Rosen MR, Fisch C, Hoffman BF, Danilo P, Lovelace DE, Knoebel SB: Can accelerated AV junctional escape rhythms be explained by delayed afterdepolarizations? *Am J Cardiol* 1980;45:1272–1282.
65. Rosen MR, Reder RF: Does triggered activity have a role in the genesis of cardiac arrhythmias? *Ann Intern Med* 1981;94:794–801.

PART II

Experimental Arrhythmias

Chapter **7**

Ventricular Arrhythmias in Experimental Subacute Myocardial Infarction

Nabil El-Sherif, MD, William B. Gough, PhD, Mark Restivo, PhD, and Mohamed Boutjdir, PhD

Following coronary artery occlusion, the area that was originally perfused becomes ischemic. Due to diffusion and collateral circulation, irreversible cell damage spreads from the central zone into the border zone.[1,2] Most infarctions have one or more areas of necrosis surrounded by ischemic border zones. Purkinje and ventricular muscle fibers in the ischemic zone develop abnormal electrophysiologic properties and generate ventricular arrhythmias at various stages following infarction. Since the classic experiments on the dog heart by Harris,[3] it is known that ventricular arrhythmias after coronary artery occlusion occur in two distinct phases. The first phase corresponds to the acute phase of ischemia and lasts until 15 to 30 min after coronary occlusion; the second starts 4 to 8 h after occlusion, peaks 1 to 2 days postinfarction, and usually subsides by the third day. The early phase of ventricular arrhythmias is more serious, can degenerate into rapid ventricular tachycardia and ventricular fibrillation, and has been attributed to reentrant excitation in ischemic myocardium.[4–6]

The second phase, which is more benign, consists of spontaneous multiform ventricular rhythms at about the same rate as the sinus rhythm. These rhythms arise from surviving subendocardial Purkinje fibers overlying the infarction.[4,7,8] In vitro studies of these surviving subendocardial Purkinje fibers have suggested two mechanisms that may be responsible for the activity: (a) abnormal automaticity[7,9] and (b) triggered activity.[10] However, the preponderance of evidence favors the second mechanism.

El-Sherif and associates[11–13] have shown that during the second phase of spontaneous ventricular rhythms, as well as following the subsidence of this phase, fast ventricular tachyarrhythmias and ventricular fibrillation could be induced by programmed electrical stimulation of the heart. These tachyarrhythmias, similar to those in the first phase, were attributed to reentrant excitation in ischemic myocardial border zones. Other investigators have shown that ventricular tachyarrhythmias can also be induced by programmed electrical stimulation in the chronic phase of canine myocardial infarction.[14]

In this chapter, recent progress in the understanding of the two major electrophysiologic mechanisms of ventricular arrhythmias in the subacute phase of myocardial infarction, that is, triggered ventricular rhythms and reentrant ventricular rhythms will be reviewed.

TRIGGERED VENTRICULAR RHYTHMS IN THE SUBACUTE PHASE OF MYOCARDIAL INFARCTION

Triggered Activity Studied In Vitro

Delayed afterdepolarizations (DADs) giving rise to triggered activity were recorded in depolarized ischemic Purkinje fibers from a 1-day-old-canine infarction.[10] The amplitude and rate of rise of the DADs were a function of both the cycle length (Fig. 7.1A) and the number of impulses in a stimulated train. A critically timed premature impulse may be followed by a DAD that triggers activity (Fig. 7.1B). When triggered activity was initiated

655 Avenue of the Americas, New York, NY 10010
Current Topics in Cardiology

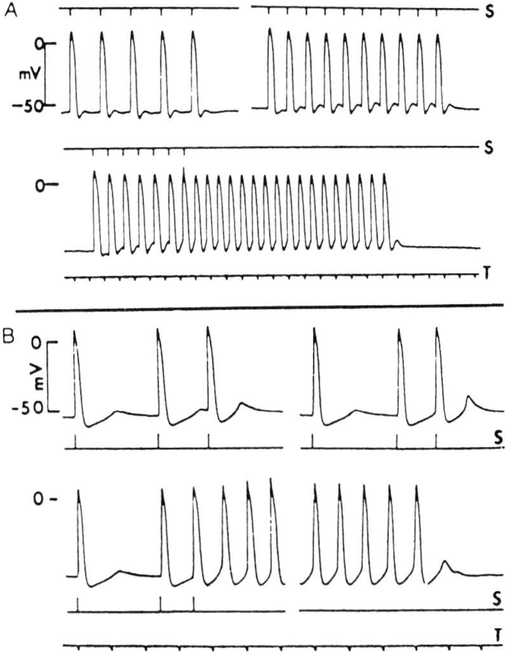

FIGURE 7.1 Triggered activity arising from delayed afterdepolarizations in endocardial preparations obtained from 1-day-old canine infarction. **A** and **B.** Transmembrane recordings from Purkinje cells in the ischemic zone from two different preparations. **A.** The preparation was stimulated at cycle lengths of 2,000 ms, 1,200 ms, and 1,000 ms. Reduction of the cycle length of stimulation resulted in an increase of the amplitude of the afterdepolarization that reached threshold and initiated a run of triggered activity in the lower recording. The triggered rhythm terminated following a subthreshold delayed afterdepolarization. **B.** The effect of premature stimulation on the amplitude of the delayed afterdepolarization is demonstrated. The preparation was paced at a basic cycle length of 2,500 ms. The coupling interval of the premature stimulus was shortened from 1,500 ms to 1,200 ms to 1,000 ms. This resulted in an increase of the amplitude of the afterdepolarization that reached threshold following the short coupling interval (CI) and initiated a triggered rhythm. The rhythm terminated following a subthreshold afterdepolarization. S = the timing of stimulation; T = time scale, representing 1-s intervals. *Reprinted from El-Sherif et al.,*[10] *with permission.*

by normal automaticity in Purkinje fibers, a subthreshold DAD reached threshold and initiated an action potential (AP). The result was extrasystolic groupings (that is, bigeminal and trigeminal rhythms) or sustained triggered activity (Fig. 7.2). Entrance and exit block around sites of triggered activity is not uncommon in ischemic subendocardial preparations. The presence of entrance block around a site of triggered activity that is able to exit, at least intermittently, to the rest of the preparation (or the ventricles) can result in a parasystolic rhythm.

Abnormal automatic or triggered activity had been described only in Purkinje fibers until the recent report of Dangman and colleagues,[15] which suggests that ischemia has the capacity to produce spontaneous activity in both Purkinje fibers and ventricular muscle.

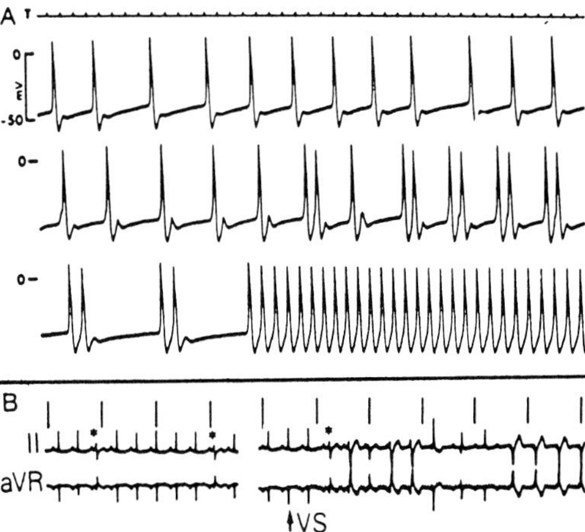

FIGURE 7.2 Extrasystolic rhythm due to triggered activity arising from delayed afterdepolarizations. **A.** Transmembrane recording from a Purkinje fiber in the ischemic zone of an endocardial preparation from a 1-day-old canine infarction. The upper tracing shows a background slow automatic rhythm. Each automatic AP is followed by a low-amplitude delayed afterdepolarization. In the middle tracing, the amplitude of the delayed afterdepolarization gradually increased, reached threshold, and triggered a single AP, resulting in a trigeminal rhythm followed by a bigeminal rhythm. In the lower tracing, a sustained triggered activity occurred when the subthreshold afterdepolarization that followed the single triggered AP reached threshold potential. *Modified from El-Sherif et al.,*[10] *with permission.* **B.** Electrocardiographic tracing obtained from a dog 1 day following ligation of the left anterior descending artery. The tracing shows sinus tachycardia and extrasystolic ventricular ectopic beats with late fixed coupling (marked by *). Slight slowing of the sinus rhythm by vagal stimulation (VS) revealed the presence of a multiform ventricular rhythm. The extrasystolic beats and the multiform ventricular rhythm may be the result of triggered activity arising from delayed afterdepolarizations in ischemic Purkinje fibers surviving the infarct.

Dependence of Triggered Activity on Diastolic Potentials

Rhythmic activity in ischemic endocardial preparations is initiated in depolarized ischemic Purkinje fibers having a maximum diastolic potential averaging -59 ± 9 mV.[10] Superfusion of an ischemic preparation causes recovery of the reduced diastolic potential toward normal resting potentials.[8,16] During this recovery, sustained activity gives way to nonsustained activity and subthreshold DADs. At potentials between -80 and -85 mV, triggered activity and DADs were no longer observed.[10] Therefore, it appeared that sustained and nonsustained activity were dependent on membrane potential (V_m).

Dresdner and associates[16] have shown that intracellular potassium ($[K^+]_i$) activity was reduced after 24 h of ischemia by 50.4 mM from 112 $\pm$ 19.8 mM. This K^+ loss was accompanied by a loss of the K^+ equilibrium potential (E_k). However, the loss of K^+ activity accounted for only about a one-half reduction of the diastolic potential and, therefore, additional factors must be considered. The intracellular sodium ($[Na^+]_i$) activity of these ischemic Purkinje fibers was elevated to 15.6 ± 6.0 from 9.4 ± 2.6 ($p < 0.005$). Thus, the large loss of K^+ was not matched by an equivalent gain of Na^+. This gain of $[Na^+]_i$ would produce an elevation or overload of $[Ca^{2+}]_i$. During prolonged superfusion in vitro, both K^+ activity and the E_k returned toward normal, thereby providing an explanation for the improved diastolic potentials measured after significant superfusion.[13] The Na^+ activity also decreased as the maximum diastolic potentials became more negative. Partial inhibition of the Na/K pump observed during ischemia may account for part of the K^+ loss and consequent depolarization.[17]

Transmembrane recordings of sustained triggered activity and abnormal automaticity have the same appearance. The dependence of triggered activity on the magnitude of diastolic potential was studied by the application of constant depolarizing or hyperpolarizing current.[18] During sustained activity (maximum diastolic potential -61 ± 7 mV) hyperpolarizing current decreased the DADs, rendered their amplitude subthreshold, and terminated triggered activity. During the quiescence caused by constant hyperpolarizing current, a stimulated train of action potentials (APs) produced DADs. Decreasing the current permitted augmented DADs. In quiescent preparations (resting potentials -68 ± 7 mV), a train of stimulated action potentials was followed by subthreshold DADs. Depolarizing current increased the DAD amplitude. Sufficient depolarization caused triggered activity to occur (Fig. 7.3). To exclude depolarization-induced automaticity, constant currents were applied without a previous train of stimuli. Neither DADs nor triggered activity was evoked.

The main finding of this study was that in ischemic Purkinje fibers, 1 day postinfarction, there is a graded response of DADs to depolarizing currents and depolarized diastolic potentials.[18] On this basis, one can consider that the transition from nonsustained triggered activity to a sustained rhythm can be due to an enhancement of the abnormal DAD accompanying postinfarction depolarization. Similarly, a sustained rhythm can become nonsustained as a consequence of hyperpolarization accompanying recovery from ischemia. The transition from threshold to subthreshold and vice versa can result from only a 1 mV change in diastolic potential. One need not invoke a separate, intrinsic oscillatory mechanism, such as abnormal automaticity, to explain the transition from sustained to nonsustained rhythms in the same heart or in isolated preparation 1 day postinfarction. In addition, the DADs may become suprathreshold to the extent that interventions such as overdrive pacing,[19] changes in temperature,[19,20] or pharmacologic inhibitors[19,21] may be unable to render these DADs subthreshold and thereby terminate the sustained triggered activity. Caution must be applied when one assumes that such interventions might distinguish sustained triggered activity from abnormal automaticity.

Ionic Mechanisms

The ionic mechanisms that produce DADs have not been studied with voltage-clamp or patch-clamp techniques in ischemic Purkinje fibers. The basic hypothesis for the mechanism by which charge is carried by the transient inward current that causes the DAD is derived from studies of Purkinje and ventricular muscle cells that have been overloaded with calcium (Ca^{2+}) in a variety of ways. As recently reviewed,[22] there are two proposed mechanisms. The first, originally proposed by Kass and Tsien,[23] postulated a nonspecific cation channel, activated by a phasic rise in $[Ca^{2+}]_i$, which had significant permeability to Na^+, K^+, and Ca^{2+}. The inward current was thought to be carried predominantly by Na^+ with some Ca^{2+} contribution. The second proposed mechanism for transient inward current (i_{ti}) is the electrogenic Na/Ca exchange pump driven by the transmembrane electrochemical gradient for Na^+ and Ca^{2+}.[24]

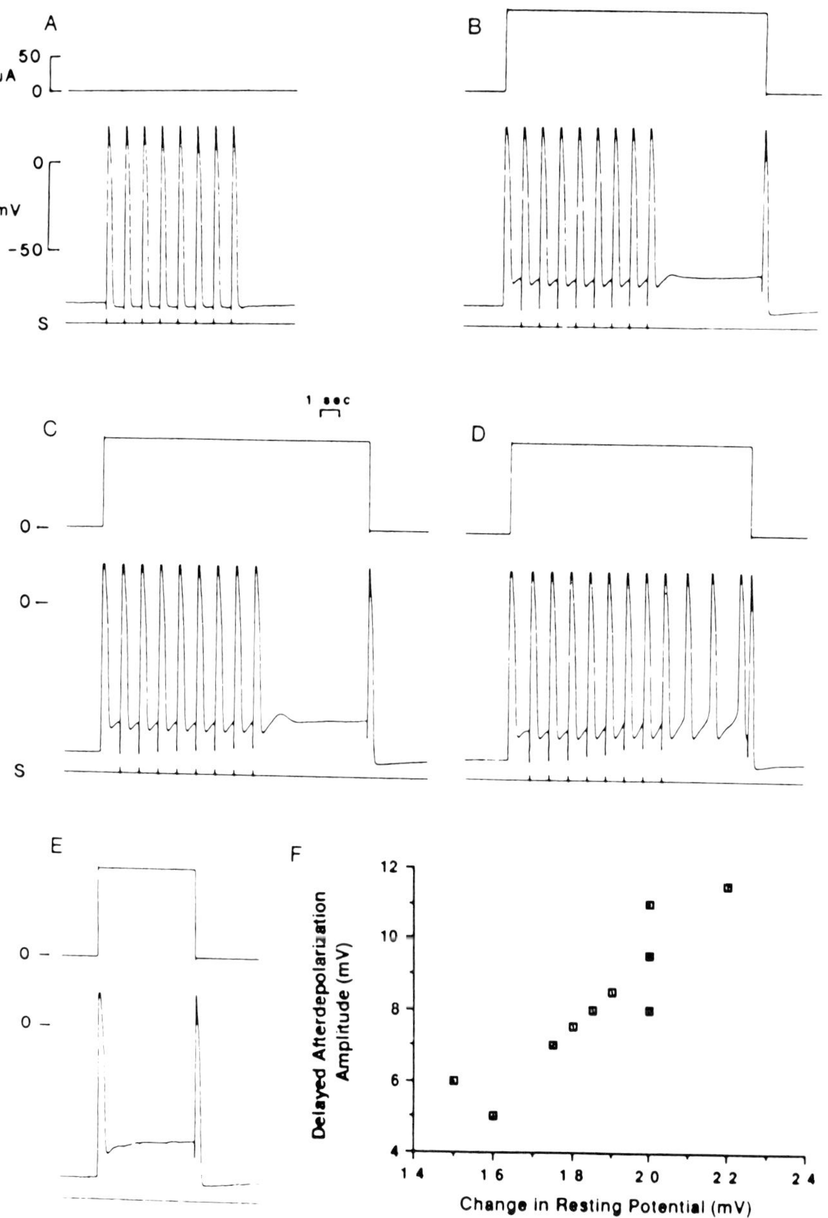

FIGURE 7.3 Effects of long depolarizing current on delayed afterdepolarization. Recordings were obtained from a canine endocardial preparation 1 day postinfarction. **A.** Control. A train of eight paced beats (cycle length = 1 s) produced no measurable delayed afterdepolarization. **B** and **C.** During long graded depolarizing current pulses initiated before stimulation, delayed afterdepolarization gradually increased in amplitude. **D.** Delayed afterdepolarization following the eighth-beat induced triggered activity that ceased as soon as current was terminated. **E.** Depolarizing current equal in amplitude to that used in **D** produced neither a delayed afterdepolarization nor depolarization-induced automaticity. **F.** Amplitude of delayed afterdepolarization after the eighth driven beat is plotted as a function of change in resting potential. *Reprinted from Gough and El-Sherif,*[18] *with permission.*

Clues to the mechanism that may be operative in ischemic Purkinje fibers have been derived by varying ionic concentrations and using specific inhibitors. An elevation of $[K^+]_o$ from 4 to 6 mM caused suppression of DADs.[25] Reduction of $[NA^+]_o$ caused initially an enhancement of the DAD and a more rapid sustained activity.

Elevation of $[Ca^{2+}]_o$ increased the amplitude of subthreshold DADs and enhanced their ability to trigger activity.[118]

The effects of ionic blockers may be direct or indirect. Tetrodotoxin (TTX) has been shown to suppress DADS.[26] It may directly block the i_{ti} or it may reduce the i_{Na} during the upstroke of the AP, thereby reducing $[Na^+]_i$. A decreased $[Na^+]_i$ reduces $[Ca^{2+}]_i$, thereby reducing the amount of phasic Ca^{2+} release. Ca^{2+} channel blockers have been shown to reduce the rate of depolarization and amplitude of the DAD.[27] This could be a direct effect on the DAD or an indirect effect on the upstroke of the APs initiated at reduced diastolic potentials. Less Ca^{2+} entering the myoplasm during the AP would reduce the Ca^{2+} overload of the cell and, thereby, the amplitude of the DAD. Caffeine, at high concentrations (10 mM), and ryanodine, at micromolar concentrations, have been shown to reduce phasic Ca^{2+} release from the sarcoplasmic reticulum. Boutjdir and associates[28] have shown that both agents suppress sustained triggered activity, thereby implying that this activity in ischemic fibers requires a phasic release of Ca^{2+} from the sarcoplasmic reticulum. When the release of Ca^{2+} is inhibited, neither sustained nor nonsustained triggered activity occurs.

The Na/Ca exchanger has been reported to be inhibited by the anthracycline antibiotic doxorubicin,[29] which suppressed DADs and triggered activity both in digitalis-toxic Purkinje fibers and in 2-day ischemic subendocardial Purkinje fibers.[21] However, abnormal automaticity or sustained rhythms were not suppressed by doxorubicin in either of these preparations. Thus, both the direct and indirect evidence suggests that Ca^{2+} overload is fundamental to the genesis of the DAD and that the Na/Ca exchanger or a nonspecific cation channel may be responsible for the ion-carrying mechanism underlying the DAD.

Actions of Antiarrhythmic Agents

The effects of antiarrhythmic drugs on sustained triggered activity has been studied both in vitro and in vivo. Ethmozin was first shown to suppress ventricular arrhythmias occurring 1 day postinfarction.[30] Ethmozin slows but does not abolish normal automaticity. However, it abolishes barium-induced automaticity, terminating it with a subthreshold DAD.[31] Sustained activity, 24 h postinfarction, was terminated with ethmozin. The termination occurred with a subthreshold DAD. The combination of lidocaine and ethmozin was used as a test for differentiating sustained triggered activity from abnormal automaticity. If sustained activity was suppressed by ethmozin, it had to be either triggered activity or abnormal automaticity,[19] but if it was suppressed by lidocaine, it was either normal automaticity or triggered activity. Therefore, if a preparation was suppressed by both ethmozin and lidocaine, the underlying mechanism was concluded to be triggered activity. The problem faced by this matrix is that the level of lidocaine is crucial, because elevated lidocaine will suppress even abnormal automaticity. If, as our recent study[18] suggests, there is a voltage dependence of the DAD, then when it is raised to suprathreshold amplitudes that maintain a sustained rhythm, the level of lidocaine necessary to render it subthreshold may be in the same concentration range as that required to suppress abnormal automaticity.

Ca^{2+} channel blockers were shown to inhibit triggered activity either by directly suppressing DADs or by inducing exit block around sites of triggered activity (Figs. 7.2 to 7.4).[10,27,33] Ouabain at a concentration that has no toxic effect on normal Purkinje fibers may enhance arrhythmias in ischemic Purkinje fibers by increasing the magnitude of DAD and enhancing triggered activity.[34] It was postulated that the common action of ischemia and digitalis causing an increase in $[Ca^{2+}]_i$ may work synergistically to cause DAD and triggered activity in Purkinje fibers. This may represent the mechanism responsible for increased susceptibility to digitalis toxicity in patients with ischemic heart disease.

The increased sensitivity of ischemic Purkinje fibers to the toxic effects of type III antiarrhythmic drugs was first shown in a study of clofilium.[35] During the course of in vitro superfusion, triggered activity ceased as the maximum diastolic potential improved. In these recovering Purkinje fibers at a resting potential of -77 ± 5 mV, clofilium produced oscillatory early afterdepolarizations at concentrations that had no such effect in normal Purkinje fibers at a resting potential of -86 ± 3 mV (Fig. 7.4). These early afterdepolarizations were able to trigger activity in the adjacent myocardium. The preferential sensitivity of ischemic fibers to type III drugs was

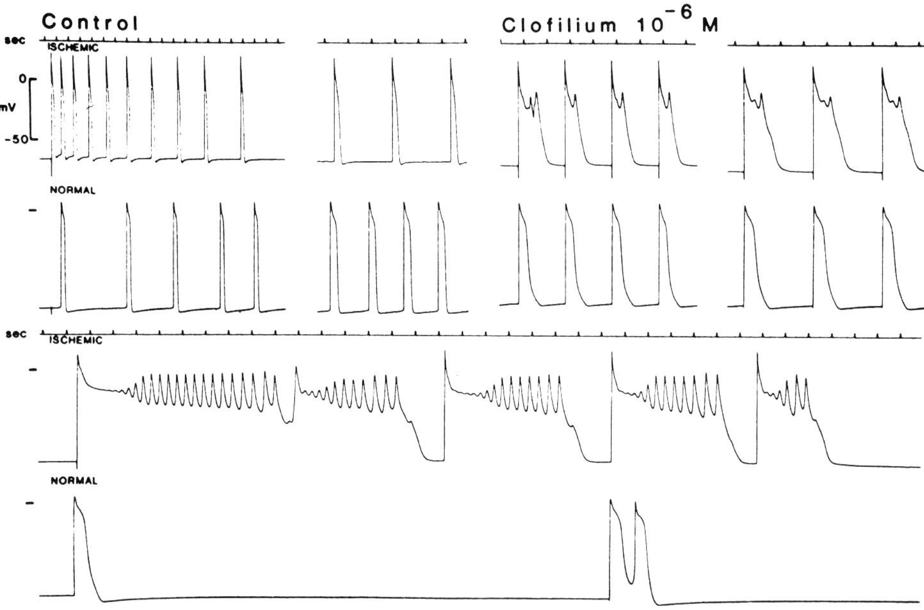

FIGURE 7.4 Transmembrane APs recorded from normal Purkinje fiber and ischemic subendocardial Purkinje fiber 1 day after infarction, demonstrating triggered activity from delayed afterdepolarizations in the ischemic fiber when perfused with clofilium. Under control conditions in Tyrode solution, both fibers showed diastolic depolarization. In the ischemic preparation, triggered activity from delayed afterdepolarizations was observed. One stimulus evoked an AP that was followed by a delayed afterdepolarization, which in turn initiated spontaneous AP formation. The triggered activity started with a cycle length of 900 ms and gradually increased to a cycle length of approximately 3 s. After further superfusion, only a slow pacemaker activity persisted at a cycle length of 2.5 s. After introducing clofilium into the perfusate at 10^{-6} M and pacing at a cycle length of 2 s, early afterdepolarizations (EADs) occurred during the plateau of the action potential of the ischemic Purkinje fiber at a CI of 500 ms. The normal AP was prolonged only in phases 2 and 3 and showed no early or delayed afterpotentials. At a driven cycle length of 3 s, repolarization of the ischemic cell was more prolonged and was accompanied by EADs. When the preparation was permitted to remain quiescent for periods of >30 s (bottom tracing), EADs developed in the ischemic fiber at a low V_m (in the range of -15 to -35 mV) and continued with a cycle length of 300 to 500 ms, thereby extending the duration of the AP to beyond 10 s. Generally, the V_m of the nadir of the EADs became progressively greater during the plateau until repolarization was completed or reexcitation occurred. The normal Purkinje fiber showed slowing of late phase 3 repolarization consistent with EADs at high V_m that gave rise to a single triggered AP. *Reprinted from Gough and El-Sherif,*[35] *with permission.*

also observed with D-sotalol and bretylium.[36] The mechanism of this differential sensitivity may be due to the differences arising from the reduced maximum diastolic potential in ischemic Purkinje fibers. Snyders and Katzung[37] have reported that clofilium inhibits the delayed K^+ current (i_K) in isolated myocytes, and Carmeliet[38] found a similar effect due to D-sotalol. Dresdner and coworkers[16] have shown that these Purkinje fibers have reduced $[K^+]_i$ and reduced E_K. These conditions could reduce K^+ permeability and driving force to the level that might permit oscillations at plateau potentials.

Triggered Activity Studied In Vivo

Studies to demonstrate or disprove the occurrence of triggered rhythms postinfarction in vivo have relied on being able to record the initiation or termination of such triggered rhythms. During a sustained rhythm, interventions have been tried that had successfully suppressed triggered activity in vitro. The first approach[39] suppressed the sustained ventricular rhythms with the Ca^{2+} channel blocker verapamil, after β-adrenergic blockade. These spontaneous ventricular rhythms were shown later to be modulated by adrenergic stimulation,[40] and so it is not surprizing that β-adrenergic blockade plus a Ca^{2+} channel blocker successfully suppressed these arrhythmias. Once the sustained rhythm was suppressed, it could be reinitiated by one or more stimulated beats. Isochronal mapping studies excluded a circus movement reentry and showed a focal origin of activity from the Purkinje fibers that had survived the infarct.[41] Either intrave-

nous catecholamines or intravenous Ca^{2+} facilitated the induction of triggered ventricular rhythms. The assumption was that the mechanism of the rhythm induced after Ca^{2+} channel blockade was the same as that before the blockade.

The ability to record DADs in dogs with small apical infarction was accomplished by Hariman and colleagues,[42] using the technique they had used to record diastolic potentials in the sinus node, namely, the high-gain, low-pass filtered electrogram. Using multiple electrodes, they found that the point of earliest activation had a negative diastolic slope and an upstroke preceding each ventricular complex. Upon termination of a nonsustained rhythm, multiple rhythmic, slow, negative diastolic potentials were recorded before the quiescence. These potentials were consistent with subthreshold DADs. The results of these studies suggest that in vivo–triggered rhythms exist postinfarction in the ventricles.

Reentrant Ventricular Rhythms in the Subacute Phase of Myocardial Infarction

In 1977, El-Sherif and associates[11–13] made the observation that in dogs that survived the initial stage of myocardial infarction arrhythmias, and studied 3 to 5 days postinfarction, reentrant ventricular rhythms occurred spontaneously but were more commonly induced by programmed electrical stimulation. The anatomic and electrophysiologic substrates for the reentrant rhythms were later characterized in a series of reports.[14,43–53] These studies have shown that reentrant excitation occurred around zones (arcs) of functional conduction block. The arcs were attributed to ischemia-induced spatially nonhomogeneous lengthening of refractoriness. Sustained reentrant tachycardia was found to have a figure-eight activation pattern whereby clockwise and counterclockwise wavefronts were oriented around two separate arcs of functional conduction block. The two circulating wavefronts coalesced into a common wavefront that conducted slowly between the two arcs of block. Using reversible cooling, reentrant excitation could be successfully terminated only from localized areas along the common reentrant wavefront.[45]

Anatomic and Electrophysiologic Substrates of Reentrant Excitation

After left anterior descending coronary artery ligation in dogs, blood flow is reduced more in the subendocardium, and resistance to flow in the infarcted tissue causes a redistribution of flow in the epicardial layers. Combined with the enlargement of collateral vessels, this results in sufficient flow to the epicardium that usually survives.[54] Although the geometry of the infarction varies in different experiments, pathologic studies consistently reveal a layer of surviving epicardial tissue overlying the core of necrotic myocardium. The epicardial layer varies in thickness from a few cells to a few millimeters, as verified histologically. The surviving epicardial layer is generally wedge shaped, with more depth at the border than at the central portion of the infarction. Although the surviving epicardial layer appears microscopically intact, the myocardial blood flow to this layer is reduced.[54]

Intracellular recordings from the surviving "ischemic" epicardial layer show cells with variable degrees of partial depolarization, reduced AP amplitude, and decreased upstroke velocity.[55–57] Full recovery of responsiveness frequently outlasts the AP duration (APD), reflecting the presence of post-repolarization refractoriness.[55,56] In these cells, premature stimuli could elicit graded responses over a wide range of coupling intervals. Slowed conduction, Wenckebach periodicity, and 2 : 1 or higher degrees of conduction block could be easily induced by fast pacing or premature stimulation (Fig. 7.5). Isochronal mapping studies have shown that both the arcs of functional conduction block and the slow activation wavefronts of the reentrant circuit develop in the surviving electrophysiologically abnormal epicardial layer overlying the infarction.

The ionic changes induced by ischemia that explain abnormal transmembrane APs of myocardial cells in the subacute phase of myocardial infarction have not been fully explored. Some studies suggest that ischemic transmembrane APs may be generated by a depressed fast Na^+ channel. This was based on experiments that showed that ischemic cells are sensitive to the depressant effect of the fast channel blocker TTX, but not to the slow channel blocker methoxyverapamil (D600).[55,56] The depression of fast channel in ischemia can be only partly explained by cellular depolarization, because the depression is usually out of proportion to the depolarization of the resting potential. The Na/K pump may be depressed in surviving ischemic myocardial cells, leading to $[Na^+]_i$ loading.[58] This can diminish the electrochemical driving force for the i_{Na}.

Ultrastructural changes of the sarcolemmal membrane, as well as the effects of products released by ischemia, including lysophosphoglycerides,[59] have been implicated. Abnormal

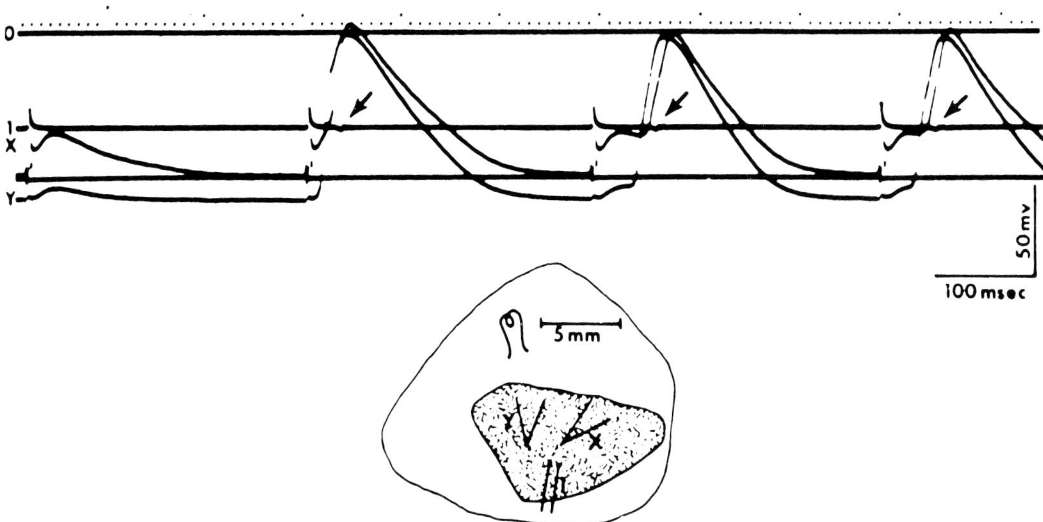

FIGURE 7.5 Recordings from a dog with 3-day-old infarction illustrating AP characteristics in ischemic epicardium. The sketch of the preparation shows two intracellular recordings (X and Y) and a close bipolar recording (1) from the infarction zone (hatched area). Ischemic cells had decreased upstroke velocity, reduced AP amplitude, and a variable degree of partial depolarization. The two cells were recorded 5 mm apart in the infarction zone, but showed significant difference in their resting potential. The resting potential of the Y cell was only slightly reduced (−80 mV), but it still had a poor AP. The preparation was stimulated at a cycle length of 290 ms, which resulted in a Wenckebach-like conduction pattern. Note that the pacing cycle length exceeded the AP duration of the two cells, suggesting that refractoriness extended beyond the completion of the AP (that is, post-repolarization refractoriness). *Reprinted from El-Sherif and Lazzara,*[55] *with permission.*

membrane properties of ischemic myocardial cells may not be the only cause for slowed conduction and block in the surviving ischemic epicardial layer. Electrical uncoupling and increase of extracellular resistance after ischemia have also been suggested.[60] Ischemia-induced increase in $[Ca^{2+}]_i$ and low pH may increase the gap-junctional resistance of the intercalated disc.[61]

Another factor considered by some authors is the anisotropic structure of the surviving epicardial layer.[62] The epicardial muscle fibers are closely packed together and arranged parallel to each other in a direction generally perpendicular to the left anterior descending artery. Conduction in the direction along the long axis of myocardial fibers is more rapid than in the transverse direction.[63,64] The slower conduction in the transverse direction is due to higher axial resistivity, which may be partly explained by fewer and shorter intercalated discs in a side-to-side direction.[64] The normal uniform anisotropic conduction properties of the epicardial layer may be altered further following ischemia.

It was suggested that the site of conduction block of premature stimuli in the ischemic epicardial layer may be determined by its anisotropic properties (that is, premature stimuli block along the long axis of epicardial muscle fibers).[62] Restivo and associates[52] have shown that functional conduction block of premature stimuli in the ischemic epicardial layer is due to abrupt and discrete change in refractoriness. The spatially nonuniform refractory distribution occurs both along and across fiber direction, in the same manner as the arcs of conduction block.[52]

Epicardial Activation Patterns of Reentrant Excitation Induced By Premature Stimulation

One to five days postinfarction, reentrant rhythms could be induced in the canine heart by one or more premature stimuli (S2S3) during regular cardiac pacing (S1) at relatively long cycle lengths (Fig. 7.6). Isochronal activation maps during S1 usually show relatively fast conduction over the epicardial surface of the infarction. The introduction of S2 results in the development of an arc of unidirectional conduction block forcing the activation wavefront to travel around the two

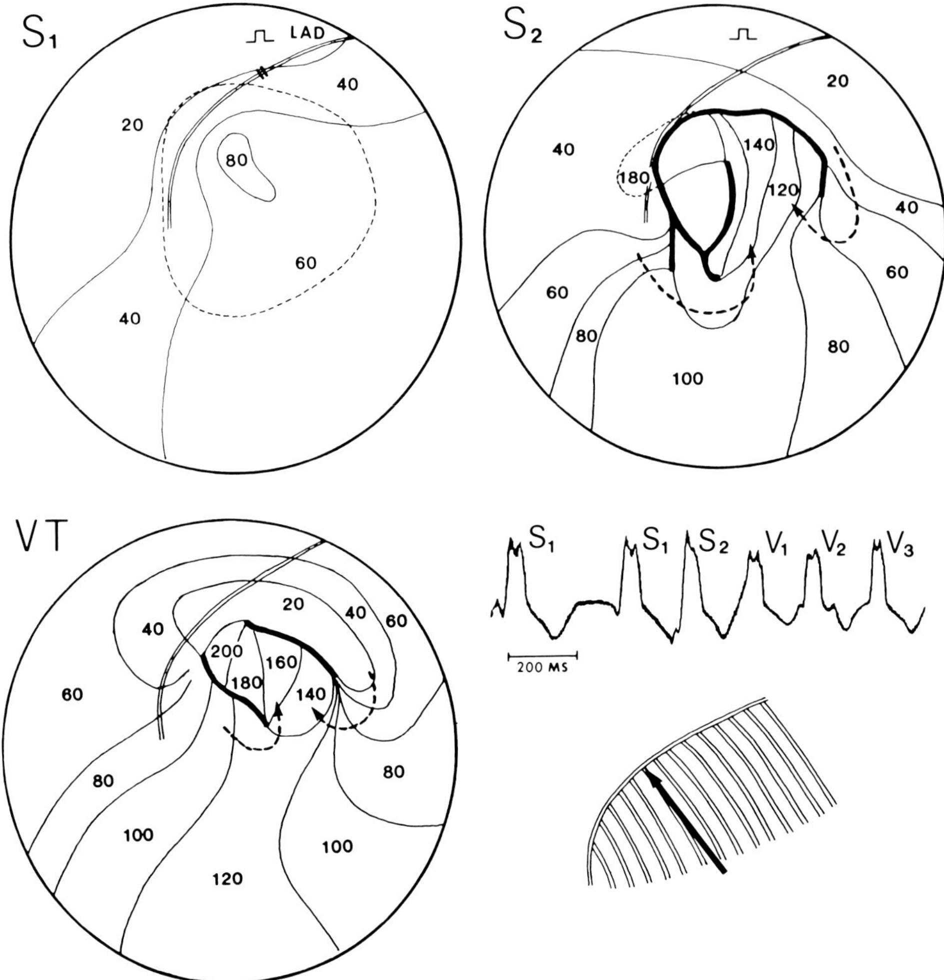

FIGURE 7.6 Epicardial isochronal activation maps during a basic ventricular stimulated beat (S1), initiation of reentry by a single premature stimulus (S2), and sustained monomorphic reentrant ventricular tachycardia (VT). A representative electrocardiogram is shown in the lower right panel. The recordings were obtained from a dog 4 days postligation of the left anterior descending artery (LAD). Site of ligation is represented by a double bar. In this and subsequent maps, epicardial activation is displayed as if the heart is viewed from the apex located at the center of the circular map. The perimeter of the circle represents the AV junction. The outline of the epicardial ischemic zone is represented by the dotted line. Activation isochrones are drawn at 20 ms intervals. Arcs of functional conduction block are represented by heavy solid lines and are depicted to separate contiguous areas that are activated at least 40 ms apart. During S1, the epicardial surface was activated within 80 ms with the latest isochrone located in the center of the ischemic zone. S2 resulted in a long continuous arc of conduction block within the border of the ischemic zone. The activation wavefront circulated around both ends of the arc of block and coalesced at the 100 ms isochrone. The common wavefront advanced within the arc of block before reactivating an area on the other side of the arc at the 180 ms isochrone to initiate the first reentrant cycle. During sustained VT, the reentrant circuit had a figure-eight activation pattern in the form of a clockwise and counterclockwise wavefront around two separate arcs of functional conduction block. The two wavefronts joined into a common wavefront that conducted between the two arcs of block. The sites of the two arcs of block during sustained VT were different to a varying degree from the site of the arc of block during the initiation of reentry by S2 stimulation. The lower right panel illustrates the orientation of myocardial fibers in the surviving ischemic epicardial layer perpendicular to the direction of the LAD. The arrow represents the longitudinal axis of propagation of the slow common reentrant wavefront during a sustained figure-eight activation pattern, which is oriented parallel to fiber orientation and perpendicular to the nearby LAD segment. *Reprinted from El-Sherif N: Electrophysiology of ventricular arrhythmias in myocardial ischemia and infarction, in El-Sherif N, Samet P (eds):* Cardiac Pacing and Electrophysiology; *1990, pp 18–56.*

ends of the arc. The arc of conduction block is functional in nature and does not exist during S1 stimulation.

The length of the arc of conduction block and the degree of slow conduction distal to the arc are crucial factors for the creation of a reentrant circuit. A premature beat that successfully initiates reentry results in a longer arc of conduction block and/or slower conduction compared with one that fails to induce reentry. When a single premature stimulus (S2) fails to initiate reentry, the introduction of a second premature stimulus (S3) may be necessary. The S3 usually creates a longer arc of conduction block and/or slower conduction around the arc. The slower activation wavefront travels around a longer, more circuitous route, thus providing more time for refractoriness to expire along the proximal side of the arc of unidirectional block. Reexcitation of this site will initiate reentry. The beat that initiates the first reentrant cycle, whether it is an S2 or an S3, results in a continuous arc of conduction block. The activation front circulates around both ends of the arc of block and rejoins at the distal side of the arc of block before breaking through the arc to reactivate an area proximal to the block. This results in splitting of the initial single arc of block into two separate arcs.

Subsequent reentrant activation continues with a figure-eight activation pattern, whereby two circulating wavefronts advance in clockwise and counterclockwise directions, respectively, around two arcs of conduction block. During monomorphic reentrant tachycardia, the two arcs of block and the two circulating wavefronts remain fairly stable. The two arcs of functional conduction block are usually oriented parallel to the long axis of the epicardial muscle fibers.[53,62,65] On the other hand, during a polymorphic reentrant rhythm, both arcs of block and the circulating wavefronts can change their geometric configurations while maintaining their synchrony. Reentrant activation spontaneously terminates when the leading edge of both reentrant wavefronts encounters refractory tissue and fails to conduct. This results in coalescence of the two arcs of block into a single arc and termination of reentrant activation.[44]

The majority of reentrant circuits in the canine postinfarction model develop in the surviving epicardial layer and can be viewed as having an essentially two-dimensional configuration. However, some reentrant circuits were identified in intramyocardial[47] or subendocardial locations.[41] The latter location is of special interest because it may be comparable to reentrant circuits described in the surviving subendocardial muscle layer in the heart of patients with chronic myocardial infarction.[66,67] This suggests that, depending on the particular anatomic features of the infarction and the geometrical configuration of ischemic surviving myocardium, reentrant circuits can be located in epicardial, subendocardial, or intramyocardial zones.[14,47,68]

Interruption of Figure-Eight Reentrant Circuit

As established by Mines,[69] the criteria for proving the presence of circulating excitation are: (a) an area of unidirectional block must be demonstrated; (b) the movement of the excitatory wave should be observed to progress through the pathway, to return to its point of origin, and then to follow again the same pathway; and (c) "the best test for circulating excitation is to cut through the ring at one point. If impulses continue to arise in the cut ring, circus movement as a cause can be ruled out."

El-Sherif and coworkers used reversible cooling and/or cryoablation of localized area of the epicardial surface of the reentrant circuit to fulfill Mines's criteria for proving the presence of circulating excitation and to identify the critical site along the reentrant pathway at which interruption of reentrant activation could be successfully accomplished.[45] These studies demonstrated that a figure-eight reentrant activation could be successfully interrupted when cooling or cryoablation was applied to the part of the common reentrant pathway immediately proximal to the zone of earliest reactivation (Fig. 7.7). At this site, the common reentrant wavefront is usually narrow and is surrounded on each side by an arc of functional conduction block. On the other hand, localized cooling at the site of earliest reactivation commonly failed to interrupt reentry. The common reentrant wavefront usually broke through the arc of functional conduction block and reactivated other sites close to the original reactivation site without necessarily changing the overall reentrant activation pattern.

Role of Spatial Nonhomogeneous Lengthening of Refractoriness in the Initiation of Reentry

In the surviving ischemic epicardial layer, refractoriness was found to be prolonged in a spatially nonuniform manner.[49] The pattern of refractoriness resembled concentric rings of isorefractoriness which increased in a monotonic fashion from the normal zone toward the center of the ischemic zone (Fig. 7.8). The disparity of refractoriness per unit distance was more marked

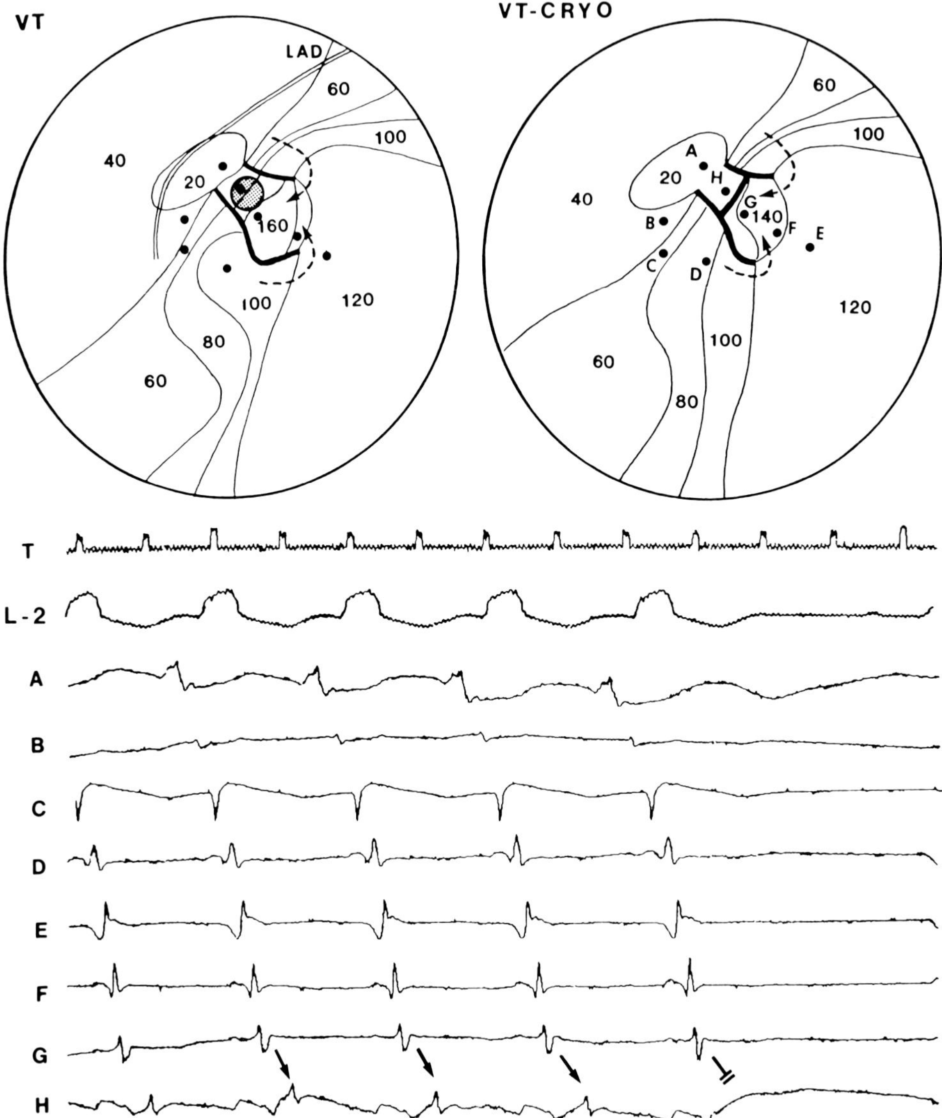

FIGURE 7.7 Interruption of a figure-eight reentrant tachycardia in the epicardial layer overlying 4-day-old canine infarction by cryothermal techniques. The control activation map is shown on the left (VT) and the map of the last reentrant beat prior to termination on the right (VT-CRYO). Selected epicardial electrograms are shown on the bottom. The position of the cryoprobe is represented by the shaded circle. The reentrant circuit was interrupted by reversible cooling of the distal part of the common reentrant wavefront (site H). During control the conduction time between the proximal electrode site G and the more distal site H was 33 ms. Prior to termination of the tachycardia, an incremental beat-to-beat increase of the conduction time between sites G and H occurred, associated with equal increases in the tachycardia cycle length. When conduction block developed between the two sites, the reentrant circuit was terminated and electrogram H recorded an electrotonic potential but no local activation potential. This was represented on the isochronal map by an arc of conduction block (heavy solid line) that joined the two separate arcs of conduction block into one. *Reprinted from Assadi et al.,*[53] *with permission.*

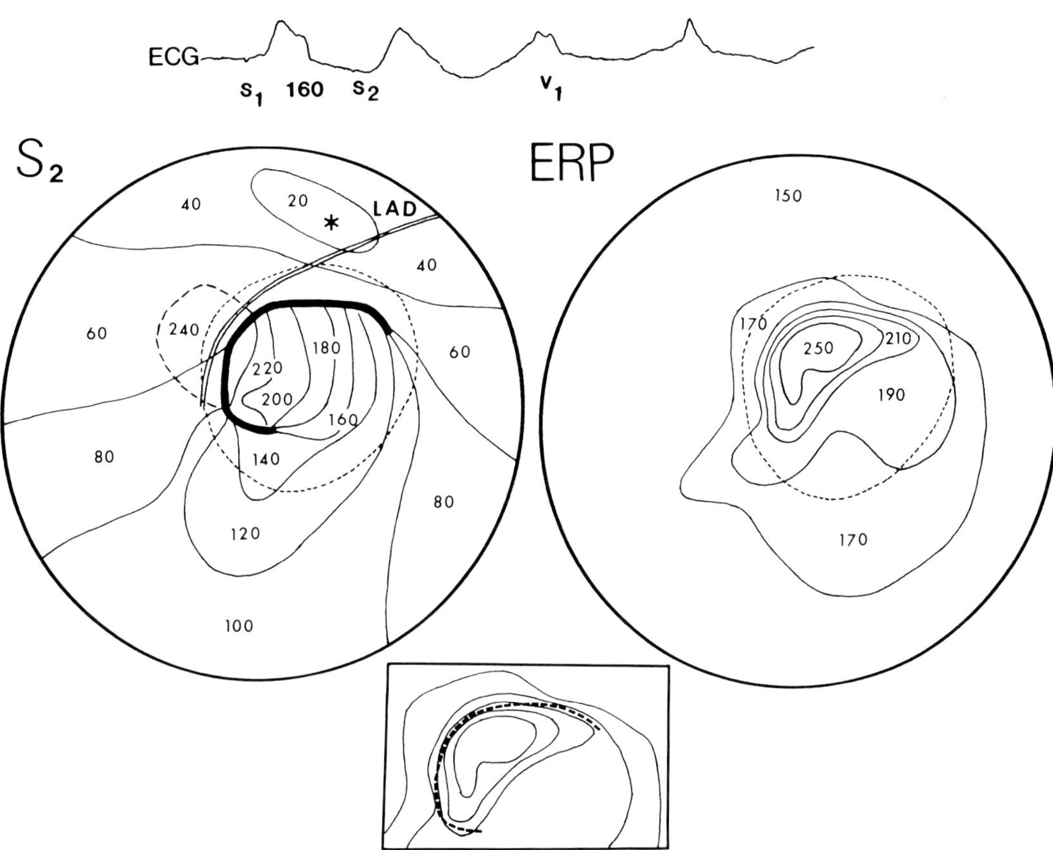

FIGURE 7.8 Correlation of isochronal maps of reentrant activation and refractory distribution in the epicardial surface from a 4-day-old infarction in the dog. The electrocardiogram on top shows that a single premature stimulus, S2, at a coupling interval of 160 ms initiated a reentrant rhythm. The epicardial activation map of S2 is shown on the left, while the refractory map of S1 as encountered by S2 is shown on the right and labeled ERP for effective refractory period. Both maps were drawn at 20 ms isochrones. The border of the ischemic zone is outlined on both maps by the dotted line. The refractory map shows a nonuniform refractory distribution with ERPs of 150 to 170 ms located in the normal right and left ventricular epicardium, while the longest ERP of 250 ms was located in the center of the ischemic region. The dispersion of refractoriness was 100 ms with concentric isochrones of refractoriness producing a graded increase in ERP going from the border zone toward the center of the ischemic zone. The steepest dispersion of refractoriness occurred inside the septal and basal borders of the infarction. The arc of functional conduction block encountered by S2 developed between adjacent sites of short and long refractoriness, with the sites of longer refractoriness being distal to the arc of block. This is shown in the inset at the bottom of the figure in which the arc of block (represented by a heavy dotted line) was superimposed on the refractory isochronal map. Note that both disparate refractoriness and the functional arc of conduction block occurred parallel as well as perpendicular to the long axis of epicardial muscle fibers. *Reprinted with permission from El-Sherif et al.: Electrophysiology of ventricular arrhythmias in myocardial ischemia and infarction, in El-Sherif N, Samet P (eds):* Cardiac Pacing and Electrophysiology; *WB Saunders, 1990, pp 18–56.*

along the septal border of the ischemic zone, resulting in more crowded refractory isochrones. The arc of functional conduction block induced by premature stimulation was found to occur along the steep gradient of refractoriness. The length and location of the arc depended on the degree of prematurity of the extrastimulus (S1S2 interval).

When a high-density electrode plaque (1 mm interelectrode distance) was used to record the patterns of activation and refractoriness, functional conduction block was found to occur ab-

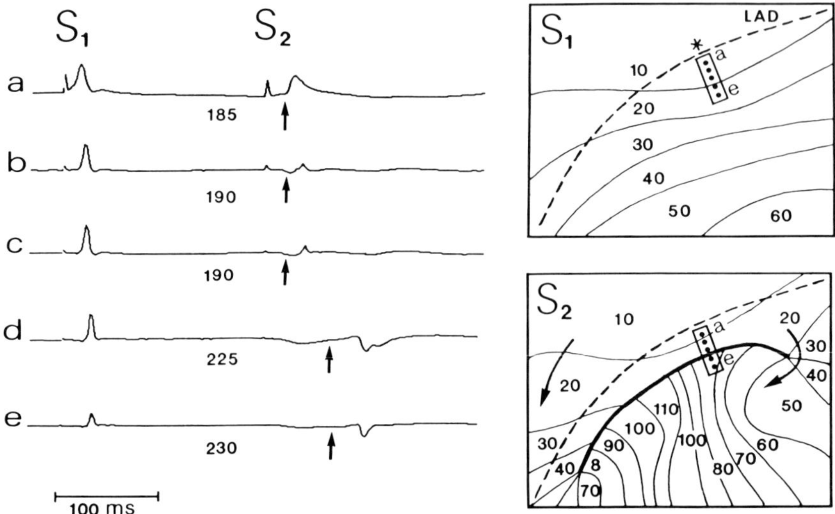

FIGURE 7.9 High-resolution determination of spatial refractory gradients and their relationship to the arc of functional conduction block from a 4-day-old canine infarction. A high-density bipolar electrode plaque with 1 mm interelectrode spacing was positioned on the epicardial surface at the site of the arc of block induced by premature stimulation (S2) as determined from an earlier low-resolution sock electrode array. The plaque was oriented with the electrode rows perpendicular to the arc. The figure illustrates five bipolar electrograms recorded successively at 1-mm distance (a to e). The values of the effective refractory period in milliseconds (ms) at each site are shown. The arrows indicate the end of the effective refractory period relative to S1 activation at each site. The S1 and S2 activation maps are shown on the right. The asterisk on the S1 map denotes the site of stimulation. During S1, sites a to e were activated sequentially within a 12 ms interval (conduction velocity of 42 cm/s). During S2, conduction between sites a and c was relatively slow compared to S1. Conduction block developed abruptly between sites c and d. Sites d and e were activated 65 ms later by the wave front that circulated in a clockwise direction around one end of the arc of block. The site of conduction block coincided with a 35 ms abrupt increase in the effective refractory period between sites c and d. Note that the arc of block was parallel to the left anterior descending artery (LAD) represented by the dashed line. *Reprinted from Restivo et al.,*[52] *with permission.*

ruptly (within 1 mm distance) and the activation wavefront prior to block did not show decremental conduction. The site of conduction block correlated with an abrupt increase of refractoriness by 10 to 85 ms over a 1-mm distance. Electrograms obtained 1 mm distal to the site of conduction block usually revealed an electrotonic deflection synchronous with the activation potential proximal to block (Fig. 7.9).

The role of differential refractoriness and fiber orientation in the formation of the arc of functional conduction block was also examined in the same study.[52] Abrupt increase in refractoriness was found both along fiber axis (27 ± 9 ms/mm) and across fiber axis (14 ± 7 ms/mm). Although the difference along the fiber axis was greater, the arc of functional conduction block occurred in both orientations in the presence of an abrupt change in refractoriness. The study suggests that in the ischemic ventricle, an arc of functional block occurs as a consequence of differentially graded refractoriness and can be independent of fiber orientation.

Sustained Reentry Orients Around Continuous Arcs of Functional Conduction Block

Several investigators[53,62,65] have shown that during sustained figure-eight monomorphic reentrant tachycardia, the two arcs of functional conduction block around which the reentrant wavefronts circulate are usually oriented parallel to the long axis of the epicardial muscle fibers. Some authors have suggested that these areas represent an apparent or a pseudo block and are in fact due to very slow and possibly discontinuous conduction across the myocardial fibers.[62] In this case, reentrant activation may be oriented around a small central region of functional block rather than a long line of block. Electrograms recorded from these sites had long duration and

were fractionated, a characteristic that was shown to result from activation transverse to myocardial fiber long axis.[70] These conclusions, however, were based on relatively low-resolution recordings (2.5 mm interelectrode distance).[62]

Restivo and associates[52] have analyzed close bipolar electrograms obtained at high resolution (1-mm interelectrode distance) from sites of the arcs of block during sustained stable reentry. Electrograms recorded at each side of the line of block showed two distinct deflections: one represented local activation and the other an elec-

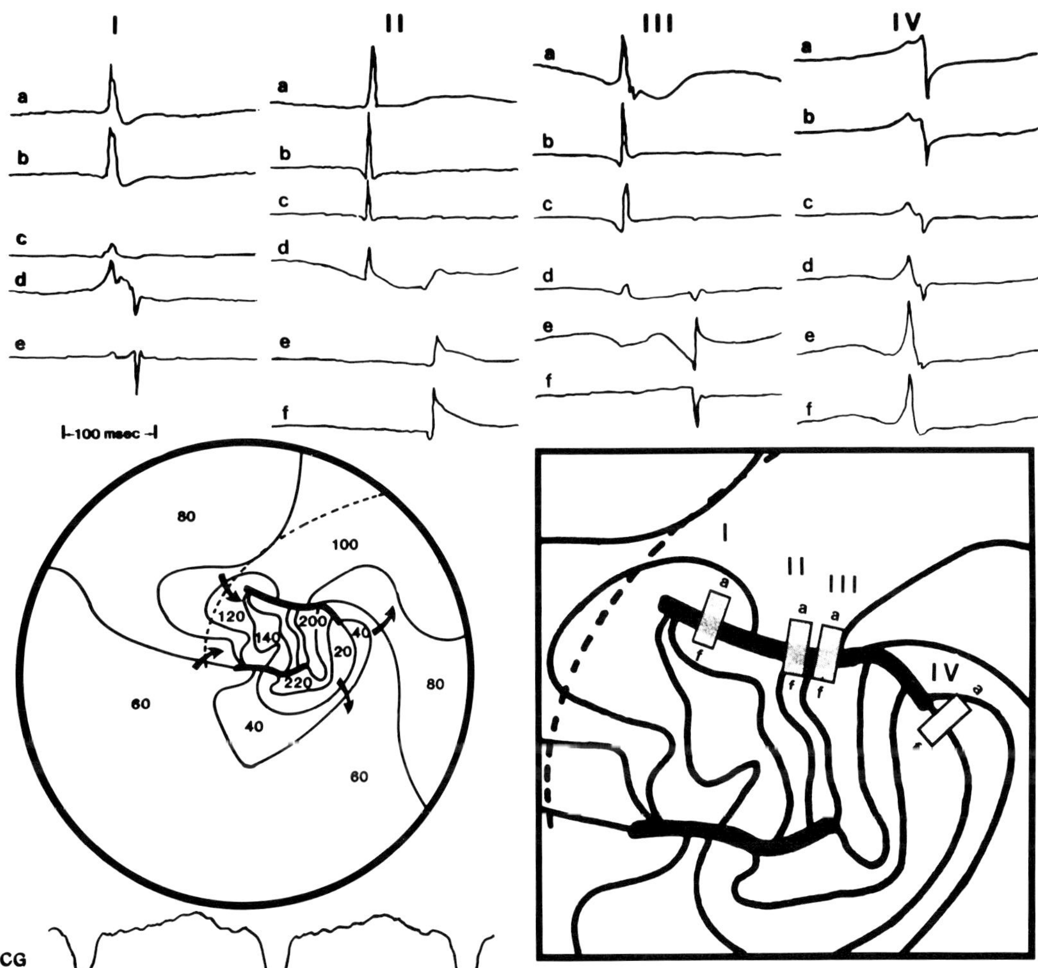

FIGURE 7.10 Recordings of high-density bipolar electrode array at multiple locations (I to IV) along a continuous arc of functional conduction block during sustained monomorphic ventricular tachycardia (top panel). A polar projection of isochronal activation map is shown in the lower left panel. Arrows indicate wave-front direction during sustained ventricular tachycardia. Both arcs are oriented approximately parallel to the longitudinal axis of the epicardial muscle fibers. A portion of this activation map is shown in the lower right panel. The shaded rectangles represent the column for each array location. Electrograms recorded in proximity to the arc of block show split electrograms composed of two discrete potentials separated by a variable isoelectric interval; one deflection represents local activation, and the other deflection is an electrotonic potential reflecting activation recorded 1 mm away. The interval between the two deflections is greatest at the center of the arc (locations II and III) where the difference in isochronal activation, by whole-ventricle mapping, is greatest. The interval between the two deflections is less as the reentrant impulse circulates around the end of the arc (location I). The electrographic characteristics of functional conduction block are the same at location IV, which indicates that the arc of functional conduction block was longer than that predicted by the whole-ventricle mapping technique. *Reprinted from Restivo et al.,*[52] *with permission.*

trotonus corresponding to activation recorded 1 mm away. Both deflections were separated by a variable isoelectric period that correlated with the isochronal difference across the arc. In recordings obtained from the center of the arc, local activation and electrotonus were separated by 90 to 110 ms. This interval successively decreased toward both ends of the arc (Fig. 7.10). These observations provide evidence that circus movement reentry was sustained around a continuous arc of abrupt functional conduction block (7 to 25 mm long) and not by very slow conduction across fibers. Although refractoriness could not be measured during sustained reentry, the electrogram configurations reflecting conduction block were similar to those obtained during functional conduction block induced by premature stimulation across a refractory gradient. This suggests that disparate refractoriness along the line of block rather than anisotropic properties of the epicardial layer may be responsible for sustained reentrant excitation.

CONCLUSION

Considerable progress has been made in the last decade toward understanding of the electrophysiological alterations and the mechanisms of ischemia-related ventricular arrhythmias. There is now convincing evidence that reentrant excitation in ischemic myocardium underlies the spontaneous and often lethal ventricular tachyarrhythmias seen in the acute phase of myocardial infarction as well as ventricular tachyarrhythmias induced by programmed stimulation in the subacute and, possibly, chronic phases. The role of non-reentrant mechanisms, possibly triggered activity in postocclusion and, particularly, in reperfusion ventricular arrhythmias is undergoing much investigation.

Triggered activity arising from delayed afterdepolarizations in ischemic Purkinje fibers, and possibly abnormal automaticity in ischemic depolarized cardiac cells, may explain some of the spontaneous multiform ventricular rhythms seen in the subacute phase of myocardial infarction and perhaps in the chronic phase as well. With the advent of new and sophisticated techniques applicable to both human and animal studies, the future holds promise for improved clinical management of postinfarction lethal ventricular arrhythmias.

ACKNOWLEDGMENT

This research was supported in part by National Institutes of Health grants HL36680 and HL31341 and the Veterans Administration Medical Research Funds.

REFERENCES

1. Reimer KA, Low JE, Rasmussen MM, Jennings RB: The wavefront phenomenon of ischemic cell death 1. Myocardial infarct size vs. duration of coronary occlusion in dogs. *Circulation* 1977,56:786–794.
2. Fenoglio JJ, Karagueuzian HS, Friedman PL, Albola A, Wit AL: Time course of infarct growth toward the endocardium after coronary occlusion. *Am J Physiol* 1979;236:H356–H370.
3. Harris AS: Late development of ventricular ectopic rhythms following experimental coronary occlusion. *Circulation* 1950;1:1318–1325.
4. Scherlag BJ, El-Sherif N, Hope RR, Lazzara R: Characterization and localization of ventricular arrhythmias due to myocardial ischemia and infarction. *Circ Res* 1974;35:372–383.
5. El-Sherif N, Scherlag BJ, Lazzara R: Electrode catheter recordings during malignant ventricular arrhythmias following experimental acute myocardial ischemia. Evidence for reentry due to local conduction delay and block in ischemic myocardium. *Circulation* 1975;51:1003–1014.
6. Janse MJ, Van Cappelle FJL, Morsink H, Kleber AG, Wilms-Schopman F, Cardinal R, D'Alnoncourt CN, Durrer D: Flow of "injury" current and patterns of excitation during early ventricular arrhythmias in acute regional myocardial ischemia in isolated porcine and canine hearts. Evidence for two different arrhythmogenic mechanisms. *Circ Res* 1980;47:151–165.
7. Friedman PL, Stewart JR, Wit AL: Spontaneous and induced cardiac arrhythmias in subendocardial Purkinje fibers surviving extensive myocardial infarction in dogs. *Circ Res* 1973;33:612–625.
8. Lazzara R, El-Sherif N, Scherlag BJ: Electrophysiologic properties of canine Purkinje cells in one-day-old myocardial infarction. *Circ Res* 1973;33:722–734.
9. Hoffman BF, Rosen MR: Cellular mechanisms of cardiac arrhythmias. *Circ Res* 1981;49:1–15.
10. El-Sherif N, Gough WB, Zeiler RH, Mehra R: Triggered ventricular rhythms in one-day-old myocardial infarction in the dog. *Circ Res* 1983;52:566–579.
11. El-Sherif N, Scherlag BJ, Lazzara R, Hope RR: Reentrant ventricular arrhythmias in the late myocardial infarction period 1. Conduction characteristics in the infarction zone. *Circulation* 1977;55:686–701.
12. El-Sherif N, Hope RR, Scherlag BJ, Lazzara R: Reentrant ventricular arrhythmias in the late myocardial infarction period 2. Patterns of initiation and termination of reentry. *Circulation* 1977;55:702–719.
13. El-Sherif N, Lazzara R, Hope RR, Scherlag BJ: Reentrant ventricular arrhythmias in the late myocardial infarction period 3. Manifest and concealed extrasystolic grouping. *Circulation* 1977;56:225–234.
14. Garan H, Fallon JT, Rosenthal S, Ruskin JN: Endocardial, intramural, and epicardial activation patterns during sustained monomorphic ventricular tachycardia in late canine myocardial infarction. *Circ Res* 1987;60:879–896.
15. Dangman KH, Dresdner KP, Zaim S: Automatic and triggered impulse initiation in canine subepicardial ventricular muscle cells from border zone of 24-hour transmural infarcts. *Circulation* 1988;78:102–103.
16. Dresdner P, Kline RP, Wit AL: Intracellular K^+ activity, intracellular Na^+ activity and maximum dia-

stolic potential in canine subendocardial Purkinje fibers from one day old infarcts. *Circ Res* 1987;60:122–132.
17. Dresdner KP, Hannah NS, Kline RP, Wit AL: Na^+/K^+ pump failure in canine cardiac Purkinje fibers surviving an infarct. *Circulation* 1988;78:II-414(abstract).
18. Gough WB, El-Sherif N: Dependence of delayed afterdepolarizations on diastolic potentials in ischemic Purkinje fibers. *Am J Physiol* 1989;257:H770–H777.
19. Le Marec H, Dangman KH, Danilo P, Rosen MR: An evaluation of automaticity and triggered activity in canine heart one to four days after myocardial infarction. *Circulation* 1985;71:1124–1236.
20. Mugelli A, Cerbai E, Amerini S, Visentin S: The role of temperature on the development of oscillatory afterpotentials and triggered activity. *J Mol Cell Cardiol* 1986;18:1313–1316.
21. Le Marec H, Spinelli W, Rosen MR: The effects of doxorubicin on ventricular tachycardia. *Circulation* 1986;74:881–889.
22. January CP, Fozzard HA: Delayed afterdepolarizations in heart muscles: Mechanisms and relevance. *Pharmacol Rev* 1988;30:219–227.
23. Kass R, Tsien RW, Weingart R: Ionic basis of transient inward current induced by strophanthidin in cardiac Purkinje fibers. *J Physiol (Lond)* 1978;281:208–226.
24. Nobel D: The surprising heart: A review of recent progress in cardiac electrophysiology. *J Physiol (Lond)* 1984;353:1–50.
25. Gough WB, Zeiler RH, El-Sherif N: Basis for reduced transmembrane potentials associated with triggered activity in ischemic subendocardial Purkinje fibers. *Circulation* 1982;66:II-156(abstract).
26. El-Sherif N, Zeiler RH, Gough WB: Effects of catecholamines, verapamil and tetrodotoxin in triggered automaticity in canine ischemic Purkinje fibers. *Circulation* 1980;62:III-281(abstract).
27. Gough WB, Zeiler RH, El-Sherif N: Effects of diltiazem on triggered activity in canine one day old infarction. *Cardiovasc Res* 1984;18:339–343.
28. Boutjdir M, El-Sherif N, Gough WB: The effects of caffeine and ryanodine on delayed afterdepolarizations and sustained rhythmic activity in one day old infarction of the dog. *Circulation* 1990;81:1393–1400.
29. Carioni P, Vallani F, Carfoli E: The cardiotoxic antibiotic doxorubicin inhibits Na^+/Ca^{++} exchange of dog heart sacrolemmal vesicles. *FEBS Lett* 1981;30:184.
30. Danilo P, Langan WB, Rosen MR, Hoffman BF: Effects of the phenothiazine analog, EN-313, on ventricular arrhythmias in the dog. *Eur J Pharmacol* 1977;45:127–139.
31. Dangman KH, Hoffman DF: Antiarrhythmic effects of automaticity and abolition of triggering. *J Pharmacol Exp Ther* 1983;227:578–586.
32. Allen JD, Brennan FJ, Wit AL: Actions of lidocaine on transmembrane potentials of subendocardial Purkinje fibers surviving in infarcted canine hearts. *Circ Res* 1978;43:470–481.
33. Gough WB, Zeiler RH, El-Sherif N: Effects of nifedipine on triggered activity in one-day-old myocardial infarction in dogs. *Am J Cardiol* 1984;53:303–306.
34. Hariman RJ, Zeiler RH, Gough WB, El-Sherif N: Enhancement of triggered activity in ischemic Purkinje fibers by ouabain: A mechanism of increased susceptibility to digitalis toxicity in myocardial infarction. *J Am Coll Cardiol* 1985;5:672–679.
35. Gough WB, Hu D, El-Sherif N: Effects of clofilium on ischemic subendocardial Purkinje fibers one day post-infarction. *J Am Coll Cardiol* 1988;11:431–437.
36. Gough WB, El-Sherif N: The differential response of normal and ischemic Purkinje fibers, to clofilium, d-sotalol, and bretylium. *Cardiovas Res* 1989; 23:554–559.
37. Synders DJ, Katzung BG: Clofilium reduces the plateau potassium current in isolated cardiac myocytes. *Circulation* 1985;72:III-233.
38. Carmeliet E: Electrophysiologic and voltage clamp analysis of the effects of sotalol on isolated cardiac muscle and Purkinje fibers. *J Pharmacol Exp Ther* 1985;232:817–825.
39. El-Sherif N, Gough WB, Zeiler RH, Mehra R: Ventricular rhythms in one day old canine infarction are due to triggered activity. *Circulation* 1982;66:II-357(abstract).
40. Martins JB: Autonomic control of ventricular tachycardia: Sympathetic neural influence on spontaneous tachycardia 24 hours after coronary occlusion. *Circulation* 1985;72:933–942.
41. El-Sherif N, Mehra R, Gough WB, Zeiler RH: Ventricular activation patterns of spontaneous and induced ventricular rhythms in canine one-day-old myocardial infarction. Evidence for focal and reentrant mechanisms. *Circ Res* 1982;51:152–166.
42. Hariman RJ, Holtzman R, Gough WB, Mehra R, Gomes JAC, El-Sherif N: In vivo demonstration of delayed afterdepolarization as a cause of ventricular rhythms in one day old infarction. *J Am Coll Cardiol* 1984;3:478(abstract).
43. El-Sherif N, Smith RA, Evans K: Canine ventricular arrhythmias in the late myocardial infarction period. 8. Epicardial mapping of reentrant circuits. *Circ Res* 1981;49:255–265.
44. Mehra R, Zeiler RH, Gough WB, El-Sherif N: Reentrant ventricular arrhythmias in the late myocardial infarction period. 9. Electrophysiologic-anatomic correlation of reentrant circuits. *Circulation* 1983;67:11–24.
45. El-Sherif N, Mehra R, Gough WB, Zeiler RH: Reentrant ventricular arrhythmias in the late myocardial infarction period. Interruption of reentrant circuits by cryothermal techniques. *Circulation* 1983;8:644–656.
46. El-Sherif N, Mehra R, Gough WB, Zeiler RH: Reentrant ventricular arrhythmias in the late myocardial infarction period. 11. Burst pacing versus multiple premature stimulation in the induction of reentry. *J Am Coll Cardiol* 1984;4:295–304.
47. El-Sherif N, Gough WB, Zeiler RH, Hariman R: Reentrant ventricular arrhythmias in the late myocardial infarction period. 12. Spontaneous versus induced reentry and intramural versus epicardial circuit. *J Am Coll Cardiol* 1985;6:124–132.
48. El-Sherif N: The figure 8 model of reentrant excitation in the canine postinfarction heart, in Zipes DP, Jalife J (eds): *Cardiac Electrophysiology and Arrhythmias*. New York, Grune & Stratton, 1985, pp 365–378.
49. Gough WB, Mehra R, Restivo M, Zeiler RH, El-Sherif N: Reentrant ventricular arrhythmia in the late myocardial infarction period in the dog. 13. Correlation

of activation and refractory maps. *Circ Res* 1985;57:432–442.
50. El-Sherif N, Gough WB, Restivo M: Reentrant ventricular arrhythmias in the late myocardial infarction period. 14. Mechanisms of resetting, entrainment, acceleration, or termination of reentrant tachycardia by programmed electrical stimulation. *PACE* 1987;10:341–371.
51. Restivo M, Gough WB, El-Sherif N: Reentrant ventricular rhythms in the late myocardial infarction period: Prevention of reentry by dual stimulation during basic rhythm. *Circulation* 1988;77:429–444.
52. Restivo M, Gough WB, El-Sherif N: Ventricular arrhythmias in the subacute myocardial infarction period. High resolution activation and refractory patterns of reentrant rhythms. *Circ Res* 1990;66:1310.
53. Assadi M, Restivo M, Gough WB, El-Sherif N: Reentrant ventricular arrhythmias in the late myocardial infarction period. 17. Correlation of activation patterns of sinus and reentrant ventricular tachycardia. *Am Heart J* 1990;119:1014–1024.
54. Hirzel HO, Nelson GR, Sonnenblick EH, Kirk ES: Redistribution of collateral blood flow from necrotic to surviving myocardium following coronary occlusion in the dog. *Circ Res* 1976;39:214–222.
55. El-Sherif N, Lazzara R: Reentrant ventricular arrhythmias in the late myocardial infarction period. 7. Effects of verapamil and D-600 and role of the "slow channel." *Circulation* 1979;60:605–615.
56. Lazzara R, Scherlag BJ: The role of the slow current in the generation of arrhythmias in ischemic myocardium, in Zipes DP, Bailey JC, Elharrar V (eds): *The Slow Inward Current and Cardiac Arrhythmias*. The Hague, Martinus Nijhoff, 1980, pp 399–416.
57. Ursell PC, Gardner PI, Albala A, Fenoglio JJ Jr, Wit AL: Structural and electrophysiological changes in the epicardial border zone of canine myocardial infarcts during infarct healing. *Circ Res* 1985;56:436–451.
58. Schwartz A, Wood JM, Allen JC, Bornet E, Entman ML, Goldstein MA, Sordahl LZ, Suzuki M, Lewis RM: Biochemical and morphologic correlates of cardiac ischemia. I. Membrane system. *Am J Cardiol* 1973;32:46–61.
59. Sobel BE, Corr PB, Robinson AK, Goldstein RA, Witkowski FX, Klein MS: Accumulation of lysophosphoglycerides with arrhythmogenic properties in ischemic myocardium. *J Clin Invest* 1978;62:546–553.
60. Spear JF, Michelson EL, Moore EN: Reduced space constant in slowly conducting regions of chronically infarcted canine myocardium. *Circ Res* 1983;52:176–185.
61. Page E, Shibata Y: Permeable junctions between cardiac cells. *Annu Rev Physiol* 1981;43:431–441.
62. Dillon S, Allessie MA, Ursell PC, Wit AL: Influences of anisotropic tissue structure on reentrant circuits in the epicardial border zone of subacute canine infarcts. *Circ Res* 1988;63:182–206.
63. Clerc L: Directional differences of impulse spread in trabecular muscle from mammalian heart. *J Physiol (Lond)* 1976;255:335–346.
64. Spach M, Miller WT, Geselowitz DB, Barr RC, Kootsey JM, Johnson EA: The discontinuous nature of propagation in normal canine cardiac muscle: Evidence for recurrent discontinuities of intracellular resistance that affect the membrane currents. *Circ Res* 1981;48:39–54.
65. Cardinal R, Vermeulen M, Shenasa M, Roberge F, Page P, Heile F, Savard P: Anisotropic conduction and functional dissociation of ischemic tissue during reentrant ventricular tachycardia in canine myocardial infarction. *Circulation* 1988;77:1162–1176.
66. Fenoglio JJ Jr, Pham TD, Harken AH, Horowitz LN, Josephson ME, Wit AL: Recurrent sustained ventricular tachycardia: Structure and ultrastructure of subendocardial regions where tachycardia originates. *Circulation* 1983;68:518–533.
67. Harris L, Downar E, Mickleborough L, Shaikh N, Parsons I, Chen T, Gray G: Activation sequence of ventricular tachycardia: Endocardial and epicardial mapping studies in the human ventricle. *J Am Coll Cardiol* 1987;10:1040–1047.
68. Kramer JB, Saffitz JE, Witkowski FX, Corr PB: Intramural reentry as a mechanism of ventricular tachycardia during evolving canine myocardial infarction. *Circ Res* 1985;56:736–754.
69. Mines GR: On circulating excitations in heart muscles and their possible relation to tachycardia and fibrillation. *Trans R Soc Can* 1914(ser 3, sect IV)8:43–52.
70. Spach MS, Dolber PC: Relating extracellular potentials and their derivatives to anisotropic propagation at a microscopic level in human cardiac muscle: Evidence for uncoupling of side to side fiber connections with increasing age. *Circ Res* 1986; 56:356–371.

Chapter **8**

Slow Conduction in Experimental Arrhythmias: Mechanisms and Role in Reentry

Benjamin J. Scherlag, PhD, Eugene Patterson, PhD, Edward J. Berbari, PhD, and Ralph Lazzara, MD

THE REENTRY THEORY: HISTORICAL BACKGROUND

During the years 1887 to 1889, when Waller recorded the first electrocardiogram in the dog and then in man,[1] McWilliam formulated the concept of reentry as a basic mechanism for arrhythmia formation.[2] By visual observation of the amphibian heart in the prefibrillatory and fibrillatory state, he concluded that this arrhythmia was not due to "rapid spontaneous discharges" associated with the then widely held view of the hyperirritable ventricle. Instead, he proposed another "cause of continued and rapid movement during prefibrillatory states" (theory of reentry). It was not until the enlightening studies of Alfred Mayer during 1903 to 1908[3,4] that the characteristics of reentry were formulated. Using the subumbrella of the jellyfish, Mayer could follow, by visual observation, the wave of contraction that swept around the flat ring of excitable tissue (Fig. 8.1). If an area of injury were placed at 10 o'clock, a stimulus starting at this point would enter and block (unidirectionally) in the damaged zone while the normal tissue would conduct the impulse at relatively rapid rate until it reached the zone of slow conduction moving in a clockwise direction. By proceeding slowly through the damaged zone, the tissue just ahead would have recovered its excitability and reentry and reexcitation would allow another and another circuit to occur for variable periods of time. To preclude the possibility that the initial provoked stimulus became a focus of ectopic activity, Mayer cut the continuously active rings at any site and observed that the wave of excitation would come to that point and stop, thus causing the entire ring to become quiescent. Mines in 1914 showed the same phenomena in rings of cardiac muscle and provided the impetus for others to promulgate the theory of reentry as a mechanism for atrial and ventricular tachycardia and flutter.[5]

Using these simple models, various investigators demonstrated that the following features characterized reentrant arrhythmias: (a) unidirectional block; (b) slow conduction; (c) circus movement that could be documented, for example, by mapping; and (d) cessation of the spontaneous activation by cutting or otherwise interrupting the circuit.

For the next 70 years, the proponents of the ectopic focus theory and the reentry theory continually battled to bring their respective positions to the forefront. It proved particularly difficult to demonstrate reentry in the mammalian heart because electrical activation was too rapid to be monitored visually and circuits that were endocardial or intramural were inaccessible and thus not demonstrable with the then available technologies. In 1943, Harris and Guevera-Rojas attempted to resurrect the basic observation of McWilliam in their own studies of the mechanism of ventricular arrhythmias caused by abrupt ligation of the left anterior descending coronary artery in the dog.[6] The McWilliam hypothesis

655 Avenue of the Americas, New York, NY 10010
Current Topics in Cardiology

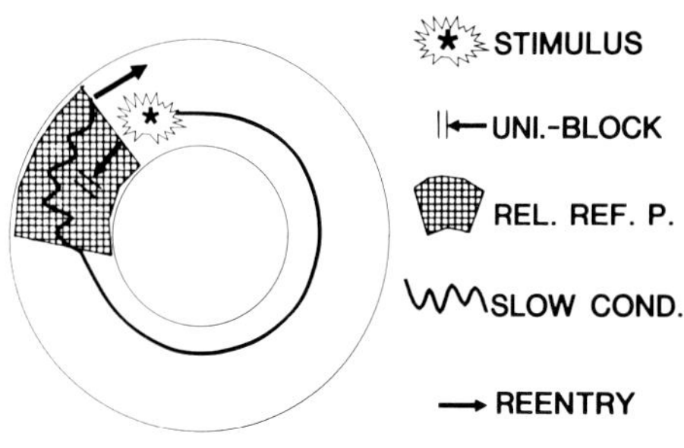

FIGURE 8.1 Schematic representation of the reentry phenomenon as demonstrated by Mayer in the jellyfish subumbrella. An evoked stimulus at the edge of a damaged zone (hatched area) was conducted into the damaged zone and blocked (unidirectional block) (II), whereas the impulse conducted, unimpeded, in the other direction until entering the damaged zone through which it conducted slowly (~ slow cond.) until reaching the blocked zone, which now was in its relatively refractory period (REL. REF. P). Upon reaching the fully recovered tissue, the impulse could reexcite and make successive circuits (→ REENTRY).

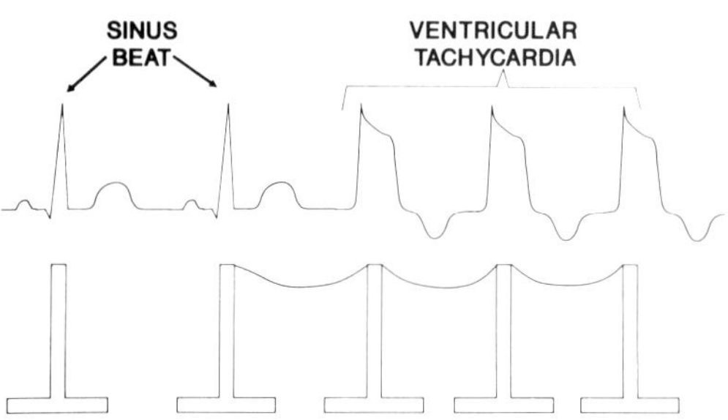

FIGURE 8.2 Cartoon demonstrating the McWilliam hypothesis of reentry. The first two sinus beats are represented (bottom) by bridge supports without a connecting electrical span. Slow conduction during the last sinus beat outlasts the T wave, thus allowing a spanning electrical connection between these two beats and a similar continuous spanning of the successive beats of the subsequent ventricular tachycardia.

they sought to test is displayed schematically as a cartoon in Figure 8.2. Their postulation was that a reentrant loop would require that the first ectopic ventricular beat resulted from continuous electrical activity bridging the interval from the previous sinus beat (depicted here by the second and third bridge supports connected by a span), and that the successive beats of the monomorphic ventricular tachycardia are similarly connected by continuous electrical activity (spans between successive ventricular ectopic beats). During ischemia, Harris and Guevera-Rojas were not able to record continuous activation because many of their electrograms showed marked diminution and apparent inexcitability. Some sites showed activation in every other beat.

The actual recording of such bridging activity was made some 30 years later in similar studies in dogs with coronary artery occlusion. Waldo and Kaiser,[7] using multiple electrogram recordings from the left ventricular epicardial surface, found similar diminution and apparent loss of electrical activity within 3 min after ligation of a major coronary artery (Fig. 8.3, first panel). However, unlike their predecessors, these workers, having higher sensitivity amplifiers available to them, could "turn up the gain" and thus record low-level, fractionated activity that bridged the interval between a supraventricular beat and a ventricular ectopic beat and between successive ectopic beats (Fig. 8.3, second, third, and fourth beats of the right-hand panel). The "continuous electrical activity" shown by Waldo and Kaiser was the demonstration that a circumscribed area of ventricles could be activated for a period that lasted longer than the entire normal ventricular depolarization. Moreover, this prolonged acti-

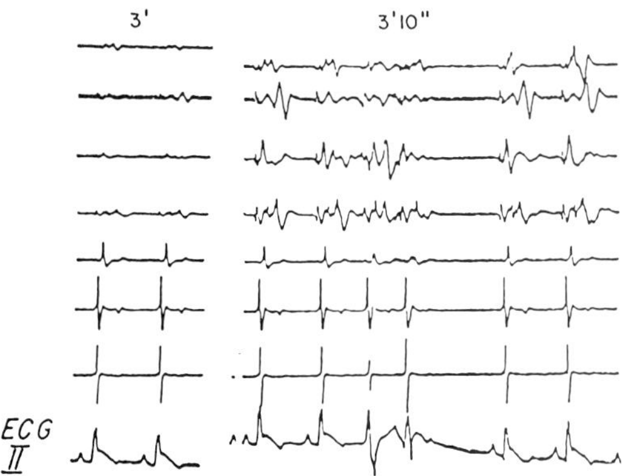

FIGURE 8.3 First recordings of continuous electrical activity associated with reentrant beats in a dog heart during acute myocardial ischemia. At 3 min after ligation of the left anterior descending coronary artery, four of the seven electrodes recording from the ischemic heart show marked diminution of activation, compared to the electrogram from the normal zone (last electrogram). Note the S-T elevation on the Lead II ECG recording. By increasing the gain of the amplifiers, the marked fractionation and delays of the diminished electrograms can be seen with electrical connections from sinus to ventricular ectopic beats and between successive ectopic beats. *Reprinted from Waldo and Kaiser,*[7] *with permission.*

vation was closely linked to ectopic impulse formation presumably due to reentry.

THE DEVELOPMENT OF METHODOLOGY TO DOCUMENT REENTRY IN THE MAMMALIAN HEART

Other workers, using multiple electrode recordings from the dog heart subjected to acute myocardial ischemia, corroborated the findings of diminished, fractionated, and delayed electrograms that bridged the gap between sinus beats and ectopic ventricular beats and between successive ventricular ectopic beats.[8,9] The major problem with these studies was the consistency in recording such evidence for reentrant activation. Scherlag et al.[9] found that in 20% of cases it was possible to record a sequence of local electrogram changes consonant with a reentrant mechanism. The problem was caused by the lack of an adequate number of electrode sites and the marked heterogenous effects of ischemia on the small number of recorded electrograms.

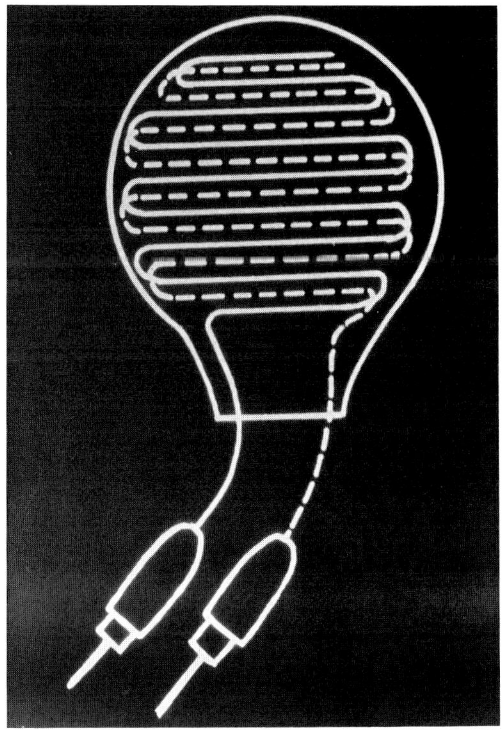

FIGURE 8.4 A diagram of a composite electrode. Two bipolar pairs of wires (solid and dashed lines) are sewn into a paper tape matrix. When applied to the epicardial surface, the bipolar pairs comprise a multicontact electrode over a relatively large area.

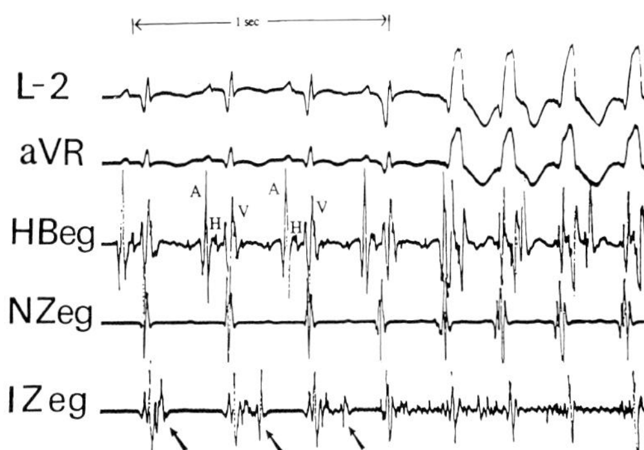

FIGURE 8.5 Spontaneous onset of sustained ventricular tachycardia in a 4-day-old infarcted dog heart. Traces, from top, are: Standard ECG Lead II (L-2) and aVR; His bundle electrogram (HBeg); composite electrograms recorded from the epicardial surface of the left ventricle overlying the uninfarcted, normal zone (NZeg) and the infarct zone (IZeg). At a constant sinus rate there is progressive fractionation in the viable, but sick, IZ leading to delay beyond the T wave in the fourth sinus beat. The electrical connection between the last sinus beat, the first ventricular ectopic beat, and the subsequent ectopic beats of the ventricular tachycardia is the hallmark of a reentrant mechanism. *Reprinted with permisison from Scherlag: The early arrhythmic phase of myocardial ischemia, in Befeler B (ed.):* Selected Topics in Cardiac Arrhythmias; *Futura Publishing, Mt. Kisco, NY, 1980, Chapter 5.*

One attempt to address this problem was the use of the composite electrode (Fig. 8.4) by which a large surface of the ischemic epicardium was contacted by a bipolar electrode, that is, a pair of wires with multiple contact sites. The major advantage of the composite electrode is the ability to record from a large area using a single amplifier channel.[10–14] However, its limitations lie in the lack of local resolution of electrical events and restriction of its field of view to the surface it contacts, for example, subepicardium. The recordings made from the composite electrode, consonant with reentrant activity, are shown in Figure 8.5. In a 4-day-old infarcted dog heart, composite electrodes were placed on the anterior epicardial surface overlying the infarct zone (IZ) and on the posterior epicardial surface in a normal, noninfarcted zone (NZ). During sinus rhythm (first four beats), there is a progressive delay and fractionation of the terminal portion of the composite electrogram overlying the infarct zone (arrows). In contrast, note the stability of the composite electrogram in the normal zone

with isoelectric diastolic intervals. When the fractionated activity extends beyond the T wave, in the infarct zone, thus encountering nonrefractory myocardium, (after the fourth beat) reentry can occur with bridging electrical activity connecting the last sinus beat and first ectopic ventricular beat. Note the continuation of the bridging electrical activity in the subsequent beats of ventricular tachycardia. The composite electrogram in the normal zone is concurrently silent during these interectopic periods. This is the actual demonstration described in the theoretic scenario shown in Figure 8.2 (above).

The fact that such recordings are not artifacts of the composite electrogram is attested to by the recording of a similar sequence of events using electrode catheters in the clinical setting of myocardial infarction. Josephson et al.[15] used programmed stimulation to induce sustained ventricular tachycardia in patients with myocardial infarcts and ventricular aneurysms. In Figure 8.6, A and B, during ventricular pacing, V_1 to V_1 = 700 ms, a premature ventricular stimulus, V_2, was coupled to the last V_1 by 310 and 300 ms, respectively. The magnified inserts clearly show the fractionation and delay recorded from an

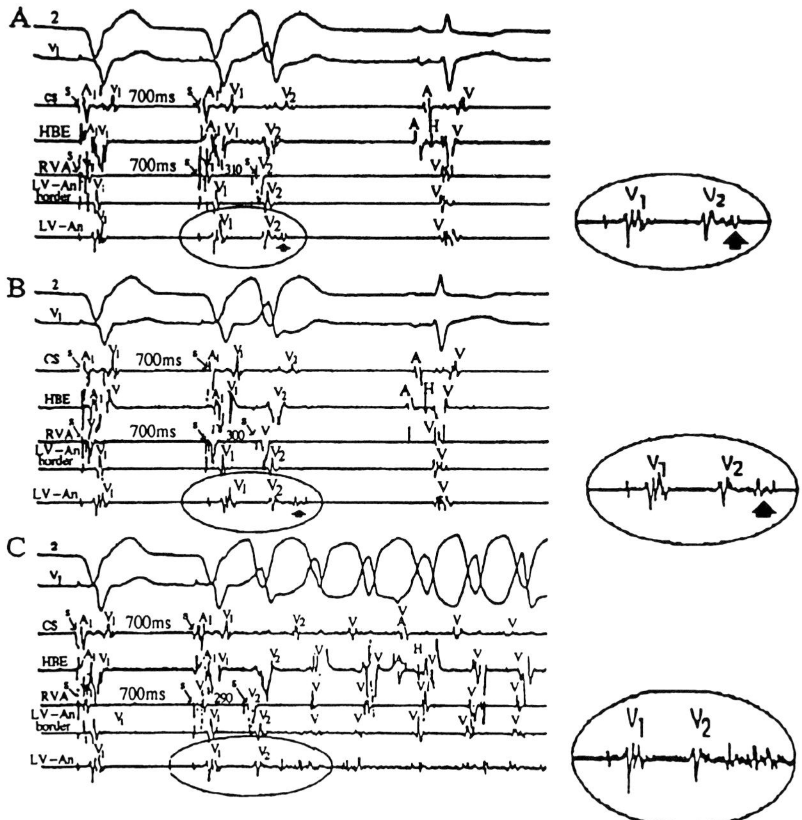

FIGURE 8.6 Continuous electrical activity and sustained ventricular tachycardia induced in the heart of a patient with chronic myocardial infarction and left ventricular aneurysm. Traces, from top: ECG leads II (2) and V_1, electrode catheter recordings from the coronary sinus (CS), the His bundle (HBE), the right ventricular apex (RVA), and two from the endocardial surface of the left ventricle at the border of the aneurysm (LV-An border) and from within the aneurysm (LV-An). After a series of ventricular paced beats (V_1) at a constant pacing interval of 700 ms, a premature stimulus (V_2) is introduced into the right ventricle. As the coupling of the premature stimulus to the last paced beat decreases from 310 ms (**A**) to 300 ms (**B**) to 290 ms (**C**) electrical activation within the left ventricle (LV-An) fractionates progressively (see magnified inserts and arrows) so that in **C** the electrical activity extends beyond the previous T wave and initiates a sustained monomorphic ventricular tachycardia. During the tachycardia continuous electrical activity can be seen bridging the interval between successive ectopic beats. *Modified from Josephson et al.,*[15] *with permission.*

electrode catheter contacting the subendocardial surface of the left ventricle within the aneurysm (LV-AN). In Figure 8.6C, when the V_1 to V_2 interval is decreased to 290 ms, the fractionated activity extended beyond the preceding T wave and was associated with the initiation of sustained ventricular tachycardia with continuous electrical activity bridging the interectopic intervals during the tachycardia. Nearby or distant electrodes do not show similar low-level fractionated activation.

More commonly, the interectopic electrical activity has been shown to have a specific reproducible character from beat to beat during any given sustained monomorphic ventricular tachycardia.[16–19] Figure 8.7 shows the typical recording of continuous electrical activity from a composite electrode overlying the epicardial border zone (IZ epi) in the infarcted dog heart. On the basis of physiologic and pharmacologic interventions, alterations of this reproducible pattern of continuous electrical activity could be induced (an example of which is shown, below, in Fig. 8.8). Such changes in the interectopic waveform allowed a subdivision of the entire interval into an initial potential (IP), a middiastolic potential (MDP), and an exit potential (exit). The initial potential starts immediately after the end of the preceding ventricular depolarization (QRS in the ECG lead), whereas the exit potential immediately precedes the onset of the next ventricular depolarization. Note the vertical dotted lines in Figure 8.7 depicting the beginning and end of the interectopic interval.

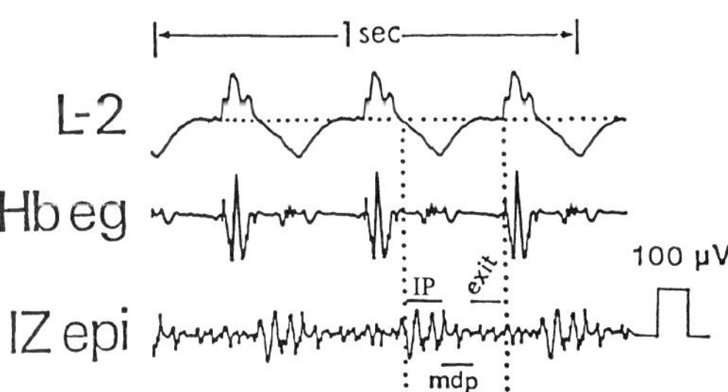

FIGURE 8.7 Recording of continuous electrical activation during sustained monomorphic ventricular tachycardia in the 4-day infarcted dog heart. The traces from above are: Lead II ECG (L-2); His bundle electrogram (Hb eg); and a composite electrode recording from the epicardial surface of the left ventricle overlying the infarct zone (IZ epi). The interectopic interval designated by the period between the vertical dotted lines was characterized by regular, reproducible patterns of electrical activation seen only in the composite electrogram from the IZ epi. This activity was divided into an initial potential (IP), a middiastolic potential (mdp), and an exit potential (exit). See text for further discussion. *Reprinted from Brachmann et al.,[17] with permission.*

The rationale for dividing this continuous electrical activity into components was based on the role that these potentials played in the initiation, maintenance, and termination of experimental forms of sustained, monomorphic ventricular tachycardia.[17] In Figure 8.8A, such a tachycardia with a cycle length of 340 ms is shown. The interectopic intervals contain a reproducible pattern of continuous electrical activity subdivided as described above into initial (IP), interectopic (MIP), and exit potentials (note magnified inset) with caps and arrows denoting each type. In Figure 8.8B, after procainamide (10 mg/kg) was administered intravenously, the cycle length of the tachycardia increased to 425 ms. This slowing was associated with marked diminution of the mid-interectopic potential and delay within this compartment (asterisk, Fig. 8.8B). Note the relative lack of action of the drug on the initial and, particularly, the exit potentials. In Figure 8.8C, after an additional dose of procainamide (total dose = 20 mg/kg), the tachycardia terminated. An attempt to reinduce with rapid atrial pacing resulted in continuous electrical activation between the last paced atrial beat and the first ectopic beat. Both the exit potential and the QRS configuration identify the first ventricular ectopic beat as similar to the preceding ventricular tachycardia. Essentially the same initial and exit potentials corroborate that the two ectopic beats belong to a markedly slowed tachycardia (cycle length, 600 ms) before the sinus pacemaker usurps the heart's rhythm (last beat, Fig. 8.8C).

This example and others[19,20] clearly indicate that there is a relatively slow conducting portion of the reentrant circuit during sustained monomorphic ventricular tachycardia and that this is a critical or vulnerable part of that circuit. During tachycardia, depolarization of the mass of the ventricles takes 100 to 120 ms (the QRS complex) whereas to depolarize a much smaller mass of tissue overlying the infarct on the anterior left ventricle takes 120 ms (Fig. 8.8A) during the mid-interectopic interval. It may require as much as 480 ms (the mid-interectopic interval in Fig. 8.8C) to traverse this slow-conducting portion completely. The size, configuration, and localization of the slow conduction zone has been variously described by a number of basic investigators using multiplexed (from 64 to more than 200) electrode recordings from hearts with myocardial ischemia or infarction. These data were acquired and analyzed by a computer, which then constructed maps of the electrical activation pat-

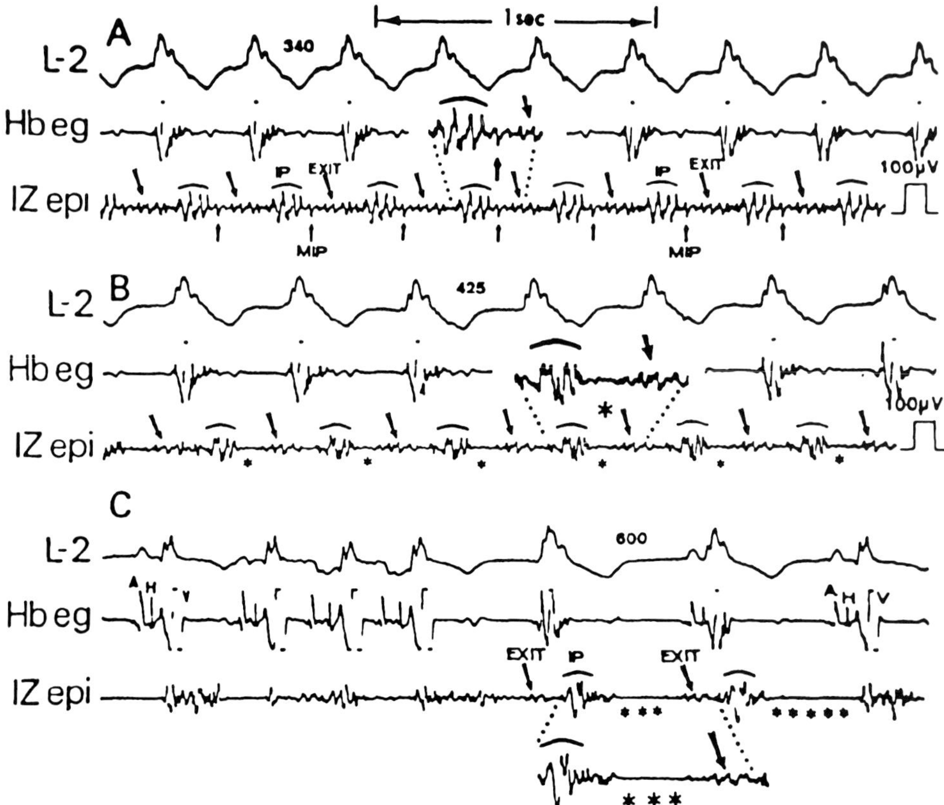

FIGURE 8.8 Effect of procaineamide on sustained ventricular tachycardia (**A**). Control state of sustained ventricular tachycardia in the 4-day-old infarcted dog heart. Initial, mid-interectopic potential, and exit potentials are designated by arcs and arrows in each of the interectopic intervals, as well as in the magnified insets that show reproducible continuous electrical activity from beat to beat. **B.** After 10 mg/kg procainamide given intravenously, the cycle length of the tachycardia increases from 340 ms (**A**) to 425. Note the essentially unchanged initial and exit potentials but the marked diminution of the middiastolic potential. **C.** After 20 mg/kg of procainamide caused termination of the tachycardia, three rapidly paced atrial beats caused bridging electrical activity and two beats of the previous tachycardia with such a long cycle length (***, 600 ms) that the sinus rate captures and extinguishes the rhythm. See text for further details. *Reprinted from Brachmann et al.,*[17] *with permission.*

terns during initiation, maintenance, and termination of the inducible sustained tachycardias.

Garan and Ruskin constructed isochronous maps of a radial spread of activation arising from a small area on the subendocardium in the infarcted dog heart.[19] The center, from which activation spread, was characterized by recording from a close (2-mm) bipolar electrogram that showed continuous electrical activity. Once a stable monomorphic tachycardia was established, the interectopic activation was reproducible from cycle to cycle (Fig. 8.9, lowest panel).

On the other hand, El-Sherif and his associates[21,22] mapped reentrant circuits on the epicardium overlying an infarct in the dog heart and showed that the circuit aligned as a double loop or figure-eight around two arcs of functional conduction block. In Figure 8.10 a typical isochronous map is shown on the left with selected electrograms on the right during sustained monomorphic ventricular tachycardia. The selected electrograms depict continuous electrical activity bridging the successive interectopic intervals of the ventricular tachycardia. Note that electrograms 31, 19, and 23, both spatially and temporally, represent initial, middiastolic, and exit potentials, respectively.[23] In other studies Gessman et al.[20] and El-Sherif and his associates[24] showed that cryoprobes placed at the areas corresponding to the middiastolic potential (electrode 19, Fig. 8.10) could consistently terminate the tachycardia. Local cooling of exit sites (for

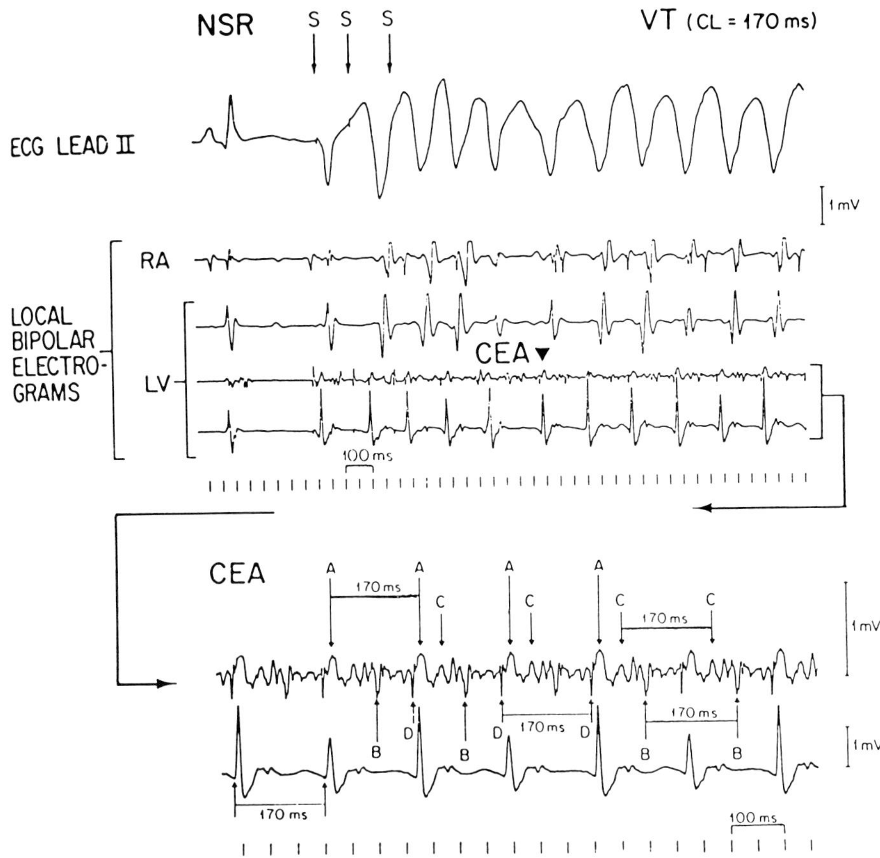

FIGURE 8.9 Organized continuous electrical activity (CEA) recorded from a local bipolar electrogram in the infarcted dog heart. The top panel shows the initiation of sustained ventricular tachycardia, which become stable in the last part of the record. Along with a standard ECG lead II are local (2-mm) bipolar recordings from the right atrium (RA) and three left ventricular sites showing relatively normal electrograms in two and a fractionated and depressed electrogram (middle). During sustained tachycardia, CEA starts at the arrowhead and is continuous to the end. The bottom panel is an amplified view of the last two local electrograms showing the reproducible CEA compared to clear interectopic intervals from an adjacent normal recording site. *Reprinted with permission from Garan and Ruskin. Localized reentry: Mechanism of induced sustained ventricular in canine model of recent myocardial infarction. J Clin Invest 1984;74:377–392.*

example, site 23, Fig. 8.10) produced slowing of the ventricular tachycardia, but did not terminate the arrhythmia.[24]

Although many of the reports from El-Sherif et al.[22–26] and others[27] have emphasized the double-loop form of the reentrant circuit with activation fronts circulating around two separate arcs of block, others have shown that single-loop reentrant circuits can exist around either a single line of block or a fulcrum, that is, a central fractionated potential within the line of block.[28] Thus, in the double-loop circuit, slow conduction exists within the common pathway made up of the coalescent wavefronts circulating around two arcs of block[22,24–26] or around a single activation barrier[28,29] in a vortical form of reentry. The length of this region of slow conduction may be 2 to 4 cm, whereas in the microcircuits described by Garan, Fallon, and Ruskin[19] and more recently detailed by Zuanetti et al. the area at the center of slow conduction may be smaller than 2 mm (Fig. 9 in ref. 30).

Conduction velocity has been measured in reentrant circuits in dog hearts with subacute infarction (first week). This "slow conduction" can be quite variable particularly in regard to whether the activation wave front is moving in a parallel or transverse direction to longitudinal fiber orientation. Dillon et al.[31] found that conduction velocities were nonuniform in reentrant

circuits that were aligned in accordance with the anisotropic tissue properties in the epicardial border zone of the infarcted dog heart. The estimated conduction velocities parallel to longitudinal fiber orientation were relatively normal at 0.5 to 1.0 meter/s; whereas, at the pivot points, that is, where the wave front turned around the line of conduction block, conduction velocity transverse to longitudinal fiber orientation was 0.1 to 0.4 meter/s (Fig. 15 in ref. 31). These investigators also detected even slower conduction through the line of conduction block, 0.04 to 0.07 meter/s.

The basis of this slow conduction may be due to multiple alterations of normal conduction properties secondary to ischemic damage. Relatively few studies have been carried out in multicellular preparations taken from the epicardial border zone overlying the infarction. The electrophysiological properties of these border zones in the 1- to 4-day-old infarcted heart show marked heterogeneity with some cells showing near-normal properties, that is, resting membrane potential, upstroke velocity, and action potential duration (APD), with others showing marked depression of these electrophysiological features, particularly in response to rate changes.[32–34] It has been suggested that derangement of ionic mechanisms, such as deranged sodium (Na^+) channel function may characterize the cells in the surviving epicardial rim in the acute and subacute stages of myocardial infarction.[35,36]

On the other hand, Gardner et al. found that in the epicardial border, 2 weeks to 18 months after infarction, normal action potentials were found within areas showing fractionated potentials. Their responses to rate were not tested. These areas did, however, show structural changes consisting of widely dispersed groups of surviving cells. "Distortion of the cells by connective tissue with a resulting decrease in intracellular connections might increase resistance to current flow and thereby slow conduction."[37] These authors, thus, proposed that another factor causing slow conduction could be "slow electrotonic transmission across high resistance or inexcitable gaps."[37] Models of this form of slow conduction have been elegantly demonstrated by Anzelevitch and Moe.[38,39]

Another related factor that may contribute to slow conduction is the alteration of the normal anisotropic properties of cardiac tissue induced by ischemic damage. The phenomenon of anisotropy and its role in discontinuous conduction has been recently reviewed by Spach and Dolber.[40] Several studies have directly tested the importance of anisotropy and its alteration by ischemic damage as it contributes to the reentry phenomenon.[28–31,41] Some controversy exits as

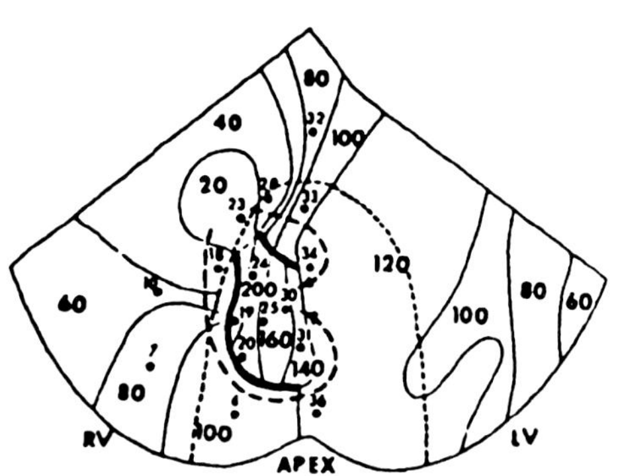

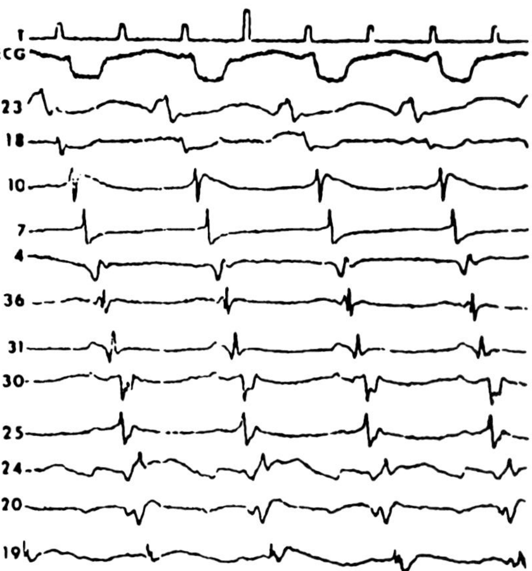

FIGURE 8.10 Double-loop or figure-eight model of reentry in the infarcted dog heart. An isochronal activation map from surviving epicardium is shown at the left, in which activation wave fronts, indicated by curved arrows, proceed around two arcs of block (heavier dark lines) during sustained monomorphic ventricular tachycardia. At the right are selected electrograms from sites numbered on the map. See text for details. *Reprinted from El-Sherif,[23] with permission.*

to the significance of the anisotropic factor as a basis for reentry and further studies are required to resolve this issue.[42]

CLINICAL CORRELATES OF BASIC EXPERIMENTAL FINDINGS

The development of catheter ablation techniques for ventricular tachycardias in the clinical setting has made the identification of slow conduction zones more than an academic exercise. Although Josephson and his associates initially described the clinical counterpart of continuous electrical activity during ventricular tachycardia and showed its relation to the slow conduction portion of the reentrant circuit,[15,43] this group later focused on the "earliest recorded discrete or fragmented ventricular electrogram" preceding the onset of the QRS as the site of "origin" of the tachycardia.[44] They stated that "continuous electrical activity can occasionally be recorded . . . [but]more often . . . presystolic activity can be recorded."[44] Mapping with an electrode catheter in the left ventricle, particularly in the area of a large aneurysm, guided the surgical excision of these arrhythmogenic sites.[45] However, from a practical viewpoint a large endocardial resection was performed that precluded the determination of the critical relationship between the local site recording presystolic activity and the clinical tachycardia.[45]

More recently, catheter ablative procedures, rather than surgery, have been used in an attempt to "excise" the symptomatic ventricular tachycardia. Fitzgerald et al. reported that delivery of DC shocks to ventricular sites whose electrograms showed presystolic activity during tachycardia was less effective in terminating the arrhythmia than shocks delivered at sites showing middiastolic potentials. Figure 8.11 shows a sustained ventricular tachycardia during which ECG lead V_1, a left and right ventricular electrogram, were recorded by electrode catheters in contact with the endocardial surfaces of the heart. In the top panel, a reproducible presystolic potential starting 60 ms before the QRS was consistently recorded. Shocks delivered at this site failed to terminate the tachycardia. When the catheter was repositioned to record a reproducible middiastolic potential (arrows), and similar shocks delivered, the tachycardia was terminated.

Figure 8.12 schematically summarizes the results found by these investigators.[46] Of the attempts to ablate 20 ventricular tachycardias, the success rate was 88% (7/8) when shocks were

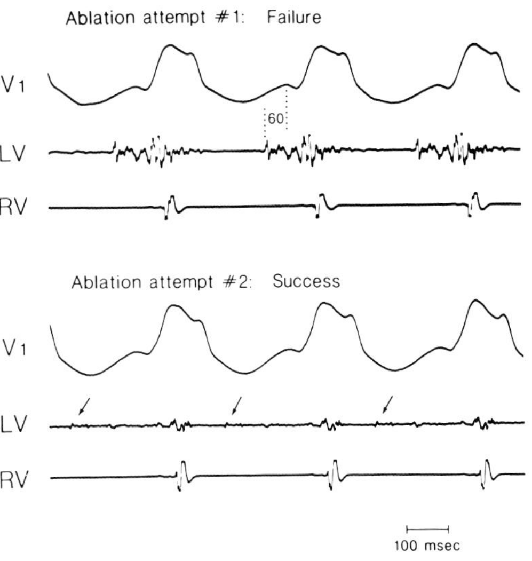

FIGURE 8.11 Electrogram recordings from two different ventricular sites showing presystolic (upper panel) and middiastolic (lower panel) fractionated activity during sustained ventricular tachycardia in a patient with chronic myocardial infarction. When DC shocks were delivered at site of presystolic activation (60 ms prior to QRS onset) tachycardia was not terminated. When similar DC shocks were delivered to the site showing low-level, reproducible potentials, which occurred in middiastole, ablation was successful. *Reprinted from Fitzgerald et al.,*[46] *with permission.*

delivered at sites showing middiastolic activity, that is, the slow-conducting portions of the reentrant circuit. This activity was directly associated with initiation, termination, and resetting by delivery of extra stimulation. On the other hand, 12 attempts to ablate tachycardia by delivering similar shocks at sites of early ventricular activation, that is, exits from the slow conduction zones, resulted in only a 25% success rate. Similar results had been reported earlier both in the experimental[20] and clinical settings.[47]

In regard to follow-up studies on a group of patients subjected to ablation procedures via an electrode catheter, a recent study by Trappe et al.[48] reported significant differences in recurrence rates depending on the areas ablated. The schematic representation of the results is shown in Figure 8.13. A total of 44 patients with sustained ventricular tachycardia were treated. DC ablation through electrode catheters was performed to terminate tachycardia by shocks delivered to sites of early ventricular activation, that is, presystolic or exit sites, in 32 patients and at sites showing middiastolic potentials, that is,

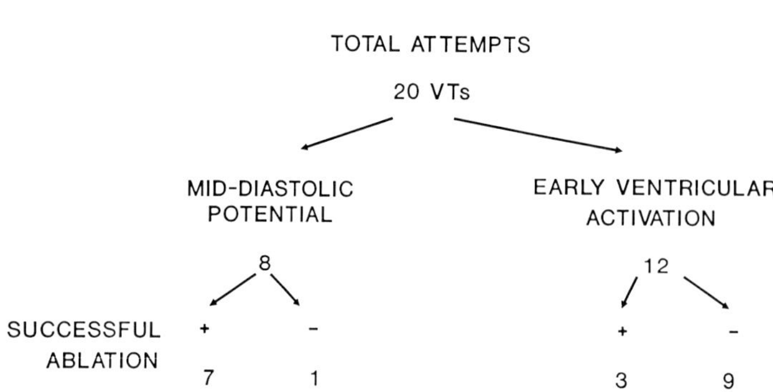

FIGURE 8.12 Schematic summary of success (+) and failure (−) in terminating sustained ventricular tachycardia in patients with myocardial infarction. Seven of 8 tachycardias were ablated by DC shocks delivered at sites showing a reproducible middiastolic potential during sustained tachycardia, whereas only 3 of 12 successful terminations were achieved when sites with presystolic or early ventricular activation were ablated. Data were taken from Fitzgerald et al.[46]

slow conduction zones, in 12. There was a 25% recurrence rate in the latter and a significantly greater recurrence rate of 81% in the former group.

As Josephson and his associates originally observed, the ability to find middiastolic or continuous electrical activity is less likely than finding presystolic activity during ventricular tachycardia in the clinical situation. However, it should be noted (Figs. 8.6 and 8.11) that low-gain recordings are uniformly used in clinical studies. This widely accepted methodology is thought to minimize detection of distal activity at local electrode catheter recording sites. A recent report from Downar et al.[49] provides evidence that increasing the sensitivity of electrogram recordings may allow recognition of low-level continuous electrical activity (Figs. 6 and 7 in ref. 48). To avoid the possibility that distant activity may be recorded from high-gain electrograms, closely adjacent sites can be monitored at comparable gains.

In any event, it seems apparent that targeting of localized ablation sites of slow conduction will play an important role in the effective treatment of patients with life-threatening ventricular tachycardias. The influence of basic experimental studies has again served to direct the course of clinical advances in the management and treatment of serious disease states in man.

ACKNOWLEDGMENTS

We are grateful for the contributions of our colleagues Drs. Glenn Kabell and Johannes Brachmann, and our technical staff, consisting of Marjo Carpenter, Betty Orange, and LaVonna Blair. We especially acknowledge the assistance of Mrs. Pam Tomey for the production of this and many of the other manuscripts cited herein.

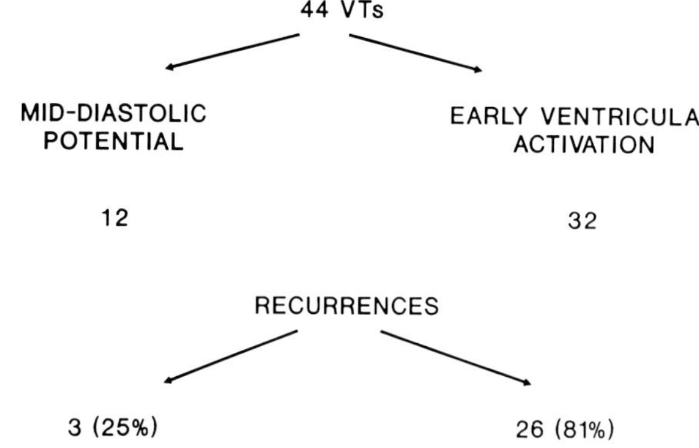

FIGURE 8.13 A similar diagrammatic summary from another study of patients with myocardial infarction. Of 44 inducible, sustained ventricular tachycardia there were 3 recurrences of the arrhythmia (25%) when middiastolic potentials were ablated (total 12), whereas there was an 81% recurrence rate when DC ablation, via catheter, was aimed at early ventricular activation sites during tachycardia (total 32). Data were taken from Trappe et al.[48]

REFERENCES

1. Waller AD: On the electromotive changes connected with the beat of the mammalian heart, and of the human heart in particular. *Phil Trans B* 1889;180:169.
2. McWilliam JA: Fibrillar contraction of the heart. *J Physiol (Lond)* 1887;8:296–310.
3. Mayer AG: Rhythmical pulsation in Scyphomedusae. *Carnegie Inst Washington Pub* 1906;47.
4. Mayer AG: Rhythmical pulsation in Scyphomedusae. II. Carnegie Inst. Washington Papers Tortuga Lab. 1:113–131. *Carnegie Inst Washington Pub* 1908;102, Part VII.
5. Mines GR: On circulating excitations in heart muscle and their possible relation to tachycardia and fibrillation. *Trans Roy Soc Canada* Series 3, Section IV, 1914;8:43–52.
6. Harris AS, Guevera-Rojas A: The initiation of ventricular fibrillation due to coronary occlusion. *Exp Med Surg* 1943;1:105–121.
7. Waldo AL, Kaiser GA: Study of ventricular arrhythmias associated with acute myocardial infarction in the canine heart. *Circulation* 1973;47:1222–1228.
8. Boineau JP, Cox JL: Slow ventricular activation in acute myocardial infarction. A source of reentrant premature ventricular contractions. *Circulation* 1973;48:702–713.
9. Scherlag BJ, El-Sherif N, Hope R, et al.: Characterization and localization of ventricular arrhythmias resulting from myocardial ischemia and infarction. *Circ Res* 1974;35:372–383.
10. Williams DO, Scherlag BJ, Hope R, et al.: The pathophysiology of malignant ventricular arrhythmias during acute myocardial infarction. *Circulation* 1974;50:1163–1172.
11. El-Sherif N, Scherlag BJ, Hope RR, et al.: Reentrant ventricular arrhythmias in the late myocardial infarction period. I. Conduction characteristics in the infarction zone. *Circulation* 1977;55:686–702.
12. El-Sherif N, Hope RR, Scherlag BJ, et al.: Reentrant ventricular arrhythmias in the late myocardial infarction period. II. Patterns of initiation and termination of reentry. *Circulation* 1977;55:702–719.
13. El-Sherif N, Lazzara R, Hope RR, et al.: Reentrant ventricular arrhythmias in the late myocardial infarction period. III. Manifest and concealed extrasystolic grouping. *Circulation* 1977;56:225–234.
14. El-Sherif N, Lazzara R: Reentrant ventricular arrhythmias in the late myocardial infarction period. 7. Effect of verapamil and D-600 and the role of the "slow channel." *Circulation* 1979;60:605–615.
15. Josephson ME, Horowitz LN: An electrophysiologic approach to the therapy of recurrent sustained ventricular tachycardia. *Am J Cardiol* 1979;43:631.
16. Kabell G, Scherlag BJ, Hope RR, et al.: Patterns of interectopic activation recorded during pleomorphic ventricular tachycardia after myocardial infarction in the dog. *Am J Cardiol* 1982;49:56–62.
17. Brachmann J, Kabell G, Scherlag BJ, et al.: Analysis of interectopic activation patterns during sustained ventricular tachycardia. *Circulation* 1983;67:449.
18. Gessman LJ, Agarwal JB, Endo T, et al.: Localization and mechanisms of ventricular tachycardia by ice mapping 1 week after the onset of myocardial infarction in dogs. *Lab Invest* 1983;68:657–666.
19. Garan H, Fallon JT, Ruskin JN: Sustained ventricular tachycardia in recent canine myocardial infarction. *Circulation* 1980;62:980–987.
20. Gessman LJ, Endo T, Egan T, Egan J, Gallagher JD, Hastie R, Maroko PR: Dissociation of the site of origin from the site of cryotermination of ventricular tachycardia. *PACE* 1983;6:1293–1305.
21. El-Sherif N, Smith RA, Evans K: Canine ventricular arrhythmias in the late myocardial infarction period. 8. Epicardial mapping of reentrant circuits. *Circ Res* 1981;49:255–264.
22. Mehra R, Zeiler RH, Gough WB, El-Sherif N: Reentrant ventricular arrhythmias in the late myocardial infarction period. 9. Electrophysiologic-anatomic correlation of reentrant circuits. *Circulation* 1983;67:11–24.
23. El-Sherif N: Reentry revisited. *PACE* 1988;11:1358–1368.
24. El-Sherif N, Mehra R, Gough WB, et al.: Reentrant, ventricular arrhythmias in the late myocardial infarction period. Interruption of reentrant circuits by cryothermal techniques. *Circulation* 1983;8:644–656.
25. El-Sherif N, Mehra R, Gough WB, Zeiler RH: Reentrant ventricular arrhythmias in the late myocardial infarction period. Interruption of reentrant circuits by cryothermal techniques. *Circulation* 1983; 68:644–656.
26. El-Sherif N, Gough WB, Restivo M: Reentrant ventricular arrhythmias in the late myocardial infarction period. 14. Mechanisms of resetting, entrainment, acceleration or termination of reentrant tachycardia by programmed electrical stimulation. *PACE* 1987;10:341–371.
27. Cardinal R, Savard P, Carson DL, Pagé P: Mapping of ventricular tachycardia induced by programmed stimulation in canine preparation of myocardial infarction. *Circulation* 1984;70:136–148.
28. Wit AL, Dillon S, Ursell PC: Influences of anisotropic tissue structure on reentrant ventricular tachycardia, in Brugada P, Wellens HJJ (eds): *Cardiac Arrhythmias: Where To Go From Here?* Mount Kisco, NY, Futura Publishing, 1987, pp 27–50.
29. Dillon SM, Allessie MA, Ursell PC, et al.: Influences of anisotropic tissue structure on reentrant circuits in the epicardial border zone of subacute canine infarcts. *Circ Res* 1988;63:182–206.
30. Zuanetti G, Hoyt RH, Corr PB: Beta-adrenergic-mediated influences on microscopic conduction in epicardial regions overlying infarcted myocardium. *Circ Res* 1990;67:284–302.
31. Dillon SM, Allessie MA, Ursell PC, Wit AL: Influences of anisotropic tissue structure on reentrant circuits in the epicardial border zone of subacute canine infarcts. *Circ Res* 1988;63:182–206.
32. Lazzara R, El-Sherif N, Vallone T, Scherlag BJ: Conduction in depressed cardiac muscle. *Circulation* 1975;II:704.
33. Lazzara R, Hope RR, El-Sherif N, Scherlag BJ: Effects of lidocaine on hypoxic and ischemic cardiac cells. *Am J Cardiol* 1978;41:872–879.
34. Spear JF, Michelson EL, Moore EL: Reduced space constant in slowly conducting regions of chronically infarcted canine myocardium. *Circ Res* 1983;53:176–185.
35. Lazzara R, El-Sherif N, Hope RR, Scherlag BJ: Ventricular arrhythmias and electrophysiological con-

sequences of myocardial ischemia and infarction. *Circ Res* 1978;42:740–749.
36. Lazzara R, Scherlag BJ: Electrophysiologic basis for arrhythmias in ischemic heart disease. *Am J Cardiol* 1984;53:1B–7B.
37. Gardner PI, Ursell PC, Fenoglio JJ Jr, et al.: Electrophysiologic and anatomic basis for fractionated electrograms recorded from healed myocardial infarcts. *Circulation* 1985;72:596–611.
38. Antzelevitch C, Moe GK: Electrotonically mediated delayed conduction and reentry in relation to "slow response" in mammalian ventricular conducting tissue. *Circ Res* 1981;49:1129.
39. Antzelevitch C, Moe GK: Electrotonic inhibition and summation of impulse conduction in mammalian Purkinje fibers. *Am J Physiol* 1983;245:H42.
40. Spach M, Dolber PC: The relation between discontinuous propagation in anisotropic cardiac muscle and the "vulnerable period" of reentry, in Zipes DP, Jalife J (eds): *Cardiac Electrophysiology and Arrhythmias*. New York, Grune & Stratton, 1985, Chapter 28.
41. Cardinal R, Vermeulen M, Shenasa M, et al.: Anisotropic conduction and functional dissociation of ischemic tissue during reentrant ventricular tachycardia in canine myocardial infarction. *Circulation* 1988;77:1162–1176.
42. Restivo M, Gough WB, El-Sherif N: Ventricular arrhythmias in the subacute myocardial infarction period: High-resolution activation and refractory patterns of reentrant rhythms. *Circ Res* 1990;66:1310–1327.
43. Josephson ME, Horowitz LN, Farshidi A, Spear JF, Kastor JA, Moore EN: Recurrent sustained ventricular tachycardia. 2. Endocardial mapping. *Circulation* 1978;57:440–448.
44. Josephson ME, Horowitz LN, Waxman HL, et al.: Role of catheter mapping in evaluation of ventricular tachycardia, in Josephson ME (ed): *Ventricular Tachycardia: Mechanisms and Management*. Mount Kisco, NY, Futura Publishing, 1982, Chapter X.
45. Josephson ME, Horowitz LN: An electrophysiologic approach to the therapy of recurrent sustained ventricular tachycardia. *Am J Cardiol* 1979;43:631.
46. Fitzgerald DM, Friday KJ, Yeung Lai Wah JA, et al.: Electrogram patterns predicting successful catheter ablation of ventricular tachycardia. *Circulation* 1988;77:806–814.
47. Morady F, Scheinmann MM, DiCarlo LA Jr, et al.: Catheter ablation of ventricular tachycardia with intracardiac shocks: Results in 33 patients. *Circulation* 1987;75:1037.
48. Trappe HJ, Klein H, Eberhard S: Where to ablate ventricular tachycardia? *Circulation* 1988;78:II-300 (abstract).
49. Downar E, Harris L, Mickleborough LL, et al.: Endocardial mapping of ventricular tachycardia in the intact human ventricle: Evidence for reentrant mechanisms. *J Am Coll Cardiol* 1988;11:783–791.

Chapter **9**

Dispersion of Refractoriness in Experimental and Clinical Arrhythmias

Borys Surawicz, MD, Chien-Sun Kuo, MD, and Andras Varro, MD, PhD

Dispersion of refractoriness is a measure of nonhomogeneous recovery of excitability in a given mass of cardiac tissue. In the normal atrial or ventricular myocardium, the dispersion of refractoriness parallels the dispersion of repolarization because the excitability is strictly proportional to the duration of repolarization (Fig. 9.1). However, in the diseased, depolarized, and pharmacologically altered myocardium, or in the normal slow channel-dependent structures, for example, atrioventricular (AV) node or sinoatrial (SA) node, the recovery of excitability may lag behind the completion of repolarization, that is, post-repolarization refractoriness.

The dispersion of repolarization is determined by the differences in the activation times and the differences in the durations of action potentials (APs).[1] Since the duration of ventricular APs tends to decrease as the ventricular activation proceeds,[2] it can be assumed that the total dispersion at the end of ventricular repolarization probably does not exceed the duration of the QRS complex. Another reasonable assumption, which is supported by experimental findings,[3] is that if the activation time differences remain unchanged, the dispersion of repolarization will be greater at slow than at fast heart rates.

Dispersion of ventricular repolarization was measured in dogs using six simultaneously recorded monophasic action potentials (MAP), five on the left and one on the right ventricular surface.[4,5] During atrial pacing, the dispersion was predominantly due to the MAP duration differences, and during ventricular pacing, it was due predominantly to the activation time (AT) differences.[4] The dispersion produced by the earliest ventricular premature complex (VPC) during atrial pacing was greater than during ventricular pacing, due mainly to an increase in the MAP duration difference. After a second VPC, which was the earliest propagated response elicited by a stimulus of 2 ms duration and up to 10 times diastolic threshold strength, dispersion increased further due to increase in both the MAP and the AT differences. No spontaneous arrhythmia was induced by single or double VPCs in any of the dogs. This shows that normal ventricular myocardium can tolerate safely a fivefold increase in the dispersion of repolarization.

The site of ventricular stimulation, and by implication the site of origin of the premature impulse, may affect the dispersion of repolarization. However, the studies in humans and experimental animals produced conflicting results. In human subjects,[6] effective refractory periods (ERP) were shorter when the pacing and the testing sites were at the same location than when the sites were separated. In contrast, Abildskov[7] and Burgess et al.[8] observed longer refractory periods when driving and testing were performed at the same site than when the sites were separated. These investigators found that testing at a distant site with fusion of activation fronts shortened the ERP, and attributed this effect of changed activation sequence to the electrotonic interaction. The difference in ERP dependent of the site of stimulation averaged about 5 ms in normal myocardium and about 19 ms in the ischemic myocardium.[8]

When the myocardial cells are well coupled, differences in action potential duration (APD) lessen due to electrotonic interaction between neighboring cells. However, when the cells are uncoupled, such as in a setting of abnormal an-

655 Avenue of the Americas, New York, NY 10010
Current Topics in Cardiology

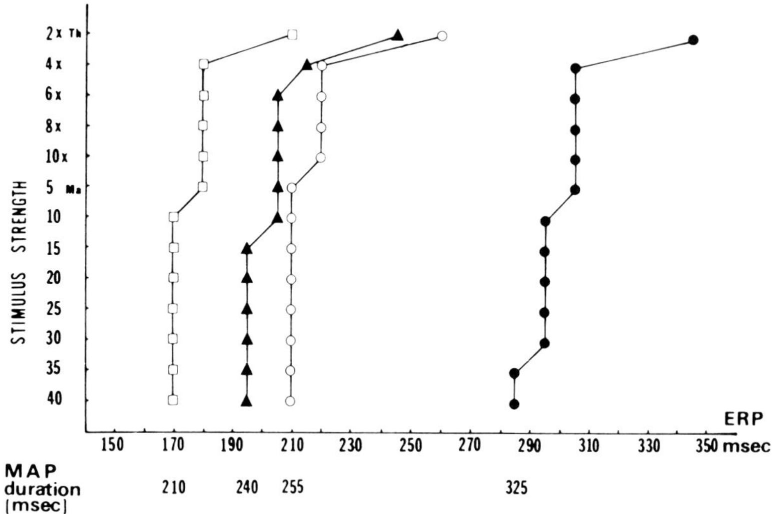

FIGURE 9.1 Strength–interval curves and duration of monophasic action potential (MAP) recorded at a site within 5 mm from the stimulation site during control, hypothermia, and regional warm blood perfusion. The effective refractory period (ERP) at each strength of stimulus is plotted against the strength of stimulus. The division on the ordinate is not a proportional scale, and the strength of the stimulus is expressed in milliamperes (mA) in the lower portion of the ordinate and in the multiples of the threshold of diastolic excitability (Th) in the upper portion when the stimulus strength ranged from twice to 10 times Th. The MAP durations corresponding to each strength–interval curve are shown at the bottom of the figure. The 275 and 255 ms MAPs were recorded during control, the 325 ms MAP during hypothermia, and MAPs of 240 and 210 ms during regional warm blood perfusion. *Reprinted from Kuo et al.,*[31] *with permission.*

isotropic impulse propagation, APD differences may be expected to increase, which can result in increased dispersion of repolarization. In the computer model simulating a tiny sheet of ventricular myocardium (perhaps such as one overlying myocardial infarction), abnormal anisotropy was simulated by an increase in axial resistance.[9] This produced both changes in activation and an increased dispersion of APD. The changes in activation were pacing site-dependent, and were associated with pacing site-dependent APD dispersion. In this model, the magnitude of increased APD dispersion due to simulated nonuniform anisotropy was less than 10 ms.[9]

In human subjects, dispersion of refractoriness measured at three right ventricular sites was greater during ventricular pacing than during atrial pacing.[6] The dispersion of QT intervals in precordial maps in 50 normal subjects (average QT of 384.1 ms) ranged from 34 to 88 ms.[10] In another study, QT dispersion averaged 89 ms in normal subjects and 155 ms in 14 patients with prolonged QT (average QT of 482 ms).[11] The dispersion of MAPs at different endocardial sites of the right ventricle was up to 40 ms in normal subjects, and from 100 to 270 ms in 10 patients with long QT and a history of torsade de pointes.[12]

In a recent study, recovery of excitability at multiple endocardial sites in the left ventricle was determined in three groups of patients.[13] In seven persons with no heart disease and no arrhythmia, the dispersion of ventricular refractoriness averaged 40 ms and the mean dispersion of total recovery time was 52 ms. In six patients with previous myocardial infarction and sustained ventricular tachycardia, the mean endocardial activation time was significantly prolonged, the dispersion of refractoriness was not significantly different from control, and the mean dispersion of total recovery times was 90 ms. In five patients with long QT intervals (average 528 ms), the dispersion of left ventricular refractoriness averaged 87 ms and the dispersion of total recovery times 114 ms.[13]

ROLE OF DISPERSION OF REPOLARIZATION IN EXPERIMENTAL ARRHYTHMIAS

In studies of dogs, Han, Moe, and coworkers firmly established an association between nonuniform recovery of excitability and lowered ven-

tricular fibrillation threshold (VFT).[14,15] The mechanism by which the increases in dispersion facilitate the induction of ventricular fibrillation is difficult to unravel because the electric stimulation employed in the measurements of VFT induces a number of complex changes in electrophysiologic properties of cardiac fibers, including altered polarizations, automaticity, and refractoriness.[16]

The most likely mechanism of arrhythmia facilitated by dispersion of repolarization is reentry, that is, a phenomenon requiring a unidirectional block and a defined conduction pathway.[17] The ingredients that are most often represented in the experimental or clinical models of arrhythmias attributed to reentry include nonuniform conduction, nonuniform excitability, and nonuniform refractoriness. These three electrophysiologic properties tend to be mutually interdependent, so that any intervention aimed at changing one of them is likely to result in concomitant changes of the other two; hence, in many models of experimental arrhythmias, it is difficult to identify the independent contribution of the dispersion of repolarization to the initiation and maintenance of ventricular arrhythmias. Occasionally, however, the experimental design of an arrhythmia model does permit the assessment of the possible role of dispersion of repolarization in the genesis of arrhythmias.

An example of an experimental model in which dispersion of repolarization is considered to play a crucial role in the maintenance of arrhythmia is the circus movement tachycardia induced by Allessie et al.[18,19] in isolated segments of rabbit left atrial tissue. In this preparation, comprising an area of about 15 mm^2, the intrinsic difference between the shortest and the longest AP was about 30 ms,[18] and the circus movement tachycardia of up to 30 min duration was initiated when the premature impulse was conducted only in one direction and was blocked in other directions, while the propagation lasted sufficiently long to allow the restoration of excitability in the area where the impulse was initiated.[18]

The premature stimuli initiating tachycardia (1 ms duration and four times threshold strength) were always delivered on a border between two areas of different refractory periods and propagated in the direction of the shorter ERP, while being blocked in the direction of the longer ERP. Using varying carbalmylcholine concentrations that produced graded ERP shortening, these investigators established that the minimal difference in the ERP for the production of unidirectional block was 11 to 16 ms; thus, the site with the long duration of AP formed the site of unidirectional block and served as a barrier around which the impulse was spreading. The authors pointed out that if the "island" of block was small, the impulse returned rapidly, that is, before the excitability at the site of block had been reestablished; hence, the circus movement depended not only on the magnitude of dispersion of refractory periods but on the dimensions of the site of block created by the long ERP. In this model, the intrinsic dispersion within the small piece of atrial tissue was sufficient to produce a unidirectional block and create the circus movement tachycardia following a single premature stimulus. Also, the dimension of the site of local block was critical, and the greater the dispersion, the wider the range of premature stimuli that induced unidirectional block.[18] The revolution time was proportional to the refractory period[19] and was not determined by the conduction block, as in another type of circus movement around an anatomic obstacle. The inexcitability in the center of tissue mass was caused not by depolarization but by continuous invasion of ineffective stimuli resulting in collision of centripetal wavelets. The leading circle formed a tight fit where the "head" of excitation was biting into the tail. The dimension of the circle was determined by the electrophysiologic properties of the tissue, and was not fixed but could change with changes in conduction and excitability. No excitable gap exists for invading stimuli but the evolution may be short-circuited by an impulse crossing the center.[18]

A similar "leading circle" type of reentry without fixed obstacle has been suggested as a mechanism of atrial flutter induced by acetylcholine and atrial stimulation in dogs.[20] The authors of this study[20] speculated that the same mechanism may operate in the so-called type II atrial flutter, which is not influenced by rapid atrial pacing in humans.

The unique manner of arrhythmia induction in the experiments of Allessie et al.[18,19] could not be duplicated in the dog ventricle in situ, where one or two ventricular premature stimuli failed to produce arrhythmia, even when the dispersion of ventricular repolarization averaged about 150 ms and the difference between MAPs averaged 44 ms.[5]

Dispersion also played an important role in ventricular arrhythmias elicited within 3 to 5 days after ligation of the left anterior descending coronary artery in dogs. In the model of a 3- to 8-day-old myocardial infarction[21] in which sustained reentrant ventricular tachycardia could be reproducibly initiated, marked disparities in effective and relative refractory periods were observed within area of infarction at sites separated by 1 to 2 mm.[21]

In a similar model, El-Sherif and coworkers[22,23] correlated the isochronal maps of the ERP with the activation maps. Their studies showed that refractoriness was prolonged in the surviving epicardial layer in a spatially nonuniform manner with steep gradients of refractoriness between contiguous zones. In the normal hearts, the dispersion of ERP averaged 30 ms, and in the infarcted hearts it was 160 ms. The conduction block occurred at the site of the most marked gradient of refractoriness, that is, where the difference in refractoriness at adjacent sites spaced at 5 to 10 mm was ⩾ 20 ms.[22,23] Single reentry was initiated by a premature complex originating in the region of shortest refractory period and conducted toward areas with longer refractory periods.[22] In a more recent study[24] in the same preparation of subacute infarction, epicardial surface activation maps were constructed using high-density bipolar electrograms spaced at 1 mm. During initiation of reentry by premature stimulation, conduction block occurred abruptly within 1 mm, and without prior decrement of the impulse. This was correlated with abrupt increases in effective refractory periods of 10 to 120 ms duration (average 27 ± 24 ms) within 1 mm or less. The abrupt change in refractoriness was not dependent on differences in excitability or the orientation of myocardial fibers, and therefore the observed continuous areas of functional conduction block that are necessary for both the initiation and the maintenance of reentrant activation were due to abrupt changes in refractoriness within a distance of ⩽1 mm. A difference of 10 ms in effective refractory periods between sites spaced 1 mm apart was sufficient for the occurrence of functional conduction block. It is not known whether prolonged refractoriness in this preparation was due to lengthening of APD or to post-repolarization refractoriness.[24]

In coronary-perfused cat left ventricles,[25] acute ischemia resulted in increased APD differences and increased difference in post-repolarization refractoriness between endocardial and epicardial cells. Increased dispersion was associated with the occurrence of spontaneous arrhythmias and facilitation of arrhythmia induction. Also, in a model of acute ischemia superimposed on healed myocardial infarction in isolated coronary-perfused cat ventricles, spontaneous rapid ventricular tachycardia was observed in four of eight preparations during the last 20 to 30 min of ischemia but not in any nonischemic preparations. The arrhythmia was associated with increased differences in AP characteristics and increased dispersion of refractoriness.[26]

Dispersion of refractoriness appeared to play a role also in the early arrhythmias after coronary ligation in dogs. In one such study,[27] the arrhythmias occurring within 2 to 8 min after coronary artery ligation were associated with increased dispersion of excitability, refractoriness, and conduction time within the ischemic zone. Arrhythmias occurring later, that is, 11.5 to 23 min after ligation, were associated with partial recovery of excitability, conduction, and refractoriness but the dispersion of refractoriness still persisted.

In contrast to postocclusion arrhythmias, the dispersion appeared to play no more than a subordinate role in the model of coronary reperfusion. In the study of Naimi et al.,[28] the authors measured the dispersion of the ERP in the electrograms recorded with both surface and plunge electrodes in the ischemic and the nonischemic myocardium. The control dispersion among epicardial electrodes was 15.5 ms and among transmural electrodes, 12.9 ms. After ligation of the left anterior descending coronary artery for 3 to 5 min, both values increased to an average of 44.3 ms and 24.8 ms, respectively. After the release of ligation lasting 5 min, the corresponding dispersion values averaged 37.9 ms and 32.5 ms, and after the release of ligation lasting 15 min, both dispersion values decreased to 20.0 ms; yet, the authors reported that the reperfusion arrhythmias were more frequent after the longer than after the shorter periods of ligation.[28] The reported results suggest that other factors, for example, nonhomogeneous conduction and excitability, did play a more important role in the genesis of reperfusion arrhythmia than the dispersion of repolarization.

The role of dispersion refractoriness was investigated by Chen et al.[29] in a model of reentry induced by strong premature stimuli applied in an isolated portion of canine right ventricular infundibulum. In this model, arrhythmias elicited by strong stimuli acting on normal tissue were not dependent on preexisting dispersion of refractoriness, and the regions with the greatest nonhomogeneity in the dispersion of refractoriness did not correspond to the sites of unidirectional block after premature stimulation. Reentrant arrhythmias were associated with large differences in activation times at two adjacent myocardial regions. In this model, cells were depolarized directly by field stimulation and not by impulse conduction, and unidirectional block occurred at the onset of activation. The stimulus was sufficiently strong to capture cells that were partially refractory, and the field stimulation induced graded responses.

In the animal model employed by Geddes et

al.,[30] surface cooling was used to prolong conduction time and warming to shorten recovery time. Reentrant activity could not be elicited by shortening recovery times alone, but in the presence of hot-spot reentrant activity occurred during a longer portion of the cardiac cycle and at less cold temperatures than those required to induce reentry in the absence of focal shortening of recovery time.[30]

MODELS OF DISPERSION OF REPOLARIZATION

In the experimental model of Kuo et al.,[31] the arrhythmogenic effect of dispersion of repolarization was studied using MAPs recorded with suction electrodes. Simultaneous recording of MAPs permits direct measurement of the contribution of the AT differences to the dispersion of repolarization of a propagated impulse. Six MAPs on the ventricular surface and 12 local electrograms in 23 open-chest anesthetized dogs were recorded simultaneously. To induce dispersion caused by changes in MAP duration, the myocardial temperature was altered by a combination of general hypothermia and selective coronary artery perfusion with heated blood.[4] The shunted blood passed through a coil immersed in a 45°C bath. The blood temperature was regulated by varying the length of the coil immersed in the bath. This temperature range was 38 to 43°C at the outlet of the shunt immediately before entering the left anterior descending coronary artery. By cooling and warming the blood rather than the cardiac surface, large transmural temperature gradients were avoided and the representative regions with the extremes of long and short MAPs were sampled.

When regional warm blood perfusion (RWBP) was added to hypothermia, MAP shortened in the perfused area and either did not change or slightly shortened in the region not perfused by the warm blood. The maximum dispersion during atrial pacing increased nearly 10-fold, that is, 111 ± 16 ms of which about 80% was due to increased difference in MAP durations. Ventricular arrhythmia did not occur spontaneously during hypothermia combined with RWBP. In each dog, when maximum dispersion reached a certain critical value, however, an early premature ventricular stimulus applied at the site in which MAP was shortened induced repetitive ventricular responses progressing to ventricular fibrillation. The stimulus strength of the premature response that induced arrhythmia ranged from 0.4 to 40 mA (average 9.5 ± 11 mA).

The following evidence suggests that the induction of arrhythmia was dependent on critical dispersion[31]: (a) In 16 dogs, hypothermia was maintained at a constant temperature, while blood temperature during RWBP was increased stepwise to produce graded increases in dispersion. In these 16 dogs, the critical maximum dispersion during atrial pacing averaged 112 ± 15 ms (range 94 to 145 ms). (b) In four dogs, the effects of ventricular premature stimulation were tested during RWBP, both with and without hypothermia. During hypothermia combined with RWBP, the critical maximum dispersion values associated with induction of ventricular fibrillation were 105, 115, 120, and 125 ms. After defibrillation, RWBP was continued, but hypothermia was withdrawn. This decreased the respective values of maximum dispersion to 60, 85, 65, and 60 ms, and made it no longer possible to induce ventricular arrhythmia in any of the four dogs. (c) In the presence of constant hypothermia and RWBP, the atrial pacing rate was increased in four dogs. During pacing at a cycle length of 510 ± 79 ms (range 450 to 580 ms), maximum dispersion was 103 ± 5 ms (range 100 to 110 ms). When the pacing cycle length was shortened by 50 to 130 ms to an average 423 ± 39 ms (range 380 to 450 ms), dispersion decreased to 86 ± 9 ms (range 75 to 95 ms) ($p < 0.05$). This decrease occurred because the shorter MAP within the 170 to 205 ms range shortened less than the longer MAPs within the 245 to 295 ms range. The average shortening of the short MAPs was 14 ms (188 ± 20 ms versus 174 ± 17 ms), and the long MAPs, 31 ms (266 ± 22 ms versus 235 ± 15 ms). The maximal difference in AT did not change (32 ± 8 ms at slow pacing rate and 32 ± 8 ms at fast pacing rate). In each dog, ventricular fibrillation was induced during pacing at a slower rate but not during pacing at a faster rate.

The effects of the site of ventricular stimulation were compared in six dogs in the presence of critical dispersion of 118 ± 16 ms. In each dog, ventricular fibrillation was induced during stimulation at the site at which the MAP duration was short, but no arrhythmia could be induced in the presence of the same dispersion during stimulation at the site at which the MAP duration was long[31] (Fig. 9.2). This study showed that the propagation of a premature impulse from the site with short duration of MAP to the sites with long duration of MAP was slower than the propagation from the site with long duration of MAP to the sites of short MAP duration. Thus, the AT difference during propagation from the site with long MAP duration averaged 51 ms, and during propagation from the site with short MAP duration, 100 ms. The corresponding values of maximum dispersion during premature stimulation averaged 127 ms and 176 ms, respectively.

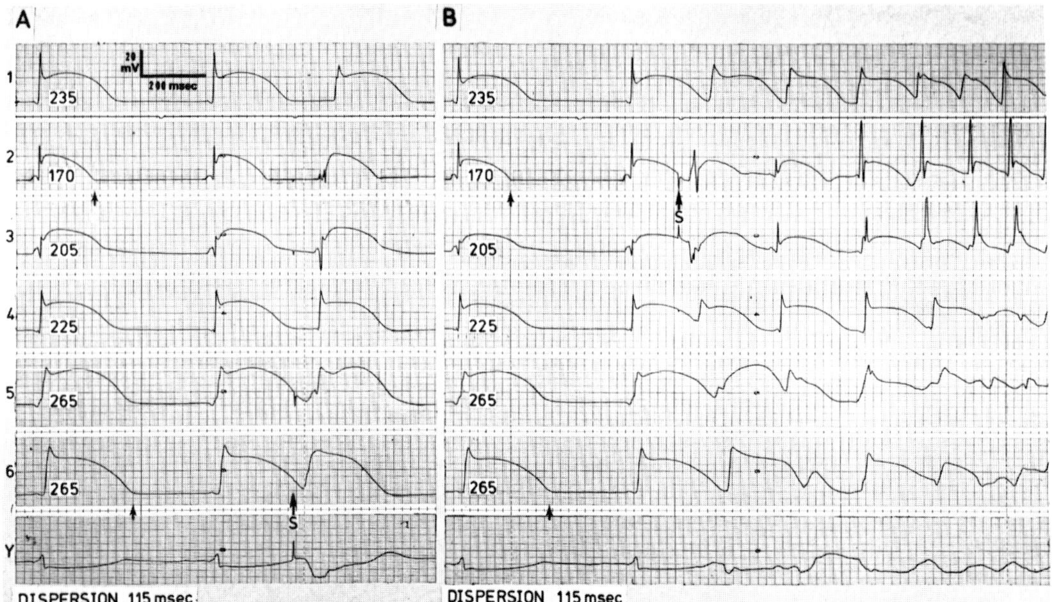

FIGURE 9.2 Effect of site of ventricular premature stimulation (VPS) on the induction of ventricular arrhythmia in the presence of critical dispersion. **A.** VPS at the lateral site of the left ventricle. **B.** VPS at the paraseptal site. The stimulus in **A** represents the strongest (40 mA) and the most premature stimulus that elicited a premature ventricular complex. Ventricular arrhythmia was induced in **B** but not in **A**. Symbols are the same as in Figure 9.1. *Reprinted from Kuo et al.,*[31] *with permission.*

The site of the earliest activation of the first spontaneous ventricular complex of tachycardia was compared with that of the premature ventricular complex (PVC) that induced this spontaneous activity. For this, adequate records were available in 18 episodes of ventricular fibrillation induced in 13 dogs. In 15 of these episodes, the sites of the earliest activation of the first spontaneous ventricular complex and of the preceding stimulus-induced PVC differed from each other.[31] In 6 episodes, the earliest activation of the first spontaneous complex was recorded in the perfused area; in 6, in the border region; and in the remaining 3 episodes, in the region not perfused with warm blood.

Three possible mechanisms of arrhythmia facilitated by increased dispersion of repolarization were considered: focal reexcitation, reentry facilitated by a conduction from an area with a long refractory period to an area with a short refractory period, and reentry facilitated by a conduction from an area with a short refractory period to an area with a long refractory period, that is, a unidirectional block caused by prolonged duration of repolarization. In the presence of critical dispersion ranging from 95 to 145 ms, the differences between the duration of the longest and shortest MAPs averaged 99 ms. Juxtaposition of two such APs might be expected to create a potential difference during repolarization and generate an excitatory current of a sufficient magnitude to reexcite the fiber with the shorter APD.[32] This phenomenon was postulated in a small strand of Purkinje fibers.[33] Such a mechanism, however, would be expected to initiate a spontaneous activity, which was not observed in our study. Most likely, a sharp temperature gradient between the cold and warm areas was prevented by an abundant intracoronary collateral communication in canine hearts and by passive heat conduction in the myocardium. Also, electrotonic interaction could have blunted the sharp potential gradients between the areas with long and short MAP durations.[34] The lack of spontaneous ventricular arrhythmia also decreased the probability that arrhythmia was due to enhanced automaticity of cardiac Purkinje fibers, which could be expected in the warmed myocardium.

The use of a strong premature stimulus required consideration of local reexcitation due to stimulation during the vulnerable period. In this study, however, the stimulus strength was not critical, because stimuli of the same or greater strength did not induce ventricular arrhythmia during control, hypothermia, or RWBP alone.

The initiation of ventricular arrhythmia by a single premature stimulus at the sites distal to the site of stimulation favors the mechanism of reentry, or triggered automaticity, and one cannot distinguish between these two possibilities.[35] Compatible with reentry was the presence of independent activation fronts at nonadjacent anatomic locations at the onset of arrhythmia. The arrhythmia could be induced only by premature stimulus applied at the site of short MAP durations.

The inability to induce arrhythmia by stimulating the site with long duration of MAP ruled out the possibility that reentrant arrhythmia was initiated by an early reexcitation at the sites with short MAPs. Because the site of premature stimulation had no significant influence on the MAP duration difference during control or during hypothermia and RWBP, the increase in dispersion during propagation of ventricular premature complex was due solely to the increase in the AT difference. When this increase was added to the preexisting large MAP duration difference, a critical magnitude of dispersion was achieved. The lack of a comparable increase in the AT difference during stimulation from the site with long MAP prevented such a degree of dispersion (Fig. 9.3). The authors assumed that when the increase in dispersion of repolarization reached a critical level, propagation of premature impulses originating from the area with a short MAP encountered a block in the area with a long MAP and created conditions favorable for reentry. It was concluded that the large dispersion of repolarization is arrhythmogenic because it creates an environment that facilitates the development of a conduction delay required to induce a sustained arrhythmia.

In the model of Kuo et al.,[31] the premature stimuli initiating arrhythmia either did not change or decreased the MAP duration differences. This suggests that the large MAP duration difference in this model played a crucial role only in the process of initiation but not in the process of maintenance of arrhythmia.

In another animal model in which large increase of dispersion of regional QT intervals was induced by intracoronary injections of Reno-

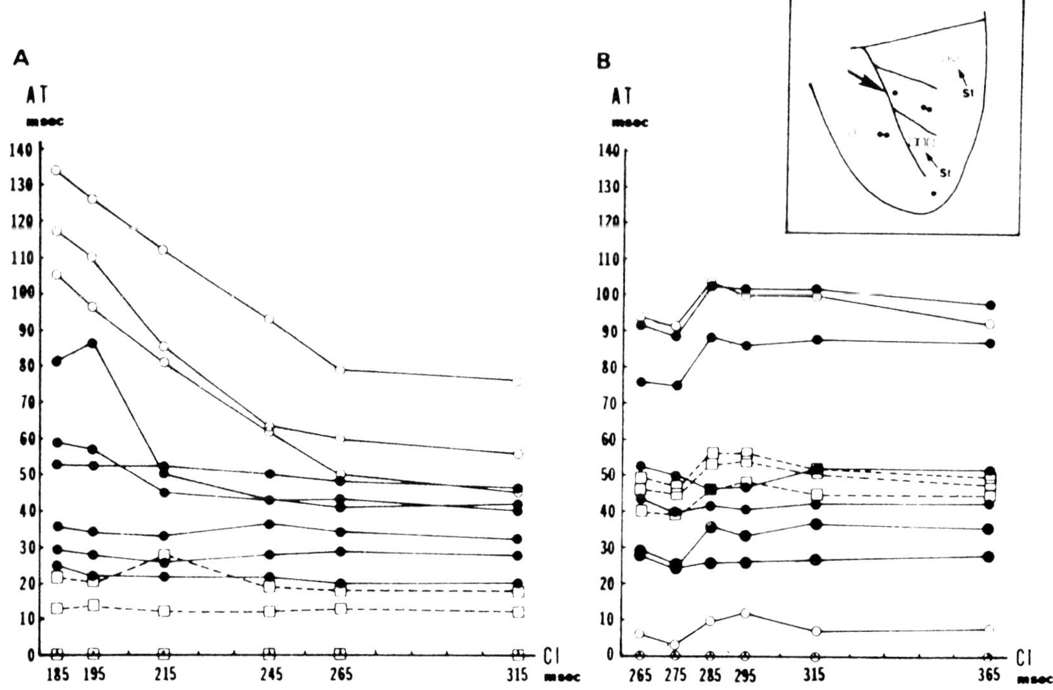

FIGURE 9.3 Results of an experiment showing the activation times (AT) at 12 recording sites during ventricular premature stimulation (VPS) at the paraseptal site (**A**) and at the lateral site (**B**) at different coupling intervals (CI) in the presence of critical dispersion. The diagram in the inset shows the locations of recording sites. Squares = perfused area; solid circles = border area; open circles = nonperfused area; small arrows = site of VPS (St); large arrow = site of coronary cannulation. *Reprinted from Kuo et al.,*[31] *with permission.*

grafin 76, ventricular fibrillation occurred both spontaneously and after premature stimulation at the site of normal regional QT, that is, in the non-Renografin-perfused area.(36) The origin of the spontaneous premature complex initiating the arrhythmia was not established, but the arrhythmia was attributed to reentry caused by slow conduction in incompletely repolarized myocardium.(36) In this animal model, administration of Ca^{2+}, which attenuated the local QT lengthening, prevented the induction of ventricular fibrillation. This finding is of some interest because the increase in $[Ca^{2+}]$ may be expected to enhance afterdepolarizations which have been postulated to precipitate arrhythmia in dogs with prolonged APDs.(36)

In another model, increased dispersion of refractoriness was produced by distention of the left ventricle in an isolated perfused rabbit heart.(37) In this model, increased dispersion was associated with facilitation of induction of ventricular arrhythmia by programmed stimulation.

In yet another recent study in anesthetized dogs, an area of the right ventricle was warmed by a directed light beam.(38) Sequences of activation and dispersion of refractoriness were measured from 40 epicardial and 24 endocardial sites in and surrounding the warmed area. Warming increased the dispersion of epicardial ERP from an average of 26 ms to an average of 93 ms, and that of endocardial ERP from an average of 20 ms to an average of 36 ms. Low-amplitude programmed stimulation that had failed to initiate arrhythmia during control conditions initiated reentrant arrhythmia using one or two premature stimuli in 7 of 10 dogs. Conduction block occurred at the border of the warmed and the normal myocardium, and the reentry occurred near the stimulation site situated within the warmed area. In the remaining three dogs, conduction block was not observed but the spread of excitation suggested that the arrhythmia elicited by programmed stimulation represented macroreentry utilizing the bundle branches and the His-Purkinje system.(38)

CLINICAL ARRHYTHMIAS DEPENDENT ON DISPERSION OF REPOLARIZATION

Few studies have examined the role of dispersion of refractoriness in the genesis of reentrant arrhythmias in man. In one study of patients with hypertrophic cardiomyopathy, ventricular tachycardia or ventricular fibrillation was inducible in 8 of 18 patients, and the inducibility of these arrhythmias was associated with increased dispersion of refractoriness between right ventricular outflow tract and right ventricular apex.(39)

The dispersion of ventricular repolarization is probably of critical importance in the genesis of arrhythmias in patients who lack other identifiable factors that might contribute to electrical nonhomogeneity. An example of such condition is the congenital long QT syndrome (LQTS). The only demonstrable electrophysiologic abnormality in such hearts is an increased dispersion of refractoriness, which in the absence of conduction disturbances is probably due to abnormal differences in ventricular APDs. This assumption is supported by studies of ERP, MAP durations, and the R on T phenomenon. Except for the changes in ventricular repolarization, the electrocardiogram of patients with LQTS is usually normal and there is no clinical evidence of valvular, myocardial, or coronary heart disease. In several autopsy studies, the myocardium appeared both grossly and histologically normal. However, in one series, a consistent finding was focal neuritis and neural degeneration within the SA node, AV node, His bundle, and ventricular myocardium. Others reported degeneration of Purkinje fibers or fibrosis of the conduction system.(40)

The characteristic arrhythmia associated with LQTS is torsade de pointes. The torsade de pointes or ventricular fibrillation in patients with LQTS is frequently precipitated by a ventricular premature complex interrupting the T wave, characteristically after a cycle with a long RR interval. Congenital LQTS has been attributed to a dysfunction of autonomic nervous system.(40) However, the precise nature of such dysfunction is not known. Similarly, the mechanism of acquired neurogenic LQTS associated with subarachnoid hemorrhage, cryohypohysectomy, truncal vagotomy, and poisoning with organophosphorus insecticides is not known.(40) However, in several of these conditions, for example, intracranial hemorrhage and organophosphorus poisoning, torsade de pointes is precipitated by the R on T depolarization which in the presence of long QT interval is a marker of increased dispersion of refractoriness.(40)

The three most common causes of torsade de pointes in clinical practice are: treatment with class IA and class III antiarrhythmic drugs, hypokalemia, and severe bradycardia. The arrhythmia in these conditions usually is precipitated by R on T depolarization that is preceded by a long RR interval (Fig. 9.4). These circumstances suggest increased dispersion of refractoriness.

Among patients treated with antiarrhythmic drugs, the largest number of cases of torsade de pointes has been attributed to quinidine, which

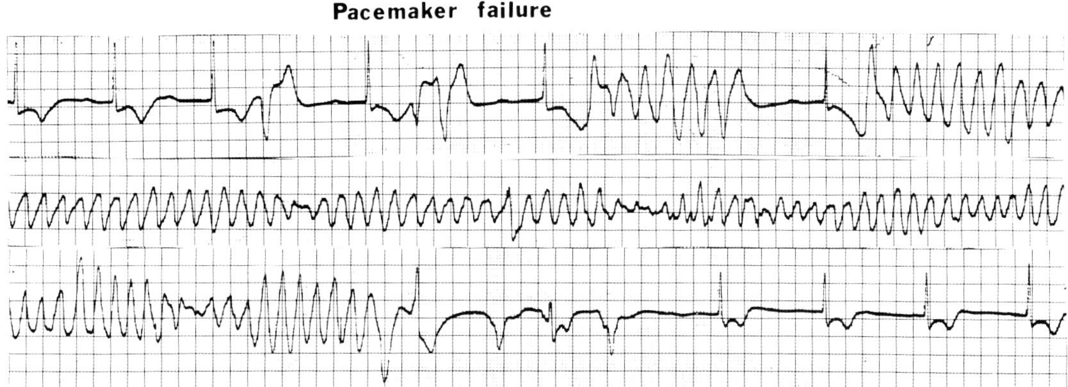

FIGURE 9.4 Ventricular premature complexes (VPC) with varying coupling intervals (CI) in a patient with a complete AV block and AV junctional escape rhythm. Continuous strip of lead II. Q-T interval = 0.54 s. In the upper strip, the duration of CI of the VPC decreased progressively: 0.56 s (4th complex); 0.52 s (6th complex); 0.50 s (9th complex); and 0.48 s (16th complex). Note the long pauses preceding complexes followed by VPC. The VPC with the shortest coupling is followed by ventricular fibrillation and torsade de pointes continuing through the middle and part of the bottom strip.

is the oldest antiarrhythmic agent in the category of Na^+ channel-blocking drugs. In the quinidine-treated patients, the incidence of torsade is about 1.5% per year.[41] The arrhythmia tends to occur within 1 week of initiation of treatment and in one study of 31 patients, approximately two-thirds had prolonged QT while off quinidine.[42] Other aggravating factors include: the presence of structural heart disease, hypokalemia, and abrupt rate slowing. No correlation has been established between the plasma concentration of quinidine, or dihydroquinidine, and the subsequent development of torsade.

Torsade de pointes has been reported in patients treated with both procainamide and its metabolite, nor-acetyl procainamide (NAPA). Of the two compounds, NAPA causes greater lengthening of the QT interval. It has been suggested that patients receiving procainamide, who have higher NAPA concentrations, are either fast acetylators or have impaired renal function, and that such individuals are at higher risk of developing arrhythmia.[43]

In patients treated with disopyramide, torsade de pointes tends to occur in the early stages of treatment but there is no correlation between its occurrence and the dose of the drug. In patients receiving amiodarone, torsade has been reported during various states of treatment, including the loading phase.

Torsade de pointes also occurs during treatment with antiarrhythmic drugs that have no effect on the QRS duration but prolong the QTc interval, for example, the β-adrenergic blocker sotalol, the Ca^{2+} channel blocker bepridil, and the drug combining both effects, for example, prenylamine. In the small subset of susceptible patients treated with antiarrhythmic drugs causing QTc lengthening, ventricular tachyarrhythmias tend to occur within the first few days of treatment when plasma drug concentrations are within the therapeutic range. In such cases arrhythmias are attributed to an idiosyncratic reaction, perhaps signifying latent LQTS and increased dispersion of repolarization.

Since in the majority of treated patients proarrhythmic effects do not occur, it may be assumed that the lengthening of repolarization induced by these drugs causes no important increase in dispersion of repolarization. Two studies support this assumption. In one, quinidine prolonged the ERP and the MAP duration in dogs but did not increase dispersion of repolarization measured at several right ventricular sites.[44] In another, procainamide prolonged QT and ERP in human subjects but did not alter dispersion measured at three right ventricular sites.[45]

In hypokalemia, QT interval is not prolonged when it can be measured, but when the repolarization is markedly distorted and T wave merges with U wave, QT is no longer measurable. Although no direct relation has been established

between the amplitude of the T + U wave and torsade de pointes, hypokalemia is a well-known risk factor for this arrhythmia either alone or in association with other factors. Lengthening of the terminal portion of repolarization by hypokalemia is expected to prolong the relative refractory period and increase dispersion of repolarization.

Torsade de pointes has been reported in a variety of conditions associated with QT lengthening, for example, after contrast injections into coronary artery, in patients with hypothyroidism or anorexia nervosa, during treatment with "liquid protein" diets or other fad weight-reducing diets, during therapeutic starvation, and in the presence of marked bradycardia. Marked dispersion of repolarization was documented in patients with permanent bradycardia, attacks of ventricular tachycardia, and long QT interval.(46)

The most effective treatment of torsade de pointes is an increase in heart rate by pacing. This results in shortening of QT, although frequently without shortening of QTc. The increased rate of pacing shortened the dispersion of repolarization in dogs(31) and in humans.(6,13) In one reported case of a patient with torsade de pointes and long QT, the dispersion of refractoriness measured at three right ventricular sites was 100 ms before and 35 ms during atrial pacing.(6)

In the absence of pacing facilities, isoproterenol has been used to terminate the arrhythmia. The presumed mechanism of isoproterenol action is an increase in heart rate and a decrease in dispersion of repolarization.(14) Also, bretylium tosylate, which has been useful in the treatment of torsade de pointes, decreased the dispersion of refractoriness in dogs in which the dispersion had been increased by intravenous quinidine administration.(47)

DISPERSION DURING PROPAGATION OF AN EARLY PREMATURE IMPULSE

It will be recalled that the dispersion of repolarization is determined by the differences in activation times and APD. The activation time difference may be expected to increase during slow-impulse propagation. The APD difference may be expected to increase when a slowly conducted impulse, that is, an early premature impulse, follows a normally conducted impulse. Under these circumstances each successive premature depolarization will occur at a progressively longer interval after the preceding repolarization, that is, a progressively longer diastolic interval (DI). Because the duration of the premature AP increases with increasing duration of the preceding DI, the increased range of DI will result in increased APD difference.

For any given range of preceding DI, the magnitude of APD differences depends on the characteristics of restitution of APD. We studied the kinetics of restitution of APD duration in canine Purkinje and ventricular muscle fibers in vitro, and the restitution of ventricular MAP durations in anesthetized dogs, and found in both types of studies that the kinetics of restitution were not changed by the duration of basic AP. In Purkinje fibers, we compared the restitution following the basic APD of 350 ms and 500 ms,(48) and in dog ventricle within an APD range of 170 to 290 ms (unpublished studies of Kuo, Atarashi, Munakata, and Surawicz). In the Purkinje fibers, the restitution followed a course determined by a sum of two exponentials, and in the ventricular fibers the restitution followed a single exponential course.

The lack of change in kinetics with changing basic APD leads to the conclusion that the premature dispersion of repolarization increases with lengthening of basic APD, and with increasing basic dispersion of APD. We also found that the relation between MAP and the preceding DI during electrical restitution favors an increased dispersion of repolarization when the early premature impulse propagates from an area with short MAP to an area with long MAP; conversely, a dispersion of repolarization will be decreased when the premature impulse propagates from an area with long MAP to an area with short MAP.

Our studies suggest that restitution of APD needs to be considered as a potential mechanism for influencing the perpetuation of reentrant ventricular arrhythmias dependent on dispersion of repolarization. Class I antiarrhythmic drugs are known to alter both the conduction and the APD, and a decrease in dispersion has been considered as a possible mechanism of their antiarrhythmic action.(49–51)

We studied the effect of six class I antiarrhythmic drugs,(52) that is, quinidine (5 μg/ml), disopyramide (10 μg/ml), procainamide (30 μg/ml), flecainide (4 μg/ml), lidocaine (4 μg/ml), and mexiletine (4 μg/ml), on the durations of the basic action potential (APD_b) at a cycle length of 500 ms and on the premature APD (APD_t) elicited at progressively increasing DIs in canine Purkinje fibers. The difference between APD_t elicited at a DI of 100 ms and the earliest APD_t elicited at the onset of ERP was defined as the range of APD_t. In controls, this range was 98 ± 1.8 ms ($n = 59$). Disopyramide and procainamide did not change the range significantly, but the other four drugs

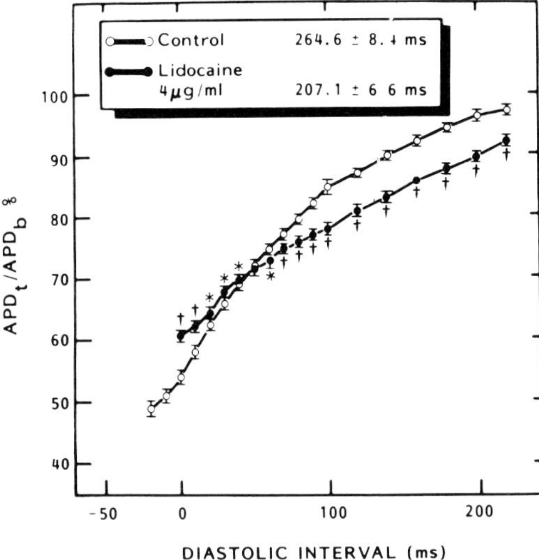

FIGURE 9.5 Restitution of action potential duration (APD) in dog Purkinje fibers, before (control, n = 12) and during superfusion with lidocaine (n = 12). On the ordinate, APD of the premature APD (APD_t) as percentage of basic APD (APD_b) at the basic cycle of 500 ms. The values of APD at basic cycle length are shown in the inset. The significance of the differences between the points corresponding to the same diastolic interval is expressed by + for $p < 0.01$ and by * for $p < 0.05$. Note that lidocaine decreases the range of APD_t predominantly because of longer ERP and slower kinetics of restitution. *Reprinted from Varro et al.,*[52] *with permission.*

decreased it significantly ($p < 0.01$) as follows: quinidine by 50.2%, lidocaine by 60.2% (Fig. 9.5), mexiletine by 61.6%, and flecainide by 61.4% (Fig. 9.6).

The range of APD_t during the first 100 ms of electrical restitution was determined also in canine ventricular muscle fibers[53] in the presence of lidocaine (4 and 8 μg/ml), mexiletine (8 μg/ml), flecainide (2 and 4 μg/ml), procainamide (50 μg/ml), quinidine (10 μg/ml), and disopyramide (10 μg/ml). The drug effects on the characteristics of AP at the basic cycle lengths of 500 and 1,000 ms were similar to those reported previously. At control conditions, the range of premature APDs was 40 ms at cycle of 1,000 ms and 30 ms at cycle of 500 ms. It was decreased 26 to 52% by lidocaine, mexiletine, and flecainide, not changed significantly by procainamide, and increased 17 to 53% by quinidine and disopyramide. The range of APD_ts at control levels and in the presence of drugs in ventricular muscle was smaller than in Purkinje fibers at the same or lower drug concentrations.

Unlike the Purkinje fibers, the range of APD_t in the ventricular muscle fibers was influenced chiefly by APD at basic cycle. The lengthening of the APD_t range by quinidine and disopyramide suggests that these drugs may be expected to increase dispersion of repolarization during the propagation of an early premature impulse, and that this increase may be greater at long than at short BCL. This applies also to other drugs that prolong AP duration, for example, bretylium tosylate, sotalol, bepridil, or amiodarone.

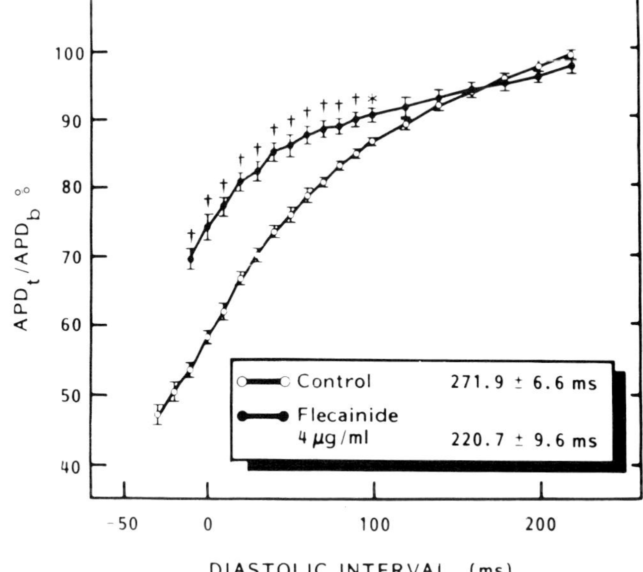

FIGURE 9.6 Restitution of APD in dog Purkinje fibers before (control, n = 8) and during superfusion with flecainide (n = 8). Abbreviations and explanations as in Figure 9.5. Note that flecainide decreases the range of APD_t, predominantly because of longer ERP and shift of restitution curve toward longer APD_t/APD_b values. *Reprinted from Varro et al.,*[52] *with permission.*

CONCLUSIONS

Dispersion of refractoriness plays an important role in several models of experimental arrhythmias, for example, acute myocardial ischemia, and a 3- to 4-day-old myocardial infarction. Several studies suggest that the underlying dispersion of refractoriness facilitates induction of arrhythmias when the premature impulse originating in the area of shorter refractoriness blocks during the propagation toward areas of longer refractoriness.

In several conditions with prolonged repolarization, the characteristic arrhythmia is torsade de pointes. The three most common causes of torsade in clinical practice are: treatment with class IA and class III antiarrhythmic drugs, hypokalemia, and severe bradycardia.

The characteristics of electrical restitution mandate an increase in dispersion of refractoriness during propagation of an early premature impulse. Our studies of antiarrhythmic drugs in vitro suggest that antiarrhythmic drugs can both decrease and increase premature dispersion of repolarization in Purkinje and ventricular muscle fibers.

REFERENCES

1. Kuo CS, Reddy CP, Munakata K, Surawicz B: Arrhythmias dependent predominantly on dispersion of repolarization, in Zipes DP, Jalife J (eds): *Cardiac Electrophysiology and Arrhythmias*. Orlando, FL, Grune & Stratton, 1985, pp 277–285.
2. Autenrieth G, Surawicz B, Kuo CS: Sequence of repolarization on the ventricular surface in the dog. *Am Heart J* 1975;89:463–469.
3. Han J, Millett D, Chizzonitti B, et al.: Temporal dispersion of recovery of excitability in atrium and ventricle as a function of heart rate. *Am Heart J* 1966;71:481–487.
4. Kuo CS, Amlie JP, Munakata K, Surawicz B: Dispersion of monophasic action potential durations and activation times during atrial pacing, ventricular pacing, and ventricular premature stimulation in canine ventricles. *Cardiovasc Res* 1983;17:152–161.
5. Kuo CS, Atarashi H, Reddy CP, Surawicz B: Dispersion of ventricular repolarization and arrhythmia: Study of two consecutive ventricular premature impulses. *Circulation* 1985;72:370–376.
6. Friehling TD, Kowey PR, Shechter JA, Engel TR: Effect of site of pacing on dispersion of refractoriness. *Am J Cardiol* 1985;55:1339–1343.
7. Abildskov JA: Effects of activation sequence on the local recovery of ventricular excitability in the dog. *Circ Res* 1976;38:240–247.
8. Burgess MJ, Steinhaus BM, Spitzer KW, Green LS: Effects of activation sequence on ventricular refractory periods of ischemic canine myocardium. *J Electrocardiol* 1985;18:323–330.
9. Lesh MD, Pring M, Spear JF: Cellular uncoupling can unmask dispersion of action potential duration in ventricular myocardium. *Circ Res* 1989;65:1426–1440.
10. Mirvis DM: Spatial variation of QT intervals in normal persons and patients with acute myocardial infarction. *J Am Coll Cardiol* 1985;5:625–631.
11. Sylvien JC, Horacek BM, Spencer CA, Klassen GA, Montague TJ: QT interval variability on the body surface. *J Electrocardiol* 1984;17:179–188.
12. Bonatti V, Rolli A, Botti G: Recording of monophasic action potentials of the right ventricle in long QT syndromes complicated by severe ventricular arrhythmias. *Eur Heart J* 1983;4:168–179.
13. Vassallo JA, Cassidy DM, Kindwall KE, Marchlinski FE, Josephson ME: Nonuniform recovery of excitability in the left ventricle. *Circulation* 1988;78:1365–1372.
14. Han J, Garcia DeJalon PD, Moe GK: Adrenergic effects on ventricular vulnerability. *Circ Res* 1964;14:516–524.
15. Han J, Moe GK: Nonuniform recovery of excitability in ventricular muscle. *Circ Res* 1964;14:44–60.
16. Surawicz B: Ventricular fibrillation. *Am J Cardiol* 1971;27:268–289.
17. Mines GR: On circulating excitations in heart muscle and their possible relation to tachycardia and fibrillation. *Trans R Soc Can* 1914;8:43–52.
18. Allessie MA, Bonke FIM, Schopman FJG: Circus movement in rabbit atrial muscle as a mechanism of tachycardia. II. The role of nonuniform recovery of excitability in the occurrence of unidirectional block, as studied with multiple microelectrodes. *Circ Res* 1976;39:168–177.
19. Allessie MA, Bonke FIM, Schopman FJG: Circus movement in rabbit atrial muscle as a mechanism of tachycardia. III. The "leading circle" concept: A new model of circus movement in cardiac tissue without the involvement of an anatomical obstacle. *Circ Res* 1977;41:9–18.
20. Allessie MA, Lammers WJEP, Bonke IM, Hollen J: Intraatrial reentry as a mechanism for atrial flutter induced by acetylcholine and rapid pacing in the dog. *Circulation* 1984;70:123–135.
21. Michelson EL, Spear JF, Moore EN: Electrophysiologic and anatomic correlates of sustained ventricular tachyarrhythmias in a model of chronic myocardial infarction. *Am J Cardiol* 1980;45:583–590.
22. Gough WB, Mehra R, Restivo M, Zeller RH, El-Sherif N: Reentrant ventricular arrhythmias in the late myocardial infarction period in the dog. 13. Correlation of activation and refractory maps. *Circ Res* 1985;57:432–442.
23. Restivo M, Gough WB, El-Sherif N: Reentrant ventricular rhythms in the late myocardial infarction period: Prevention of reentry by dual stimulation during basic rhythm. *Circulation* 1988;77:429–444.
24. Restivo M, Gough WB, El-Sherif N: Ventricular arrhythmias in the subacute myocardial infarction period. *Circ Res* 1990;66:1310–1327.
25. Kimura S, Bassett AL, Kohya T, Kozlovskis PL, Myerburg RJ: Simultaneous recording of action potentials from endocardium and epicardium during ischemia in the isolated cat ventricle: Relation of temporal electrophysiologic heterogeneities to arrhythmias. *Circulation* 1986;74:401–409.
26. Kimura S, Bassett AL, Cameron JS, Huikuri H, Kozlovskis PL, Myerburg RJ: Cellular electrophysiological changes during ischemia in isolated, coro-

nary-perfused cat ventricle with a healed myocardial infarction. *Circulation* 1988;78:401–406.
27. Horacek T, Neumann S, Budden M, Meesman W: Nonhomogeneous electrophysiological changes and the bimodal distribution of early ventricular arrhythmias during acute coronary artery occlusion. *Bas Res Cardiol* 1984;79:649–667.
28. Naimi S, Avitall B, Mieszale J, Levine HJ: Dispersion of effective refractory period during abrupt reperfusion of ischemic myocardium in dogs. *Am J Cardiol* 1977;39:407–412.
29. Chen PS, Wolf PD, Dixon EG, et al.: Mechanism of ventricular vulnerability to single premature stimuli in open-chest dogs. *Circ Res* 1988;62:1191–1209.
30. Geddes J, Burgess MJ, Millar K, Abildskov JA: Accelerated repolarization as a factor in reentry—Stimulation of the electrophysiology of acute myocardial infarction. *Am Heart J* 1974;88:61–68.
31. Kuo CS, Munakata K, Reddy CP, Surawicz B: Characteristics and possible mechanism of ventricular arrhythmia dependent on the dispersion of action potential durations. *Circulation* 1983;67:1356–1367.
32. Kleber AG, Janse MJ, Va Capelle FJL, et al.: Mechanisms and time course of S-T and T-Q segment changes during acute regional myocardial ischemia in pig heart determined by extracellular and intracellular recordings. *Circ Res* 1978;42:603–613.
33. Sano T, Sawanobori T: Abnormal automaticity in canine Purkinje fibers focally subjected to low external concentrations of calcium. *Circ Res* 1972;31:158–164.
34. Mendez C, Moe GK: Some characteristics of transmembrane potentials of AV nodal cells during propagation of premature beats. *Circ Res* 1966;19:993–1010.
35. Wit AL, Cranefield PF, Gadsby DC: Triggered activity, in Zipes DP, Bailey JC, Elharrar V (eds): *The Slow Inward Current and Cardiac Arrhythmias*. The Hague, Martinus Nijhoff, 1980, pp 437–454.
36. Murdock DK, Euler DE, Becker DM, Murdock JD, Scanlon PJ, et al.: Ventricular fibrillation during coronary angiography: An analysis of mechanisms. *Am Heart J* 1985;109:265–273.
37. Reiter MJ, Synhorst DP, Mann DE: Electrophysiological effects of acute ventricular dilatation in the isolated rabbit heart. *Circ Res* 1988;62:554–562.
38. Kubota I, Lux RL, Burgess MJ, Abildskov JA: Activation sequence at the onset of arrhythmias induced by localized myocardial warming and programmed premature stimulation in dogs. *J Electrocardiol* 1988;21:345–354.
39. Watson RM, Schwartz JL, Maron JM, Tucker E, Rosing DR, Josephson ME: Inducible polymorphic ventricular tachycardia and ventricular fibrillation in a subgroup of patients with hypertrophic cardiomyopathy at high risk for sudden death. *J Am Coll Cardiol* 1987;10:761–774.
40. Surawicz B, Knoebel SB: Long QT: Good, bad or indifferent. *J Am Coll Cardiol* 1984;4:398–413.
41. Roden DM, Woosley RL, Primm RK: Incidence and clinical features of the quinidine-associated long QT syndrome: Implications for patient care. *Am Heart J* 1986;111:1088–1093.
42. Bauman JL, Bauernfeind RA, Hoff JV, et al.: Torsade de pointes due to quinidine: Observations in 31 patients. *Am Heart J* 1984;107:425–430.
43. Stratmann HG, Walter KE, Kennedy HL: Torsade de pointes associated with elevated N-acetylprocainamide levels. *Am Heart J* 1985;109:375–377.
44. Brugada J, Sassine A, Escande D, Masse C, Puech P: Effects of quinidine on ventricular repolarization. *Eur Heart J* 1987;8:1340–1345.
45. Schechter JA, Caine R, Friehling T, Kowey PR, Engel TR: Effect of procainamide on dispersion of ventricular refractoriness. *Am J Cardiol* 1983;52:279–282.
46. Solti F, Szatmary L, Renyi-Vamos F Jr, Szabo Z: Pacemaker therapy for the treatment of the long QT syndrome associated with long-lasting bradycardia and ventricular tachycardia. Electrophysiological characteristic and therapy. *Cor Vasa* 1987;29:428–435.
47. Inoue H, Toda I, Nozaki A, Matsuo H, Sugimoto T: Effects of bretylium tosylate on inhomogeneity of refractoriness and ventricular fibrillation threshold in canine hearts with quinidine-induced long QT interval. *Cardiovasc Res* 1985;10:655–660.
48. Elharrar V, Surawicz B: Cycle length effect on restitution of action potential duration in dog cardiac fibers. *Am J Physiol* 1983;244:H782–H792.
49. Kupersmith J, Antman EM, Hoffman BF: In vivo electrophysiological effects of lidocaine in canine acute myocardial infarction. *Circ Res* 1975;36:84–91.
50. Kus T, Sasyniuk BI: Electrophysiological actions of disopyramide phosphate on canine ventricular muscle and Purkinje fibers. *Circ Res* 1975;37:844–854.
51. Myerburg RJ, Bassett AL, Epstein K, Gaide MS, Kozlovskis P, Wong SS, Castellanos A, Gelband M: Electrophysiological effects of procainamide in acute and healed experimental ischemic injury of cat myocardium. *Circ Res* 1982;50:386–393.
52. Varro A, Elharrar V, Surawicz B: Effect of antiarrhythmic drugs on the premature action potential duration in canine cardiac Purkinje fibers. *J Pharmacol Exp Ther* 1985;233:304–311.
53. Varro A, Saitoh H, Surawicz B: Effects of antiarrhythmic drugs on premature action potential duration in canine ventricular muscle fibers. *J Cardiovasc Pharmacol* 1987;10:407–414.

Chapter **10**

Ventricular Fibrillation: Update

Eric E. Johnson, MD, and Raymond E. Ideker, MD, PhD

Ventricular fibrillation, frequently preceded by ventricular tachycardia, is the most common cause of sudden cardiac death[1] and remains a major cause of mortality in the United States.[2] The best way to devise effective therapy for the prevention of ventricular fibrillation and sudden cardiac death is to understand the basic mechanisms by which it occurs. The purpose of this paper is to provide an update on ventricular fibrillation by reviewing pertinent experimental studies from over the last 5 to 10 years. For a complete past history and further background information on ventricular fibrillation, the reader is referred to three previous reviews: two by Surawicz[3,4] and one by Gettes.[5]

To limit the scope of this chapter, an emphasis has been placed on the mechanisms of ventricular fibrillation as delineated by recent advances in cardiac mapping. With simultaneous multichannel cardiac mapping, the basic mechanisms responsible for the initiation and maintenance of ventricular fibrillation are being better defined and have been shown to include both reentrant and nonreentrant mechanisms. Yet even with the improvements in mapping techniques, there are still many unanswered basic questions regarding the exact mechanisms and location of critical electrophysiologic events responsible for the initiation and propagation of ventricular fibrillation. Once these have been answered it then will be possible to characterize more precisely the biochemical, morphologic, and neurologic variables responsible for this electrical abnormality. Thus, cardiac mapping is but one of many tools for studying ventricular fibrillation, yet it is a very important tool because it is able to pinpoint the site and mechanism of origin of ventricular fibrillation so that other techniques, for example, biochemical analysis, identification of receptor levels, and determination of refractory periods, can be focused on this site.

COMPUTER-ASSISTED MAPPING OF VENTRICULAR FIBRILLATION

Whether ventricular fibrillation is initiated by a reentrant or a nonreentrant mechanism, it is thought to be maintained by multiple, disorganized, wandering wavelets that follow constantly changing reentrant pathways. The idea that the number of wavelets during the first minute is small is supported by several pieces of evidence. Wiggers, on the basis of cinematographic studies, stated that ventricular fibrillation cannot be described adequately as asynchronous contraction of individual myocardial fibers, but rather that incoordination and asynchronism first involve comparatively large sections of myocardium.[6] These large sections progressively decrease in size and increase in number over the first 5 min of ventricular fibrillation. By dissection of the heart into pieces of various sizes, Garrey showed that a critical mass of myocardium must be present for ventricular fibrillation to persist.[7] In dogs, the critical mass is about one-fourth of the total ventricular mass. Ventricular fibrillation is sustained only in larger hearts, such as in humans and dogs, and is frequently difficult to induce in smaller hearts, such as in frogs.[6] Finally, frequency analysis of humans and dogs demonstrates a power spectrum with a well-defined peak and its higher harmonics suggest some degree of organized activation during ventricular fibrillation.[8] This evidence suggests that ventricular fibrillation is sufficiently organized to be mapped with closely spaced electrodes if the electrodes do not cause excessive damage to the tissue. This has been recognized for decades.[9] Until recently, the technical problems of recording simultaneously from many electrodes made such studies extremely difficult. However, with advances in digital hardware,[10] electrode construction,[11] and identification of activation during ventricular fibrillation,[12] it is now possible

655 Avenue of the Americas, New York, NY 10010
Current Topics in Cardiology

to map the activation sequences during the initiation and maintenance of ventricular fibrillation as well as during the transition from ventricular tachycardia to ventricular fibrillation from hundreds of electrodes placed directly on and in the ventricles and recorded simultaneously. Several groups of investigators have begun to use such techniques to map the sequences of activation during fibrillation, which will help aid in the determination of the electrophysiologic mechanisms responsible for ventricular fibrillation.

Moe's Multiple Wavelet Hypothesis

Moe, based on the previous work of Garrey, Mines, and Lewis, revived the theory of reentry as a mechanism of fibrillation. Through a series of astute observations, he developed the well-known multiple wandering wavelet hypothesis of fibrillation.(13) At that time, technology had not yet advanced to allow multiple electrode recordings of the heart to be made; thus, Moe and his colleagues developed a mathematical computer model in which the heart was represented as an electrophysiologically inhomogenous two-dimensional sheet of hexagonal cells.(13) Using this model, he demonstrated multiple wandering wavelets that exhibited a self-sustained turbulent activity resembling fibrillation. The computer model studies strengthened the probability of the multiple wavelet hypothesis as a mechanism for fibrillation in the heart.

Sano and Scher, in 1964, were among the first to use cardiac mapping techniques to study fibrillation in vivo.(14) They placed 36 bipolar electrodes in the dog atria in situ. Atrial fibrillation was induced by giving single electrical shocks and its onset and recovery were recorded. Usually about 20 leads were successfully recorded. They proposed that the initiating mechanism was ectopic impulse formation but that reentry was probably the mechanism responsible for its maintenance.

Allessie et al., in a series of in vivo experiments using isolated Langendorff-perfused canine hearts, confirmed Moe's multiple wavelet theory as a basis of atrial fibrillation.(15) Using two solid egg-shaped multiple electrodes, each composed of 480 electrodes, they were able to reconstruct excitation of the atria during stable atrial fibrillation. One multiple electrode was placed in the right atrium and the other in the left atrium. During one episode of atrial fibrillation, the right atrium would be mapped, whereas during a second episode of fibrillation, initiated a couple of minutes later, the activation of the left atrium would be mapped. They demonstrated multiple wandering wavelets as the basis for the continuity of impulse conduction during fibrillation. These wavelets were due to intraatrial reentry of the leading circle type.(16) Allessie et al. interpreted activation patterns in one atrium that appeared to be "foci" of new impulses as in reality sites of endocardial breakthrough of impulses coming from the other atrium. However, they did note that it was possible that the presence of a normal or abnormal pacemaker may sometimes succeed in supporting the continuation of multiple wandering wavelets and may be necessary for the perpetuation of atrial fibrillation. In the chance occurrence of simultaneous extinction of all activation fronts, the generation of an impulse either by the sinus node or some abnormal pacemaker shortly after cancellation of the fibrillatory wavelets almost certainly will restart the arrhythmia. Thus, it is possible that both explanations for fibrillation, that is, multiple rapidly firing ectopic foci and multiple reentry, may act together in maintaining chronic atrial fibrillation.(15)

Mechanisms of Ischemia-Induced Ventricular Fibrillation

Janse et al., studying the flow of injury current and patterns of excitation during early ventricular arrhythmias in acute regional myocardial ischemia, recorded 60 extracellular electrograms simultaneously from epicardial and intramural sites of the left ventricle of isolated porcine and canine hearts during the first 15 min after occlusion and subsequent reperfusion of the left anterior descending artery.(17) During ventricular fibrillation, fragmentation of wavefronts occurred, and multiple wandering wavelets following tortuous paths were seen on the epicardial surface. Circus movements were seldom completed; when they were, their diameter was small (0.5 cm). These data suggested that two mechanisms were responsible for the arrhythmias in early ischemia: one, a "focal" mechanism located at the normal side of the ischemic border, possibly induced by injury currents, and, two, reentry in ischemic myocardium. However, since mapping was limited to only a portion of the ventricles, a definitive assessment of the underlying mechanism could not be made.

Ideker et al. demonstrated a period of organized epicardial activation during the transition to ventricular fibrillation induced by reperfusion following acute ischemia, recording from 27 epicardial electrodes spaced over both ventricles of the canine heart after a 15-min occlusion of the proximal circumflex artery and subsequent reperfusion.(18) Analysis of the initial 1.5

to 2.5 s of the transition from sinus rhythm or ventricular tachycardia to fibrillation revealed that ventricular activation occurred in an orderly, rapidly repeating sequence in all hearts. Each cycle of activation broke through to the epicardium near the border of the ischemic-reperfused region and passed across the nonischemic portion of the ventricles to the opposite side of the heart as a single, organized wavefront. The rate increased during the onset of fibrillation, as indicated by the decreasing time between successive activations. Concurrently, the duration of each cycle increased because of slowing conduction, as shown by the decreasing distance between isochrones. As the conduction velocity decreased and the rate increased, the interval between the end of one cycle of activation and the start of the next cycle decreased. Soon, the duration of each cycle grew longer than the interval between successive epicardial breakthroughs for consecutive cycles. Thus, before one activation front had terminated over the right ventricle, the next activation front had broken through to the epicardium in the ischemic-reperfused region. The overlap increased progressively during subsequent cycles, so that in some cases, three successive activation fronts were present simultaneously on the epicardium.[19] This demonstrates that although the body surface electrocardiogram appears chaotic, there is actually organization and periodicity in the underlying cardiac activation sequence. Due to the limited number of electrodes, the wide spacing between the epicardial electrodes, and the lack of intramyocardial and subendocardial electrodes, the actual mechanism by which activation was initiated during the transition to fibrillation could not be determined.

Figure-Eight Reentry

El-Sherif et al. demonstrated circus movement reentry in the surviving electrophysiologically abnormal epicardial layer overlying canine infarction. In two studies, they looked at isochronal maps of ventricular activation during spontaneous ventricular rhythms and pacing-induced ventricular tachyarrhythmias in dogs 1 and 4 days, respectively, after myocardial infarction.[20,21] Sixty-four electrograms from the entire epicardial surface and from selected endocardial and intramural sites were recorded. Spontaneous ventricular rhythms had a focal origin from the surviving subendocardial Purkinje network underlying the infarction and showed frequent shift of the pacemaker site. Fast ventricular tachyarrhythmias were consistently induced by bursts of rapid ventricular pacing or programmed premature stimulation and had a tendency to degenerate into ventricular fibrillation. Pacing-induced rhythms were due to reentrant activation that developed mainly in the surviving, electrophysiologically abnormal, epicardial layer overlying the infarction. The last stimulated beat that initiated reentry resulted in a continuous arc of functional conduction block and two slowly circulating activation fronts around both ends of the arc of block. The activation fronts rejoined on the distal side of the arc of block before breaking through the arc of block to form two separate arcs of block. Reentrant activation subsequently continued as two synchronous circuits that conducted in clockwise and counterclockwise directions, respectively, in a so-called figure-eight reentry pattern (Fig. 10.1).[21,22]

Three-Dimensional Mapping of Ischemic Induced Ventricular Arrhythmias

The electrophysiologic mechanisms responsible for malignant ventricular arrhythmias associated with reperfusion of ischemic myocardium have been further delineated using a computerized three-dimensional mapping system, with simultaneous eight-level transmural recordings from 232 bipolar sites throughout the entire feline heart in a series of experiments by Pogwizd and Corr.[23,24] In six cats, regional ischemia was induced for 10 min by occlusion of the left anterior descending coronary artery, followed by reperfusion.[23] After 10 min of ischemia, just before reperfusion, total ventricular activation time during sinus rhythm was significantly delayed. Ventricular tachycardia occurred within 15 s after reperfusion and in three animals culminated in ventricular fibrillation.

In 75% of cases of nonsustained ventricular tachycardia, initial activation occurred in the subendocardium, at the border of the reperfused zone, via a mechanism not involving reentry, as determined by the fact that continuous activation was not apparent and the time from the end of the sinus beat to the beginning of ventricular tachycardia was not associated with any intervening depolarizations. In the remaining 25% of cases of nonsustained ventricular tachycardia, initiation of the ventricular tachycardia resulted from intramural reentry in the subendocardium adjacent to the site of delayed midmyocardial activation from the preceding sinus beat. Unlike tachycardias initiated by a nonreentrant mechanism, reentrant tachycardias demonstrated marked delay in the activation of the preceding sinus beat that was comparable to that during early ischemia without reperfusion, resulting in

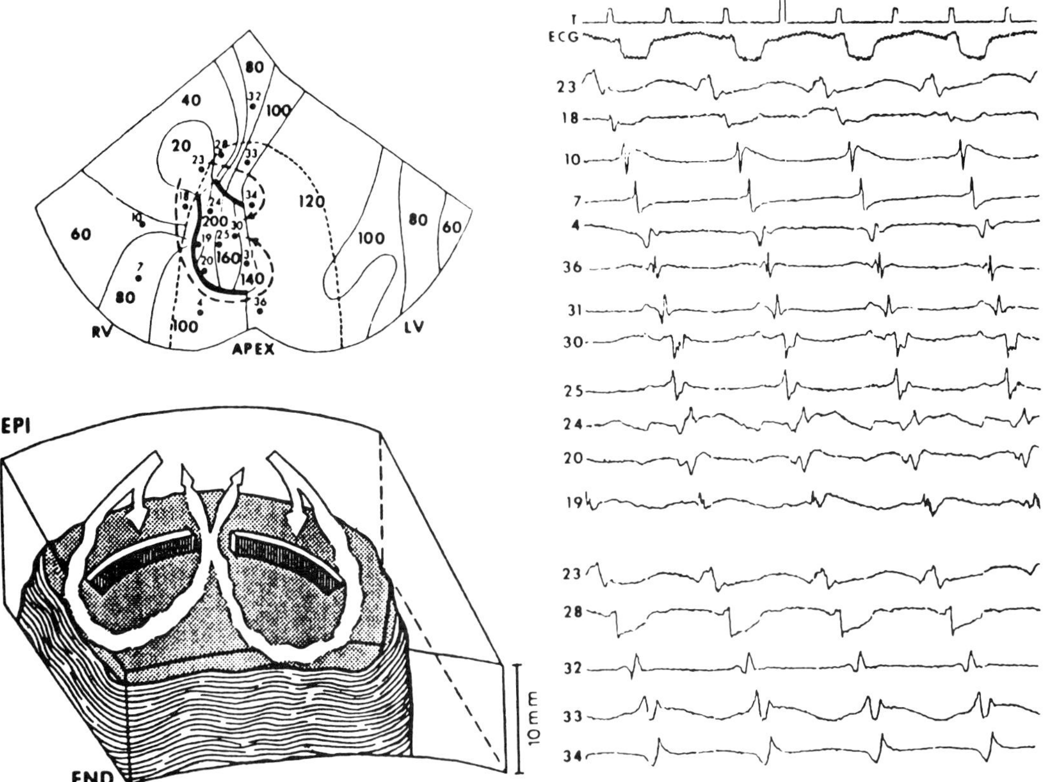

FIGURE 10.1 Figure-eight model of reentry. Isochronal activation map during monomorphic reentrant ventricular tachycardia. Recordings were obtained from a dog 4 days after ligation of the left anterior descending coronary artery. Activation isochrones are drawn at 20-ms intervals. The reentrant circuit has a characteristic figure-eight activation pattern whereby two circulating wave fronts advance in clockwise and counterclockwise directions, respectively, around two zones (arcs) of conduction block (represented by heavy solid lines). The epicardial surface is depicted as if the ventricles were folded out after a cut was made from the crux to the apex. The right panel shows selected simultaneous electrograms recorded along the two arcs of functional conduction block and the common reentrant wave front and depicts the presence of diastolic bridging between reentrant beats. A three-dimensional diagrammatic illustration of the ventricular activation pattern during the reentrant tachycardia is shown at the lower left corner of the figure. In this experimental model, reentrant activation occurs in the surviving thin epicardial layer overlying the infarction. RV = right ventricle, LV = left ventricle, EPI = epicardium, END = endocardium, T = time lines at 100-ms intervals. *Reprinted from El-Sherif,*[22] *with permission.*

continuous activation to initiate the first beat of the tachycardia in the subendocardium.

Maintenance of ventricular tachycardia during reperfusion occurred by nonreentrant mechanisms as well as by intramural reentry, with most cases of ventricular tachycardia involving both mechanisms. Ventricular tachycardia leading to ventricular fibrillation was initiated in the subendocardium at the border of the reperfused zone by a nonreentrant mechanism and was maintained by both nonreentrant and reentrant mechanisms, at times in combination in the same beat. During the transition from ventricular tachycardia to ventricular fibrillation, nonreentrant mechanisms arising both in the subendocardium and in the subepicardium led to very rapid acceleration of the tachycardia, resulting in enhanced functional block and further conduction delay, with the total activation time of the transition beats exceeding the coupling interval (CI) of the tachycardia to give overlapping cycles of activation as discussed previously. This subsequently led to the development of multiple small reentrant circuits and multiple simultaneous wavefronts characteristic of ventricular fibrillation. Thus, ventricular tachycardia during re-

perfusion of ischemic myocardium is most commonly initiated by a nonreentrant mechanism, although intramural reentry can contribute. Ventricular fibrillation occurs through a rapid nonreentrant acceleration of the tachycardia. The nature of this nonreentrant excitation remains to be elucidated but may involve an abnormal form of automaticity or triggered activity.

Pogwizd and Corr also studied the mechanisms underlying the development of ventricular fibrillation during early myocardial ischemia without reperfusion.[24] Using the same mapping system as noted above, occlusion of the proximal left anterior descending coronary artery led to ventricular tachycardia which degenerated into ventricular fibrillation in 1 to 5 min in four of fifteen animals. Initiation of ventricular tachycardia leading to ventricular fibrillation occurred by intramural reentry in three of four animals. In one case, the mechanism responsible for the initiation of ventricular tachycardia could not be assigned. Maintenance of the ventricular tachycardia that led to ventricular fibrillation was due primarily to intramural reentry involving multiple activation sites in and around the border region of the ischemic zone. Nonreentrant mechanisms, arising in the subendocardium and subepicardium, also contributed to the maintenance of ventricular tachycardia. The transition from ventricular tachycardia to ventricular fibrillation was due exclusively to intramural reentry with initiation of the reentrant beats in the subendocardium and, occasionally, the subepicardium. Acceleration of the tachycardia by intramural reentry, along with very rapid and inhomogenous recovery of excitability, led to increased functional block and conduction delay. As a result, the total activation time for a given beat exceeded the CI for that beat and led to multiple reentrant circuits and multiple simultaneous activations characteristic of ventricular fibrillation. Thus, the initiation and maintenance of ventricular tachycardia leading to ventricular fibrillation during early ischemia is due to intramural reentry, although nonreentrant mechanisms also contribute. However, the development of ventricular fibrillation is due to continued intramural reentry and rapid recovery of excitability.

Note that the mechanism underlying the transition from ventricular tachycardia to ventricular fibrillation during early ischemia differs from that during subsequent reperfusion. The transition to ventricular fibrillation after reperfusion is due to acceleration by nonreentrant mechanisms arising in the subepicardium; in contrast, acceleration of the tachycardia by a nonreentrant mechanism was not observed during the transition to ventricular fibrillation during ischemia without reperfusion. However, once acceleration of the tachycardia occurred, the development of ventricular fibrillation during ischemia and reperfusion was similar.

The Stimulus-Induced Critical Point

The reentrant mechanism for the initiation of ventricular fibrillation by electrical stimulation during the vulnerable period of the cardiac cycle has been thought to be due to nonuniform dispersion of refractoriness. Refractoriness or recovery is dispersed nonuniformly when it changes at different rates in different places and changes rapidly over a small distance so that one site remains refractory long after an adjacent site has recovered. An electrical stimulus delivered prematurely during the vulnerable period is thought to give rise to an activation front that propagates away from the site of stimulation in all directions and then blocks unidirectionally when it reaches regions that have not yet recovered excitability. It continues to propagate through adjacent regions that are more recovered, however, and later circles back to excite the blocked regions that have since had time to recover. These later activation fronts may propagate into the tissue that was excited soon after the premature stimulus and that has had time to recover, causing reentry and ventricular fibrillation (Fig. 10.2).

Although nonuniform dispersion of recovery or refractoriness probably plays a role in the mechanism of the initiation of ventricular fibrillation during ischemia, investigators from Duke University have demonstrated that it is not a requirement for the electrical stimulation of fibrillation during the vulnerable period; rather, refractoriness may be dispersed uniformly, that is, changing approximately the same amount over a given distance throughout the region, and still initiate ventricular fibrillation.[25–27]

In the first of three studies,[25] recordings were made simultaneously from 117 epicardial electrodes in a 30-mm by 30-mm region of the anterior right ventricle in 14 dogs. A train of 10 regularly spaced S1 stimuli was given simultaneously through a row of eight pacing wires on one side of the mapped region, creating uniform, parallel activation fronts. Recovery periods were determined individually at 24 to 44 sites after the tenth S1. The refractory periods were found to to be similar at all sites, so that recovery was dispersed uniformly across the mapped region, with earliest recovery near the row of S1 pacing electrodes (Fig. 10.3).

After verification that both activation iso-

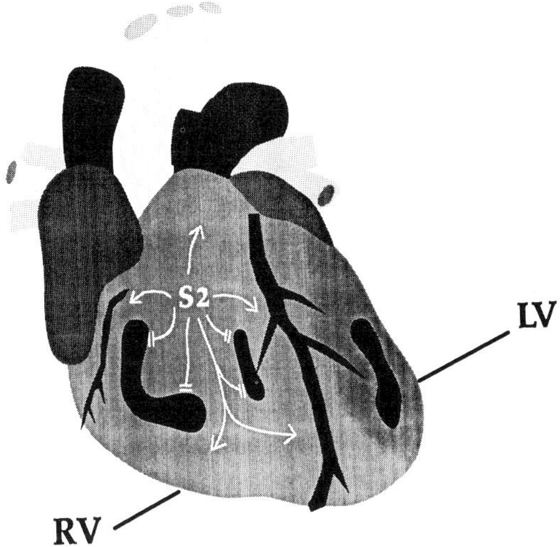

FIGURE 10.2 Nonuniform dispersion of refractoriness and reentry. Refractoriness or recovery is dispersed nonuniformly when it changes at different rates in different places and changes rapidly over a small distance so that one site remains refractory long after an adjacent site has recovered. An electrical stimulus delivered prematurely during the vulnerable period is thought to give rise to an activation front that propagates away from the site of stimulation in all directions and then blocks unidirectionally when it reaches regions that have not yet recovered excitability. It continues to propagate through adjacent regions that are more recovered, however, and later circles back to excite the blocked regions that have since had time to recover. These later activation fronts may propagate into the tissue that was excited soon after the premature stimulus and that has had time to recover, causing reentry and ventricular fibrillation.

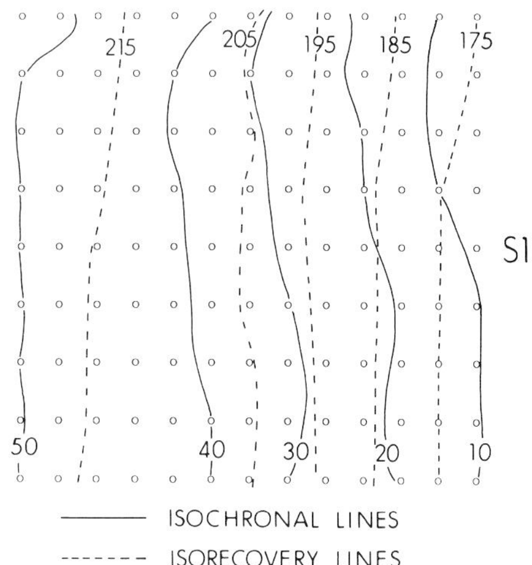

FIGURE 10.3 S1 activation and isorecovery patterns. The row of eight epicardial pacing wires was tied together as a single source (S1) and paced at 10 mA. Approximately parallel isochronal lines (solid lines) are created by S1 pacing, with conduction velocities of 0.5 to 0.7 m/s. The recovery periods (dashed lines) were calculated at 32 electrode sites evenly spaced across the array for this example. The refractory periods were similar at all electrode sites (166 ± 3 ms), indicating minimal inhomogenous refractoriness. The mean epicardial fiber orientation under the array in this case was 24 ± 5 degrees with respect to the horizontal. *Reprinted from Frazier et al.,*[25] *with permission.*

chrones and isorecovery lines were uniform and parallel, 25- to 250-V monophasic S2 shocks were given prematurely following the tenth S1 to scan the recovery period. The shocks were given through a mesh electrode that spanned one side of the mapped region. This side of the mapped region was adjacent to and at a right angle to the side to which the S1 was delivered, so that the potential gradient electric field of the shock was perpendicular to the isorecovery lines. The potential gradient indicates how rapidly the transmembrane potential changes with distance. It is usually estimated in the heart by recording potentials from electrodes spaced known distances apart and dividing the difference in potential at adjacent electrodes by the distance between the electrodes. The potential at a site must always be measured with respect to some reference potential and will differ for different reference potentials. However, the potential difference between two myocardial electrodes will be the same no matter where the reference electrode is placed. The potential gradient at each point is a vector quantity since the potential at a site can change at different rates in different directions away from the site. Isogradient lines were found to be parallel with highest gradients near the S2 mesh electrode.

Reentrant activation leading to ventricular fibrillation was induced by the S2 shock at some CIs in all animals. Circus reentry in a leading circle activation pattern was created by the shock (Fig. 10.4). Earliest activation after the shock was not in the region adjacent to the mesh S2 electrode where the shock field was the greatest, but instead, the earliest poststimulus propagation occurred distant from the S2 electrode at the edge of the directly excited region on the opposite side of the mapped region. The "critical point" is that point where critical values of the S2 field strength and the refractoriness intersect and about which reentry is created. The critical point of the shock potential gradient field, at which conduction prop-

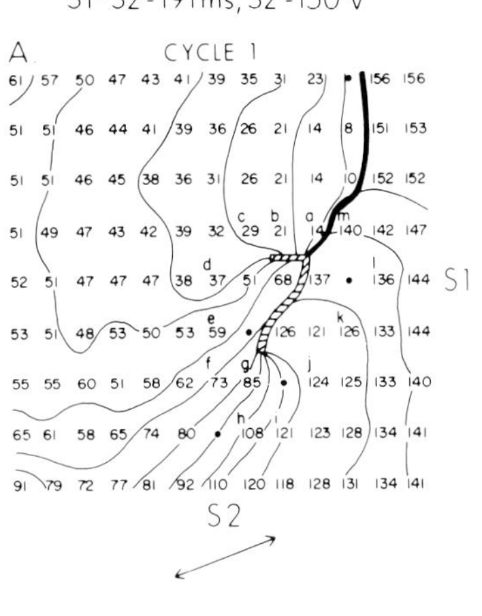

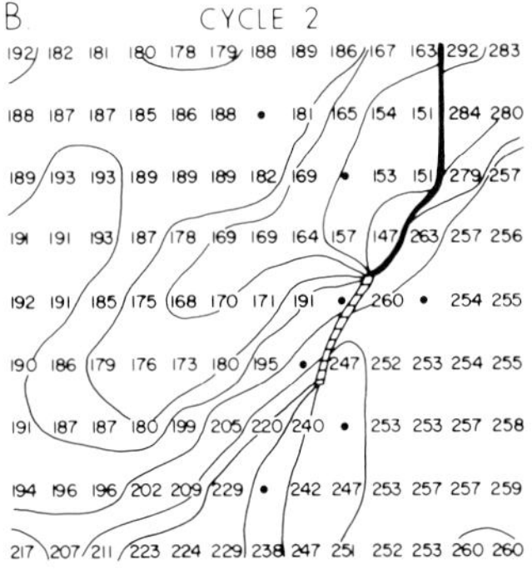

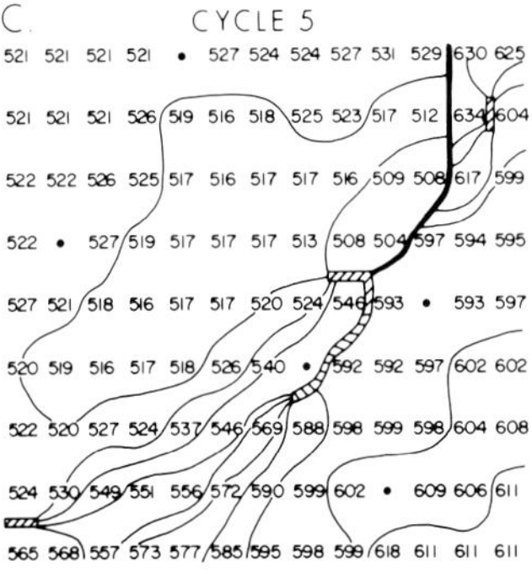

FIGURE 10.4 Activation patterns for perpendicular isorefractory and isogradient lines. The patterns of activation following S1 pacing from the right and S2 shock from the bottom are shown in **A** (cycle 1), **B** (cycle 2), and **C** (cycle 5). Activation times, shown in milliseconds, are measured from the start of the 3-ms S2 shock. Solid dots represent sites of inadequate recordings. The solid line represents the transition between successive activation maps and is called a "frame line." It is required because each static isochronal map can only show a single cycle of a continuous dynamic reentrant circuit. The hatched line represents a zone of conduction block and is called a "block line." Isochrones are at 10-ms intervals. The double-headed arrow for this figure represents the mean epicardial fiber orientation in the area of conduction block, in this case 21° with respect to the horizontal. **A.** The initial activation pattern following the S2 shock at an S1–S2 interval of 191 ms and an S2 strength of 150 V. The hatched area indicates the region assumed to be directly excited by the S2 shock field. Earliest postshock activation occurs distant from the S2 site, with no early activation wave fronts conducting away from the directly excited region located between the S2 site and the critical point, that is, the point where the activation front blindly ends at the junction between the frame line and the block line. A counterclockwise reentrant circuit is formed around the region containing the critical point and the line of block. The potential gradient equals 5.8 V/cm and the preshock interval equals 171 ms at the critical point (critical refractory period = 169 ms). **B.** The second cycle of the reentrant pattern. No slow conduction occurs at the frame line between the first and second reentrant cycles. The activation pattern is similar to the first cycle except that the circuit time is decreased to 122 ms. **C.** The fifth cycle of the reentrant pattern. The reentrant pattern is similar to the previous cycles; however, the total circuit time has further decreased to 101 ms. *Reprinted from Frazier et al.,*[25] *with permission.*

agated away from the directly excited border where potential gradients were weaker than this point and did not conduct away from the border where gradients were stronger than this point, averaged approximately 5 V/cm. Activation circled around this point to form a circus reentrant pattern.

Reentry was clockwise when the row of S1 electrodes was at the right and the S2 mesh electrode was at the top of the mapped region; it was also clockwise but spatially shifted 180° in phase when the S1 electrodes were at the left and the S2 mesh electrode was at the bottom of the mapped region. Reentry was counterclockwise when the row of S1 electrodes was at the left and the S2 electrode was at the bottom, and was also counterclockwise but spatially shifted 180° in-phase when the S1 electrodes were at the left and the S2 electrode was at the top of the mapped region (Fig. 10.5). The location of the rotor could be moved predictably by changing either the strength or the timing of the S2.

Thus, reentry was not solely a function of intrinsic differences in refractoriness throughout

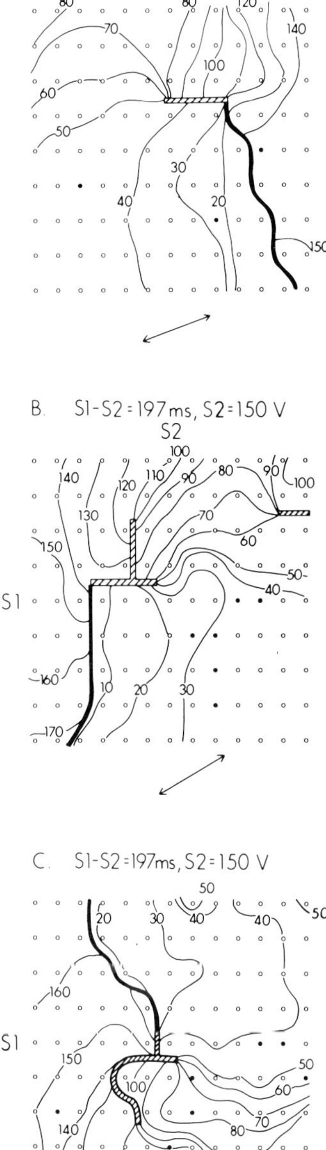

FIGURE 10.5 Effect of locations of S1 and S2 on direction of reentrant circuits. **A.** The first cycle of reentry following a 150-V S2 at an S1–S2 interval of 190 ms is shown with S2 at the top and S1 toward the septum. Earliest activation occurs distant from the S2 site, with activation wave fronts then conducting around a line of block with slow conduction near the S2 site. A clockwise reentrant circuit is formed, versus the counterclockwise circuit with S2 at the bottom (Fig. 10.4A). In addition, the reentrant circuits in Figures 10.4A and 10.5A differ in phase by approximately 180°, as earliest activation occurs at the top of the array for S2 at the bottom (Fig. 10.4A) and at the bottom of the array for S2 at the top (Fig. 10.5A). The potential gradient equals 5.4 V/cm and the preshock interval equals 172 ms at the critical point. In **B** and **C,** earliest activation occurs distant from the S2 site, conducts around a line of block with a region of slow conduction near the S2 site, and forms a reentrant circuit, the same type of pattern seen in Figure 10.4A. In **B** (S2 of 150 V, S1–S2 of 197 ms) with S2 at the top and S1 at the RV side, a counterclockwise reentrant circuit is formed, the same as for Figure 10.4A (septal S1 with bottom S2). The two reentrant patterns, however, differ in phase by approximately 180°. The potential gradient equals 5.2 V/cm and the preshock interval equals 173 ms (critical refractory period = 171 ms) at the critical point. In **C** (S2 of 150 V, S1–S2 of 197 ms), with S2 at the bottom and S1 at the RV side, a clockwise reentrant circuit is formed, the same as for Figure 10.5A (septal S1 with top S2). These two reentrant patterns, however, also differ in phase by approximately 180°. The potential gradient equals 5.9 V/cm and the preshock interval equals 169 ms (critical refractory period = 170 ms) at the critical point. *Reprinted from Frazier et al.,*[25] *with permission.*

the mapped region, but rather the initiation and location of reentry was a function of the interaction of refractoriness with the strength of the shock field. Activation did not conduct away from the S2 electrode in all directions and later block in a region where there was a large change in refractoriness, as predicted by the nonuniform dispersion of recovery hypothesis. Instead, conduction block leading to reentry and ventricular fibrillation was caused by the interaction of a uniformly dispersed change in refractoriness, forming a critical point around which activation fronts rotated. Therefore, nonuniform dispersion of refractoriness was not the mechanism for the electrical induction of ventricular fibrillation in this study. This does not rule out the likelihood that nonuniform dispersion of refractoriness may be the mechanism for naturally occurring ventric-

ular fibrillation, such as that caused by acute ischemia, or for ventricular fibrillation induced by low-voltage stimuli, that is, twice diastolic threshold, and for S1 and S2 given from the same site.

In a second study,[26] S1 and S2 were delivered from point sources 1 cm apart in the right ventricular outflow tract in seven dogs. An array of 40 plunge needles was used for transmural recording. An S1 was given just to the left of the needle array, while S2 was given from the center. Earliest activation following S2 was not at the site of S2 stimulation but was at a site between the S1 and S2 stimulation sites. Activation fronts spread toward the S1 site where the tissue was more recovered and also circled around both sides of an arc of block near the S2 site to form a figure-eight reentrant pattern. The activation fronts coalesced to excite the region around the S2 site last, and if the difference in times between the early site and the S2 site was large, reentered the tissue toward the S1 site (Fig. 10.6). Two critical points are present, one above and the other below the S2 site, where a critical level of potential gradient induces a pair of mirror-image rotors (figure-eight reentry).

The critical point mechanism predicts that as S2 strength is increased so that the critical level of potential gradient is created at a greater distance from the S2 electrode, the two critical points should move away from the S2 site and thus farther apart. This prediction was tested in a third study,[27] in which S1 was delivered to the base of either the right or left ventricle, while S2 was delivered through defibrillation electrodes on the left ventricular apex and right atrium in seven dogs. Two mirror-image rotors of reentry were created on opposite sides of the ventricles, presumably where two critical points were created by the large premature S2 shock.

Thus, in all three studies, a region directly excited by the S2 field was present. The border of this directly excited region was dependent upon both the strength of the S2 field as well as the stage of refractoriness of the cells. Where the S2 potential gradient was weak (under approximately 5 V/cm for 3 ms square monophasic waveforms), the border of the directly excited region was in cells that were only mildly refractory and, after the S2 stimulus, an activation front conducted away from this border through the tissue that was not directly excited by the S2 field. Where the S2 potential gradient was higher, myocardium that was more refractory was directly excited, and activation fronts did not conduct away from the border of the directly excited region, creating a zone of temporary unidirectional block. The cause for this block may have been

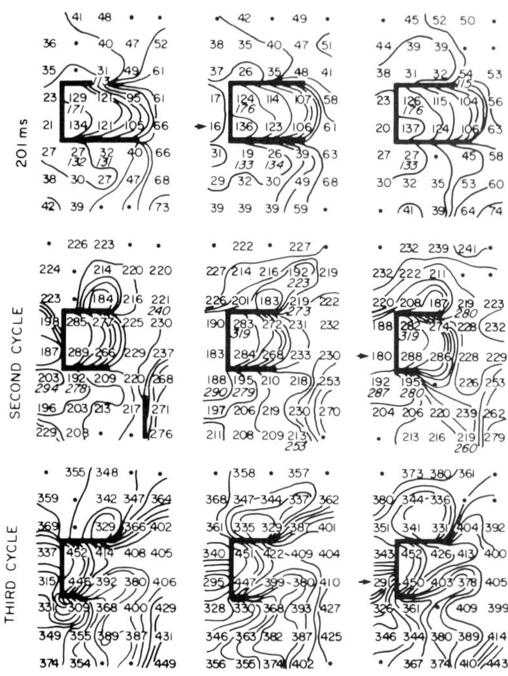

FIGURE 10.6 Figure-eight reentry at the onset of ventricular fibrillation. Activation times for the first three cycles of ventricular fibrillation are shown timed from a 5-ms S2 of 50 mA given prematurely with a CI of 201 ms. The three maps from left to right in each panel represent the subendocardial, midmyocardial, and subepicardial layers of recording electrodes. Each number gives the activation time at an electrode site in milliseconds after the S2 stimulus, which was given to the endocardium from a point electrode at the center of the recording array. S1 stimuli were given to the endocardium from a point electrode at the left side of the array. Small closed circles represent bad recording electrodes. The heavy black bar in the center of the mapped region represents the frame line. The small arrows indicate the earliest recorded site of activation for each cycle. *Reprinted from Chen et al.,[25] with permission.*

prolongation of refractoriness by the S2 stimulus electric field.[28] This situation, in which, after the S2, activation spread away from one portion of the directly excited region but not another, created an activation front that ends blindly in the tissue, and led to reentry and ventricular fibrillation. The type of reentry (leading circle or figure-eight) depended upon the spatial relationship of the S2 isogradient field to the distribution of refractoriness, and hence depended upon the

location of the S1 and S2 sites. These types of reentrant patterns were predictable based on the different shape and intersections of the isogradient and isorefractory lines.[29,30]

Comparison of Activation Patterns during Ventricular Tachycardia and Ventricular Fibrillation

Although ventricular fibrillation is thus thought to arise from reentry caused by functional block, until recently ventricular tachycardia was thought to be caused by reentry around anatomic blocks of areas of patchy necrosis and scar tissue from myocardial ischemia and infarction interspersed with small bundles of viable myocardium.[31] Slow conduction and variable degrees of conduction block, conditions necessary for reentry to occur, have been demonstrated in ischemic myocardium.[32–36] Although these factors may be responsible for ventricular tachycardia in some cases, recent experimental evidence has demonstrated that they are not necessities for ventricular tachycardia to occur. Spach and coworkers have shown that reentry can occur because of the anisotropy in the safety factor for conduction along and across fibers.[37] Results from several groups suggest that functional block may also be responsible for ventricular tachycardia. Waldo and coworkers[38] have shown in patients that regions of slow conduction during reentry do not exhibit slow conduction during normal sinus rhythm.

Ventricular tachycardia is normally induced by electrical stimulation in which a small S2 (< 2 mA) is given outside the infarct so that a premature activation front conducts into the infarct and blocks, leading to reentry. Ventricular fibrillation is normally induced by giving a large S2 (>20 mA) during the vulnerable period. As noted previously, El-Sherif's group[21] and Wit's group,[39] using programmed electrical stimulation from electrodes outside a 4-day-old, nonreperfused infarct, were able to maintain ventricular tachycardia in a figure-eight reentry pattern in a thin sheet of viable spared myocardium in the epicardium over the infarct. As seen in Chen's study noted above,[26] for a large S2 stimulus given in normal myocardium during the vulnerable period with S1 given at another site, ventricular fibrillation is initiated by figure-eight reentry near the S2 site.

To test the hypothesis that a large S2 given during the vulnerable period over a 4-day-old infarct may induce ventricular tachycardia and not ventricular fibrillation, 10 dogs underwent 30 min of partial and 90 min of complete left anterior descending coronary artery occlusion followed by reperfusion.[40] Four days later, 10 S1 stimuli were given to right or left ventricular sites while the vulnerable period was scanned with an S2 stimulus given to the center of the region of epicardial sparing over the infarct. The S2 strength was increased until ventricular tachycardia or ventricular fibrillation was induced. Sustained ventricular tachycardia in a figure-eight reentry pattern was induced repeatedly from 23 S1 sites with a mean S2 of 39 mA, with ventricular fibrillation being obtained from 10 S1 sites with a mean S2 of 56 mA. Increasing S2 strength above the lowest strength that initiated ventricular tachycardia up to 100 mA yielded only ventricular tachycardia, not ventricular fibrillation. The mean difference in the arrhythmia cycle lengths for the initial six cycles was 153 ± 33 ms for ventricular tachycardia and 110 ± 8 ms for ventricular fibrillation ($p < 0.001$). Histologically, the mean transmural extent of infarction was 80% in five dogs with ventricular tachycardia from all S1 sites, 60% in three dogs with ventricular tachycardia from some and ventricular fibrillation from other S1 sites, and 20% from two dogs with ventricular fibrillation from all S1 sites. After 20 to 60 s of sustained ventricular tachycardia in some animals, the ventricular tachycardia degenerated into ventricular fibrillation. Thus, the mechanism of ventricular tachycardia and ventricular fibrillation can be the same, that is, figure-eight reentry, and can be induced in the same way, that is, by giving a large S2 stimulus during the vulnerable period; while the infarct lowers the ventricular fibrillation threshold for large S2 stimuli given outside the infarcted region, the infarct has a protective effect for large S2 stimuli given over the center of the infarct during the vulnerable period, causing slower rotational velocity and increased stability of rotors.

ION CHANNELS IN VENTRICULAR FIBRILLATION

The cellular events responsible for the initiation and maintenance of ventricular fibrillation associated with ischemia remain unclear. Several investigators have recorded transmembrane action potentials (APs) during ventricular fibrillation.[41–43] Akiyama, using a motion-compensated micropippette holder, collected data on the transmembrane AP of in situ ventricular cells during ventricular fibrillation induced in 14 dogs by 30 min of ischemia followed by reperfusion and compared them with those recorded from the same areas during regular sinus rhythm prior to occlusion of the left anterior descending coronary artery.[41] Most of the APs of the ventricular

areas had a low resting potential, low overshoot potential, and shortened AP duration and were similar to those of the slow-channel type recorded from the sinoatrial (SA) or atrioventricular (AV) node, and from the cardiac muscle exposed to high external potassium (K^+) and adrenergic agents or to sodium (Na^+)-free and calcium (Ca^{2+})-rich solutions. Akiyama demonstrated that epicardial application of verapamil suppressed those APs during ventricular fibrillation, whereas tetrodotoxin (TTX), a fast Na^+ channel blocker, did not affect the APs. The mechanism responsible for converting these cells from fast-channel type to slow-channel type is unknown, but it is postulated that sustained ischemia depolarizes the membrane potential (V_m) beyond the threshold level of fast Na^+ channels, and some yet unidentified factors, probably including elevated local levels of catecholamines and extracellular K^+, restore the excitability of ischemic fibers by enhancing inward currents flowing through slow channels. The transmembrane potential recordings obtained 1 min after the onset of ventricular fibrillation in this study appeared similar to those obtained in two comparable studies by Czarnecka and Downar.[42,43] During the first 10 min of ventricular fibrillation, however, in contrast to Downar's study, little inhomogeneity among the subepicardial ventricular cells from the reperfused area was encountered.[41]

It has also been demonstrated that Ca^{2+} channel antagonists decrease vulnerability to ventricular fibrillation during ischemia.[44–48] Clusin et al. suggested that Ca^{2+}-mediated ionic currents participate in ischemic ventricular fibrillation.[49] To study the role of the Ca^{2+} ion and Ca^{2+} channel availability played in the initiation and maintenance of ventricular fibrillation, Merillat and coworkers[50] developed an isolated, nonischemic, Langendorff-perfused rabbit heart preparation in which sustained ventricular fibrillation could be induced either by alternating current (AC), or by Na/K pump inhibition (10 μM ouabain or K^+-free perfusate), and which allowed changes in perfusate composition. AC stimulation or Na/K inhibition always initiated ventricular fibrillation. Ca^{2+} channel blockade by verapamil or nitrendipine uniformly inhibited the initiation of ventricular fibrillation in both models. During Na/K pump inhibition the following was noted: one, ventricular fibrillation was prevented by Ca^{2+} channel blockade, despite evidence of Ca^{2+} overload, and, two, abolition of spontaneous sarcoplasmic reticulum–generated cytosolic Ca^{2+} oscillations by ryanodine or Na^+ channel blockade with TTX did not prevent ventricular fibrillation initiation. Lowering extracellular Ca^{2+} concentration ($[Ca^{2+}]_o$) to 80 μM uniformly prevented the initiation of ventricular fibrillation by Na/K pump inhibition, but not by AC stimulation.

Ventricular fibrillation maintenance was also studied using the following techniques: reduction in perfusate Ca^{2+}, blockade of Ca^{2+} channels, or electrical defibrillation. Decreasing the perfusate Ca^{2+} to 80 μM resulted in defibrillation during ventricular fibrillation whether AC or Na/K pump inhibition induced. Verapamil or nitrendipine also resulted in defibrillation regardless of the initiation method. Electrical defibrillation was successful only in AC-induced ventricular fibrillation. These experiments demonstrate that sustained ventricular fibrillation can be studied in rabbit hearts during nonischemic Langendorff perfusion. The data suggest that increases in slow-channel Ca^{2+} flux, as opposed to increases in cytosolic Ca^{2+} per se or activation of the Na^+ channel, are necessary for the initiation and maintenance of ventricular fibrillation. The data, however, do not exclude an important role for cytosolic Ca^{2+} in the modulation of ventricular fibrillation characteristics, for example, heterogenous variations in intracellular Ca^{2+} concentration ($[Ca^{2+}]_i$) may lead to heterogeneity in depolarization and changes in cell-to-cell coupling progressing to areas of locally propagated wavefronts.[50]

ROLE OF THE AUTONOMIC NERVOUS SYSTEM IN VENTRICULAR FIBRILLATION

There is a multitude of experimental evidence to implicate the autonomic nervous system in the development of cardiac arrhythmias and in the genesis of sudden cardiac death.[51,52] In a study by Kolman et al.,[53] stimulation of the cardiac sympathetic nerves alone increased the vulnerability to ventricular fibrillation. Vagal stimulation alone had no significant effect on the vulnerability to ventricular fibrillation. However, when the same vagal stimulation was given in the presence of tonic sympathetic stimulation, the vagal activity substantially attenuated the increased vulnerability produced by the sympathetic stimulation. There is also evidence that sympathetic–parasympathetic interactions regulate myocardial electrical stability in conscious animals.[54] Transmural infarction causes both sympathetic and parasympathetic imbalance in regions distal to the infarct.[55] This autonomic heterogeneity may contribute to arrhythmogenesis. β-Adrenergic receptor blockade decreases the incidence of sudden cardiac death following

myocardial infarction[56]; this effect may be due in part to lessening this autonomic heterogeneity.[57] Stimulation of the central nervous system in normal pigs causes ventricular fibrillation, and the pathway has been mapped all the way from the frontal cortex, through the brain stem and the sympathetic nerves, to the heart.[55] Several lines of experimental evidence suggest that myocardial ischemia results in the accumulation of arrhythmogenic metabolites.[58] These findings suggest that accumulating metabolites and/or ions may be critical in the evolution of the electrophysiologic changes during ischemia and may provide the substrate for the generation of malignant ventricular arrhythmias during ischemia.[52] Increases in catecholamines may increase these arrhythmogenic metabolites.[59]

FREQUENCY ANALYSIS OF VENTRICULAR FIBRILLATION

Frequency analysis of ventricular fibrillation demonstrates that there is often a clear dominant frequency with a narrow band-width and a peak in the power spectrum around 9 to 12 Hz.[8,60] Carlisle et al. used fast Fourier transform analysis to study ventricular fibrillation induced by several different methods in dogs.[61] In their study, the dominant frequency at the body surface of ventricular fibrillation induced by acute occlusion of the left anterior descending coronary artery was initially 12 Hz, which then fell rapidly to 5 to 6 Hz by 120 s. The recordings from the limb leads were similar for ischemic ventricular fibrillation, reperfusion ventricular fibrillation, and electrically induced ventricular fibrillation in the presence of ischemia. Fibrillation recorded from the endocardium of the heart initially showed a dominant frequency similar to that recorded at the body surface, but there was no significant fall in frequency over 3 min. The precise cause of the fall in frequency in the limb leads after 1 to 2 min of continued fibrillation is uncertain. However, this is consistent with a previous study by Worley et al. in which direct bipolar recordings from the left ventricular endocardium showed that activation rates in the endocardium continued at a rapid rate for many minutes, while activation rates in the myocardium and epicardium decreased markedly with time.[62]

This decline in frequency of fibrillation may be of clinical significance. The success rate in the treatment of ventricular fibrillation has been shown to vary inversely with the downtime, that is, the duration of time between the onset of ventricular fibrillation and the initiation of cardiopulmonary resuscitation and transthoracic defibrillation.[63] Downtime has been assessed from the amplitude of the standard electrogram with the amplitude inversely proportional to the downtime.[64] Dzwoncyzk et al. developed and tested an algorithm in pigs using the median frequency of the power spectrum to track the decrease in frequency with the increase in time of fibrillation.[65] They were able to predict the duration of ventricular fibrillation with an average error of -0.86 min, thus identifying a signal processing tool that may some day be useful in the prehospital treatment of ventricular fibrillation.

CONCLUSION

This paper reviews many of the experimental studies on ventricular fibrillation from over the last 5 to 10 years. With recent advances in cardiac mapping, we have an improved understanding of the basic mechanisms of ventricular fibrillation. Both reentrant and nonreentrant mechanisms have been demonstrated to play a role in the initiation and maintenance of ventricular fibrillation. The mechanism underlying the transition from ventricular tachycardia to ventricular fibrillation during early ischemia differs from that during reperfusion. The transition to ventricular fibrillation after reperfusion is due to acceleration by nonreentrant mechanisms; in contrast, acceleration of the tachycardia by a nonreentrant mechanism was not observed during the transition to ventricular fibrillation during ischemia without reperfusion. Once acceleration of the tachycardia occurred, the development of ventricular fibrillation during ischemia and reperfusion was similar.

The reentrant mechanism for the initiation of ventricular fibrillation by electrical stimulation during the vulnerable period has been thought due to nonuniform dispersion of refractoriness; however, it has been shown that ventricular fibrillation may be initiated electrically in the presence of uniform dispersion of refractoriness. Conduction block leading to reentry and ventricular fibrillation can be caused by the interaction of a uniformly dispersed change in refractoriness with the strength of a shock field which forms a critical point around which activation fronts rotate. This does not rule out the likelihood that nonuniform dispersion of refractoriness may be the mechanism for naturally occurring ventricular fibrillation, such as that caused by acute ischemia or for ventricular fibrillation induced by low-voltage stimuli.

Our knowledge of the cellular events responsible for the initiation and maintenance of ventricular fibrillation has increased with evi-

dence supporting the importance of the slow channel. Data suggest that increases in slow channel Ca^{2+} flux, as opposed to increases in cytosolic Ca^{2+} per se or activation of the Na^{+} channel, are necessary for the initiation and maintenance of ventricular fibrillation.

Although we have learned much about ventricular fibrillation over the last few years, there is still much to learn. Since ventricular fibrillation and sudden cardiac death remain a major cause of mortality, work should continue in this area. We hope, in the future, what is learned in experimental studies will be applied in the clinical setting to improve our ability to identify patients at risk and to prevent occurrences and recurrences of ventricular fibrillation.

ACKNOWLEDGMENTS

This work was supported in part by the National Institutes of Health research grants HL-42760, HL-44066, HL-28429, and HL-33637, and the National Science Foundation Engineering Research Center grant CDR-8622201.

REFERENCES

1. Luna AB, Coumel P, Leclercq JF: Ambulatory sudden cardiac death: Mechanisms of production of fatal arrhythmia on the basis of data from 157 cases. *Am Heart J* 1989;117:151–159.
2. Gillum RF: Sudden coronary death in the United States. *Circulation* 1989;79:756–765.
3. Surawicz B: Ventricular fibrillation. *Am J Cardiol* 1971;28:268–287.
4. Surawicz B: Ventricular fibrillation II: Progress report since 1971. *Clin Prog Pacing Electrophysiol* 1984;2:395–419.
5. Gettes LS: Ventricular fibrillation, in B Surawicz, CP Reddy, EN Prystowsky (eds): *Tachycardias*. Boston, Martinus Nijhoff, 1984, pp 37–54.
6. Wiggers CJ: The mechanism and nature of ventricular defibrillation. *Am Heart J* 1940;20:399.
7. Garrey WE: The nature of fibrillatory contractions of the heart—Its relation to tissue mass and form. *Am J Physiol* 1914;33:397–414.
8. Herbschleb JN, Heethaar RM, Tweel L, Meijler RL: Frequency analysis of the ECG before and during ventricular fibrillation. *Proc Computers Cardiology* 1980;365–368.
9. Moe GK, Harris AS, Wiggers CJ: Analysis of the initiation of fibrillation by electrographic studies. *Am J Physiol* 1941;134:473–492.
10. Smith WM, Wharton JM, Blanchard SM, Wolf PD, Ideker RE: Direct cardiac mapping, in Zipes DP, Jalife J (eds): *Cardiac Electrophysiology, From Cell to Bedside*. Philadelphia, WB Saunders, 1990, pp 849–58.
11. Mastrototaro JJ, Pilkington TC, Ideker RE, Massoud HZ: Thin-film multielectrode arrays for potential gradient measurements in the heart. *Proc 10th Annual Conf of the IEEE Engineering in Medicine and Biology Society*, 1988;10:90.
12. Cabo C, Wharton JM, Simpson EV, Ideker RE, Smith WM: Use of coherence in activation detection during ventricular fibrillation. *Proc 11th Annual Conf of the IEEE Engineering in Medicine and Biology Society*, 1989;11:1733–1734.
13. Moe GK, Rheinboldt WC, Abildskov JA: A computer model of atrial fibrillation. *Am Heart J* 1964;67:200–220.
14. Sano T, Scher AM: Multiple recording during electrically induced atrial fibrillation. Circ Res 1964;14:117–125.
15. Allessie MA, Lammers WJEP, Bonke FIM, Hollen J: Experimental evaluation of Moe's multiple wavelet hypothesis of atrial fibrillation, in Zipes DP, Jalife J (eds): *Cardiac Electrophysiology and Arrhythmias*. Orlando, FL, Grune & Stratton, 1985, pp 265–275.
16. Allessie MA, Bonke FIM, Schopman FJG: Circus movement in rabbit atrial muscle as a mechanism of tachycardia. III. The "leading circle" concept: A new model of circus movement in cardiac tissue without the involvement of an anatomical obstacle. *Circ Res* 1977;41:9–18.
17. Janse MJ, van Capelle FJL, Morsink H, et al.: Flow of "injury" current and patterns of excitation during early ventricular arrhythmias in acute regional myocardial ischemia in isolated porcine and canine hearts: Evidence for two different arrhythmogenic mechanisms. *Circ Res* 1980;47:151–165.
18. Ideker RE, Klein GJ, Harrison L, et al.: The transition to ventricular fibrillation induced by reperfusion following acute ischemia in the dog: A period of organized epicardial activation. *Circulation* 1981;63:1371–1379.
19. Ideker RE, Bardy GH, Worley SJ, German LD, Smith WM: Patterns of activation during ventricular fibrillation, in Josephson ME, Wellens HJJ (eds): *Tachycardias: Mechanisms, Diagnosis, Treatment*. Philadelphia, Lea & Febiger, 1984, pp 519–536.
20. El-Sherif N, Mehar R, Gough WB, Zeiler RH: Ventricular activation patterns of spontaneous and induced ventricular rhythms in canine one-day-old myocardial infarction: Evidence for focal and reentrant mechanisms. *Circ Res* 1982;51:152–166.
21. El-Sherif N, Gough WB, Restivo M: Reentrant ventricular arrhythmias in the late myocardial infarction period: 14. Mechanisms of resetting, entrainment, acceleration, or termination of reentrant tachycardia by programmed electrical stimulation. *PACE* 1987;10:341–371.
22. El-Sherif N: Reentry revisited. *PACE* 1988;11:1358–1368.
23. Pogwizd SM, Corr PB: Electrophysiologic mechanisms underlying arrhythmias due to reperfusion of ischemic myocardium. *Circulation* 1987;76:404–426.
24. Pogwizd SM, Corr PB: Mechanisms underlying the development of ventricular fibrillation during early myocardial ischemia. *Circ Res* 1990;66:672–695.
25. Frazier DW, Wolf PD, Wharton JM, Tang ASL, Smith WM, Ideker RE: Stimulus-induced critical point: Mechanism for the electrical initiation of reentry in normal canine myocardium. *J Clin Invest* 1989;83:1039–1052.
26. Chen P-S, Wolf PD, Dixon EG, et al.: Mechanism of ventricular vulnerability to single premature stimuli in open chest dogs. *Circ Res* 1988;62:1191–1209.
27. Shibata N, Chen P-S, Dixon EG, et al.: Influence of

shock strength and timing on induction of ventricular arrhythmias in dogs. *Am J Physiol* 1988;255:H891–H901.
28. Kao CY, Hoffman BF: Graded and decremental response in heart muscle fibers. *Am J Physiol* 1958;194:187–196.
29. Ideker RE, Frazier DW, Krassowska W, et al.: Experimental evidence for autowaves in the heart, in Jalife J (ed): *Mathematical Approaches to Cardiac Arrhythmias*. New York, New York Academy of Sciences, 1990, pp 208–218.
30. Winfree AT: *When Time Breaks Down: The Three-Dimensional Dynamics of Electrochemical Waves and Cardiac Arrhythmias*. Princeton, NJ, Princeton University Press, 1987.
31. Michelson EL, Spear JF, Moore EN: Electrophysiologic and anatomic correlates of sustained ventricular tachyarrhythmias in a model of chronic myocardial infarction. *Am J Cardiol* 1980;45:583–590.
32. Boineau JP, Cox JL: Slow ventricular activation in acute myocardial infarction: A source of re-entrant premature ventricular contractions. *Circulation* 1973;48:702–713.
33. Williams DO, Scherlag BJ, Hope RR, El-Sherif N, Lazzara R: The pathophysiology of malignant ventricular arrhythmias during acute myocardial ischemia. *Circulation* 1974;50:1163–1172.
34. Elharrar V, Foster PR, Jirak TL, Gaum WE, Zipes DP: Alterations in canine myocardial excitability during ischemia. *Circ Res* 1977;40:98–105.
35. Lazzara R, El-Sherif N, Hope RR, Scherlag BJ: Ventricular arrhythmias and electrophysiological consequences of myocardial ischemia and infarction. *Circ Res* 1978;42:740–749.
36. Janse MJ, Kléber AG: Electrophysiological changes and ventricular arrhythmias in the early phase of regional myocardial ischemia. *Circ Res* 1981;49:1069–1081.
37. Spach MS, Dolber PC, Heidlage JF: Influence of the passive anisotropic properties on directional differences in propagation following modification of the sodium conductance in human atrial muscle: A model of reentry based on anisotropic discontinuous propagation. *Circ Res* 1988;62:811–832.
38. Okumura K, Olshansky B, Henthorn RW, Epstein AE, Plumb VJ, Waldo AL: Demonstration of the presence of slow conduction during sustained ventricular tachycardia in man: Use of transient entrainment of the tachycardia. *Circulation* 1987;75:369–378.
39. Dillon SM, Allessie MA, Ursell PC, Wit AL: Influences of anisotropic tissue structure on reentrant circuits in the epicardial border zone of subacute canine infarcts. *Circ Res* 1988;63:182–206.
40. Kavanagh KM, Kabas JS, Rollins DL, Melnick SB, Smith WM, Ideker RE: Epicardial mapping of ventricular tachycardia induced by a large premature stimulus over nontransmural infarcts. *J Am Coll Cardiol* 1990;15:123A(abstract).
41. Akiyama T: Intracellular recording of in situ ventricular cells during ventricular fibrillation. *Am J Physiol* 1981;140:H465–H471.
42. Czarnecka M, Lewartowski B, Prokopczuk A: Intracellular recording from the *in situ* working dog heart in physiological conditions and during acute ischemia and fibrillation. *Acta Physiol Pol* 1973;24:331–337.
43. Downar E, Janse MJ, Durrer D: The effect of acute coronary artery occlusion on subepicardial transmembrane potentials in the intact porcine heart. *Circulation* 1977;56:217–224.
44. Thandroyen FT: Protective action of calcium channel antagonist agents against ventricular fibrillation in the isolated perfused rat heart. *J Mol Cell Cardiol* 1982;14:21–32.
45. Raeder EA, Verrier RL, Lown B: Protective effect of tiapamil against ventricular fibrillation during coronary artery occlusion. *Am Heart J* 1986;111:878–882.
46. Clusin WT, Buchbinder M, Ellis AK, Kernoff RS, Giacomini JC, Harrison DC: Reduction of ischemic depolarization by the calcium channel blocker diltiazem: Correlation with improvement of ventricular conduction and early arrhythmias in the dog. *Circ Res* 1984;54:10–20.
47. Cavero I, Boudot J-P, Feuvray D: Diltiazem protects the isolated rabbit heart from the mechanical and ultrastructural damage produced by transient hypoxia, low-flow ischemia and exposure to Ca^{++}-free medium. *J Pharmacol Exp Ther* 1983;226:258–268.
48. Jonkman FAM, Boddeke HWGM, van Zwieten PA: Protective activity of calcium entry blockers against ouabain intoxication in anesthetized guinea pigs. *J Cardiovasc Pharmacol* 1986;8:1009–1013.
49. Clusin WT, Bristow MR, Karagueuzian HS, Katzung BG, Schroeder JS: Do calcium-dependent ionic currents mediate ischemic ventricular fibrillation? *Am J Cardiol* 1982;49:606–612.
50. Merillat JC, Lakatta EG, Osamu H, Guarnieri T: A role of calcium and the calcium channel in the initiation and maintenance of ventricular fibrillation. *Circ Res* 1990;67:1115–1123.
51. Zipes DP, Levy MN, Cobb LA, et al.: Task force 2: Sudden cardiac death. Neural-cardiac interactions. *Circulation* 1987;76:I-202–I-207.
52. Corr PB, Pitt B, Natelson BH, et al.: Task Force 3: Sudden cardiac death. Neutral-chemical interactions. *Circulation* 1987;76:I-208–I-214.
53. Kolman BS, Verrier RL, Lown B: The effect of vagus nerve stimulation upon vulnerability of the canine ventricle. *Circulation* 1975;52:578–585.
54. Verrier RL, Lown B: Behavioral stress and cardiac arrhythmias. *Ann Rev Physiol* 1984;46:155–176.
55. Skinner JE: Regulation of cardiac vulnerability by the cerebral defense system. *J Am Coll Cardiol* 1985;5:88B–94B.
56. Gillis AM, Clusin WT: The role of β-adrenergic receptor antagonists in the prevention of sudden cardiac death. *Comp Ther* 1984;10:60–67.
57. Kammerling JM, Green FJ, Watanabe AM, et al.: Transmural myocardial infarction produces denervation supersensitivity of refractoriness in noninfarcted areas apical to the infarction. *Circulation* 1987;76:383–393.
58. Corr PB, Gross RW, Sobel BE: Amphipathic metabolites and membrane dysfunction in ischemic myocardium. *Circ Res* 1984;55:135–154.
59. Gross RW, Sobel BE: Lysophosphatidylcholine metabolism in the rabbit heart. *J Biol Chem* 1982;257:6702–6708.
60. Carlisle EJF, Allen JD, Bailey A, et al.: Fourier analysis of ventricular fibrillation and synchronization of DC countershocks in defibrillation. *J Electrocardiol* 1988;21:337–343.

61. Carlisle EJF, Allen JD, Kernohan WG, Anderson J, Adgey AAJ: Fourier analysis of ventricular fibrillation of varied aetiology. *Eur Heart J* 1990;11:173–181.

62. Worley SJ, Swain JL, Colavita PG, Smith WM, Ideker RE: Development of an endocardial-epicardial gradient of activation rate during electrically induced, sustained ventricular fibrillation in the dog. *Am J Cardiol* 1985;55:813–820.

63. Yakaitis RW, Ewy A, Otto CW, Taren DL, Moon TE: Influence of time and therapy on ventricular defibrillation in dogs. *Crit Care Med* 1980;8:157–163.

64. Weaver WD, Cobb LA, Dennis D, Ray R, Hallstrom AP, Copass MK: Amptitude of ventricular fibrillation waveform and outcome after cardiac arrest. *Ann Intern Med* 1985;102:53–55.

65. Dzwonczyk R, Brown CG, Werman HA: The median frequency of the ECG during ventricular fibrillation: Its use in an algorithm for estimating the duration of cardiac arrest. *IEEE Trans Biomed Eng* 1990; 37:640–646.

PART III

Clinical Arrhythmias

Chapter **11**

Reentrant Supraventricular Arrhythmias: Update

Agustin Castellanos, MD, Alberto Interian, Jr., MD, Liaqat Zaman, MD, and Robert J. Myerburg, MD

In this chapter we will deal only with those supraventricular tachycardias in which the reentry circuits are located in the atrioventricular (AV) node and in the atria. Other reentrant supraventricular tachycardias where the ventricles are essential components of the reentry circuit, such as circus movement tachycardias due to accessory pathways of the Kent bundle type, or Mahaim fibers, are dealt with elsewhere in this book. Because AV nodal reentrant tachycardias have been discussed exhaustively in various publications, we will only emphasize the recently gained knowledge and our own views.

AV NODAL REENTRANT TACHYCARDIAS OF THE SHORT RP–LONG PR TYPE

Mechanisms

These tachycardias are referred to as the usual or common type of AV nodal reentrant tachycardias. They require all components of classical reentry, functional, and/or anatomical dissociation into a fast pathway (FP) and a slow pathway (SP).[1–5]

In the type of tachycardia under consideration, the SP has shorter effective refractory period than the FP. Reentry is initiated when a premature atrial impulse arriving at the AV node encounters the refractory FP, thus being able to be conducted with delay through the SP, and on reaching a final or distal common pathway (FCP), is capable of traversing the SP retrogradely. On its journey back to the atria, the impulse reenters anterogradely the now no longer refractory FP. Perpetuation of this sequence leads to sustained tachycardia (Fig. 11.1). In the surface electrocardiogram (ECG), a prolonged PR interval and a short or absent RP interval are seen (Figs. 11.1 and 11.2). The relationship between QRS and P^- is determined by the relative times for anterograde conduction through the His-Purkinje system and the retrograde conduction through the FP.[1] Since the difference between these conduction times is small, the time between the onset of atrial (A) and ventricular (V) activation ranges from a negative value (A before V) to a VA interval of about 60 ms. This means that atrial activation usually occurs either within the ventricular complexes or slightly after its termination, therefore simulating a late ventricular deflection (Figs. 11.1 and 11.2).[1–7]

Classically, the tachycardia incorporates all components of the reentry circuit, namely, upper common pathway, FP, SP, and FCP.[1–7] In some instances, the electrophysiologic studies support the concept of intranodal location of the upper common pathway. The strongest evidence is the existence of a 2 : 1 HA block during tachycardia (Fig. 11.3).[8,9] Less compelling is the transient dissociation of the low septal right atrial electrograms from the tachycardia circuit in the presence of an unchanged tachycardia cycle length.[10,11] For example, in Figure 11.4, the impulses produced by S6 and S7 (as recorded in the low septal right atrium) occur at the pacing cycle length of 365 ms while the tachycardia persists at its intrinsic rate of 400 ms. Although such a phenomenon reflects septal right atrial–tachycardia dissociation, it may only indicate that the site from which the atrial deflection of the His-bundle electrographic lead was recorded did not constitute an essential link of the circuit, presumably by not reflecting the precise site of anterograde entry into the AV node.

Another suggestive finding is the appearance, while the atria are paced incrementally, of

655 Avenue of the Americas, New York, NY 10010
Current Topics in Cardiology

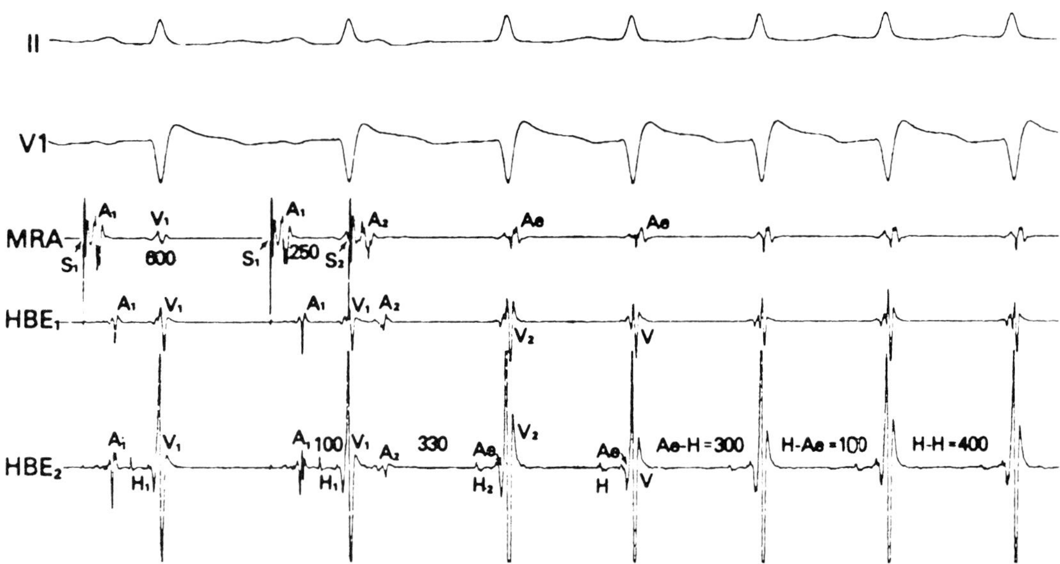

FIGURE 11.1 Atrioventricular (AV) nodal reentrant tachycardia of the usual type initiated by a premature atrial beat (A2) during pacing with the extrastimulus technique. MRA = mid-right atrium; HBE = His-bundle electrographic lead; A_1 and A_2 = atrial electrograms produced by driving, and premature stimuli; Ae = retrograde electrograms occurring during tachycardia; H = His-bundle deflections. While the tachycardia was present, the atrial electrograms were inscribed during the narrow ventricular (V) complexes. In this, and all figures, values are expressed in milliseconds.

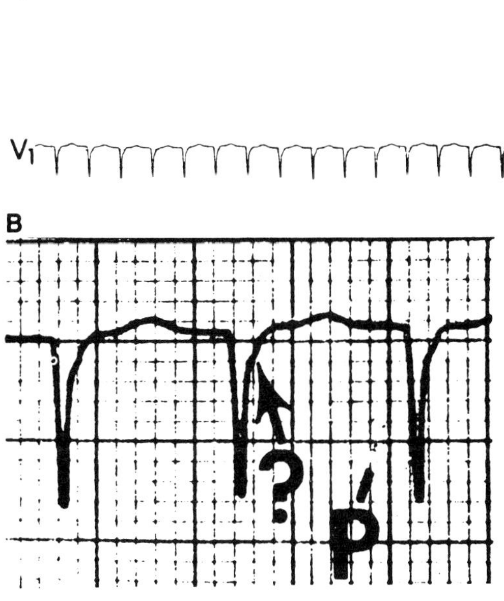

FIGURE 11.2 Surface leads II and VI recorded during AV nodal reentrant tachycardia at a rate of approximately 180/min. What appears as a small notch at the end of the ventricular complex (marked by the arrow at bottom) probably represents retrograde P wave.

AV nodal Wenckebach block at cycle lengths longer than those of the tachycardia.[6,11,12] The latter can be assumed to indicate that during reentry the impulse bypasses not only the atrium but also the more refractory portions of the node encountered during atrial pacing. Yet, a valid argument against this hypothesis is the possibility of a different input into the AV node during atrial pacing than during tachycardia. Additional proposed evidence is the finding that Wenckebach cycle lengths are longer during incremental pacing than during pacing at a constant interval after the preceding QRS complex (R-to-stimulus or RP interval), or more explicitly when atrial stimulation at a constant RP interval mimics the situation existing during tachycardia.[13]

Conversely, there is also evidence favoring a supranodal atrial location of the upper common pathway. The relatively gross catheter techniques may not be able to detect a small atrial rim above the AV nodal tissue.[14] In addition, two distinct inputs into the AV node have been reported both in animals and in man during atrial pacing from different atrial sites. Microelectrode studies performed in the isolated rabbit heart by Janse et al. revealed two distinct inputs into the AV node that were responsible for the occurrence of AV nodal echoes and reentry tachycardia.[15] This was corroborated by other investigators

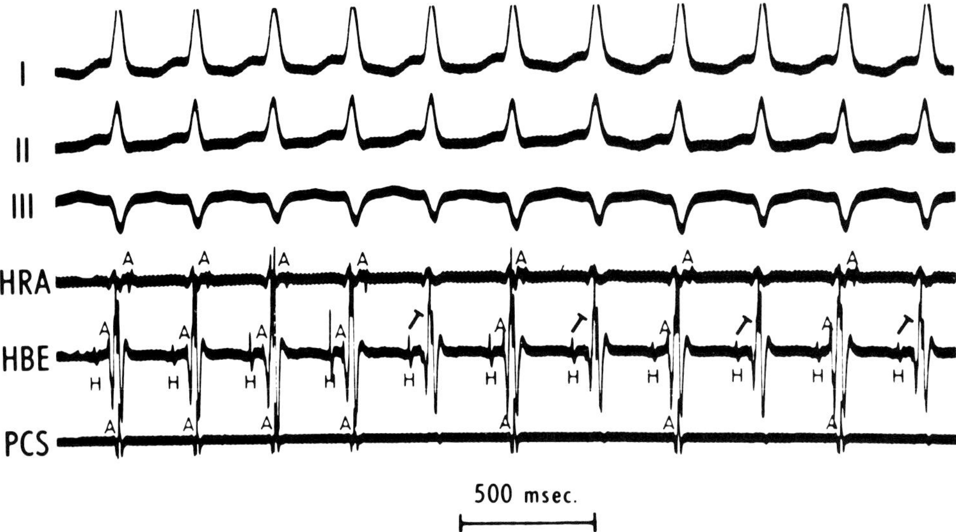

FIGURE 11.3 AV nodal reentrant tachycardia of the usual type showing sudden 2 : 1 HA block. The latter suggests that the atria are not always a necessary part of the reentry circuit, which means that the upper common pathway may have an intranodal location. HRA = high right atrium; PCS = proximal coronary sinus. *Reprinted from Portillo et al.,*[8] *with permission.*

who observed that experimental block at either input site could result in reentrant tachycardia.[16,17] Batsford et al. noted that in 50% of patients with normal PR intervals pacing from coronary sinus pacing but not from high right atrium produced shortening of the AH interval and of AV nodal refractoriness.[18] Similarly, in patients with short PR intervals, Aranda et al. demonstrated the existence of, not two, but several, inputs into the AV node.[19]

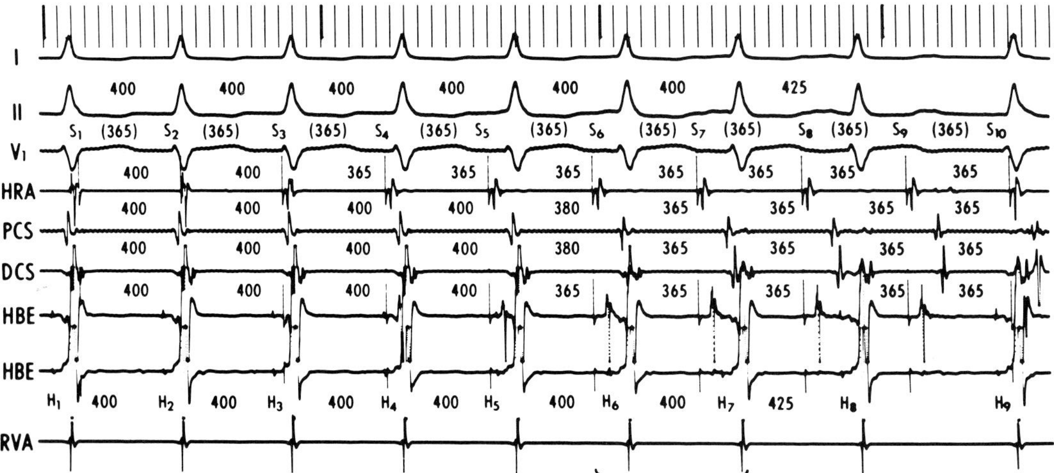

FIGURE 11.4 Transient dissociation between septal right atrial electrogram and AV nodal reentrant reciprocating tachycardia during overdrive atrial pacing. S = pacemaker stimuli. Note that A deflections in the HBE lead produced by S5, S6, and S7 occur at the pacing cycle length (365 ms), whereas the tachycardia persists at the unchanged cycle length of 400 ms. Thus, these atrial electrograms are not related to the immediately following ventricular electrograms. Tachycardia terminates presumably because S7 is blocked in the slow pathway. Alternatively, S7 may have been conducted with a long AH interval and S8 blocked. DCS = distal coronary sinus; RVA = right ventricular apex. *Reprinted from Portillo et al.,*[8] *with permission.*

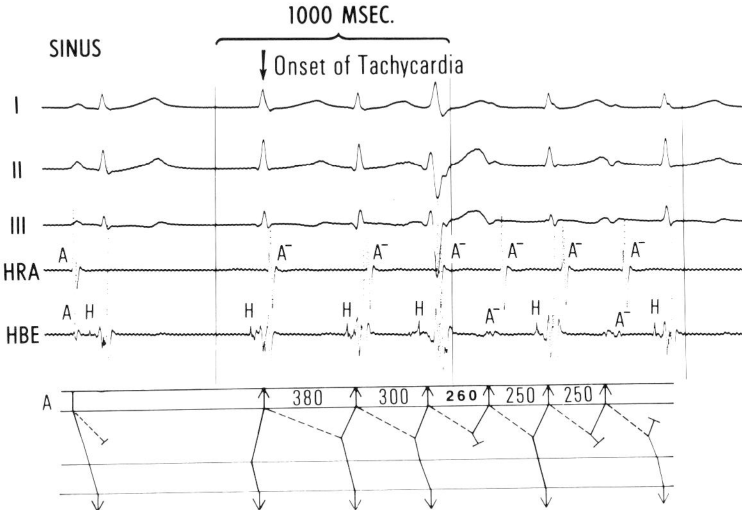

FIGURE 11.5 Spontaneously initiated AV nodal reentrant tachycardia of the usual type showing 2 : 1 block above the His bundle, presumably in an intranodal final common pathway.

Using the more refined technique of perinodal mapping during intracardiac surgery of AV nodal reentrant tachycardias, Ross et al. found during arrhythmia two different sites of retrograde atrial activation which presumably corresponded to the previously mentioned separate inputs.[20] This included anterior sites close to the atrial septum and posterior sites in the vicinity of the coronary sinus os. By means of strategically placed incisions in these atrial regions, Ross et al. and others were able selectively to prevent AV nodal reentry while preserving transnodal function.[20–25] Considering the relatively macroscopic techniques used by these and other investigators, the results favor the existence of an extranodal upper common pathway because it appears improbable that one could selectively obliterate an intranodal pathway while leaving the remaining AV nodal conduction unmodified.[21–25] Similar results have been obtained by some authors using the gross techniques of catheter ablation, presumably because of the same mechanism.[26–28]

An intranodal location of the FCP can be more readily accepted. The presence of the FCP was convincingly demonstrated by the experimental studies of Moe et al. and by the occurrence of second degree AV block during episodes of AV nodal reentrant tachycardia (Fig. 11.5).[9,29] According to Schamroth, the existence of a FCP is also inferred by the concealed reciprocal conduction of an AV nodal impulse to and discharge of the AV nodal pacemaker in some cases of reciprocal rhythm.[30]

Electrophysiologic Studies

Electrophysiologic studies are performed on patients with clinically recurrent sustained or incessant narrow QRS tachycardias to determine the mechanism of the arrhythmia, to guide pharmacologic treatment, and to evaluate the efficacy of stimulation techniques for tachycardia termination, catheter ablation, or surgical treatment.

Although dual AV conduction pathways are demonstrated during atrial stimulation in only

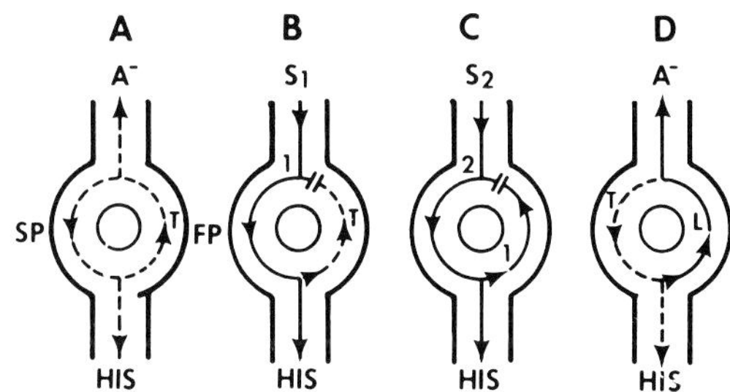

FIGURE 11.6 Diagrammatic representation of entrainment of AV nodal reentry tachycardia of the usual type during overdrive atrial stimulation. A^- = retrograde atrial impulse emerging from the upper common pathway; FP = fast pathway; SP = slow pathway; S1 = impulse produced by the first stimulus; S2 = impulse produced by the second stimulus; T = tachycardia impulse; L = impulse produced by the last stimulus. *Reprinted from Portillo et al.,[8] with permission.*

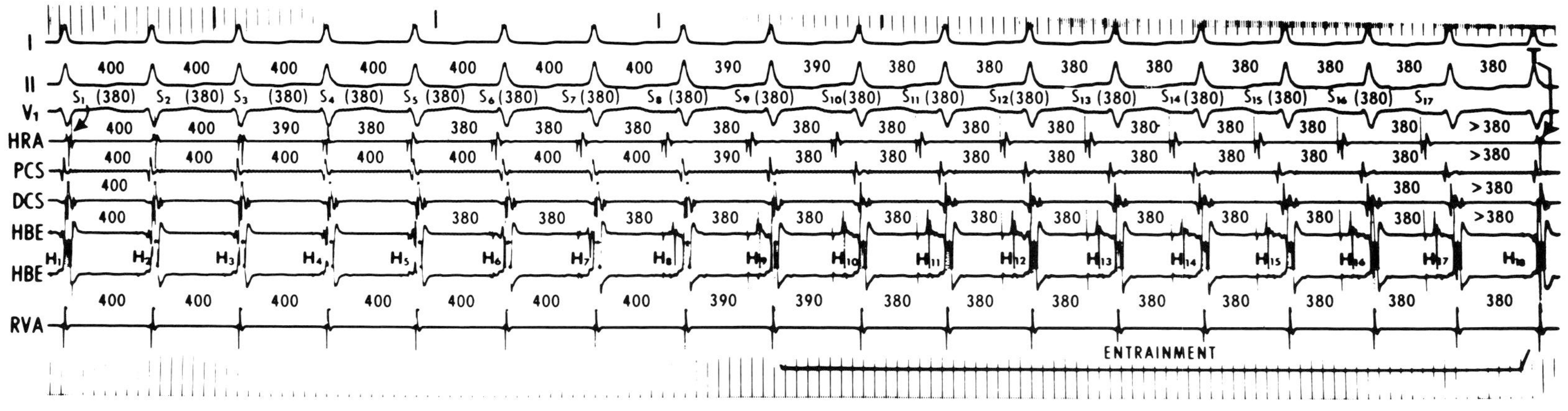

FIGURE 11.7 Entrainment of AV nodal reentry tachycardia at a cycle length of 400 ms of the usual type while pacing from high right atrium (HRA) at a slightly shorter cycle length of 380 ms. Arrow after first HRA electrogram points at first stimulus delivered in this run. S17 was the last paced stimulus. Entrainment occurred after S8 because thereafter: (a) atrial, His-bundle, and ventricular cycle lengths decreased, that is, becoming similar to the pacing cycle length; and (b) the tachycardia resumed after cessation of stimulation (last arrow). The fact that the interval between H17 and H18 is equal to the paced cycle length indicates that H18 and the last ventricular complex were indeed produced by S17. Note that H18 was followed by retrograde atrial electrograms having the same sequence as those of the tachycardia because it met no (paced) anterograde atrial impulse with which to collide. Consequently, the intervals between these retrograde atrial electrograms and the anterograde electrograms produced by S17 were longer than the pacing cycle length. *Reprinted from Portillo et al.,*[8] *with permission.*

about 70% of the patients (Fig. 11.1), failure to demonstrate dual AV nodal pathways does not rule out reentry as the underlying mechanism.

Tachycardias can be initiated by atrial pacing with the extrastimulus technique or by incremental atrial stimulation, and much less frequently by ventricular stimulation. The VA intervals are usually shorter than 60 ms when the A is measured in the His-bundle electrographic lead or in any other atrial leads. Recording V-high right atrial intervals shorter than 90 ms tends to exclude retrograde conduction through an accessory pathway.[31]

Most tachycardias are terminated by the induction of anterograde block in the SP with carotid sinus massage, verapamil, adenosine, digoxin, and β-blockers. The functional behavior of FP is more intriguing because the arrhythmia is usually not terminated retrogradely by the same maneuvers and drugs. It has to be pointed out that assessment of carotid sinus massage effect can be done only during ventricular pacing when the tachycardia is present as the occurrence of anterograde delay or block does not allow the assessment of the effect of vagal maneuvers on FP conduction. Moreover, retrograde conduction usually does not show the expected increase in conduction time with incremental or premature ventricular stimulation. Intravenous procainamide selectively blocks retrograde conduction. De-

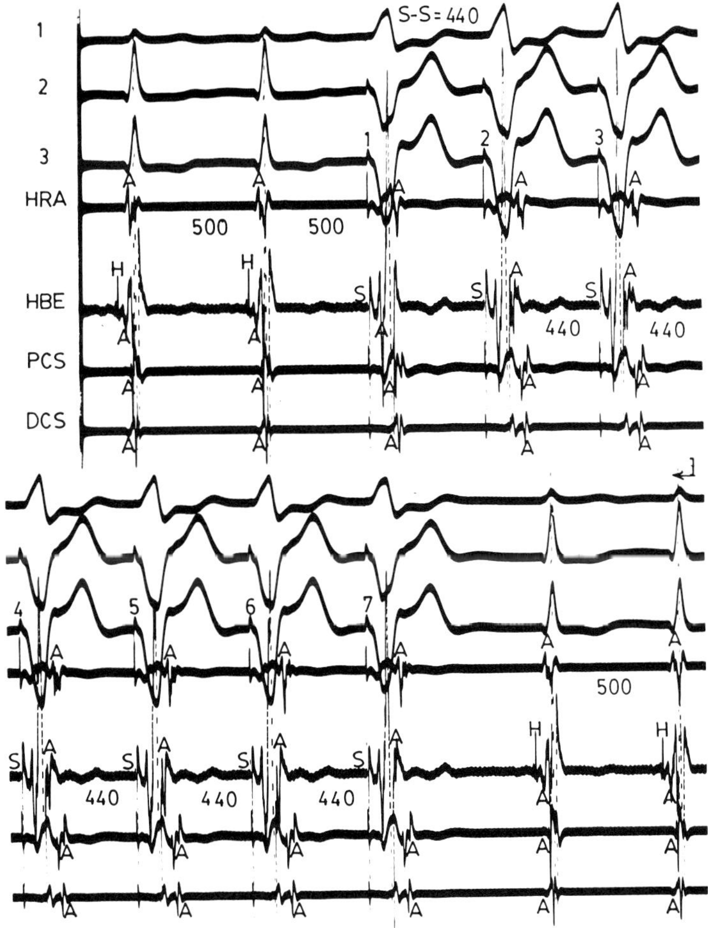

FIGURE 11.8 Entrainment of AV nodal reentrant tachycardia (cycle length of 500 ms) of the usual type by overdrive ventricular stimulation at a shorter cycle length (440 ms). During entrainment, intervals between the retrograde atrial electrograms equaled the pacing cycle length. In contrast to atrial entrainment in Figure 11.7 the interval between the last paced (seventh) QRS complex and the first (nonpaced) QRS complex of the reinitiated tachycardia was longer than the cycle lengths of stimulation and tachycardia. This phenomenon most likely was a consequence of concealed pacing-induced relative refractoriness that became manifest upon cessation of stimulation by slight anterograde (AH) delay occurring either in the slow pathway or in the final common pathway.

spite these findings, most electrophysiologists do not consider the FP as an anatomically distinct either partial or total AV nodal bypass tract.

Entrainment

In clinical electrophysiologic studies it is customary to accept that if a tachycardia can be entrained, then reentry is the underlying mechanism.[32,33] Classically, during entrainment, which is best performed at stimulation rates faster than those of the tachycardia, paced impulses engage the circuit in such a way that the tachycardia can resume at its original rate upon the cessation of pacing.

The mechanisms proposed to explain entrainment during overdrive atrial pacing are depicted in Figure 11.6.[8] Figure 11.6A shows that during tachycardia the impulse propagates anterogradely through the SP and retrogradely through the FP. The upper common pathway (UCP) leading into, or constituting, the atria, as well as the FCP, are not essential components of the circuit. In panel B, the first anterograde impulse (S1), initiated by the first stimulus, collides with the retrograde impulse of the last tachycardia beat (T) in the superior part of the FP and extinguishes the original episode of tachycardia without preventing its capacity to resume, as will be mentioned later. S1 is also conducted anterogradely via the SP to penetrate the FP retrogradely in an unsuccessful attempt to continue the tachycardia. In panel C, S1 reaches the superior part of the FP where it collides with S2, the anterograde impulse produced by the second stimulus. Therefore, S1 cannot complete the circuit. S2 also propagates anterogradely through the SP. This sequence of events continues with successive stimuli until the impulse (L) (produced by the last stimulus) in panel D arrives retrogradely at the superior part of the FP and, finding no anterograde impulse with which to collide because pacing had been stopped, can reenter anterogradely through the SP which allows the tachycardia to resume. Examples of entrainment during atrial pacing and ventricular pacing are shown in Figures 11.7 and 11.8, respectively. The findings are described in the corresponding legends.

AV NODAL REENTRANT TACHYCARDIAS OF THE LONG RP–SHORT PR TYPE

These tachycardias, which are usually referred to as the unusual types of AV nodal reentry tachycardias, result in a surface ECG pattern that can be present also in circus movement tachycardias using a posteroseptal accessory pathway with long VA conduction times, and in intraatrial tachycardias.[34–39] While the arrhythmia is present, the impulse is conducted anterogradely through the FP (because the SP has a longer refractory period) with small or insignificant AV nodal delay (Fig. 11.9). Despite the fact that the arrhythmia can start late in the cycle, a small but nevertheless progressive increase in AH conduction times is noted as the coupling intervals shorten. Retrograde conduction takes place via the SP (Fig. 11.9).

If the H2A2 intervals during ventricular stimulation with the extrastimulus technique show an inverse relationship with the estimated H1H2 intervals, one can assume that the retrograde pathway is connected (directly or through the FCP) to the His bundle. This is an important finding because physiologists consider that such ECG pattern can be also produced when retrograde conduction occurs through a concealed nodoventricular (Mahaim) tract, bridging the intranodal SP and the right ventricle.[38] It should be emphasized that the AV nodal reentrant tachycardia under consideration is believed to be uncommon,[34] and there are no extensive studies of the reentry circuit. According to Brugada et al., sustained reentry tachycardia showing a shorter PR than RP interval most likely favors a circuit in which retrograde conduction occurs through an accessory pathway with long VA conduction time.[34] But this needs reevaluation.

The following considerations have to be entertained:

(a) In the type of circus movement tachycardia involving an accessory pathway, premature ventricular impulses delivered at a time when the His bundle is refractory (or without interfering with the corresponding anterograde H deflection) can preexcite the atria with the same retrograde atrial sequence indicating that the ventricular impulse reaches the atria through an extra-Hisian, extra-AV nodal pathway.[34] This does not happen in the unusual types of AV nodal reentry tachycardias in which very premature ventricular impulses can sometimes reach the atria, usually when the tachycardia rates are slower, and the retrograde impulse activates the His bundle before the latter is depolarized anterogradely by the incoming tachycardia atrial impulse (Fig. 11.9).[39]

(b) Propagation through posteroseptal accessory pathways results in retrograde atrial sequence during which the coronary sinus region is depolarized before the low septal right atrium. However, Sung and Waxman observed that during sudden transition from retrograde FP to SP conduction, coronary sinus ostial atrial activation sometimes occurs before that of the low sep-

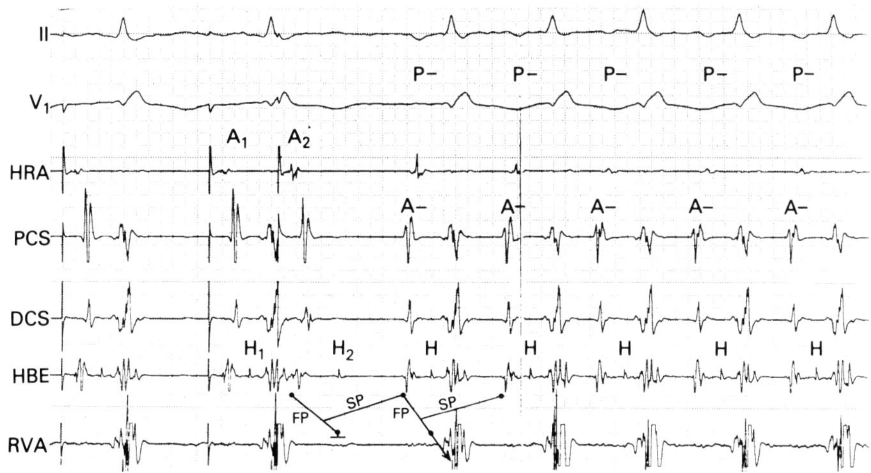

FIGURE 11.9 Initiation and perpetuation of AV nodal tachycardia of the unusual type. A2 (which starts the tachycardia) is conducted anterogradely through the fast pathway (as indicated diagrammatically) and on reaching the lower common pathway returns via the slow pathway. In addition, this impulse, after activating the His bundle (H2), is unable to reach the ventricles which therefore cannot be part of the reentry circuit. Note that during tachycardia, retrograde atrial activation is first detected in the proximal coronary sinus. The patient also had right bundle branch block.

tal right atrium.[40] The same can happen in the unusual type of AV nodal reentrant tachycardia (Fig. 11.9).

(c) Spontaneous, vagal- or drug-induced second-degree AV nodal block excludes participation of an accessory pathway in the reentry circuit (Fig. 11.9).[34] However, second-degree AV nodal block can be seen with either AV nodal reentrant tachycardia or with intraatrial tachycardia.[34] Then the differential diagnosis has to be made by finding an identical retrograde atrial activation sequence during ventricular stimula-

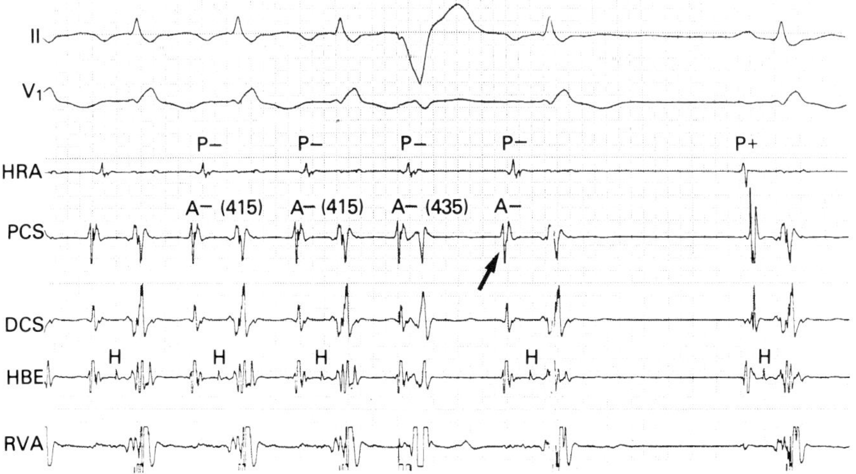

FIGURE 11.10 Termination of a run of unusual AV nodal reentry tachycardia in the same patient as in Figure 11.9. Retrograde conduction time became longer after the early induced ventricular ectopic beat since the corresponding A–A– interval increased from 415 to 435 ms. This could have reflected atrial capture with delay, in which case the retrograde activation patterns produced by the latter would have been identical to that of the tachycardia. This finding would exclude an intraatrial tachycardia. Subsequently, the last retrograde impulse was blocked in the slow pathway, effectively terminating the tachycardia. An alternate explanation is the spontaneous occurrence of an abrupt second-degree block in the slow pathway.

tion and during tachycardia because it would be exceptional for the atrial focus to be located precisely at the site at which the retrograde impulse emerges from the AV node (Fig. 11.10).

INTRAATRIAL TACHYCARDIAS

Most of the few reports dealing with tachycardias originated and perpetuated in the atria have included small numbers of cases.[7,41–45] This does not mean that atrial tachycardias are clinically rare since it shows only that few such patients have been studied by invasive methods, presumably because the tachycardias are transient and not recurrent or incessant.[44] Several mechanisms of atrial tachycardias have been reported and, although criteria have been given for different diagnoses,[41–46] differentiation can be difficult and sometimes all that can be said is that they are "focal" in origin.[47]

Intraatrial Reentrant Tachycardias

According to Wellens, intracardiac studies performed in patients with recurrent monomorphic atrial tachycardias suggest a reentry mechanism.[45] An important criterion is initiation by premature atrial impulses occurring during the relative atrial refractory period, and the onset not related to critical AV nodal delay (Fig. 11.11).[7,46] Direct relationship between PR interval and atrial rate, P waves different from sinus P waves, and occurrence of spontaneous or induced vagal second-degree AV nodal block does not exclude mechanisms other than reentrant atrial tachycardias.[7,46] More than half of the patients with intraatrial reentrant tachycardias reported by Coumel et al.[48] and Haines and DiMarco[46] had more than one P-wave morphology and different cycle lengths which suggested the presence of more than one tachycardia type.

Classically, diagnosis of intraatrial reentrant tachycardias has been made by their initiation and termination by one or several artificial premature atrial beats delivered in succession.[7,46] However, experimental studies have shown that fast atrial rhythms resulting from triggered activity due to delayed afterdepolarization may also be initiated by atrial beats.[49] Also, in some preparations, triggered activity due to afterdepolarizations can remain subthreshold beat after beat for long periods only reaching threshold, and thus becoming manifest, after a premature beat.[49] On the other hand, automatic rhythms are not initiated by premature beats.[7,41,42]

According to Cranefield and Aronson, a fundamental difference between automatic fibers and triggerable fibers may be related to the fact that automatic fibers, even when quiescent, are on the brink of rhythmic activity and can become spontaneously active in response to small, spontaneous fluctuations in current and voltage that almost always occur across any excitable membrane.[49] In contrast, triggerable fibers are not ordinarily poised on the verge of spontaneous rhythmic activity and may remain quiescent even when the current and voltage "noise" of the membrane are much enhanced.[49] They do not develop spontaneous activity unless triggered by extraneous impulses.[49]

When an intraatrial tachycardia starts spontaneously during sinus rhythm, the exact mechanism may not be identified exclusively by the mode of initiation (Fig. 11.12). However, the response to drugs may be helpful in differentiating between reentry and triggered intraatrial tachycardias, since verapamil may inhibit triggered activity and terminate the arrhythmia whereas it only causes AV nodal block in the intraatrial reentrant tachycardias.[46] Although a single case of termination of "focal" intraatrial tachycardia by adenosine triphosphate has been reported,[50] clinical experience with adenosine suggests that this drug is most valuable in differentiation between intraatrial tachycardias and AV nodal reentry tachycardias by virtue of terminating the latter and not the former.[51] Milchak and Greenberg reported a patient with the unusual type of AV nodal reentry tachycardia in whom carotid sinus massage induced second-degree AV block while adenosine terminated the arrhythmia.[52] Such a response suggests that acetylcholine and aden-

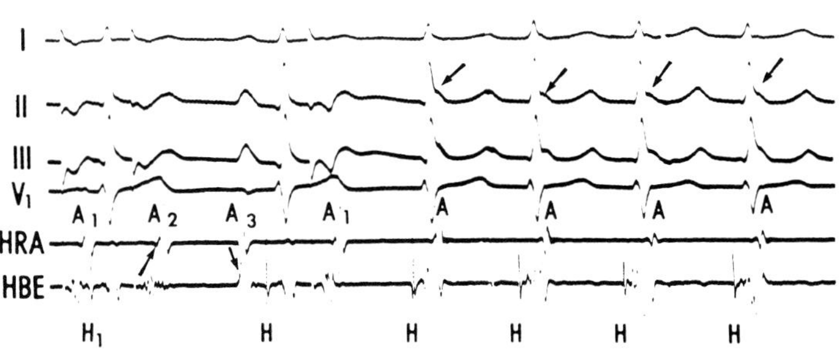

FIGURE 11.11 Initiation of intraatrial reentrant tachycardia (rate, 140/min) by an atrial impulse delivered to the coronary sinus (second A1). During the tachycardia, the P waves (arrows), which seem to be positive in leads II and III, were conducted with delay through the AV node.

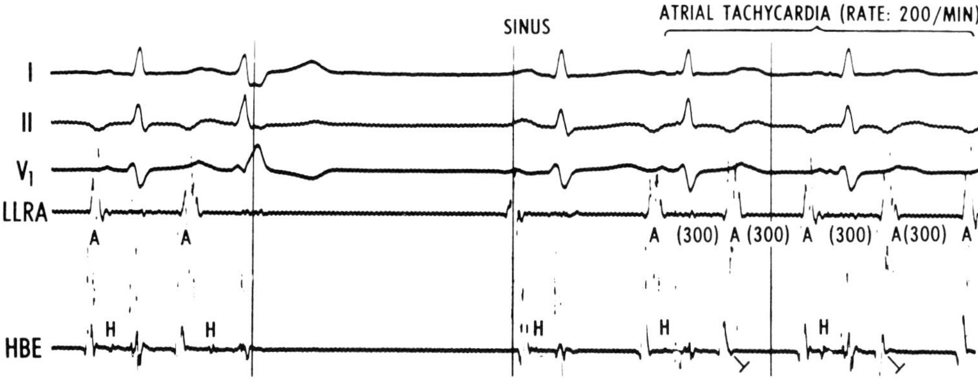

FIGURE 11.12 Spontaneous initiation of intraatrial tachycardia (at a rate of 200/min) with 2 : 1 AV (AH) block. The beginning of the figure shows the end of a previous episode, which had a slightly slower rate (170/min) and a Wenckebach type of AV (AH) block. The exact mechanism of this tachycardia could not be determined because it could not be consistently initiated or terminated by atrial or ventricular stimulation techniques. Furthermore, the ventricular premature beats that were able to activate the atria produced a different sequence than that occurring during tachycardia.

osine affect different regions of the human AV node.[52]

The automatic atrial tachycardias are neither initiated nor terminated by premature atrial beats, their onset is independent of intraatrial or AV nodal delay, and they frequently show a so-called warming-up phenomenon.[7,41,42]

Criteria for entrainment are difficult to demonstrate when dealing with intraatrial reentrant tachycardias.[46] Low amplitude and location in the cycle of P waves make it difficult to demonstrate progressive fusion.[46] Also, differentiation between concealed entrainment and overdrive suppression of an automatic tachycardia may not be possible since both show a post-overdrive "pacing pause."

Addendum

While this book was in press, McGuire et al. published a very pertinent article (*Circulation* 1991; 83:1222–1246).

REFERENCES

1. Akhtar M: Supraventricular tachycardias, in Josephson ME, Wellens HJJ (eds): *Tachycardias: Mechanisms, Diagnosis, Treatment.* Philadelphia, Lea & Febiger, 1984, pp 137–170.
2. Denes P, Wu D, Amat-y-Leon F, Dhingra R, Wyndham CR, Rosen K: The determinants of atrioventricular nodal re-entrance with premature atrial stimulation in patients with dual A-V nodal pathways. *Circulation* 1977;56:253–259.
3. Wu D, Denes P, Amat-y-Leon F, et al.: Clinical electrocardiographic and electrophysiologic observations in patients with paroxysmal supraventricular tachycardia. *Am J Cardiol* 1978;41:1045–1051.
4. Josephson ME: Paroxysmal supraventricular tachycardia: An electrophysiologic approach. *Am J Cardiol* 1978;41:1123–1126.
5. Goldreyer BN, Bigger JT Jr: Site of reentry in paroxysmal supraventricular tachycardia in man. *Circulation* 1971;43:15–26.
6. Miller JM, Rosenthal ME, Vassallo JA, Josephson ME: Atrioventricular nodal reentrant tachycardia: Studies on upper and lower "common pathways." *Circulation* 1987;75:930–940.
7. Josephson ME, Seides SF: *Clinical Cardiac Electrophysiology: Techniques and Interpretations.* Philadelphia, Lea & Febiger, 1979, p 147.
8. Portillo B, Mejias J, Leon-Portillo N, Zaman L, Myerburg RJ, Castellanos A: Entrainment of atrioventricular nodal reentrant tachycardias during overdrive pacing from high right atrium and coronary sinus. *Am J Cardiol* 1984;53:1570–1576.
9. Wellens HJJ, Wesdorp JC, Duren DR, Lie KR: Second degree block during reciprocal atrioventricular nodal tachycardia. *Circulation* 1976;53:595–604.
10. Ko PT, Naccarelli GV, Gulamhusein S, Prystowsky EN, Zipes DP, Klein GJ: Atrioventricular dissociation during paroxysmal junctional nodal tachycardia. *PACE* 1981;4:670–676.
11. Josephson ME, Kastor JA: Paroxysmal supraventricular tachycardia: Is the atrium a necessary link? *Circulation* 1976;54:430–435.
12. Akhtar M, Damato AN, Ruskin JN, et al.: Antegrade and retrograde conduction characteristics in three patterns of paroxysmal atrioventricular junctional reentrant tachycardia. *Am Heart J* 1978;95:22–42.
13. Levy MN, Martin PJ, Edelstein J, Goldberg LB: The AV nodal Wenckebach phenomenon as a positive feedback mechanism. *Prog Cardiovasc Dis* 1974;16:601–613.
14. Sharma AD, Yee R, Guiraudon GM, Klein GJ: AV nodal reentry—Current concepts and surgical

treatment, in Zipes DP, Rowlands DJ (eds): *Progress in Cardiology*. Philadelphia, Lea & Febiger, 1988, pp 129–145.

15. Janse MJ, van Capelle FJL, Anderson RH, Touboul P, Billette J: Electrophysiology and structure of the atrioventricular node of the isolated rabbit heart, in Wellens HJJ, Lie KI, Janse MJ (eds): *The Conduction System of The Heart*. Philadelphia, Lea & Febiger, 1976, pp 296–315.
16. Mazgalev T, Dreifus LS, Bianchi J, Michelson EL: The mechanism of AV junctional reentry: Role of the atrioventricular junction. *Anat Rec* 1981;201:179–188.
17. Iinuma H, Dreifus LS, Mazgalev T, Price R, Michelson EL: Role of the perinodal region in atrioventricular reentry: Evidence in an isolated rabbit heart preparation. *J Am Coll Cardiol* 1983;2:465–473.
18. Batsford WP, Akhtar M, Caracta AR, Josephson ME, Seides SF, Damato AN: Effect of atrial stimulation site on the electrophysiological properties of the atrioventricular node in man. *Circulation* 1974;50:283–292.
19. Aranda JM, Castellanos A, Moleiro F, Befeler B: Effects of the pacing site on A-H conduction and refractoriness in patients with short P-R intervals. *Circulation* 1976;53:33–39.
20. Ross DL, Johnson DC, Denniss AR, Cooper MJ, Richards DA, Uther JB: Curative surgery for atrioventricular junctional ("AV nodal") reentrant tachycardia. *J Am Coll Cardiol* 1985;6:1383–1392.
21. Ross DL, Johnson DC, Koo CC, et al.: Surgical treatment of supraventricular tachycardia without the WPW syndrome: Current indications, techniques, and results, in Brugada P, Wellens HJJ (eds): *Cardiac Arrhythmias: Where To Go From Here?* Mt. Kisco, NY, Futura Publishing, 1987, pp 591–603.
22. Guiraudon GM, Klein GJ, Sharma AD, Yee R: Use of old and new anatomic, electrophysiologic, and technical knowledge to develop operative approaches to tachycardia, in Brugada P, Wellens HJJ (eds): *Cardiac Arrhythmias: Where To Go From Here?* Mt Kisco, NY, Futura Publishing, 1987, pp 639–652.
23. Guiraudon GM, Klein GJ, Sharma AD, Yee R: Pathological insights gained by direct surgical approach to AV nodal reentrant tachycardias. *J Clin Invest Med* 1987;10:C75.
24. Pritchett ELC, Anderson RW, Benditt DG, et al.: Reentry within the atrioventricular node: Surgical cure with preservation of atrioventricular conduction. *Circulation* 1979;60:440–446.
25. Klein GJ, Guiraudon GM, Perkins DG, et al.: Controlled cryothermal injury to the AV node: Feasibility for AV node modification. *PACE* 1985;8:630.
26. Gallagher JJ, Svenson RH, Kasell JH: Catheter technique for closed-chest ablation of the atrioventricular conduction system. *N Engl J Med* 1982;306:194–200.
27. Scheinman MM, Morady F, Hess DS, Gonzalez R: Catheter induced-ablation of the atrioventricular junction to control refractory supraventricular arrhythmias. *JAMA* 1982;248:851–855.
28. Scheinman MM, Evans-Bell GT, et al.: Catheter electrocoagulation of the atrioventricular junction: A report of the percutaneous mapping and ablation registry, in Fontaine G, Scheinman MM (eds): *Ablation in Cardiac Arrhythmias*. Mt. Kisco, NY, Futura Publishing, 1987.
29. Moe G, Mendez C: The physiologic basis of reciprocal rhythm. *Prog Cardiovasc Dis* 1966;8:461–482.
30. Schamroth L: *The Disorders of Cardiac Rhythm*, 2d ed., vol. 1. London, Blackwell Scientific Publication, 1980, p 247.
31. Benditt DG, Pritchett ELC, Smith WM, Gallagher JJ: Ventriculoatrial intervals: Diagnostic use in paroxysmal supraventricular tachycardia. *Ann Intern Med* 1979;91:161–166.
32. Waldo AL, Plumb VJ, Arciniegas JG, et al.: Transient entrainment and interruption of the atrioventricular bypass pathway type of paroxysmal atrial tachycardia. A model for understanding and identifying reentrant arrhythmias. *Circulation* 1983;67:73–83.
33. Waldo AL, Plumb VJ, Arciniegas JG, James TN: Overdrive pacing of A-V bypass pathway type PAT — A model for demonstrating and understanding reentrant rhythms. *Circulation* 1980;62(suppl III):III-47(abstract).
34. Brugada P, Bar FWHM, Vanagt EJ, Friedmman PL, Wellens HJJ: Observations in patients showing A-V junctional echoes with a shorter P-R than R-P interval. Distinction between intranodal reentry or reentry using an accessory pathway with a long conduction time. *Am J Cardiol* 1981:48:611–622.
35. Coumel P: Junctional reciprocating tachycardias. The permanent and paroxysmal forms of AV nodal reciprocating tachycardias. *J Electrocardiol* 1975;8:79–90.
36. Guarnieri T, German LD, Gallagher JJ: The long R-P′ tachycardias. *PACE* 1987;10:103.
37. Castellanos A, Myerburg RJ: The wide electrophysiologic spectrum of tachycardias having R-P intervals longer than the P-R intervals. *PACE* 1987; 10:1382–1384.
38. Sung RJ, Tau DY: Electrophysiologic characteristics of accessory connections: An overview, in Benditt DG, Benson DW (eds): *Cardiac Preexcitation Syndromes. Origins, Evaluation and Treatment*. Boston, Martinus Nijhoff Publishing, 1988, pp 165–200.
39. Wu D, Denes P, Wyndham C, Amat-y-Leon F, Dhingra RC, Rosen KM: Demonstration of dual atrioventricular nodal pathways utilizing a ventricular extrastimulus in patients with atrioventricular nodal reentrant paroxysmal supraventricular tachycardia. *Circulation* 1975;52:789–798.
40. Sung RJ, Waxman HL: Physiological shift of retrograde atrioventricular nodal output mimicking retrograde accessory pathway conduction. *Circulation* 1979;60(suppl II):II-191(abstract).
41. Goldreyer, BN, Gallagher JJ, Damato AN: The electrophysiologic demonstration of atrial ectopic tachycardia in man. *Am Heart J* 1973;85:205–215.
42. Scheinman MM, Basu D, Hollenberg M: Electrophysiologic studies in patients with persistent atrial tachycardia. *Circulation* 1974;50:266–273.
43. Gillette PC, Garson A Jr: Electrophysiologic and pharmacologic characteristics of automatic ectopic atrial tachycardia. *Circulation* 1977;56:571–575.
44. Castellanos A, Fernandez P, Luceri RM, Myerburg RJ: Mechanisms of atrial tachycardias, in Attuel P, Coumel P, Janse MJ (eds): *The Atrium in Health and Disease*. Mt. Kisco, NY, Futura Publishing, 1989, pp 263–269.
45. Wellens HJJ: Antiarrhythmic drug treatment of atrial tachyarrhythmias, in Attuel P, Coumel P, Janse MJ (eds): *The Atrium in Health and Disease*. Mt. Kisco, NY, Futura Publishing, 1989, pp 233–238.

46. Haines DE, DiMarco JP: Sustained intraatrial reentrant tachycardia: Clinical, electrocardiographic and electrophysiologic characteristics and long-term follow-up. *J Am Coll Cardiol* 1990;15:1345–1354.
47. Seals AA, Lawrie GM, Magro S, et al.: Surgical treatment of right atrial focal tachycardia in adults. *J Am Coll Cardiol* 1988;11:1111–1117.
48. Coumel P, Flammang D, Attuel P, Leclercq JF: Sustained intraatrial reentrant tachycardia: Electrophysiologic study of 20 cases. *Clin Cardiol* 1979;2:167–178.
49. Cranefield PF, Aronson RS: *Cardiac Arrhythmias: The Role of Triggered Activity and Other Mechanisms*. Mt. Kisco, NY, Futura Publishing, 1988, pp 238,239,622.
50. Pereleman MS, Krikler DM: Termination of focal atrial tachycardia by adenosine triphosphate. *Br Heart J* 1987;58:520–530.
51. DiMarco JP, Sellers TD, Lerman BB, Greenberg ML, Berne RM, Belardinelli L: Diagnostic and therapeutic use of adenosine in patients with supraventricular tachyarrhythmias. *J Am Coll Cardiol* 1985; 6:417–425.
52. Michak MA, Greenberg MC: Discordant effects of carotid sinus massage and intravenous adenosine in atypical (fast-slow) atrioventricular nodal reentry tachycardia. *PACE* 1989;12:1903–1909.

Chapter **12**

Nonreentrant Supraventricular Tachycardias

Pratap C. Reddy, MD, and James W. Cox, Jr., MD

A number of mechanisms for the development of supraventricular tachycardia (SVT) have been identified in isolated cell preparations and animal models.[1-6] However, their precise role in clinical SVT remains uncertain. The three major suggested mechanisms of SVT are reentry, abnormal automaticity, and triggered automaticity due to late afterdepolarizations.[7] Most clinical SVTs are attributed to reentry; however, in clinical practice, accurate categorization of the arrhythmia mechanism frequently is difficult and the true incidence of SVTs due to each of these mechanisms is unknown.

Tachycardias that are inducible by programmed stimulation are thought to be due to reentry and those not inducible are ascribed to enhanced or abnormal automaticity. However, since patients with asymptomatic tachycardias or tachycardias noninducible by programmed stimulation are not referred for electrophysiologic studies, it is possible that the incidence of nonreentrant SVT has been underestimated. Supraventricular tachycardias due to reentry mechanism are described elsewhere in this book. In this chapter, our current knowledge of the specific SVTs attributed to abnormal or triggered automaticity is reviewed and the electrophysiologic criteria used to differentiate various mechanisms of SVT are described.

Nonreentrant SVTs include: (a) unifocal ectopic automatic atrial tachycardia, (b) ectopic atrial tachycardia with atrioventricular (AV) block, (c) multifocal atrial tachycardia, (d) junctional ectopic tachycardia, and (e) nonparoxysmal AV junctional tachycardia. In most cases, these tachycardias are attributed to abnormal automaticity, and triggered automaticity is implicated in only isolated cases of clinical SVT.

AUTOMATIC ATRIAL TACHYCARDIA

Atrial tachycardia due to abnormal automaticity may be transient, recurrent, or persistent. The more common transient or recurrent forms are found in patients with chronic lung disease, or those with cardiomyopathy with or without congestive heart failure; they are self-limited, lasting only for a few seconds to minutes. In contrast, chronic persistent atrial tachycardia is rare. It tends to occur in patients with no structural heart disease and is frequently disabling. This tachycardia has been described under such diverse names as ectopic auricular or atrial tachycardia, persistent supraventricular or atrial tachycardia, and chronic atrial tachycardia.[8] Although the existence of this tachycardia has been recognized for over 50 years, a review in 1974 by Scheinman et al.[9] found only 22 cases of persistent automatic atrial tachycardia in the literature. Although the precise incidence of automatic atrial tachycardia is not known, it is believed that in adult patients with SVT this form constitutes 2 to 5% of all SVTs.[10] Among children with SVT, the incidence of automatic atrial tachycardia appears to be higher. Garson and Gillette[11-14] diagnosed automatic atrial tachycardia in 11% of children with SVT.

Clinical Features

Unlike reentrant SVT, which occurs in paroxysms and is characterized by abrupt onset and termination, automatic atrial tachycardia tends to be incessant or permanent with sinus rhythm being either absent or present only for brief periods.[9,14,15] In various reported cases, tachycar-

655 Avenue of the Americas, New York, NY 10010
Current Topics in Cardiology

dia persisted from 3 months to 43 years and the reported age of onset varied from 1 month to 36 years.[12,14,16] In adult patients with automatic atrial tachycardia, structural heart disease is infrequently present. In children, diagnosis is usually made before the age of 6 years and a family history of SVT is frequently present.[11,14] While most children with persistent automatic atrial tachycardia have no underlying structural heart disease, a few have congenital or valvular heart disease, or cardiomyopathy.[11–14] In one reported case of incessant automatic atrial tachycardia, the tachycardia was due to an atrial tumor.[17] In most reported cases of persistent automatic atrial tachycardia, congestive heart failure, cardiac enlargement, and impaired ventricular contractile function were present either at the time of initial presentation or occurred sometime during follow-up.[11,14,18] However, in several cases, heart failure was absent during follow-up periods varying from 3 to 43 years.[14,16,18] In the study of Garson and Gillette,[11,12] congestive heart failure was present in 9 of 11 children with automatic atrial tachycardia, and both cardiomegaly and depressed left ventricular ejection fraction were present in 6 patients. Scheinman and coworkers[9] reported a patient with mitral regurgitation and incessant atrial tachycardia in whom the tachycardia and congestive heart failure persisted even after mitral valve replacement. During the past 13 years, one of us (P.C.R.) has followed 3 patients with chronic persistent atrial tachycardia of 1 to 37 years duration and no apparent structural heart disease. In each of these 3 patients, cardiomegaly with depressed left ventricular ejection fraction or congestive heart failure or both appeared during the follow-up period. The presence or absence of congestive heart failure may depend on the duration and the rate of tachycardia and the presence or absence of AV block. Whereas congestive heart failure is common in patients with persistent atrial tachycardia, it is rare in patients with the repetitive form of atrial tachycardia. In the series of Morgan and Nadas,[19] congestive heart failure occurred in 3 of 6 patients with persistent atrial tachycardia, but in none of those with the repetitive form of atrial tachycardia. Cerebrovascular accidents, presumably secondary to a cerebral embolus, have been reported in a small percentage of patients with the permanent form of automatic atrial tachycardia.[18]

ECG Characteristics

In patients with the recurrent form of automatic atrial tachycardia, an atrial premature complex arising late in the cardiac cycle interrupts sinus rhythm and initiates tachycardia[20] (Fig. 12.1). The P-wave configuration of tachycardia initiating atrial complex is identical to the P waves of succeeding complexes, suggesting a common origin of both the initiating atrial complex and the subsequent complexes of ectopic tachycardia[20] (Fig. 12.1). In contrast to the reentrant SVT, the PR interval of the tachycardia-initiating complex usually is not prolonged. After the onset, the cycle length of tachycardia first shortens progressively (warm-up phenomenon), but once established, the PP intervals do not vary by more than 50 ms unless an exit block from the tachycardia focus is present.[8,21] During tachycardia, the P wave is always visible and its axis is abnormal (Fig. 12.2). The rate of tachycardia is usually slower than that in patients with reentrant tachycardia, but rates as high as 270 bpm have been reported.[8,11,20] The rate of tachycardia tends to vary significantly from day to day and

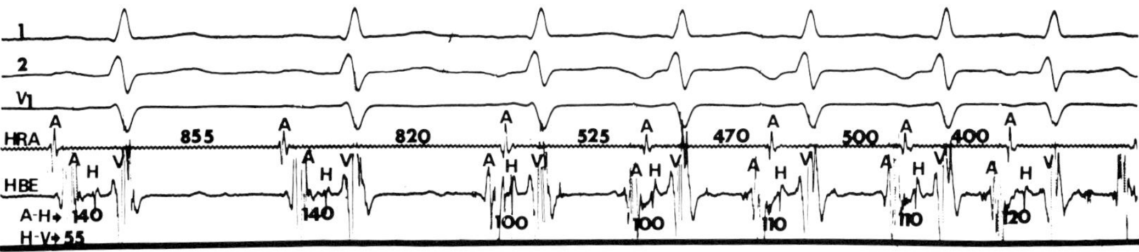

FIGURE 12.1 Initiation of automatic atrial tachycardia. The tracing shows ECG leads I, II, V1, and intracardiac electrograms from high right atrium and the His bundle. The first two complexes are sinus with normal sequence of atrial activation. A premature atrial complex appearing late in the cycle (third complex) initiates tachycardia. Note that: (a) the sequence of atrial activation in the tachycardia initiating premature complex is different from that of the sinus complex and similar to that of subsequent complexes, (b) initiation of tachycardia is not dependent on AH interval prolongation, and (c) cycle length of tachycardia gradually shortens ("warm up") before stabilization. *Reprinted from Reddy,[8] with permission.*

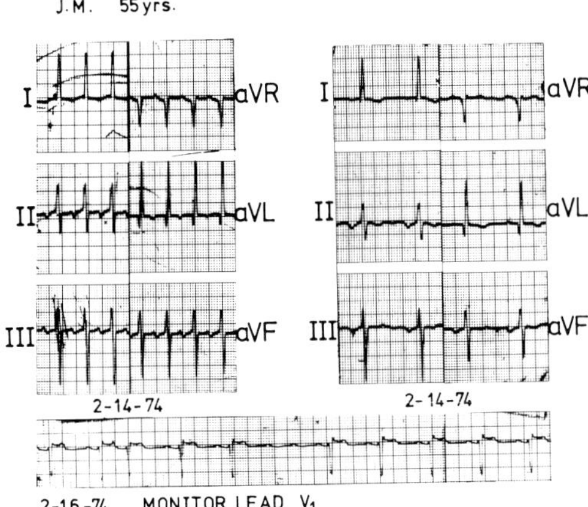

FIGURE 12.2 Automatic atrial tachycardia in a 55-year-old man with thyrotoxicosis showing fluctuation of tachycardia rate due to exit block from the tachycardia focus and the occurrence of AV block. ECGs recorded at different times on the same day (2-14-74) show rates of 150 and 75, the lower rate most likely is due to 2 : 1 exit block from the tachycardia focus. A monitor lead recorded 2 days later (2-16-74) during sleep shows AV block. *Reprinted from* Reddy CP, *with permission.*[76]

from hour to hour, being influenced by many physiologic factors including posture, exercise, emotion, and time of the day[9,14,19] (Fig. 12.2). During sleep, the rate of tachycardia slows and AV block may occur or, when already present, may become more pronounced[8,9,19] (Fig. 12.2). During exercise, the rate of tachycardia increases and the AV block disappears or becomes less pronounced.[14,18] Atropine and isoproterenol may increase and phenylephrine and carotid sinus stimulation may decrease the rate of tachycardia.[11,14,19,22] The latter maneuvers may also produce AV block, but they do not terminate tachycardia, and sinus rhythm does not emerge even when the rate of tachycardia slows.

Clinical Electrophysiology

In most patients with reentrant SVT, the tachycardia can be initiated by programmed atrial stimulation or by rapid pacing, and the initiation depends on a critical coupling interval of premature atrial complex which, in turn, propagates with a critical degree of conduction delay.[23] In patients with SVT due to abnormal automaticity, initiation of tachycardia requires a premature atrial complex, but the initiation does not depend on a critical coupling of the premature atrial complex or on the achievement of critical conduction delay in the premature complex[20] (Fig. 12.1). In automatic atrial tachycardia, the atrial activation sequence of the tachycardia-initiating complex is identical to that of the succeeding complexes of tachycardia, whereas in reentrant SVT, the atrial activation sequence of the tachycardia-initiating complex differs from that of the succeeding complexes of tachycardia.[8,20] During tachycardia, programmed atrial extrastimuli introduced at long coupling intervals (CIs) result in a fully compensatory pause in both reentrant and automatic SVTs. However, at short CIs, the response of automatic SVT differs from that of reentrant SVT, in that the former becomes reset without cessation of tachycardia (Fig. 12.3) while the latter may stop with resumption of sinus rhythm. In this regard, the ectopic atrial focus behaves like the sinoatrial node. If an entrance block into the ectopic focus is present, however, a fully compensatory pause without reset will be observed at all CIs[20] (Fig. 12.3). In patients with reentrant SVT, the tachycardia can be terminated by overdrive pacing. In contrast, in patients with automatic atrial tachycardia, overdrive pacing captures the atrium but upon cessation of pacing, tachycardia immediately resumes without an intervening sinus complex[9,20] (Fig. 12.4). In some instances, overdrive pacing may slow the rate of atrial tachycardia or suppress the tachycardia focus transiently with emergence of sinus rhythm for a very brief period.

Diagnosis

Diagnosis of automatic atrial tachycardia is important because if left untreated, the arrhythmias may precipitate left ventricular dysfunction and congestive heart failure. More importantly, in patients with automatic atrial tachycardia, a permanent cure can be achieved by surgical or electrical ablative techniques.[24,25] Diagnosis of automatic atrial tachycardia is made by analysis of resting and ambulatory electrocardiograms. When analysis of resting and ambulatory electrocardiograms alone is not sufficient, electrophysiologic studies with intracardiac recordings may be required to define the tachycardia mechanism. Unlike the AV reentrant SVT, which cannot be sustained in the presence of AV block, automatic atrial tachycardia can perpetuate in the presence of AV block because AV node is not an integral part of the tachycardia circuit.

Clinical Course

A characteristic of automatic atrial tachycardia is its resistance to treatment including DC countershock. Drug therapy is frequently ineffective

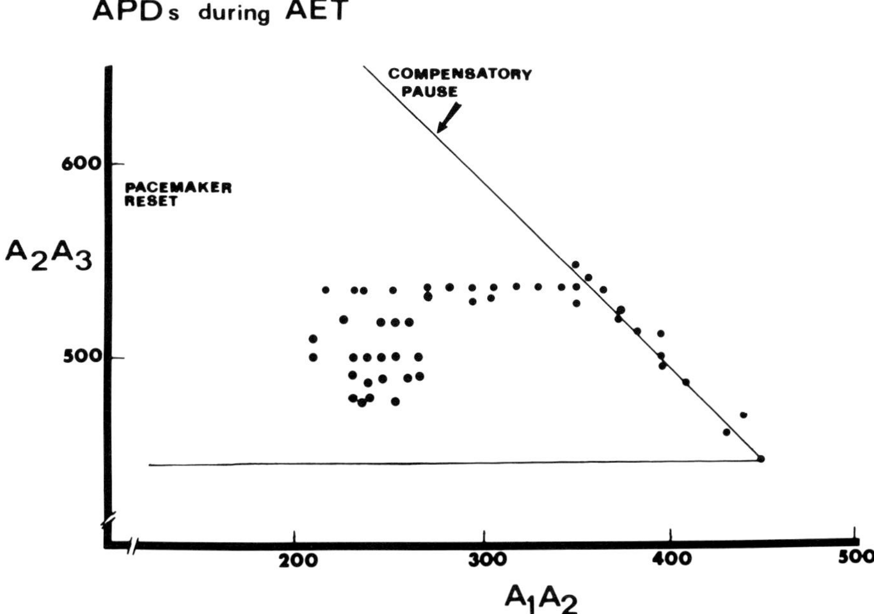

FIGURE 12.3 Response to the introduction of premature atrial extrastimuli during tachycardia. The effect of atrial premature depolarizations (APDs) (A2) introduced during automatic tachycardia (AET) is shown. The interval following the APD (A2–A3 in milliseconds) is plotted as a function of the CI of the APD (A1–A2 in milliseconds). The diagonal line indicates a fully compensatory pause following the APD. The area inside the diagonal line labeled "pacemaker reset" indicates the range of values that might be expected if A2 reset the atrial cycle with a variable degree of suppression of the ectopic focus. Note that as the CI of A2 is progressively decreased, all points fall within the area of pacemaker reset. *Reprinted from Goldreyer et al.,*[20] *with permission.*

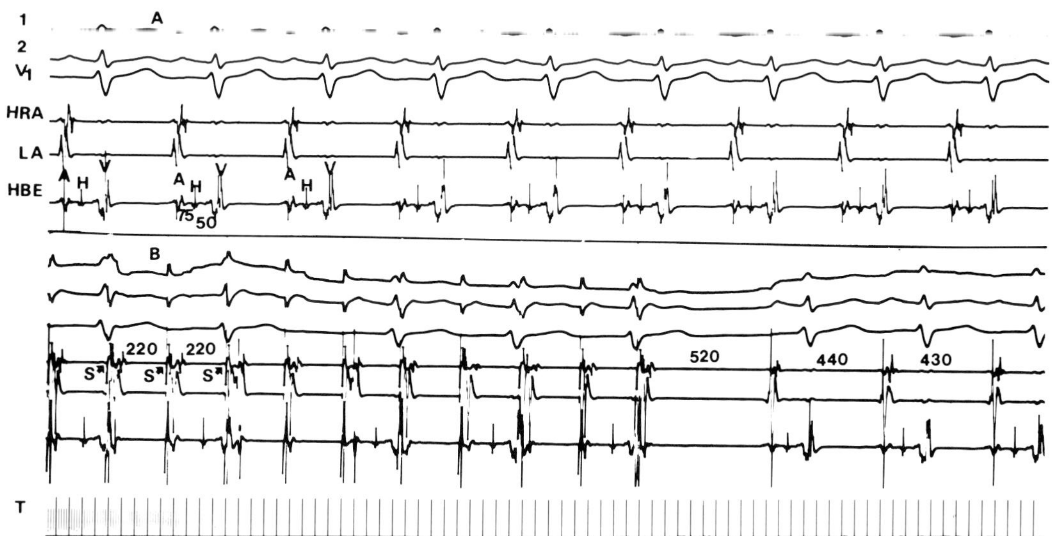

FIGURE 12.4 Effect of overdrive pacing. ECG leads I, II, V1, and electrograms from the high right atrium, left atrium, and His bundle are shown in both panels. **A.** Abnormal atrial activation during tachycardia. **B.** Overdrive atrial pacing at a cycle length of 220 ms fails to terminate tachycardia and upon termination of pacing ectopic atrial tachycardia promptly returns. *Reprinted from* Reddy,[76] *with permission.*

in maintaining sinus rhythm. In the series of children with automatic atrial tachycardia reported by Garson and Gillette,[11–13] automatic atrial tachycardia either did not convert to sinus rhythm or recurred in 10 of 11 patients after aggressive drug therapy or countershock. During follow-up for several years, however, tachycardia disappeared in 2 patients and could be controlled with drugs in 4 other patients. These observations suggest that ectopic foci responsible for tachycardia may cease to function spontaneously. They may also account for the rarity of this tachycardia in adults. Wider application of surgical and catheter ablation of ectopic atrial foci may be expected to improve significantly the prognosis of patients with persistent automatic atrial tachycardia.[24,25]

ATRIAL TACHYCARDIA WITH AV BLOCK

In the past, atrial tachycardia with AV block was described as paroxysmal atrial tachycardia with block. However, recently, the adjective "paroxysmal" has been omitted because the tachycardia lacks both the abrupt onset and the termination that are characteristic of paroxysmal SVT.

Clinical Features

Atrial tachycardia with AV block was first described by Sir Thomas Lewis in 1909.[26] Heyl[27] was the first to attribute this tachycardia to digitalis intoxication. Subsequent description of this tachycardia in patients with serious underlying heart disease with or without digitalis therapy was provided by Barker et al.[28] The importance of this arrhythmia as a serious manifestation of digitalis intoxication became recognized in 1959 by Lown and associates.[29,30]

In their study of 88 patients with 112 episodes of atrial tachycardia with block, digitalis intoxication was the cause of arrhythmia in 73% of patients and tachycardia was precipitated most often by concomitant potassium (K^+) depletion caused by excessive diuresis, hemodialysis, vomiting, diarrhea, cortisone therapy, or calcium (Ca^{2+}) ingestion.[30] Although atrial tachycardia with block is a serious manifestation of digitalis toxicity, it represents only about 10% of the total digitalis-induced arrhythmias.

Despite the close association between atrial tachycardia with block and digitalis toxicity, it should be emphasized that in some cases, this arrhythmia may not be due to digitalis toxicity but to serious underlying heart disease. Atrial tachycardia with block has been reported in normal young people, in patients with rheumatic or ischemic heart disease, and in the presence of severe hypokalemia alone.[31,32] Morgan and Breneman[33] reported 15 cases of atrial tachycardia with block; in 9 of these the arrhythmia could not be attributed to excessive use of digitalis glycosides. According to Soffer[34] and Dreifus et al.,[31] the incidence of atrial tachycardia with block as a manifestation of digitalis toxicity has declined and the arrhythmia is much less common than nonparoxysmal AV junctional tachycardia. After an extensive review of the literature, Fisch and Knoebel[35] stated that probably one-half or more of the cases of atrial tachycardia with block may be due to patient's underlying heart disease and not to digitalis and that some cases of reported atrial tachycardia with block may represent atrial flutter in which the rate of flutter waves is slower than usual. They also believe that atrial tachycardia with block is a less common manifestation of digitalis toxicity than nonparoxysmal AV junctional tachycardia.

ECG Characteristics

Atrial tachycardia begins with a sudden change in the configuration of the P wave. Atrial rate gradually increases over a period of several seconds with gradual prolongation of PR interval.[29,30,35] One-to-one AV conduction is usually present at the onset of tachycardia but as the atrial rate increases, AV Wenckebach block supervenes.[35] Although Wenckebach block is most common, high-degree or complete AV block with escape rhythm may be present in a small percentage of patients. The site of block, presumably, is within the AV node. Simultaneous occurrence of atrial tachycardia and AV junctional tachycardia resulting in AV dissociation has been reported.[35] The tachycardia is characterized by diminutive, upright P waves separated by an isoelectric line in all leads and an atrial rate of less than 250 bpm. Since P waves are small, they may be discernible only in the right precordial leads (Fig. 12.5). The PP interval may be constant or vary by as much as 0.12 s and the PR interval is usually prolonged.

Diagnosis

Differentiation of atrial tachycardia with block from sinus tachycardia and slow atrial flutter with varying block may be difficult. In sinus tachycardia, the amplitude of P waves is greater, the PR interval is usually within normal limits, and carotid sinus pressure may transiently slow the rate without change of the PR interval. In atrial tachycardia with block, the P waves are of lower amplitude, PR interval is usually prolonged, and

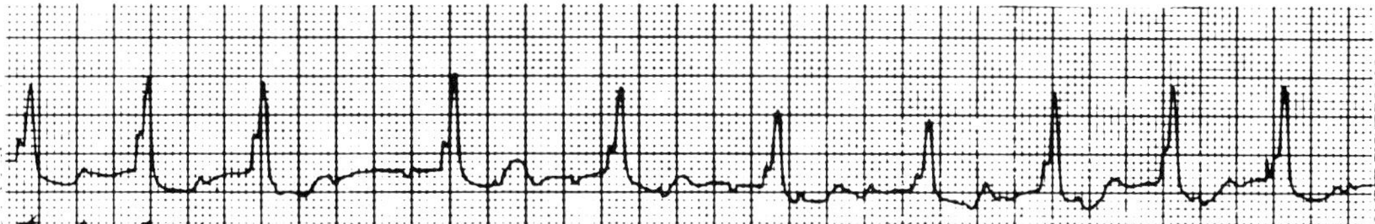

FIGURE 12.5 Monitor lead from a patient with digitalis toxicity showing atrial tachycardia mostly with 2 : 1 atrioventricular block. The irregularity of PP intervals is probably due to exit block from the ectopic focus.

carotid sinus pressure produces AV block, but does not influence the atrial rate. In atrial flutter, the atrial rate usually exceeds 250 bpm, PP interval is regular, no isoelectric baseline is present, and carotid sinus pressure may increase AV block but does not terminate the arrhythmia.

Clinical Course

The prognosis of atrial tachycardia with block depends on the etiology of the arrhythmia and the underlying heart disease. In many patients, prognosis is poor because of the serious underlying heart disease and because of the failure to recognize the relationship between digitalis toxicity and the arrhythmia. Lown and coworkers[29,30] reported a 50% death rate when the arrhythmia was caused by digitalis intoxication and a 75% death rate when it was accompanied by K^+ loss. Mortality was found to be 35% even when digitalis was stopped and appropriate treatment instituted. Dreifus et al.[31] reported an overall mortality of 35%, and a mortality of 100% when digitalis was continued.

MULTIFOCAL ATRIAL TACHYCARDIA

Multifocal atrial tachycardia (MAT) is a SVT characterized by P waves of differing morphology.[36] It is generally assumed that the P waves of varying morphology arise from multiple foci within the atria. However, the possibility that P waves originate from a single focus and that their differing morphology is due to intraatrial conduction disturbances cannot be ruled out.[8,37] Earlier descriptions of MAT attributed the arrhythmia to enhanced automaticity of atrial pacemakers, but recent evidence suggests that MAT may be due to triggered activity of atrial fibers caused by hypoxia, increased circulating catecholamines, and direct mechanical effects produced by changes in atrial pressure and volume.[38–40]

Clinical Features

MAT has been described in all age groups, but most frequently in very ill, elderly patients, particularly those with heart or respiratory failure associated with chronic pulmonary disease.[36,40] The tachycardia may be transient, recurrent, or continuing for months to years. A paroxysmal form of MAT with AV conduction disturbance has been described in young patients.[41] The mean age of patients in most studies was greater than 70 years with a preponderence of men. MAT is most frequently precipitated by an exacerbation of chronic pulmonary disease caused by an acute cardiac decompensation or an infection.[36,40] Although acute or chronic pulmonary disease and ischemic heart disease are the two most frequent conditions associated with MAT, this tachycardia has been reported in a number of other conditions, including carcinoma, pulmonary embolism, valvular heart disease, renal failure, and electrolyte disorders such as hypokalemia, hypomagnesemia, and hyponatremia.[8,36,40] There is a high incidence of MAT in patients recovering from surgery. In one study, 9 of 32 patients with MAT were convalescents from major surgery.[40] The postoperative courses of some of these patients were complicated by infection, pulmonary embolism, and congestive heart failure. Mulitfocal atrial tachycardia has been described as "prefibrillatory rhythm," suggesting a close relationship between MAT and atrial fibrillation.[42,43] Mulitfocal atrial tachycardia may evolve into atrial fibrillation or appear after conversion of atrial flutter or fibrillation. In one study, MAT evolved into atrial fibrillation in 53% of patients.[44] The incidence of this arrhythmia in outpatients has not been studied adequately but in hospitalized patients, the reported incidence varied from 0.13 to 0.38%.[45] In patients 75 years old and older, MAT was the second most commonly diagnosed SVT, exceeded only by atrial fibrillation.[45]

ECG Characteristics

The electrocardiogram in MAT is characterized by three or more morphologically distinct P waves, varying PP, PR, and RR intervals, and an isoelectric baseline between P waves[36,45] (Fig. 12.6). Some of the P waves may be conducted to the ventricles with aberration and some may be

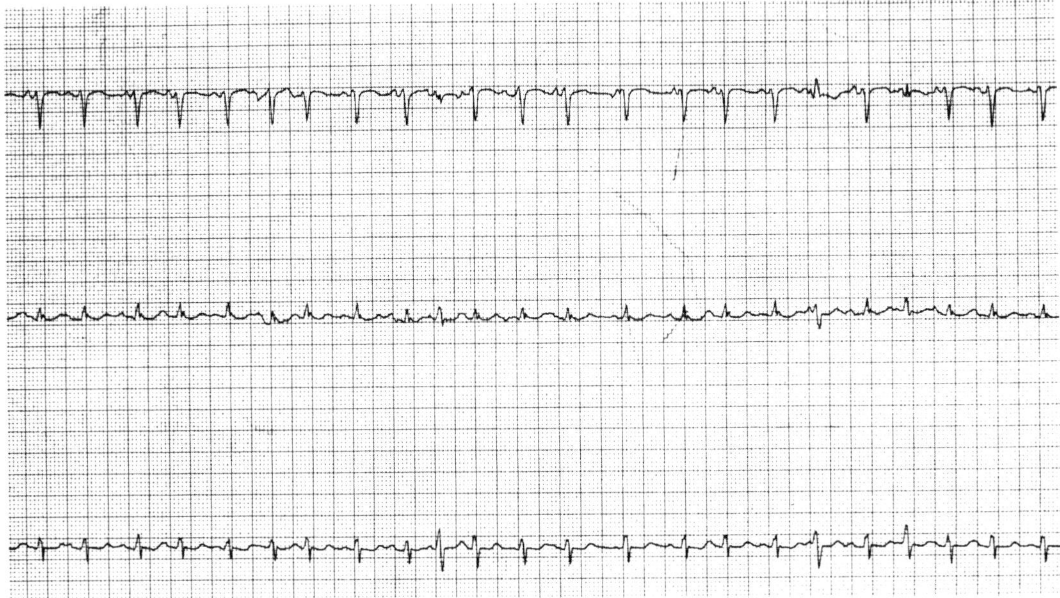

FIGURE 12.6 Multifocal atrial tachycardia. The tracing shows leads V1, V2, and V6, and was recorded in a patient with chronic pulmonary disease and acute respiratory failure. Note P waves of differing morphology with irregular PP and RR intervals and an occasional QRS complex with aberrant conduction.

nonconducted. The ventricular rate is usually between 100 and 150 bpm, but rates as high as 250 bpm have been reported.[8,45] The tachycardia is clinically indistinguishable from atrial fibrillation and the diagnosis can be made only by an electrocardiogram. The arrhythmias may be misdiagnosed when P waves are inconspicuous on the surface ECG. In such cases, intraatrial electrogram recordings or recordings from an esophageal lead may be needed to differentiate MAT from atrial fibrillation.[40] In one retrospective study of hospitalized patients with MAT, the arrhythmia was correctly diagnosed at the bedside in only 22% of patients.[40]

Clinical Course

In patients with respiratory failure, congestive heart failure, or infection, the presence of MAT implies poor prognosis and is accompanied by a high mortality rate. The average in-hospital mortality rate in all studies that investigated this relation was 46%, with rates as high as 60% in some series.[8,36,40,41] The mortality, however, is not due to MAT but to the underlying disease. Very few studies have examined the long-term survival rates in patients with MAT. In one study of 50 patients with MAT, the survival time from the first observed episode of MAT ranged from 1 day to 55 months, with an average duration of 14 months.[40]

JUNCTIONAL ECTOPIC TACHYCARDIA

Junctional ectopic tachycardia (JET) is an uncommon form of SVT first described in infants in 1976.[46] Since then, JET has been described in children of all ages and in adults[47–51]; however, the severest form of this tachycardia occurs in infants and children with uncorrected congenital cardiac defects. When JET occurs before the age of 6 months, it is described as a "congenital" arrhythmia.[51] JET is generally thought to be due to abnormal automaticity in the His-bundle region.[50] However, some investigators have suggested that JET may be due to slow channel-dependent triggered activity.[52]

Clinical Features

Junctional ectopic tachycardia is known to occur at all ages. However, the clinical features and prognosis of congenital JET appear to be different from those of JET that occur in older children and adults.[50,51] Congenital JET typically is incessant with either brief or absent periods of intervening sinus rhythm between the attacks.[46,47,51] In a recent study of 26 infants with congenital JET, the age at which diagnosis was made varied from birth to 6 months, with the majority of the patients having symptoms before 4 weeks of age.[51] In 3 infants, a history of intra-

uterine tachycardia was diagnosed in vitro. Family history of JET was present in 50% of children. At the time of diagnosis, 16 of 26 children had varying degrees of congestive heart failure. The occurrence of congestive heart failure was not related to the age of the child but to the rate of tachycardia.(48–51) The heart rate in children with congestive heart failure was significantly greater than in those without congestive heart failure.(51) Noninvasive evaluation of cardiac function by echocardiography in asymptomatic children with JET frequently reveals dilation and decreased shortening fraction of the left ventricle.(48–51)

The clinical manifestations of JET in older children and adults share many features with congenital JET, but they also differ in several aspects.(48,50) In adults with JET, structural heart disease is infrequently present and the tachycardia occurs sporadically, being separated by long intervals of sinus rhythm. Several attacks may occur during a day lasting from several seconds to hours.(50) Junctional ectopic tachycardia is highly sensitive to changes in sympathetic tone. Exercise and stress frequently precipitate tachycardia, presumably by a reflex increase in sympathetic tone. Adults and older children with JET are frequently asymptomatic, but significant symptoms, including syncope, have been reported.(48,50)

ECG Characteristics

JET usually has a normal QRS morphology except when rate-dependent aberrant conduction is present (Fig. 12.7). The rate of tachycardia varies from 110 to 420 bpm, with faster rates occurring in the congenital form of JET. In adults and older children, the tachycardia is irregular at a slower rate (Fig. 12.7), but in congenital JET, the tachycardia tends to be regular at a faster rate. The irregularity and cycle-length alternation of tachycardia observed in some patients with JET remains unexplained but may be due, at least in some cases, to Wenckebach exit block from the ectopic focus. AV dissociation is the rule with intermittent sinus captures.

Electrophysiology

Intracardiac recordings with His-bundle electrograms determine the origin of this tachycardia in the AV junction; that is, the QRS complexes are always preceded by a His-bundle deflection with

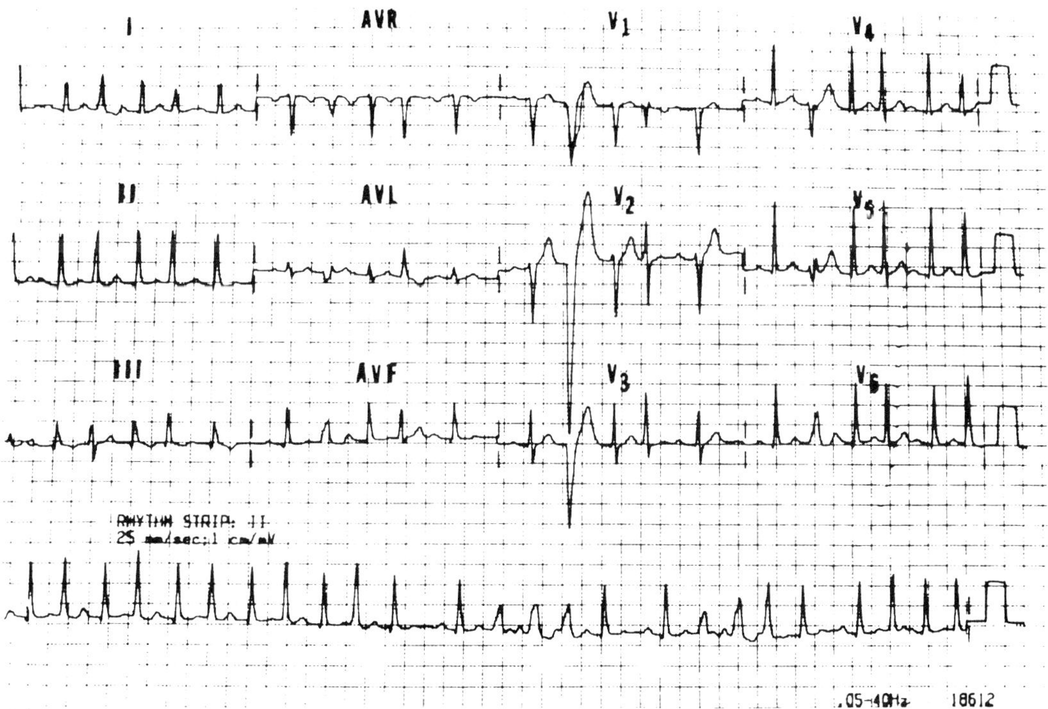

FIGURE 12.7 Junctional ectopic tachycardia. The 12-lead ECG and lead II rhythm strip from a patient with junctional ectopic tachycardia. The tachycardia is irregular with a rate of approximately 160 bpm, and occasional aberrant ventricular conduction. There is atrioventricular dissociation with intermittent sinus captures. *Reprinted from Ruder et al.,*(50) *with permission.*

a normal or prolonged HV interval.[48,50] Atrial or ventricular programmed stimulation does not initiate tachycardia. During spontaneous tachycardia neither atrial and ventricular overdrive pacing nor programmed stimulation causes acceleration or termination of tachycardia.[48,50,51] During atrial pacing, AV conduction is always present, but during ventricular pacing, retrograde conduction to the atria is rarely present.[48,50] The response to isoproterenol can be blunted or prevented by β-adrenergic blocking agents. The presence of AV dissociation during tachycardia and the inability to initiate or terminate the tachycardia by pacing techniques rule out reentry and suggest abnormal automaticity as the mechanism of JET.

Diagnosis

In most cases, an electrocardiogram is sufficient to make the diagnosis of JET. JET can be easily differentiated from the more common nonparoxysmal junctional tachycardia because the latter is usually associated with digitalis toxcity, cardiac surgery, or inferior wall myocardial infarction. Also, nonparoxysmal AV junctional tachycardia tends to be regular with a slower rate. When the tachycardia is irregular, an erroneous diagnosis of atrial fibrillation or multifocal atrial tachycardia can be made. However, the presence of dissociated, unimorphic P waves facilitates the correct diagnosis.

Clinical Course

The clinical course of congenital JET is malignant. In infants, the rate of JET is very rapid, and the arrhythmia usually does not respond to any treatment, including amiodarone, resulting in congestive heart failure and, not infrequently, in the child's death.[48,51] In a multicenter study of 26 patients with congenital JET, the overall mortality was 35%.[51] In these patients, digitalis had no beneficial effect, propranolol was rarely useful, and class I antiarrhythmic agents had no significant effect.[48,49,51]

Adults and older children with JET are frequently asymptomatic, have a benign prognosis, and treatment with β-blockers, Ca^{2+} channel blockers, digoxin, and class I antiarrhythmic agents alone or in various combinations is frequently successful in suppressing tachycardia or reducing its rate.[50] In congenital JET, amiodarone appears to be the most useful agent in suppressing or controlling the rate of tachycardia. Catheter or surgical ablation of AV junction may be useful in patients with JET and congestive heart failure unresponsive to medical treatment.

NONPAROXYSMAL ATRIOVENTRICULAR JUNCTIONAL TACHYCARDIA

Nonparoxysmal AV junctional tachycardia has been attributed to accelerated impulse generation in the His bundle or contiguous structures with the intrinsic property of automaticity.[53] However, recent evidence suggests that the mechanism of triggered automaticity is due to delayed afterdepolarizations.[54] The exact site of origin of this tachycardia cannot be determined with certainty. On the basis of their studies on the effect of verapamil in dogs with AV junctional rhythm produced by SA nodal ablation, Hariman and Chen[55] postulated that the N region of the AV junction was the site of origin of this arrhythmia. In their study, verapamil decreased the rate of junctional rhythm. In clinical studies, verapamil has produced variable effects with either no change, decrease, or increase in the rate of AV junctional rhythm.[56–58] It is possible that in some cases this tachycardia may originate in a small group of atrial fibers, the mass of which is insufficient to generate a P wave that can be recorded on the electrocardiogram.[59] In such cases, diligent mapping of atria during electrophysiologic study may disclose its atrial origin.[58]

Clinical Features

Nonparoxysmal AV junctional tachycardia seldom occurs in the absence of digitalis toxicity, inferior or posterior myocardial infarction, cardiac surgery, viral or rheumatic carditis, or hypokalemia.[60–62] In one study, heart disease was present in 28 of 30 patients with nonparoxysmal AV junctional tachycardia, and in more than half of the patients the tachycardia was due to digitalis toxicity.[60] Dreifus et al.[63] examined 11,000 consecutively recorded electrocardiograms and found 41 cases of nonparoxysmal AV junctional tachycardia. The tachycardia was due to digitalis intoxication in 16 patients, cardiac surgery in 12 patients, and acute myocardial infarction or hypokalemia in 13 patients. The clinical course of this arrhythmia is usually benign and self-limiting. However, in patients with significant preexisting hemodynamic impairment, loss of atrial contribution may result in decreased cardiac output and significant symptoms.

ECG Features

The electrocardiogram usually shows a narrow QRS complex, but may occasionally show aberrant conduction (Fig. 12.8). In either case, the

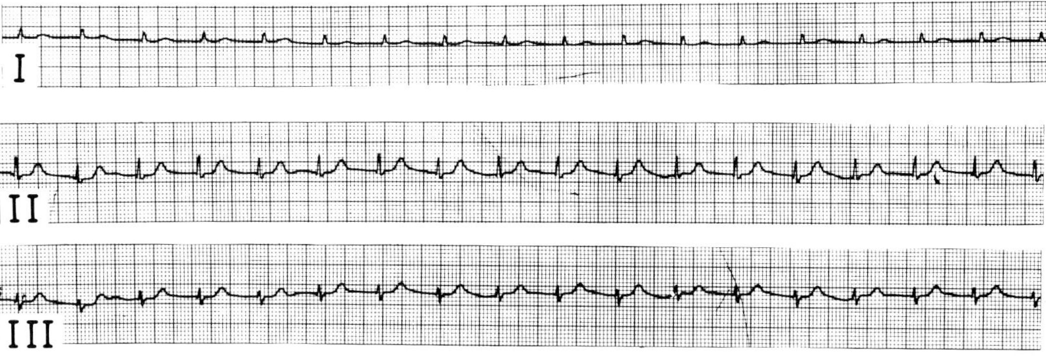

FIGURE 12.8 Nonparoxysmal AV junctional tachycardia. The ECG showing leads I, II, and III was recorded in a patient recovering from open heart surgery. The tachycardia shows a regular rhythm with a rate of 100 bpm. P waves are not apparent on the ECG, but intracardiac electrogram recordings (not shown) revealed the presence of atrial activity within the QRS complex.

QRS is very similar to that in sinus rhythm and the rate of tachycardia is usually 70 to 130 bpm (Fig. 12.8). When the rate of tachycardia is less than 100 bpm, it is described as accelerated AV junctional rhythm. Because the conditions precipitating nonparoxysmal AV junctional tachycardia often impair AV nodal conduction, AV dissociation is frequently present. Retrograde conduction to the atria is present in less than 20% of patients, and the P wave may be inscribed before, during, or after the QRS complex. When the P wave is before the QRS, the PR interval is less than 0.10 s and when the P wave follows the QRS, the RP interval does not exceed 0.20 s. When AV dissociation is present, atria may be controlled by sinus rhythm, atrial tachycardia, atrial flutter, or atrial fibrillation.[64–66] In some cases, atria and ventricles may be depolarized by two separate AV junctional pacemakers, resulting in double AV junctional tachycardia.[60] Depending upon the severity of impairment of AV nodal conduction, the AV dissociation may be complete with no conduction across the AV node in both anterograde and retrograde direction or incomplete with intermittent capture of the ventricles by the atrial rhythm. Similarly, the AV junctional focus may intermittently capture the atria, depending upon the relationship between the atrial (P wave) and ventricular (QRS complex) events.

Diagnosis

The diagnosis of nonparoxysmal AV junctional tachycardia is rarely difficult. The gradual onset and termination, slower rate of tachycardia, frequent AV dissociation, and presence of associated clinical conditions help to differentiate nonparoxysmal AV junctional tachycardia from other types of SVT. When clinical evaluation and analysis of electrocardiogram are inconclusive, electrophysiologic studies may be helpful.

Clinical Course

Nonparoxysmal AV junctional tachycardia is a benign arrhythmia in which the prognosis depends on the underlying etiology. The tachycardia resolves following correction of the underlying cardiac disorder or discontinuation of digitalis. In patients with acute myocardial infarction, the arrhythmia lasts only for a few days and does not indicate unfavorable outcome. In rheumatic fever, AV junctional tachycardia may be the only manifestation of cardiac involvement, and resolves with clinical recovery.[61] In patients after cardiac surgery, AV junctional tachycardia usually disappears within 1 to 5 days and is of no prognostic significance.[60]

Triggered Automaticity as a Mechanism of SVT

In 1974, Cranefield and Aronson[67] suggested that clinical arrhythmias may be due to triggered automaticity produced by delayed afterdepolarizations and could induce clinical arrhythmias. Triggered activity due to delayed afterdepolarizations can be readily induced in canine,[68] simian,[69] and human[70] mitral valves superfused with norepinephrine and in human atrial fibers obtained from dilated and diseased atria.[71] However, the relevance of these findings to clinically observed SVT has not been established because of the absence of specific criteria to diagnose this mechanism. The inability to confirm or to exclude the role of triggered automaticity in clinical SVT is due in part to the similarity of many electrophysiologic characteristics of trig-

gered rhythms to those of reentrant rhythms. Supraventricular tachycardias due to both mechanisms can be initiated by spontaneous or induced premature complexes or by rapid pacing.

Based on their knowledge of the cellular electrophysiologic characteristics of delayed afterdepolarization, Rosen et al.(54) devised rules expected from clinical arrhythmias due to triggered automaticity. They tested these rules in 55 cases of accelerated AV junctional escape rhythm and concluded that the characteristics of these rhythms were similar to those expected of triggered rhythms.(54) In their study, the escape intervals and numbers of escape complexes were directly influenced by the cycle length of the dominant rhythm and by the CI of the premature complex preceding the escape complex.

In a study of 330 patients with SVT who underwent programmed stimulation studies, Wellens et al.(72) found only 6 patients who fulfilled one or more criteria for triggered automaticity (Table 12.1). Goren et al.(73) reported a case of repetitive AV junctional tachycardia, the characteristics of which resembled those of triggered rhythms in vitro. Tachycardia could be initiated by rapid atrial pacing, but not by atrial or ventricular extrastimuli; initiation did not depend on achievement of critical conduction delay, and was facilitated by sympathetic stimulation and prevented by propranolol.

Some SVTs may exhibit characteristics of both abnormal automaticity and triggered automaticity. In a case of incessant automatic atrial tachycardia reported by Moro et al.,(74) atrial fibers from the ectopic focus excised during a curative operation exhibited abnormal automaticity and triggered activity with both early and delayed afterdepolarizations and after isoproterenol infusion. Fujiki et al.(75) reported a case of ectopic automatic atrial tachycardia in which the electrophysiologic characteristics were compatible with triggered automaticity. After intravenous verapamil administration, these characteristics were absent and arrhythmia characteristics became compatible with abnormal automaticity.

Criteria Used to Differentiate Various Mechanisms of SVT During Programmed Stimulation

Although significant progress has been made in our understanding of electrophysiologic mechanisms that can cause SVT, we still cannot relate specific mechanisms to specific clinical SVTs with certainty. However, several criteria based on the observations made in the electrophysiology laboratory during initiation, perpetuation, and termination of tachycardia can be helpful in the differential diagnosis of various mechanisms (see Table 12.1). These criteria are discussed in detail elsewhere in this book. It should be emphasized that in many cases not all of these criteria can be evaluated, and that even detailed electrocardiographic, electrophysiologic, or pharmacologic studies may fail to distinguish reentry from automaticity. For instance, induction of tachycardia by programmed stimulation suggests reentry; inability to induce tachycardia by this method does not rule out reentry.

The inverse relation between the coupling interval of premature complex initiating tachycardia and the interval between premature complex and the first complex of tachycardia, listed in Table 12.1, appears to be clinically most useful in differentiating reentry from triggered automaticity. Overdrive acceleration cannot be accepted as specific because it can also occur in reentrant tachycardias. Similarly, before one accepts evidence based on suppression by verapamil, it is necessary to rule out the role of the AV node in initiation or perpetuation of tachy-

TABLE 12.1 Electrophysiologic Criteria Used to Differentiate Reentrant from Nonreentrant Supraventricular Tachycardia

Criteria	Reentry	Abnormal Automaticity	Triggered Automaticity
Initiation by premature complex	+	−	+
Initiation by rapid pacing	+	−	+
Termination by premature complex	+	−	+
Termination by overdrive pacing	+	−	+
Overdrive acceleration	?	−	+
Inverse relation between the coupling interval of tachycardia initiating premature complex and the interval between the premature complex and the first complex of tachycardia	+	−	−
Delayed conduction of tachycardia initiating premature complex	+	−	−

cardia. A more accurate diagnosis of tachycardia mechanisms must await the development of more specific clinical criteria, and, perhaps, more reliable extracellular recording techniques for detection of automatic activity.

REFERENCES

1. Allessie MA, Bonke FIM, Schopman F: Circus movement in rabbit atrial muscle as mechanism of tachycardia. *Circ Res* 1973;33:54–62.
2. Wit AL, Hoffman BF, Cranefield PF: Slow conduction and reentry in the ventricular conducting system. I. Return extrasystole in canine Purkinje fibers. *Circ Res* 1972;30:1–10.
3. Wit AL, Cranefied PF, Hoffman BF: Slow conduction and reentry in the ventricular conducting system. II. Single and sustained circus movement in networks of canine and bovine Purkinje fibers. *Circ Res* 1972;30:11–22.
4. Antzelevitch C, Jalife J, Moe GK: Characteristics of reflection as a mechanism of reentrant arrhythmias and its relationship to parasystole. *Circulation* 1980;61:182–191.
5. Wit AL, Cranefield PF: Triggered activity in cardiac muscle fibers of the simian mitral valve. *Circ Res* 1976;38:85–98.
6. Frame LH, Hoffman BF: Mechanisms of tachycardia, in Surawicz B, Reddy CP, Prystowsky EN (eds): *Tachycardia*. Boston, MA, Martinus Nijhoff Publishing, 1984, pp 7–36.
7. Akhtar M, Tchou PJ, Jazayeri M: Mechanisms of clinical tachycardias. *Am J Cardiol* 1988;61:9A–19A.
8. Reddy CP: Supraventricular ectopic tachycardias due to mechanisms other than reentry, in Surawicz B, Reddy CP, Prystowsky EN (eds): *Tachycardia*. Boston, MA, Martinus Nijhoff Publishing, 1984, pp 173–183.
9. Scheinman MM, Basu D, Hollenberg M: Electrophysiologic studies in patients with persistent atrial tachycardia. *Circulation* 1974;50:266–273.
10. Wu D, Denes P, Amat-y-Leon F, Dhingra R, Wyndham CRC, Bauernfeind R, Latif P, Rosen KM: Clinical electrocardiographic and electrophysiologic observations in patients with paroxysmal supraventricular tachycardia. *Am J Cardiol* 1978;41:1045–1051.
11. Garson A Jr: Supraventricular tachycardia, in Gillete P, Garson A (eds): *Pediatric Cardiac Dysrhythmias*. New York, Grune & Stratton, 1981, pp 177–255.
12. Garson A Jr, Gillette PC: Electrophysiologic studies of supraventricular tachycardia in children. I. Clinical-electrophysiologic correlations. *Am Heart J* 1981;102:233–250.
13. Garson A Jr, Gillette PC: Electrophysiologic studies of supraventricular tachycardia in children. II. Prediction of specific mechanism by noninvasive features. *Am Heart J* 1981;102:383–388.
14. Gillette PC, Garson A: Electrophysiological and pharmacologic characteristics of automatic ectopic atrial tachycardia. *Circulation* 1977;56:571.
15. Dolara A, Pozzi L: Persistent supraventricular tachycardia. *Am J Cardiol* 1965;16:449–451.
16. Weiss HB, McGuire J: Ectopic tachycardia auricular in origin of unusual duration. *Am Heart J* 1936;12:585–591.
17. Ross BA, Crawford FA Jr, Whitman V, Gillette PC: Atrial automatic tachycardia due to an atrial tumor. *Am Heart J* 1988;115:660–610.
18. Miller R, Perelman JS: Chronic auricular tachycardia with unusual response to change in posture. *Am Heart J* 1945;29:555–569.
19. Morgan CL, Nadas AS: Chronic ectopic tachycardia in infancy and childhood. *Am. Heart J.* 1964;67:617–627.
20. Goldreyer BN, Gallagher JJ, Damato AN: The electrophysiologic demonstration of atrial ectopic tachycardia in man. *Am Heart J* 1973;85:205–215.
21. Josephson ME, Kastor JA: Supraventricular tachycardia: Mechanism and management. *Ann Intern Med* 1977;87:348–358.
22. Claiborne TS: Auricular tachycardia with auriculoventricular block of 12 years duration in 16 year girl. *Am Heart J* 1950;39:444–450.
23. Wellens HJJ: Value and limitations of programmed electrical stimulation of the heart in the study and treatment of tachycardias. *Circulation* 1978;57:845–853.
24. Wyndham CRC, Arnsdorf MR, Levitsky S, Smith TC, Dhingra RC, Denes P, Rosen KM: Successful surgical excision of a focal paroxysmal atrial tachycardia. *Circulation* 1980;62:1365.
25. Hendry PJ, Packer DL, Anstadt MP, Plunket MD, Lowe JE: Surgical treatment of automatic atrial tachycardias. *Ann Thorac Surg* 1990;49:253–260.
26. Lewis T: Observations upon flutter and fibrillation. IV. Impure flutter: Theory of circus movement. *Heart* 1920;7:293.
27. Heyl AF: Auricular paroxysmal tachycardia caused by digitalis: Report of case. *Ann Intern Med* 1932;5:858.
28. Barker PS, Wilson FN, Johnston FD, Wishard SW: Auricular paroxysmal tachycardia with auriculoventricular block. *Am Heart J* 1943;25:765.
29. Lown B, Marcus F, Levine HD: Digitalis and atrial tachycardia with block. *N Engl J Med* 1959;260:301.
30. Lown B, Wyatt NF, Levine HD: Paroxysmal atrial tachycardia with block. *Circulation* 1960;21:129.
31. Dreifus LS, McKnight EH, Katz M, Likoff: Digitalis intolerance. *Geriatrics* 1963;18:494.
32. Freiermuth LJ, Jick S: Paroxysmal atrial tachycardia with atrioventricular block. *Am J Cardiol* 1958;1:584.
33. Morgan WL, Breneman GM: Atrial tachycardia with block treated with digitalis. *Circulation* 1962;25:787.
34. Soffer A: The changing clinical picture of digitalis intoxication. *Arch Intern Med* 1962;104:442.
35. Fisch C, Knoebel SB: Recognition and therapy of digitalis toxicity. *Prog Cardiovasc Dis* 1970;13:71.
36. Chung EK: Appraisal of multifocal atrial tachycardia. *Br Heart J* 1971;33:500–504.
37. James TN, Sherf L: Specialized tissues and preferential conduction in the atria of the heart. *Am J Cardiol* 1971;28:414–427.
38. Levine JH, Michael JR, Guarmieri T: Treatment of multifocal atrial tachycardia with verapamil. *N Engl J Med* 1985;312:21–25.
39. Wit AL, Rosen MR: Cellular electrophysiology of cardia arrhythmias, in Josephson ME, Wellens HJJ (eds): *Tachycardias: Mechanisms, Diagnosis and Treatment*. Philadelphia, Lea & Febiger, 1984.
40. Scher DL, Asura EL: Multifocal atrial tachycardia: Mechanisms, clinical correlates and treatment. *Am Heart J* 1989;118:574–580.

41. Omori Y: Repetitive multifocal paroxysmal atrial tachycardia with cyclic Wenckebach phenomenon under observation for thirteen years. *Am Heart J* 1971;32:527–530.
42. Berlinerblau R, Feder W: Chaotic atrial rhythm. *J Electrocardiol* 1972;5:135–144.
43. Kones RJ, Phillips JH, Hersh J: Mechanism and management of chaotic atrial mechanism. *Cardiology* 1974;59:92–101.
44. Phillips J, Spano J, Burch G: Chaotic atrial mechanism. *Am Heart J* 1969;78:171–179.
45. Shine KI, Kastor JA, Yurchak PM: Multifocal atrial tachycardia: Clinical and electrophysiologic features in 32 patients. *N Engl J Med* 1968;279:344–349.
46. Nouaille J, Kachaner J, Ribiere M, Toumiex MC: Tachycardies focales hissiennes congenitales. *Arch Mal Coeur* 1976;69:899.
47. Brechenmacher C, Coumel P, James TN: Intractable tachycardia in infancy. *Circulation* 1976;53:377.
48. Garson A, Gillette PC: Junctional ectopic tachycardia in children: Electrocardiography, electrophysiology and pharmacologic response. *Am J Cardiol* 1979;44:298.
49. Gillette PC, Garson A, Porter CJ, Ott O, McVey P, Zinner A, Blair H: Junctional automatic tachycardia: New proposed treatment by transcatheter His bundle ablation. *Am Heart J* 1983;106:619.
50. Ruder MA, Davis JC, Eldar M, Abbott JA, Griffin JC, Seger JJ, Scheinman MM: Clinical and electrophysiologic characterization of automatic junctional tachycardia in adults. *Circulation* 1986;73:930–937.
51. Villian E, Vetter VL, Garcia JM, Herre J, Cifarelli A, Garson A: Evolving concepts in the management of congenital junctional ectopic tachycardia: A multitcenter study. *Circulation* 1990;81:1544–1549.
52. Santinelli V, Chiariello M, Condovelli M: Junctional ectopic tachycardia and verapamil. *Chest* 1984;85:121–122.
53. Hecht HH, Kossmann CE: Atrioventricular and intraventricular conduction: Revised nomenclature and concepts. *Am J Cardiol* 1973;31:232–244.
54. Rosen MR, Fisch C, Hoffman BF, Danilo P Jr, Lovelace DE, Knoebel SB: Can accelerated atrioventricular junctional escape rhythms be explained by delayed afterdepolarizations? *Am J Cardiol* 1980;45:1272–1284.
55. Hariman RJ, Chen CM: Recording of diastolic slope from the junctional area in dogs with junctional rhythm. *Circulation* 1983;68:636–643.
56. Hariman RJ, Gomes JAC, El-Sherif N: Recording of diastolic slope with catheters during junctional rhythm in man. *Circulation* 1984;69:485–491.
57. Schwartz JB, Nielsen AP, Griffin JC: Concentration dependent enhancement of junctional pacemaker activity by verapamil in man. *Circulation* 1985;71:450–457.
58. Santinelli V, DePaola M, Smimmo D, Turco P, Condorelli M: Further observations on junctional rhythm. Role of verapamil. Letter to the editor. *J Am Coll Cardiol* 1986;8:255–256.
59. Zipes DP, Guam WE, Genetos BC, Glassman RD, Noble RJ, Fisch C: Atrial tachycardia without P waves masquerading as an AV junctional tachycardia. *Circulation* 1977;55:253–260.
60. Pick A, Dominguez P: Nonparoxysmal AV nodal tachycardia. *Circulation* 1957;16:1022.
61. Zimdahl W: Rheumatic fever in young adults. *Br Heart J* 1952;14:70–76.
62. Surawicz B: Tachycardia, in Conn HF (ed): *Current Therapy*. Philadelphia, WB Saunders, 1976, pp 187–194.
63. Dreifus LS, Bartolucci G, Likoff W: Nodal tachycardia: Etiology and therapy. *Circulation* 1960;22:741.
64. Chung EK, Thomas J: Arrhythmias caused by digitalis toxicity. *Geriatrics* 1965;10:1006–1016.
65. Igarashi M: Electrocardiogram in potassium depletion. II. Atrial and nodal tachycardia in digitalis intoxication. *Jpn Circ J* 1963;27:476.
66. Irons GV, Orgain ES: Digitalis induced arrhythmias and their management. *Prog Cardiovasc Dis* 1966; 8:539–569.
67. Cranefield PF, Aronson RS: Initiation of sustained rhythmic activity by single propagated action potentials in canine cardiac Purkinje fibers exposed to sodium-free solution or to ouabain. *Circ Res* 1974;34:477–481.
68. Wit AL, Fenoglio JJ JR, Wagner BM, Bassett AL: Electrophysiological properties of cardiac muscle in the anterior mitral valve leaflet and the adjacent atrium in the dog: Possible implications for the genesis of atrial dysrhythmias. *Circ Res* 1973;32:731–745.
69. Wit AL, Cranefield PF: Triggered activity in cardiac muscle fibers of the simian mitral valve. *Circ Res* 1976;38:85–98.
70. Wit AL, Fenogli JJ Jr, Hordof AJ, Reemstsma K: Ultrastructure and transmembrane potentials of cardiac muscle in the human anterior mitral leaflet. *Circulation* 1979;59:1284–1292.
71. Hordof AJ, Edie R, Malm JR, Hoffman BF, Rosen MR: Electrophysiologic properties and response to pharmacologic agents of fibers from diseased human atria. *Circulation* 1976;54:774–779.
72. Wellens HJJ, Brugada P, Banagt EJDM, Ross DL, Bar FW: New studies with triggered automaticity, in Harrison DC (ed): *Cardiac Arrhythmias: A Decade of Progress*. Boston, GK Hall Medical, 1981, pp 601–610.
73. Goren C, Santucci BA, Bucheleres HG, Denes P: Chronic recurrent ectopic junctional tachycardia resembling triggered automaticity in mitral valve prolapse syndrome. *Am Heart J* 1981;101:504–507.
74. Moro C, Rufilanchas JJ, Tamargo J, Novo L, Martinez J: Evidence of abnormal automaticity and triggering activity in incessant ectopic atrial tachycardia. *Am Heart J* 1986;116:550–552.
75. Fujiki A, Yoshida S, Mizumaki K, Sasayama S: Atrial automatic tachycardia with two different types of electrophysiologic characteristics. *Am Heart J* 1989;117:699–701.
76. Reddy CP, Arnett JD: Automatic atrial tachycardia and nonparoxysmal atrioventricular junctional tachycardia. In Horowitz LN(ed): Current management of arrythmias. Philadelphia, PA, B.C. Decker, Inc. 1991, pp 67–73.

Chapter **13**

Differential Diagnosis of Narrow QRS Tachycardia

Hein J. J. Wellens, MD

Careful analysis of the 12-lead electrocardiogram (ECG) during a supraventricular tachycardia (SVT) can give important information about the site of origin and mechanism of the arrhythmia. This information has become available by analyzing the ECG during tachycardia after programmed stimulation of the heart and intracardiac recordings have established the type of SVT.[1]

In this chapter the emphasis will be upon the value of the 12-lead electrocardiogram during a narrow QRS tachycardia. While references will be given to the use of intracardiac recordings to prove the site of origin of the tachycardia. I believe that for the practicing cardiologist it is much more important to be able to diagnose the different types of SVT correctly on the 12-lead ECG, and when necessary combined with the results of the physical examination.

CLASSIFICATION OF THE DIFFERENT TYPES OF SVT

Table 13.1 gives a classification of the different types of SVT and their manifestations during physical examination.

The ECG in SVT

Figures 13.1 and 13.2 schematically illustrate the site of origin or mechanism of five types of SVT and their ECG during tachycardia. As shown, the location of the P wave in relation to the QRS complex is of great importance in diagnosing the site of origin of SVT.[2] The ECG during tachycardia should be searched carefully for P waves. As will be shown, recognition of the P wave is often facilitated by comparing the ECG during tachycardia with the ECG recorded after termination of tachycardia. The most useful tracing is a multi-lead ECG recorded at the time of termination of tachycardia either by carotid sinus massage or drug administration.

Atrial Tachycardia

All SVT originating in the atrium logically have atrial activation preceding ventricular activation. In atrial tachycardia the configuration of the P wave depends upon its origin in the atrium. This is demonstrated in Figure 13.3. In a series of patients with atrial tachycardia not related to digitalis intoxication, careful mapping of the atria during the arrhythmia revealed that they can occur anywhere in the atrium. Two types of atrial tachycardia should be recognized. One, the so-called paroxysmal type, has a sudden onset and offset (Fig. 13.4). The other, the permanent or incessant type (Fig. 13.5), is present more than half of the day. This distinction is of therapeutic interest. The paroxysmal type of atrial tachycardia can be easily managed in most cases with drugs like digitalis, calcium antagonists, β-blocking agents, and class I antiarrhythmic agents. The incessant form is much more difficult to control with drug therapy, although some success has been reported with class IC and class III agents. The mechanisms of these two types of atrial tachycardia appear to be different. In most cases, the paroxysmal form can be easily initiated and terminated during programmed stimulation, which suggests reentry or triggered activity as its basic mechanism.

The incessant form of atrial tachycardia cannot be initiated and terminated during programmed electrical stimulation of the heart. This suggests abnormal automaticity as the mechanism of the arrhythmia. Recognition of the incessant form of atrial tachycardia is important, not only because continuous tachycardia may

TABLE 13.1 Clinical Findings and Blood Pressure Measurements in Different Types of Supraventricular Tachycardia

Tachycardia	Pulse	Neck Veins	Blood Pressure	Loudness First Heart Sound
Sinus tachycardia	Regular	No abn.	Constant	Constant
Atrial tachycardia	Regular	No abn.	Constant	Constant
Atrial flutter	a. Regular during 2:1 AV conduction	Flutter waves	a. Constant during regular pulse	Constant
	b. irregular during changing AV conduction		b. Changing during irregular pulse	Changing
Atrial fibrillation	Irregular	Irregular pulsation	Changing	Changing
AV nodal tachycardia	Regular	"Frog" sign	Constant	Changing
Circus movement tachycardia (using concealed accessory pathway)	Regular	"Frog" sign	Constant	Changing

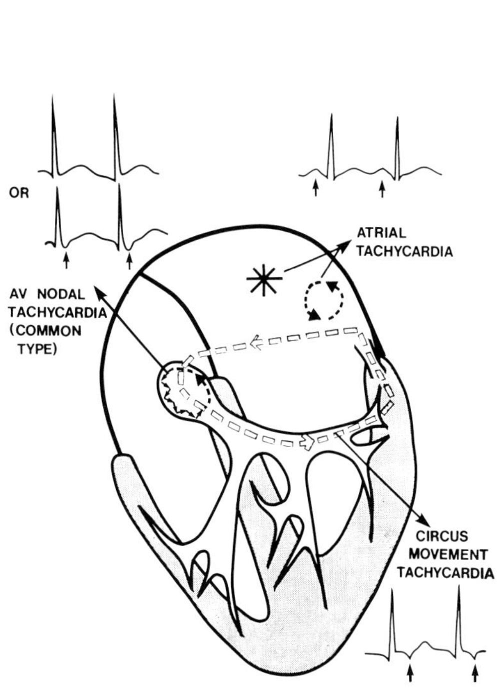

FIGURE 13.1 Three types of supraventricular tachycardia and the relation between QRS and P wave during tachycardia. Note that in atrial tachycardia P precedes QRS; P occurs simultaneously with QRS in the common type of AV nodal tachycardia and follows QRS in circus movement tachycardia using a fast-conducting accessory AV pathway.

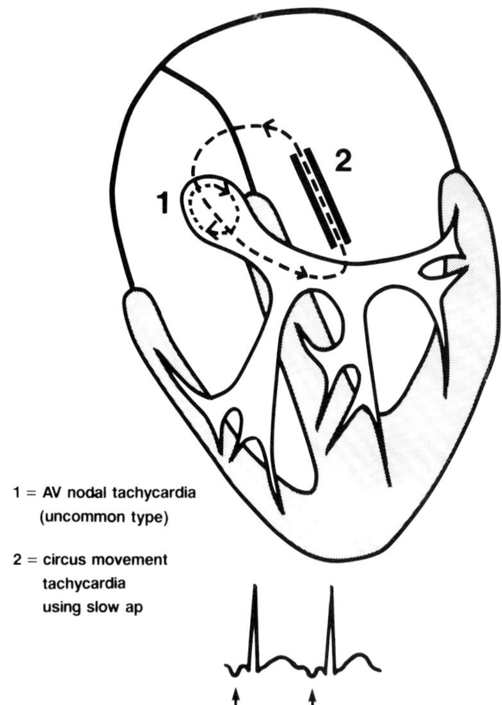

FIGURE 13.2 Diagrammatic representation of two unusual types of supraventricular tachycardia both resulting in an ECG during tachycardia showing a long RP interval. The uncommon type of AV nodal tachycardia is rarely sustained. The circus movement tachycardia using an accessory AV pathway with long conduction times for retrograde conduction is frequently incessant and may lead to a dilated cardiomyopathy.

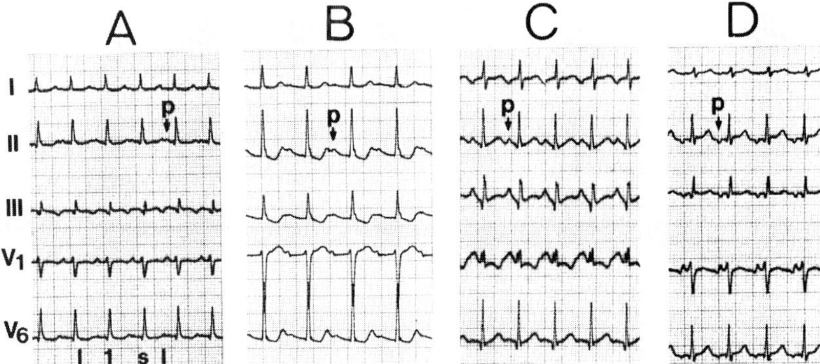

FIGURE 13.3 Four panels showing five surface leads in four different patients with atrial tachycardia. Note the difference in P-wave configuration related to the site of impulse formation in the atrium.

lead to dilated cardiomyopathy, but also because surgical therapy or electrical ablation of the arrhythmia focus should be considered.

A special form of atrial tachycardia is caused by digitalis intoxication. As shown in Figure 13.6 and described elsewhere,[3] this type of atrial tachycardia has several characteristic features, including an atrial wave form that suggests an origin close to the sinus node area. Another feature is the phenomenon of ventriculophasic behavior of the P waves: The PP interval around a QRS complex is shorter than the PP interval not enclosing a QRS complex. Although it has been demonstrated that delayed afterdepolarizations play a role in the genesis of digitalis-induced arrhythmias, no information is currently available in support of such a mechanism as the cause of clinically occurring digitalis-induced atrial tachycardia.

In atrial tachycardia the relation between atrial and ventricular events depends upon the properties of the AV node. Therefore, 1 : 1 AV conduction, Wenckebach or higher degree of AV conduction block can be seen. Figure 13.7 shows the effect of carotid sinus massage in a patient with incessant atrial tachycardia.

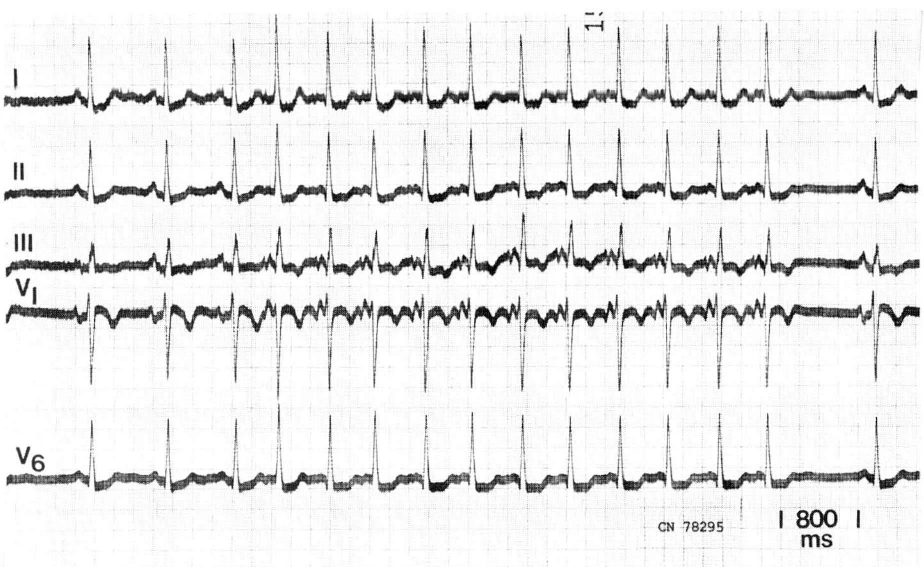

FIGURE 13.4 Example of a paroxysmal atrial tachycardia. Note the onset of the arrhythmia after three conducted sinus beats with a P wave that precedes the QRS complex but has a different configuration from the sinus P waves.

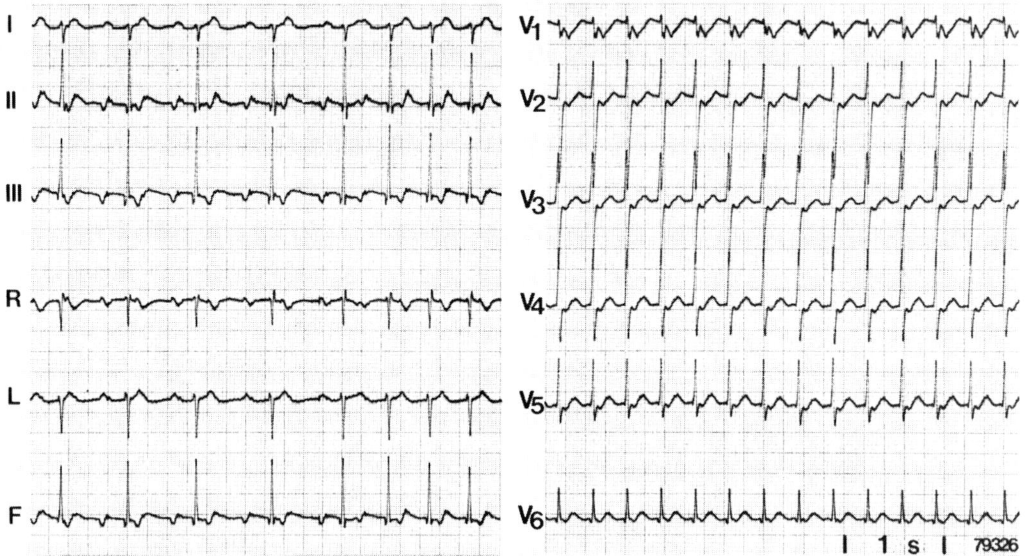

FIGURE 13.5 Example of an incessant atrial tachycardia. This patient, initially showing 2 : 1 and later 1 : 1 AV conduction, had been continuously in tachycardia for 12 years and presented with a picture of a dilated cardiomyopathy.

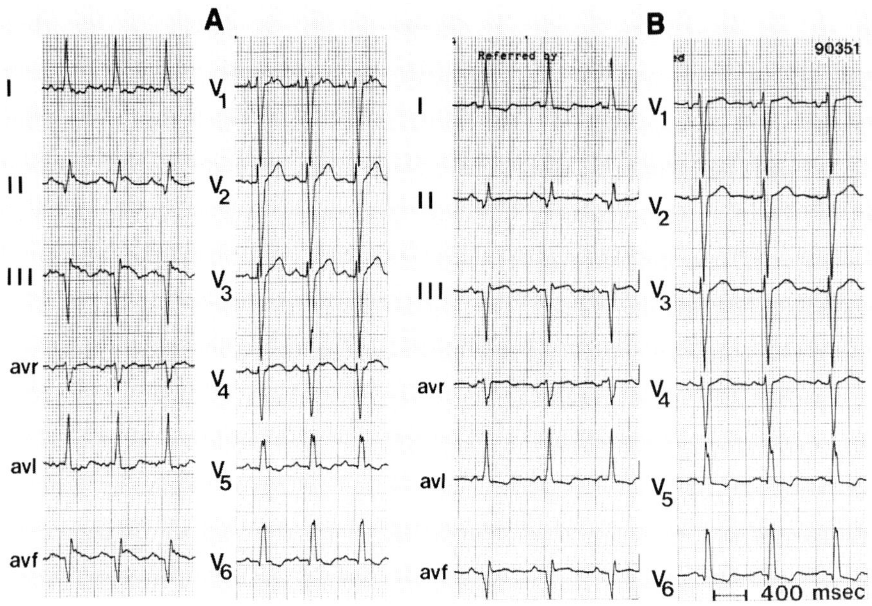

FIGURE 13.6 A. Atrial tachycardia with 2 : 1 AV conduction in digitalis intoxication. Note upright P waves in II, III, and AVF during atrial tachycardia, and the phenomenon of ventriculophasic behavior of the P–P interval (see text). **B.** The same patient after digitalis intoxication has subsided.

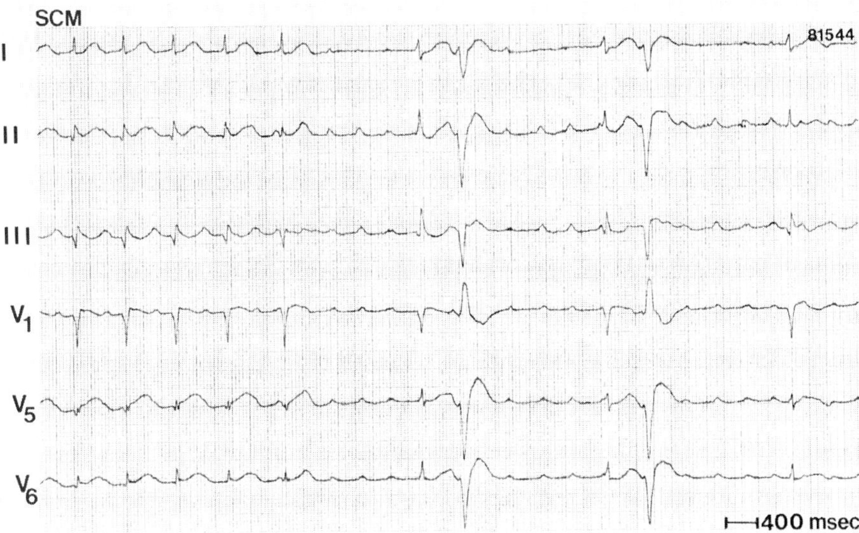

FIGURE 13.7 Effect of carotid sinus massage in incessant atrial tachycardia. By creating AV block the atrial complexes can easily be recognized.

Atrial Flutter

During atrial flutter there is a rapid regular rhythm at the atrial level (250 to 350/min). Observations both in the tissue bath and in patients suggest reentry as the most likely mechanism of the arrhythmia.[4–10] Electrocardiographically there is no isoelectric interval between two successive flutter waves. The ventricular rate during atrial flutter depends upon the conductive properties of the AV node. Especially during 2 : 1 AV conduction, recognition of flutter as the underlying mechanism can be difficult in the patient presenting with a regular narrow tachycardia and a ventricular rate of approximately 150/min. As shown in Figures 13.8 and 13.9, carotid sinus massage facilitates making the correct diagnosis.

Atrial Fibrillation

This arrhythmia is the SVT that is electrocardiographically the most easy to recognize. Because of the rapid irregular atrial rate, the ventricular rhythm is completely irregular unless complete AV block or an AV junctional tachycardia is present.

Atrioventricular Nodal Tachycardia

The reproducible initiation and termination of paroxysmal AV nodal tachycardia by programmed stimulation of the heart suggests reentry as the underlying mechanism. This is supported by the finding of "dual" AV nodal conduction in many of these patients.[11] The common type of paroxysmal AV nodal tachycardia typically shows simultaneous activation of the atrium and the ventricle during the arrhythmia[12] (see Fig. 13.1). Anterograde conduction in the AV node during tachycardia is considered to occur over a slowly conducting pathway and retrograde conduction over a rapidly conducting pathway. The uncommon type of paroxysmal AV nodal tachycardia is characterized by a P wave that follows the QRS complex, the mechanism being anterograde AV nodal conduction over a rapid pathway and retrograde conduction over a slow pathway (see Fig. 13.2). There has been debate for many years about the participation of atria and ventricles in the reentry circuit.[13,14] Evidence of the continuation of AV nodal tachycardia has been presented in the case of conduction block to the atria or ventricles and in the presence of atrial fibrillation, which suggests a true intranodal location of the reentry circuit.[15,16] Recent surgical interventions, however, have again questioned the participation of portions of the atrium in the tachycardia circuit.[17–19] The work of Ross et al.[18] especially favors such a situation in reentrant AV nodal tachycardia. These investigators successfully operated on patients by making a discrete dissection around the AV node.

Accelerated AV junctional impulse formation (of a nonparoxysmal form) may occur in ischemia, inflammation, postcardiac surgery, and digitalis intoxication. The exact site of origin in the AV junction (bundle of His?) and the mechanism of these tachycardias are not known.

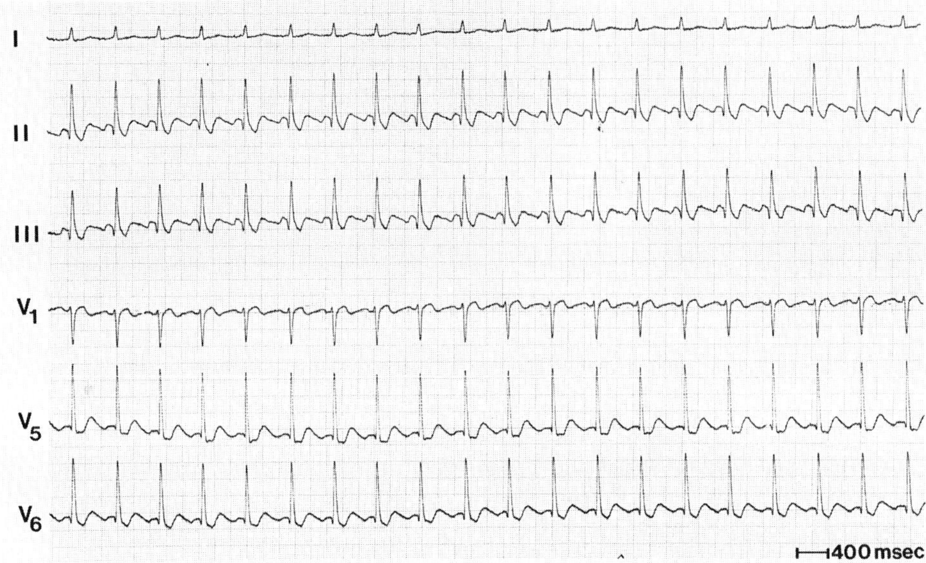

FIGURE 13.8 Supraventricular tachycardia with a ventricular rate of 130/min

In the common form of AV nodal tachycardia the P wave usually occurs simultaneously with the QRS complex. When the P wave is located in the last part of the QRS complex, pseudo S waves in leads II, III, and AVF and pseudo R waves in V1 may result (Fig. 13.10). Rarely, the P wave is located at the beginning of the QRS leading to a pseudo Q wave (Fig. 13.11).

Occasionally during AV nodal tachycardia there is 2 : 1 conduction from the AV node to the ventricle.(15) As shown in Figure 13.12, this results in a very characteristic electrocardiogram with the P wave (which is negative in II, III, and AVF, positive in V1, and negative in V3 to V6) exactly in between two QRS complexes (the other P wave occurring simultaneously with the QRS complex).

In patients with AV nodal tachycardia, ven-

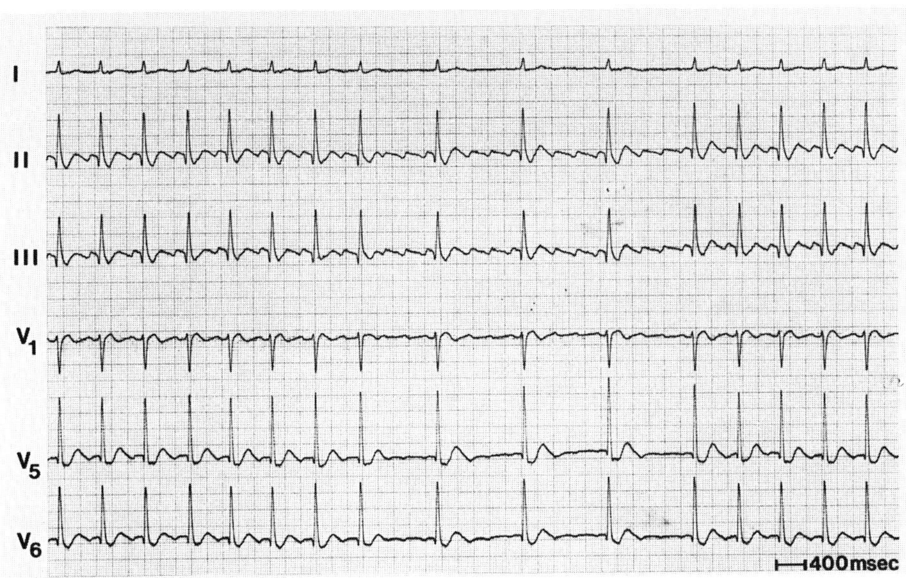

FIGURE 13.9 Carotid sinus massage reveals that atrial flutter is the underlying rhythm at the atrial level.

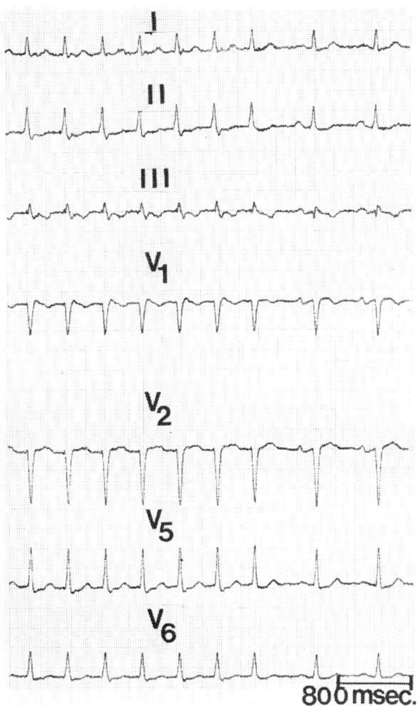

FIGURE 13.10 Example of an AV nodal tachycardia that ends spontaneously and allows a comparison of the QRS–T during tachycardia and during sinus rhythm. Note that the atrial activation during tachycardia results in pseudo S waves in leads II and III and a pseudo incomplete right bundle branch block pattern in lead V1.

tricular premature beats can be very helpful in identifying the P wave (Fig. 13.13).

Circus Movement Tachycardia

Epicardial mapping[20] and electrophysiologic investigations[21,22] have shown that accessory connections between atrium and ventricle frequently participate in tachycardia circuits. Sophisticated intracardiac stimulation techniques and recordings are required to demonstrate the location and participation of these extra connections.

An interesting subgroup of patients with SVT are those using a so-called concealed accessory pathway. These connections conduct the impulse in only a ventriculoatrial direction. They are often present in patients referred for evaluation of their SVT. There are two groups of concealed accessory AV pathways (see Figs. 13.1 and 13.2). The most common type of concealed accessory pathway is the one that conducts the impulse rapidly from ventricle to atrium. This results in an RP interval shorter than the PR interval during tachycardia. The group using a slowly conducting accessory pathway (see Fig. 13.2) is small. Atrial activation typically follows the QRS complex with an RP interval that is longer than the PR interval. This type of SVT must be differentiated from a low atrial tachycardia and an AV nodal tachycardia of the uncommon type.[23] Figure 13.14 illustrates schematically the two types of circus movement tachycardia using a concealed accessory AV pathway. It is of interest that during carotid sinus massage the circus movement tachycardia using a fast-conducting accessory pathway usually terminates because of block in the AV node, while the circus movement tachycardia using a slow-conducting accessory pathway frequently stops because of block in the slow pathway (Fig. 13.15). The latter finding suggests that such pathways have AV nodal properties.

In patients with a circus movement tachycardia using retrogradely an accessory pathway, clues about the atrial location of the pathway can usually be obtained from the configuration of the

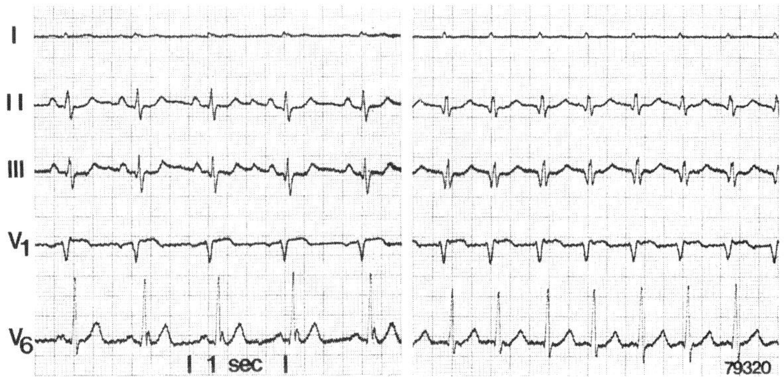

FIGURE 13.11 The ECG on the right recorded during AV nodal tachycardia shows that atrial activation results in pseudo Q waves in leads II and III. Compare the ECG during sinus rhythm with the one taken during tachycardia to recognize the effect of the P wave on QRS configuration.

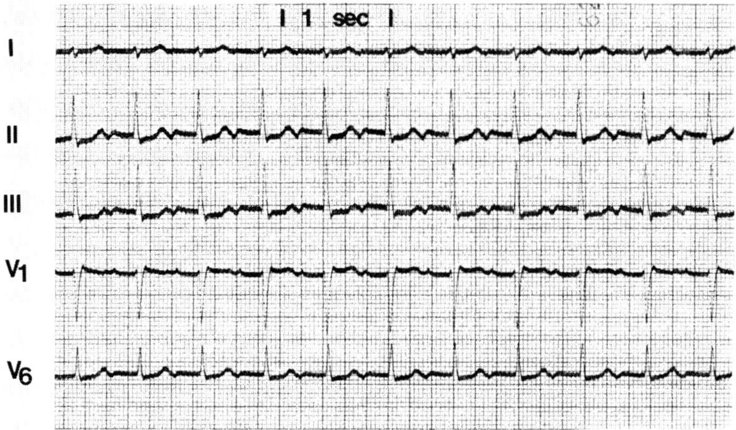

FIGURE 13.12 Example of AV nodal tachycardia with 1 : 1 conduction to the atrium and 2 : 1 conduction to the ventricle. This results in P waves, negative in II and III, located exactly in between two QRS complexes, the other P wave being hidden in the QRS complex.

P wave during tachycardia. A left lateral insertion results in negative P waves in lead I and frequently upright P waves in leads III and AVF.(24) To have the best information about the configuration of the P wave during tachycardia, careful comparison with the ECG during sinus rhythm is essential!

An important pointer to the diagnosis of a circus movement tachycardia is the finding of QRS alternation during SVT (Fig. 13.16). As we have described, alternating changes in the QRS complex during a narrow QRS tachycardia are highly suggestive of a circus movement tachycardia using an accessory pathway for ventriculoatrial conduction.(25) It is important to stress that QRS alternation as a clue to a circus

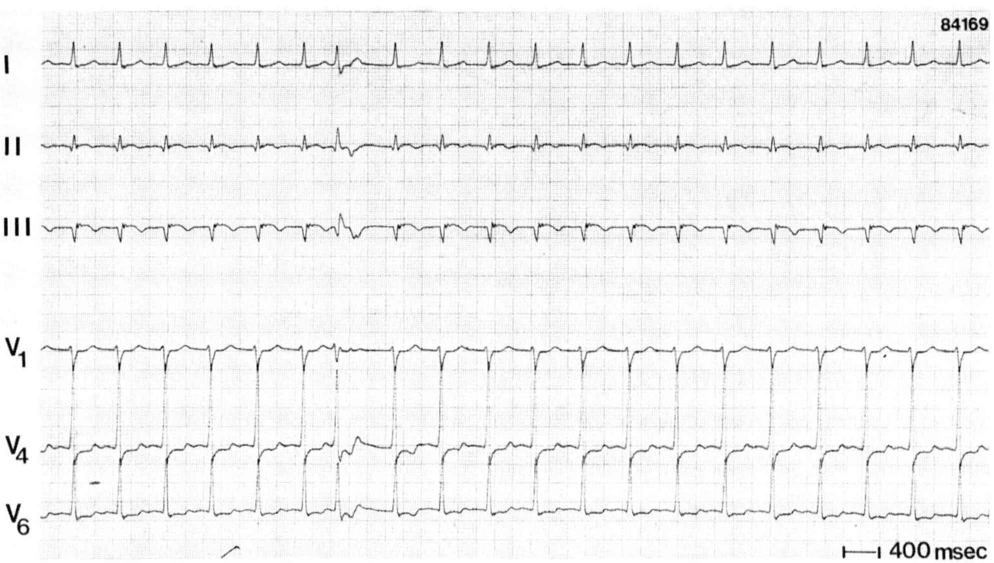

FIGURE 13.13 Example of the value of a ventricular premature beat in recognizing the P wave in AV nodal tachycardia. Note that the ventricular premature beat exposes a P wave, negative in leads II and III, which occurs exactly at the time of the expected QRS of AV nodal tachycardia if no ventricular premature beat had occurred.

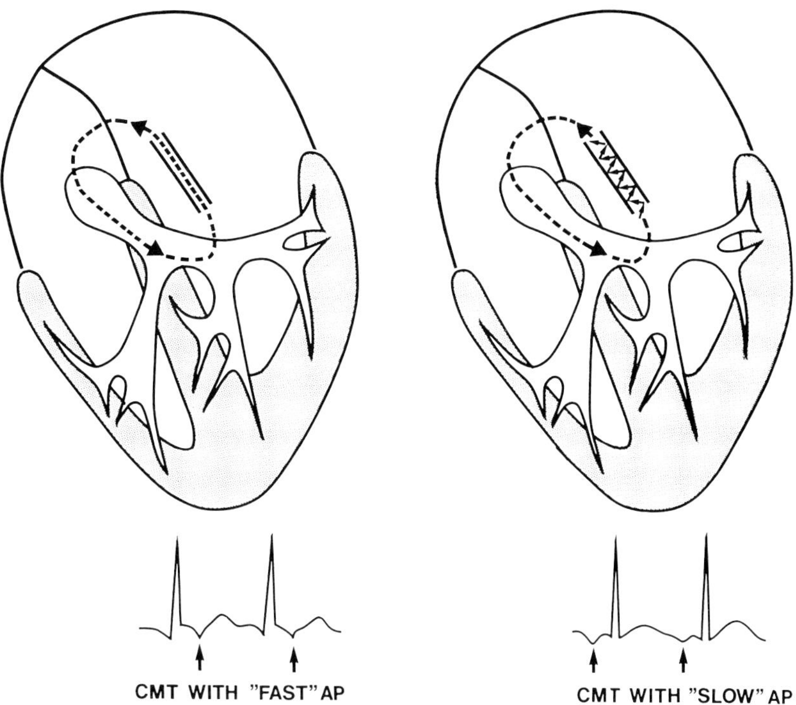

FIGURE 13.14 Schematic representation of a circus movement tachycardia using a "concealed" fast-conducting (left panel) or slow-conducting (right panel) accessory AV pathway. The corresponding ECGs during tachycardia with their characteristic RP/PR ratio are shown below.

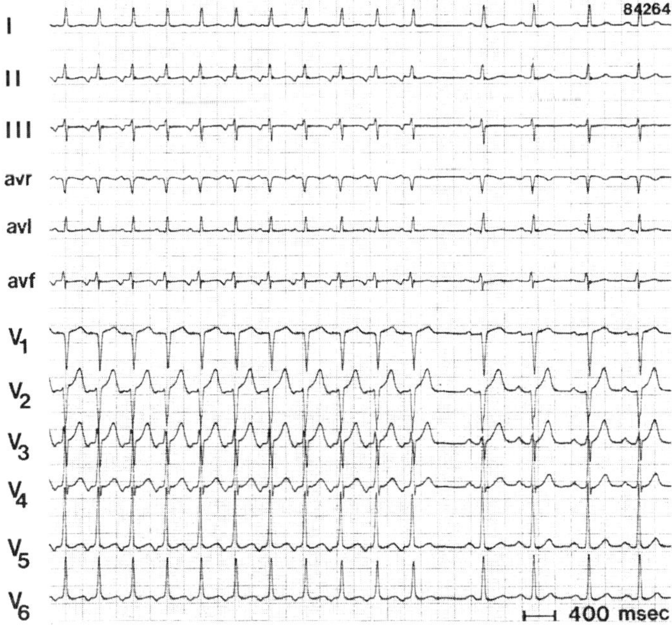

FIGURE 13.15 Termination of a circus movement tachycardia using a slow-conducting accessory pathway by carotid sinus massage. Note that block occurs in retrograde conduction over the accessory pathway.

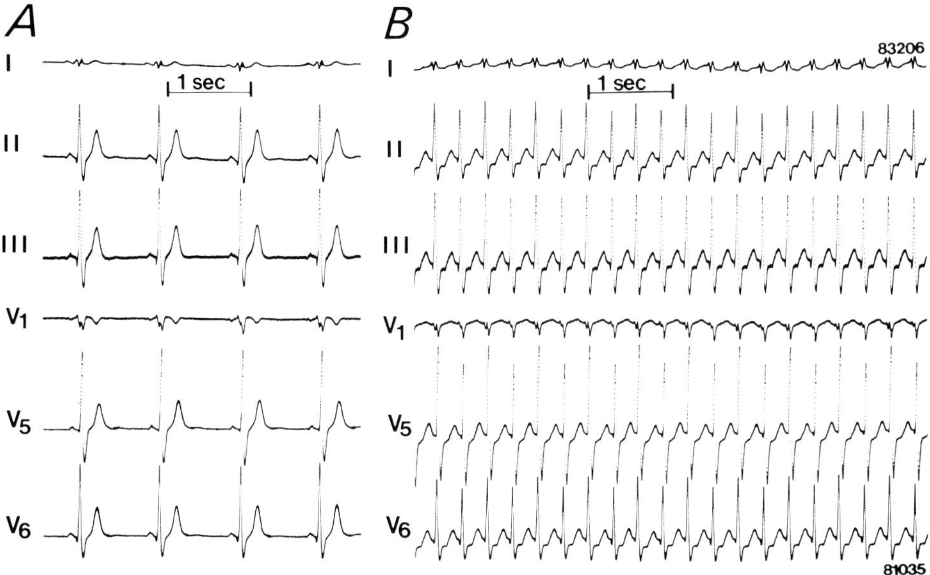

FIGURE 13.16 Example of electrical alternans of the QRS complex during a circus movement tachycardia using a "concealed" accessory AV pathway. Note that QRS alternation is best seen in leads II, V5, and V6. **A.** ECG during sinus rhythm.

movement tachycardia can only be used when present more than 5 s after the start of tachycardia. Changes in QRS configuration are common at the start of any SVT because of the sudden acceleration in ventricular rate leading to different degrees of intraventricular conduction.

The Practical Approach

As shown in Figure 13.17, when confronted with an SVT, a stepwise approach usually leads to the correct analysis of the electrocardiogram.[26] The first step is to identify the presence of AV block or its creation by carotid sinus massage. Table 13.2 summarizes the findings on carotid sinus massage in different types of tachycardia. If no AV block occurs spontaneously or cannot be induced, the ECG during tachycardia should be carefully screened for QRS alternation. Presence suggests a circus movement tachycardia using an accessory AV pathway for retrograde conduction (do not forget the 5-s rule!). The next step is the location of the P wave during tachycardia in relation to the QRS complex. The last step, the analysis of the P-wave axis in the frontal and horizontal planes, gives information about the site of origin of atrial activation during tachycardia. This four-step systematic approach usually allows the correct identification of the type of SVT.

TABLE 13.2 Findings on Carotid Sinus Massage in Different Types of Tachycardia

Tachycardia	Effect of Carotid Sinus Massage
Sinus tachycardia	Gradual and temporary slowing in heart rate
Atrial tachycardia	
Paroxysmal form	a. Cessation of tachycardia b. No effect
Incessant form	a. Temporary slowing because of increase in AV block b. No effect
Atrial flutter	a. Temporary slowing because of increase in AV block b. Transformation into atrial fibrillation c. No effect
Atrial fibrillation	a. Temporary slowing because of increase in AV block b. No effect
AV nodal tachycardia	a. Cessation of tachycardia b. No effect
Circus movement tachycardia (using concealed accessory pathway)	a. Cessation of tachycardia b. No effect

STEPS IN DIAGNOSIS OF NARROW QRS TACHYCARDIA (QRS < 0,12 SEC)

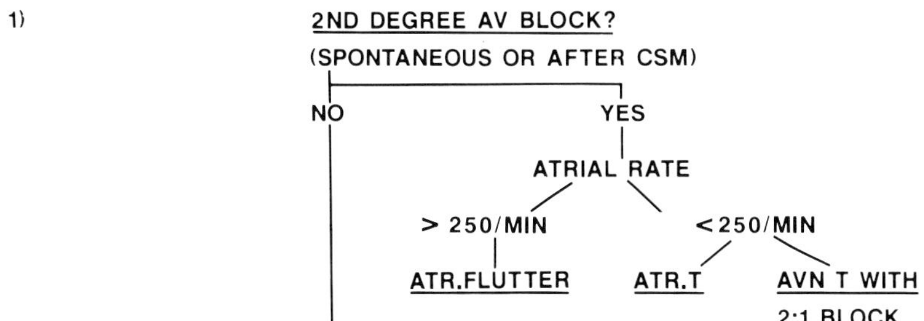

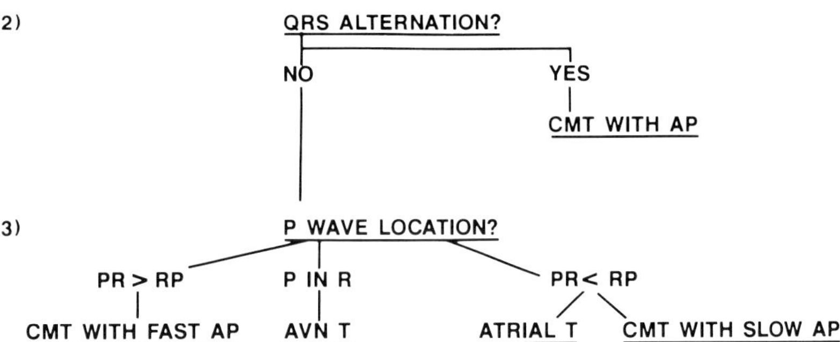

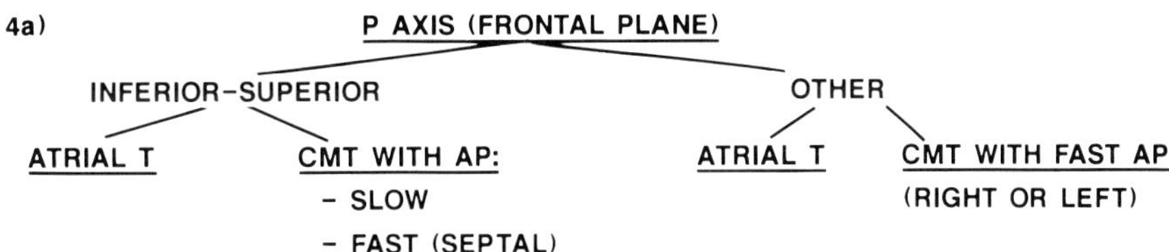

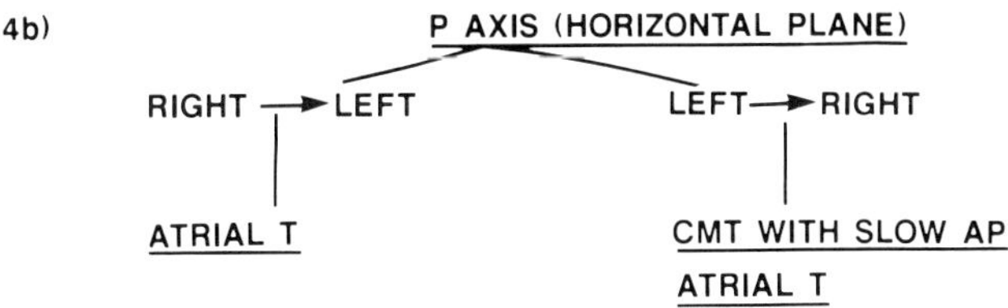

FIGURE 13.17 Steps in the diagnosis of narrow QRS tachycardia (QRS < 0.12 s). (See text for details.)

REFERENCES

1. Wellens HJJ: The electrocardiogram 80 years after Einthoven. *J Am Coll Cardiol* 1986;7:484–491.
2. Farré J, Wellens HJJ: The value of the electrocardiogram in diagnosing, site of origin and mechanism of supraventricular tachycardia, in Wellens HJJ, Kulbertus H (eds): *What's New in Electrocardiography*. The Hague, Martinus Nijhoff, 1981, pp 131–171.
3. Wellens HJJ: The electrocardiogram in digitalis intoxication, in Yu PN, Goodwin JF (eds): *Progress in Cardiology 5*. Philadelphia, Lea & Febiger, 1976, pp 271–290.
4. Lewis T, Drury AN, Iliescu CC: A demonstration of circus movement in clinical flutter of the auricles. *Heart* 1921;8:341–345.
5. Puech P, Latour M, Grolleau R: Le flutter et ses limites. *Arch Mal Coeur* 1970;63:116–120.
6. Boineau JP, Mooney CR, Hudson RD, Hughes DG,

Erchin RA Jr, Wylds AC: Observations on re-entrant excitation pathways and refractory period distributions in spontaneous and experimental atrial flutter in the dog, in Kulbertus HE (ed): *Re-entrant Arrhythmias: Mechanisms and Treatment.* Baltimore, University Park Press, 1977, pp 72–98.

7. Waldo AL, Maclean WAH, Karp RB, Kouchoukos NT, James TN: Entrainment and interruption of atrial flutter with atrial pacing; studies in man following open heart surgery. *Circulation* 1977;56:737–744.
8. Wellens HJJ: *Electrical Stimulation of the Heart in the Study and Treatment of Tachycardias.* Baltimore, University Park Press, 1971, pp 27–39.
9. Allessie MA, Bonke FIM, Schopman FJG: Circus movement in rabbit atrial muscle as a mechanism of tachycardia. III. The "leading circle" concept: A new model of circus movement in cardiac tissue without the involvement of an anatomic obstacle. *Circ Res* 1977;41:9–18.
10. Klein GJ, Guiraudon GM, Sharma AD, Milstein S: Demonstration of macro re-entry and feasibility of operative therapy in the common type of atrial flutter. *Am J Cardiol* 1986;57:587–591.
11. Denes P, Wu D, Dhingra RC, Chuguimia R, Rosen KM: Demonstration of dual AV nodal pathways in patients with paroxysmal supraventricular tachycardia. *Circulation* 1973;48:549–555.
12. Wellens HJJ, Durrer D: The role of an accessory atrioventricular pathway in reciprocal tachycardia. *Circulation* 1975;52:58–72.
13. Mignone RH, Wallace AS: Ventricular echoes. Evidence for dissociation of conduction and re-entry within the AV node. *Circ Res* 1966;19:638–645.
14. Schuilenburg RM, Durrer D: Further observations on the ventricular echo phenomenon elicited in the human heart. Is the atrium part of the echo pathway? *Circulation* 1972;45:629–638.
15. Wellens HJJ, Wesdorp JC, Düren DR, Lie KI: Second degree block during reciprocal atrioventricular nodal tachycardia. *Circulation* 1976;53:595–599.
16. Schamroth L: *The Disorders of Cardiac Rhythm.* Oxford, Blackwell Scientific, 1980.
17. Pritchett ELC, Anderson RW, Benditt DG, Kasell J, Harrison L, Wallace AG, Sealy WC, Gallagher JJ: Reentry within the atrioventricular node: Surgical cure with preservation of atrioventricular conduction. *Circulation* 1979;60:440–446.
18. Ross DL, Johnson DC, Denniss AR, Cooper MJ, Richards DA, Uther JB: Curative surgery for atrioventricular junctional ("AV nodal") re-entrant tachycardia. *J Am Coll Cardiol* 1985;6:1383–1392.
19. Holman WL, Ikeshita M, Lease JG, Ferguson TB, Lofland GK, Cox JL: Alteration of antegrade atrioventricular conduction by cryoablation of periatrioventricular nodal tissue. *J Thorac Cardiovasc Surg* 1984;88:67–75.
20. Durrer D, Roos JP: Epicardial excitation of the ventricles in a patient with Wolff Parkinson White syndrome (type B). *Circulation* 1967;35:15–22.
21. Durrer D, Schoo L, Schuilenburg RM, Wellens HJJ: The role of premature beats in the initiation and termination of supraventricular tachycardia in the Wolff Parkinson White syndrome. *Circulation* 1967;36:644–662.
22. Gallagher JJ, Pritchett ELC, Sealy WC: The preexcitation syndromes. *Prog Cardiovasc Dis* 1978;20:285–327.
23. Brugada P, Farré J, Green M, Heddle B, Roy D, Wellens HJJ: Observations in patients with supraventricular tachycardia having a P R interval shorter than the R-P interval. Differentiation between atrial tachycardia using an accessory pathway with long conduction times. *Am Heart J* 1984;107:556–570.
24. Puech P, Grolleau R, Cinca J: Reciprocating tachycardia using a latent left-sided accessory pathway. Diagnostic approach by conventional ECG, in Kulbertus H (ed): *Mechanisms and Treatment.* Lancaster, MTP, 1977, p 117.
25. Green M, Heddle B, Dassen W, Wehr M, Abdollah H, Brugada P, Wellens HJJ: The value of QRS alternation in diagnosing the site of origin of narrow QRS supraventricular tachycardia. *Circulation* 1983;68:368–373.
26. Bär FWHM, Brugada P, Dassen WRM, Wellens HJJ: Differential diagnosis of tachycardia with narrow QRS complex (shorter than 0.12 sec). *Am J Cardiol* 1984;54:555–560.

Chapter **14**

What's New in Atrial Flutter

Albert L. Waldo, MD

INTRODUCTION

Atrial flutter was described by Jolly and Ritchie in 1911.[1] Despite recognition of this supraventricular tachycardia for so very long, until recently little was known about its mechanism, and techniques in diagnosis and management had changed little over the years. The mechanism was thought to be either due to a single focus firing rapidly or due to some form of reentry. The diagnosis of atrial flutter remained largely within the province of the 12-lead electrocardiogram. Acute treatment of atrial flutter had been to use a digitalis preparation to slow and control the ventricular response rate until atrial fibrillation or sinus rhythm was produced, and to use either quinidine or procainamide to prevent recurrence of the rhythm.

As a result of recent advances, there is a considerably better understanding of the mechanism of atrial flutter. In addition, some old techniques to diagnose atrial flutter have been significantly refined and some new ones have evolved. And, beginning with the advent of DC cardioversion in the 1960s,[2,3] significant improvements in therapy of atrial flutter have been developed. The past 15 years have seen the addition of β-blockers and calcium (Ca^{2+}) channel blockers as an adjunct to, but on occasion in lieu of, digitalis therapy in an effort to control ventricular response rate to atrial flutter.[4–6] And there are now still new therapies of atrial flutter that have evolved more recently. This chapter will describe more fully the recent advances in the understanding, diagnosis, and treatment of atrial flutter.

UPDATE—MECHANISM OF ATRIAL FLUTTER

It is now generally accepted that atrial flutter is due to a reentrant mechanism. This was not always the case. Sir Thomas Lewis first thought atrial flutter was due to a single focus firing rapidly.[7] Others subsequently presented data to support the notion that a single focus firing rapidly can produce atrial flutter. This evidence came from studies in canine models in which substances such as aconitine or delphinine were applied to a localized site on the atria and produced atrial flutter,[8–10] or from atrial pacing applied in patients at rates equal to the typical atrial flutter rate which then mimicked atrial flutter.[11] There were also some data obtained during surgery in patients by Prinzmetal et al.[12] and Wellens et al.[13] that also were interpreted as indicating that atrial flutter could emanate from a single focus firing rapidly.

Although Lewis first thought that atrial flutter was due to a single focus firing rapidly, he subsequently changed his mind, based on what would now be considered scanty mapping studies of atrial flutter induced in a canine model. Because the atrial flutter episodes induced by Lewis and his colleagues were rather brief, the number of sites from which atrial electrograms could be obtained were few. Therefore, much extrapolation had to be made regarding sequence of atrial activation during the induced atrial flutter. Nevertheless, on the basis of their studies in experimental animals and also on vector analysis of electrocardiograms in man,[14,15] Lewis et al.

655 Avenue of the Americas, New York, NY 10010
Current Topics in Cardiology

concluded that atrial flutter was the result of circus movement in the atria. Subsequently, primarily after the studies of Rosenbleuth and Garcia-Ramos[16] in which a lesion was created between the venae cavae that occasionally extended into a portion of the free wall of the right atrium, it was demonstrated that atrial flutter could indeed be due to an intraatrial reentrant mechanism. It had been assumed, and to some degree demonstrated, that the reentry circuit was around the obstacle created by the atrial lesion in this model. However, recently it has been shown that in the presence of a lesion similar to that created by Rosenbleuth and Garcia-Ramos, atrial flutter could be explained by circus movement around the tricuspid valve annulus.[17]

Perhaps a clinical counterpart of the latter experimental atrial flutter preparation that relies on incisional lesions in the right atrium now exists in patients who have undergone repair of some congenital lesions requiring considerable incision and suturing of the atria. Such procedures include the Mustard procedure and the Fontan procedure. There is a remarkably high incidence of atrial flutter over both the short and long term in these patients after open heart surgery,[18,19] and it is very likely that the large incisions made in the free wall of the right atrium and also in the intraatrial septum, which are a necessary part of the corrective surgery, in essence create in patients the condition of the experimental animal and thereby provide an explanation for why atrial flutter is so often seen after these operations.

The recent studies of Allessie et al.,[20,21] using a Langendorf-perfused canine atrial preparation, have shown that atrial flutter due to reentry of the leading circle type can occur at numerous places in the atria, and can occur in the absence of an anatomic obstacle. In addition, three different atrial flutter models in the canine heart recently have been studied using multisite (multiplexing) mapping techniques, and have been shown to be due to reentry in the right atrial free wall. One model is a spontaneous example of atrial flutter in a dog studied by Boineau et al.[22] Another is the model of right atrial enlargement provided by the creation of tricuspid regurgitation and banding of the pulmonary artery by Boyden.[23] A third is the pericarditis model of atrial flutter model described by our laboratory.[24,25] The location of the reentry circuit in these preparations is especially of interest, as these models are thought to have clinical counterparts.

Studies of atrial flutter in patients have been limited by the inherent difficulties of catheter mapping techniques and the difficulty of intraoperative mapping during open heart surgery. Nevertheless, as summarized by Rytand,[26] catheter mapping studies (limited) performed in the 1950s and 1960s in humans that used esophageal leads and selected intraatrial leads, usually correlating them with electrocardiographic recordings, supported the notion that atrial flutter is an intraatrial reentrant rhythm in which activation travels in a caudocranial direction in the left atrium and in a craniocaudal direction in the right atrium. Then, studies during cardiac catheterization by Puech and colleagues[27,28] suggested that the entire duration of the atrial flutter cycle length could be explained by activation of the right atrium alone, and that atrial flutter could be explained by circus movement in the right atrium. Limited mapping studies of atrial flutter intraoperatively performed by the group in London, Ontario, also are consistent with the reentrant circuit limited to the right atrium.[29]

The demonstration of transient entrainment and interruption of atrial flutter by rapid atrial pacing in patients[30] is more recent strong evidence to support the notion that atrial flutter is an intraatrial reentrant rhythm, with the reentrant circuit having an excitable gap.[31] More recent studies using cardiac catheterization techniques by Cosio et al.[32,33] and from our laboratory[34,35] have again been consistent with the notion that atrial flutter is due to a reentrant circuit with an excitable gap in the right atrium, with the reentrant excitation wave front seeming to travel up the atrial septum and down the free wall of the right atrium. The considerable limitations of catheter mapping make it difficult to be confident about the precise location of the reentrant circuit in the right atrium. However, data from these studies,[32–35] as well as studies from animal models[21,25,36,37] have shown that the portions of the reentrant circuit are characterized by recording fractionated electrograms, which represent areas of slow conduction, and double potentials, which are markers for the center of the reentry circuit around which the impulse circulates. Using the principles of transient entrainment and the above markers of slow conduction, we have shown that an area of slow conduction in the atrial flutter reentry circuit of patients is located inferiorly and posteriorly in the right atrium,[35] and the group in Lyon, France, has demonstrated that the area of slow conduction can be localized and ablated, providing a cure of the atrial flutter (see following).[38]

In sum, it is clear that classical atrial flutter is due to reentry with an excitable gap. The reentrant circuit must still be more clearly defined

and characterized, but appears to be limited to the right atrium, with an important area of slow conduction inferior in the right atrium. Whether the reentrant circuit involves the entire right atrium or only a portion of it remains to be clarified. In addition, it is possible that some of the atrial flutter in children following correction of complicated congenital heart disease requiring various long incisions in the right atrium may be due to a reentrant circuit around the tricuspid valve annulus.

UPDATE ON THE DIAGNOSIS OF ATRIAL FLUTTER

Types of Atrial Flutter

We recently have provided evidence that there are two types of atrial flutter: type I (classical) and type II (very rapid) (Fig. 14.1).[39] Three lines of evidence led to this distinction. First, type I atrial flutter can always be influenced by rapid atrial pacing from sites high in the right atrium, whereas type II atrial flutter cannot.[39] This observation was the initial marker that led to the separation of two types of atrial flutter. Parenthetically, if type II atrial flutter is due to leading-circle-type reentry as has been suggested,[21] it is understandable that rapid atrial pacing will not interrupt this rhythm because in leading-circle-type reentry there is only a very small, difficult-to-penetrate excitable gap.[40] Second, type II atrial flutter was observed to convert spontaneously to type I atrial flutter in a stepwise fashion.[36] Third, when rapid atrial pacing from a high right atrial site was used in an effort to interrupt type I atrial flutter, on several occasions type II atrial flutter was present after termination of the rapid pacing.[39]

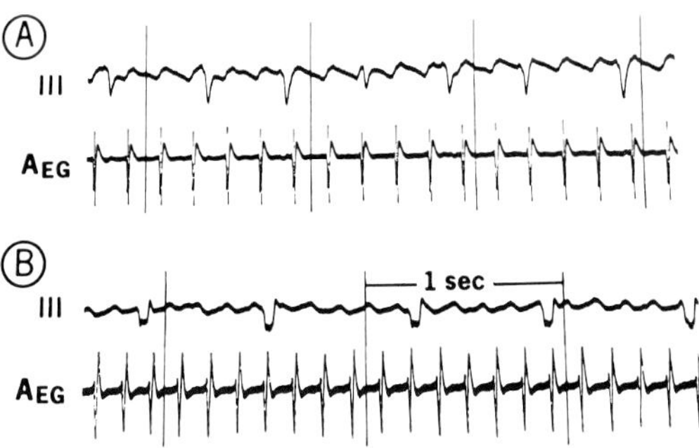

FIGURE 14.1 **A** and **B**. The simultaneous recording of electrocardiogram lead III and a bipolar arterial electrogram (AEG) during type I atrial flutter with an atrial rate of 296 bpm (**A**) and type II atrial flutter with a rate of 420 bpm (**B**). In each example, note the constant beat-to-beat cycle length, polarity, morphology, and amplitude of recorded atrial electrogram signal characteristic of atrial flutter. (See text for discussion.) *Modified from Wells et al., with permission.*[39]

Types I and II atrial flutter are further differentiated in terms of range of atrial rates. Type I atrial flutter occurs over a range of atrial rates from 240 to 340 bpm, and type II atrial flutter occurs over a range of atrial rates from 340 to 433 bpm.[39] There almost certainly is some overlap between the upper range of type I atrial flutter and the lower range of type II atrial flutter. Similarly, there is almost certainly some flexibility in the upper and lower range of atrial rates of each of these types of atrial flutter. However, common to both types of atrial flutter is the remarkably constant beat-to-beat atrial cycle length, first noted by Lewis[41] and recently confirmed by recording atrial electrograms over long periods.[39] This, along with the constant beat-to-beat morphology and polarity of recorded atrial electrograms, differentiates atrial fibrillation from type I atrial flutter.[42]

Diagnosis of Suspected Atrial Flutter

The demonstration of classic saw-toothed flutter waves in the electrocardiogram, particularly in leads II, III, and AVF, remains the standard for the diagnosis of type I atrial flutter. If the diagnosis is suspected but not clear from an electrocardiogram recording, several diagnostic options have recently been refined or have evolved.

In patients following open heart surgery, temporary atrial epicardial wire electrodes should be used to record atrial electrograms either simultaneously or sequentially with an electrocardiogram. This should establish the diagnosis readily and definitively (Fig. 14.2).[43] For other patients, if a simple vagal maneuver to produce increased atrioventricular (AV) block transiently while recording the electrocardiograms does not work so that the diagnosis is still unclear, any of the following maneuvers may be used: (a) placement of an esophageal lead to record atrial electrograms; (b) use of a pharmacologic agent (edrophonium, adenosine, esmolol, or verapamil) to increase AV conduction transiently; or (c) per venous placement of a temporary endocardial bipolar electrode catheter. Use of an esophageal lead is really an old technique rediscovered and refined. Although edrophonium is an old drug, adenosine, esmolol, and verapamil are newer antiarrhythmic agents. And, use of an endocardial catheter electrode is an old, reliable invasive technique to record atrial electrograms.

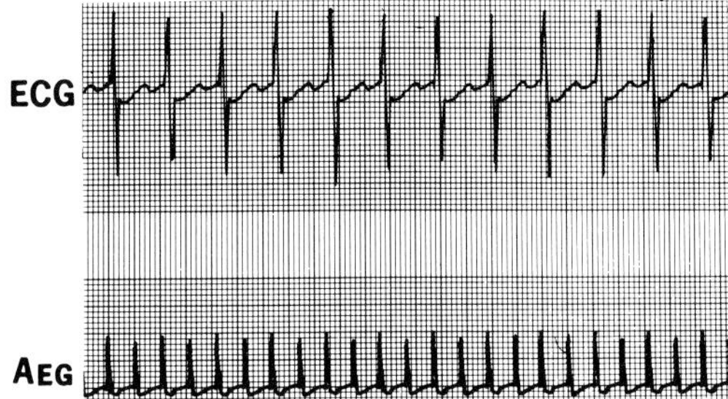

FIGURE 14.2 Monitor electrocardiogram lead recorded simultaneously with a bipolar atrial electrogram (AEG) during an episode of type I atrial flutter. The atrial rate is 280 bpm, and there is 2 : 1 AV conduction. Note that atrial complexes are not readily discernible from the electrocardiogram alone, but the bipolar atrial electrogram clearly establishes the nature of atrial activation and its relationship to ventricular activation. *Reprinted from Waldo et al.,*[43] *with permission.*

And finally, it is occasionally possible and useful to establish the diagnosis of atrial flutter by more indirect techniques. This includes echocardiography (flutter waves seen on the mitral valve echoes), signal averaged ECG, or jugular venous pulsations.

UPDATE IN THE TREATMENT OF ATRIAL FLUTTER

Acute Treatment of Type I Atrial Flutter

Once the diagnosis of atrial flutter is established, three courses of acute therapy are available: DC cardioversion, antiarrhythmic drug therapy, or rapid atrial pacing to interrupt the atrial flutter and restore sinus rhythm.

DC Cardioversion

There is not much new in DC cardioversion therapy of atrial flutter. However, recent clinical studies using an automated, impedance-based method of delivered energy adjustment show that current-based cardioversion, in which delivered energy automatically takes into account the patient's own impedance, provides a safer and more likely effective means of performing successful cardioversion with minimal delivered energy on the first shock.[44] When such cardioverters become commercially available, this mode of DC cardioversion likely will become the method of choice.

Drug Therapy

Regarding interruption of atrial flutter with an antiarrhythmic drug, the old form of treatment was aggressive administration of a digitalis preparation, usually intravenously, until conversion of the rhythm either to atrial fibrillation or to sinus rhythm. It has long been known that administration of a type I antiarrhythmic agent would slow the atrial flutter rate, but was very unlikely to interrupt atrial flutter. Thus, unlike the use of lidocaine or procainamide for ventricular tachycardia, there has been no drug that given intravenously can be reasonably expected to convert atrial flutter to sinus rhythm. However, there have been recent reports that indicate that administration of intravenous flecainide has a reasonable likelihood of restoring sinus rhythm.[45]

Atrial Pacing

Much has been learned about rapid pacing to interrupt atrial flutter.[31] First, the atria have to be paced at a critical rate for a critical duration of time. If the atria are paced at rates faster than the spontaneous rate of the atrial flutter but slower than a critical pacing rate, the atrial flutter will only be transiently entrained. Thus, following termination of pacing at such rates, the atrial flutter will return promptly at its previous spontaneous rate. However, when the critically rapid pacing rate is achieved, atrial flutter can always be interrupted. In addition, even when pacing at the appropriate pacing rate, our studies have shown that there is a critical duration of pacing at that rate before the atrial flutter is interrupted. As a result, we usually recommend pacing for at least 15 s, with longer durations sometimes required to interrupt type I atrial flutter.

The reason for the critical pacing rate and critical duration of pacing at the critical rate is well understood by considering the principles of transient entrainment of a reentrant rhythm by rapid pacing.[30,31,46] During transient entrainment of atrial flutter, a wave front from each pacing impulse enters into the excitable gap of the reentrant circuit and travels in two directions: antidromically, that is, in the opposite direction of the circulating wave front of the spontaneous tachycardia, and orthodromically, that is, in the same direction as the circulating wave front of the spontaneous tachycardia (Fig. 14.3). Each antidromic wave front collides with the orthodromic wave of the previous beat. However, each orthodromic wave front continues the tachycardia, resetting it to the pacing rate. When one stops pacing at a rate that only entrains the tachycardia, the last orthodromic wave front

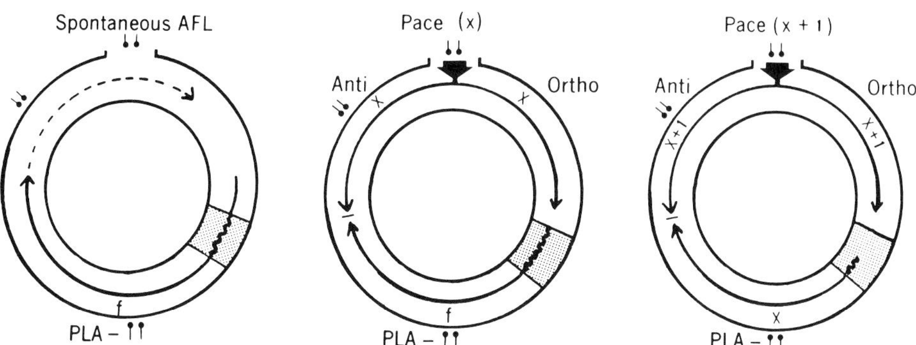

FIGURE 14.3 Diagrammatic representation of the reentry circuit during spontaneous atrial flutter and during transient entrainment of the atrial flutter. The left diagram shows the reentry circuit during spontaneous type I (classical) atrial flutter (AFL). f = circulating wavefront of the atrial flutter. The middle diagram shows the introduction of the first pacing impulse (X) during rapid pacing from a high atrial site during the atrial flutter. The large arrow indicates the entry of the pacing impulse into the reentry circuit, whereupon it is conducted orthodromically (Ortho) and antidromically (Anti). The antidromic wave front of the pacing impulse (X) collides with the previous beat, in this case the circulating wave front of the spontaneous atrial flutter (f), resulting in an atrial fusion beat. This, in effect, terminates the atrial flutter. However, the orthodromic wave front from the pacing impulse (X) continues the tachycardia, resetting it to the pacing rate. The right diagram shows the introduction of the next pacing impulse (X + 1) during rapid pacing from the same high atrial site. The large arrow again indicates the entry of the pacing impulse into the reentry circuit, whereupon it is conducted orthodromically and antidromically. Once again, the antidromic wave front from the pacing impulse (X + 1) collides with the orthodromic wave front of the previous beat, in this case, it is the orthodromic wave front of the previous paced beat (X), resulting in an atrial fusion beat; and the orthodromic wave front from the pacing impulse (X + 1) continues the tachycardia, resetting it to the pacing rate. In this and subsequent diagrams of transient entrainment and interruption of type I atrial flutter, the arrows indicate the direction of spread of the impulses, the serpentine line indicates slow conduction through a presumed area of slow conduction (stippled region) in the reentry circuit, and the black dots with tails indicate bipolar electrodes at the high atrial pacing site, the posterior–inferior portion of the left atrium (PLA), and another atrial site. *Reprinted from Waldo et al., with permission*, in Surawicz B (ed): Tachycardias. The Hague, Martinus Nijhoff, 1984, p 213.

from the pacing impulse will travel around the reentry circuit unopposed (there is no subsequent antidromic pacing wave front with which to collide), resulting in resumption of the atrial flutter at its previous spontaneous rate (Fig. 14.4). This mechanism, universal for transient entrainment of any tachycardia due to reentry with an excitable gap, explains why termination of pacing at a rate that simply entrains the tachycardia will fail to interrupt it.[30,31,46]

To interrupt the tachycardia, pacing must occur at a rate sufficiently fast to produce block not only of the antidromic wave front of the pacing impulse, but also of the orthodromic wave front of the same pacing impulse during the same beat (Fig. 14.5). Direct data from simultaneous multisite mapping studies in a canine atrial flutter model[37] and indirect data from studies in patients[35] indicate that when atrial flutter is interrupted by rapid atrial pacing, the orthodromic wave front from the pacing impulse blocks in an area of slow conduction in the reentrant circuit. When such block occurs, the atrial flutter reentrant circuit is no longer present. Termination of pacing, therefore, will demonstrate that atrial flutter is no longer present.

It is recommended that rapid atrial pacing to interrupt atrial flutter be performed from the high right atrium because the interruption of atrial flutter is heralded by the appearance of positive atrial complexes in the inferior leads (II, III, AVF) of the ECG (Fig. 14.6). During rapid atrial pacing from the site at rates that transiently entrain rather than interrupt atrial flutter, the atrial complexes in the ECG will be fusion beats (Figs. 14.3 and 14.6), such that they will not have the usual positive polarity in these leads that normally occurs when overdrive pacing of a sinus rhythm is performed from this site. With the interruption of atrial flutter by atrial pacing, because the atrial complexes no longer are fusion beats but rather the expected complexes during overdrive pacing of a sinus rhythm, atrial complexes in leads II, III, and AVF will become pos-

itive (Fig. 14.6). In short, the appearance of positive atrial complexes in these leads indicates that the atrial flutter is no longer being transiently entrained, but rather that it has been interrupted.

In sum, when using rapid atrial pacing techniques to interrupt atrial flutter, when the atria are paced at rates faster than the spontaneous rate, the atrial flutter may not be interrupted even though atrial capture has been obtained. This phenomenon, called transient entrainment,[30,31,46] should not be considered evidence that rapid atrial pacing will be unsuccessful. Rather, it provides evidence that pacing at a more rapid rate is required to interrupt the atrial flutter.

Last, it has recently been shown that the simultaneous administration of a type IA antiarrhythmic agent increases the efficacy of conversion of atrial flutter to sinus rhythm with rapid atrial pacing.[47,48] In the absence of a type IA an-

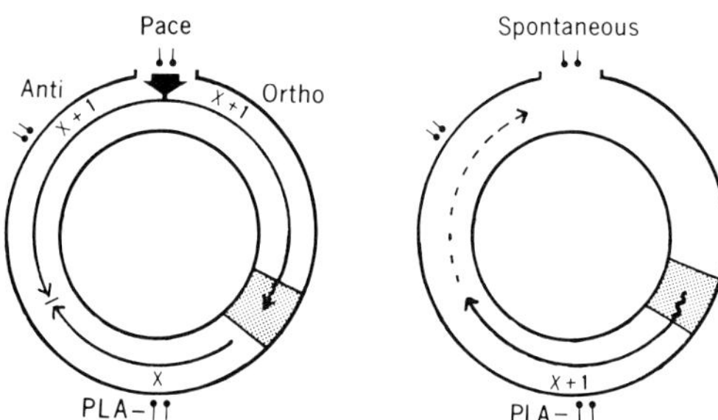

FIGURE 14.4 Diagrammatic representation of the reentry circuit during termination of rapid atrial pacing of atrial flutter (initiated diagrammatically in Fig. 14.3) with resulting resumption of the spontaneous tachycardia. In the left diagram,the large arrow indicates the last pacing impulse from the high atrial pacing site entering into the reentry circuit, whereupon it is conducted orthodromically (Ortho) and antidromically (Anti). The antidromic wave front from the pacing impulse (X + 1) collides with the orthodromic wave front of the previous beat (X), resulting in an atrial fusion beat, and the orthodromic wave front from the pacing impulse (X + 1) continues the tachycardia, resetting it to the pacing rate. In the right diagram, the orthodromic wave front of the last paced beat (X + 1) is unopposed by an antidromic wave front because there is no subsequent pacing impulse. Therefore, no atrial fusion beat occurs despite transient entrainment at the PLA recording site. This last entrained orthodromic wave front continues the tachycardia (dashed lines), which resumes at its previous spontaneous rate. *Reprinted from Waldo et al., with permission.*

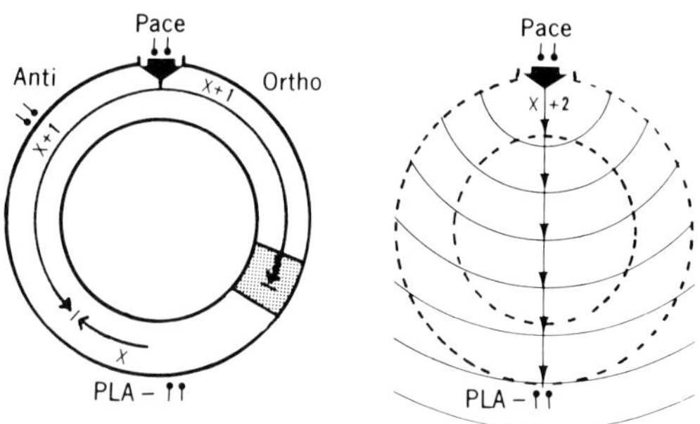

FIGURE 14.5 Diagrammatic representation of the interruption of atrial flutter during rapid atrial pacing. In the left diagram, the large arrow indicates the pacing impulse from the atrial pacing site entering into the reentry circuit, whereupon it is conducted orthodromically (Ortho) and antidromically (Anti). The antidromic wave front (X + 1) collides with the orthodromic wave front from the previous beat (X), resulting in an atrial fusion beat. The orthodromic wave front (X + 1) also blocks, presumably in the area of slow conduction, so that the tachycardia is no longer continued. Thus, because the antidromic and orthodromic wave fronts of the same pacing impulse have blocked during the same beat, the tachycardia has been interrupted. Note that the block of both the antidromic and orthodromic wave fronts of the same pacing impulse is associated with absence of conduction of one pacing impulse to the posterior–inferior left atrial recording site (PLA). In the right diagram, the large arrow indicates the next pacing impulse (X + 2) from the same atrial pacing site as in the left diagram. The dashed circle represents the reentry circuit that had been present during both the previous period of spontaneous atrial flutter and transient entrainment of the atrial flutter. Because the atrial flutter has been interrupted by the previous pacing impulse (X + 1), the reentry circuit is no longer present. Thus, the sequence of atrial activation during the next pacing impulse (X + 2) is the same as one would expect during overdrive pacing of a sinus rhythm. Therefore, the PLA recording site will be activated from a different direction (the activation wave front from the pacing impulse no longer first engages the reentry circuit) and with a shorter conduction time (because the activation wave front does not have to engage the reentry circuit before activating the PLA site, there is no longer a functional area of slow conduction through which it must pass) than during previous periods of transient entrainment from the same pacing site. The isochrones are drawn to indicate the general direction of spread of the impulse and are not meant to imply that the sequence of atrial activation from the pacing site to the PLA recording is radial. *Reprinted from Waldo et al., with permission.*

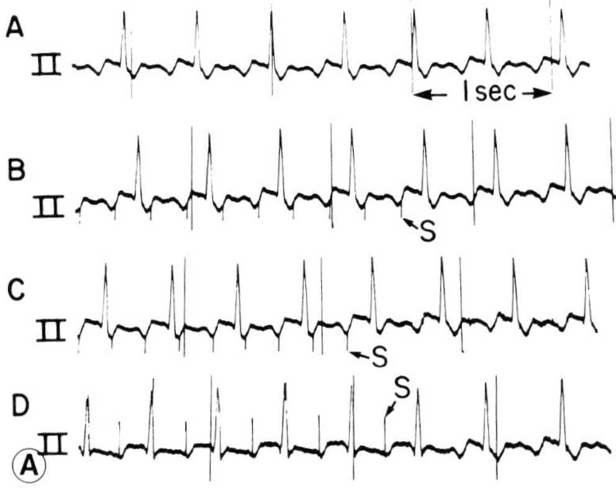

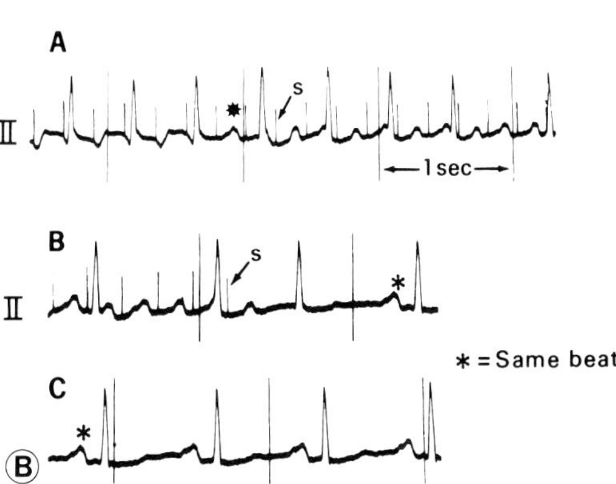

FIGURE 14.6 A. ECG lead II recorded from a patient with classical (type I) atrial flutter (atrial cycle length of 264 ms) (**A**) and at the end of 30 s of rapid atrial pacing from a high right atrial site at a cycle length of 254 ms (**B**), at a cycle length of 242 ms (**C**) and at a cycle length of 232 ms (**D**). The atrial flutter was transiently entrained at each pacing rate. Note that when comparing the morphology of the atrial complexes during atrial pacing at each cycle length, especially comparing the atrial complexes during atrial pacing in **D** with those in **A** and **B,** progressive fusion has occurred. S = stimulus artifact. Time lines are at 1-s intervals. *Modified from Waldo et al.*[30] **B.** In part A, ECG lead II recorded in the same patient as in **A** during high right atrial pacing from the same site at a cycle length of 224 ms. Note that with the seventh atrial beat in this tracing, and after 22 s of atrial pacing at a constant rate, the atrial complexes suddenly become positive (*). In part B, ECG lead II recorded at the termination of atrial pacing in the same patient. Note that with abrupt termination of pacing, sinus rhythm occurs. In part C, the first beat in this panel (*) is identical with the last beat in part B (*). S = stimulus artifact. Time lines are at 1-s intervals. *Modified from Waldo et al.*[30]

tiarrhythmic agent, there is an important incidence of induction of atrial fibrillation. While this rhythm often is transient, spontaneously reverting to sinus rhythm, in many instances, the atrial fibrillation will revert back to atrial flutter, or the atrial fibrillation will persist, requiring DC cardioversion. Thus, it is also recommended that a type I antiarrhythmic agent be administered prior to atrial pacing whenever possible.

Esophageal pacing. It has been shown recently that the atria can be paced from the esophagus.[49] Atrial pacing via the esophagus can also be used to interrupt atrial flutter. Since atrial pacing via the esophagus tends to be a bit painful (usually causing a stinging sensation), pacing for shorter periods, usually less than 15 s at a time, is recommended. In addition, rapid pacing should be initiated after first demonstrating a good atrial electrogram recording with only a minimal ventricular deflection. And then, pacing at a sufficiently slow rate should be initiated so that if ventricular capture is inadvertently obtained, there will be no adverse effects. In addition, the pacing stimulus duration should be at least 9 to 10 ms, and the stimulus strength should initially be at least 15 mA. This is because threshold for atrial capture via the esophagus is generally quite high, and it is not unusual for the stimulus strength to be greater than 20 mA, even up to 30 mA.

Deliberate precipitation of atrial fibrillation. For those patients in whom type I atrial flutter recurs despite interruption by rapid atrial pacing, it has recently been shown that continuous rapid atrial pacing to precipitate and sustain atrial fibrillation may be very useful on a temporary basis, particularly in patients following open heart surgery, until pharmacologic control of the atrial flutter is achieved.[50] In such instances, atrial fibrillation usually is a more desirable rhythm than the continuation of type I atrial flutter because the atrial fibrillation is almost always associated with a slower ventricular response rate than during atrial flutter. Also, in almost all instances, the ventricular response rate to atrial fibrillation can be easily controlled with digitalis or, occasionally, with a Ca^{2+} channel blocker or a β-blocker.

Update on Chronic Treatment of Atrial Flutter

Drug Therapy

Paroxysmal atrial flutter may be quite difficult to suppress clinically. Standard treatment is to administer a type IA agent in an effort to pre-

vent recurrence. When a type IA antiarrhythmic agent is ineffective or cannot be administered for any reason (for example, adverse effects), recent data indicate that one of the IC agents may be quite effective and well tolerated.[51] However, should atrial flutter recur, because the atrial flutter rate may be remarkably slowed, it is desirable to be sure that sufficient AV block will be present to prevent 1 : 1 AV conduction. Thus, medication that increases AV conduction time (digitalis, Ca^{2+} channel blockers, or β-blockers) probably ought to be administered with these agents. In addition, type III antiarrhythmic agents (amiodarone or sotalol) likewise may be quite effective. However, there are very few data concerning use of sotalol in this circumstance, and potential amiodarone toxicity is a well-described concern.

Permanent Antitachycardia Pacing

Although the published series of patients who have received antitachycardia pacemakers to treat recurrent atrial flutter have been small, they have nevertheless recently been shown to be effective in interrupting recurrent atrial flutter in properly selected patients.[52] These patients usually still need chronic treatment with antiarrhythmic drug therapy. Nevertheless, the ability to interrupt atrial flutter promptly when it recurs should provide safe and effective treatment. A potential problem with the use of a permanent antitachycardia pacemaker is the precipitation of atrial fibrillation, usually transient. If precipitation of atrial fibrillation is clinically unacceptable, this technique is probably best avoided.

Catheter Ablation Techniques

Three catheter ablation techniques have recently been used successfully to treat chronic or recurrent atrial flutter. One, His-bundle ablation to create high-degree AV block, thereby preventing the rapid ventricular response rate to atrial flutter, is now a generally accepted technique.[53] For patients who are intolerant of antiarrhythmic drug therapy or in whom atrial flutter with an unacceptably rapid ventricular response rate occurs despite antiarrhythmic drug therapy, producing third-degree AV block or high-degree AV block provides a successful form of therapy without the need for any more antiarrhythmic agents. Of course, it does include the need for a permanent pacemaker system, but there are a host of very acceptable pacemakers available for use, VVIR, DDD, or DDDR. The second type of catheter ablation therapy is AV node modification.[54] Although this technique, currently investigational, has been used primarily to treat AV nodal reentrant tachycardia, it may be used to modify AV nodal conduction such that the ventricular response rate to atrial flutter is acceptably slow or easily controlled with drugs that affect AV conduction such as digitalis, Ca^{2+} channel blockers, or β-blockers.

The third mode of catheter ablation therapy is also currently investigational. It involves identification of an apparent critical area of slow conduction in the atrial flutter reentry circuit. When this area is mapped and identified during cardiac catheterization, it can be ablated. In a small series of patients published to date, this technique has provided apparent successful therapy.[38] Clearly, the latter two catheter ablation techniques await further study before the associated short- and long-term efficacy rates and potential adverse effects can be better appreciated.

Surgical Therapy

Acceptable surgical therapy awaits more definitive understanding of the nature of the reentrant circuit of atrial flutter in patients, and a clearer understanding of which surgical intervention(s) (ablation, incision, excision, and so forth) might prove effective when applied appropriately. Guiraudon et al.[55] have recently reported on three operated patients in whom intraoperative mapping showed an area of slow conduction or a gap in activation during atrial flutter between the coronary sinus surface and the tricuspid annulus. Cryoablation of the region successfully prevented recurrent atrial flutter in two, but the third developed symptomatic atrial fibrillation. Clearly, this surgical series is too small to project the future, but the similarities between these data and the ablation data of Saoudi et al.[38] seem related and promising.

One special case should be mentioned. Atrial flutter that occurs in patients who have the Wolff-Parkinson-White syndrome can be its usual troublesome self or, when associated with 1 : 1 AV conduction, may be life-threatening. For reasons that are not understood, ablation of the accessory AV connection is initially always associated with elimination of the atrial flutter (true for atrial fibrillation as well[56]). Therefore, in this group of patients, surgical ablation or catheter ablation of the accessory AV connection should be considered as a therapeutic option (it is clearly indicated when 1 : 1 AV conduction of atrial flutter over an accessory AV connection has been demonstrated[57]), as it will likely effect a cure.

SUMMARY

The diagnosis of atrial flutter should always be possible using either old or new techniques. Likewise, interruption of atrial flutter should always be possible using pacing or DC cardioversion

techniques. Long-term antiarrhythmic drug suppression of atrial flutter may be a clinical problem, but availability of newer antiarrhythmic agents holds promise for finding an effective regimen, perhaps in concert with an antitachycardia pacemaker. Ablation techniques may hold the most promise, as very early data suggest cure of atrial flutter using these techniques may be possible.

ACKNOWLEDGMENTS

This work was supported in part by grant RO1 HL38408 from the National Institutes of Health, National Heart, Lung, and Blood Institute, Bethesda, MD; a Research Initiative Award from the Northeast Ohio Affiliate of the American Heart Association, Cleveland, OH; and a grant from the Wuliger Foundation, Cleveland, OH.

REFERENCES

1. Jolly WA, Ritchie WJ: Auricular flutter and fibrillation. *Heart* 1911;2:177–221.
2. Morris JJ, Kong Y, North WC, McIntosh HD: Experience with "cardioversion" of atrial fibrillation and flutter. *Am J Cardiol* 1964;14:94–100.
3. Castellanos A, Lemberg L, Gosselin A, Fonseca UJ: Evaluation of countershock treatment of atrial flutter. *Arch Intern Med* 1965;115:426–433.
4. Zipes DP: Management of cardiac arrhythmia: Pharmacological, electrical and surgical techniques, in Braunwald E (ed): *Heart Disease: A Textbook of Cardiovascular Medicine*. Philadelphia, WB Saunders, 1988, p 634.
5. The Esmolol Research Group: Intravenous esmolol for the treatment of supraventricular tachyarrhythmia: Results of a multi center, baseline-controlled safety and efficacy study of 160 patients. *Am Heart J* 1986;112:498–505.
6. Plumb VJ, Karp RB, Kouchoukos NT, Zorn GO Jr, James TN, Waldo AL: Verapamil therapy of atrial fibrillation and atrial flutter following open heart surgery. *J Thorac Cardiovasc Surg* 1982;83:590–596.
7. Lewis T: *Clinical Disorders of the Heart Beat*. New York, Paul B. Hoeher, 1918, p 74.
8. Scherf D: Studies on auricular tachycardia caused by aconitine administration. *Proc Exp Biol Med* 1947;64:233.
9. Scherf D, Blumenfeld S, Taner D, Yildiz M: Cardiac arrhythmias provoked by focal application of delphinine. *Arch Kreisl Forsch* 1960;33:4.
10. Hayden WG, Hurley EJ, Rytand DA: The mechanism of canine atrial flutter. *Circ Res* 1967;20:496–505.
11. Rosen K, Lau SH, Damato AN: Simulation of atrial flutter by rapid coronary sinus pacing. *Am Heart J* 1969;78:635–642.
12. Prinzmetal M, Corday E, Oblath RW, Kruger AG, Brill IC, Fields J, Kennamer SR, Osborne JA, Smith LA, Sellers AL, Flieg W, Finston E: Auricular flutter. *Am J Med* 1951;11:410–430.
13. Wellens HJJ, Janse MJ, Van Dam RTh, Durrer D: Epicardial excitation of the atria in a patient with atrial flutter. *Br Heart J* 1971;33:233–237.
14. Lewis T, Feil HS, Stroud WD: Observations upon flutter and fibrillation. Part II. The nature of auricular flutter. *Heart* 1920;7:191–245.
15. Lewis T, Drury AN, Iliescu CC: A demonstration of circus movement in clinical flutter of the auricles. *Heart* 1921;8:341–389.
16. Rosenbleuth A, Garcia-Ramos J: Studies on flutter and fibrillation. II. The influence of artificial obstacles on experimental auricular flutter. *Am Heart J* 1947;33:677–684.
17. Frame LH, Page RL, Hoffman BF: Atrial reentry around an anatomic barrier with a partially refractory excitable gap. A canine model of atrial flutter. *Circ Res* 1986;58:495–511.
18. Flinn CJ, Wolff GS, Dick M II, Campbell RM, Borkat G, Cast A, Hordof A, Hougen TJ, Kavey R, Kugler J, Liebman J, Greenhouse J, Hees P: Cardiac rhythm after the Mustard operation for complete transposition of the great arteries. *N Engl J Med* 1984; 310:1635–1638.
19. Garson A Jr, Bink-Boelkens M, Hesslein PS, Hordoff AJ, Keane JF, Neches WH, Porter CJ: Atrial flutter in the young: A collaborative study of 380 cases. *J Am Coll Cardiol* 1985;6:871–878.
20. Allessie M, Lammers W, Smeets J, Bonke F, Hollen J: Total mapping of atrial excitation during acetylcholine-induced atrial flutter and fibrillation in the isolated canine heart, in Kulbertus HE, Olsson SB, Schlepper M (eds): *Atrial Fibrillation*. Molndal, Sweden, AB Hassell, 1982, pp 44–59.
21. Allessie MA, Lammers WJEP, Bonke FIM, Hollen J: Intraatrial reentry as a mechanism for atrial flutter induced by acetylcholine in rapid pacing in the dog. *Circulation* 1984;70:123–135.
22. Boineau JP, Schuessler RB, Mooney CR, Miller CB, Wylds AC, Hudson RD, Borremans JM, Brockus CW: Natural and evoked atrial flutter due to circus movement in dogs. *Am J Cardiol* 1980;45:1167–1187.
23. Boyden PA: Activation sequence during atrial flutter in dogs with surgically induced right atrial enlargement. I. Observations during sustained rhythms. *Circ Res* 1988;62:596–608.
24. Pagé P, Plumb VJ, Okumura K, Waldo AL: A new model of atrial flutter. *J Am Coll Cardiol* 1986;8:872–879.
25. Okumura K, Plumb VJ, Pagé PL, Waldo AL: Atrial activation sequence during atrial flutter in the canine pericarditis model and its effects on the polarity of the flutter wave in the electrocardiogram. *J Am Coll Cardiol* 1991;17:509–518.
26. Rytand DA: The circus movement (entrapped circuit wave) hypothesis of atrial flutter. *Arch Intern Med* 1966;65:125.
27. Puech P: *L'activité électrique auriculaire normal et pathologique*. Paris, Masson & Cie, 1956, p 214.
28. Puech P, Latour H, Grolleau R: Le flutter et ses limites. *Arch Mal Coeur* 1970;61:116–144.
29. Klein GJ, Guiraudon GM, Sharma AD, Milstein S: Demonstration of macroreentry and feasibility of operative therapy in the common type of atrial flutter. *Am J Cardiol* 1986;57:587–591.
30. Waldo AL, MacLean, WAH, Karp RB, Kouchoukos NT, James TN: Entrainment and interruption of atrial flutter with atrial pacing: Studies in man following open heart surgery. *Circulation* 1977;56:737–745.
31. Waldo AL, Carlson MD, Biblo LA, Henthorn RW: The transient entrainment concept in understanding atrial flutter, in Touboul P, Waldo AL (eds): *Atrial Arrhythmias: Current Concepts and Management*. St. Louis, CV Mosby 1990; p 210.

32. Cosio FG, Arribas F, Palacios J, Tascon J, Lopez-Gil M: Fragmented electrograms and continuous electrical activity in atrial flutter. *Am J Cardiol* 1986; 57:1309–1314.
33. Cosio FG, Arribas F, Barbero JM, Kallmeyer C, Giocolea A: Validation of double spike electrograms as markers of conduction delay or block in atrial flutter. *Am J Cardiol* 1988;61:775–780.
34. Olshansky B, Okumura K, Henthorn RW, Waldo AL: Characterization of double potentials in human atrial flutter: Studies during transient entrainment. *J Am Coll Cardiol* 1990;15:833–841.
35. Olshansky B, Okumura K, Hess PG, Waldo AL: Demonstration of an area of slow conduction in human atrial flutter. *J Am Coll Cardiol* 1990;16:1634–1648.
36. Shimizu A, Nozaki A, Rudy Y, Waldo AL: The onset of induced atrial flutter in the canine pericarditis model. *J Am Coll Cardiol* 1991;17:1223–1234.
37. Shimizu A, Nozaki A, Rudy Y, Waldo AL: Multiplexing studies of the effects of rapid atrial pacing on the area of slow conduction during atrial flutter in the canine pericarditis model. *Circulation* 1991;83:938–944.
38. Saoudi N, Atallah G, Kirkorian G, Touboul P: Catheter ablation of the atrial myocardium in human type I atrial flutter. *Circulation* 1990;81:762–771.
39. Wells JL Jr, MacLean WAH, James TN, Waldo AL: Characterization of atrial flutter. Studies in man after open heart surgery using fixed atrial electrodes. *Circulation* 1979;60:655–673.
40. Allessie MA, Bonke FIM, Schopman FJG: Circus movement and rapid atrial muscle as a mechanism of tachycardia. III. The "leading circle" concept: A new model of circus movement in cardiac tissue without the involvement of an anatomic obstacle. *Circ Res* 1977;41:9–18.
41. Lewis T: Observations upon flutter and fibrillation. I. The regularity of clinical auricular flutter. *Heart* 1920;7:127–130.
42. Wells JL Jr, Karp RB, Kouchoukos NT, MacLean WAH, James TN, Waldo AL: Characterization of atrial fibrillation in man. Studies following open heart surgery. *PACE* 1978;3:426–438.
43. Waldo AL, MacLean WAH, Cooper TB, Kouchoukos NT, Karp RB: Use of temporarily placed epicardial atrial wire electrodes for the diagnosis and treatment of cardiac arrhythmias following open heart surgery. *J Thorac Cardiovasc Surg* 1978;76:500–505.
44. Kerber RE, Martins JB, Kienzle MG, Constantin L, Olshansky B, Hopson M, Charbonnier F: Energy, current, and success in defibrillation and cardioversion: Clinical studies using automated impedance-based method of energy adjustment. *Circulation* 1988;77:1038–1046.
45. Hellestrand KJ: Intravenous flecainide acetate for supraventricular tachycardias. *Am J Cardiol* 1988;62:16D–22D.
46. Waldo AL, Plumb VJ, Arciniegas JG, MacLean WAH, Cooper TB, Priest MF, James TN: Transient entrainment and interruption of A-V bypass pathway type paroxysmal atrial tachycardia. A model for understanding and identifying reentrant arrhythmias in man. *Circulation* 1982;67:73–83.
47. Camm J, Ward D, Spurrell R: Response of atrial flutter to overdrive atrial pacing and intravenous disopyramide phosphate singly and in combination. *Br Heart J* 1980;44:240–247.
48. Olshansky B, Okumura K, Hess PG, Henthorn RW, Waldo AL: Use of procainamide with rapid atrial pacing for successful conversion of atrial flutter to sinus rhythm. *J Am Coll Cardiol* 1988;11:359–364.
49. Crawford W, Plumb VJ, Epstein AE, Keeg GN: Prospective evaluation of transesophageal pacing for the interruption of atrial flutter. *Am J Med* 1989;86:663–667.
50. Waldo AL, MacLean WAH, Karp RB, Kouchoukos NT, James TN: Continuous rapid atrial pacing to control recurrent or sustained supraventricular tachycardias following open heart surgery. *Circulation* 1976;54:245–250.
51. Anderson JL, Gilbert LM, Alpert BL, Henthorn RW, Waldo AL, Bhandari AK, Hawkinson RW, Pritchett ELC: Prevention of symptomatic recurrence of paroxysmal atrial fibrillation in patients initially tolerating antiarrhythmic therapy. *Circulation* 1989;80:1557–1570.
52. Barold SS, Wyndham CRC, Kappenberger LL, Abinader EG, Griffin JC, Falkoff MD: Implanted atrial pacemakers for paroxysmal atrial flutter: Long term efficacy. *Ann Intern Med* 1987;107:144–149.
53. Scheinman MM: Catheter techniques for ablation of supraventricular tachycardia. *N Engl J Med* 1989; 320:460–461.
54. Huang SK: Radio-frequency catheter ablation of cardiac arrhythmias: Appraisal of an evolving therapeutic modality. *Am Heart J* 1990;118:1317–1323.
55. Guiraudon GM, Klein GJ, Sharma AD, Yee R: Surgical alternatives for supraventricular tachycardias. *Am J Cardiol* 1989;64:92J–96J.
56. Sharma AD, Klein GJ, Guiraudon GM, Milstein S: Atrial fibrillation in patients with Wolff-Parkinson-White syndrome: Incidence after surgical ablation of the accessory pathway. *Circulation* 1985;72:161–169.
57. Waldo AL, Akhtar M, Benditt VG, Brugada P, Camm AJ, Gallagher JJ, Gillette PC, Klein GJ, Levy S, Scheinman MM, Wellens HJJ, Zipes DP: Appropriate electrophysiologic study and treatment of patients with the Wolff-Parkinson-White syndrome. *J Am Coll Cardiol* 1988;11:1124–1129.

Chapter **15**

Atrial Fibrillation: The Blind Man's Elephant

Frits L. Meijler, MD, FACC, and
Fred H.M. Wittkampf, MD

This paper seeks to explain that most puzzling of all the forms of irregularity of the heart, where the heart is never regular in its action, where seldom or never two beats of the same character follow one another.

This is the opening sentence of Mackenzie's paper in 1904 in the *British Medical Journal*[1] in which he described an arrhythmia of the heart that we call "atrial fibrillation" today.

Definition

Atrial fibrillation has been defined by a WHO/ISC Task Force as "an irregular, disorganized electrical activity of the atria. P-waves are absent and the baseline consists of irregular wave forms which continuously change in shape, duration, amplitude and direction. In the absence of advanced or complete AV block, the resulting ventricular response is totally irregular (random)."[2] This is an electrocardiographic description that is quite useful for routine, daily clinical practice, but which has limited meaning for understanding the mechanism(s) involved. When one takes a closer look at the above-mentioned definition, one notices that the electrocardiographic diagnosis of atrial fibrillation requires a number of anatomical and functional conditions. The anatomical components of the definition are: the atria, the atrioventricular (AV) node, and the ventricles. The functional components are: atrial electrical behavior, AV nodal function, and the ventricular response.

Diagnosis

The clinical picture of atrial fibrillation consists of a number of separate entities, and each student of this frequently occurring arrhythmia will have to take this into account. Emphasis may be put on one of the anatomical and/or physiological ingredients of the definition. In the clinical setting of atrial fibrillation the ventricular response must be random[2] to conclude that the atria are fibrillating. To the naked eye, the differential diagnosis between atrial flutter and atrial fibrillation may be quite difficult[3] and to be certain, a mathematical analysis of the ventricular response would be desirable. A random ventricular rhythm requires normal AV nodal/junctional function, whether one accepts concealed conduction[4,5] as the mechanism that scales the atrial impulses or believes that electrotonic modulation of an AV nodal pacemaker protects the ventricles against the high rate of the fibrillating atria.[6]

Since atrial fibrillation has many facets to be considered, looking at only one of them may lead to the "blind man's elephant syndrome." One may actually reach conclusions that, once the whole animal is seen, will not meet the test of the real-life situation. We will consider each component of atrial fibrillation separately, ultimately trying to come up with a picture of the whole elephant.

Clinical Relevance

The clinical and social relevance of atrial fibrillation surpasses all other arrhythmias. Selzer[7] called atrial fibrillation "the grandfather of all cardiac arrhythmias," an expression that may be paraphrased by stating that atrial fibrillation is the cardiac arrhythmia of grandfathers and grandmothers, since it is a frequent finding in old age. More than 50% of 1,212 patients in Godtfredsen's series with atrial fibrillation[8] were over 70 years of age. Moreover, it is a well-known

655 Avenue of the Americas, New York, NY 10010
Current Topics in Cardiology

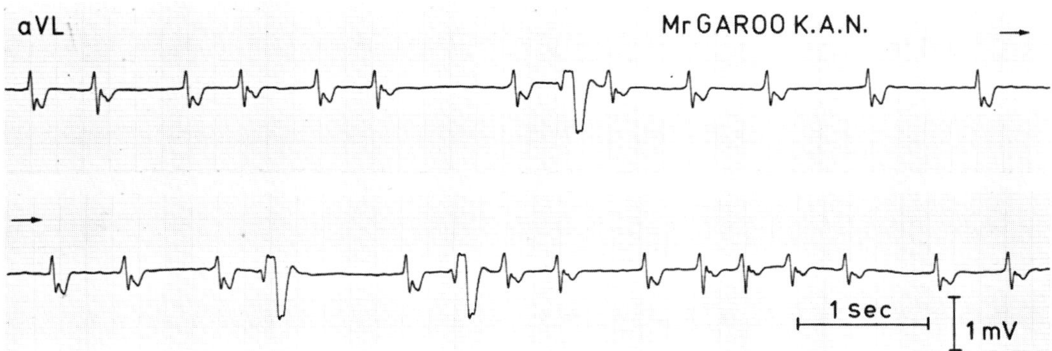

FIGURE 15.1 Atrial fibrillation in a kangaroo—the Ashman phenomenon.[117]

complication of mitral valvular disease, hyperthyroidism, and coronary artery disease.[8–11] It is frequently seen in the intensive care units, especially after cardiac surgery.[12] There is sustained atrial fibrillation, paroxysmal atrial fibrillation, and lone atrial fibrillation; all have in common a random ventricular response. However, in case of atrial fibrillation and complete AV block or a ventricular tachycardia, the ventricular rhythm may be regular, at least not randomly irregular, while the atria fibrillate. Atrial fibrillation occurs not only in humans but in large dogs (over 20 kg),[13] and horses.[14–16] We have observed one case in a kangaroo (Fig. 15.1), and have been looking for it in elephants[17] and whales.[18]

HISTORICAL REMARKS

In 1850, Hoffa and Ludwig[19] described atrial fibrillation in an animal experiment. What is called atrial fibrillation in patients today was discovered as "pulsus irregularis perpetuus" by Hering in 1903,[20] although Scherf and Schott[21] mention the discovery of atrial fibrillation by Vulpian in 1874. Mackenzie in his paper in 1904[1] did not link "the cause of continuous irregularity of the heart" to atrial or auricular fibrillation, as it was called at the beginning of our century.

> The history of the recognition of fibrillation of the auricles will impress you with the dimness of our eyes and the opacity of the obstacles which embarrass our vision. You will know how blind we have been to things which once seen, are so apparent.

This statement by Sir Thomas Lewis in 1912[22] is no less appropriate than the opening sentence of Mackenzie's article in 1904.[1] Mackenzie originally attributed the continuous irregularity of the heart to what he called a "nodal rhythm" because of the absence of a wave in the jugular pulse. He still may have been right, albeit for the wrong reasons, as will be explained later. Five years later, Lewis produced evidence that the irregularity of the heart probably results from fibrillation of the auricles, but this was after the introduction of the electrocardiograph.[23] Final proof had to wait until 1912, when Lewis observed that horses with complete irregularity of the heart had chronic fibrillation of the auricles.[24] After this, atrial fibrillation was established as an important and frequently occurring arrhythmia.[25,26] If Mackenzie, like Lewis,[22] had used the electrocardiograph, he would have recognized that during atrial fibrillation there is no atrial arrest as seems to be the case when we observe only the jugular pulse.

THE ROLE OF THE ATRIA IN ATRIAL FIBRILLATION

The random pattern of the ventricular response during atrial fibrillation points to an electrical behavior of the atria that almost certainly is random[27,28] because whatever the mechanism in the AV node that reduces the fast atrial rate to a much slower ventricular rate, it is unlikely that the AV node would be able to transform a nonrandom atrial input into a random ventricular response. Moreover, we and others have shown that in patients with atrial fibrillation who have Wolff-Parkinson-White syndrome and conduction through the bypass the ventricular rhythm is also random.[28,29] Thus, it seems fair to conclude that the cause of atrial fibrillation is an erratic electrical behavior of the atrial myocardium, resulting in a random sequence of atrial impulses of varying amplitude, which reach the AV node from random directions.[30,31]

Electrical Inhomogeneity

Garrey[32] in 1914 probably was the first to relate the nature of fibrillatory contractions of the heart to tissue mass and form. A solid basis for our understanding of (especially atrial) fibrillation has been provided by Moe and his colleagues.[33–35] Using computer simulation, they analyzed and defined the factors that cause the transition of the organized pattern of normal atrial excitation into self-sustaining, chaotic electrical behavior of the fibrillating atria. Moe and coworkers discovered that fibrillatory activity depends on (a) the total number of cells involved and (b) the electrical properties of those cells.

The large number of cells and unstable electrical properties (inhomogeneity) control the excitation process of the atria. Large atria will fibrillate much easier than small atria with a smaller number of cells and with the same or similar cellular electrical properties.

The Atrial Electrogram

Puech et al.[36] and others[27–29] have attempted to study the electrical activity of the atria in patients with atrial fibrillation. Using electrode-catheters, they recorded uni- and bipolar electrograms from different sites in the fibrillating atria and concluded from the appearance of the recorded electrograms the presence of atrial fibrillation. They also observed that in different patients atrial fibrillation may look different, while in one patient different atrial electrograms were recorded from different sites in the right atrium. We performed signal analysis of the atrial electrogram to learn more about the electrical behavior of the atria in patients with atrial fibrillation.[28] Analysis of the intervals between the zero crossings of the electrical signal by means of autocorrelograms showed absence of correlation between successive intervals. In other words, the atrial rhythm, assuming that the zero crossing method delivers a rhythm that reliably represents the rhythm of the fibrillating atria, is random with a rate between 300 and 600 episodes per minute.[28,36–38] However, not only the sequence of the recorded signals displays the erratic electrical behavior of the fibrillating atria, but also the shape and the amplitude of the signals do not seem to show any repetition either.

Unipolar electrograms do not reveal information about the direction(s) of atrial excitation during fibrillation. Bipolar recordings show continuously changing patterns of atrial excitation, which implies that there is no organized spread of excitation in the atria as observed during sinus or any other atrial rhythm.[39] Signal analysis of ventricular electrograms and electrocardiograms during ventricular fibrillation has demonstrated local areas of electrical activity that give an illusion of wave fronts traveling from one site to another.[40] We may assume that also during atrial fibrillation local areas of electrical activity are present. The work of Allessie and his group[41,42] has produced further evidence that multiple so-called wandering wavelets are the basis of atrial fibrillation. However, as clearly demonstrated,[43] experimentally induced atrial fibrillation may not be representative of sustained atrial fibrillation in humans, dogs, and horses.

We recapitulate the electrical behavior of the atria during atrial fibrillation, as follows[44]: (a) The rhythm at any site of the atria is random with a rate of approximately 300 to 600/min. (b) The electrical activity is restricted to more or less defined areas—in other words, there is no spread of activation. (c) The recorded signals vary continuously in amplitude, shape, duration, and direction.

We conclude that atrial signals reaching the AV node during fibrillation have complex patterns that differ in different patients or in the same patient under different circumstances. Drugs like digitalis,[45] quinidine, amiodarone, and others also affect the electrical behavior of the fibrillating atria, and thus the pattern of the signals that reach and surround the AV node.

AV NODAL FUNCTION IN ATRIAL FIBRILLATION

Since Engelmann[46] at the end of previous century introduced the word *Leitung* for the functional or anatomical connection between atria and ventricles, the coupling between ventricular and atrial excitation and contraction has been described in terms of conduction. *Leitung* may be translated as "connection" (cable or wire) or indeed "conduction"—the transport of electrons as later expressed by Hoffman and Cranefield in their cable theory for AV nodal function.[47]

Concealed Conduction

Engelmann was the first to observe and describe what is called today concealed conduction. While studying the effect of atrial extrasystoles on the contractions of the ventricles of isolated frog hearts, he noticed that every effective atrial contraction, even if it did not elicit a ventricular systole, prolonged the subsequent AV interval.

Langendorf[4] in 1948 observed the same phenomenon in clinical electrocardiograms and named the phenomenon "concealed conduction." According to Fisch,[48] concealed conduc-

tion is incomplete conduction coupled with an unexpected behavior of the subsequent impulse.

Hoffman and Cranefield[49] introduced the concept of decremental conduction to explain AV nodal delay during sinus rhythm. Concealed conduction and decremental conduction were recognized as the basic mechanisms by which the AV node slows the ventricular rate during atrial fibrillation.[50]

Moore[51] and Mazgalev et al.[52] showed concealed conduction using microelectrodes in isolated rabbit preparations. Short RR intervals were associated with lesser or absent concealment, whereas several atrial impulses were concealed during long cycles. There can be no doubt about the essential role of the AV node to slow down the rate of the fibrillating atria. This is best demonstrated in patients with the Wolff-Parkinson-White syndrome, atrial fibrillation, and conduction through the accessory pathway[28,53] in whom ventricular rates may reach values of 300/min and can produce ventricular fibrillation and sudden death.[54,55]

The AV Node as a Biological Oscillator

For many years there was little reason to doubt the Langendorf–Moore concept of concealed conduction as the slowing mechanism of the AV node. However, Grant[56] and James and his group[57,58] brought forward alternative mechanisms that could explain AV nodal function without viewing the AV conduction system as an electrical cable with a high-resistance site in the AV node. Grant[56] suggested that the AV node, like the sinus node, may behave like a relaxation oscillator and thus be modulated by outside influences.[59,60] In 1983, Cohen et al.[61] published a model that characterized the behavior of the AV node as an equivalent cell in terms of a hypothetical transmembrane potential—in other words, as an electrotonically modulated pacemaker. This model was attractive because it could easily explain the characteristics of the ventricular rhythm in patients with atrial fibrillation, without necessarily adopting the concealed conduction theory. In 1986, Van der Tweel et al.[62] presented experimental evidence that the function of the canine AV node could be described as a periodically perturbed, biological oscillator.

The Compensatory Pause

Langendorf,[63] Pritchett et al.[64] and others[65,66] demonstrated that after ventricular extrasystoles the ventricular cycle was lengthened also in the presence of atrial fibrillation.[63] Langendorf called this phenomenon the compensatory pause in atrial fibrillation. It was believed to be caused by lengthening of the AV node refractory period due to retrograde concealed conduction into the AV node of the spontaneous or artificially evoked ventricular extrasystole. However, Moore and Spear[67] and Akhtar and coworkers[68,69] showed that properly timed retrograde concealed conduction facilitates rather than slows AV nodal anterograde conduction.

Renewed doubt in the validity of concealed conduction theory as an explanation for the ventricular rate and rhythm in atrial fibrillation originated from the work of Wittkampf et al.[70] They found that ventricular pacing at intervals almost twice as long as the shortest spontaneous RR intervals in patients with atrial fibrillation and normal AV conduction could block all anterograde conduction, resulting in a pacemaker rhythm without any anterogradely conducted QRS complexes (Fig. 15.2). These observations cannot be explained by the classical AV nodal "filter" theory, despite the defense of concealed conduction by Dreifus and Mazgalev[71] in an editorial that accompanied Wittkampf's paper.[70]

In further studies, Wittkampf and coworkers,[72,73] using induced ventricular extrasystoles, obtained further evidence that in patients with atrial fibrillation, concealed anterograde conduction at different levels in the AV node is unlikely. The apparent ability to reset a random discharge cycle in the AV node by ventricular extrasystoles suggests that the distal side of a weakly coupled area inside the AV node behaves as a pacemaker for the ventricular rhythm during atrial fibrillation.

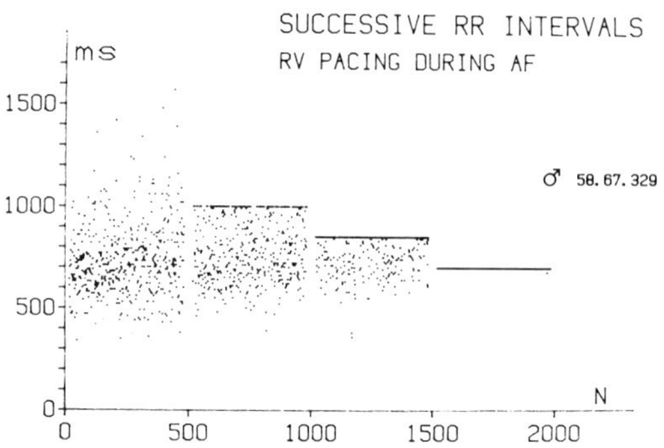

FIGURE 15.2 Successive RR intervals in a patient with atrial fibrillation before (first 500 cycles) and during pacing of the right ventricle with a pacing interval of 1,000 ms, 850 ms, and 700 ms (cycles 500–2,000). At a pacing interval of 700 ms (last 500 cycles) all anterograde "conduction" ceased and the rhythm became regular. *Reprinted from Wittkampf et al.,*[70] with permission.

Scaling in the AV Junction?

Scaling of atrial impulses in the AV junction has been widely accepted as the cause of the random and relatively slow ventricular response in patients with atrial fibrillation.[29,30] Wittkampf et al.[74,75] studied ventricular rhythm in patients with atrial fibrillation under various circumstances known to modulate AV nodal conduction characteristics. With a fixed scaling factor (ratio between atrial and ventricular rates), the relative variability of the atrial and ventricular cycle length should relate to the square root of the scaling factor.[76] This principle is the basic mechanism of the atomic clock. However, a linear relationship between variability and ventricular cycle length was found with an almost constant relative variability of 22%. In any scaling process in which the cumulative effect of randomly irregular impulses of randomly varying magnitude would lead to ventricular activation, individual variations would partly compensate each other and thus tend to stabilize the rhythm at longer average cycle lengths when on the average more impulses are required to generate the next ventricular depolarization. Thus, a proportional increase in absolute variability is not to be expected if longer intervals are derived from randomly irregular short intervals by such a process of scaling.

When a more frequent invasion of the AV node by atrial impulses (for instance during digitalis treatment) would be responsible for a slower ventricular rate, the average scaling factor would be more than proportionally higher in patients with a long average ventricular cycle length than in those with a relatively fast ventricular rate under similar circumstances. This would have to result in even greater reduction of relative variability at slow ventricular rates. Figure 15.3 shows that this is not the case. Due to the effect of digitalis treatment, average cycle length has increased. At the same time the width of the histogram (standard deviation) also has increased. In atrial fibrillation, the irregularity of the ventricular rhythm increases linearly with the ventricular rate. In other words, all patients with atrial fibrillation have about the same irregularity of their ventricular rhythm irrespective of their ventricular rate. An unusual and odd filtering process has to be assumed to explain such a linear relationship between average cycle length and standard deviation. This suggests to us that scaling as traditionally conceived almost certainly does not take place in the AV junction. We realize that these findings disagree with our former ideas,[30] the views of Dreifus and Mazgalev,[71] and the conclusions derived from the experimental findings of Moore,[51] Janse,[77] and others[78,79] in in vitro preparations.

Alternative Explanation

The linear relationship between average cycle length and absolute variability at different randomly irregular atrial and ventricular rhythms suggests that randomly irregular fibrillatory atrial impulses are not conducted through the AV junction but rather impart to cells within the AV junction a similarly irregular behavior, but on a slower time scale. When all intervals between atrial impulses are multiplied by a constant factor, then both average cycle length and standard deviation are multiplied by the same factor, and consequently the relative variability remains constant. The relationship observed by us seems to be the expression of an intrinsic property of AV nodal cells which is responsible for both the average ventricular cycle length and the standard deviation during atrial fibrillation. Although these findings and this hypothesis are in agreement with the AV nodal pacemaker theory, they do not support the mechanism of phase 4 electrotonic modulation that would also result in a decrease of relative irregularity with increasing ventricular cycle length.[6,61,80] Our findings could be explained by electrotonic inhibition and summation in AV nodal cells,[81,82] a mechanism requiring further experimental support. We restipulate the characteristics of AV nodal function during atrial fibrillation as follows: (a) complete blocking of all anterograde conduction during right ventricular pacing at pacing intervals that are twice as long as the spontaneous short RR intervals; (b) the reset of the AV nodal activation sequence following ventricular extrasystoles; and (c) the absence of scaling of the atrial impulses as the cause of the slower ventricular rate.

This suggests to us that the classical concept of propagation through the AV junction by concealed and decremental conduction in the AV node resulting in scaling down of the atrial rate does not explain the ventricular rate and rhythm in patients with atrial fibrillation. A region within the AV node seems to determine ventricular rate and rhythm modulated by the inefficient electrical impulses of the fibrillating atria.[60]

It should be mentioned that also during sinus rhythm the AV node may not conduct the well-organized atrial impulse in a generally accepted manner. This is apparent from the mismatch between the heart size and the PR interval.[83,84]

As early as 1913, Waller[85] drew attention to the correlation between the size of the animal and the "aurioculoventricular" interval. Clark in 1927[86] was struck by the fact that the PR interval varies so little in different animals. There is, paradoxically, a comparatively short PR interval in hearts of large animals. Since the heart weight is closely and linearly related to the body weight,[87,88] the latter can be substituted by the

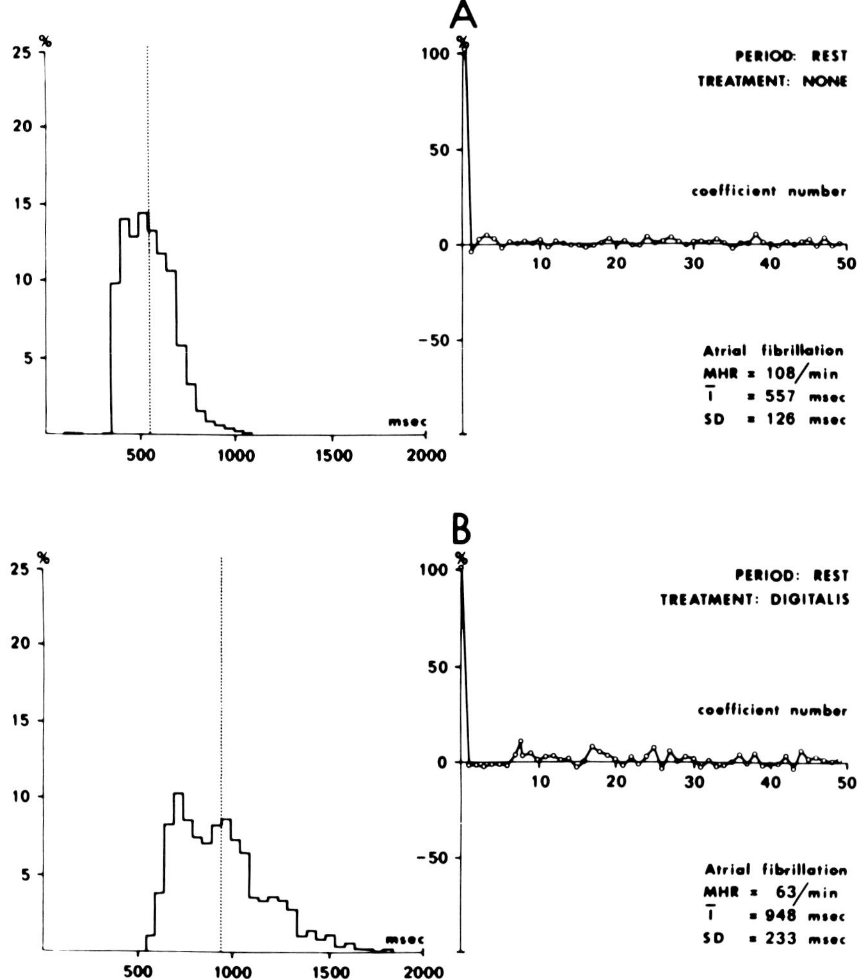

FIGURE 15.3 Histogram and autocorrelogram of the RR intervals of a patient with atrial fibrillation at rest without (**A**) and with (**B**) digitalis treatment. The autocorrelogram is unchanged; the ventricular rhythm remains random, despite change in form and shift to the right of the histogram. *Reprinted from Bootsma et al.,*[30] with permission.

former and the mismatch between the PR interval and the heart weight becomes evident.[83]

The length of the AV conduction system can be approximated by the third root of heart weight.[80] The relationship between PR interval and the third root of heart weight is shown in Figure 15.4. The S-shape of this relationship indicates a relatively long AV delay in small mammals and a relatively short AV delay in large mammals. Assuming a more or less constant conduction velocity in the His-Purkinje system,[89] although differences of some orders of magnitude may be also present,[90,91] the contribution of the AV node to the total AV delay (PR interval) in small mammals is relatively large, and relatively small in large mammals such as horses, elephants, and whales. From Kawamura's work[92] it became apparent that small mammals have relatively large AV nodes, whereas large mammals have relatively small ones. James[60] made it plausible that the AV nodes in large mammals contain more fibrous tissue (thus fewer P cells) than the AV nodes in small mammals. These two findings may offer at least in part a morphological substrate for the mismatch between the PR interval and the body (heart) size. If the conduction through the AV system were to take place as conceived by Hoffman and Cranefield,[47] the relationship between PR interval and heart size would be almost certainly different.

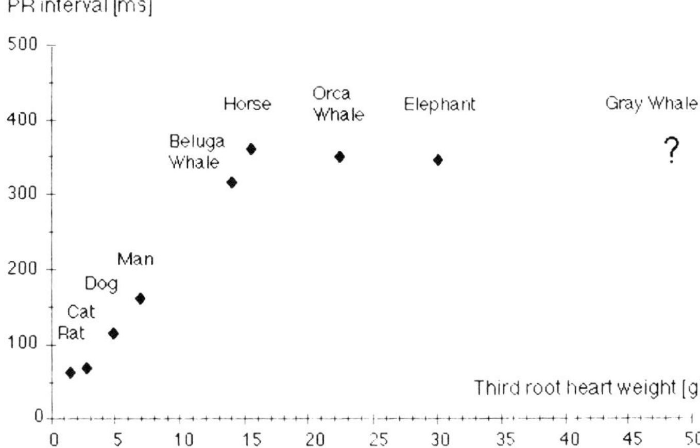

FIGURE 15.4 Relationship between PR interval and the third root of heart weight. Values obtained from Altman and Dittmer.[118]

THE VENTRICULAR RHYTHM IN ATRIAL FIBRILLATION

The most striking clinical feature of atrial fibrillation is the irregularity of the pulse. Although the complete irregularity of the heart was discovered and established long ago, the advent of computers enabled the ventricular rhythm to be analyzed more accurately and its mathematical properties to be established. The introduction of sophisticated time sequence analyzing techniques in the early 1950s stimulated several groups of investigators to study the irregular ventricular rhythm in atrial fibrillation.[30,93–96] The results of these studies were sometimes conflicting and created confusion and even controversies.[97,98] The differences in the results were probably due to, or at least influenced by, the choice of the statistical techniques, the selection of patients, and the use of induced (artificial) atrial fibrillation in atria that would otherwise not have fibrillated. Moreover, high-frequency atrial stimulation was employed in healthy dogs to simulate atrial fibrillation and to obtain irregular ventricular rhythms that looked like the rhythms obtained at the bedside, but may not have represented real-life situations.

Human Patients

In a study published in 1970,[30] we used serial autocorrelograms and histograms to analyze the ventricular rhythm of patients with sustained atrial fibrillation before and during digitalis treatment at rest and during exercise. The results of this study led to the conclusion that the ventricular rhythm in patients with uncomplicated atrial fibrillation is random where randomness is defined as a time series having mutually independent intervals. In Figure 15.3, reproduced from this paper, the histogram and autocorrelogram of the RR intervals of a patient with atrial fibrillation is shown before (A) and during (B) digitalis treatment. Despite a shift to the right and significant changes in the shape of the histogram due to digitalis treatment, the autocorrelogram was not affected. All correlation coefficients remained at zero before and during digitalis treatment, and thus the ventricular rhythm remained random. The same was true for the situations before and during exercise, namely, there was a change in the histogram, but the ventricular rhythm remained random. Different interventions and different drugs like digitalis, quinidine, verapamil, and β-blockers change the histogram but do not affect the random pattern of the ventricular rhythm in human patients.

Horses and Dogs

In a subsequent study[43] we demonstrated that dogs with true atrial fibrillation on the basis of mitral incompetence also had a random ventricular rhythm. Atrial fibrillation induced by rapid atrial stimulation, however, although resulting in an irregular ventricular rhythm, did not create randomness of the successive RR intervals. Horses are known to develop atrial fibrillation, usually on the basis of some form of myocardial disease.[15] It is of considerable interest that horses with atrial fibrillation at rest, with or without digitalis treatment, have signs of periodicity in their autocorrelogram (Fig. 15.5).[99] This periodicity can be abolished by atropine and quinidine, drugs that cause an increase in ventricular rate. From these observations we concluded that a considerable drop in blood pressure caused by

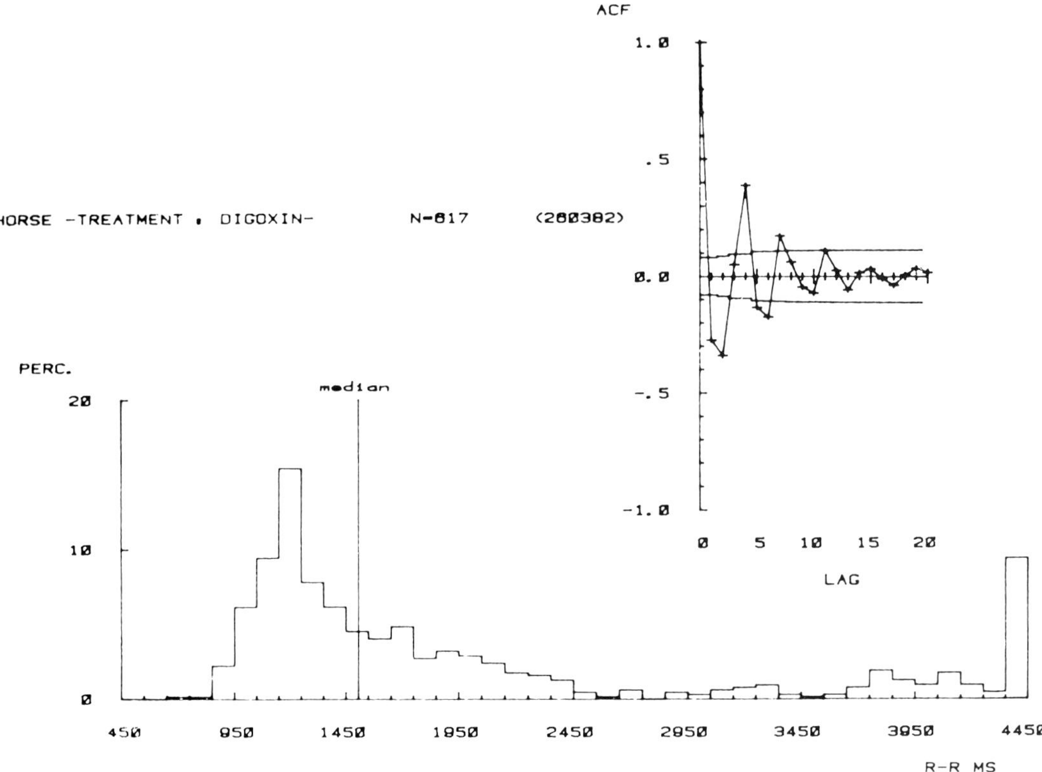

FIGURE 15.5 Histogram (left) and serial autocorrelogram (right upper corner) of a horse with atrial fibrillation during digitalis treatment. Note the large number of much longer intervals (4,450 ms or more) and the strong periodicity in the autocorrelogram. *Reprinted from Meijler et al.*,[99] with permission.

very long RR intervals (up to 5 s) in horses with atrial fibrillation creates autonomic nervous interference with AV nodal propagation properties through baroreceptor triggering.[99] Our overall conclusion was that the cause of ventricular irregularity resides somewhere else than in the AV conduction system, and we offered the hypothesis that randomly spaced atrial impulses of random strength reaching the AV node from random directions are responsible for the renewal process of the RR interval sequence.[30] This opinion was strengthened by the fact that the fast ventricular rhythm is also random in patients with Wolff-Parkinson-White syndrome and atrial fibrillation and conduction through the bypass,[28] while patients and dogs with ventricular fibrillation during heart–lung bypass turned out to have a random atrial rhythm; again randomness is defined as a time series without mutually dependent intervals.[100]

We have concluded that the essential role of the AV node in atrial fibrillation is restricted to the process of changing the rapid atrial rhythm into a slower ventricular rhythm, probably by some form of electrotonic modulation of an AV nodal pacemaker.

DELIRIUM CORDIS

Apart from the lack of insight into the AV nodal function, another baffling puzzle is the regulation of cardiac output during atrial fibrillation, especially in the presence of valvular disease. It is well known that atrial fibrillation may be totally unnoticed by the patient who may be unaware of an irregular heart beat, and may not be limited in daily physical activities. Often there are no signs of heart failure; cardiac output may be adequate and blood pressure remain normal, even during severe exertion.[50] Nevertheless, atrial fibrillation was discovered not only because the pulse was irregular but also because the arterial pulsations differed in form and size. This led Hering[20] and also Wenckebach[101] to the introduction of the term "delirium cordis." It is fascinating to learn how at the beginning of this century, that is, before the introduction of the electrocardiograph, the irregularity and inequality of the

pulse occupied the minds of the pioneers in cardiology.[1,20,22,101,102]

The Frank–Starling Mechanism

In 1915, Einthoven and Korteweg[102] demonstrated that the amplitude of the pulse wave was related to the length of the preceding heart period. In the following decades, the relationship between duration of the cardiac cycle and variations in hemodynamics in patients with atrial fibrillation was explained by the effect of varying degrees of diastolic filling on the contractile force of the heart (Frank–Starling mechanism).[103,104] In 1960, Braunwald and associates[105] reasoned that it was "unlikely that the duration of the preceding diastole, per se, determines the characteristics of the subsequent ventricular contraction" but that "Starling's law of the heart operates in patients with mitral stenosis and atrial fibrillation on a beat to beat basis."

In 1968 we showed[106] that in isolated Langendorff-perfused rat hearts in which Starling's law cannot operate, an induced random rhythm creates the same variability of ventricular contractions as observed in patients with atrial fibrillation (Fig. 15.6). In agreement with Braunwald et al.,[105] we found in patients with atrial fibrillation and mitral stenosis that left ventricular end diastolic pressure and dimensions may be closely related to the duration of the diastolic pause. However, in patients with atrial fibrillation but without mitral stenosis, the left ventricle attains maximum volume early in diastole and remains constant and independent of the duration of the diastole. Left ventricular beat-to-beat variations are thus unaffected by preload.[107–109]

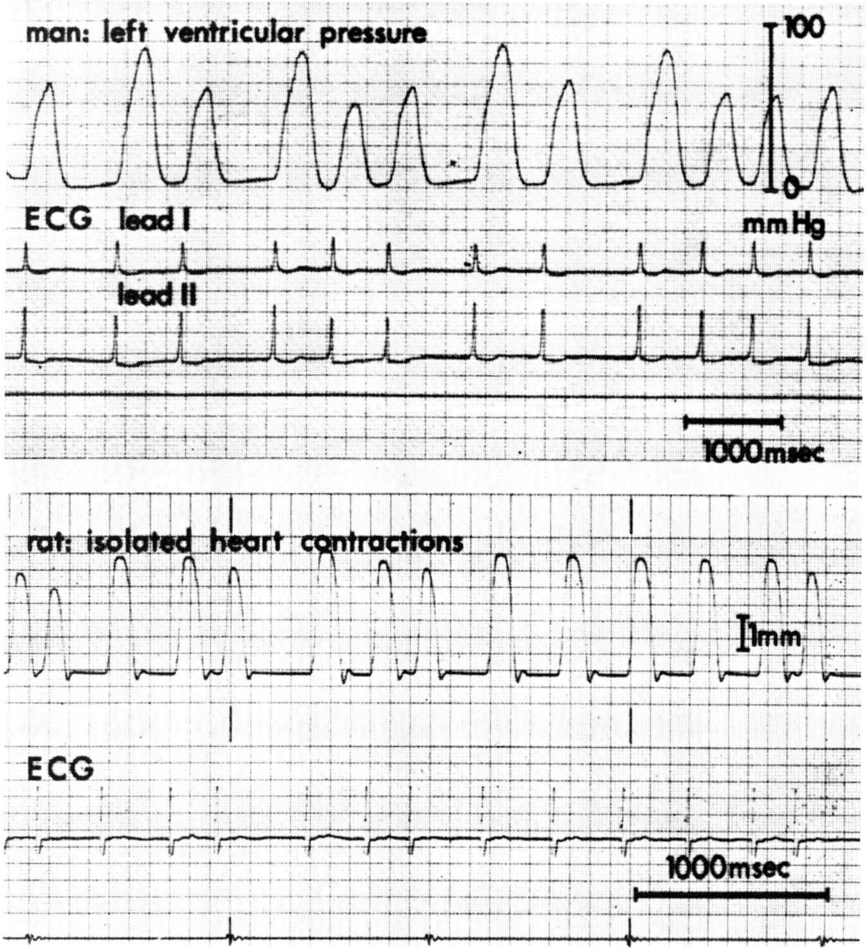

FIGURE 15.6 Left ventricular pressure of a patient with atrial fibrillation (upper curve) and contractions of an isolated rat heart (lower curve) stimulated with a (ventricular) rhythm from a patient with atrial fibrillation. (For further details, see text). *Reprinted from Meijler et al.,*[106] with permission.

Postextrasystolic Potentiation

During atrial fibrillation, left ventricular contractions are mainly dependent on cycle length. Brooks in 1960[110] introduced the term "post-extrasystolic potentiation," a physiological phenomenon that describes the strengthening effect of a short RR interval on the subsequent ventricular contraction. The opposite is also true: a longer-than-average interval depotentiates or weakens the subsequent contraction.[111] The weakening after long intervals is less pronounced than the potentiation after short intervals.

The RR interval histogram in atrial fibrillation (Fig. 15.3) shows that there are more shorter-than-average intervals than longer-than-average ones. The result of this is that beat-to-beat variations in atrial fibrillation are positively related to the preceding RR interval and inversely related to the pre-preceding interval(s).[106,112] The effect of RR interval duration on contractile behavior of the heart is quite complicated but can be accounted for by the potentiation and depotentiation effects.[113,114]

When we consider hemodynamic parameters like stroke volume or arterial blood pressure in relation to preceding RR intervals, it should be realized that long RR intervals diminish aortic impedance, while ventricular contractions after a short RR interval may face a high aortic impedance.[115] This, of course, tends to strengthen the first coefficient of any RR interval/hemodynamic parameter correlation.

Despite the evolving knowledge of the relation between stroke volume and the preceding RR intervals,[108,113,116] we still do not fully understand the regulation of the circulation in patients with atrial fibrillation at rest and during exercise and the effects of changes in ventricular rates induced by drugs such as digitalis, verapamil, β-blockers, or quinidine. Since in general cardiac output tends to be adequate within a wide range of ventricular rates,[50] one is tempted to assume a loose association between cardiac output and ventricular rate and rhythm in patients with atrial fibrillation.

We conclude that beat-to-beat variations (delirium cordis) in patients with atrial fibrillation can be explained satisfactorily on the basis of potentiation and depotentiation of myocardial contractility and cycle-length-dependent aortic impedance. However, the maintenance of an adequate cardiac output and blood pressure is still inadequately explained.

FINAL REMARKS

Although atrial fibrillation is the most frequently occurring arrhythmia, we still do not know how the AV node transforms fast irregular atrial rhythm into a random ventricular rhythm at a rate compatible with life. We also do not know how in patients with atrial fibrillation cardiac output is regulated and adequate circulation is maintained. After almost 100 years of research into the riddles of atrial fibrillation, the picture of our metaphoric elephant is still opaque, and the dimness of our vision still obscures the intervening obstacles.[22]

CONCLUSION

Atrial fibrillation is a commonly observed cardiac arrhythmia that shows the following characteristics:

(a) Fibrillation of the atria is caused by electrical inhomogeneity of the atrial myocardial cells.

(b) Transformation of a rapid atrial rate into a slower ventricular rate that is compatible with maintenance of life. Recent findings make it unlikely that concealed and/or decremental conduction in the AV junction takes place during atrial fibrillation.

(c) Random ventricular rhythm is created by an erratic electrical behavior of the atria. The atrial impulses impart to the AV node, possibly by electrotonic modulation, a random activation sequence that is much slower than the rate of the atrial impulses.

(d) While ventricular rhythm is random in atrial fibrillation, the ventricular contractions correlate up to the second (and sometimes third) coefficient with preceding RR intervals. This correlation is based on postextrasystolic potentiation and depotentiation following long intervals.

ACKNOWLEDGMENT

This work was supported by the Wijnand M. Pon Foundation, Leusden, The Netherlands.

REFERENCES

1. Mackenzie J: The inception of the rhythm of the heart by the ventricle as the cause of continuous irregularity of the heart. *Br Med J* 1904;March 5:529–536.
2. Robles de Medina EO, Bernard R, Coumel Ph, Damato AN, Fisch Ch, Krikler D, Mazur NA, Meijler FL, Mogensen L, Moret P, Pisa Z, Wellens HJJ (WHO/ISFC Task Force): Definition of terms related to cardiac rhythm. *Am Heart J* 1978;95:796–806.
3. Robles de Medina EO, Meijler FL: Atrial flutter–atrial fibrillation: Is a distinction clinically necessary? *Practical Cardiol* 1981;7:77–88.
4. Langendorf R: Concealed A-V conduction: The effect of blocked impulses on the formation and conduction of subsequent impulses. *Am Heart J* 1948;35:542–552.
5. Langendorf R, Pick A, Katz LN: Ventricular response

in atrial fibrillation. Role of concealed conduction in the AV junction. *Circulation* 1965;32:69–75.

6. Meijler FL, Fisch C: Does the AV node conduct? *Br Heart J* 1989;61:309–315.
7. Selzer A: Atrial fibrillation revisited. *N Engl J Med* 1982;306:1044–1045.
8. Godtfredsen J: Atrial fibrillation. Etiology, course and prognosis. A follow-up study of 1212 cases. Copenhagen, University of Copenhagen, 1975; p 39 (thesis).
9. Brill IC: Auricular fibrillation; the present status with a review of the literature. *Ann Intern Med* 1937;10:1487–1502.
10. Åberg H: Atrial fibrillation. *Acta Med Scand* 1968;184:425–431.
11. Alpert JS, Petersen P, Godtfredsen J: Atrial fibrillation: Natural history, complications, and management. *Ann Rev Med* 1988;39:41–52.
12. Lauer MS, Eagle KA, Buckley MJ, DeSanctis RW: Atrial fibrillation following coronary artery bypass surgery. *Prog Cardiovasc Dis* 1989;31:367–378.
13. Bohn FK, Patterson DF, Pyler RL: Atrial fibrillation in dogs. *Br Vet J* 1971;127:485–496.
14. Deegen E, Buntenkotter S: Behaviour of the heart rate of horses with auricular fibrillation during exercise and after treatment. *Equine Vet* 1976;8:26–29.
15. Deem DA, Fregin GF: Atrial fibrillation in horses: A review of 106 clinical cases, with consideration of prevalence, clinical signs, and prognosis. *J Am Vet Med Assoc* 1982;180:261–265.
16. Meijler FL, Van der Tweel I, Herbschleb JN, Heethaar RM, Borst C: Lessons from comparative studies of atrial fibrillation in dog, man and horse, in Zipes DP, Jalife J (eds): *Cardiac Electrophysiology and Arrhythmias: Mechanisms & Management. Part 8. Electrocardiography.* New York, Harcourt Brace Jovanovich Publ, Grune & Stratton, 1985, pp 489–493.
17. Meijler FL, Van der Tweel LH: De elektrocardiogrammen van 10 olifanten en van de orka in Harderwijk. *Ned T Geneeskd* 1986;130:2344–2348.
18. Meijler FL: An off beat whale hunt. *Br Med J* 1989;299:1563–1565.
19. Hoffa M, Ludwig C: Einige neue Versuche über Herzbewegung. *Zeitschr f rat Med* 1850;9:107–144.
20. Hering HE: Analyse des pulsus irregularis perpetuus. *Prager Med Wochenschr* 1903;28:377–381.
21. Scherf D, Schott A: *Extrasystoles and Allied Arrhythmias.* Chicago, Wm Heinemann, 1973, p 400.
22. Lewis T: A lecture on evidences of auricular fibrillation treated historically. *Br Med J* 1912; 2663(January 13):57–60.
23. Katz LN, Hellerstein HK: Electrocardiography, in Fishman AP, Richards DW (eds): *Circulation of the Blood. Men and Ideas.* Baltimore, Williams & Wilkins, 1982, pp 265–351.
24. Lewis T: Irregularity of the heart's action in horses and its relationship to fibrillation of the auricles in experiment and to complete irregularity of the human heart. *Heart* 1911/12;3:161–171.
25. McMichael J: History of atrial fibrillation 1628–1819. Harvey–De Senac–Laënnec. *Br Heart J* 1982;48:193–197.
26. Kannel WB, Abbott RD, Savage DD, McNamara PM: Epidemiologic features of chronic atrial fibrillation. *N Engl J Med* 1982;306:1018–1022.
27. Slocum J, Sahakian A, Swiryn S: Computer discrimination of atrial fibrillation and regular atrial rhythms from intra-atrial electrograms. *PACE* 1988;11:610–621.
28. Meijler FL, Van der Tweel I, Herbschleb JN, Hauer RNW, Robles de Medina EO: Role of atrial fibrillation and atrioventricular conduction (including Wolff-Parkinson-White syndrome) in sudden death. *J Am Coll Cardiol* 1985;5:17B–22B.
29. Kirsh JA, Sahakian AV, Baerman JM, Swiryn S: Ventricular response to atrial fibrillation: Role of atrioventricular conduction pathways. *J Am Coll Cardiol* 1988;12:1265–1272.
30. Bootsma BK, Hoelen AJ, Strackee J, Meijler FL: Analysis of R-R intervals in patients with atrial fibrillation at rest and during exercise. *Circulation* 1970;41:783–794.
31. Hoffman BF, Bigger JT: Digitalis and allied glycosides, in Gilman AG, Goodman LS, Gilman A (eds): *The Pharmacological Basis of Therapeutics*, 6th ed. New York, Macmillan, 1980, pp 729–760.
32. Garrey WE: The nature of fibrillatory contraction of the heart.—Its relation to tissue mass and form. *Am J Physiol* 1914;33:397–414.
33. Moe GK, Rheinboldt WC, Abildskov JA: A computer model of atrial fibrillation. *Am Heart J* 1964;67:200–220.
34. Moe GK: On the multiple wavelet hypothesis of atrial fibrillation. *Arch Int Pharmacodyn* 1962;140:183–188.
35. Moe GK, Abildskov JA: Atrial fibrillation as a self-sustaining arrhythmia independent of focal discharge. *Am Heart J* 1959;58:59–70.
36. Puech P, Grollean R, Rebuffat G: Intra-atrial mapping of atrial fibrillation in man, in Kulbertus HE, Olsson SB, Schlepper M (eds): *Atrial Fibrillation.* Sweden, AB Hässle, 1982; pp 94–108.
37. Åberg H: Atrial fibrillation. III. A study of fibrillatory wave using a new technique. *Acta Soc Med Uppsala* 1969;74:17–27.
38. Drury AN, Brow GR: Observations relating to the unipolar electrical curves of heart muscle with especial reference to the mammalian auricle. *Heart* 1925/26;12:321–369.
39. Van der Kooi MW, Durrer D, Van Dam RTh, Van der Tweel LH: Electrical activity in sinus node and atrioventricular node. *Am Heart J* 1956;51:684–700.
40. Herbschleb JN, Van der Tweel I, Meijler FL: The apparent repetition frequency of ventricular fibrillation. *Comp in Cardiol, IEEE Comp Soc* 1982;249–252.
41. Allessie MA, Lammers WJEP, Bonke FIM, Hollen J: Experimental evaluation of Moe's multiple wavelet hypothesis of atrial fibrillation, in Zipes DP, Jalife J (eds): *Cardiac Electrophysiology and Arrhythmias: Mechanisms & Management. Part 8. Electrocardiology.* New York, Harcourt Brace Jovanovich Publ, Grune & Stratton, 1985, pp 265–275.
42. Kirchhof CJHJ, Allessie MA, Bonke FIM: The sinus node and atrial fibrillation. *Ann NY Acad Sci* 1990; 591:166–177.
43. Strackee J, Hoelen AJ, Zimmerman ANE, Meijler FL: Artificial atrial fibrillation in the dog; an artifact? *Circ Res* 1971;28:441–445.
44. Meijler FL: Atrial fibrillation: A new look at an old arrhythmia (editorial). *J Am Coll Cardiol* 1983;2:391–393.
45. Meijler FL: An "account" of digitalis and atrial fibrillation. *J Am Coll Cardiol* 1985;5:60A–68A.
46. Engelmann ThW: Beobachtungen und Versuche am suspendierten Herzen. Zweite Abhandlung. Ueber

die Leitung der Bewegungsreize im Herzen. *Pflügers Arch* 1894;56:149–202.
47. Hoffman BF, Cranefield PF: *Electrophysiology of the Heart.* New York, McGraw-Hill, 1960, pp 23 ff.
48. Fisch C: *Electrocardiography of Arrhythmias.* Philadelphia/London, Lea & Febiger, 1990, p 1.
49. Hoffman BF, Cranefield PF: *Electrophysiology of the Heart.* New York, McGraw-Hill 1960, pp 156–162.
50. Bigger TJ: Mechanism and diagnosis of arrhythmias, in Braunwald E (ed): *Heart Disease. A Textbook of Cardiovascular Medicine*, Vol. I. Philadelphia/London, WB Saunders, 1980, p 658.
51. Moore EN: Observations on concealed conduction in atrial fibrillation. *Circ Res* 1967;21:201–208.
52. Mazgalev T, Dreifus LS, Bianchi J, Michelson EL: Atrioventricular nodal conduction during atrial fibrillation in rabbit heart. *Am J Physiol* 1982; 243:H754–H760.
53. Wellens HJJ, Durrer D: Wolff-Parkinson-White syndrome and atrial fibrillation. Relation between refractory period of accessory pathway and ventricular rate during atrial fibrillation. *Am J Cardiol* 1974;34:777–782.
54. Dreifus LS, Haiat R, Watanabe Y, Arriage J, Reitman N: Ventricular fibrillation. A possible mechanism of sudden death in patients with Wolff-Parkinson-White syndrome. *Circulation* 1971;43:520–527.
55. Boineau JP, Moore EN: Evidence for propagation of activation across an accessory atrioventricular connection in types A and B pre-excitation. *Circulation* 1970;41:375–397.
56. Grant RP: The mechanism of A-V arrhythmias. With an electronic analogue of the human A-V node. *Am J Med* 1956;20:334–344.
57. Katholi CR, Urthaler F, Macy J Jr, James TN: A mathematical model of automaticity in the sinus node and AV junction based on weakly coupled relaxation oscillators. *Comp Biomed Res* 1977;10:529–543.
58. Urthaler F, Katholi CR, Macy J, James TN: Mathematical relationship between automaticity of the sinus node and the AV junction. *Am Heart J* 1973;86:189–195.
59. Van der Tweel LH, Meijler FL, Van Capelle FJL: Synchronization of the heart. *J Appl Physiol* 1973; 34:283–287.
60. James TN: Automaticity in the atrioventricular junction, in Rosen MR, Janse MJ, Wit AL (eds): *Cardiac Electrophysiology: A Textbook.* Mt. Kisco, NY, Futura Publ, 1990, chapter 2.3.
61. Cohen RJ, Berger RD, Dushane TE: A quantitative model for the ventricular response during atrial fibrillation. *IEEE Trans Biom Eng* 1983;BME-30:769–780.
62. Van der Tweel I, Herbschleb JN, Borst C, Meijler FL: Deterministic model of the canine atrio-ventricular node as a periodically perturbed, biological oscillator. *J Appl Cardiol* 1986;1:157–173.
63. Langendorf R: Aberrant ventricular conduction. *Am Heart J* 1951;41:700–707.
64. Pritchett LC, Smith WM, Klein GJ, Hammill SC, Gallagher JJ: The "compensatory pause" of atrial fibrillation. *Circulation* 1980;62:1021–1025.
65. Neuss H, Golling F-R, Thormann J, Weissmüller P, Kindler M: Regularisierung der Kammerintervalle bei Vorhofflimmern—elektrophysiologische Befunde zum zugrundeliegenden Mechanismus. *Zeitschr Kardiol* 1984;73:106–112.
66. Morady F, Dicarlo LA, Krol RB, De Buitleir M, Baerman JM: An analysis of post-pacing R-R intervals during atrial fibrillation. *PACE* 1986;9:411–416.
67. Moore EN, Spear JF: Experimental studies on the facilitation of AV conduction by ectopic beats in dogs and rabbits. *Circ Res* 1971;29:29–39.
68. Akhtar M, Damato A, Batsford WP, Ruskin JN, Ogunkelu JB: A comparative analysis of antegrade and retrograde conduction patterns in man. *Circulation* 1975;52:766–779.
69. Lehmann MH, Mahmud R, Denker S, Soni J, Akhtar M: Retrograde concealed conduction in the atrioventricular node: Differential manifestations related to level of intranodal penetration. *Circulation* 1984;70:392–401.
70. Wittkampf FHM, De Jongste MJL, Lie KI, Meijler FL: Effect of right ventricular pacing on ventricular rhythm during atrial fibrillation. *J Am Coll Cardiol* 1988;11:539–545.
71. Dreifus LS, Mazgalev T: "Atrial paralysis": Does it explain the irregular ventricular rate during atrial fibrillation? *J Am Coll Cardiol* 1988;11:546–547.
72. Wittkampf FHM, De Jongste MJL, Meijler FL: Atrioventricular nodal response to retrograde activation in atrial fibrillation. *J Cardiovasc Electrophysiol* 1990;1:437–447.
73. Wittkampf FHM, De Jongste MJL, Meijler FL: Competitive anterograde and retrograde atrioventricular junctional activation in atrial fibrillation. *J Cardiovasc Electrophysiol* 1990;1:448–456.
74. Wittkampf FHM, Robles de Medina EO, Strackee J, Meijler FL: Scaling in atrial fibrillation? in: Wittkampf FHM *Atrioventricular Nodal Transmission in Atrial Fibrillation.* State University Utrecht, the Netherlands, 1991, pp 89–104 (thesis).
75. Wittkampf FHM, Robles de Medina EO, Strackee J, Meijler FL: Conservation of cycle length variability in atrial fibrillation. *Science* (submitted).
76. Armitage P, Berry G: *Statistical Methods in Medical Research.* Oxford, Blackwell Sci Publ, 1987, p 90.
77. Janse MJ: Influence of the direction of the atrial wave front on AV nodal transmission in isolated hearts of rabbits. *Circ Res* 1969;25:439–449.
78. Billette J: Atrioventricular nodal activation during premature stimulation of the atrium. *Am J Physiol* 1987;252:H163–H177.
79. Billette J, Janse MJ, Van Capelle FJL, Anderson RH, Touboul P, Durrer D: Cycle-length-dependent properties of AV nodal activation in rabbit hearts. *Am J Physiol* 1976;231:1129–1139.
80. Meijler FL, Janse MJ: Morphology and electrophysiology of the mammalian atrioventricular node. *Physiol Rev* 1988;68:608–647.
81. Antzelevitch C, Moe GK: Electrotonic inhibition and summation of impulse conduction in mammalian Purkinje fibers. *Am J Physiol* 1983;245:H42–H53.
82. Chialvo DR, Jalife J: Non-linear dynamics of cardiac excitation and impulse propagation. *Nature* 1987;330:749–752.
83. Meijler FL: Atrioventricular conduction versus heart size from mouse to whale. *J Am Coll Cardiol* 1985;5:363–365.
84. Meijler FL: Comparative aspects of the dual role of the human atrioventricular node. *Br Heart J* 1986;55:286–290.
85. Waller AD: Cardiology and cardiopathology. *Br Med J* 1913;2:375–376.
86. Clark AJ: Conduction in the heart of mammals, in *Comparative Physiology of the Heart.* Cambridge, UK, Cambridge University Press, 1927, pp 49–51.

87. Schmidt-Nielsen K. *Scaling. Why is Animal Size so Important?* Cambridge, UK, Cambridge University Press, 1984, pp 126–130.
88. Prothero J: Heart weight as a function of body weight in mammals. *Growth* 1979;43:139–150.
89. Durrer D, Janse MJ, Lie KI, Van Capelle FJL: Human cardiac electrophysiology, in Dickinson CJ, Marks J (eds): *Developments in Cardiovascular Medicine.* Cambridge, MTP Press, 1978, pp 53–75.
90. Pressler ML: Cable analysis in quiescent and active sheep Purkinje fibres. *J Physiol* 1984;352:739–757.
91. Pressler ML: Membrane properties of the cardiac conduction system: comparative aspects. *Proc Royal Neth Acad Sci* 1990;93:477–487.
92. Kawamura K: Size of the atrioventricular node in mammals. *Proc Royal Neth Acad Sci* 1990;93:431–435.
93. Brody DA: Ventricular rate patterns in atrial fibrillation. *Circulation* 1970;41:733–735.
94. Horan LG, Kistler JC: Study of ventricular response in atrial fibrillation. *Circ Res* 1961;9:305–311.
95. Goldstein RE, Barnett GO: A statistical study of the ventricular irregularity of atrial fibrillation. *Comp Biomed Res* 1967;1:146–161.
96. Braunstein JR, Franke EK: Autocorrelation of ventricular response in atrial fibrillation. *Circ Res* 1961;9:300–303.
97. Urbach JR: The ventricular response to atrial fibrillation. *J Am Coll Cardiol* 1984;3:1105–1106.
98. Urbach JR, Grauman JJ, Straus SH: Quantitative methods for the recognition of atrioventricular junctional rhythms in atrial fibrillation. *Circulation* 1969;39:803–817.
99. Meijler FL, Kroneman J, Van der Tweel I, Herbschleb JN, Heethaar RM, Borst C: Nonrandom ventricular rhythm in horses with atrial fibrillation and its significance for patients. *J Am Coll Cardiol* 1984;4:316–323.
100. Scher AM, Heethaar RM, Zimmerman ANE, Meijler FL: Atrial rhythm during ventricular fibrillation in the dog. *Circ Res* 1976;38:41–45.
101. Wenckebach KF: *Die Arhytmie als Ausdruck bestimmter Funktionsstörungen des Herzens*, 1st ed. Leipzig, Wilhelm Engelmann Verlag, 1903, p 181.
102. Einthoven W, Korteweg AJ: On the variability of the size of the pulse in case of atrial fibrillation. *Heart* 1915;6:107–120.
103. Frank O: Zur Dynamik des Herzmuskels. *Zeitschr f Biologie* 1895;32:370–447.
104. Starling EH: *The Lineacre Lecture on the Law of the Heart.* London, Longmans, Green, 1915.
105. Braunwald E, Frye RL, Aygen MM, Gilbert JW: Studies on Starlings law of the heart. III. Observations in patients with mitral stenosis and atrial fibrillation on the relationships between left ventricular end diastolic segment length, filling pressure, and the characteristics of ventricular contraction. *J Clin Invest* 1960;39:1874–1884.
106. Meijler FL, Strackee J, Van Capelle FJL, Du Perron JC: Computer analysis of the RR interval–contractility relationship during random stimulation of the isolated heart. *Circ Res* 1968;22:695–702.
107. Van Dam IM, Van Zwieten G, Vogel JA, Meijler FL: Left ventricular (diastolic) dimensions and relaxation in patients with atrial fibrillation. *Eur Heart J* 1980;1(suppl A):149–156.
108. Van Dam I: Left ventricular dimensions during atrial fibrillation. Utrecht, The Netherlands, State University of Utrecht, 1988 (thesis).
109. Iwase M, Aoki T, Maeda M, Yokota M, Hayashi H: Relationship between beat to beat interval and left ventricular function in patients with atrial fibrillation. *Int J Card Imaging* 1988/89;3:217–226.
110. Brooks C, Hoffman BF, Suckling EE, Orias O: *Excitability of the Heart.* New York, Grune & Stratton, 1960, pp 325–327.
111. Meijler FL, Nieuwendijk ES, Durrer D: Physiological basis of paired stimulation potentiation, in Cranefield PF, Hoffman BF (eds): *Paired Pulse Stimulation of the Heart.* New York, Rockefeller University Press, 1968, pp 65–71.
112. Rawles JM: A mathematical model of left ventricular function in atrial fibrillation. *Int J Biomed Comput* 1988;23:57–68.
113. Blinks JR, Koch-Weser J: Analysis of the effects of changes in rate and rhythm upon myocardial contractility. *J Pharmacol Exp Ther* 1961;134:373–389.
114. Kuijer PJP, Van der Werf T, Meijler FL: Postextrasystolic potentiation without a compensatory pause in normal and diseased hearts. *Br Heart J* 1990;63:284–287.
115. Gibson DG, Broder G, Sowton E: Effect of varying pulse interval in atrial fibrillation on left ventricular function in man. *Br Heart J* 1971;33:388–393.
116. Meijler FL, Kuijer PJP, Van Dam-Koopman IM, Heethaar RM, Van der Werf T: Clinical relevance of post-extrasystolic potentiation. *Mayo Clinic Proc* 1982;57:34–40.
117. Ashman R, Byer E: Aberration in the conduction of premature, supraventricular impulses. *J LA State Med Soc* 1946;8:62–65.
118. Altman PL, Dittmer DS: *Biological Handbooks: Respiration and Circulation.* Bethesda MD, FASEB, 1971.

Chapter **16**

Differential Diagnosis of "Wide QRS Tachycardia"

Charles Fisch, MD

The difficulties attending the differential diagnosis of "wide QRS tachycardia" due to supraventricular tachycardia with aberration and one due to ventricular tachycardia (VT) were recognized during the early days of electrocardiography. In 1920, Lewis, analyzing records with wide QRS tachycardia, wrote,

> The paroxysm would seem at the first blush to be of ventricular origin, for the ventricular complexes are anomalous and the notch towards the end of the complex in leads II and II probably represents an abnormal auricular systole. Yet this origin is not certain, for an alternative interpretation is equally plausible, namely, that the paroxysm is in reality auricular; the view being that the ventricle is responding to the preceding auricular impulse after an increased conduction interval, and that the excitation wave takes an aberrant course in the ventricle; aberration is known to be a frequent phenomenon in patients who are the subjects of paroxysmal tachycardia. It is impossible to decide the exact origin of a paroxysm of the kind illustrated in the present figure unless its first or last beat is recorded [Fig. 16.1].[1]

In 1990, 80 years after Lewis, and despite numerous studies, the problems connected with the differential diagnosis of wide QRS tachycardia are still with us.[2–9] Although the criteria proposed by Wellens and associates are widely recognized and accepted by many,[10–13] their specificity has been questioned by others.[12] Despite this difference of opinion, our ability to differentiate the causes of wide QRS tachycardia has improved considerably. It is generally agreed that an accurate differential diagnosis based on the 12-lead electrocardiogram (ECG) is possible in approximately 90% of the patients.

The conditions that may manifest a wide QRS tachycardia include: (a) ventricular tachycardia, (b) supraventricular tachycardia with preexisting bundle branch block, (c) supraventricular tachycardia with rate-dependent aberration, and (d) supraventricular tachycardia with conduction over an anomalous pathway (W-P-W).

VENTRICULAR TACHYCARDIA

Although some criteria have been shown to be more helpful than others, none, with the exception of captures and fusions, approach a 100% specificity.

Some criteria favor VT but lack sufficient specificity to permit a definitive diagnosis. These include (a) marked left-axis deviation, (b) a QRS duration of 0.14 s or greater except when recorded in presence of electrolyte disturbances such as hyperkalemia, or in the course of administration of class I antiarrhythmia drugs, or during supraventricular tachycardia with preexisting bundle branch block, or W-P-W with conduction through the anomalous pathway.

As indicated, the classical criteria for diagnosis of VT, with a specificity approaching 100%, are captures and fusions (Fig. 16.2). Their value, however, is limited by a sensitivity of only approximately 5%. Similarly atrioventricular (AV) dissociation has a relatively high degree of specificity for VT but a sensitivity of only about 25% (Fig. 16.2). Its sensitivity is limited by retrograde, ventriculoatrial conduction (Figs. 16.3 and 16.4) approaching 50% and failure to recognize the sinus P waves in 25%. In a rare instance of AV dissociation the QRS complex originates in the junction and conducts aberrantly, and a differential diagnosis is nearly impossible.

With the advent of His-bundle electrocardiography, a definitive diagnosis of the cause of wide QRS tachycardia became possible and a number of investigators proposed criteria for the

Published 1991 by Elsevier Science Publishing Co., Inc.
655 Avenue of the Americas, New York, NY 10010
Current Topics in Cardiology

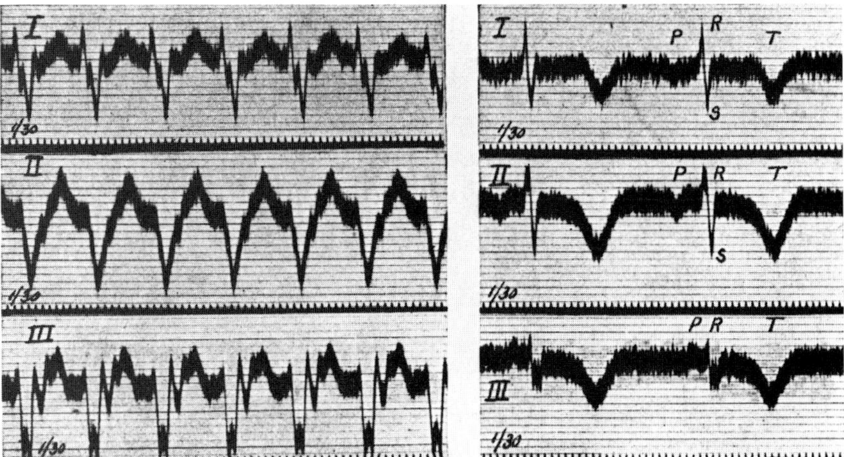

FIGURE 16.1 This figure is most likely the earliest illustration of the difficulty of differential diagnosis of supraventricular tachycardia with aberration and ventricular tachycardia. *Reprinted from Lewis,*[1] *with permission.*

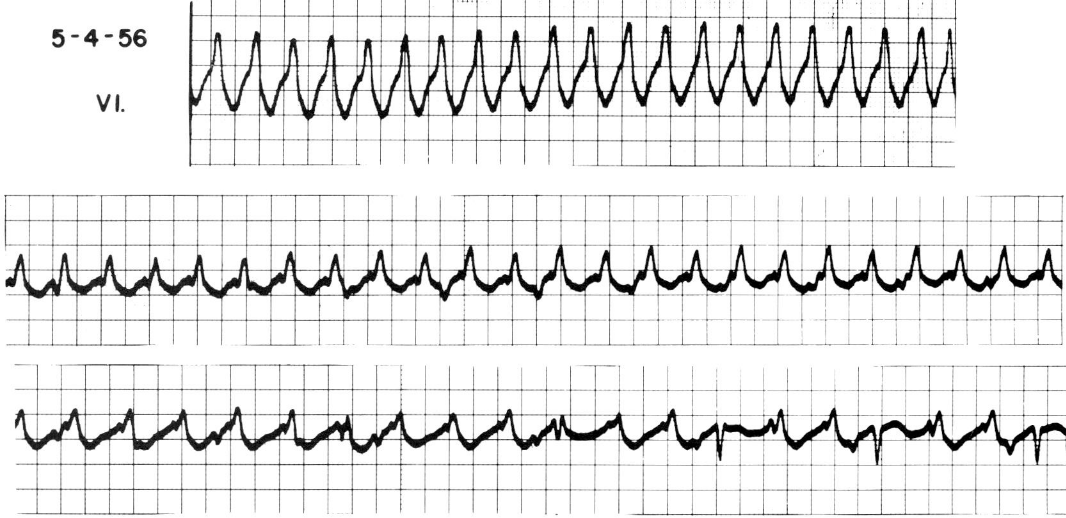

FIGURE 16.2 This figure illustrates ventricular tachycardia with fusion and capture complexes. The tracings were recorded in the course of administration of quinidine with gradual slowing of the heart rate. Although the monomorphic RBBB QRS in the top trace and the AV dissociation in the middle trace support a diagnosis of VT, a definitive diagnosis may still not be possible. In the middle trace, the possibility of a junctional tachycardia with aberration and AV dissociation cannot be excluded, although the specificity of the AV dissociation for VT approaches 90%. The captures and fusions recorded in the bottom trace permit a definitive diagnosis of VT. *Reprinted from Fisch and Pinsky, with permission.*

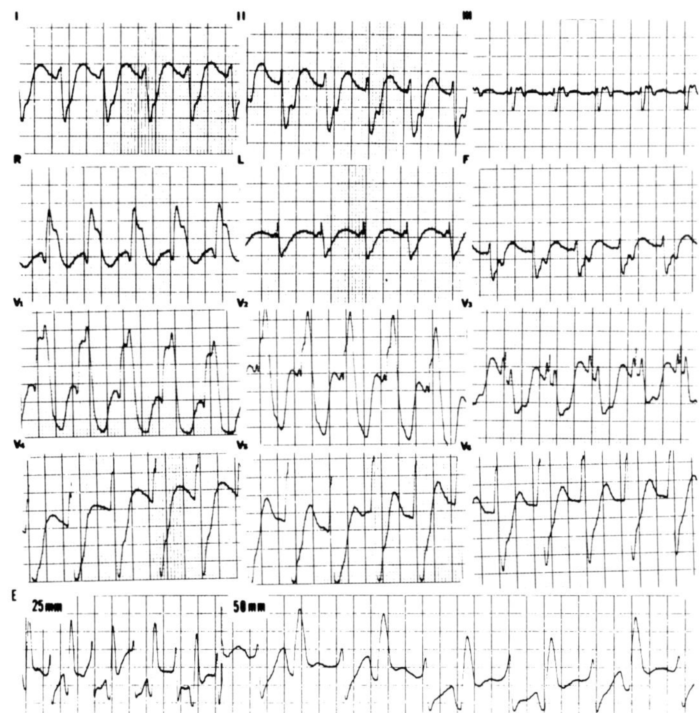

FIGURE 16.3 This figure illustrates a RBBB wide QRS tachycardia with 1 : 1 retrograde conduction. The axis, the duration, and the configuration of the QRS in lead V1 with the initial Q wave and the notch high on the upstroke of the R wave are highly supportive of VT. The esophageal lead (E), recorded at a 25- and 50-mm speed, indicates presence of 1 : 1 retrograde conduction. *Reprinted from Fisch,*[9] *with permission.*

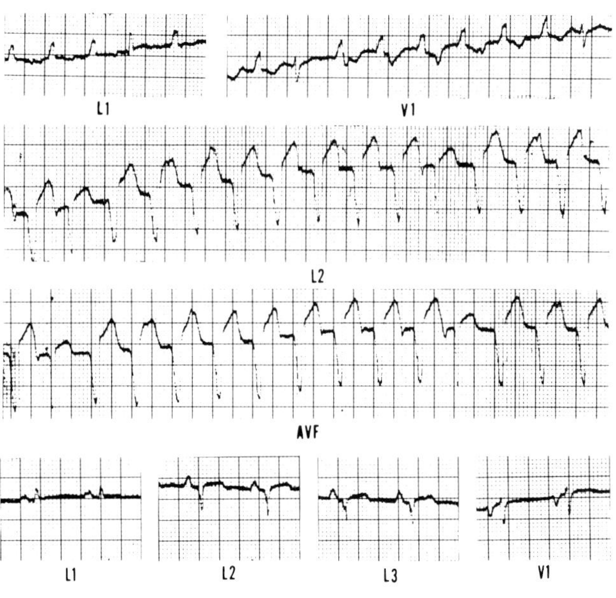

FIGURE 16.4 This figure, recorded from a patient with an acute inferior myocardial infarction, illustrates VT with retrograde Wenckebach conduction and ventricular reciprocation. The dominant rhythm is wide QRS tachycardia with an R-R interval of 400 ms. The successively longer R-P interval and the gradually more distinct inverted P waves are best seen in the middle of lead AVF. Following the longest R-P interval a ventricular reciprocation inscribes a normal QRS complex, a capture. The R-R interval of the capture is 40 ms shorter than the R-R interval of the wide QRS tachycardia. The ventricular capture with normal QRS is diagnostic of VT and excludes preexisting bundle branch block and rate-dependent aberration. *Reprinted from Fisch,*[9] *with permission.*

differential diagnosis of the wide QRS tachycardia based on the 12-lead electrocardiogram. Because the diagnosis was not aided by either right or left BBB, the attention was directed to the specific QRS pattern, largely in leads V1 and V6.

With RBBB wide QRS tachycardia, a monophasic (R) or biphasic (qR, QR, RS) QRS complex in lead V1 is highly specific for VT. In lead V6, QRS complexes with rS, QS, qR configuration are highly specific of VT (Figs. 16.3 and 16.5).

With LBBB wide QRS tachycardia, the qR, QS, or QRS pattern in V6 is highly specific for VT. It has also been suggested that in V1 an initial R wave of 30 ms or greater in duration, presence of a slur on the downstroke of the S wave, and an interval from the onset of the R wave to the nadir of the S wave of 70 ms or longer favor the diagnosis of VT[13] (Fig. 16.6).

QRS complexes that are positive in all the precordial leads, positive concordance, or negative in all the precordial leads, negative concordance, are highly specific for VT (Figs. 16.7 and 16.8). Positive concordance, while highly suggestive of VT, can be recorded with W-P-W (Fig. 16.9).

SUPRAVENTRICULAR TACHYCARDIA WITH ABERRATION

As indicated earlier, supraventricular tachycardia with preexisting BBB (Figs. 16.10, 16.11) or with rate-dependent aberration (Fig. 16.3) or W-P-W (Fig. 16.9) may simulate VT, and in fact a differential diagnosis of the wide QRS tachycardia may not be possible.

Preexisting BBB does not alter the sensitivity and specificity of the different markers of VT (Fig. 16.14),[14] because it does not interfere with ventricular excitation induced by an ectopic ven-

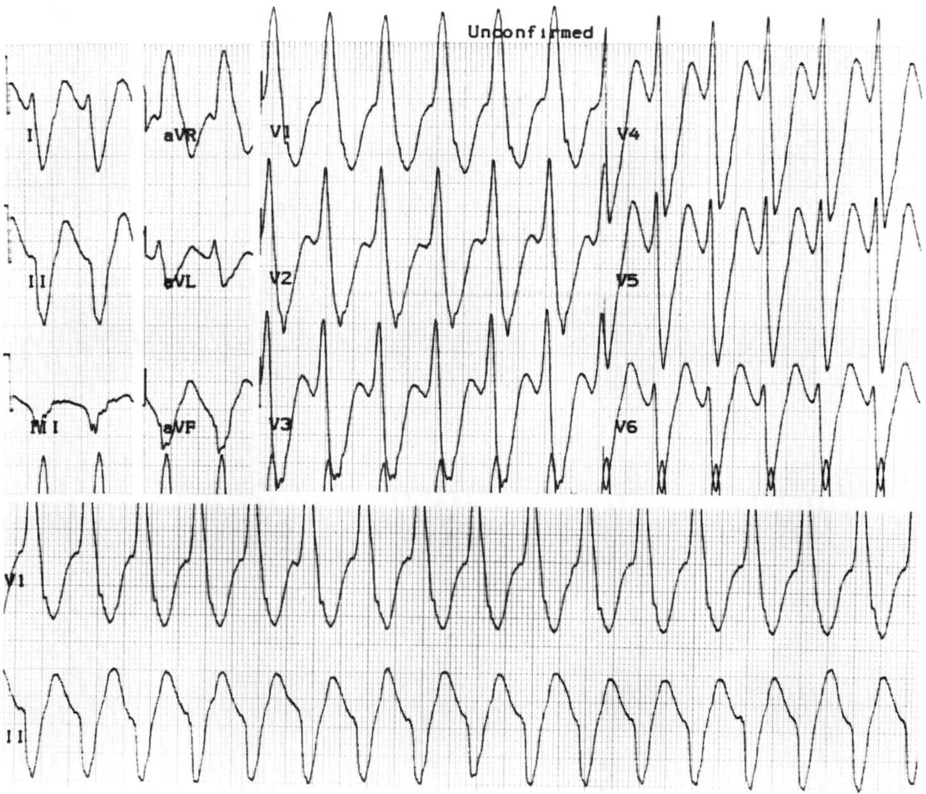

FIGURE 16.5 Wide QRS tachycardia with RBBB pattern. The monophasic QRS in lead V1 coupled with the rS pattern in lead V6 are highly specific for VT.

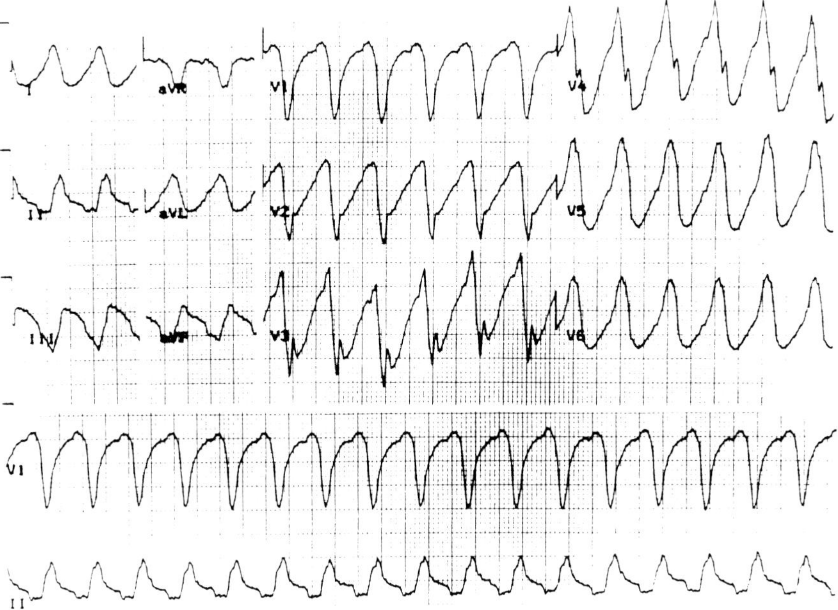

FIGURE 16.6 Wide QRS tachycardia with LBBB pattern. The width of the QRS of 200 ms, the R wave 100 ms in duration, and an interval from the onset of the R wave to the nadir of the S wave of 160 ms in lead V1 are all supportive of VT.

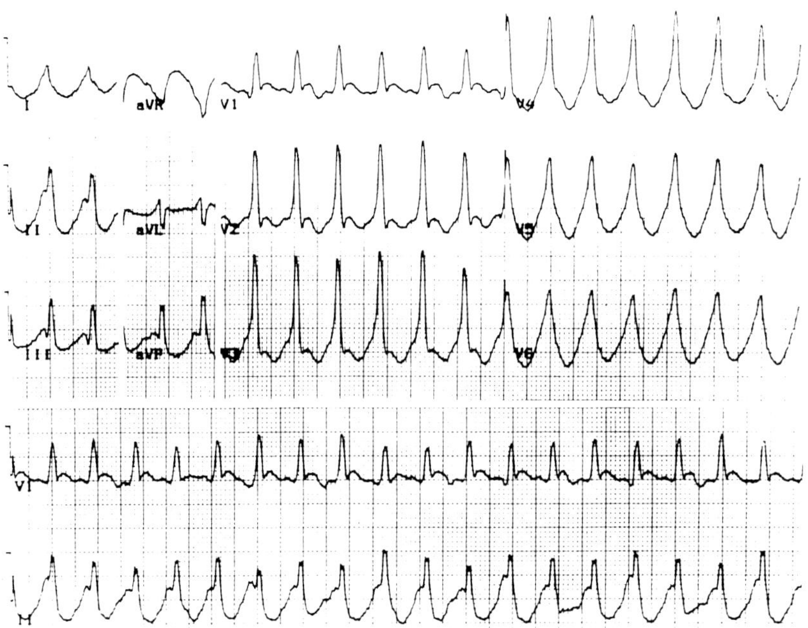

FIGURE 16.7 Wide QRS tachycardia with positive concordance of the precordial QRS complexes. Although highly specific for VT, such concordance may be seen with W-P-W (see Fig. 16.9).

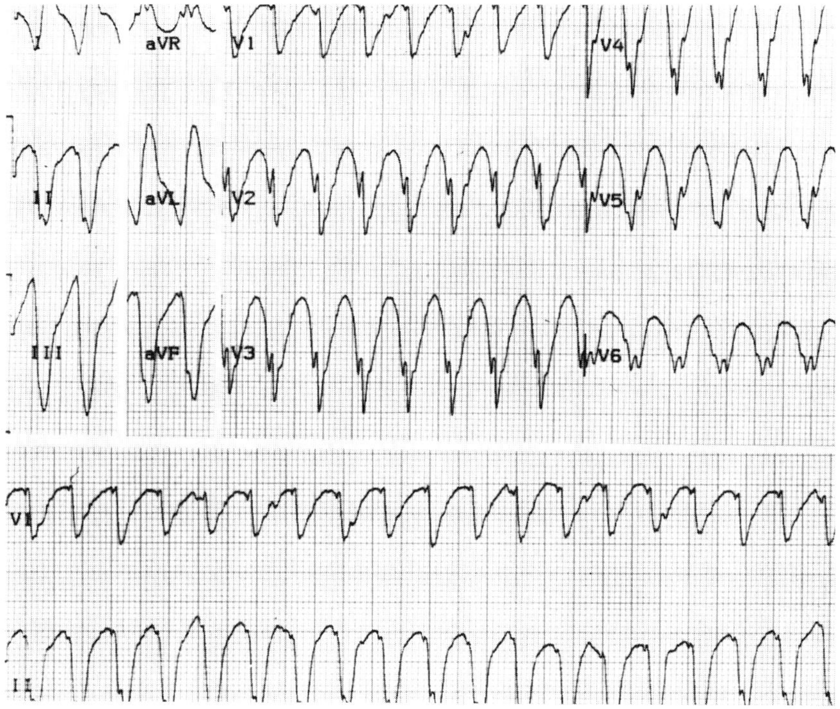

FIGURE 16.8 Wide QRS tachycardia with negative concordance. The negative concordance, the marked left-axis deviation, and the width of the QRS of 200 ms are highly specific for VT. The Q wave in lead 1 localizes the site of impulse formation to the anterior wall near the septum.

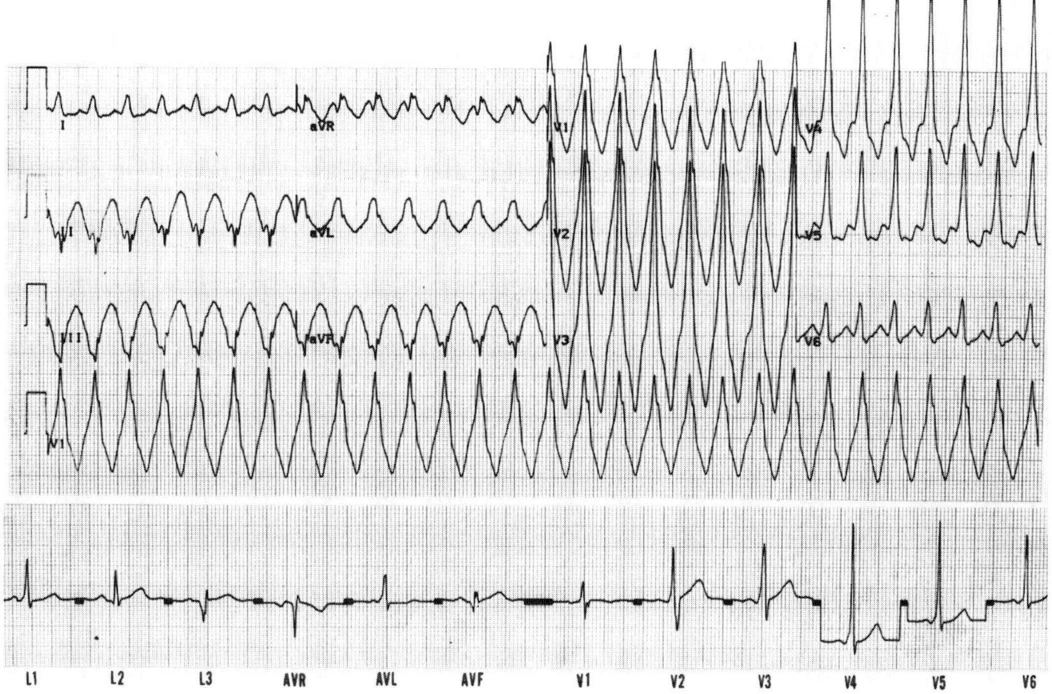

FIGURE 16.9 Wide QRS tachycardia with a positive concordance due to conduction through an accessory pathway. Although positive concordance is highly specific for VT, it may be recorded with W-P-W. *Reprinted from Fisch,[9] with permission.*

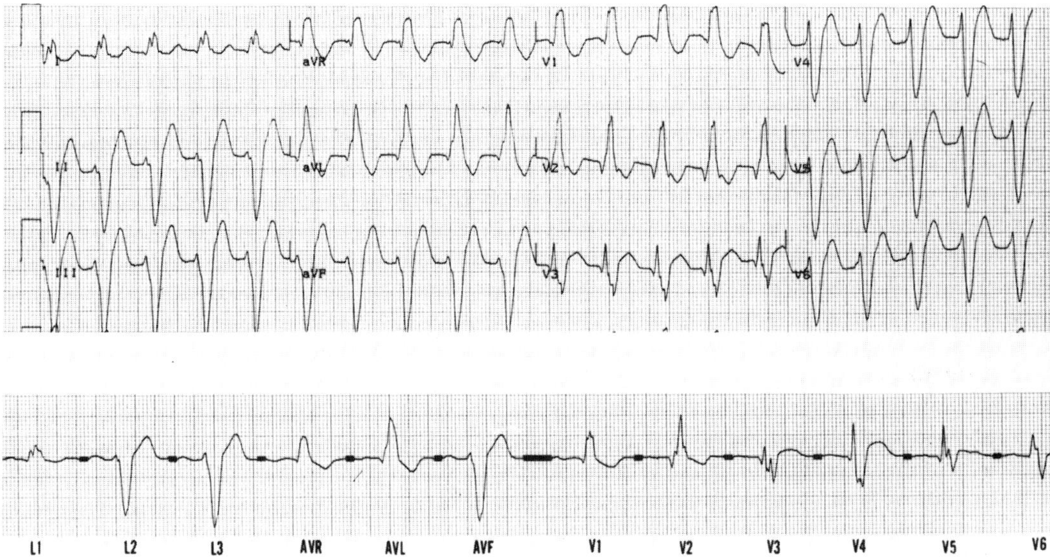

FIGURE 16.10 Wide QRS tachycardia due to supraventricular tachycardia, probably sinus, with preexisting "masquerading" bundle branch block. The latter is indicated by RBBB pattern in the precordial leads and LBBB in the limb leads. The width of the QRS of 200 ms, the marked left-axis deviation, the qR pattern in lead V1, and rS in lead V6 are highly suggestive of VT. A definitive diagnosis would be impossible were it not for the evidence of fixed bundle branch block noted in the bottom trace.

tricular pacemaker. However, a rapid supraventricular rhythm with a preexisting BBB or a rapid rate with aberration may make a differential diagnosis of wide QRS tachycardia virtually impossible (Figs. 16.10 and 16.11).

In presence of a wide QRS tachycardia, a triphasic QRS pattern in V1 and qRS in V6, while not excluding VT, do favor aberration. Other triphasic QRS patterns in lead V1 are recorded equally often during VT and supraventricular tachycardia with aberration (Fig. 16.12).

A wide QRS tachycardia due to ventricular activation along an accessory pathway may be difficult or impossible to differentiate from VT because the ventricular activation in both the VT and W-P-W originates outside the normal conduction system. In fact, as indicated earlier, the pattern of positive concordance considered highly specific for VT can be recorded in W-P-W with a left posterior pathway (Fig. 16.9).

Supraventricular tachycardia activating the ventricles along the Mahaim fibers, the nodoventricular connection, inscribe a LBBB pattern with left-axis deviation which may be difficult if not impossible to differentiate from VT. Similarly, if AV dissociation is associated with a wide QRS originating in the AV junction, a differentiation from VT is impossible.

Rapid rate in presence of depressed intraventricular conduction may not be distinguishable from VT. These include, among others, metabolic disturbances such as hyperkalemia, the depressing effect of antiarrhythmic drugs, and intramyocardial delay due to pathologic changes.

SUMMARY

(a) In the presence of wide QRS tachycardia, a differential diagnosis between VT and aberration based on the 12-lead ECG is possible in about 90% of the patients.

(b) Criteria with a high specificity but a low sensitivity include captures, fusions, and AV dissociation.

(c) Marked left-axis deviation and a QRS duration greater than 0.14 s favor VT.

(d) With RBBB wide QRS tachycardia, a monophasic (R) or biphasic (qR, QR, RS) QRS complex in lead V1 is highly specific for VT. Similarly, in lead V6, rS, QS, and qR patterns favor VT.

(e) With LBBB wide QRS tachycardia, a qR or QS pattern in lead V6 is highly specific for VT. In lead V1, an initial R greater than 30 ms in duration, slur on downstroke of the S

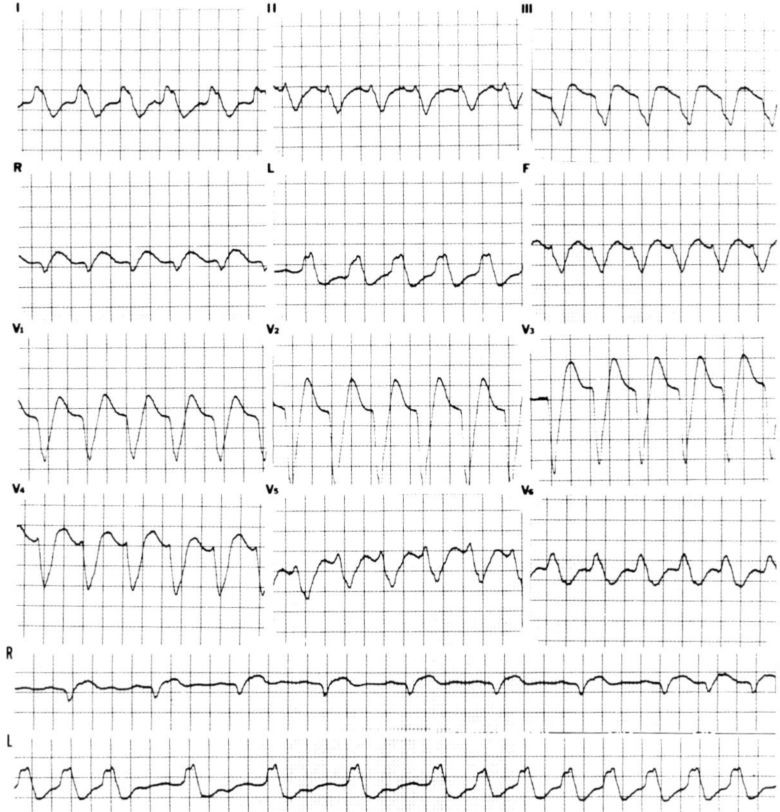

FIGURE 16.11 Wide QRS tachycardia due to atrial tachycardia with preexisting LBBB. Although the QRS duration of nearly 0.20 s and the marked left-axis deviation suggest VT, the 2 : 1 AV conduction noted in the bottom two traces clearly indicate a preexisting LBBB as the cause of the wide QRS tachycardia. Were it not for the episodes of 2 : 1 AV conduction a differential diagnosis would not be possible. *Reprinted from Fisch,*[9] *with permission.*

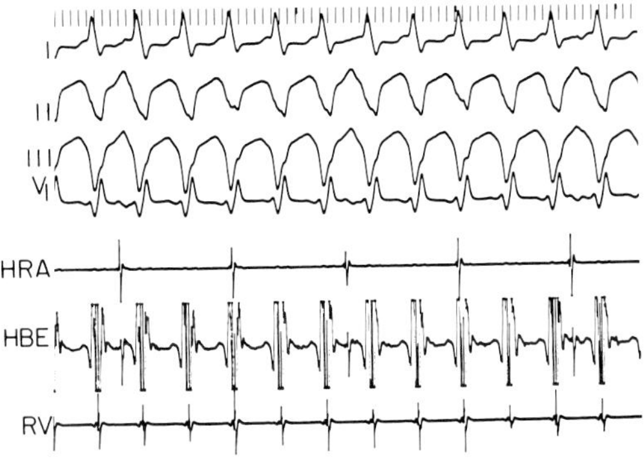

FIGURE 16.12 Wide QRS tachycardia with a QRS of approximately 125 ms and an rsR QRS pattern in lead V1 strongly suggests a supraventricular origin of the tachycardia. However, the intracardiac recording including a high right atrial (HRA), a His bundle (HBE), and a right ventricular (RV) electrogram illustrates AV dissociation and ventricular complexes without a preceding His bundle, thus confirming the diagnosis of VT. The AV dissociation, although highly specific for VT, may occasionally accompany a junctional tachycardia with aberration and may make a differential diagnosis impossible. *Reprinted from Fisch,*[9] *with permission.*

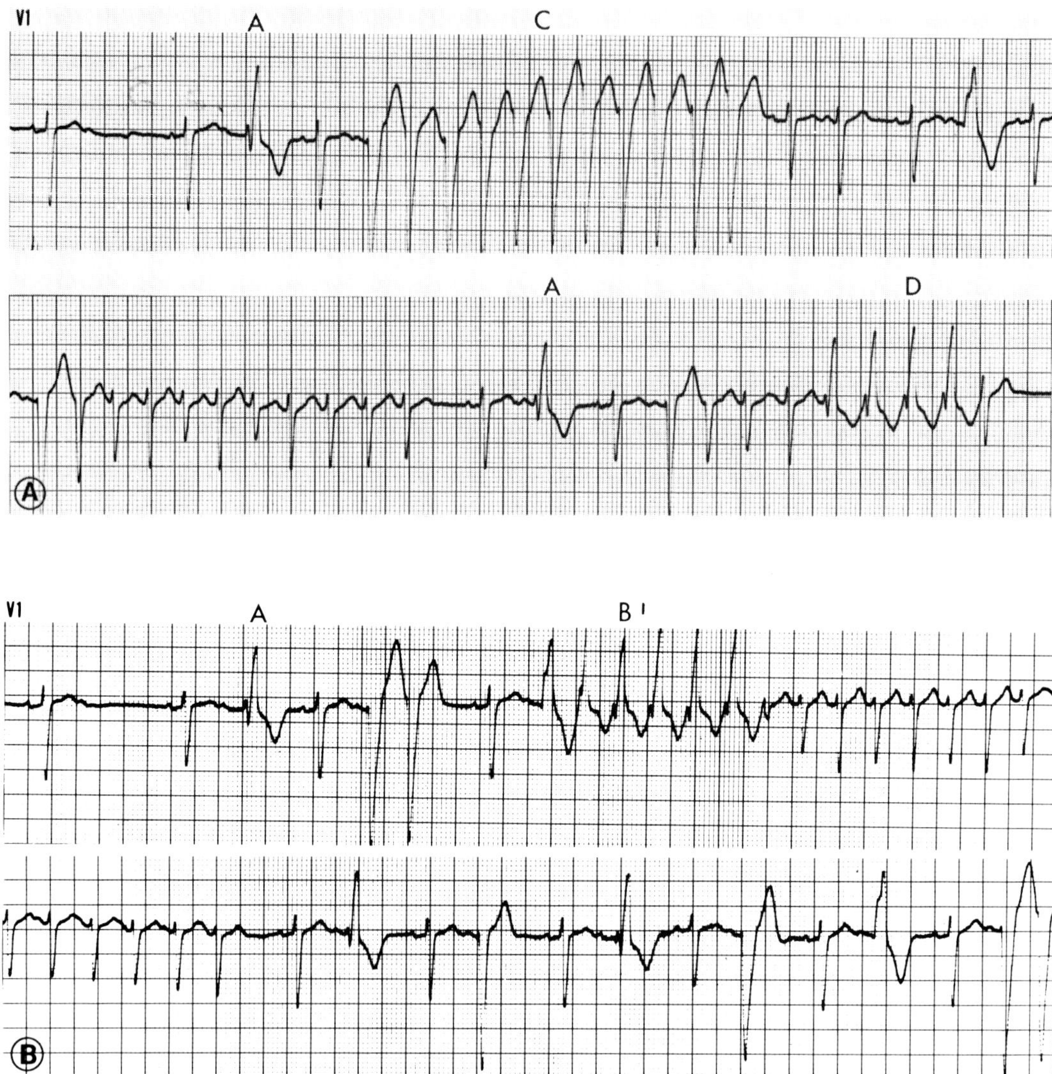

FIGURE 16.13 A and **B**. These figures, sections of a continuous tracing, illustrate the different manifestations of aberration with increasing heart rate. These include RBBB caused by the Ashman phenomenon **(A)**, persistence of RBBB aberration once initiated by Ashman phenomenon **(B)**, acceleration-dependent LBBB **(C)**, RBBB **(D)**, and a bigeminal rhythm with alternation of R and LBBB caused by concealed transseptal conduction **(E)**. *Reprinted from Fisch,*[9] *with permission.*

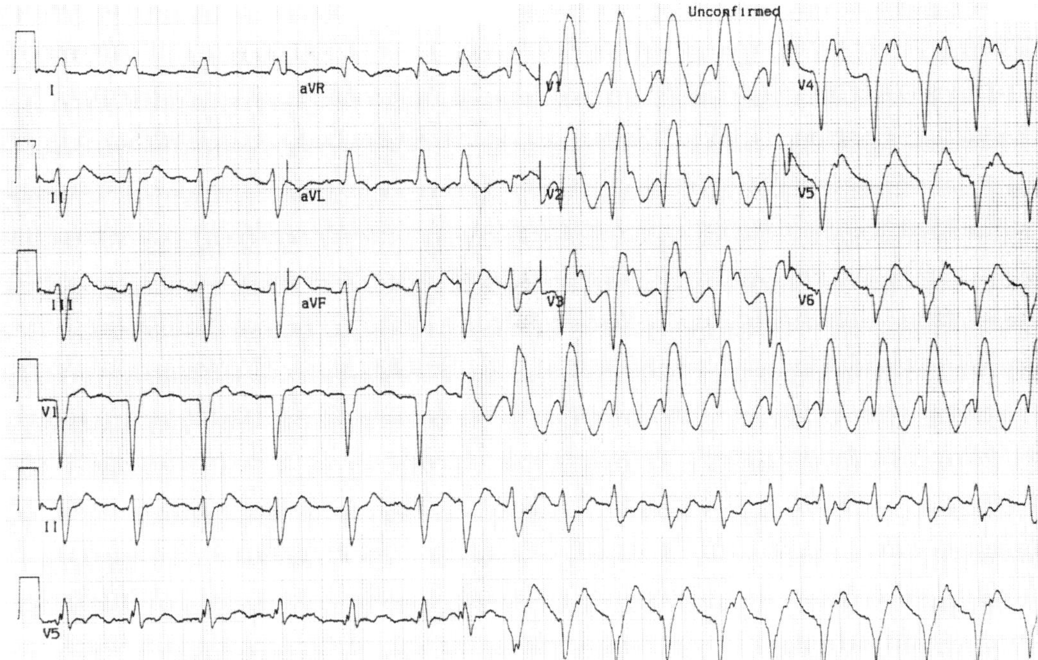

FIGURE 16.14 This figure illustrates that preexisting bundle branch block does not alter or affect the evidence of VT and does not interfere with a differential diagnosis of wide QRS tachycardia. The first six QRS complexes indicate a sinus rhythm with LBBB and left anterior fascicular block. These are followed by a PVC and VT. The diagnosis of the latter is supported by the QRS of 0.20 s in duration a qR and QS pattern in leads V1 and V6, respectively.

wave, and duration from onset of QRS to nadir of the S wave greater than 70 ms similarly favor VT.

(f) QRS complex of 0.12 s or shorter, especially when there is an association of a triphasic QRS pattern in lead V1, favors aberration.

REFERENCES

1. Lewis T: *The Mechanism and Graphic Registration of the Heart Beat,* 2d ed. London, Shaw and Sons, 1920, p 259.
2. Pick A, Langendorf R: Differentiation of supraventricular and ventricular tachycardia. *Prog Cardiovasc Dis* 1960;2:391.
3. Kistin AD: Problems in differential diagnosis of ventricular arrhythmias from supraventricular arrhythmia with abnormal QRS. *Prog Cardiovasc Dis* 1966;8:1.
4. Sandler IA, Mariott HJL: The differential morphology of anomalous ventricular complexes of RBBB type in lead V1: Ventricular ectopy versus aberration. *Circulation* 1965;31:551.
5. Marriott HJL, Sandler IA: Criteria, old and new, for differentiating between ectopic ventricular beats and aberrant ventricular conduction in the presence of atrial fibrillation. *Prog Cardiovasc Dis* 1966;9:18.
6. Marriott HJL: Differential diagnosis of supraventricular and ventricular tachycardia. *Geriatrics* 1970;25:91.
7. Benditt DG, Pritchett ELC, Gallagher JJ: Spectrum of regular tachycardias with wide QRS complexes in patients with accessory atrioventricular pathways. *Am J Cardiol* 1978;42:828.
8. Dongas J, Lehmann MH, Mahmud R, Denker S, Soni J, Akhtar M: Value of preexisting bundle branch block in the electrocardiographic differentiation of supraventricular from ventricular origin of "wide QRS tachycardia." *Am J Cardiol* 1985;55:717.
9. Fisch C: *Electrocardiography of Arrhythmias.* Philadelphia, Lea & Febiger, 1990, pp 47,94.
10. Wellens HJJ, Bar FWHM, Lie KI: The value of the electrocardiogram in the differential diagnosis of a tachycardia with a widened QRS complex. *Am J Med* 1978;64:27.
11. Wellens HJJ, Bar FWHM, Vanagt EJ, Brugada P, Farre J: The differentiation between ventricular tachycardia and supraventricular tachycardia with aberrant conduction: The value of the electrocardiogram? in Wellens HJJ, Kulbertus HE (eds): *What's New in Electrocardiography?* The Hague, Martinus Nijhoff, 1981, p 184.
12. Akhtar M, Shenasa M, Tchou PJ, Jazayeri M, Caceres J: Role of surface electrocardiogram in the diag-

nosis of wide QRS complex tachycardia: A reappraisal of surface electrocardiographic criteria, in Zipes DP, Rowlands DJ (eds): *Progress in Cardiology*. Lea & Febiger, 1988, p 101.

13. Kindwall K, Brown J, Josephson ME: Electrocardiographic criteria for ventricular tachycardia with wide complex left bundle branch block. *Am J Cardiol* 1988;61:1279–1283.
14. Kremers S, Black WH, Wells PH, Solodyna M: Effect of preexisting bundle branch block on the electrocardiographic diagnosis of ventricular tachycardia. *Am J Cardiol* 1988;62:1208–1212.

Chapter **17**

Preexcitation Update

Andrea Natale, MD, George J. Klein, MD, and Raymond Yee, MD

The classical preexcitation syndrome is that described by Wolff, Parkinson, and White in 1930[1] as an electrocardiogram (ECG) consisting of "functional bundle branch block and a short PR interval occurring in otherwise healthy young people with paroxysm of tachycardia." We now appreciate that the anatomic substrate in the great majority of patients with this syndrome is the accessory atrioventricular pathway. Electrophysiological testing is able to characterize this pathway and establish mechanisms of tachycardia while a broad range of pharmacologic and nonpharmacologic therapies have become available. The last decade has witnessed some clarification of variant forms of preexcitation, the consolidation of surgery as an important and readily available option, the recording of accessory pathway potentials to understand further the physiology of the pathway, and, finally, the emergence of transvenous ablation techniques.

THE NODOVENTRICULAR ACCESSORY PATHWAY

Since Mahaim and Winston[2] described a large connection between the node of Tawara and the myocardial septum in man, providing a direct anatomical continuity between these two structures, the presence of nodoventricular fibers has been invoked to explain electrocardiographic and electrophysiologic phenomena compatible with this entity. The electrophysiologic findings in a patient felt to have a nodoventricular Mahaim were first reported by Wellens in 1971[3] and subsequently by others.[4–6] However, most of the evidence supporting the presence of this pathway as causing the clinical syndrome has been based on deductive reasoning using electrophysiologic data with only sparse clinical anatomic correlations.[7,8] The usual clinical features suggesting a nodoventricular pathway are as follows: The baseline ECG is usually normal or may show subtle preexcitation (Fig. 17.1A). With incremental right atrial pacing or atrial extrastimulus testing, the PR interval prolongs as the QRS becomes more preexcited and a left bundle branch block morphology emerges. Intercavitary recordings with progressively premature atrial extrastimuli show the AH interval lengthening and displacement of the His deflection into the ventricular electrogram (Fig. 17.2). Unlike classical accessory pathways, the PR interval progressively lengthens with increasing atrial prematurity and this decremental behavior led to the postulate that the pathway was continuous with the atrioventricular (AV) node. The presenting tachycardia is usually preexcited tachycardia with left bundle branch block morphology (Fig. 17.1B). Nodoventricular pathways frequently coexist with other AV pathways and dual AV node physiology.[8–10] Hence, many tachycardia mechanisms are possible, including antidromic tachycardia, AV nodal reentry, and preexcited tachycardia utilizing a second accessory pathway as part of the circuit. The nodoventricular pathway may be a critical part of the tachycardia mechanism or may be activated in a bystander fashion when it is not included in the main circuit (Fig. 17.3).

The anatomical substrate for these clinical presentations has been considered to be the Mahaim fiber arising from the right side of the AV node and joining the right ventricle, even though most of the clinical phenomena could be explained by an accessory AV pathway with decremental properties. Observations by several laboratories have now clarified the location of the pathway involved with the preceding clinical electrophysiologic features.[11–14] These observations have been made with attempts to eliminate the nodoventricular pathway by catheter ablation[15,16] and operatively.[11–14] Here, attempts

655 Avenue of the Americas, New York, NY 10010
Current Topics in Cardiology

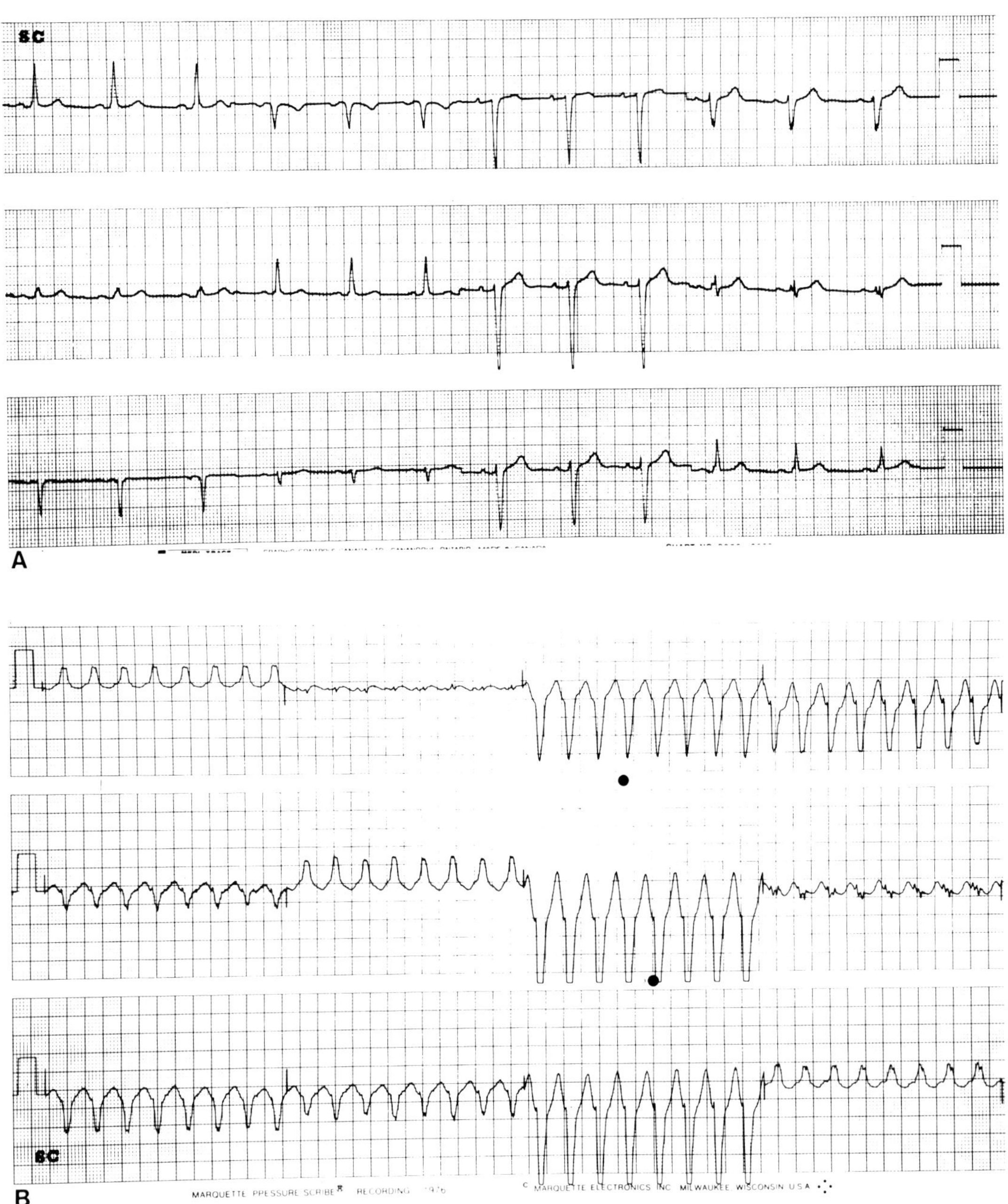

FIGURE 17.1 A 12-lead electrocardiogram in a patient with a nodoventricular Mahaim fiber. **A.** During normal sinus rhythm, the QRS is essentially normal with subtle preexcitation. **B.** During preexcited tachycardia (rate 180 bpm), the QRS shows left bundle branch block aberration with a leftward axis.

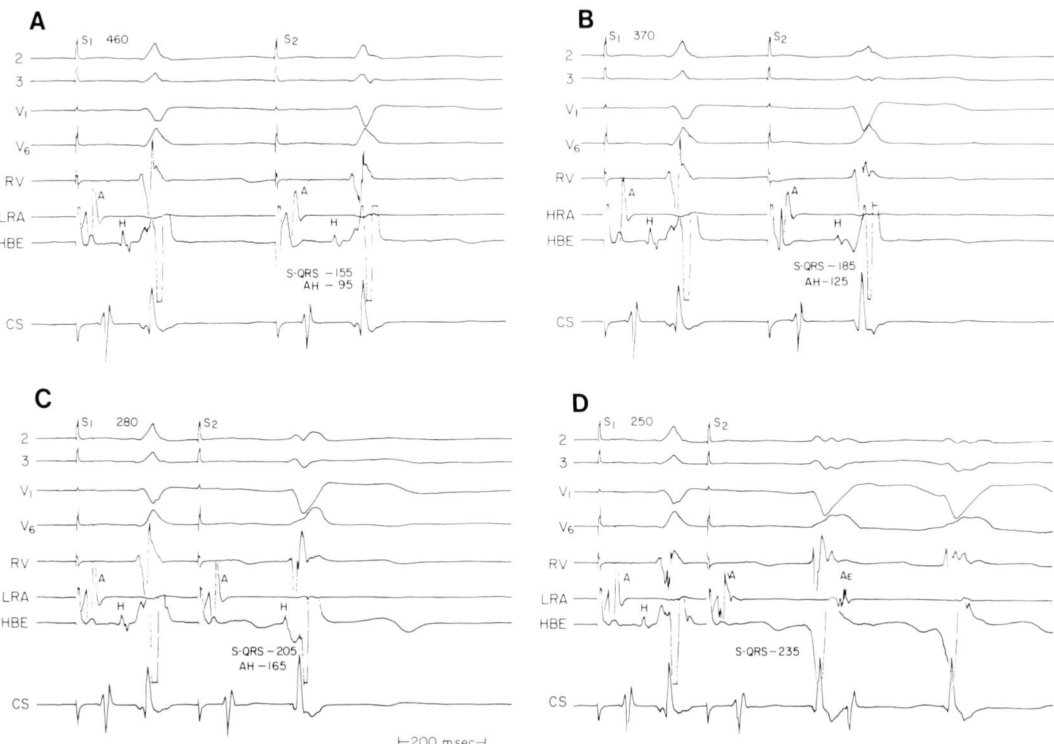

FIGURE 17.2 Intracavitary tracing during premature atrial extrastimuli (S2) introduced at progressively shorter coupling intervals (CIs) in a patient with a nodoventricular pathway. **A,B,** and **C.** As the atrial extrastimulus becomes more premature, the AH interval lengthens and the His deflection merges with the QRS complex, which becomes progressively preexcited. At the same time, the S–QRS interval also prolongs or "decrements." **D.** At a critical CI, a reentrant beat is induced with a frankly preexcited QRS complex. The anterograde limb of this reentrant cycle is the accessory pathway and the retrograde limb is the normal AV conduction system. CS = coronary sinus, HBE = His bundle electrogram, LRA = low right atrial electrogram, RV = right ventricular electrogram; 2, 3, V1, V6 = electrocardiographic leads.

to eliminate the pathway by AV node destruction have uniformly led to AV nodal block and persistence of preexcitation. These data have led to the view that the true nodoventricular fiber must indeed be a rare entity and the majority of patients with the clinical presentation suggesting this have an atypical accessory pathway with the following characteristics:

(a) The pathway is atrioventricular and exhibits anterograde decremental properties.

(b) The pathway is in the right anterior free wall location (right atrial to right ventricular pathway).

(c) Conduction over the pathway is unidirectional (anterograde only).

(d) The ventricular insertion site of the pathway is generally not at the AV ring as typical accessory pathways. Rather it inserts more apically into the right ventricle or more probably into the right bundle branch system (Fig. 17.4).

The histology of the pathway is not clear. Anderson and associates[17] reported islands of AV node tissue in several hearts along the AV groove. The most frequent location of this tissue was in the right anterior atrioventricular annulus, where virtually all of these accessory connections have been described. Similarly, Guiraudon and associates[18] reported the finding of cells typical of AV node (transitional cells, nodal cells, P cells) in a right atrial block of tissue removed at the site of the atrial insertion of the Mahaim pathway in a patient undergoing operative ablation. This raises the possibility that this type of accessory pathway may be a "displaced" or accessory AV node.

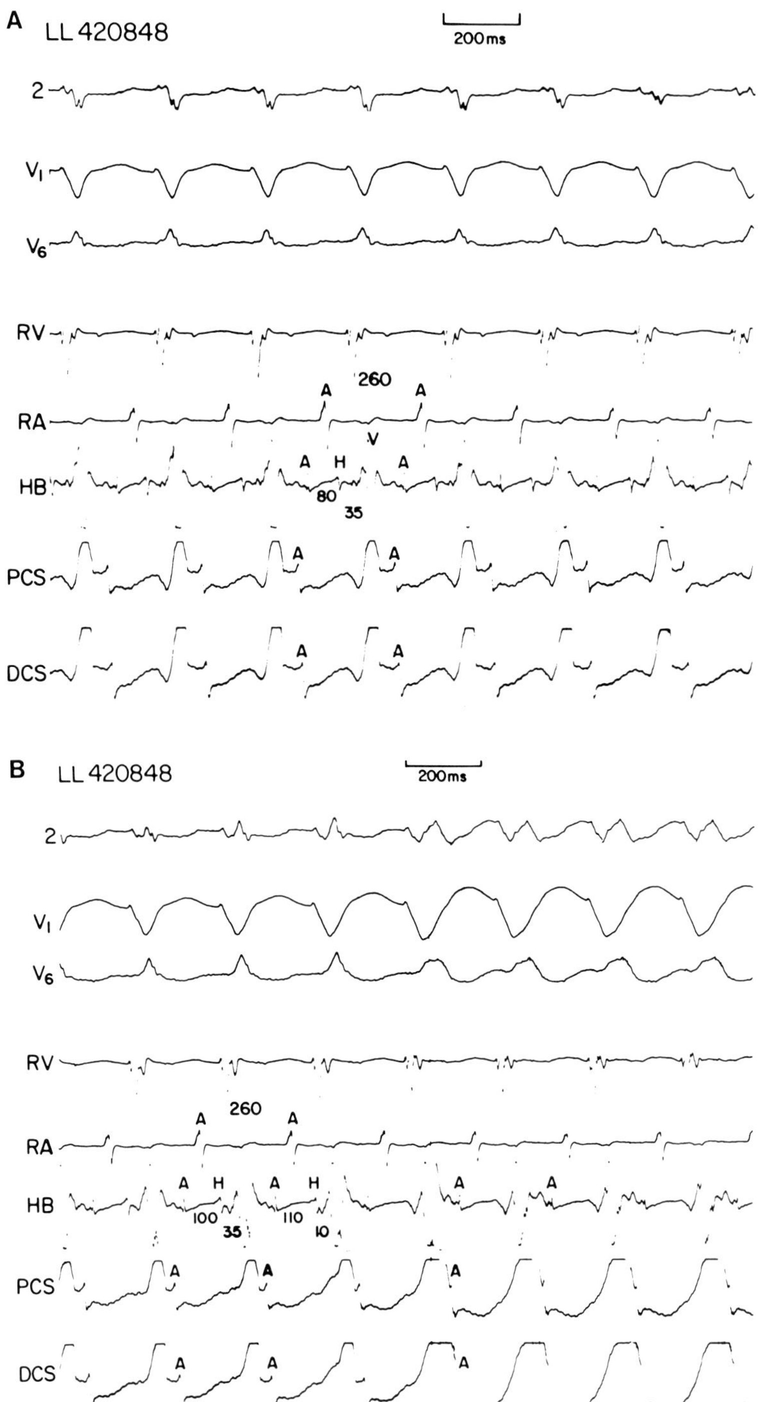

FIGURE 17.3 Intracavitary recording from a patient with a nodoventricular pathway. **A.** A slightly preexcited tachycardia with the anterograde limb consisting of the normal AV conduction system and the nodoventricular accessory pathway (fusion) and the retrograde limb being a second posteroseptal accessory pathway. **B** is continuous and shows prolongation of the AH, increasing preexcitation, and loss of the His deflection. The cycle length of tachycardia and the retrograde activation sequence remain unchanged. This tachycardia utilizes the nodoventricular (in reality, a right-sided atrioventricular) pathway as the anterograde limb and a posteroseptal pathway as the retrograde limb of a reentrant circuit. The AV node is a "bystander" for anterograde conduction. DCS = distal coronary sinus, HB = His bundle electrogram, PCS = proximal coronary sinus, RA = right atrial electrogram, RV = right ventricular electrogram, 2, V1, V6 = electrocardiographic leads.

NAME: ML NUMBER: 377819

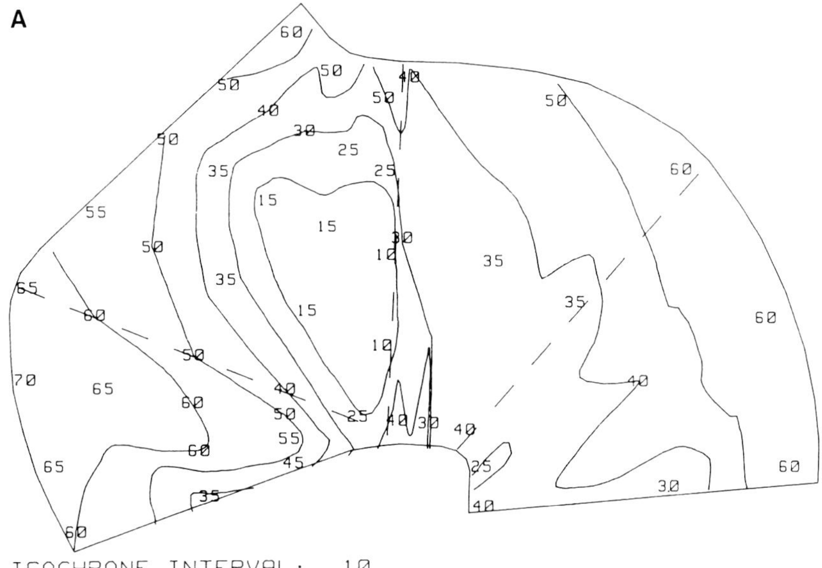

NAME: ML NUMBER: 377819

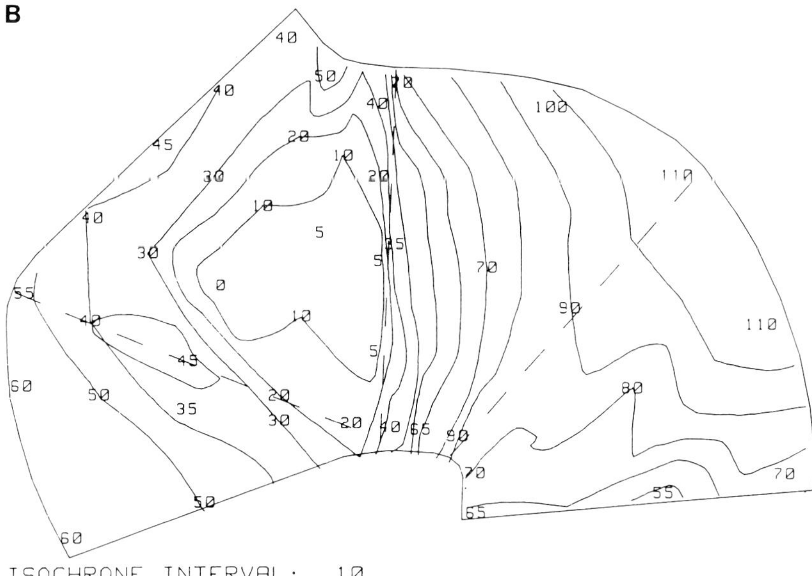

FIGURE 17.4 Intraoperative epicardial mapping in a patient with a nodoventricular Mahaim fiber. The heart is displayed as if cut along the posterior ventricular septum and flattened. The numbers represent activation times relative to the onset of polarization. **A.** During normal sinus rhythm, earliest epicardial activation occurs over the low anterior right ventricle at the terminus of the right bundle branch. **B.** During preexcited tachycardia, earliest epicardial activation occurs at the same area, suggesting ventricular insertion of the pathway into the right bundle branch system.

Although much remains to be learned, the clinical importance of these observations focuses on operative or ablation management of these patients that can be performed with little or no risk of AV nodal block since the pathway is remote from the AV node.

PERMANENT FORM OF JUNCTIONAL RECIPROCATING TACHYCARDIA

Coumel and coworkers[19] described a unique supraventricular tachycardia with distinctive electrocardiographic and clinical features and named it permanent junctional reciprocating tachycardia, or PJRT. They observed this tachycardia predominantly in infants and children and noted that it was incessant or nearly incessant. Its initiation was usually related to a critical shortening of the atrial cycle length without PR interval prolongation. During tachycardia, the atrioventricular relationship was 1 : 1 with inverted P waves observed in the inferior leads (Fig. 17.5A). Much has been learned since Coumel et al. originally described this syndrome. Coumel et al. originally felt that the mechanism of tachycardia was AV nodal reentry that was atypical in that the retrograde limb had a long conduction time. Gallagher and Sealy[20,21] later suggested that the retrograde limb was an accessory AV pathway, and a number of other investigators have now shown that the mechanism of PJRT is usually AV reentry with retrograde conduction over a posteroseptal accessory pathway with decremental properties. This pathway exhibits properties usually characteristic of AV nodal tissue showing shortening of conduction time with catecholamines, atropine, and exercise and prolongation of conduction time with vagal tone, digoxin, and β-blockers.[10] During AV reciprocating tachycardia, earliest retrograde atrial activation occurs at the orifice of the coronary sinus. The critical feature distinguishing AV reentry from AV nodal reentry occurs when premature ventricular stimuli are introduced during the tachycardia at the time when the His bundle is refractory. These either advance or delay[22] the next atrial electrogram or terminate tachycardia. Conduction over these atypical accessory pathways is generally unidirectional (retrograde only), but anterograde preexcitation with a typical posteroseptal pattern has been observed in some patients (Fig. 17.5B), especially after His-bundle ablation.[23] The clinical presentation is also considerably more varied and eclectic than originally described by Coumel and coworkers. Arrhythmia can occur in all age groups, including the occasional geriatric patient. Patients may be entirely asymptomatic, may have palpitations, may have effort intolerance, or may present with symptoms attributable to left ventricular dysfunction or frank congestive heart failure. Patients may develop severe and irreversible congestive heart failure with incessant tachycardia without an awareness of rapid heart beating or other symptoms for a long period of time.

While it is clear that long-standing tachycardia can lead to congestive heart failure, the mechanism is not well understood, and it is not obvious why some patients develop cardiomyopathy and others with apparently similar tachycardia rates and duration have normal ventricular function. Nonetheless, the control of tachycardia can result in a dramatic improvement of ventricular function even in patients with very severe left ventricular dysfunction.[24–27] The mechanism of improvement has been related to the restoration of high-energy phosphates believed to be depleted with incessant tachycardia.[28]

The substrate for this unusual type of accessory pathway is not known. Recently, Critelli and associates[29] examined an accessory pathway in a patient with PJRT who was dying of coronary artery disease. They found a group of tiny atrial fibromuscular bundles originating from the lower rim of the coronary sinus orifice and descending through the fat of the AV groove in a sinuous manner to insert into the top of the posterior intraventricular septum. The accessory connection consisted of working muscle fiber with nothing to suggest AV nodal tissue. The fibers exhibited interstitial fibrosis, and sharp angulations were apparent, especially at the ventricular insertion of the pathway. Thus, it is possible that the behavior and decremental properties of this pathway are related to pathology and abnormal geometry in a conventional accessory pathway rather than to structural resemblance to the AV nodal tissue.[30] A study by Lerman and coworkers[31] showed that some of these accessory pathways were sensitive to adenosine but not to verapamil, suggesting that they consisted of partially depolarized working myocardial fibers with depressed sodium (Na^+) channels rather than with functioning calcium (Ca^{2+}) channels. In spite of the appearance of new, antiarrhythmic drugs, pharmacological therapy is frequently disappointing, requiring in many cases operative therapy or catheter ablation of the pathway. Patients with normal ventricular function may be treated conservatively and followed for evidence of impaired left ventricular function in a manner similar to the follow-up of asymptomatic patients with chronic volume overload due to aortic insufficiency.

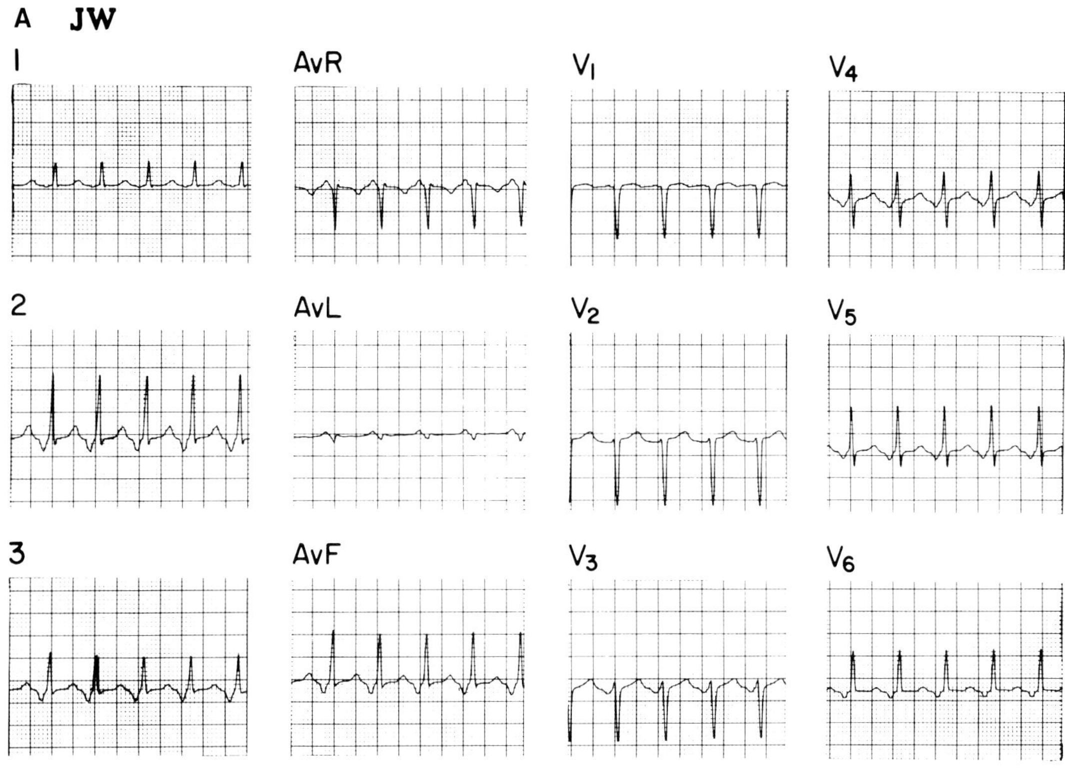

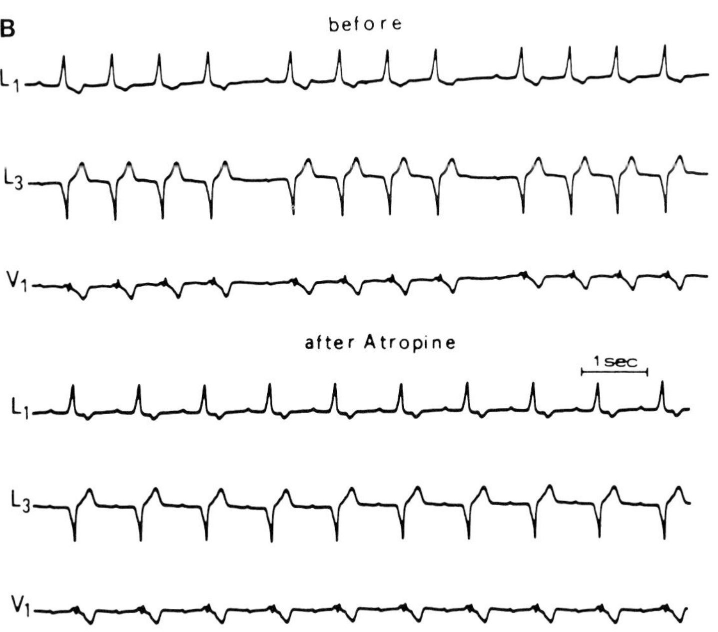

FIGURE 17.5 A. A 12-lead electrocardiogram showing the typical appearance of PJRT. **B.** After AV node ablation in another patient with PJRT, manifest preexcitation with Wenckebach AV block is due to conduction over the posteroseptal pathway. After the administration of atropine, 1 : 1 AV conduction with first-degree block is present. **B** *reprinted from Gallagher,*[10] *with permission.*

DIRECT RECORDING FROM THE ACCESSORY PATHWAY

Most of the information that we have on accessory pathway location and properties is based on recording atrial or ventricular activation near the site of the accessory pathway and making inferences about accessory pathway location and function based on these recordings. The recording of the accessory pathway potential provides a more direct way for localizing the accessory pathway and uncovering its physiological behavior. Recording of the accessory pathway potential is occasionally seen during routine electrophysiological testing and has been reported.[32,33] However, Jackman has recently described electrophysiological techniques to allow consistent recording of the accessory pathway potential in a prospective way. The success of the recordings is based on closer interelectrode distance on the recording catheter in the range of 1 to 2 mm and the use of low recording gain with a very clean signal free of artifact. The use of electrodes oriented perpendicularly to the AV anulus may also improve the ability to record the signal[34] although there is no uniform agreement on this.[35,36] Recording of accessory pathway potentials will allow a more direct analysis of accessory pathway physiology and may provide new insight into the action of antiarrhythmic drugs on accessory pathway function. For example, it is now possible to localize the site of block of an impulse to the atrial insertion of the accessory pathway or the ventricular insertion of the accessory pathway,[37] and it has been suggested that the site of block for left-sided accessory pathways is at the ventricular insertion whereas the site of block for right-sided pathways is at the atrial insertion of the accessory pathway. The sites of block do not change with pharmacological intervention. Additionally, direct recording has provided finer resolution for localizing the accessory pathway, which may contribute importantly to the success of catheter ablative procedures (Fig. 17.6). Jackman suggested that many accessory pathways have an oblique course and exhibit branching.[38,39] The direct recording of the accessory pathway has been an important achievement and will continue to provide important information relating to the localization and function of the accessory pathway.

OPERATIVE THERAPY OF THE WOLFF-PARKINSON-WHITE SYNDROME

The previous two decades have seen a steady evolution in the operative management of the Wolff-Parkinson-White syndrome first described by Cobb and coworkers in 1968.[40] Operative ther-

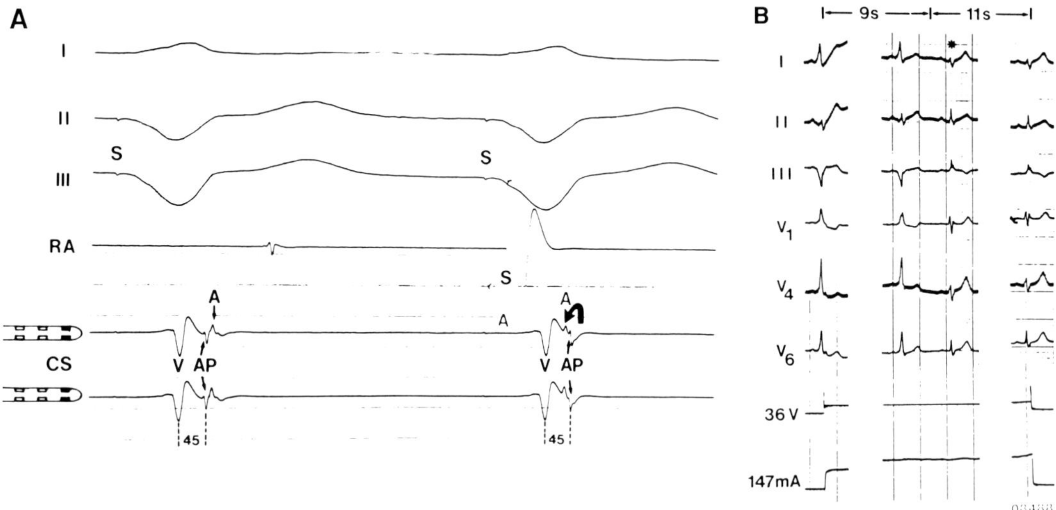

FIGURE 17.6 Recording of accessory pathway potentials. **A.** Endocardial recording during right ventricular pacing showing three distinct potentials indicative of ventricular (V), accessory pathway (AP), and atrial (A) activation. The V–AP conduction time is 45 ms. Evidence for accessory pathway, not atrial, origin of the second potential (AP) is given by a properly timed atrial extrastimulus (S) which advances the atrial potential (curved arrow) without changing timing and morphology of the AP potential. **B.** In the same patient, radiofrequency energy caused conduction block in the left-sided accessory pathway leading to normalization of the PR interval and the QRS complex (asterisk). CS = coronary sinus; RA = right atrium. *Reprinted from Kuck et al., with permission.*

apy is now relatively widely available, with excellent results obtained by centers doing a sufficient volume of this work. The operation has changed from one used as a last resort to one that is an acceptable and even preferred alternative to a lifelong commitment to antiarrhythmic drugs.

Two operative approaches currently are widely used. The first is the endocardial approach initially developed by Sealy under which the atria are opened to provide an endocardial approach. Cardioplegia is used for left-sided accessory pathways and may be used for right-sided accessory pathways at the discretion of the operator. An incision is made on the atrial side of the anulus fibrosis at the site of the accessory pathway to obtain access to the AV fat pad. The AV fat pad is then dissected bluntly from its atrial and ventricular insertion. This interrupts the accessory pathway that traverses the fat pad at that site (Fig. 17.7A).

The other widely used newer approach is

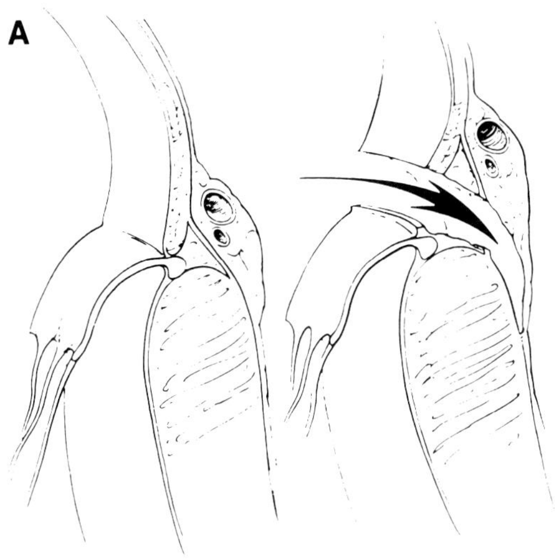

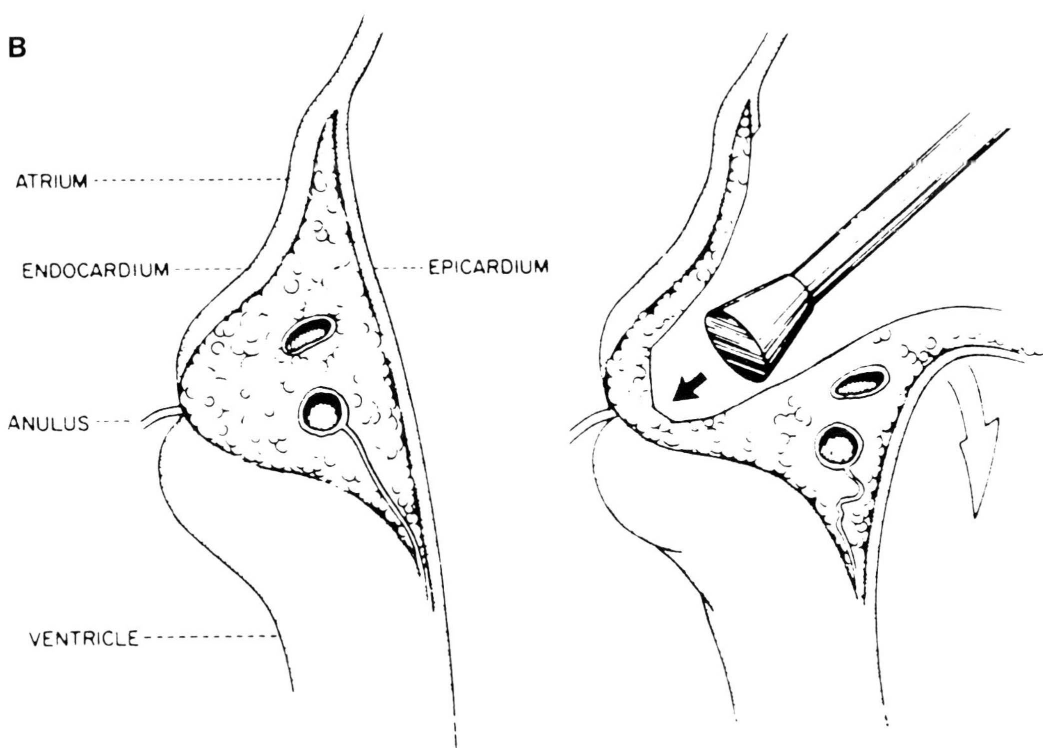

FIGURE 17.7 Operative techniques for accessory pathway ablation. **A.** The endocardial technique for accessory pathway interruption. A supraanular incision from the endocardial surface of the atrium allows access to the AV fat pad. **B.** The epicardial approach to accessory pathway ablation. The fat pad and vascular contents are dissected away and the cryoprobe is applied to this region. **A** *reprinted from Lawrie, used with permission.*

from the epicardial surface.[41] This approach does not require cardiopulmonary bypass or cardioplegia for the great majority of accessory pathways. The AV fat pad and its vascular contents are dissected from the atrial wall and AV anulus in the area of interest. Generally, this results in a sudden cessation of accessory pathway conduction as it is interrupted in its course through the fat pad. The operation is finished with the application of a cryosurgical lesion on the AV anulus to ensure that an accessory pathway adherent to the anulus is ablated (Fig. 17.7B). Occasionally, pathways are endocardial and these can be dealt with by atriotomy or by insertion of a cryoprobe through a purse string and subsequent endocardial cryoablation.

The epicardial approach has the advantage of not requiring cardiopulmonary bypass or cardioplegia, and it also enables continuous assessment of the electrograms preoperatively, intraoperatively, and immediately after the ablation procedure has been performed. It allows for immediate mapping of a second accessory pathway after a first accessory pathway is eliminated. Finally, the bypass time required for a concomitant cardiac procedure is not added to that required for ablation of the accessory pathway which is done on the normothermic, beating heart.

In spite of differences in individual preferences, proponents of each approach report excellent results with an expected primary success rate of more than 95% and a mortality rate of less than 1%.[42,43] Recent studies[44] have documented the advantages of operative therapy versus a long-term commitment to medical therapy. Surgically treated patients enjoy a better quality of life and more active life-style with fewer arrhythmias and fewer visits for medical therapy. Operative therapy also compares favorably in cost when compared to a long-term commitment to antiarrhythmic drugs and ongoing medical attention.[45]

CATHETER ABLATION OF ACCESSORY PATHWAYS

The concept of permanent ablation of an arrhythmia focus or part of the conduction system began when Scheinman et al.[46] and Gallagher et al.[47] demonstrated the feasibility of nonoperative ablation of the AV node His bundle by a catheter technique. The original and still most widely used technique involves the use of a high-energy DC shock delivered through the pole of an electrode catheter positioned in the area of interest. The catheter is guided to the area of interest fluoroscopically and by electrograms recorded at the ablating catheter. This technique was applied to accessory pathways in the right atrioventricular ring,[48–51] the orifice of the the coronary sinus for posteroseptal accessory pathways,[50,52,53] and the distal coronary sinus for left lateral accessory pathways.[54–56] High-energy shock in the distal coronary sinus has been largely abandoned because of low success rate and a serious risk of rupture of the coronary sinus. However, countershock at the orifice of the coronary sinus for interruption of posteroseptal pathways has met with reasonable success and a relatively low complication rate. Morady and coworkers[57] reported permanent posteroseptal accessory pathway ablation using high-energy countershock in the orifice of the coronary sinus in 67% of 48 patients with posteroseptal pathways during a mean follow-up of 26 ± 19 months. One patient had a ruptured coronary sinus that was attributed in part to delivery of the ablation shock too deep within the coronary sinus. Others[50,53] have reported similar results, including the occurrence of inadvertent AV block in 1 patient. Because of the potential for coronary sinus rupture, some authors[53,58] recommend performing the procedure with full operative standby support. Warin et al.[51] recently described the use of high-energy ablation in a series of 70 patients including many with left-sided pathways. The ablating catheter for the left-sided pathway was positioned at the endocardial ventricular insertion of the accessory pathway via the left ventricular cavity and not the coronary sinus.

Although encouraging results have been demonstrated with these high-energy ablation techniques, the trauma is relatively large and maintains the potential for atrial rupture and excessive myocardial necrosis. Other investigators have demonstrated that altering the shape[59] and polarity[60,61] of the discharge and geometry of the electrode[62] can increase the amount of current delivered without the potentially destructive excess arcing and barotrauma.

Direct current ablation has the advantage of producing a relatively large lesion that increases the probability of successful ablation. On the other hand, the method is potentially more destructive and more prone to serious complications. Alternate energy sources are being considered of which radiofrequency energy is the most promising and most frequently used. The current is very similar to that used in a standard cautery or coagulation surgical unit. Application of this energy endocardially is painless and less likely to create complication by inadvertent damage to surrounding structures. On the other hand, the discrete size of a lesion necessitates exact localization and the maintenance of good contact between the radiofrequency electrode and the

myocardium. Early experiences with radiofrequency ablation for accessory pathways[63-65] have been encouraging (Fig. 17.6). Current refinements to catheter technology and technique[66-69] are expected to continue to improve the efficacy of this particular procedure. The procedure is readily applicable to right-sided accessory pathways and septal pathways. Left free wall accessory pathways can be approached via the coronary sinus or via the left ventricle.[70,71] The latter approach uses energy directed between the left ventricular catheter and the coronary sinus catheter at the site of the pathway and probably results in ablation of the pathway closer to its left ventricular insertion site.

The current "gold standard" for nonpharmacological ablation of accessory pathways is operative therapy with a long and excellent track record for safety and efficacy. Ablation has a long way to go to achieve the confidence that operative therapy has earned but will nonetheless probably supplant operative therapy in the future.

The last decade has witnessed substantial progress in our understanding of the Wolff-Parkinson-White syndrome. Variant forms of preexcitation have been more clearly defined, operative therapy has matured as an important therapy, and ablation is on the horizon. The future will witness refinement in catheter ablation techniques and clearer understanding of preexcitation variants.

ACKNOWLEDGMENTS

This work was supported by the Heart and Stroke Foundation of Ontario, Toronto, Ontario. Dr. Klein is a Distinguished Research Professor of the Heart and Stroke Foundation of Ontario, Toronto, Ontario. We wish to thank Sheila Doan for preparation of the manuscript.

REFERENCES

1. Wolff L, Parkinson J, White PD: Bundle branch block with short P-R interval in healthy young people prone to paroxysmal tachycardia. *Am Heart J* 1930;5:685–704.
2. Mahaim I, Winston MR: Recherches d'anatomic comparee et du pathologie experimentale sur les connexions hautes du faisceau de His-Tawara. *Cardiologia* 1941;5:189–260.
3. Wellens HJJ: The preexcitation syndrome, in Wellens HJJ (ed): *Electrical Stimulation of the Heart in the Study or Treatment of Tachycardias.* Baltimore, University Park Press, 1971, pp 97–109.
4. Ward DE, Camm AJ, Spurrel RAJ: Ventricular preexcitation due to anomalous nodoventricular pathways. Report of 3 patients. *Eur J Cardiol* 1979;9:111–127.
5. Touboul P, Vexler RM, Chatelain MT: Reentry via Mahaim fibers as a possible basis for tachycardia. *Br Heart J* 1978;40:806–811.
6. Gallagher JJ, Smith WM, Kasell JH, Benson DW Jr, Sterba R, Grant AO: Role of Mahaim fibres in cardiac arrhythmias in man. *Circulation* 1981;64:176–189.
7. Gmeiner R, Ng CK, Hammer I, Becker AE: Tachycardia caused by an accessory nodoventricular tract: A clinico-pathologic correlation. *Eur Heart J* 1984;5:233–242.
8. Motté G, Brechenmacher C, Davy JM, Belhassen B: Association de fibres nodo-ventriculaires et atrioventriculaires a l'origine de tachycardies reciproques. *Arch Mal Coeur* 1980;73:737–746.
9. Bardy GH, German LD, Packer DL, Coltorti F, Gallagher JJ: Mechanism of tachycardia using a nodoventricular Mahaim fiber. *Am J Cardiol* 1984;54:1140–1141.
10. Gallagher JJ: Variants of preexcitation: Update 1984, in Zipes DP, Jalife J (eds): *Cardiac Electrophysiology and Arrhythmias.* New York, Grune & Stratton, 1985, pp 419–427.
11. Klein GJ, Guiraudon GM, Kerr CR, Sharma AD, Yee R, Szabo T, Yeung LW: "Nodoventricular" accessory pathway: Evidence for a distinct accessory atrioventricular pathway with atrioventricular node-like properties. *J Am Coll Cardiol* 1988;5:1035–1040.
12. Gillette PC, Garson A Jr, Cooley DA, McNamara DG: Prolonged and decremental antegrade conduction properties in right anterior accessory connections: Wide QRS antidromic tachycardia of left bundle branch block pattern without Wolff-Parkinson-White configuration in sinus rhythm. *Am Heart J* 1982;103:66–74.
13. Gallagher JJ, Selle JG, Sealy WC, Fedor JM, Svenson RH, Zimmern SH, Cox JC: Surgical interruption of nodoventricular Mahaim fibers with preservation of normal A-V conduction. *J Am Coll Cardiol* 1986; 7:133A(abstract).
14. Ross DL, Denniss AR, Johnson DC, Sadick N, Richards DA, Uther JB: Further observation on nodoventricular fibres: Anatomic localizataion and electrophysiology. *Circulation* 1989;80(suppl II):II-432 (abstract).
15. Bhandari A, Morady F, Shen EN, Schwarts AB, Botvinick E, Scheinman MM: Catheter-induced His bundle ablation in a patient with reentrant tachycardia associated with a nodoventricular tract. *J Am Coll Cardiol* 1984;4:611–616.
16. Ellenbogen KA, O'Callaghan WG, Colavita PG, Packer DL, Gilbert MR, German LD: Catheter atrioventricular junction ablation for recurrent supraventricular tachycardia with nodoventricular fibers. *Am J Cardiol* 1985;55:1227–1229.
17. Anderson RH, Becker AE, Wenink ACG: The development of the conducting tissues, in Roberts NK, Gelband H (eds): *Cardiac Arrhythmias in the Neonate, Infant and Child.* New York, Appleton-Century-Crofts, 1977, p 23.
18. Guiraudon CM, Guiraudon GM, Klein GJ: "Nodal ventricular" Mahaim pathway: Histological evidence for an accessory atrioventricular pathway with AV node-like morphology. *Circulation* 1988;78(suppl II):II-40(abstract).
19. Coumel P, Cabrol C, Fabiato A, et al.: Tachycardie permanent par rhythm reciproque. 1. Preuves du diagnostic par stimulation auriculaire et ventriculaire. *Arch Mal Coeur* 1967;60:1830–1864.
20. Gallagher JJ, Sealy WC: The permanent form of junctional reciprocating tachycardia: Further elucidation of the underlying mechanism. *Eur J Cardiol* 1978;8:413–430.

21. Gallagher JJ, Sealy WC: The permanent form of junctional reciprocating tachycardia in an asymptomatic adult: Further evidence of an accessory ventriculoatrial nodal structure. *Eur J Cardiol* 1981;102:282.
22. Bardy GH, Packer DL, German LD: Paradoxycal delay in accessory pathway conduction during long R-P′ tachycardia following interpolated premature ventricular depolarizations. *Am J Cardiol* 1985;55:1223–1225.
23. Critelli G, Perticone G, Coltorti F, et al.: Antegrade slow bypass conduction following closed chest ablation of the His bundle in permanent junctional reciprocating tachycardia. *Circulation* 1983;67:687–692.
24. Guarnieri T, Sealy WC, Kasell JH, German LD, Gallagher JJ: The nonpharmacologic management of the permanent form of junctional reciprocating tachycardia. *Circulation* 1984;69:269–277.
25. O'Neill BJ, Klein GJ, Guiraudon GM, Yee R, Fujimura O, Boahene A, Sharma AD: Results of operative therapy in the permanent form of junctional reciprocating tachycardia. *Am J Cardiol* 1989;63:1074–1079.
26. Packer DL, Bardy GH, Worley SJ, Smith MS, Cobb FR, Coleman RE, Gallagher JJ, German LD: Tachycardia-induced cardiomyopathy: A reversible form of left ventricular dysfunction. *Am J Cardiol* 1986;57:563–570.
27. McLaran CJ, Gersh BJ, Sugrue DD, Hammill SC, Seward JB, Holmes DR: Tachycardia-induced dysfunction: a reversible phenomenon. *Br Heart J* 1985;53:323–327.
28. Kloner RA, DeBoer LWV, Darsee JR, Ingwall JS, Braunwald E: Recovery from prolonged abnormalities of canine myocardium salvaged from ischemic necrosis by coronary reperfusion. *Proc Natl Acad Sci USA* 1981;78:7152–7156.
29. Critelli G, Gallagher JJ, Monda V, Coltorti F, Scherillo M, Rossi L: Anatomic and electrophysiologic substrate of the permanent form of junctional reciprocating tachycardia. *J Am Coll Cardiol* 1984;4:601–610.
30. Spach MS: The electrical representation of cardiac muscle based on discontinuities of atrial resistivity at a microscopic and macroscopic level. A basis for saltatory propagation in cardiac muscle, in Carvalho AP, Hoffmann BF, Lieberman M (eds): *Normal and Abnormal Conduction in the Heart. Biophysics, Physiology, Pharmacology and Ultrastructure.* Mt. Kisco, NY, Futura Publishing, 1982, pp 145–153.
31. Lerman BB, Greenberg M, Overholt ED, Swerdlow CD, Smith RT, Sellers TD, DiMarco JP: Differential electrophysiologic properties of decremental retrograde pathways in long RP tachycardia. *Circulation* 1987;76:21–31.
32. Jackman WM, Friday KJ, Scherlag BJ, Dehning MM, Schechter E, Reynolds DW, Olson EG, Berbari EJ, Harrison LA, Lazzara R: Direct endocardial recording from an accessory atrioventricular pathway: Localization of the site of block, effect of antiarrhythmic drugs, and attempt at nonsurgical ablation. *Circulation* 1983;68:909–916.
33. Prystowsky EN, Browne KF, Zipes DP: Intracardiac recording by catheter electrode of accessory pathway depolarization. *J Am Coll Cardiol* 1983;1:468–470.
34. Jackman WM: New catheter techniques for recording accessory AV pathway activation, in Benditt DG, Benson DW (eds): *Cardiac Preexcitation Syndromes: Origins Evaluation and Treatment.* Boston, Martinus Nijhoff Publishing, 1986, pp 413–434.
35. Winters SL, Gomes AJ: Intracardiac electrode catheter recordings of atrioventricular bypass tracts in Wolff-Parkinson-White syndrome: Techniques, electrophysiologic characteristics and demonstration of concealed and decremental propagation. *J Am Coll Cardiol* 1986;7:1392–1403.
36. Kuck KH, Kunze KP: Recording technique of accessory pathway potentials. Comparison of catheters with different electrode configurations. *Circulation* 1986;74(suppl II):II-302(abstract).
37. Kuck KH, Geiger M: Accessory pathway conduction block: Concordance of spontaneous, stimulation-induced and drug-induced site of block. *Circulation* 1987;76(suppl IV):IV-136(abstract).
38. Jackman WM, Friday KJ, Fitzgerald DM, Bowman AJ, Yeung-Lai-Wah JA, Lazzara R: Localization of left free-wall and posteroseptal accessory atrioventricular pathways by direct recording of accessory pathway activation. *PACE* 1989;12(part II):204–214.
39. Jackman WM, Friday KJ, Yeung-Lai-Wah JA, Fitzgerald DM, Beck B, Bowman AJ, Steltzer P, Harrison L, Lazzara R: New catheter technique for recording left free-wall accessory atrioventricular pathway activation. *Circulation* 1988;78:598–610.
40. Cobb FR, Blumenschein SD, Sealy WL, et al.: Successful surgical interruption of the Bundle of Kent in a patient with Wolff-Parkinson-White syndrome. *Circulation* 1968;38:1018.
41. Klein GJ, Guiraudon GM, Perkins DG, Jones DL, Yee R, Jarvis E: Surgical correction of the Wolff-Parkinson-White syndrome in the closed heart using cryosurgery: A simplified approach. *J Am Coll Cardiol* 1984;3:405–409.
42. Guiraudon GM, Klein GJ, Sharma AD, Yee R: Surgical alternatives for supraventricular tachycardias. *Am J Cardiol* 1989;64:92J–96J.
43. Cain ME, Cox JL: Surgical treatment of supraventricular tachyarrhythmias, in Platia EV (ed): *Management of Cardiac Arrhythmias—The Nonpharmacologic Approach.* Philadelphia, JP Lippincott, 1987, pp 304–309.
44. Pressley JC, German LD, Packer DL, Lowe JE, Burks BD, Shander GS, Prystowsky EN: Comparisons of baseline characteristics and long-term quality of life in 540 Wolff-Parkinson-White patients treated with pharmacologic vs. surgical therapy. *J Am Coll Cardiol* 1988;11:110A(abstract).
45. Lezaun R, Brugada P, Talajic M, Trappe JH, Della Bella P, Sternick E, Penn O, Wellens HJJ: Medical vs. surgical treatment of patients with accessory atriventricular pathways. *Circulation* 1987;76(suppl IV):IV-499(abstract).
46. Scheinman MM, Morady F, Hess DS, Gonzalez R: Catheter induced ablation of the atrioventricular junction to control refractory supraventricular arrhythmias. *JAMA* 1982;248:851–855.
47. Gallagher JJ, Svenson RH, Kasell JH: Catheter techniques for closed-chest ablation of the atrioventricular conducting system. *N Engl J Med* 1982;306:194–200.
48. Ruder MA, Davis JC, Eldar M, Scheinman MM: Effects of electrode catheter shock delivered near the tricuspid annulus in dogs. *J Am Coll Cardiol* 1986; 7:7A(abstract).
49. Kunze KP, Kuck KH: Transvenous ablation of accessory pathway in patients with incessant atrio-

ventricular tachycardia. *Circulation* 1984;70(suppl II):II-412(abstract).
50. Ruder MA, Mead RH, Gaudian V, Buch WS, Smith NA, Winkle RA: Transvenous catheter ablation of extranodal accessory pathways. *J Am Coll Cardiol* 1988;11:1245–1253.
51. Warin JF, Haissaguerre M: Fulguration of accessory pathway in any location: Report of seventy cases. *PACE* 1989;12(part II):215–218.
52. Morady F, Scheinman MM, Winston SA, DiCarlo L, Davis JC, Griffin JC, Ruder M, Abbott JA, Eldar M: Efficacy and safety of transcatheter ablation of posteroseptal accessory pathways. *Circulation* 1985;72:170–177.
53. Bardy GH, Ivey TD, Coltorti F, Stewart RB, Johnson G, Greene HL: Developments, complications and limitations of catheter-mediated electrical ablation of posterior accessory atrioventricular pathways. *Am J Cardiol* 1988;61:309–316.
54. Brodman R, Fisher JD: Evaluation of a catheter technique for ablation of accessory pathways near the coronary sinus using a canine model. *Circulation* 1983;67:923–929.
55. Fisher JD, Brodman R, Kim SG, Matos JA: Nonsurgical Kent bundle ablation via the coronary sinus in patients with the Wolff-Parkinson-White syndrome. *Circulation* 1982;66(suppl II):II-375(abstract).
56. Fisher JD, Brodman R, Kim SG, Matos JA, Brodman E, Wallerson D, Waspe LE: Attempted nonsurgical electrical ablation of accessory pathways via the coronary sinus in the Wolff-Parkinson-White syndrome. *J Am Coll Cardiol* 1984;4:685–694.
57. Morady F, Scheinman MM, William HK, Griggin JC, Dick II M, Herre J, Kadish AH, Langberg J: Long-term results of catheter ablation of a posteroseptal accessory atrioventricular connection in 48 patients. *Circulation* 1989;79:1160–1170.
58. Packer DL, Prystowsky EN: Wolff-Parkinson-White syndrome: Further progress in evaluation and treatment, in Zipes DP, Rowlands DJ (eds): *Progress in Cardiology.* Philadelphia, Lea & Febiger, 1988, pp 147–179.
59. Coltorti F, Rackson MM, Hanson K, Greene HL, Ivey TD: Current waveform modulation to avoid plasma-arcing and barotrauma with catheter mediated electric discharges. *Circulation* 1985;72(suppl III):III-474(abstract).
60. Boyd EG, Holt P: Haematological effects of the ablation technique: A comparison between anodal and cathodal delivery. *J Am Coll Cardiol* 1986; 7:131A(abstract).
61. Bardy GH, Coltorti F, Ivey TD, Alferness C, Rackson MM, Hansen K, Stewart R, Greene HL: Some factors affecting bubble formation with catheter-mediated defibrillator pulses. *Circulation* 1986;73:525–538.
62. Cunningham D, Rowland E, Rickards A: A new non-arcing electrode for catheter ablation—Design and trial performance. *Circulation* 1987;76(suppl IV):IV-405(abstr).
63. Roman CA, Friday KJ, Wang X, Calame JD, Dyer JW, Moulton KP, Lazzara R, Jackman WM: Ablation of single and multiple accessory pathway with radiofrequency current. *Circulation* 1989;80(suppl II):II-323(abstract).
64. Kuck KH, Kunze KP, Geiger M, Schluter M: Radiofrequency ablation of accessory atrioventricular pathways. *Circulation* 1989;80(suppl II):II-323(abstract).
65. Borggrefe M, Budde T, Podczeck A, et al.: High frequency alternating current ablation of an accessory pathway in humans. *J Am Coll Cardiol* 1987;10:576–582.
66. Langberg JJ, Chin MC, Lee MA, Scheinman MM: Radiofrequency ablation; the effect of electrode size on lesion area in vivo. *Circulation* 1989;80(suppl II):II-323 (abstract).
67. Blouin LT, Marcus FI: The effect of electrode design on the efficiency of delivery of radiofrequency energy to cardiac tissue in vitro. *PACE* 1989;12(part II):136–143.
68. Wittkampf FH, Hauer RN, Robles de Medina EO: Radiofrequency ablation with a cooled porous catheter. *J Am Coll Cardiol* 1988;11:17A(abstract).
69. Haines DE, Watson DD, Halperin C: Monitoring electrode tip temperature during radiofrequency fulguration of ventricular myocardium is strongly predictive of lesion size. *Circulation* 1987;76(suppl IV):IV-406(abstract).
70. Jackman WM, Kuck KH, Naccarelli GV, et al.: Radiofrequency current directed across the mitral annulus with a bipolar epicardial endocardial catheter electrode configuration in dogs. *Circulation* 1988;78:1288–1298.
71. Kuck KH, Schluter M, Geiger M, Kunze KP: Attempted ablation of left sided accessory pathways using radiofrequency current and an epicardial-endocardial electrode configuration. *Circulation* 1988;78(suppl II):II-40(abstract).

Chapter **18**

Long QT Syndromes

J. A. Abildskov, MD and Larry S. Green, MD

The QT interval of the electrocardiogram is the only easily available clinical measure of the duration of ventricular repolarization. That process is affected by various physiologic factors and pathologic conditions and evidence of its duration has well-established medical value. Acquired prolongation of the QT interval is a nonspecific but useful manifestation of several cardiac and extracardiac conditions. Examples are ischemic heart disease, myocarditis, hypokalemia, neurologic disease, and drug effects such as those of quinidine or procainamide. The significance of acquired QT interval prolongation is highly dependent on the nature of the responsible condition. It ranges from diagnostic value, suggesting the presence of disease, to prognostic value and therapeutic implications as a marker of quinidine or other drug effects.

Idiopathic QT interval prolongation occurs in association with congenital deafness in the Jervell Lange-Nielson syndrome and with normal hearing in the Romano Ward syndrome. Both syndromes are characterized by syncope and a high incidence of sudden death due to ventricular tachyarrhythmias. Electrocardiographic and clinical cardiac manifestations are similar in the syndromes.

This chapter will review the physiologic basis of the QT interval and the cardiac manifestations of both the acquired and idiopathic syndromes as well as their possible mechanisms. Many findings suggest a role for the autonomic nervous system in the idiopathic syndromes and this evidence will be discussed. Other possible mechanisms, including that of afterdepolarizations in the initiation of tachyarrhythmias, will also be discussed.

PHYSIOLOGIC BASIS OF THE QT INTERVAL

The QT interval extends from the onset of the QRS complex to the end of the T wave in the electrocardiogram. It is often measured as the interval in a particular electrocardiographic lead; however, when multiple simultaneous leads are available, the earliest QRS onset to the latest T wave completion in whatever leads these occur may be a more accurate measure of the duration of repolarization. However it is measured, it does not reflect the duration of recovery at particular cardiac sites. The interval extends from the electrocardiographic evidence of the onset of excitation to the latest evidence of completion of recovery, regardless of the cardiac location of these events.

The onset of the QRS complex can usually be identified with considerable accuracy, but termination of the T wave is sometimes ambiguous and limits precision of the measurement. It is frequently necessary to select a particular lead for measurement on the basis of having the clearest T-wave termination. When a U wave can be distinguished, it should not be included in the measurement but it is likely that some apparent T wave terminations include U waves and this is one of the ambiguities of measurement.

Effect of Heart Rate

One of the important physiologic factors affecting QT interval duration is heart rate. It is clear that the interval decreases as rate increases but it is difficult to separate effects of rate from other factors with intrinsic effects on the duration of ventricular repolarization. Exercise, for example, increases heart rate but changes of the QT interval may be due to altered autonomic effects and other factors in addition to rate. Various relations such as those in the Bazett formula have been proposed to define the normal QT interval at various rates.[1] According to the Bazett formula, the QT interval varies with the square root of the cycle length. The normal QT interval in seconds equals the square root of the cycle length in seconds times a constant (k) determined by Bazett as 0.37 in men and 0.40 in women. Thus, at a heart rate of 60 per minute, the cycle length is 1 s and the normal QT interval for men is $0.37\sqrt{1}$ or 0.37

655 Avenue of the Americas, New York, NY 10010
Current Topics in Cardiology

s. Various modifications of this formula as well as other relations have been proposed to identify the normal QT interval at various heart rates. It is unlikely, however, that any single relation is applicable to all conditions and it seems necessary to accept some ambiguities in definition of the normal QT interval as well as those associated with determination of T-wave endpoint.

Cardiac and ECG Lead Factors

In addition to difficulties in the measurement and definition of normal QT intervals, some features of its cardiac origin and their expression in ECG leads affect the QT interval. In particular, decreased duration of ventricular repolarization can sometimes result in paradoxical prolongation of the QT interval. As with other electrocardiographic quantities, the QT interval is influenced by the responsible cardiac events and also the characteristics of ECG leads. The events responsible for the QT interval include both ventricular excitation and recovery. Excitation is expressed in the QRS complex, which is the result of currents between excited and resting myocardium. Similarly, the T wave is the result of currents between myocardium in various stages of repolarization. Normal excitation proceeds from endocardium to epicardium and apex to base in the free ventricular walls and bidirectionally from right and left sides of the interventricular system. The multidirectional spread of excitation results in cancellation of the electrocardiographic expression of parts of the excitation process, so QRS complex onset in a particular lead may not reflect the onset of ventricular excitation. Similarly, cancellation of the ECG effects of repolarization occurs and termination of the T wave may not reflect completion of repolarization. The normal sequence of repolarization is grossly opposite to that of normal excitation due to intrinsically longer recovery properties in regions of early excitation.[2] Since excitation and recovery are processes of opposite polarity, this sequence leads to T waves of the same polarity as major portions of the QRS complex. The cardiac origin of T waves, however, differs importantly from that of QRS complexes. At a particular moment, excitation has a particular cardiac location or limited number of locations while repolarization is widely distributed with much or all of the ventricular muscle in various stages of that process. Since the sequence of recovery is inversely related to that of excitation, the two processes result in similar boundaries and the simultaneous presence of all boundaries during much of recovery increases the degree of cancellation of ECG effects.[3] Cancellation can be expected to be highest during late portions of normal recovery when the remaining boundaries between various stages of the process are subendocardial and tend to surround the ventricular cavities. Objective evidence that late portions of normal ventricular repolarization are not expressed in the T wave is furnished by the observation that prolongation of the QT interval with ectopic ventricular excitation is only partially accounted for by increased QRS duration.[4] This is illustrated in Figure 18.1. The additional QT duration represents the expression of recovery whose ECG effects must have been canceled after normal excitation. Therefore, it is possible to prolong the QT interval by shortening repolarization in a region whose recovery previously cancelled the ECG effects of other regions. Such paradoxical prolongation of the QT interval was noted after

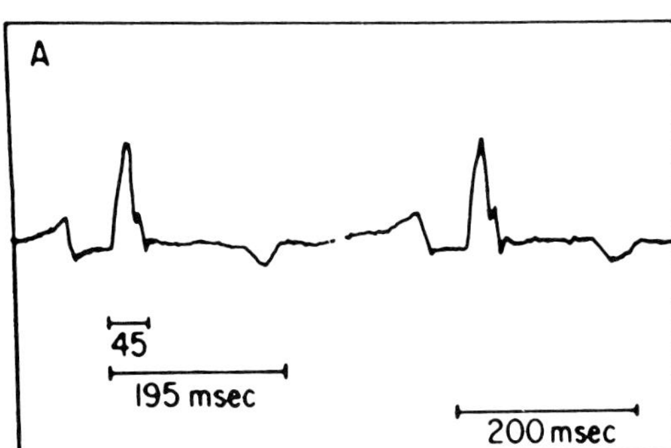

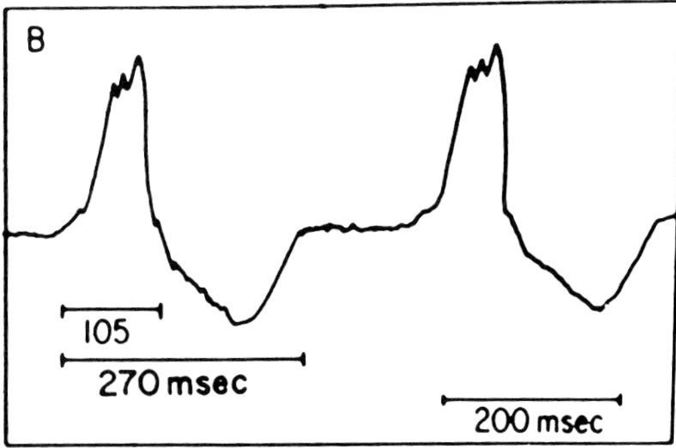

FIGURE 18.1 Effect of epicardial ventricular pacing near the pulmonary conus on the QT interval. The QRS interval is 60 ms and the QT interval 75 ms longer than those associated with supraventricular activation. Other ventricular pacing sites also resulted in greater prolongation of the QT than of the QRS, suggesting the abnormal activation orders exposed ECG evidence of ventricular repolarization not apparent in the T wave following normal excitation. *Reprinted from Abildskov,[4] with permission.*

left stellate ganglion stimulation and was part of the rationale for left stellectomy in the treatment of long QT syndromes.[5,6] It was subsequently shown, however, that QT prolongation after left stellate stimulation was transient and was not unique for the left stellate ganglion.[4] Transient prolongation also occurs after right-sided sympathetic stimulation and after bolus injections of catecholamines. These findings make it unlikely that excessive left-sided sympathetic activity is the mechanism of QT prolongation in the idiopathic long QT syndromes. However, left stellectomy has proven to be effective therapy for the prevention of tachyarrhythmias. The sometimes paradoxical behavior of the QT interval does add still another difficulty to the use of the interval to estimate duration of ventricular repolarization. In particular, the probable mechanism of such behavior suggests at least the theoretic possibility that prolongation of recovery might occur in regions where ECG effects are canceling and fail to result in QT interval prolongation. Although cancellation of ECG effects may affect QT interval duration, especially in the case of transient changes, long-lasting prolongation of the interval strongly suggests prolonged ventricular repolarization. In the case of idiopathic long QT syndromes, ECG body surface mapping studies suggest that the prolongation of repolarization is regional, not global.[5]

The U Wave

Both the physiologic basis and the role of the U wave in the long QT syndromes are uncertain. The wave has been variously attributed to late repolarizing portions of ventricular myocardium, repolarization of the Purkinje system, mechanical factors, and afterdepolarizations. Whatever their origin, prominent U waves are often associated with prolongation of the QT interval although evidence from body surface ECG mapping indicates that it is the actual QT interval rather than the QU interval, which is prolonged in the idiopathic long QT syndromes.[7] Afterdepolarizations, however, which may be the basis of the U wave, may be the initiating agency for tachyarrhythmias in the long QT syndromes.

CLINICAL FEATURES OF LONG QT SYNDROMES

Ventricular Arrhythmias

The ventricular tachycardia that occurs in the setting of a long QT interval often has the appearance of alternating sharp peaks and rounded valleys that periodically change axis by 180°. This twisting morphology has been given the name of torsade de pointes.[8] Confusion in use of this term has arisen in part because this morphology of ventricular tachycardia is not unique to long QT syndrome patients and the ventricular tachycardia that occurs in this group of patients is not always of this morphology. The term "torsade de pointes" as used in this chapter refers to ventricular tachycardia that occurs in the setting of a prolonged QT interval, usually after slowing of the intrinsic heart rate and usually after pauses. Electrocardiograms of such patients often show abnormal T-wave morphology and/or prominent U waves. An example of the tachycardia that occurs in the long QT syndromes is shown in Figure 18.2. The example rhythm strip is continuous and shows both the "twisting" morphology and the initiation of the arrhythmia following a pause.

Acquired Long QT Syndrome

As already noted, acquired long QT syndrome has been documented to occur with a variety of initiating factors including drugs, neurologic diseases, electrolyte abnormalities, dietary deficiencies, and disease of the cardiac conduction system leading to complete heart block. In nearly all cases, removal of the offending agent or increasing the heart rate by pacing or infusion of catecholamines provides control of the frequently associated ventricular arrhythmias.[33] The latter characteristic distinguishes the acquired syndromes from the idiopathic or congenital long QT syndromes which are distinctly associated with increased adrenergic tone and often respond to β-blocker therapy.

The drug most frequently associated with acquired long QT syndrome is quinidine, although the other class IA antiarrhythmic drugs, procainamide and disopyramide, appear to have an equal propensity to cause the syndrome. The importance of the additional factors of hypomagnesemia or hypokalemia in the initiation of quinidine-induced arrhythmias is uncertain since serum levels of potassium (K^+) and magnesium (Mg^{2+}) may not reflect myocardial levels, but hypokalemia and hypomagnesemia with quinidine therapy are felt by most experts to enhance the risk of ventricular arrhythmias including torsade de pointes. Documentation of normal electrolyte levels in patients with quinidine-induced arrhythmias at least suggests the drug effect may be independent of electrolyte abnormalities. The serum level of quinidine has varied widely in reported cases of drug-induced long QT syndrome and the arrhythmias cannot be considered as an overdose effect.[9,10] Although prediction of patients at risk with quinidine therapy is not pos-

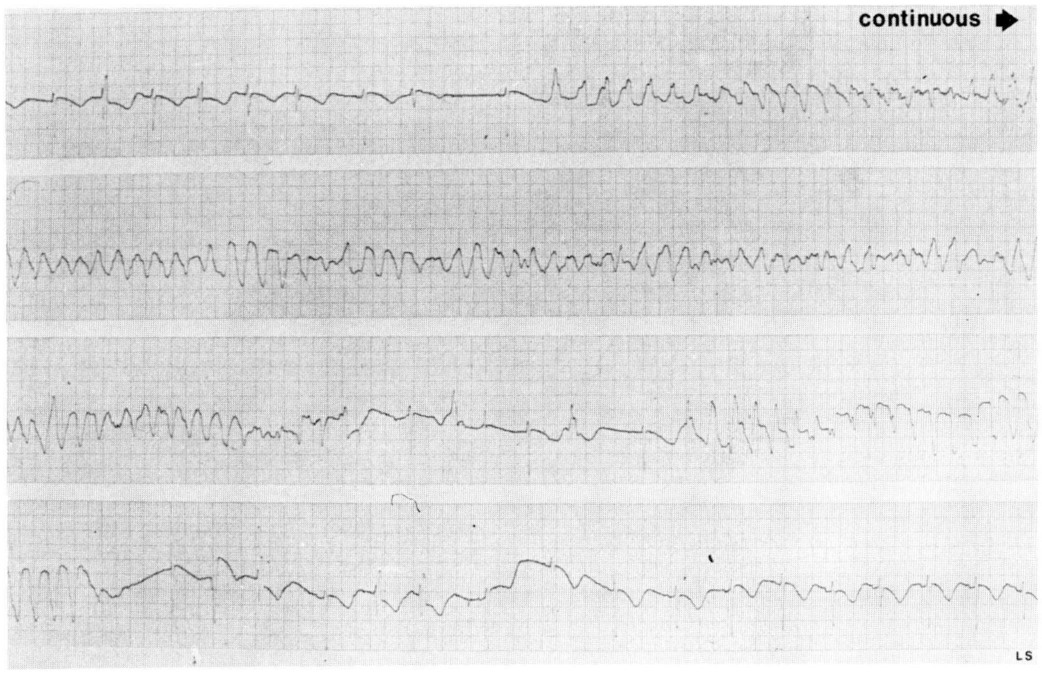

FIGURE 18.2 The above continuous rhythm was recorded from a patient with idiopathic long QT interval syndrome. The sinus rate slowed and the arrhythmia began following a pause. The "twisting" morphology is illustrated as well as spontaneous termination of the ventricular tachycardia.

sible if a normal QT interval is present, a variety of studies from different groups suggests that over half of patients that develop torsade de pointes with quinidine do so within the first 3 or 4 days of therapy.[11–13] Prolongation of the QT interval as an effect of quinidine therapy is well known and some studies have proposed that prolongation of the QT interval beyond 0.60 s (uncorrected) is a marker for arrhythmia risk. Several other reports, however, have documented occurrence of torsade de pointes with quinidine therapy in which there was little prolongation of the QT interval, and sensitivity of this marker of arrhythmia risk cannot be recommended. During chronic quinidine therapy, changes in drug preparation, dosage (up or down), and reinstitution of therapy after a pause have all been associated with development of torsade de pointes.[14]

Sotalol has also been associated with drug-induced prolongation of the QT interval and occurrence of torsade de pointes.[15–17] This drug is a nonselective β-blocker known to prolong action potential duration substantially. Other β-blockers have not been implicated in the long QT syndrome. Class IB drugs, which include lidocaine and its oral congeners mexiletine and tocainide, have not been clearly documented to cause long QT syndrome arrhythmias and the same is true for the class IC antiarrhythmic drugs encainide, flecainide, and indecainide, although this latter class is known to have a high incidence of proarrhythmic effects especially in the setting of poor ventricular function. Recent data suggest that amiodarone is only rarely a cause of torsade de pointes[18] and it appears to be safe from the proarrhythmic standpoint to use in patients with a history of drug-induced torsade de pointes.[19]

Numerically the next most common drug-related cause of long QT interval–associated arrhythmias involves the tricyclic and tetracyclic antidepressants. The occurrence of torsade de pointes is usually in the setting of a drug overdose as part of a suicide attempt. The arrhythmia frequently responds to parenteral Mg^{2+}, even in the setting of normal serum Mg^{2+} levels.[20] Abuse of phenothiazines in the same setting has also been associated with torsade de pointes.[21] Therapeutic use of erythromycin has also been reported to cause torsade de pointes.[22]

The marked dietary change of liquid protein diets and the eating disorder anorexia nervosa have both been documented to produce a prolonged QT interval and result in torsade de

pointes.[23,24] This has occurred in the absence of underlying heart disease, without use of antiarrhythmic drugs, and with normal electrolyte values. Numerous case reports have also documented the occurrence of torsade de pointes in patients with complete AV block as well as in patients with marked bradycardia.[25,13] Central nervous system pathology, including tumors and trauma, have resulted in prolonged QT interval syndrome.[26,27] Subarachnoid hemorrhage is the most frequent central nervous system (CNS) pathology resulting in long QT syndrome and torsade de pointes. These patients are usually responsive to temporary pacing.

Idiopathic Long QT Syndrome

Patients with classic long QT syndrome differ from those described above under acquired long QT syndrome primarily in that arrhythmias often occur in circumstances of enhanced sympathetic tone. Idiopathic long QT syndrome shares with the acquired syndrome a tendency for arrhythmias to more often occur in the setting of bradycardia and following pauses. Patients with the idiopathic syndrome are more often younger and most likely to be female. Two inherited forms of the syndrome have been well described. The Jervell Lang-Nielsen syndrome is an association between neural deafness and long QT syndrome arrhythmias. It is inherited as an autosomal recessive trait and is therefore less common. The Romano Ward long QT syndrome is inherited as an autosomal dominant trait without deafness and it is known to be carried on chromosome 11n.[28] Nonfamilial, sporadic forms of the syndrome also occur and as a consequence of the arrhythmias these patients have in the past often been misdiagnosed as seizure disorder patients. One patient in our clinic with long QT syndrome and no family history to suggest the syndrome has given birth to a child with a markedly prolonged QT interval.

Any of a variety of events can trigger arrhythmias in these patients including exercise, anger, fright, or other emotional states. All inciting states share the feature of enhanced sympathetic activity, and the circumstances causing arrhythmias can be very specific and reproducible in individual patients. One case report documents a young girl in whom torsade de pointes could reproducibly be initiated by awakening her with any type of loud sound.[29] The mortality rate in untreated patients is high and has been estimated at 71% over 15 years of follow-up in one series.[30]

Electrophysiologic study has not been helpful in management of patients in this group because arrhythmias cannot reliably be produced by programmed stimulation. The mainstay of therapy in idiopathic long QT syndrome has been treatment with β-blockade. The most experience has been with the use of propranolol but other agents have been successfully used. As noted earlier, left cervico-thoracic ganglionectomy has also been successfully employed in these patients, but treatment failures have occurred with either modality.[31] Some surgical treatment failures may be attributed to inadequate sympathectomy because the stellate ganglion and first three to five thoracic ganglia should be removed. Because of the consequences of treatment failure, some clinicians use both types of therapy in most patients with arrhythmias associated with idiopathic long QT syndrome. There is limited experience with long-term pacing therapy, but it appears to be a useful option in selected patients.[32]

OTHER MECHANISMS

The role of afterdepolarizations, specifically early afterdepolarizations or EADs, has received increasing attention in recent years as the possible mechanism of arrhythmias in both acquired and idiopathic long QT interval syndromes. As defined by Cranefield[34] and Damiano and Rosen,[35] EADs refer to an afterdepolarization that interrupts or delays repolarization. They may be caused by a reduced outward current, a delayed inward current, or both[34] and the exact mechanisms are not yet clear. A variety of mechanisms have been proposed by various investigators including enhancement of calcium currents (i_{Ca}), transient inward currents activated by elevated intracellar calcium (Ca^{2+}), abnormalities of sodium (Na^+) channel exchange mechanisms and abnormalities of K^+ conductance. Experimental studies show that EADs can occur in phase 2 or phase 3 of the AP[36] and Cranefield has pointed out that some EADs do not show a clear afterdepolarization but rather a delay of repolarization. Microelectrode studies in isolated tissue and whole-organ experiments using monophasic AP recordings have both documented that EADs can reach potentials sufficient to trigger sustained arrhythmias of the type seen in long QT syndromes. Examples of EADs are shown in Figure 18.3. In this example, EADs were recorded from canine Purkinje preparations superfused with normal Tyrode solution containing quinidine (0.5 to 1.0 μg/ml).

EADs have occurred and been modulated under a variety of conditions relevant to the clinical occurrence of arrhythmias in long QT syn-

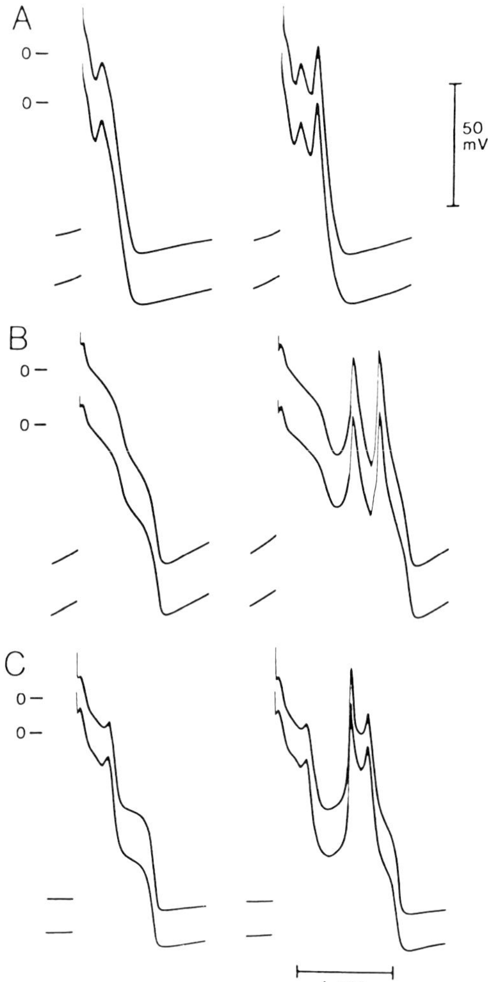

FIGURE 18.3 Tracings of two types of quinidine-induced EADs and triggered activity. Each panel shows two traces representing intracellular activity recorded from opposite ends of three different Purkinje preparations. Left panels depict responses displaying only EADs; right panels show responses manifesting triggered activity. **A.** EAD and triggered activity occurring at the plateau level only. **B.** EAD and triggered activity occurring during phase 3 only. **C.** Both types of activity occurring in the same preparation. *Reprinted from Antzelevitch, with permission.* (Used by permission of American Heart Association, Inc. Figure 4 and legend reprinted from Antzelevitch C: Quinidine-induced action potential prolongation, early afterdepolarizations, and triggered activity in canine Purkinje fibers: Effects of stimulation rate, potassium and magnesium. *Circulation* 79:677, March, 1989)[37]

drome. In isolated tissue studies EADs are generated by quinidine, cesium (Cs^{2+}) and acidosis, and amplitude of these EADs is enhanced by epinephrine, slow pacing rates, and hypokalemia. In these same experimental conditions, EADs are abolished by Mg^{2+}, faster pacing rates, and normalization of K^{+}. Cs^{2+} has been used intravenously in whole-animal preparations to produce EADs recorded from monophasic injury potentials in preparations that exhibited torsade de pointes.[37,38] Although a recent study points out several limitations of the Cs^{2+} model for long QT syndrome,[39] the model nonetheless shares many features of the clinical syndrome.

Clinical studies of patients with long QT syndrome and arrhythmias have been reported by several investigators.[40–42] Bonatti et al. have reported the monophasic injury potential recording of EADs in 10 patients with long QT syndrome, 2 with the idiopathic variety, and 8 with the acquired. Evaluation of the recordings is convincing and suggests that the recordings of afterdepolarizations are more than mechanical artifact. Schechter et al. have also reported endocardial recording of "slow waves" following endocardial T waves that correspond to U waves in Romano-Ward patients.[43]

The accumulating clinical and experimental evidence strongly suggests that afterdepolarizations play a major role in the long QT syndromes. The cellular membrane changes leading to the afterdepolarizations are not yet defined nor is the mechanism explaining why only some patients acquire long QT syndrome in response to drugs or other agents. The role of the sympathetic nervous system may be one of enhancement in the genesis of arrhythmias in the context of afterdepolarizations as they relate to the idiopathic long QT syndromes. One hypothesis advanced for the greater role of the left-sided sympathetics in the idiopathic long QT syndromes is that the increased mass of the left ventricular regions innervated by the left sympathetic chain results in a greater myocardial mass being subject to norepinephrine enhancement of afterdepolarizations.[44,45] This hypothesis and the hypothesis of afterdepolarizations as the mechanisms leading to arrhythmias in the long QT syndromes are compatible with clinical features of the syndromes.

ACKNOWLEDGMENTS

This work was supported by a grant #HL35204 from the National Heart, Lung, and Blood Institute, U.S. Public Health Service; and awards from the Richard A. and Nora Eccles Harrison Fund for Cardiovascular Research and the Nora Eccles Treadwell Foundation.

REFERENCES

1. HC. An analysis of the time relations of electrocardiograms. *Heart* 1920;7:353–370.
2. Burgess MJ, Green LS, Millar CK, Abildskov JA: The sequence of normal ventricular recovery. *Am Heart J* 1972;84:660–669.
3. Burgess MJ, Millar K, Abildskov JA: Cancellation of electrocardiographic effects during ventricular recovery. *J Electrocardiol* 1969;2(2):101–108.
4. Abildskov JA: Adrenergic effects on the QT interval of the electrocardiogram. *Am Heart J* 1976;92:210–216.
5. Abildskov JA: The prolonged QT interval syndromes, in Balsano F, Marigliano V (eds): *Sidedness in the Neurovegetative Regulation of the Cardiovascular Apparatus*. Rome, Edizioni Luigi Pozzi, 1981, pp 170–185.
6. Yanowitz F, Preston JB, Abildskov JA: Functional distribution of right and left stellate inervation to the ventricles: Production of neurogenic electrocardiographic changes by unilateral alteration of sympathetic tone. *Circ Res* 1966;18:416–428.
7. Moss AJ, McDonald J: Unilateral cervicothoracic sympathetic ganglionectomy for the treatment of the long QT interval syndrome. *N Engl J Med* 1971;285:903–907.
8. De Ambroggi L, Tocati E, Bertoni T, Monza E, Schwartz PJ: Body surface potentials during the T-U interval in patients with the idiopathic long QT syndrome, in Butrous GS, Schwartz PJ (eds): *Clinical Aspects of Ventricular Repolarization*. London, Farrand Press, 1989, pp 433–436.
9. Dessertenne F: La tachycardie ventriculaire a deux foyers opposes variables. *Arch Mal Coeur* 1966; 59:263–272.
10. Roden DM, Woosley RL, Primm RK: Incidence and clinical features of the quinidine associated long QT syndrome: Implications for patient care. *Am Heart J* 1986;111:1088–1093.
11. Kay GN, Plumb VJ, Arciniegas JG, et al.: Torsades de pointes: The long short initiating sequence and other clinical features: Observations in 32 patients. *J Am Coll Cardiol* 1983;2:806–817.
12. Jackman WM, Clark M, Friday KJ, et al.: Ventricular tachyarrhythmias in the long QT syndrome. *Med Clin North Am* 1984;68:1079–1109.
13. Keren A, Tzivoni D, Gavish D, et al.: Etiology, warning signs and therapy of torsades de pointes: A study of 10 patients. *Circulation* 1981;64:1167–1174.
14. Jackman WM, Friday KJ, Anderson JL, Aliot EM, Clark M, Lazzara R: The long QT syndromes: A critical review, new clinical observations and a unifying hypothesis. *Prog Cardiovasc Dis* 1988;31(2):115–172.
15. Keren A, Tzivoni D, Gottlieb S, et al.: Atypical ventricular tachycardia (torsades de pointes) induced by amiodarone: Arrhythmia previously induced by quinidine and disopyramide. *Chest* 1982;81:384–386.
16. Totterman KJ, Turto J, Pellinen T: Overdrive pacing as treatment of sotalol-induced ventricular tachyarrhythmias (torsades de pointes). *Acta Med Scand* 1982;668:28–33 (suppl).
17. Kuck KH, Kunze KP, Roewer N, et al.: Sotalol-induced torsades de pointes. *Am Heart J* 1984;107:179–180.
18. McKibbin JK, Pocock WA, Barrow JR, et al: Sotalol, hypokalemia, syncope and torsades de pointes. *Br Heart J* 1984;51:157–162.
19. Mattioni TA, Zheutlin TA, Dunnington C, Kehoe RF: The proarrhythmic effects of amiodarone. *Prog Cardiovasc Dis* 1989;31:439–446.
20. Mattioni TA, Zheutlin TA, Sarmiento JJ, Parker M, Lesch M, Kehoe RF: The long-term safety and efficacy of amiodarone in patients with prior drug-mediated torsade de pointes. *Ann Intern Med* 1989; 111:574–580.
21. Tzivoni D, Keren A, Cohen AM, et al.: Magnesium therapy for torsades de pointes. *Am J Cardiol* 1984;53:528–530.
22. Fowler NO, McCall D, Chou T, et al.: Electrocardiographic changes and cardiac arrhythmias in patients receiving psychotropic drugs. *Am J Cardiol* 1976;37:223–230.
23. Freedman RA, Anderson KP, Green LS, Mason JW: Effect of erythromycin on ventricular arrhythmias and ventricular repolarization in idiopathic long QT syndrome. *Am J Cardiol* 1987;59:168–169.
24. Isner JM, Roberts WC, Heymsfield SB, et al.: Anorexia nervosa and sudden death. *Ann Intern Med* 1985; 102:49–52.
25. Singh BN, Gaarder TD, Kanegae T, et al.: Liquid protein diets and torsades de pointes. *JAMA* 1978; 240:115–119.
26. Steinbrecher UP, Fitchett DH: Torsade de pointes: A cause of syncope with atrioventricular block. *Arch Intern Med* 1980;140:1223–1226.
27. Grossman MA: Cardiac arrhythmias in acute nervous system disease: Successful management with stellate ganglion block. *Arch Intern Med* 1976;136:203–207.
28. Keating M, Atkinson D, Dunn C, Timothy K, Vincent GM, Leppert M: Linkage of a cardiac arrhythmia, the long QT syndrome, and the Harvey ras-1 gene. Science 1991;252:704–706.
29. Itoh S, Munemura S, Satoh H: A study of the inheritance pattern of Romano-Ward syndrome. *Clin Pediatr* 1982;21:20–24.
30. Wellens HJJ, Vermuelen A, Durrer D: Ventricular fibrillation occurring on arousal from sleep by auditory stimuli. *Circulation* 1972;46:661–665.
31. Schwartz PJ: Idiopathic long QT syndrome. Progress and questions. *Am Heart J* 1985;109:399–411.
32. Bhandari AK, Scheinmann MM, Morady F, et al.: Efficacy of left cardiac sympathectomy in the treatment of patients with the long QT syndrome. *Circulation* 1984;70:1018–1023.
33. Eldar M, Griffin JC, Abbott JA, et al.: Permanent cardiac pacing in patients with the long QT syndrome. *J Am Coll Cardiol* 1987;10:600–607.
34. Cranefield, PF: *The Conduction of the Cardiac Impulse*. Mt. Kisco, NY, Futura Publishing, 1975, pp 9–10.
35. Damiano BP, Rosen MR: Effects of pacing on triggered activity induced by early afterdepolarizations. *Circulation* 1984;69:1013–1025.
36. Cranefield PF, Aronson RS: *Cardiac Arrhythmias: The Role of Triggered Activity and Other Mechanisms*. Mt. Kisco, NY, Futura Publishing, 1988.
37. Davidenko JM, Cohen L, Goodrow R, Antzelevitch C: Quinidine-induced action potential prolongation, early afterdepolarizations, and triggered activity in canine Purkinje fibers: Effects of stimulation rate,

potassium, and magnesium. *Circulation* 1989; 79:674–686.
38. Levine JH, Spear JG, Guarnieri T, et al.: Cesium chloride-induced long QT syndrome: Demonstration of afterdepolarizations and triggered activity in vivo. *Circulation* 1985;72:1092–1103.
39. Ben-David J, Zipes DP: Differential response to right and left stellate stimulation of early afterdepolarizations and ventricular tachycardia in the dog. *Circulation* 1988;78:1241–1250.
40. Nayebpour M, Solymoss BC, Nattel S: Cardiovascular and metabolic effects of caesium chloride injection in dogs—limitations as a model for the long QT syndrome. *Cardiovasc Res* 1989;23:756–766.
41. Gavrilescu S, Luca C: Right ventricular monophasic action potentials in patients with long QT syndrome. *Br Heart J* 1978;40:1014–1018.
42. Hartzler GO, Osborn MJ: Invasive electrophysiological study in the Jervell and Lange-Nielsen syndrome. *Br Heart J* 1981;45:225–229.
43. Bonatti V, Rolli A, Botti G: Monophasic action potential studies in human subjects with prolonged ventricular repolarization and long QT syndromes. *Eur Heart J* 1985;6:131–143.
44. Schechter E, Freeman C, Lazzara R: Afterdepolarizations as a mechanism for the long QT syndrome: Electrophysiologic studies of a case. *J Am Coll Cardiol* 1984;3:1556–1561.
45. Zipes DP: Cardiac electrophysiology: Promises and contributions. *J Am Coll Cardiol* 1989;13:1329–1352.

PART IV

Procedures to Identify Mechanisms, Prognosis, and Therapy

Chapter **19**

Exercise-Induced Ventricular Arrhythmias: Prevalence, Mechanisms, and Prognostic Implications

Paul L. McHenry, MD

Several comprehensive, prospectively designed studies of exercise-induced ventricular arrhythmias in normal or apparently healthy subjects have provided reliable estimates of arrhythmia prevalence and accurate data for assessing clinical significance based on long-term follow-up studies.[1–8] On the other hand, much of the available data dealing with the prevalence and prognostic significance of exercise-induced ventricular arrhythmias in patients with ischemic heart disease have been derived from retrospective analyses of exercise tests performed for purposes other than arrhythmia documentation.[9–14] The original study protocols did not ensure the accurate recording of all arrhythmias during and after exercise, and in this setting the on-line recognition and recording of arrhythmias tends to be biased toward those patients with frequent and complex arrhythmias.

RECOGNITION, CLASSIFICATION, AND CLINICAL CHARACTERISTICS OF EXERCISE-INDUCED VENTRICULAR ARRHYTHMIAS

A significant percentage of ventricular arrhythmias occurring during exercise testing can be missed, even when the technicians and monitoring physicians are following protocols designed for the comprehensive recording of arrhythmias. We have previously studied the accuracy of on-line ventricular arrhythmia detection in our laboratory by means of off-line reviews of taped recordings (Table 19.1). A significant percentage of both single and consecutive premature ventricular complexes were missed during on-line monitoring. There are several unique circumstances that hamper on-line arrhythmia detection during exercise testing. These include the problems of noise and artifact produced by vigorous exercise and the tendency toward fusion complexes as the exercise heart rate increases.

The prevalence, frequency, and complexity of ventricular arrhythmias during exercise vary significantly depending on whether or not patients have premature ventricular complexes at rest. In general, patients with ischemic heart disease and premature ventricular complexes at rest have a greater prevalence of ventricular arrhythmias during exercise and the arrhythmias are more frequent and complex compared to patients without arrhythmias at rest (Table 19.2). The mechanisms and prognostic significance for ventricular arrhythmias occurring during exercise may also differ, depending on whether or not they are present at rest. Most of the reported studies of the prognostic significance of exercise-induced ventricular arrhythmias in patients with ischemic heart disease have not provided a separate analysis of patients with ventricular arrhythmias at rest, and this omission may have influenced the results. An acceptable definition of patients who are free of ventricular arrhythmias at rest has not been established. In our laboratories we have defined this arbitrarily as the absence of ventricular arrhythmias on the baseline 12-lead electrocardiogram (ECG) recording and during a 2-min baseline monitoring period just prior to starting the exercise test.[1] As noted

655 Avenue of the Americas, New York, NY 10010
Current Topics in Cardiology

TABLE 19.1 Detection of Resting and Exercise-Induced Ventricular Arrhythmias During On-Line Monitoring of Exercise Tests Versus Off-Line Playback of Taped Data—338 Patients With Known or Suspected Heart Diseases

Detection Mode	Number With EIVA	Number With Consecutive PVC	
		Couplets	VT
On-line monitoring	194 (57%)	38 (11%)	8 (2%)
Tape playback	240 (71%)*	55 (16%)*	13 (4%)+

EIVA = exercise-induced ventricular arrhythmias, PVC = premature ventricular complexes, VT = ventricular tachycardia.
* $p < 0.001$, + $p < 0.01$.

TABLE 19.2 Prevalence and Types of Ventricular Arrhythmias (VA) Observed During Treadmill Exercise Testing of Coronary Heart Disease Patients With and Without Premature Ventricular Complexes (PVC) at Rest

Type of VA During Exercise	197 Patients Free of VA at Rest	79 Patients With Single PVC at Rest	Combined 276 Patients
Uniform PVC	45 (23%)	20 (25%)	65 (24%)
Multiform PVC	7 (4%)	31 (39%)	38 (14%)
Consecutive PVC	5 (3%)	8 (10%)	13 (5%)

in Table 19.2, even this liberal definition has a significant bearing on the recorded prevalence and complexity of exercise-induced ventricular arrhythmias.

The prevalence, frequency, and complexity of exercise-induced ventricular arrhythmias increase with age and with increasing exercise heart rates in both normal subjects and patients with coronary heart disease (Tables 19.3 and 19.4).[1–3,6–8] Whether this relationship of increasing ventricular arrhythmias with increasing heart rates is related directly to the increase in heart rate or to the concomitant increases in plasma catecholamine levels and β-adrenergic stimulation of the heart is not clear. As higher heart rates are reached during progressive exercise, there is also a tendency for exercise-induced ventricular arrhythmias to be suppressed in both normal subjects and patients with coronary heart disease.[1] The suppression of ventricular arrhythmias with increasing heart rates is most likely due to the shortening of the sinus R-R interval below the sinus complex–premature ventricular complex coupling interval, a form of overdrive suppression. Quite often one can observe varying degrees of fusion of the premature ventricular complexes with the sinus complexes before the ventricular premature complexes are completely suppressed. These varying effects of exercise heart rate on ventricular arrhythmias have to be considered when studying the effects of antiarrhythmic medication on exercise-induced ventricular arrhythmias. If the frequency and complexity of exercise-induced ventricular arrhythmias appear to decrease after administration of a medication that attenuates the heart rate response to exercise compared to the control state, one cannot be cer-

TABLE 19.3 Prevalence of Ventricular Arrhythmias During Maximal Exercise Testing of Indiana State Policemen

Age Group (year)	Patients Studied	All PVC	Frequent PVC*	Multiform PVC	Consecutive PVC	Heart Rate—Onset of PVC	
						<150/min	>150/min
25 to 34							
N	266	77 (29%)	22 (30%)	3 (1.1%)	7 (2.6%)	19 (24%)	58 (76%)
CVD	16	7 (44%)	4 (57%)	0	0	5 (70%)	2 (30%)
35 to 44							
N	237	81 (34%)	26 (32%)	8 (3.4%)	15 (6.7%)	20 (25%)	61 (75%)
CVD	46	25 (54%)	12 (48%)	3 (6.5%)	8 (17%)	15 (58%)	10 (42%)
45 to 54							
N	58	25 (43%)	13 (52%)	1 (1.8%)	4 (6.8%)	7 (26%)	18 (74%)
CVD	27	13 (48%)	9 (69%)	3 (11%)	6 (22%)	8 (62%)	5 (38%)

* = more than six PVC per minute during at least 1 min of exercise, N = normals, CVD = known or suspected cardiac disease.

TABLE 19.4 Cumulative Prevalance of Ventricular Arrhythmias During Progressive Increases in Exercise Heart Rate in 561 Clinically Normal Males (Age Range 25 to 54 years) Free of Ventricular Arrhythmias at Rest

Exercise Arrhythmia	Exercise Heart Rates			
	≤130/min	≤150/min	≤170/min	>170 or Recovery
Any PVC	34 (6%)	90 (16%)	157 (28%)	183 (32.6%)
Multiform PVC	1 (0.2%)	4 (0.7%)	8 (1.4%)	12 (2.1%)
Consecutive PVC	0	4 (0.7%)	16 (2.9%)	26 (4.6%)

PVC = premature ventricular complexes.

tain whether this represents a primary antiarrhythmic effect of the medication or a secondary effect resulting from the lower exercise heart rate. Suppression of ventricular arrhythmias due to attainment of a higher exercise heart rate during a serial test could also lead to misleading conclusions.

PREVALENCE AND CLINICAL SIGNIFICANCE OF EXERCISE-INDUCED VENTRICULAR ARRHYTHMIAS IN APPARENTLY HEALTHY SUBJECTS

Rare to occasional premature ventricular complexes are observed quite often during maximal or near-maximal exercise testing of apparently healthy subjects.[1–8] Both the prevalence and the complexity of exercise-induced ventricular arrhythmias increase as the exercise work load and heart rate approach maximal levels (Table 19.4). Ventricular arrhythmias sometimes appear only in the immediate recovery period following a maximal or near-maximal effort.[1–8,15,16] This well-recognized phenomenon may be related to a combination of factors that include a rise in plasma catecholamine levels upon termination of exercise[15] and a slowing of the heart rate that unmasks overdrive suppression of ventricular arrhythmias during peak exercise.[16] The prevalence and complexity of exercise-induced premature ventricular complexes also increase with age.[1–3] In our studies of asymptomatic Indiana State Policemen, the prevalence and complexity of exercise-induced ventricular arrhythmias was greater in those subjects who had nonspecific baseline ECG abnormalities, hypertension, valvular abnormalities, or suspected coronary heart disease (Table 19.3).[2] The prevalence of ventricular arrhythmia with exercise was 33% in 561 apparently healthy subjects compared to 51% in 89 subjects with known or suspected heart disease. Frequent ventricular premature complexes defined as more than 6 per minute for at least 1 min of exercise were observed in 11% of the apparently healthy subjects compared to 28% in the heart disease group. In the apparently healthy subject group, ventricular couplets were observed in 3.5% and ventricular tachycardia, defined as three or more consecutive premature ventricular complexes, in 1.1%. In the heart disease group, couplets and ventricular tachycardia were observed in 10% and 5.6%, respectively. Our laboratory has been following members of the Indiana State Police Department with serial exercise testing for 22 years and our experience has been that ventricular tachycardia of more than three in a row is uncommon, occurring in 0.5% of 5,000 exercise tests. We have never encountered sustained ventricular tachycardia requiring immediate therapy in this population.

The total prevalence of exercise-induced ventricular arrhythmias has proven to be reproducible during serial exercise testing of our Indiana State Police subjects, but the reproducibility of ventricular arrhythmias in any given subject during two consecutive tests is highly variable (Table 19.5).[3,4] The reproducibility of complex or consecutive premature ventricular complexes with consecutive tests is also poor.[3,17] The prevalence of ventricular arrhythmias detected in at least one of three consecutive tests approaches 90% in these subjects. Our long-term follow-up studies of the Indiana State Policemen have demonstrated no predictive value of exercise-induced ventricular arrhythmias for future coronary events.[18] Other studies of normal populations have also found that exercise-induced ventricular arrhythmias are of no prognostic significance for future coronary events.[5,7,8,19] Recent studies of an asymptomatic, community-dwelling population over a broad age range have reported that even complex

TABLE 19.5 Reproducibility of Exercise-Induced Ventricular Arrhythmia in Healthy Male Subjects During Two Serial Maximal Exercise Tests Performed an Average of 2.9 Years Apart (Range 1 to 4 Years)

		Prevalence of Arrhythmias		Men with Arrhythmias on	
Age Group (year)	Patients Studied	Test 1	Test 2	Test 1 Without Test 2	Test 2 Without Test 1
25 to 34	217	64 (30%)	78 (36%)	29 (45%)	46 (59%)
35 to 44	200	64 (32%)	76 (38%)	27 (42%)	45 (58%)
45 to 54	45	16 (36%)	19 (42%)	6 (38%)	11 (52%)

ventricular arrhythmias, including nonsustained ventricular tachycardia, are not predictive of future cardiovascular morbidity or mortality.(7,8)

PREVALENCE AND CLINICAL SIGNIFICANCE OF EXERCISE-INDUCED VENTRICULAR ARRHYTHMIAS IN PATIENTS WITH STABLE CORONARY HEART DISEASE

The reported prevalence of ventricular arrhythmias during exercise testing of patients with stable coronary heart disease has varied from 10 to 27%, which is considerably lower than that reported for apparently healthy subjects.(1,9–14) This seeming discrepancy is even more remarkable given the fact that most of these studies have included patients with ventricular arrhythmias at rest. The two studies that excluded patients with ventricular arrhythmias at rest were prospectively designed and both reported a 27% prevalence of exercise-induced ventricular arrhythmias.(1,12) The majority of the other studies were carried out by retrospective analyses of exercise tests performed primarily for reasons other than the comprehensive collection of arrhythmia data and infrequent or simple ventricular arrhythmias may have been missed in some patients. A major factor influencing the relatively low prevalence of ventricular arrhythmias in patients with stable coronary heart disease is the lower exercise heart rates attained by these patients. Our study(1) compared the prevalence of exercise-induced ventricular arrhythmias in 197 patients with documented coronary artery disease to that in 144 age-matched, clinically normal subjects from our Indiana State Police population. All subjects were free of arrhythmias at rest. The mean exercise heart rate for the coronary disease group was 128 per minute compared to 179 per minute for the normal group. The total prevalence of ventricular arrhythmias in the normal group was 44% compared to 27% in the coronary disease group. However, when the appearance of ventricular arrhythmias was correlated with exercise heart rate, all patients in the coronary disease group demonstrated arrhythmias at or below a heart rate of 130 per minute compared to only 6% of the normal group. Only 15% of the normal group demonstrated ventricular arrhythmias before reaching a heart rate of 150 per minute. Our study and the study of Nair et al.(12) excluded patients on β-blockers and antiarrhythmic drugs while the other cited studies(9–11,13,14) did not, and these differences may have influenced the observed prevalences of exercise-induced ventricular arrhythmias.

In our comparative study of normal subjects and patients with stable coronary heart disease, the prevalence and frequency of complex or consecutive ventricular arrhythmias was greater in the coronary disease group even when the exercise heart rate attained was not taken into account.(1) In the coronary heart disease group, the frequency and complexity of ventricular arrhythmias was greater in those patients with triple vessel coronary artery disease and/or significant left ventricular dysfunction and this correlation has been observed by other investigators.(9–11,14) We observed no significant difference in the prevalence or complexity of ventricular arrhythmias between patients with and without exercise-induced ST depression, and this has been the finding of others as well.(9,11,12,20,21) Successful coronary bypass surgery does not alter the prevalence of exercise-induced ventricular arrhythmias,(22–26) and some studies have actually found a higher prevalence after coronary bypass surgery.(24–26) There does appear to be a very small subset of patients with coronary artery disease who have ventricular arrhythmias secondary to exercise-induced ischemia, and not infrequently

they demonstrate more serious ventricular arrhythmias. Coronary revascularization can lead to an abolition or a reduction in exercise-induced ventricular arrhythmias in this subset.[27]

The suppression of resting ventricular arrhythmias during exercise is a nonspecific finding that is observed in patients with significant heart disease as well as apparently healthy subjects. Likewise, suppression of exercise-induced ventricular arrhythmia during progressive exercise as higher heart rates are attained is observed as often in patients with coronary heart disease as in normal subjects.[1]

There have been several studies of the prognostic significance of exercise-induced ventricular arrhythmias in patients with chronic stable coronary heart disease and the reported results have been conflicting.[9–14] As previously noted, many of these studies employed retrospective analysis of exercise tests performed for other purposes, and the patients were not subdivided with regard to the presence or absence of ventricular arrhythmias at rest. Table 19.6 lists the findings of five studies dealing with reasonably large numbers of patients.[9–12,14] The first three studies found no differences in the mortality rates between patients with and without exercise-induced ventricular arrhythmias during long-term follow-up. Califf et al.[9] reported a significantly higher mortality rate in patients with exercise-induced ventricular arrhythmias, and patients who had consecutive premature ventricular complexes during exercise had a higher mortality rate than did those with only single premature ventricular complexes. Exercise-induced ventricular arrhythmias added independent prognostic information to the noninvasive evaluation of the patient (history, physical examination, chest roentgenogram, electrocardiogram, and other exercise test variables) but ventricular arrhythmias made no independent contribution once the cardiac catheterization data were known. They found a close relationship between ventricular arrhythmias with exercise and the presence of left ventricular dysfunction, and they concluded that the arrhythmias may be a clinical marker for left ventricular dysfunction rather than an independent marker of prognosis. The Milis Study Group[14] also reported a significant, positive correlation between ventricular arrhythmias during exercise and the subsequent 1-year mortality rate in patients studied 6 months after myocardial infarction. Other variables predictive of subsequent mortality in these patients included ST-segment elevation with exercise, inadequate blood pressure response to exercise, and a very limited exercise capacity. These patients had no cardiac catheterization or noninvasive studies of left ventricular function but the authors observed that the exercise test variables predictive of increased mortality were frequently associated with abnormal left ventricular function.

Many studies have evaluated the prognostic significance of exercise-induced ventricular arrhythmias in patients early after acute myocardial infarction,[28–33] and several of these have reported that ventricular arrhythmias predicted a subsequent increase in cardiac morbidity and mortality. Other adverse prognostic variables identified by these studies included poor exercise capacity, inadequate blood pressure responses to exercise, a history of previous myocardial infarction, and a history of congestive heart failure. Therefore, the data suggest that exercise-induced ventricular arrhythmias in this population sometimes represent markers of significantly impaired left ventricular function, and that the occurrence of asymptomatic ventricular arrhythmias with exercise in the absence of left ventricular dysfunction may be of no clinical significance.

TABLE 19.6 Prognostic Significance of Exercise-Induced Ventricular Arrhythmias in Patients With Chronic Stable Coronary Heart Disease

Study	Number	Follow-Up (years)	Mortality (%) in Patients With EIVA	Mortality (%) in Patients No EIVA
Nair et al. (1983)[12]	280	3.9	No difference between groups	
Sami et al. (1984)[11]	1,486	4.3	29 (any PVC)	24 (NS)
Weiner et al. (1984)[10]	446	5.3	14 (any PVC)	10 (NS)
Califf et al. (1983)[9]	620	3.0	17 (single PVC)	10 (<0.05)
			25 (consecutive PVC)	10 (<0.01)
Milis Study Group (1986)	473	1.0	8 (any PVC)	3 (<0.01)

EIVA = exercise-induced ventricular arrthymias, PVC = premature ventricular complexes.

THE VALUE OF EXERCISE TESTING FOR STUDYING PATIENTS WITH KNOWN OR SUSPECTED SYMPTOMATIC VENTRICULAR ARRHYTHMIAS

The value of exercise testing for exposing ventricular arrhythmias in patients with known or suspected symptomatic arrhythmias is in general limited. In patients with episodes of spontaneous ventricular tachyarrhythmias but relatively infrequent ventricular arrhythmias between episodes, exercise testing is not a very reliable technique for inducing or documenting the arrhythmias.[20,34–38] There is a very small subset of patients in which exercise testing represents the most reliable and sometimes the only means for exposing symptomatic ventricular arrhythmias.[39–42] Some of these patients can be identified through a clinical history of exercise-related symptoms. Patients who have frequent and often complex ventricular arrhythmias between episodes of symptomatic ventricular tachyarrhythmia are more susceptible to arrhythmia induction during exercise, and in this subset serial exercise testing may be helpful in assessing the efficacy of antiarrhythmic drug therapies.[39,43,44] Some investigators have reported that exercise testing is useful in identifying those patients at increased risk after successful resuscitation from ventricular fibrillation or cardiac arrest.[39,43] However, this is again dependent upon the clinical setting and the frequency of ventricular arrhythmias between symptomatic episodes. Weaver et al.[36] studied 90 patients following resuscitation from out-of-hospital ventricular fibrillation and found no correlation between the occurrence of ventricular arrhythmias with maximal exercise testing and subsequent recurrence of cardiac arrest.

In our experience, symptom-limited exercise testing of patients who have experienced spontaneous episodes of symptomatic ventricular tachycardia or ventricular fibrillation is of very limited value for exposing these arrhythmias. This is especially true in those patients who do not demonstrate recurrent ventricular tachyarrhythmias during in-hospital electrocardiographic monitoring prior to exercise testing. We have previously reported a study comparing the induction of ventricular tachycardia by exercise testing versus programmed ventricular stimulation in patients with one or more episodes of spontaneous, symptomatic ventricular tachycardia or ventricular fibrillation.[34] Patients with a clinical history of exercise-induced ventricular tachycardia were excluded. The clinical characteristics of the 118 study patients are listed in Table 19.7. The limiting symptom with exercise testing was dyspnea or fatigue in the majority of the patients irrespective of the type of underlying heart disease or their functional classification. Over one-half of the patients had presented with sustained ventricular tachycardia or ventricular fibrillation and yet ventricular tachycardia was induced by exercise in only 26%. In the vast majority, the exercise-induced ventricular tachycardia consisted of three- or four-beat runs. None of the patients required electrical cardioversion for sustained ventricular tachyarrhythmias during exercise.

The treadmill exercise test and clinical characteristics of the 31 patients with exercise-induced ventricular tachycardia and the 87 patients without are compared in Table 19.8. The mean exercise heart rate and peak systolic blood pressure were not significantly different for the two groups. The mean exercise duration was shorter for the ventricular tachycardia group but this was influenced by the fact that the exercise test was terminated in 21 patients because of asympto-

TABLE 19.7 Clinical Characteristics of 118 Patients With Symptomatic Ventricular Tachyarrhythmias Who Were Studied With Baseline Maximal Exercise Testing and Programmed Ventricular Stimulation

Sex	
82 Men (70%), 36 Women (30%)	
Age	
Mean 50 years (range 15 to 73)	
Presenting arrhythmia	
VF	30 (26%)
Sustained VT	27 (23%)
Nonsustained VT	61 (51%)
Type of heart disease	
Coronary	60 (51%)
Cardiomyopathy	23 (20%)
Mitral valve prolapse	15 (12%)
Primary electrical	20 (17%)
Functional NYHA classification	
1	55 (47%)
2	52 (44%)
3	11 (9%)
4	0
Ventricular arrhythmias with exercise testing	
None	39 (33%)
Single PVC	29 (25%)
Paired PVC	19 (16%)
VT	31 (26%)
VF	0

VF = ventricular fibrillation, VT = ventricular tachycardia, PVC = premature ventricular complexes, NYHA = New York Heart Association.

TABLE 19.8 Treadmill Exercise Test and Clinical Characteristics For Patients With and Without Exercise-Induced Ventricular Tachycardia

	EIVT Group (N = 31)	p Value	No EIVT Group (N = 87)
Treadmill exercise data (mean)			
Peak systolic blood pressure (mm Hg)	163	(NS)	169
Maximal heart rate (bpm)	133	(NS)	138
Exercise duration (s)	378*	(<0.05)	466
Number of patients according to type of heart disease			
Coronary	15	(NS)	45
Cardiomyopathy	7	(NS)	16
Mitral valve prolapse	5	(NS)	10
Primary electrical	4	(NS)	16
Number of patients according to presenting rhythm disturbance			
Nonsustained VT	16	(NS)	45
Sustained VT	9	(NS)	18
Ventricular fibrillation	6	(NS)	24

* Exercise terminated due to VT in 21 patients.

matic ventricular tachycardia. The type of underlying heart disease and the presenting rhythm disturbance did not differ between the two groups. Table 19.9 compares the incidence of ventricular tachycardia induction by exercise testing and by programmed ventricular stimulation. Of the 31 patients (26%) with exercise-induced ventricular tachycardia, 10 (8%) were negative with programmed ventricular stimulation. Ventricular tachycardia was induced by programmed ventricular stimulation in 83 patients (70%) and in 62 (53%) of these patients the exercise test was negative. Both tests were negative in 25 patients (21%). The incidence of ventricular tachycardia induction was greater for programmed ventricular stimulation than exercise testing for all types of heart disease but the difference was significant only for the coronary heart disease patients. Programmed ventricular stimulation was significantly more effective than exercise testing for induction of ventricular tachycardia for all types of presenting arrhythmia.

The low yield of ventricular tachycardia induction by exercise testing in our study population was consistent with the finding of other investigators,[35–38] although some have reported a much better yield.[39,43,44] The different results reported by various studies are most likely related to differences in patient selection. All of our patients were hospitalized and being monitored in a telemetry unit at the time of study and patients who demonstrated ventricular tachycardia (three or more consecutive premature ventricular complexes) during a 12- to 24-h baseline period were excluded from exercise testing. Had we not excluded these patients, our yield with exercise testing would undoubtedly have been higher. However, this study clearly demonstrates that exercise testing is not an effective technique for exposing ventricular tachyarrhythmias in patients who have infrequent, spontaneous episodes of these arrhythmias.

TABLE 19.9 Comparison of Ventricular Tachycardia Induction By Exercise Testing and Programmed Ventricular Stimulation in 118 Patients

For Total Population (N = 118)	ET + Only	PVS + Only	Both +	Neither +
	10 (8%) ET Positive*	62 (53%) PVS Positive*	21 (18%)	25 (21%) p Value
By cardiac disease (no.)				
Coronary (60)	15 (25%)	55 (92%)		(<0.001)
Cardiomyopathy (23)	6 (26%)	10 (43%)		NS
Mitral valve prolapse (15)	3 (20%)	8 (53%)		NS
Primary electrical (20)	7 (35%)	10 (50%)		NS
By presenting arrhythmia (no.)				
Nonsustained VT (61)	16 (26%)	38 (62%)		(<0.001)
Sustained VT (27)	10 (37%)	25 (93%)		(<0.001)
Ventricular fibrillation (30)	5 (17%)	20 (68%)		(<0.001)

+ = positive, * = total numbers may be less than or more than 100% because some patients were in the neither + or both + groups, VT = ventricular tachycardia, ET = exercise testing, PVS = programmed ventricular stimulation.

Another potential use of exercise testing is to evaluate the efficacy of antiarrhythmic medications. However, the reproducibility of exercise-induced ventricular arrhythmias during serial, baseline testing is quite variable depending on clinical circumstances.[3,17,39,40,43–46] As one might expect, patients with frequent and complex ventricular arrhythmias at rest or during casual activities tend to have reproducible ventricular arrhythmias with serial exercise testing and in this subset of patients, exercise testing may be of value in assessing the efficacy of antiarrhythmic drug therapies. Within the group of patients with recurrent clinical episodes of symptomatic ventricular tachycardia, there is a small subset of about 10% of patients in whom exercise testing represents the most reliable and sometimes the only technique for exposing the arrhythmia.[37–42] However, before one can rely on exercise testing to evaluate efficacy of antiarrhythmic drug therapies, reproducibility of the exercise-induced ventricular tachycardia must be documented by means of serial baseline testing.[17,40,46,47] The duration of the exercise-induced ventricular tachycardia seems to have some influence on the reproducibility of the arrhythmia with serial testing.[40] However, even sustained and symptomatic ventricular tachycardias may not be reproducible in some patients. In our experience, patients with no history of symptomatic ventricular arrhythmias who exhibit an unexpected run of asymptomatic ventricular tachycardia during maximal or near-maximal exercise do not require antiarrhythmic medications if the arrhythmia cannot be reproduced with repeat testing, which is often the case.

MECHANISMS AND THERAPY OF EXERCISE-INDUCED VENTRICULAR TACHYCARDIA

Of those patients with recurrent ventricular tachyarrhythmias that are exercise-induced, only a very small percentage manifest the arrhythmias as a result of exercise-induced myocardial ischemia. Sung et al.[41,42] recently reviewed the electrophysiologic mechanisms for nonischemic, exercise-induced ventricular tachycardia. The milieu thought to be common to all of these arrhythmias is an increase in sympathetic tone and plasma catecholamines with exercise but there are three potential electrophysiologic mechanisms: (a) reentry, (b) catecholamine-sensitive automaticity, and (c) triggered activity due to delayed afterdepolarizations. These authors categorized patients with exercise-induced ventricular tachycardias into verapamil-unresponsive (group I) and verapamil-responsive (group II) based on the response of the ventricular tachycardia to intravenous verapamil. Within the verapamil-unresponsive group they described two subsets. Patients in group IA have ventricular tachycardia, which is easily induced by programmed ventricular stimulation, and the mechanism for the arrhythmia is felt to be reentry. The arrhythmias in group IB patients cannot be induced by programmed ventricular stimulation, but they can be with intravenous isoproterenol, and the mechanism for the ventricular tachycardia seems best explained by enhanced, catecholamine-sensitive automaticity.

Patients in group IA tend to have significant myocardial disease, and paroxysms of sustained, monomorphic ventricular tachycardia are inducible by programmed ventricular stimulation but not by incremental atrial or ventricular pacing. The pattern of the ventricular tachycardia on the 12-lead ECG may have either a left or a right bundle branch block morphology. The electrophysiologic mechanism is felt to be reentry related to areas of slow conduction from depressed fast sodium (Na^+) channels, and verapamil or propranolol have no effect on the ventricular tachycardia. In contrast, patients in group IB may have no structural heart disease, and the ventricular tachycardia cannot be induced by programmed ventricular or atrial stimulation. The pattern of the ventricular tachycardia usually has a left bundle branch block morphology. Ventricular tachycardia inducibility can be abolished by either intravenous procainamide (depression of phase 4 depolarization) or intravenous propranolol (β-adrenergic blockade) but not by intravenous verapamil. Patients with exercise-induced ventricular tachycardia generally do not have abnormally high levels of plasma catecholamines during exercise and the group IB patients more likely possess a low sympathetic threshold and/or an arrhythmia-prone anatomic substrate.[41,42] Although enhanced, catecholamine-sensitive automaticity is the most likely mechanism for the exercise-induced ventricular tachycardia in these patients, the authors noted that β-adrenergic blocking agents may exert other electrophysiologic effects on both ischemia- and nonischemia-induced ventricular tachycardia. Therefore, ventricular tachycardias that are triggered both by exercise and by intravenous infusion of isoproterenol may be responsive to β-adrenergic blockade irrespective of the underlying mechanism for the arrhythmia.

Patients with verapamil-responsive ventricular tachycardia (group II) are usually younger with no clinical evidence for structural heart disease. The ventricular tachycardia can be induced

by atrial or ventricular pacing as well as exercise, a critical range of cardiac cycle length is usually required, and the tachycardia is initiated by a late coupled premature ventricular complex. The ECG pattern of the ventricular tachycardia most often has a left bundle branch block morphology. This cycle length–dependent ventricular tachycardia can be terminated by overdrive suppression similar to group IA but can also be completely suppressed by intravenous verapamil. The authors felt the mechanism for this form of exercise-induced ventricular tachycardia is most likely a triggered rhythm due to delayed afterdepolarizations. When a critical range of cardiac cycle length is reached, the delayed afterpotentials reach threshold and initiate an action potential. The ionic currents believed responsible for the genesis of delayed afterpotentials are carried by Na^+ and modulated by calcium ions (Ca^{2+}). Intravenous β-adrenergic blockade and type IA antiarrhythmic agents only slow the rate of the ventricular tachycardia in these patients, while verapamil can terminate the arrhythmia or totally suppress ventricular tachycardia inducibility. An alternative mechanism for the ventricular tachycardia in this group of patients is reentry involving Ca^{2+} channel-dependent slow responses within the reentrant circuit.

The authors conclude that different mechanisms for exercise-induced ventricular tachycardias can only be inferred but the clinical and electrophysiologic characteristics of the arrhythmias in these three groups are distinctively different as are their responses to various pharmacologic agents.

Woelfel et al.[(40)] described 11 patients with exercise-induced ventricular tachycardia that was reproducible with consecutive exercise tests. They observed that patients who had sustained ventricular tachycardia or multiple episodes of ventricular tachycardia during the initial exercise test were more likely to have reproducible arrhythmias. Ventricular tachycardia inducibility was suppressed with oral β-adrenergic blocking agents in 10 of the patients, and 8 of these demonstrated a consistent relationship between a critical exercise heart rate and the onset of ventricular tachycardia. Successful therapy correlated with prevention of this critical heart rate during subsequent exercise tests. Of the 11 patients studied, 6 had no demonstrable heart disease while 5 had coronary heart disease. Those without structural heart disease had pure exercise-induced ventricular tachycardia, while those with coronary heart disease demonstrated spontaneous ventricular tachycardia as well and none had exercise-induced symptoms by history. The authors felt that the critical sinus rate–ventricular tachycardia relationship in these patients suggested a mechanism of triggered activity due to delayed afterdepolarizations, and that control of the arrhythmia with β-adrenergic blockade could be explained by a decrease in the magnitude of the delayed afterpotentials produced by slowing of the heart rate. Reported suppression of exercise-induced ventricular tachycardia by verapamil, which is known to attenuate delayed afterpotentials, is also consistent with this mechanism. However, another potential explanation is that the heart rate increase does not directly initiate ventricular tachycardia but simply identifies the exercise-induced adrenergic state necessary for tachycardia induction. The importance of β-adrenergic influences on exercise-induced tachycardia is further suggested by the response of some of these patients to isoproterenol infusion. Sung et al.[(41,42)] implicated enhanced, spontaneous discharge of a catecholamine-sensitive automatic focus as the mechanism for the isoproterenol inducible ventricular tachycardias (Sung group IB), but these authors concluded that the multiple electrophysiologic effects of β-adrenergic stimulation make it difficult to exclude either reentry or triggered activity due to afterpotentials.

The clinical, electrophysiologic, and pharmacologic characteristics of the patients reported by Woelfel et al.[(40)] seem to fit both group IB and group II of Sung's classification[(42)] and therefore this classification may be of limited clinical value. In a subsequent report, Woelfel et al.[(47)] described a very similar group of 16 patients with reproducible, exercise-induced ventricular tachycardia, 12 of whom responded to intravenous verapamil and 8 to oral verapamil. The majority of these patients did not demonstrate a critical sinus heart rate–ventricular tachycardia relationship (Sung group II), but in those who did, verapamil was effective even when the exercise heart rate exceeded the heart rate that precipitated ventricular tachycardia prior to initiation of verapamil therapy. Other clinical characteristics that identified those patients responding to verapamil included an absence of structural heart disease and monomorphic ventricular tachycardias with a left bundle branch block morphology. Verapamil was effective in some patients who also had episodes of spontaneous ventricular tachycardia, but it was usually ineffective in patients in whom tachycardias were inducible by premature ventricular stimulation (Sung group IA). The authors postulated that the mechanism for the exercise-induced ventricular tachycardia in these verapamil-responsive patients may somehow involve the slow inward Ca^{2+} current (i_{Ca}) that is inhib-

ited by verapamil, resulting in direct suppression of delayed afterdepolarizations. However, reentry or enhanced catecholamine-sensitive automaticity could not be excluded.

Lerman et al.[48] described 4 patients with structurally normal hearts and exercise-induced ventricular tachycardia that could be terminated by adenosine. The ventricular tachycardia could also be induced by isoproterenol infusion or atrial pacing. The morphology of the ventricular tachycardias was of the left bundle branch block type. The arrhythmias could also be prevented or terminated by β-adrenergic blockade, verapamil infusion, and various vagal maneuvers. In contrast, a group of 14 patients with ventricular tachycardia consistent with a reentry mechanism did not respond to adenosine, verapamil, or β-adrenergic blocking agents. The authors postulated that the mechanism for the adenosine-sensitive ventricular tachycardia was catecholamine-induced triggered activity caused by intracellular Ca^{2+} overload mediated by elevations in intracellular cyclic AMP. β-Receptor stimulation is known to induce an increase in cyclic AMP leading to an increase in intracellular Ca^{2+} and adenosine antagonizes this reaction. Verapamil, β-adrenergic blockade, and vagal maneuvers (acetylcholine) are all known to decrease slow-inward Ca^{2+} current either directly by blocking Ca^{2+} channels (verapamil) or indirectly by inhibiting production of cyclic AMP (β-adrenergic blockers, acetylcholine). The authors felt that the triggered activity could best be explained on the basis of initiation of delayed afterdepolarizations caused by overload of intracellular Ca^{2+} and mediated by catecholamines and elevations of cyclic AMP, and that the inhibition of this response is the only known electrophysiologic mechanism that could be attributed to adenosine.

Eldar et al.[49] recently reported three cases of exercise-induced double (atrial and ventricular) tachycardia. The double tachycardia could also be induced by isoproterenol infusion but not by programmed atrial or ventricular stimulation. The onset of ventricular tachycardia, which had a left bundle branch block morphology, was preceded by the onset of ectopic atrial tachycardia. Two of the patients were young, with no structural heart disease, and both had a history of syncope or sudden death. The third patient had coronary artery disease with previous myocardial infarction. Intravenous adenosine was administered to one of these patients and was effective in terminating the exercise-induced ventricular tachycardia. The ventricular tachycardia in these patients seemed to be triggered when either a sinus or atrial tachycardia reached a critical rate. One patient was treated successfully with oral propranolol, which prevented the sinus rate from reaching the critical range. The authors postulated that the most likely mechanism for these double tachycardias was triggered activity due to rate-dependent augmentation of delayed afterdepolarization. However, they could not exclude triggered activity due to catecholamine-induced automaticity. Of interest is the report of Sokoloff et al.[50] describing 10 patients with reproducible, exercise-induced ventricular tachycardia who responded to maximal intravenous β-adrenergic blockade with propranolol and plasma norepinephrine levels were significantly decreased compared with levels before β-adrenergic blockade.

Other investigators have described groups of patients with exercise or isoproterenol-induced ventricular tachycardia with a left bundle branch block morphology consistent with an origin from the right ventricular outflow tract.[51–54] Some patients had no evidence of structural heart disease while others had findings consistent with arrhythmogenic right ventricular dysplasia.[54] In the latter group, the ventricular tachycardia was often induced by programmed ventricular stimulation, but patients with and without evidence of structural heart disease often responded to β-adrenergic blockade. Therefore, one cannot exclude reentry as the mechanism for those exercise-induced ventricular tachycardias with the clinical, electrophysiologic and pharmacologic characteristics of Sung's groups IB and II.

SUMMARY

Exercise-induced ventricular arrhythmias are commonly observed in normal subjects as well as in patients with coronary heart disease. The prevalence of arrhythmias increases in both populations with age and with higher exercise heart rates. The prevalence, frequency, and complexity of exercise-induced ventricular arrhythmias is greater in individuals with premature ventricular complexes at rest. In patients with coronary heart disease, the exercise-induced ventricular arrhythmias occur at lower work loads and exercise heart rates and the frequency and complexity of the arrhythmias is greater compared to normal subjects. Some studies have found that patients with coronary heart disease and exercise-induced ventricular arrhythmias are at increased risk for subsequent coronary mortality but these studies have not excluded patients who have resting ventricular arrhythmias as well. Existing data suggest that exercise-induced ventricular arrhythmias are not independent prognostic markers but simply reflect the presence of ad-

vanced coronary artery disease and left ventricular dysfunction. Exercise-induced ventricular arrhythmias in otherwise healthy subjects are of no prognostic significance.

The clinical value of exercise testing for exposing patients prone to symptomatic ventricular tachyarrhythmias has proven to be disappointing, especially in those patients who have rare or intermittent arrhythmias at rest. In those patients who demonstrate potentially significant ventricular arrhythmias with exercise, the poor reproducibility of the arrhythmias with serial testing limits the value of exercise testing for evaluating efficacy of antiarrhythmic drug therapies. There is a small subset of about 10% of patients with recurrent, symptomatic ventricular tachyarrhythmias in whom exercise testing may be the best and sometimes the only reliable means for exposing the arrhythmia. If reproducibility of the arrhythmia can be demonstrated in these patients, exercise testing can also be used to assess efficacy of therapeutic interventions.

There is an interesting group of patients who have reproducible, exercise-induced ventricular tachycardia in the absence of ischemic heart disease. These patients tend to be relatively young, are often free of structural heart disease, and the ventricular tachycardia usually has a left bundle branch block morphology consistent with an origin from the right ventricular outflow tract area. The inducibility of the ventricular tachycardia by exercise or isoproterenol infusion is consistent with catecholamine-induced automaticity or triggered activity due to delayed afterdepolarizations. Suppression of the arrhythmias by β-adrenergic blocking agents and verapamil is also consistent with one of the above mechanisms although a reentry mechanism cannot be excluded.

ACKNOWLEDGMENTS

This work was supported in part by the Herman C. Krannert Fund; by grants HL-06308 and HL-07182 from the National Heart, Lung, and Blood Institute of the National Institutes of Health, U.S. Public Health Service; and the American Heart Association, Indiana Affiliate, Inc.

REFERENCES

1. McHenry PL, Morris SN, Kavalier M, Jordan JW: Comparative study of exercise-induced ventricular arrhythmias in normal subjects and patients with documented coronary artery disease. *Am J Cardiol* 1976;37:609–616.
2. McHenry PL, Fisch C, Jordan JW, Corya BR: Cardiac arrhythmias observed during maximal treadmill exercise testing in clinically normal men. *Am J Cardiol* 1972;29:331–336.
3. Faris JV, McHenry PL, Jordan JW, Morris SN: Prevalence and reproducibility of exercise-induced ventricular arrhythmias during maximal exercise testing in normal men. *Am J Cardiol* 1976;37:617–622.
4. Ekblom B, Hartley LH, Day WC: Occurrence and reproducibility of exercise-induced ventricular ectopy in normal subjects. *Am J Cardiol* 1979;43:35–40.
5. Froelicher VF, Thomas MM, Pillow C, Lancaster MC: Epidemiologic study of asymptomatic men screened by maximal treadmill testing for latent coronary artery disease. *Am J Cardiol* 1974;34:770–777.
6. Blackburn H, Taylor HL, Hamrell B, Buskirk E, Nicholas WC, Thorsen RD: Premature ventricular complexes induced by stress testing. Their frequency and response to physical conditioning. *Am J Cardiol* 1973;31:441–449.
7. Fleg JL, Lakatta EG: Prevalence and prognosis of exercise-induced nonsustained ventricular tachycadia in apparently healthy volunteers. *Am J Cardiol* 1984;54:762–764.
8. Busby MJ, Shefrin EA, Fleg JL: Prevalence and long-term significance of exercise-induced frequent or repetitive ventricular ectopic beats in apparently healthy volunteers. *J Am Coll Cardiol* 1989;14:1659–1665.
9. Califf RM, McKinnis RA, McNeer JF, Harrell FE, Lee KL, Pryor DB, Waugh RA, Harris PJ, Rosati RA, Wagner GS: Prognostic value of ventricular arrhythmias associated with treadmill exercise testing in patients studied with cardiac catheterization for suspected ischemic heart disease. *J Am Coll Cardiol* 1983;2:1060–1067.
10. Weiner DA, Levine SR, Klein MD, Ryan TJ: Ventricular arrhythmias during exercise testing: Mechanisms, response to coronary by-pass surgery and prognostic significance. *Am J Cardiol* 1984;53:1553–1557.
11. Sami M, Chaitman B, Fisher L, Holmes D, Fray D, Alderman E: Significance of exercise-induced ventricular arrhythmias in stable coronary artery disease: A Coronary Artery Surgery Study Project. *Am J Cardiol* 1984;1182–1188.
12. Nair CK, Aronow WS, Sketch MH, Pagano T, Lynch JD, Moss AN, Esterbrooks D, Runaco V, Ryschon K: Diagnostic significance of exercise-induced premature ventricular complexes in men and women: A four year follow-up. *J Am Coll Cardiol* 1983;5:1201–1206.
13. Udall JA, Ellestad MH: Predictive implications of ventricular premature contractions associated with treadmill stress testing. *Circulation* 1977;56:985–989.
14. Stone PH, Turi ZG, Muller JE, Parker C, Hartwell T, Rutherford JD, Jaffe AS, Raabe DS, Passamani ER, Willerson JT, Sobel BE, Robertson TL, Braunwald E, the Milis Study Group: Prognostic significance of the treadmill exercise test performance 6 months after myocardial infarction. *J Am Coll Cardiol* 1986;8:1007–1017.
15. Dimsdale JE, Hartley LH, Guiney T, Riskin JN, Greenblatt D: Post-exercise peril. Plasma catecholamines and exercise. *JAMA* 1984;251:630–632.
16. Coumel P: Rate dependence and adrenergic dependence of arrhythmias. *Am J Cardiol* 1989;64:41J–45J.
17. Sheps DS, Ernst JC, Briese FR, Lopez LV, Conde CA, Castellanos A, Myerburg RJ: Decreased frequency of exercise-induced ventricular ectopic activity in

the second of two consecutive treadmill tests. *Circulation* 1977;55:892–895.
18. McHenry PL, O'Donnell J, Morris SN, Jordan JW: The abnormal exercise electrocardiogram in apparently healthy men: A predictor of angina pectoris as an initial coronary event during long term follow-up. *Circulation* 1984;70:547–551.
19. Crow R, Prineas R, Blackburn H: The prognostic significance of ventricular ectopic beats among the apparently healthy. *Am Heart J* 1981;101:244–248.
20. Sellers TD, Beller GA, Bigson RS, Watson DD, DiMarco JP: Prevalence of ischemia by quantitative thallium-201 scintigraphy in patients with ventricular tachycardia or fibrillation inducible by programmed stimulation. *Am J Cardiol* 1987;59:828–832.
21. Reinke A, Michel D, Mathis P: Arrhythmogenic potential of exercise-induced myocardial ischemia. *Eur Heart J* 1987;8(suppl G):119–124.
22. Huikuri HV, Korhonen UR, Takkunen JT: Ventricular arrhythmias induced by dynamic and static exercise in relation to coronary artery bypass grafting. *Am J Cardiol* 1985;55:948–851.
23. Guinn GA, Mathur VS: Ambulatory, nocturnal and exercise arrhythmias in coronary artery disease: A prospective randomized study to assess the influence of aorto-coronary bypass surgery. *Am J Cardiol* 1977;39:270–277.
24. Anastassiades LC, Antonopoulos AG, Petsas AA: The effect of coronary revascularization on exercise-induced ventricular ectopic activity. *Eur Heart J* 1987;8(suppl D):75–78.
25. Michelson EL, Morganroth J, MacVaugh H: Postoperative arrhythmias after coronary artery and cardiac valvular surgery detected by long-term electrocardiographic monitoring. *Am Heart J* 1979;97:442–448.
26. Tilkian AG, Pfeifer JF, Barry WH, Lipton MJ, Hultgren HN: The effect of coronary bypass surgery on exercise-induced ventricular arrhythmias. *Am Heart J* 1976;92:707–714.
27. Rasmussen IK, Lunde PI, Lee M: Coronary bypass surgery in exercise-induced ventricular tachycardia. *Eur Heart J* 1987;8:444–448.
28. Weld FM, Chu KL, Bigger TJ, Rolnitzky LM: Risk stratification with low-level exercise testing 2 weeks after acute myocardial infarction. *Circulation* 1981;64:306–314.
29. DeBusk RF, Kraemer HC, Nash E, Berer WE, Lew H: Stepwise risk stratification soon after acute myocardial infarction. *Am J Cardiol* 1983;52:1161–1186.
30. Waters DD, Bosch X, Bouchard A, Moise A, Roy D, Pelletier G, Theroux P: Comparison of clinical variables and variables derived from a limited predischarge exercise test as predictors of early and late mortality after myocardial infarction. *J Am Coll Cardiol* 1985;5:1–8.
31. Henry RL, Kennedy GT, Crawford MH: Prognostic value of exercise-induced ventricular ectopic activity for mortality after acute myocardial infarction. *Am J Cardiol* 1987;59:1251–1255.
32. Krone RJ, Gillespie JA, Weld FM, Miller JP, Moss AJ, and Multicenter Postinfarction Research Group: Low-level exercise testing after myocardial infarction; usefulness in enhancing clinical risk stratification. *Circulation* 1985;71:80–89.
33. Saunamaki KI, Andersen JD: Post-myocardial infarction exercise testing. Clinical significance of a left ventricular function index and ventricular arrhythmias. A prospective study. *Acta Med Scand* 1985;218:271–278.
34. Evans J, Skale B, Wendle J, Heger J, McHenry P, Prystowski E: Comparison of ventricular tachycardia induction between exercise and electrophysiologic testing in patients with ventricular tachycardia. *Circulation* 1984;70:II-423.
35. Mokotoff DM, Quinones MA, Miller RR: Exercise-induced ventricular tachycardia: Clinical features, relation to chronic ventricular ectopy and prognosis. *Chest* 1980;77:10–16.
36. Weaver WD, Cobb LA, Hallstrom AP: Characteristics of survivors of exertion- and nonexertion-related cardiac arrest: Value of subsequent exercise testing. *Am J Cardiol* 1982;50:671–676.
37. Coelho A, Palileo E, Ashley W, Swiryn S, Petropoulos AT, Welch WJ, Bauernfeind RA: Tachyarrhythmias in young athletes. *J Am Coll Cardiol* 1986;7:237–243.
38. Allen BJ, Casey TP, Brodsky MA, Luckett CR, Henry WL: Exercise testing in patients with life-threatening ventricular tachyarrhythmias: Results and correlation with clinical and arrhythmia factors. *Am Heart J* 1988;116:997–1002.
39. Podrid PJ, Venditti FJ, Levine PA, Klein MD: The role of exercise testing in evaluation of arrhythmias. *Am J Cardiol* 1988;62:24H–33H.
40. Woelfel A, Foster JR, Simpson RJ, Gettes LS: Reproducibility and treatment of exercise-induced ventricular tachycardia. *Am J Cardiol* 1984;53:751–756.
41. Sung RJ, Keung EC, Nguyen N, Huycke EC: Effects of beta-adrenergic blockade on verapamil-responsive and verapamil-irresponsive sustained ventricular tachycardia. *J Clin Invest* 1988;81:688–699.
42. Sung RJ, Huycke EC, Lai W, Tseng C, Chu H, Keung E: Clinical and electrophysiologic mechanisms of exercise-induced ventricular tachyarrythmias. *PACE* 1988;11:1347–1357.
43. Graboys TB, Lampert S, Lown B: Yield of ventricular arrhythmia during exercise testing in patients with prior cardiac arrest. *Circulation* 1982;66:II 27.
44. Young DZ, Lampert S, Graboys TB, Lown B: Safety of maximal exercise testing in patients at high risk for ventricular arrhythmia. *Circulation* 1984;70:184–191.
45. Sami M, Kraemer H, DeBusk RF: Reproducibility of exercise-induced ventricular arrhythmia after myocardial infarction. *Am J Cardiol* 1979;43:724–730.
46. Saini V, Graboys TB, Twone V, Lown B: Reproducibility of exercise-induced ventricular arrhythmia in patients undergoing evaluation for malignant ventricular arrhythmia. *Am J Cardiol* 1989;63:697–701.
47. Woelfel A, Foster JR, McCallister RG, Simpson RJ, Gettes LS: Efficacy of verapamil in exercise-induced ventricular tachycardia. *Am J Cardiol* 1985;56:292–297.
48. Lerman BB, Belardinelli L, West GA, Berne RM, DiMarco JP: Adenosine-sensitive ventricular tachycardia: Evidence suggesting cyclic AMP-mediated triggered activity. *Circulation* 1986;74:270–280.
49. Eldar M, Belhassen B, Hod H, Schuger CD, Scheinman MM: Exercise-induced double (atrial and ventricular) tachycardia: A report of three cases. *J Am Coll Cardiol* 1989;14:1376–1381.
50. Sokoloff NM, Spielman SR, Greenspan AM, Rae AP, Porter RS, Lowenthal DT, Hakki AH, Iskandrian AS, Kay HR, Horowitz LN: Plasma norepinephrine in ex-

ercise-induced ventricular tachycardia. *J Am Coll Cardiol* 1986;8:11–17.

51. Ritchie AH, Kerr CR, Qi A, Yeung-Lai-Wah JA: Nonsustained ventricular tachycardia arising from the right ventricular outflow tract. *Am J Cardiol* 1989;64:594–598.
52. Palileo EV, Ashley WW, Swiryn S, Bauernfeind RA, Strasberg B, Petropoulos AT, Rosen KM: Exercise-induced provocable right ventricular outflow tract tachycardia. *Am Heart J* 1982;104:185–193.
53. Lemery R, Brugada P, Bella PD, Dugernier T, VandenDool A, Wellens HJ: Nonischemic ventricular tachycardia. Clinical course and long-term follow-up in patients without clinically overt heart disease. *Circulation* 1980;79:990–999.
54. Lemery R, Brugada P, Janssen J, Cheriex E, Dugernier T, Wellens HJ: Nonischemic sustained ventricular tachycardia: Clinical outcome in 12 patients with arrhythmogenic right ventricular dysplasia. *J Am Coll Cardiol* 1989;14:96–105.

Chapter **20**

Signal-Averaged Electrocardiography: Technique and Clinical Applications

Michael B. Simson, MD

For more than a decade, signal averaging has been applied to the electrocardiogram (ECG) to record low-amplitude potentials that are masked by noise and artifact when recorded with conventional electrocardiographic methods. One low-level waveform, the ventricular late potential, has been found to have considerable prognostic significance in detecting patients prone to ventricular tachyarrhythmias. This chapter reviews the signal-averaging methods and discusses the ventricular late potential as an example of new electrocardiographic information of clinical importance available on the body surface.

TECHNIQUE

The purpose of signal averaging is to decrease the level of noise that contaminates the ECG.[1,2] The primary sources of noise are skeletal muscle activity, electrodes, power lines, and electronic noise from amplifiers. Modern electronic equipment can minimize interference from powerlines and electronic noise, but noise from muscle activity is typically 5 to 20 μV under optimal recording conditions. Muscle noise cannot be eliminated by filtering because its frequency content is similar to high-frequency (>25 Hz) cardiac potentials.[3]

The most commonly used type of signal averaging is ensemble or temporal averaging, which averages multiple samples of a repetitive waveform. Random, nonrepeating noise cancels and is reduced.[2] The process averages an analog waveform by first measuring the voltage of highly amplified ECG frequently at 1,000 or more times per second. An interval of the ECG is identified by a computer algorithm, aligned with previous similar intervals, and added together point by point. After division by the number of cycles processed, an average value of the ECG for each discrete point in time is obtained. The averaged waveform appears smooth and continuous when plotted because the time between each point is brief, 1 ms or less, and each measurement is made with high resolution.

Several requirements must be met so that the ensemble signal-averaging technique can reduce noise effectively and give reliable results.[2,4] First, the waveform of interest must repeat so that multiple samples can be obtained. If the waveform changes over time, then ensemble signal averaging cannot be used. Second, there must be a unique feature in the ECG that can be used as a reference time so that the computer can average together appropriate points of the repeating signal. The reference time is usually derived by cross-correlation of high amplitude or rapidly moving portions of the QRS complex.[1,5–7] If the waveform of interest does not have a fixed temporal relationship with the reference time, then the averaged waveform will be smoothed, and high-frequency details of the waveform will be eliminated.[2] For example, a variation in the reference time of 1.3 ms (trigger jitter) limits the high-frequency response to <100 Hz.[2] Third, the signal of interest must be independent in time from the contaminating noise. If an artifact, such as one arising from electrode motion, repeats each cardiac cycle, then the signal-averaging process will not attentuate the artifact. Fourth, the noise must have certain prop-

655 Avenue of the Americas, New York, NY 10010

erties for the signal-averaging technique to be most effective; the noise should be unpatterned and random with a Gaussian distribution.[2] Sine wave interference from powerlines, for example, is not minimized as effectively by the technique as purely random disturbances such as muscle activity.[4]

The reduction of noise by ensemble signal averaging is, in general, proportional to the square root of the number of cycles processed.[2,4] Averaging 100 cycles will reduce noise by a factor of 10. The technique allows a low-level signal that is hidden by noise to be readily recovered. The remaining high-frequency noise level in most studies is under 1 μV.

A second form of signal averaging is spatial averaging in which potentials from 4 to 16 independent electrode pairs are summed.[8,9] The theoretical noise reduction is two to four times. The advantage of spatial averaging is that transient, beat-by-beat events can be analyzed. There is a practical limit, however, to the number of electrodes used if all are to record the same ECG vectors and, moreover, the closely spaced electrodes may record a common noise source that will not effectively cancel.[9] Most studies use ensemble signal averaging because substantial noise reduction is easier to achieve in practice.

Once the ECG is signal-averaged, it is usually high-pass filtered to reduce large-amplitude, low-frequency signal content. The rationale for high-pass filtering is that depolarization of cells generates rapid changes in membrane voltage, movement of wave fronts of activation, and high frequencies on the body's surface.[10] The plateau or repolarization phases of the action potential produced more slowly changing voltages and lower-frequency signals. The ST segment, for example, commonly contains slowly changing potentials that can exceed 100 μV or more. If displayed at high gain without high-pass filtering, microvolt signals corresponding to depolarization of small areas of the myocardium would be difficult to measure. Filter corner frequencies of 25 to 40 Hz are used in most studies.

A problem with high-pass filters of conventional design is that they create an artifact (ringing or overshoot) after a large signal ends abruptly.[6] This property impedes the detection of low-level potentials that occur after large amplitude transients, such as ventricular late potentials which exist for only a few tens of milliseconds after the QRS complex. A bidirectional digital filter free of ringing artifact was developed to study ventricular late potentials.[6] With this filter, the computer processes the signal-averaged ECG forward in time until the middle of the QRS. It then processes the signal in reverse time, beginning after the end of the T wave, until the middle of the QRS is reached. Filter artifact or ringing from the QRS complex is completely eliminated with this strategy at the expense of a discontinuity in the middle of the QRS complex.

High-pass filtering is sometimes called time domain analysis; the filter output corresponds in time with the input signal. One can obtain timing information directly from high-pass filtering, such as the duration of the filtered QRS complex. Another means to extract diagnostic high-frequency content is frequency domain analysis, usually performed by the fast Fourier transform (FFT).[7,11] In this process, an interval of input signal is mathematically decomposed into fundamental and harmonic frequencies, such as 8, 16, 24 Hz, and so forth. The fundamental frequency is the inverse of the duration of the input signal; 8 Hz for a 125-ms sample, for example. The output of the FFT is a spectrum that displays amplitude versus frequency. Timing information is lost, unless repeated transforms are performed, each displaced by small units of time (spectro-temporal maps).[11] The Fourier transform can generate errors if the endpoints of the input signal are not at the same voltage level. To avoid this, the input interval is first multiplied by a "window" function, a bell-shaped curve, that sharply attenuates the beginning and end of the input signal. The use of a window, in turn, further restricts the frequency resolution and produces complex artifacts.[11] Used carefully, both time and frequency domain analysis can give excellent results.

Orthogonal XYZ leads are used in most investigations of late potentials. When studied with body surface maps, late potentials were found to have a dipolar distribution and were well detected by orthogonal leads.[12] The filtered leads are commonly combined into a vector magnitude, $\sqrt{X^2 + Y^2 + Z^2}$, a measure that sums the high-frequency information contained in all leads. The vector magnitude is often termed the "filtered QRS complex." Various amplitude and duration measurements are made by the computer on the filtered QRS complex. Voltages are often measured by the "root mean square" method (RMS).

VENTRICULAR LATE POTENTIALS

Since 1978, multiple groups have reported evidence of small, high-frequency (>25 Hz) waveforms on signal-averaged ECGs in patients with ventricular tachycardia (VT) after myocardial infarction (MI).[6,7,13–16] The microvolt level signals, commonly termed "late potentials," are continuous with the QRS complex and persist for tens

of milliseconds into the ST segment. Late potentials correspond to delayed and fragmented ventricular activation that has been observed on epicardial and endocardial electrograms recorded from patients with VT.[14,15,17–21] The amplitude of the fragmented electrograms, which occur in only a few areas of the heart, is typically under 1 mV when directly recorded; conventional electrocardiographic technique cannot reliably detect late potentials from the body surface because they are masked by noise.

Examples of records obtained from patients without VT after MI are shown in Figure 20.1. The filtered QRS complex during sinus rhythm in patients without VT shows a rapid onset, a peak of high-frequency voltage 40 to 50 ms after QRS onset, and an abrupt decline to noise level at the end of the QRS complex. There is no signal above noise level (<1 μV) in the ST segment. The peak of the T wave contains a small amount of high-frequency signal.

The initial portion of the filtered QRS complex recorded from patients with VT after MI (Fig. 20.2) is generally similar to that from patients without VT. At the end of the filtered QRS complex, however, there is a low-amplitude waveform, a late potential, which is not present in patients without VT. The late potential is continuous with the QRS complex and corresponds to fine ripples and notches in the QRS complexes of the signal-averaged unfiltered leads. The amplitude of the late potential varies from 1 to 25 μV with 25 Hz high-pass filtering.

The late potential extends the duration of the filtered QRS complex in patients with VT after MI. In an early study, the mean filtered QRS duration was 139 ms for patients with VT versus 95 ms for patients without VT.[6] Seventy-two percent of patients with VT had a filtered QRS duration longer than 120 ms; no patient in the control group did. Patients with bundle branch block were excluded from the study.

The late potential has low amplitude. A measurement of the voltage in the last 40 ms of the filtered QRS complex was found to distinguish well between the patients with and without VT after MI.[6] The patients with VT had an average of 14.9 μV at the end of the filtered QRS complex; patients without the arrhythmia had 73.8 μV. A threshold of 25 μV separated the two groups well. Ninety-two percent of the patients with, but only 7% of patients without, VT had less than 25 μV. Another good criterion used to identify the late potential, first suggested by Denes et al., is the duration that the terminal portion of the filtered QRS complex remains below 40 μV.[16]

Late potentials are high-frequency waveforms that extend into the ST segment. When this

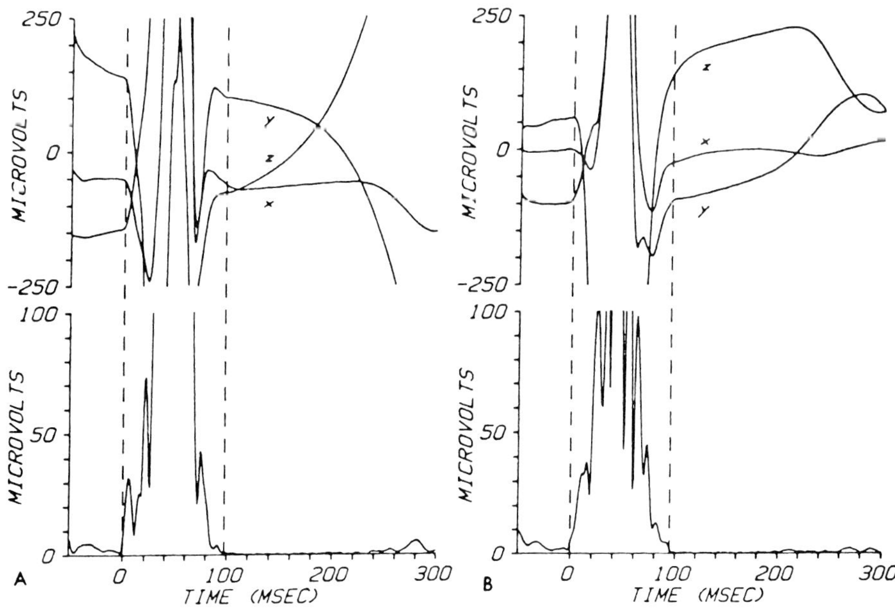

FIGURE 20.1 Signal-averaged ECGs in 2 patients who did not have ventricular tachycardia. On the top are the signal-averaged leads shown at high gain. On the bottom is the filtered QRS complex. The filtered QRS complex is less than 100 ms in duration in both, and there is a large amplitude of signal in the last 40 ms (100 and 53 μV, respectively). The patients had an inferior (**A**) and an anterior (**B**) Myocardial Infarction (MI). *Reprinted from Simson,*[93] *with permission.*

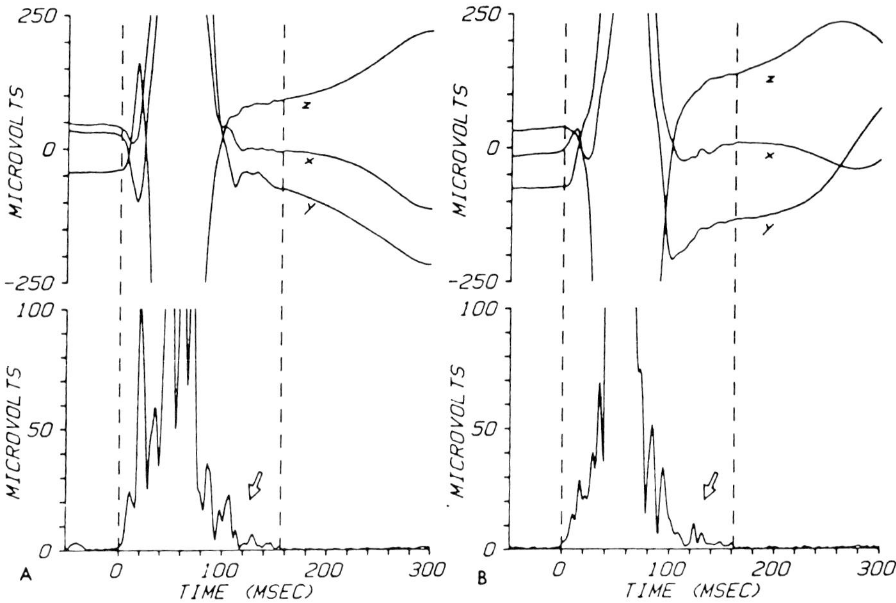

FIGURE 20.2 Signal-averaged electrocardiograms in patients with VT. For each patient the filtered QRS complex shows a longer duration and a late potential (arrow) which is not present in the filtered QRS complexes from control patients (Fig. 20.1). The voltage in the last 40 ms of the filtered QRS complex measured 3.1 and 4.3 μV, respectively. The patients had an inferior (**A**) and an anterior (**B**) infarction. *Reprinted from Simson,*[93] *with permission.*

region is analyzed by FFT, an increase in high-frequency signal content is detected.[7,11,22] Figure 20.3 shows representative findings in patients with and without VT after MI. The terminal 40 ms of the QRS complex and the ST segment were analyzed as a single unit; because a window function was applied before the FFT, the distal QRS complex contributes minimal high-frequency components. The terminal QRS and ST segments from patients with VT contained a 10- to 100-fold increase in 20- to 50-Hz components as compared to patients without VT. Cain et al. found that no frequencies above 50 Hz contributed substantially to the energy spectrum of the terminal QRS and ST segment in patients with or without VT.[22] Another study suggested that 60- to 120-Hz information may be used to identify patients with VT.[11]

Since first described by Fontaine et al. in 1977 and Uther et al. in 1978, many groups have recorded late potentials with signal averaging from patients with VT.[6,7,13–17] Patients with sustained and inducible VT after MI have abnormal signal-averaged ECGs in 79 to 92% of cases.[6,7,16,23] In contrast, 7 to 15% of patients without VT after MI and Lown class 0-1 ectopy have abnormal signal-averaged ECGs. Only 0 to 2% of healthy volunteers have abnormal signal-averaged ECGs; most studies report late potentials are not detected in normals.

The definition of a late potential and the scoring of a signal-averaged ECG as normal or abnormal is highly technique specific.[11,16,24–26] Representative criteria for 40-Hz bidirectional filtering are that a late potential exists when (a) the filtered QRS complex is longer than 114 to 120 ms, (b) there is a smaller than 20 μV of signal in the last 40 ms of the filtered QRS complex, or (c) the terminal filtered QRS complex remains below 40 μV for longer than 38 ms.[27,28] The total duration of the filtered QRS complex appears to be a critical value in patients without bundle branch block. Worley et al. compared several time and frequency domain variables from the signal-averaged ECG with multivariate analysis.[26] The filtered QRS duration was the only independent factor that separated patients with MI with and without VT. Another study emphasized that measurement of voltage alone was less accurate than measures that included the duration of the filtered QRS complex.[29] The presence of bundle branch block alters the criteria for abnormality in time domain analysis,[30] and frequency domain analysis may be preferred.[11,31] The European Society of Cardiology, the American College of Cardiology, and the American Heart Associa-

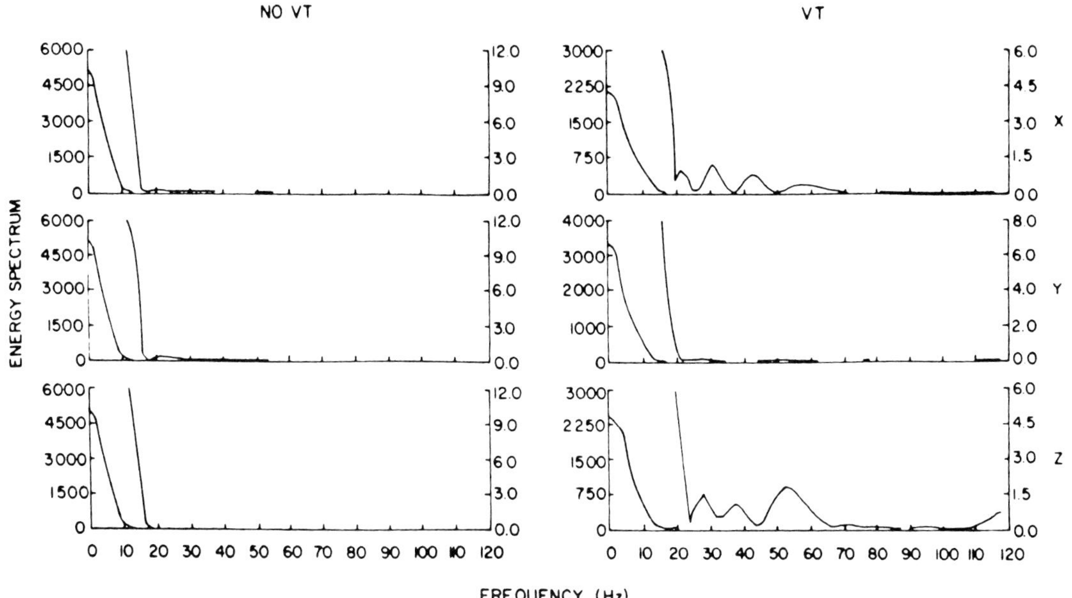

FIGURE 20.3 Examples of frequency analysis of the terminal QRS and ST segments from a patient with a prior MI and sustained VT (**right**) and from a patient with a prior MI without sustained VT (**left**). Shown in each panel are the initial (**left scale**) and magnified (**right scale**) energy versus the frequency plots of the signal-averaged X, Y, and Z ECG leads. Values for the area ratio and the magnitude ratio (mag ratio) refer to measurements of the relative contribution of frequencies between 20 and 50 Hz to the entire signal. In each lead, the patient with sustained VT contains a 10- to 100-fold greater proportion of components in the 20- to 50-Hz range compared to the corresponding values from the patient without VT. *Reprinted from Cain et al.,*[22] *with permission.*

tion convened a committee to establish guidelines for the recording and analysis of signal-averaged ECGs; the findings have been published recently in the Journal of the American College of Cardiology.

Late potentials originate from small areas of delayed and disorganized ventricular activation. Patients with VT after MI have prolonged fragmented electrograms during sinus rhythm that often last beyond the end of the QRS complex as conventionally recorded.[20,21,31–34] Studies in experimental MI have shown that delayed ventricular activation, recorded directly from ischemic or infarcted myocardium, can span diastole and that VT depends on a wave front slowly conducting over a reentrant pathway.[35–38] Josephson et al. demonstrated with endocardial recordings in man that delayed fragmented activity can outlast the surface QRS complex, that it prolongs under the stress of premature beats, and that the onset and maintenance of VT depended on continuous diastolic activity.[19] In canine experimental preparations, fragmented electrograms are recorded in areas of chronic infarcts where interstitial fibrosis caused wide separation of individual myocardial fibers.[39] A similar anatomical substrate was found in endocardium resected from patients undergoing surgery for control of VT.[40] The individual spikes of electrical activity in fragmented electrograms are thought to represent asynchronous electrical activity in separate bundles of surviving myocardium.[39] Recently, it has been shown that fragmented electrogram activity can occur when a wave front passes through regions of abrupt increases in intercellular resistance; such changes would occur in areas where muscle fibers do not have electrically tight connections.[41] Slow conduction per se does not appear to cause fragmented activity.[39,41]

The relationship between the body surface late potentials and delayed ventricular activation has been studied in patients undergoing surgical control of VT after MI.[18] Endocardial and epicardial electrogram mapping was performed along with the signal-averaged ECG. Eighty-eight percent of the fragmented and prolonged (>60 ms) electrograms were endocardial and all were

recorded within the infarct zone.[18] The fragmented electrograms began during the QRS complex and the latest fragmented electrogram for each patient ended a mean of 161 ms after QRS onset, approximately 60 ms after the high-amplitude portion of the filtered QRS complex ended. Figure 20.4 shows the time relationship between electrogram maps and the filtered QRS complex in two patients with ventricular tachycardia. In the patient with a late potential (Fig. 20.4A), the late potential correlated to delayed, fragmented electrogram activity recorded at two endocardial sites. Examples of patients with late potentials and delayed ventricular activation have been reported by several groups.[14,15,17,42–45] Studies in canine infarct models have shown a similar relationship between delayed ventricular activation and late potentials.[42,46,47]

The mapping studies suggest that fragmented electrogram activity can be detected on the body surface as a distinctive late potential only when the fragmented activity outlasts normal ventricular activation. Figure 20.4B illustrates the findings from a patient with VT after MI who did not have a late potential; the fragmented electrogram activity has a brief duration, and ends a mean of 80 ms after QRS onset during the time of normal ventricular activation. In bundle branch block, the delayed ventricular activation in sites with normal electrograms may obscure the recordings of late potentials from areas with normal delayed fragmented electrograms.[18] Fragmented electrogram activity can be recorded in patients after MI who do not have sustained VT.[18,32] The fragmented electrogram activity, however, is detected at fewer sites and has a shorter duration, too brief to outlast normal ventricular activation. In one study, the longest

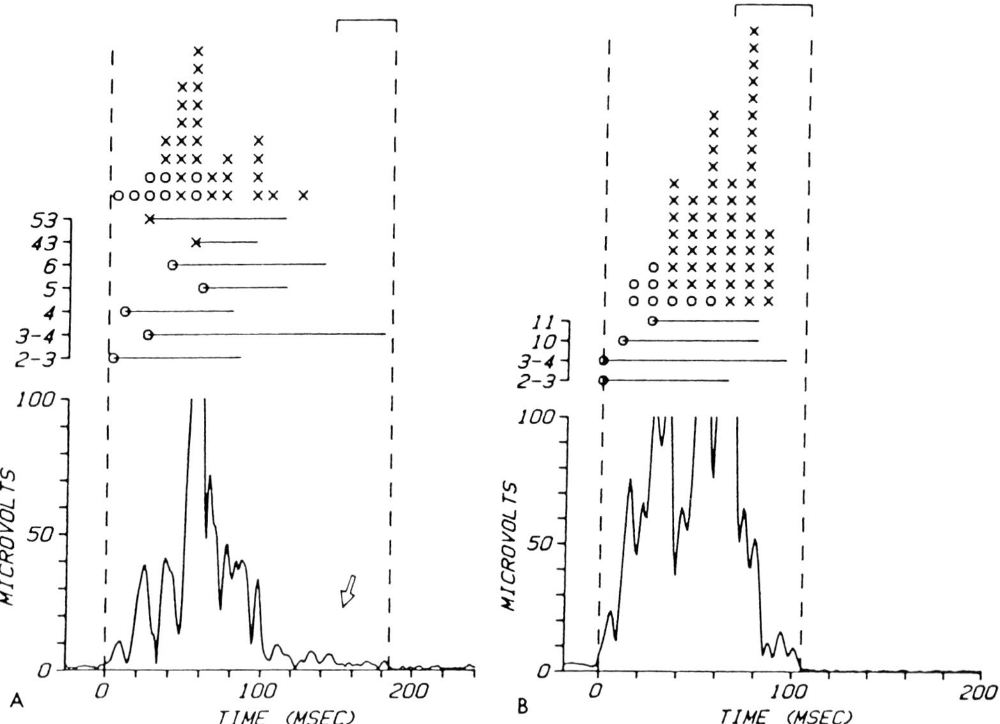

FIGURE 20.4 Examples of the time relationship between directly recorded electrograms (**top**) and the filtered QRS complex (**bottom**) in two patients with VT. The symbols on top indicate the timing of the electrograms. The **X**s are epicardial electrograms, and the **circles** are endocardial electrograms. Isolated symbols represent electrograms with normal configuration. Fragmented electrograms are indicated with a line that shows the timing and duration of the electrogram. The bracket indicates the last 40 ms of the filtered QRS complex. **A.** The patient has a late potential (arrow) which corresponds to fragmented electrogram activity recorded at two sites on the endocardium. **B.** The filtered QRS complex for this patient does not have a late potential. Four endocardial sites show fragmented activity, each of which ended less than 100 ms after the QRS onset during the activation of many sites with normal electrograms.

electrogram activity lasted a mean of 76 ms in patients without VT compared to 135 ms in patients with sustained VT.[32]

Delayed ventricular activation in patients with coronary artery disease is associated with previous myocardial infarct. A correlation between the presence of late potentials and left ventricular dysfunction would be expected and has been noted in most studies.[23,46–53] Breithardt et al. found that in patients without a history of VT or ventricular fibrillation (VF), late potentials were detected in only 9% of patients with normal left ventricular function, but were present in 46% of patients with left ventricular akinesia or aneurysm.[23] Regardless of the degree of ventricular dysfunction, however, late potentials were found more frequently and lasted longer in patients with VT or VF. Denes et al. correlated clinical, Holter recording, and angiographic findings with the signal-averaged ECG in 166 patients with coronary artery disease.[49] An abnormal signal-averaged ECG had a univariate correlation with age, prior MI, previous VT or VF, left ventricular wall motion abnormalities, and ejection fraction. Multivariate analysis, however, revealed that the only independent determinants of an abnormal signal-averaged ECG were the history of a prior MI or sustained VT or VF. In short, these findings and other studies argue that in patients with prior MI, there is a linkage between sustained VT or VF and an abnormal signal-averaged ECG which is independent of other clinical or angiographic variables.[7,49,53,54]

Recent studies suggest that patients presenting with VF have a lower prevalence of an abnormal signal-averaged ECG than patients with a history of sustained VT.[44,45,55–58] Denniss et al.[56] studied 65 patients after MI with inducible ventricular tachyarrhythmias. Patients with spontaneous and inducible VF had a shorter ventricular activation time on the signal-averaged ECG during sinus rhythm (mean 129 ms) than did patients with spontaneous VF and inducible VT (mean 158 ms) or patients with both spontaneous and inducible VT (mean 181 ms). The patients with spontaneous VT had a longer ventricular activation time as detected by epicardial mapping than did patients with spontaneous VF (mean 210 versus 192 ms). These findings agreed with a study in a canine infarct model which showed that animals with inducible VT had longer electrogram duration in the infarct zone than those with inducible VF.[57] In investigations of patients with heart disease of diverse etiology, the prevalence of late potentials was higher in patients presenting with VT (52 to 83%) as compared to patients presenting with VF (21 to 50%).[55,58] Patients with spontaneous VF after MI appear to have less conduction delay during sinus rhythm and are less amenable to detection by the signal-averaged ECG than patients presenting with sustained VT.

PROGNOSTIC VALUE OF AN ABNORMAL SIGNAL-AVERAGED ECG AFTER ACUTE MI

The prognostic significance of late potentials or an abnormal signal-averaged ECG in patients after an acute MI has been evaluated in several studies since 1983.[59] Table 20.1 presents five peer-reviewed prospective studies on a total of 954 patients.[27,28,60–62] The signal-averaged ECG was performed less than 1 month after an acute MI and the mean duration of follow-up ranged from 7.5 to 15 months. Late potentials or an abnormal signal-averaged ECG were found in 24 to 44% of patients studied soon after an acute MI. Patients with a normal signal-averaged ECG or no late potentials had an incidence of sudden cardiac death or sustained VT during the follow-up period of 0.8 to 3.5%. In contrast, patients with an abnormal signal-averaged ECG or definite late potentials (>40 ms) had an incidence of sudden death or sustained VT of 16.7 to 28.9%. One study showed a higher incidence of sudden death and sustained VT in patients with anterior wall MI as compared to those with inferior wall,[60] but the other four studies showed no difference. The management of patients after MI, although uncontrolled, probably did not bias the findings. There was a similar use of antiarrhythmic agents, β-blockers, or coronary artery bypass surgery between groups except in one study.[27] An abnormal signal-averaged ECG appeared to predict an increase in overall cardiac mortality as well as serious arrhythmic events following an acute MI; Denniss reported that patients with an abnormal signal-averaged ECG had a 10% mortality in the first year following an MI, as compared to 2% for patients with a normal signal-averaged ECG.[61]

Studies performed in the 1980s demonstrated that ventricular arrhythmias and left ventricular dysfunction are independently related to the risk of mortality and sudden cardiac death following an acute MI. Two large multicenter studies showed, for example, that patients with repetitive ventricular premature depolarizations were 2.3 to 5.6 times more likely to die during follow-up than patients without the arrhythmias.[63,64] Patients with a left ventricular ejection fraction less than 40% showed an odds ratio of dying during follow-up of 3.4 to 5.1. The risk of sudden death following an MI was also independently increased by left ventricular dysfunction and ventricular arrhythmias.[63–65]

TABLE 20.1 Prognostic Significance of the Signal-Averaged ECG in Patients After MI

Authors	n	Day of Study After MI	Follow-Up Duration (mo)		Incidence of Arrhythmia or Sudden Death (%): Sudden Death	Sustained VT
Breithardt et al.[60]	160	25.5 (median)	7.5 ± 3.3	No LP	3.8	0
			Mean ± SD	LP < 40 ms	3.2	4.8
				LP > 40 ms	11.1	16.6
					Sustained VT or VF	
Denniss et al.[61]	306	7 to 28	24	Normal	2	4
			Mean 12	Abnormal	15	19
					(1 year)	(2 years)
Gomes et al.[28]	102	10 ± 6	12 ± 6	Normal	3.5	
				Abnormal	28.9	
Kuchar et al.[27]	210	11 (mean)	6 to 24 (mean 14)	Normal	0.8	
				Abnormal	16.7	
Cripps et al.[62]	176	12 ± 19	15 ± 7	Normal	1.5	
				Abnormal	21.9	

VT = ventricular tachycardia, MI = myocardial infarction, n = number of patients, LP = late potentials, VF = ventricular fibrillation. Normal or abnormal refers to a normal or abnormal signal-averaged electrocardiogram (ECG).

An important question to be answered was whether the signal-averaged ECG provided information useful in identifying patients with lethal tachyarrhythmias that was independent of left ventricular dysfunction or ventricular ectopy. A partial answer was provided by a retrospective study which compared the findings on a signal-averaged ECG, Holter monitoring, and cardiac catheterization in 98 patients who had recurrent sustained VT and 76 patients without VT after MI.[66] Twelve variables had a univariant associated with VT; they included an abnormal signal-averaged ECG (found in 90% of the patients with VT and 30% of patients without VT), the presence of a left ventricular aneurysm, a left ventricular ejection fraction less than 40%, and frequent or complex ventricular ectopy (including nonsustained VT). Logistic regression analysis showed that only three variables provided significant independent information to identify patients with VT: an abnormal signal-averaged ECG, a peak premature ventricular contraction rate greater than 100/h, and the presence of a left ventricular aneurysm. If two or more variables were abnormal, then the probability of VT was 80 to 99%. Patients with VT were identified with a sensitivity of 81% and a specificity of 90%. When aneurysm was excluded as a variable in the analysis, then a left ventricular ejection fraction of less than 40% replaced aneurysm as a third independent variable. The study suggested that the signal-averaged ECG could be combined with other clinical information to provide a more accurate identification of patients with VT after MI. Several other investigations have demonstrated that the signal-averaged ECG provides data independent of left ventricular ejection fraction or ventricular ectopy in identifying patients with VT.[7,26–28,53,54,67]

Prospective studies by Gomes et al. and Kuchar et al. have evaluated the signal-averaged ECG, Holter monitoring, and radionuclide ventriculography as a means to predict sustained VT or sudden death after MI.[27,28] A total of 312 patients were studied for a mean of 12 to 14 months after an acute MI. The median time to develop life-threatening arrhythmias was less than 2 months. Multivariate analysis in both studies showed that the signal-averaged ECG, an ejection fraction of less than 40%, or complex ventricular ectopic activity each had independent value to predict serious arrhythmic events after an acute MI. The tests could be combined to identify more accurately high-risk patients. The combination of a signal-averaged ECG and low ejection fraction had the highest sensitivity and specificity in each study. No patient in either study with a normal signal-averaged ECG and ejection fraction greater than 40% had sustained VT or sudden death (Fig. 20.5). In contrast, 34 to 36% of patients with an abnormal signal-averaged ECG and ejection fraction of less than 40% had a life-threatening arrhythmic event or sudden death during follow-up. A combination of an abnormal signal-aver-

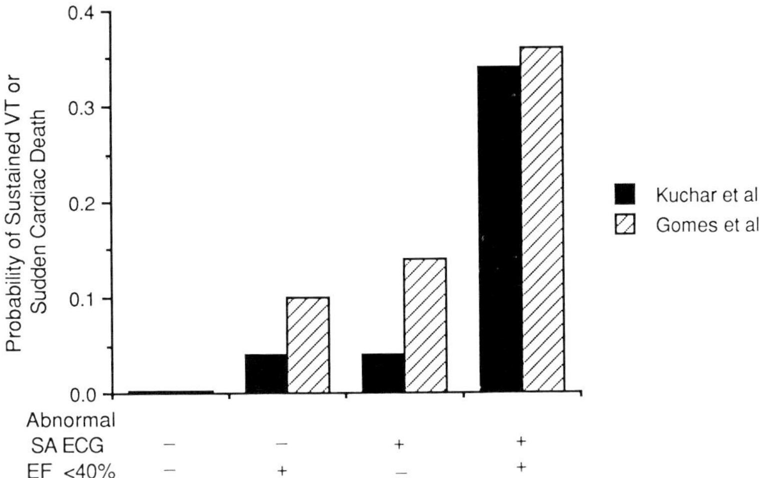

FIGURE 20.5 The probability of sustained VT or sudden cardiac death versus the finding of an abnormal signal-averaged ECG (SA ECG) and the presence of a left ventricular ejection fraction less than 40% (EF < 40%). *Reprinted from Kuchar et al.*[27] *and Gomes et al.*,[28] *with permission.*

aged ECG and a low ejection fraction provided a sensitivity of 80% and 100% and a specificity of 89% and 59% to identify patients who would have sudden death or sustained VT in the studies by Kuchar et al. and Gomes et al., respectively.[27,28]

The two studies argue that a combination of noninvasive tests, including the signal-averaged ECG, can be used to stratify patients according to risk of serious arrhythmic events after an acute MI. Patients with normal tests are at low risk for arrhythmic events, and need no further therapy. Patients with a combination of abnormal tests are at high risk for developing sustained VT or sudden death often soon, within 2 months, after an acute MI. The most effective therapy for high-risk patients is unknown. The roles of new antitachycardia devices, implantable cardioverters/defibrillators, and therapy guided by electrophysiologic stimulation tests need to be defined. A prospective study by Breithardt and Borggrefe suggests that programmed ventricular stimulation, along with the signal-averaged ECG, may be used to define the patients at highest risk for sustained VT after an acute MI; in their study, the patients at highest risk were characterized by the presence of a ventricular late potential and an inducible sustained VT at a rate less than 270 bpm.[45]

The natural history of late potentials after MI has been reported in several studies.[68–70] Late potentials can be detected as early as 3 h after the onset of chest pain and the prevalence increases in the first week after MI (from 32% to 52%).[68] In the first year after MI, the filtered QRS duration decreases slightly in patients without VT. The voltage in the terminal portions of the filtered QRS complex may normalize after 12 months in 30% of patients with late potential initially.[70] In patients without late potentials soon after MI, delayed potentials rarely developed within 12 months of an MI in the absence of reinfarction.[69]

EVALUATION OF PATIENTS WITH SYNCOPE AND NONSUSTAINED VT

Syncope may be caused by a variety of conditions, and establishing an etiology for it may be difficult and expensive.[71,72] The role of electrophysiologic stimulation tests in patients with unexplained syncope has been evaluated.[73–77] Sustained VT was induced and considered the explanation for syncope in approximately 25% of patients. Because of the low incidence of induced sustained VT, it is desirable to have a noninvasive means for identifying patients in whom sustained VT is likely to be induced during electrophysiology testing. Several reports have examined the role of the signal-averaged ECG in detecting patients with unexplained syncope in whom sustained VT can be induced (Table 20.2).[67,78–81] The cardiac disease was diverse in these studies; coronary artery disease was present in 29 to 53% and organic heart disease in 61 to 82%. Sustained VT was induced in 8 to 38% of patients. An abnormal signal-averaged ECG detected those patients who were found to have inducible sustained VT with a sensitivity of 82 to 100% and a

TABLE 20.2 Predictive Value of the Signal-Averaged ECG for Induced Sustained VT in Patients with Syncope or Nonsustained VT

Authors	Number of Patients and Presenting Arrhythmias	Number of Patients with Induced Sustained VT	Signal-Averaged ECG (%)	
			Sensitivity	Specificity
Gang et al.[78]	24 (Syncope)	9 (38%)	82	91
Lindsay et al.[67]	26 (Syncope)	5 (13%)	100	77
	12 (Nonsust VT)			
Kuchar et al.[79]	150 (Syncope)	22 (15%)*	73	89
Winters et al.[80]	34 (Syncope)	12 (35%)	82	91
Nalos et al.[81]	38 (Syncope)	3 (8%)	100	100
	24 (Nonsust VT)	7 (29%)	100	88
	5 (Sust VT)	16 (64%)	88	100
	13 (Cardiac Arrest)	3 (23%)	100	80

Nonsust VT = nonsustained VT, sust VT = sustained VT, * = Symptomatic nonsustained VT or sustained VT, either induced or detected by prolonged ECG monitoring.

specificity of 77 to 91%; Kuchar et al. used a less restrictive definition of VT (see Table 20.2) and found that an abnormal signal-averaged ECG detected 73% of the patients with VT with a specificity of 89%.[79] An abnormal signal-averaged ECG was more predictive of inducible VT than 24 h ambulatory ECG monitoring,[80] and multivariate analysis showed that the abnormal signal-averaged ECG was independent of other determinants including left ventricular ejection fraction and prior MI.[67]

A prospective study by Nalos et al. examined whether the clinical mode of presentation (syncope, sustained or nonsustained VT, or cardiac arrest) affected the utility of the signal-averaged ECG to predict the inducibility of sustained VT.[81] The sensitivity and specificity of the signal-averaged ECG was similar among groups with different presentations (Table 20.2). Multivariate analysis identified the signal-averaged ECG as the best predictor of induction of sustained VT; only the history of sustained VT resulted in a further increase in predictive accuracy. The predictive value of the signal-averaged ECG was shown to be independent of left ventricular ejection fraction, presence of ventricular aneurysm, history of MI, nonsustained VT, or syncope.[81] When late potentials were absent, then 0% or 18% of patients had an inducible sustained VT, depending on whether a history of sustained VT was absent or present, respectively. If late potentials were present, then 77 to 100% of patients were found to have inducible VT. Many studies support that signal-averaged ECG is a sensitive and specific test for the induction of sustained VT and has independent predictive value.[30,55,56,61,67,80,81] It appears to be useful as a screening test prior to electrophysiology stimulation tests to detect patients in whom sustained VT is likely to be induced. Further investigation is necessary, however. A recent report examining patients with nonsustained VT after MI showed that the signal-averaged ECG had predictive utility only in patients with inferior MIs.[82]

EVALUATION OF EFFICACY OF SURGICAL AND MEDICAL THERAPY OF VENTRICULAR TACHYCARDIA

Abnormalities on the signal-averaged ECG often disappear after successful surgical control of VT.[5,12–15,45,83,84] In 24 patients in whom VT could not be induced after endocardial excision, the filtered QRS duration shortened from a mean of 137 to 121 ms and the prevalence of late potentials decreased from 71 to 33%.[84] Loss of the late potential after surgery in 9 of 10 patients was associated with the absence of inducible VT. The prevalence of late potentials and the filtered QRS complex duration were not changed in 13 patients in whom VT remained inducible after operation. Breithardt et al. reported similar results.[45,83] Surgical control of VT may eliminate the late potential, but not invariably. The persistence of the late potential in some cases, despite successful outcome, argues that the operation need not remove all areas of delayed activation to abolish the arrhythmia.

Antiarrhythmic agents, in general, prolong the filtered QRS duration and decrease the amplitude throughout the filtered QRS complex. Late potentials are not abolished with drug ther-

apy, however.[12,14,15,47,85–87] One study found that effective drug therapy was associated with a relative decrease in the 20- to 50-Hz content of the terminal QRS and early ST segment, a change which was seen in only one of 10 unsuccessful trials.[86] Other investigations have noted no pattern of alteration in the filtered QRS complex, which indicated a successful response to antiarrhythmic drugs.[45,85,87] Further studies are needed to resolve these differences, but it does not appear at this time that the signal-averaged ECG is useful to guide antiarrhythmic drug therapy.

CARDIOMYOPATHIES

Patients with congestive cardiomyopathy have a high incidence of life-threatening ventricular arrhythmias and sudden cardiac death. The role of the signal-averaged ECG in identifying patients with sustained ventricular arrhythmias has been assessed in nonischemic dilated cardiomyopathy.[23,88,89] In one study, patients with a history of sustained VT or VF had a longer filtered QRS duration as compared to patients without sustained ventricular arrhythmias, and a lower voltage in the terminal portion of the filtered QRS complex.[88] Eighty-three percent of patients with sustained ventricular arrhythmias had an abnormal SAECG, characterized by both a long, filtered QRS duration and a late potential; only 14% of patients without sustained ventricular arrhythmias have these abnormalities. A prospective study of 54 patients with dilated cardiomyopathy (mean follow-up of 28 months) showed that the SAECG was useful for predicting sudden death, but not death from congestive heart failure.[89] Life table analysis showed that the probability of remaining free from sudden death was lower in patients with an abnormal SAECG than in patients with a normal SAECG (68% versus 100% in 3 years, $p < 0.01$). More prospective studies are needed to assess the role of the SAECG in patients with dilated cardiomyopathy.

Extremely long late potentials have been detected in arrhythmogenic right ventricular dysplasia.[5,17] Little information is available on the signal-averaged ECG in hypertrophic cardiomyopathy. One study reported the signal-averaged ECG was abnormal in 15% of cases, but there was a low correlation with high-grade ventricular arrhythmias on ambulatory monitoring.[90]

The signal-averaged ECG has been used to detect episodes of rejection after cardiac transplantation. One study showed a decrease in the amplitude of filtered QRS complex associated with histologic evidence of rejection.[91] Another study reported a relative increase in the 70- to 110-Hz content of the QRS during acute rejection.[92] These changes were not apparent on the standard ECG and offer the promise of a noninvasive means of detecting rejection after cardiac transplantation.

CONCLUSION

Signal averaging is an enhancement to electrocardiography.[93] With computer processing, microvolt-level late potentials are frequently detected in patients with VT, especially after MI. The late potentials correlate with delayed and disorganized activation in small areas of the myocardium, and they associate with an increased risk for sudden death and lethal ventricular tachyarrhythmias after MI. The information from a signal-averaged ECG is independent from other clinical factors, such as a depressed ventricular function, known to increase the risk for sudden death or VT following an acute MI. A combination of noninvasive tests, including the signal-averaged ECG, can identify a group of patients at high risk for sudden death or sustained VT following an MI. The signal-averaged ECG is also useful in detecting patients who are likely to have sustained ventricular tachyarrhythmias induced with electrophysiologic stimulation testing.

REFERENCES

1. Hombach V, Höpp HW, Braun V, et al.: The applicability of the signal averaging technique in clinical cardiology. *Clin Cardiol* 1982;5:107–124.
2. Ros HH, Koeleman ASM, Akker TJv. d: The technique of signal averaging and its practical application in the separation of atrial and His Purkinje activity, in Hombach V, Hilger HH (eds): *Signal Averaging Technique in Clinical Cardiology*. Stuttgart and New York, F.K. Schattauer Verlag, 1981, p 3.
3. Taylor TP, Macfarlane PW: Digital filtering of the e.c.g.—A comparison of low-pass digital filters on a small computer. *Med Biol Engin* 1974;12:493–502.
4. Evanich MJ, Newberry O, Partridge LD: Some limitations of the removal of periodic noise by averaging. *J Appl Physiol* 1972;33:536.
5. Abboud S, Belhassen B, Laniado S, Sadeh D: Noninvasive recording of late ventricular activity using an advanced method in patients with a damaged mass of ventricular tissue. *Electrocardiology* 1983; 16:245–252.
6. Simson MB: Use of signals in the terminal QRS complex to identify patients with ventricular tachycardia after myocardial infarction. *Circulation* 1981;64:235–242.
7. Cain ME, Ambos D, Witkowski FX, Sobel BE: Fast-Fourier transform analysis of signal-averaged electrocardiograms for the identification of patients prone to sustained ventricular tachycardia. *Circulation* 1984;69:711–720.
8. Flowers NC, Shvartsman V, Kennelly BM, Sohi GS, Horan LG: Surface recording of His-Purkinje activity

on an every-beat basis without digital averaging. *Circulation* 1981;63:948–952.
9. El-Sherif N, Mehra R, Gomes JAC, Kelen G: Appraisal of a low noise electrocardiogram. *J Am Coll Cardiol* 1983;1:456–467.
10. Plonsey R: *Bioelectric Phenomena.* New York, McGraw-Hill, 1969, pp 281–299.
11. Haberl R, Jilge G, Pulter R, Steinbeck G: Comparison of frequency and time domain analysis of the signal-averaged electrocardiogram in patients with ventricular tachycardia and coronary artery disease: Methodologic validation and clinical relevance. *J Am Coll Cardiol* 1988;12:150–158.
12. Faugere, Savard P, Nadeau RA, et al.: Characterization of the spatial distribution of late ventricular potentials by body surface mapping in patients with ventricular tachycardia. *Circulation* 1986;74:1323–1333.
13. Uther JB, Dennett CJ, Tan A: The detection of delayed activation signals of low amplitude in the vectorcardiogram of patients with recurrent ventricular tachycardia by signal averaging, in Sandor E, Julian DJ, Bell JW (eds): *Management of Ventricular Tachycardia—Role of Mexiletine.* Amsterdam-Oxford, Excerpta Medica, 1978, p 80.
14. Breithardt G, Becker R, Seipel L, et al.: Noninvasive detection of late potentials in man—a new marker for ventricular tachycardia. *Eur Heart J* 1981;2:1–11.
15. Rozanski JJ, Mortara D, Myerburg RJ, et al.: Body surface detection of delayed depolarization in patients with recurrent ventricular tachycardia and left ventricular aneurysm. *Circulation* 1981; 63:1172–1178.
16. Denes P, Santarelli P, Hauser RG, Uretz EF: Quantitative analysis of the high-frequency components of the terminal portion of the body surface QRS in normal subjects and in patients with ventricular tachycardia. *Circulation* 1983;67:1129–1138.
17. Fontaine G, Guiraudon G, Frank R, et al.: Stimulation studies and epicardial mapping in ventricular tachycardia. Study of mechanisms and selection for surgery, in Kulbertus H (ed): *Reentrant Arrhythmias.* Lancaster, MTP Press, 1977, p 334.
18. Simson MB, Untereker WJ, Spielman SR, et al.: The relationship between late potentials on the body surface and directly recorded fragmented electrograms in patients with ventricular tachycardia. *Am J Cardiol* 1983;51:105–112.
19. Josephson ME, Horowitz LN, Farshidi A: Continuous local electrical activity. A mechanism of recurrent ventricular tachycardia. *Circulation* 1978;57:659–665.
20. Wiener I, Mindich B, Pitchon R: Determinants of ventricular tachycardia in patients with ventricular aneurysms: Results of intraoperative epicardial and endocardial mapping. *Circulation* 1982;65:856–861.
21. Klein H, Karp RB, Kouchoukos NT, Zorn GL, James TN, Waldo AL: Intraoperative electrophysiologic mapping of the ventricles during sinus rhythm in patients with previous myocardial infarction. Identification of the electrophysiologic substrate of ventricular arrhythmias. *Circulation* 1982;66:847–853.
22. Cain ME, Ambos HD, Markham J, Fischer AE, Sobel BE: Quantification of differences in frequency content of signal-averaged electrocardiograms in patients with compared to those without ventricular tachycardia. *Am J Cardiol* 1985;55:1500–1505.
23. Breithardt G, Borggrefe M, Karbenn U, et al.: Prevalence of late potentials in patients with and without ventricular tachycardia: Correlation and angiographic findings. *Am J Cardiol* 1982;49:1932–1937.
24. Kelen GJ, Henkin R, Fontaine JM, El-Sherif N: Effect of analyzed signal duration and phase on the results of fast Fourier transform analysis of the surface electrocardiogram in subjects with and without late potentials. *Am J Cardiol* 1987;60:1282–1289.
25. Gomes JA, Winter SL, Steward D, Targonski A, Barreca P: Optimal bandpass filters for time-domain analysis of the signal-averaged electrocardiogram. *Am J Cardiol* 1987;60:1290–1298.
26. Worley SJ, Mark DB, Smith WM, et al.: Comparison of time domain and frequency domain variables from the signal-averaged electrocardiogram: A multivariable analysis. *J Am Coll Cardiol* 1988;11:1041–1051.
27. Kuchar DL, Thorburn CW, Sammel NL: Prediction of serious arrhythmic events after myocardial infarction: Signal-averaged electrocardiogram, Holter monitoring and radionuclide ventriculography. *J Am Coll Cardiol* 1987;9:531–538.
28. Gomes JA, Winters SL, Steward D, et al.: A new noninvasive index to predict sustained ventricular tachycardia and sudden death in the first year after myocardial infarction: Based on signal-averaged electrocardiogram, radionuclide ejection fraction and Holter monitoring. *J Am Coll Cardiol* 1987; 10:349–357.
29. Simson MB: Optimal identification of late potentials, in Santini M, Pistolese M, Alliegro M (eds): *Progress in Clinical Pacing.* Amsterdam, Elsevier Science Publishers B.V., Exerpta Medica, 1988, pp 225–238.
30. Buckingham TA, Thessen CC, Stevens RN, Redd RM, Kennedy HL: Effect of conduction defects on the signal-averaged electrocardiographic determination of late potentials. *Am J Cardiol* 1988;61:1265–1271.
31. Lindsay BD, Markham J, Schechtman KB, Ambos HD, Cain ME: Identification of patients with sustained ventricular tachycardia by frequency analysis of signal-averaged electrocardiograms despite the presence of bundle branch block. *Circulation* 1987;77:122–130.
32. Untereker WJ, Spielman SR, Waxman HL, Horowitz LN, Josephson ME: Ventricular activation in normal sinus rhythm: Abnormalities with recurrent sustained tachycardia and a history of myocardial infarction. *Am J Cardiol* 1985;55:974–979.
33. Cassidy DM, Vassallo JA, Miller JM, et al.: Endocardial catheter mapping in patients in sinus rhythm: Relationship to underlying heart disease and ventricular arrhythmias. *Circulation* 1986;73(4):645–652.
34. Vassallo JA, Cassidy D, Simson MB, et al.: Relation of late potentials to site of origin of ventricular tachycardia associated with coronary heart disease. *Am J Cardiol* 1985;55:985–989.
35. Boineau JP, Cox JL: Slow ventricular activation in acute myocardial infarction: A source of reentrant premature ventricular contraction. *Circulation* 1973;48:702–713.
36. Waldo AL, Kaiser G: A study of ventricular arrhythmias associated with acute myocardial infarction in the canine heart. *Circulation* 1973;47:1222–1228.
37. El-Sherif N, Scherlag BJ, Lazzara R, et al.: Reentrant ventricular arrhythmias in the late myocardial in-

farction period. II. Patterns of initiation and termination of reentry. *Circulation* 1977;55:702.
38. Dillon SM, Allessie MA, Ursell PC, Wit AL: Influences of anisotropic tissue structure on reentrant circuits in the epicardial border zone of subacute canine infarcts. *Circ Res* 1988;63:182–206.
39. Gardner PI, Ursell P, Fenoglio JJ Jr, Wit AL: Electrophysiologic and anatomic basis for fractionated electrograms recorded from healed myocardial infarcts. *Circulation* 1985;72(3):596–611.
40. Fenoglio JJ, Duc Pham T, Harken AH, et al.: Recurrent sustained ventricular tachycardia: Structure and ultrastructure of subendocardial regions in which tachycardia originates. *Circulation* 1983; 68(3):518–533.
41. Lesh MD, Spear JF, Simson MB: A computer model of the electrogram: What causes fractionation? *J Electrocardiol* 1989;(suppl):S69–S73.
42. Berbari EJ, Scherlag BJ, Hope RR, et al.: Recording from the body surface of arrhythmogenic ventricular activity during the ST segment. *Am J Cardiol* 1978;41:697–702.
43. Josephson ME, Simson MB, Harken AH, Horowitz LN, Falcone RA: The incidence and clinical significance of epicardial late potentials in patients with recurrent sustained ventricular tachycardia and coronary artery disease. *Circulation* 1982;66:1199–1204.
44. Kertes PJ, Glabus M, Murray A, Julian DG, Campbell WF: Delayed ventricular depolarization-correlation with ventricular activation and relevance to ventricular fibrillation in acute myocardial infarction. *Eur Soc Cardiol* 1984;5:975–983.
45. Breithardt G, Borggrefe M: Pathophysiological mechanisms and clinical significance of ventricular late potentials. *Eur Heart J* 1986;7:364–385.
46. Simson MB, Euler D, Michelson EL, et al.: Detection of delayed ventricular activation on the body surface in dogs. *Am J Physiol* 1981;241(Heart and Circ Physiol 10):H363–H369.
47. Spear JF, Richards DA, Blake GJ, Simson MB, Moore EN: The effects of premature stimulation of the His bundle on epicardial activation and body surface late potentials in dogs susceptible to sustained ventricular tachyarrhythmias. *Circulation* 1985;72:214–224.
48. Coto H, Maldonado C, Palakurthy P, Flowers NC: Late potentials in normal subjects and in patients with ventricular tachycardia unrelated to myocardial infarction. *Am J Cardiol* 1985;55:384–390.
49. Denes P, Uretz E, Santarelli P: Determinants of arrhythmogenic ventricular activity detected on the body surface QRS in patients with coronary artery disease. *Am J Cardiol* 1984;523:1519–1523.
50. Freedman RA, Gillis AM, Keren A, Soderholm-Difatte V, Mason JW: Signal-averaged electrocardiographic late potentials in patients with ventricular fibrillation or ventricular tachycardia: Correlation with clinical arrhythmia and electrophysiologic study. *Am J Cardiol* 1985;55:1350–1353.
51. Pollak SJ, Kertes PJ, Bredlau CE, Walter PF: Influence of left ventricular function on signal averaged late potentials in patients with coronary artery disease with and without ventricular tachycardia. *Am Heart J* 1985;110:747–752.
52. Zimmermann M, Adamec R, Simonin P, Richez J: Prognostic significance of ventricular late potentials in coronary artery disease. *Am Heart J* 1985;109:725.
53. Gomes JA, Horowitz SF, Millner M, et al.: Relation of late potentials to ejection fraction and wall motion abnormalities in acute myocardial infarction. *Am J Cardiol* 1987;59:1071–1074.
54. Buckingham TA, Ghosh S, Homan SM, et al.: Independent value of signal-averaged electrocardiography and left ventricular function in identifying patients with sustained ventricular tachycardia with coronary artery disease. *Am J Cardiol* 1987;59:568–572.
55. Freedman RA, Gillis AM, Keren A, Soderholm-Difatte V, Mason JW: Signal-averaged electrocardiographic late potentials in patients with ventricular fibrillation or ventricular tachycardia: Correlation with clinical arrhythmia and electrophysiologic study. *Am J Cardiol* 1985;55:1350–1353.
56. Denniss AR, Ross DL, Richards DA, et al.: Differences between patients with ventricular tachycardia and ventricular fibrillation as assessed by signal-averaged electrocardiogram, radionuclide ventriculography and cardiac mapping. *J Am Coll Cardiol* 1988;11:276–283.
57. Richards DA, Blake GJ, Spear JF, Moore EN: Electrophysiologic substrate for ventricular tachycardia: Correlation of properties *in vivo* and *in vitro*. *Circulation* 1984;69:369–381.
58. Dolack GL, Callahan DB, Bardy GH, Greene HL: Signal-averaged electrocardiographic late potentials in resuscitated survivors of out-of-hospital ventricular fibrillation. *Am J Cardiol* 1990;1102–1104.
59. Breithardt G, Broggrefe M: Recent advances in the identification of patients at risk of ventricular tachyarrhythmias: Role of ventricular late potentials. *Circulation* 1987;75:1091–1096.
60. Breithardt G, Schwarzmaier M, Borggrefe M, Haerten K, Seipel L: Prognostic significance of late ventricular potentials after acute myocardial infarction. *Eur Heart J* 1983;4:487–495.
61. Denniss AR, Richards DA, Cody DV, et al.: Prognostic significance of ventricular tachycardia and fibrillation induced at programmed stimulation and delayed potentials detected on the signal-averaged electrocardiograms of survivors of acute myocardial infarction. *Circulation* 1986;74:731–745.
62. Cripps T, Bennett D, Camm J, Ward D: Prospective evaluation of clinical assessment, exercise testing and signal-averaged electrocardiogram in predicting outcome after acute myocardial infarction. *Am J Cardiol* 1988;62:995–999.
63. Mukharji J, Rude RE, Poole WK, et al.: Risk factors for sudden death after acute myocardial infarction: Two-year follow-up. *Am J Cardiol* 1984;54:31–36.
64. Bigger JT, Fleiss JL, Kleiger R, et al.: The relationships among ventricular arrhythmias, left ventricular dysfunction, and mortality in the 2 years after myocardial infarction. *Circulation* 1984;69:250–258.
65. Bigger JT Jr: Identification of patients at high risk for sudden cardiac death. *Am J Cardiol* 1984;54:3D–8D.
66. Kanovsky MS, Falcone RA, Dresden CA, Josephson ME, Simson MB: Identification of patients with ventricular tachycardia after myocardial infarction: Signal-averaged electrocardiogram, Holter monitoring, and cardiac catheterization. *Circulation* 1984;70:264–270.

67. Lindsay BD, Ambos HD, Schechtman KB, Cain ME: Improved selection of patients for programmed ventricular stimulation by frequency analysis of signal-averaged electrocardiograms. *Circulation* 1986;73:675–683.
68. McGuire M, Kuchar D, Ganis J, Sammel N, Thorburn C: Natural history of late potentials in the first ten days after acute myocardial infarction and relation to early ventricular arrhythmias. *Am J Cardiol* 1988;61:1187–1190.
69. Denniss AR, Ross DL, Richards DA, Uther JB: Changes in ventricular activation time on the signal-averaged electrocardiogram in the first year after acute myocardial infarction. *Am J Cardiol* 1987; 60:580–583.
70. Kuchar DL, Thorburn CW, Sammel NL: Late potentials detected after myocardial infarction: Natural history and prognostic significance. *Circulation* 1986;74:1280–1289.
71. Kapoor WN, Karpf M, Wieand S, Peterson JR, Levey GS: A prospective evaluation and follow-up of patients with syncope. *N Engl J Med* 1983;309:197–204.
72. Kapoor WN, Karpf M, Maher Y, Miller RA, Levey GS: Syncope of unknown etiology: The need for a more cost-effective approach to its diagnostic evaluation. *J Am Med Assoc* 1982;247:2687–2691.
73. DiMarco JP, Garan H, Harthorne JW, Ruskin JN: Intracardiac electrophysiologic techniques in recurrent syncope of unknown cause. *Ann Intern Med* 1981;95:542–548.
74. Hess DS, Morady F, Scheinman MM: Electrophysiologic testing in the evaluation of patients with syncope of undetermined origin. *Am J Cardiol* 1982;50:1309–1315.
75. Olshansky BM, Mazuz M, Martins JB: Significance of inducible tachycardia in patients with syncope of unknown origin: A long-term follow-up. *J Am Coll Cardiol* 1985;5:216–223.
76. Morady F, Shen E, Schwartz A, et al.: Long-term follow-up of patients with recurrent unexplained syncope evaluated by electrophysiologic testing. *J Am Coll Cardiol* 1983;2:1053–1059.
77. Doherty JU, Pembrook-Rogers D, Grogan EW, et al.: Electrophysiologic evaluation and follow-up characteristics of patients with recurrent unexplained syncope and presyncope. *Am J Cardiol* 1985; 55:703–708.
78. Gang ES, Peter T, Rosenthal ME, Mandel WJ, Lass Y: Detection of late potentials on the surface electrocardiogram in unexplained syncope. *Am J Cardiol* 1986;58:1014–1020.
79. Kuchar DL, Thorburn CW, Sammel NL: Signal-averaged electrocardiogram for evaluation of recurrent syncope. *Am J Cardiol* 1986;58:949–953.
80. Winters SL, Stewart D, Gomes JA: Signal averaging of the surface QRS complex predicts inducibility of ventricular tachycardia in patients with syncope of unknown origin: A prospective study. *J Am Coll Cardiol* 1987;10:775–781.
81. Nalos PC, Gang EL, Mandel WJ, et al.: The signal-averaged electrocardiogram as a screening test for inducibility of sustained ventricular tachycardia in high risk patients: A prospective study. *J Am Coll Cardiol* 1987;9:539–548.
82. Buxton AE, Simson MB, Falcone RA, et al.: Results of signal-averaged electrocardiography and electrophysiologic study in patients with nonsustained ventricular tachycardia after healing of acute myocardial infarction. *Am J Cardiol* 1987;60:80–85.
83. Breithardt G, Seipel L, Ostermeyer J, et al.: Effects of antiarrhythmic surgery on late ventricular potentials recorded by precordial signal averaging in patients with ventricular tachycardia. *Am Heart J* 1982;104:996–1003.
84. Marcus NH, Falcone RA, Harken AH, Josephson ME, Simson MB: Body surface late potentials: Effects of endocardial resection in patients with ventricular tachycardia. *Circulation* 1984;70:632–637.
85. Simson MB, Waxman HL, Falcone R, Marcus NH, Josephson ME: Effects of antiarrhythmic drugs on noninvasively recorded late potentials, in Breithardt G, Loogen F (eds): *New Aspects in the Medical Treatment of Tachyarrhythmias.* Munich, Urban and Schwarzenberg, 1983, pp 80–86.
86. Cain ME, Ambos HD, Fischer AE, Markham J, Schectman KB: Noninvasive prediction of antiarrhythmic drug efficacy in patients with sustained ventricular tachycardia from frequency analysis of signal averaged ECGs. *Circulation* 1984;70:II–252(abstract).
87. Breithardt G, Borggrefe M, Karbenn U, Schwarzmaier J: Effects of pharmacological and non-pharmacological interventions on ventricular late potentials. *Eur Heart J* 1987;(suppl A):97–104.
88. Poll DS, Marchlinski FE, Falcone RA, Simson MB: Abnormal signal averaged ECG in nonischemic congestive cardiomyopathy: Relationship to sustained ventricular tachyarrhythmias. *Circulation* 1986;72:1308–1313.
89. Ohnishi Y, Inoue T, Fukuzaki H: Value of the signal-averaged electrocardiogram as a predictor of sudden death in myocardial infarction and dilated cardiomyopathy. *Jpn Circ J* 1990;54:127.
90. Gavaghan TP, Kelley RP, Kuchar DL, Hickie JB, Campbell TJ: The prevalence of arrhythmias in hypertrophic cardiomyopathy: Role of ambulatory monitoring and signal-averaged electrocardiography. *Aust NZ J Med* 1986;16(5):666–670.
91. Keren A, Gillis AM, Freedman RA, et al.: Heart transplant rejection monitored by signal-averaged electrocardiography in patients receiving cyclosporine. *Circulation* 1984;70(1):1124–1129.
92. Haberl R, Weber M, Reichenspurner H, et al.: Frequency analysis of the surface electrocardiogram for recognition of acute rejection after orthotopic cardiac transplantation in man. *Circulation* 1987;76:101–108.
93. Simson MB: Clinical application of signal averaging, in Zipes DP (ed): *Cardiology Clinics. Arrhythmias.* Philadelphia, WB Saunders, 1983, pp 109–119.

Chapter **21**

Programmed Electrical Stimulation: Elucidation of Arrhythmia Mechanisms

Marshall S. Stanton, MD

Physicians and physiologists have attempted to determine the mechanisms responsible for various cardiac arrhythmias for centuries. The past century has seen marked advancement since the first recording of cardiac electrical activity by Waller[1] in 1887 and, shortly thereafter, with the clinical tracings published by Einthoven.[2] Knowledge about rhythm disorders continued to accelerate with the work of Lewis,[3] and Katz, Langendorf, and Pick masterfully showed how mechanisms could be deduced from analysis of the surface electrocardiogram.[4,5] These and other developments in the history of electrocardiography have been reviewed by Fisch[6] and Cooper.[7]

A new era in the determination of arrhythmia mechanisms began with the advent of programmed electrical stimulation of the heart[8–11] and recording of activity from the bundle of His.[12,13] Durrer et al.[14] described the various mechanisms responsible for arrhythmias in a 1978 review.

Elucidation of the mechanism underlying an arrhythmia is of more than mere academic interest; there are direct implications for therapy. For instance, knowing that macroreentry within the His-Purkinje system (that is, bundle branch reentrant ventricular tachycardia) is the underlying mechanism of clinical arrhythmia allows for curative therapy by interrupting one of the limbs of the reentrant circuit by means of catheter ablation of the right bundle branch. Further, an antitachycardia pacemaker would not be chosen as therapy for an arrhythmia known to be automatic. Many such examples exist.

Current knowledge states that three general mechanisms exist for arrhythmias in man: reentry, triggered activity, and automaticity. In this chapter I will review the role of these mechanisms in arrhythmias, and the usefulness of programmed electrical stimulation in their diagnosis.

In its simplest form, a reentrant arrhythmia consists of a circular pathway of conduction that circles an anatomic[15] or functional barrier. The anatomic barrier may be scar tissue, while the functional one may be normal tissue that is refractory.[16] The best-documented arrhythmias accepted to be due to reentry are the reciprocating tachycardias involving an accessory pathway and bundle branch reentrant ventricular tachycardia.

For the arrhythmia to begin, block must occur in one direction of the circuit with the impulse proceeding in the other direction. For tachycardia to ensue, the impulse must traverse the circuit with a sufficient delay to the site of unidirectional block after that region has recovered from its refractory period. This usually means that a region of slow conduction is contained within the circuit. Studies have shown that the normal specialized ventricular conducting system may serve as a reentrant circuit.[17,18] Tachycardia begins when the impulse from a premature ventricular complex blocks in one bundle branch, conducts across the septum to the contralateral bundle branch, travels retrogradely to the bundle of His, and proceeds anterogradely down the ipsilateral bundle branch, which is now no longer refractory. This reentrant circuit is capable of maintaining sustained ventricular tachycardia.[19–22]

Reentry as a mechanism for ventricular arrhythmias was suggested by extensive ventricular mapping by Boineau and Cox[23] and by Downar et al.,[24] and recording of continuous

655 Avenue of the Americas, New York, NY 10010
Current Topics in Cardiology

electrical activity throughout diastole in patients during sustained ventricular tachycardia by Josephson et al.[25] Interruption of a portion of the reentrant circuit by cryoablation lends support to reentry as the mechanism of sustained ventricular tachycardia in patients with coronary artery disease.[26,27]

The mechanism for sinus rhythm, which usually drives the heart, is normal automaticity. Cells in a different region of the heart may, for various reasons, develop more rapid phase-4 depolarization than the dominant pacemaker. As this happens, they overtake the rhythm of the heart and a tachycardia is generated by abnormal automaticity. The cellular mechanisms involved in normal and abnormal automaticity have been reviewed.[28–30]

Triggered activity may generate extrasystoles when an afterpotential generated by an action potential (AP) attains threshold. Afterpotentials have been called delayed afterdepolarizations (DADs) when they occur during electrical diastole and early afterdepolarizations (EADs) when they occur during the plateau or rapid repolarization phases of the AP.

DADs usually appear under conditions of calcium (Ca^{2+}) overload. In vitro, the amplitude of DADs tends to increase and their coupling intervals (CIs) decrease at shorter pacing cycle lengths. Thus, clinical arrhythmias due to DADs would be expected to be induced by rapid burst pacing or premature extrastimuli following a short pacing cycle length. Evidence suggests that DADs may be the mechanism of arrhythmias in digitalis toxicity,[31–33] the congenital long QT syndrome,[34] and, in some instances, early after myocardial infarction.[35]

DADs have, as yet, not been recorded in vivo from humans. In vitro, both the amplitude and number of triggered APs from EADs increase at progressively longer pacing cycle lengths.[36] EADs are thought to be responsible for arrhythmias associated with the acquired long QT syndrome.[34,37,38] Endocardial monophasic AP recordings have shown deflections during repolarization that may represent EADs.[39] The cellular electrophysiology of EADs and DADs has been reviewed elsewhere.[40–42]

Pacing techniques and pharmacologic manipulations may be used in the electrophysiology laboratory in an attempt to distinguish between reentry, triggered activity, and automaticity as the underlying mechanisms of an arrhythmia. Table 21.1 lists the similarities and distinguishing features between the three mechanisms. Difficulty usually arises in separating reentry from triggered activity. Therefore, the bulk of this chapter is devoted to discussing the phenomenon of entrainment, which is currently the best technique for documenting high probability of reentry as the etiology of an arrhythmia.

TABLE 21.1 Characteristics of Arrhythmia Mechanisms

1. Reentry
   (a) Initiated by premature extrastimuli
   (b) May be terminated by premature extrastimuli
   (c) Reset by premature extrastimuli (may show flat or mixed response)
   (d) May be entrained by overdrive pacing
   (e) CI of first complex of tachycardia is inversely related to CI of initiating premature extrastimulus
   (f) May demonstrate warm-up
2. Triggered activity
   (a) Initiated by premature extrastimuli
   (b) May be terminated by premature extrastimuli
   (c) May be reset by premature extrastimuli
   (d) CI of first complex of tachycardia is directly related to CI of the initiating premature extrastimulus
   (e) May demonstrate warm-up
3. Automaticity
   (a) Not initiated by premature extrastimuli
   (b) Not terminated by premature extrastimuli
   (c) May be reset by premature extrastimuli
   (d) Not entrainable
   (e) May demonstrate warm-up

REENTRY

The ability to initiate or terminate an arrhythmia by premature stimulation is no longer sufficient to determine that the arrhythmia is reentrant because triggered rhythms may also demonstrate these characteristics.[40,42–44] However, it has been convincingly shown that the ability to entrain a rhythm with overdrive pacing establishes the mechanism of reentry. Work during the past decade has greatly advanced this concept. The thrust of this chapter deals with the phenomenon of entrainment. Resetting, in general, is not unique to reentrant arrhythmias. Further, certain specific types of resetting responses strongly suggest reentry. Understanding the physiology of resetting helps in the comprehension of entrainment.

Contrary to the behavior of rhythms due to triggered activity,[45–48] the CI of the first complex of reentrant tachycardia shows an inverse relationship to the prematurity of the initiating extrastimulus.[49] One reason for this is that with increasing prematurity of the inducing extrastimulus, more delay is encountered in the region

of slow conduction within the reentrant circuit. Thus, exit of the first impulse of tachycardia is delayed.

Resetting

The introduction of a single extrastimulus (S2) during tachycardia yields, if the rhythm is not terminated, a return cycle (S2X3). If S2 does not affect the arrhythmogenic focus, then the CI of the extrastimulus (X1S2) plus the return cycle after the extrastimulus (S2X3) will be twice the spontaneous tachycardia cycle length (Fig. 21.1, upper row). That is, X1S2 + S2X3 = 2(X1X1); a fully compensatory pause occurs. When the extrastimulus causes a pause that is not fully compensatory, the rhythm is said to be "reset" (Fig. 21.1, lower row). This response has been elicited in presumed reentrant arrhythmias[(50)] as well as in known nonreentrant rhythms due to automaticity[(51,52)] or triggered activity.[(36,48)]

Resetting has been studied in ventricular tachycardia using single and double ventricular extrastimuli.[(53–58)] Analysis of the response to extrastimuli during tachycardia yields information that may implicate a reentrant circuit containing an excitable gap. Kay et al.[(58)] recently analyzed the resetting response in entrainable sustained ventricular tachycardia. Resetting of the tachycardia with a single extrastimulus with a conduction time from the stimulus to the advanced electrogram that is longer than the tachycardia cycle length is termed "orthodromic resetting." This occurs when the paced stimulus traverses the tachycardia circuit including the region of slow conduction in the same direction as the spontaneous tachycardia impulse. Orthodromic resetting supports a reentrant mechanism with an excitable gap and suggests that the pacing site is proximal to the region of slow conduction.[(58,59)] Recording electrograms from critical sites is necessary to show the orthodromic resetting response and, therefore, lack of its demonstration does not exclude reentry as a possible mechanism.

The resetting response can be flat, increasing, or mixed, depending upon characteristics of the region of slow conduction in the tachycardia circuit. In the leading circle model described by Allessie et al.,[(16)] a functional barrier of refractory tissue creates the boundary of a reentrant loop. No excitable gap is present because the leading edge of the wave front directly follows its own tail of refractoriness. A single extrastimulus would always encounter tissue that is at least partially refractory when entering the circuit. Although it is possible to reset this type of tachycardia, with increasing prematurity of the extrastimulus the resetting response will change, that is, will not be flat. The manner of change is such that the return cycle (S2X3) lengthens with closer coupled premature stimuli (X1S2) and an

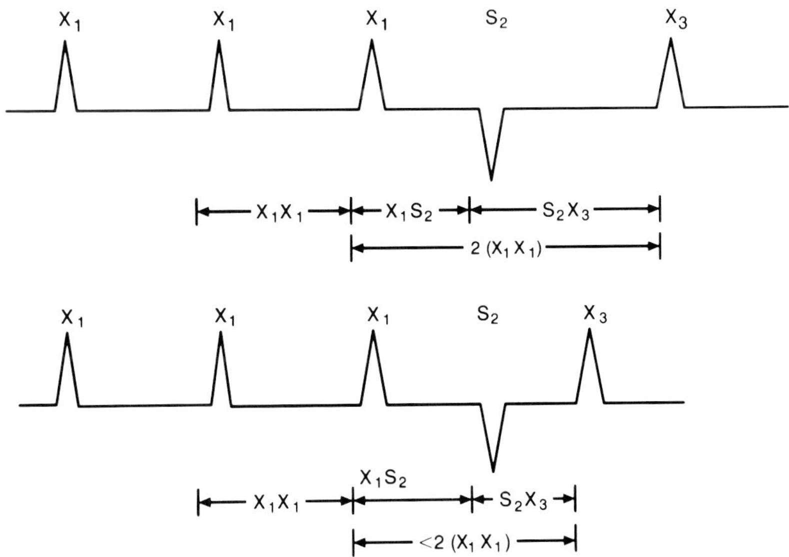

FIGURE 21.1 Response of tachycardia to a single extrastimulus (S2). The tachycardia cycle length is X1X1. The CI of the extrastimulus is X1S2. The return cycle length of the first complex of tachycardia after the extrastimulus is S2X3. In the top portion of the figure, the extrastimulus does not affect the tachycardia circuit and a compensatory pause occurs. Resetting of the tachycardia is shown in the bottom portion of the figure. See text for discussion.

increasing resetting response pattern is obtained. Reentrant circuits containing a region of fully recovered tissue within its loop at any given time could produce a flat or mixed resetting curve by entering the circuit during this period of full recovery[55] (see Fig. 21.2).

The resetting response has been also used to deduce the mechanism by which procainamide slows the rate of ventricular tachycardia (VT)[60] (Fig. 21.2). The presence of a flat or mixed resetting response during VT, the rate of which had been slowed by procainamide, suggests that the drug-induced slowing of tachycardia is not caused by longer refractoriness, that is, increased AP duration (APD), at least not in the region of the excitable gap. If refractoriness had been prolonged in the excitable gap, an increasing resetting response would occur. Also, the steeper slope of the increasing portion of the resetting response curve implies that suppression of conduction in the region of slow conduction by the drug is responsible for the tachycardia cycle length increase. These results suggest that in the studied cases procainamide slowed the rate of sustained VT by slowing conduction within the reentrant circuit.

Computer simulations have been used to assess whether the resetting response pattern and site of pacing would aid in localization of the tachycardia circuit. Resetting of tachycardia

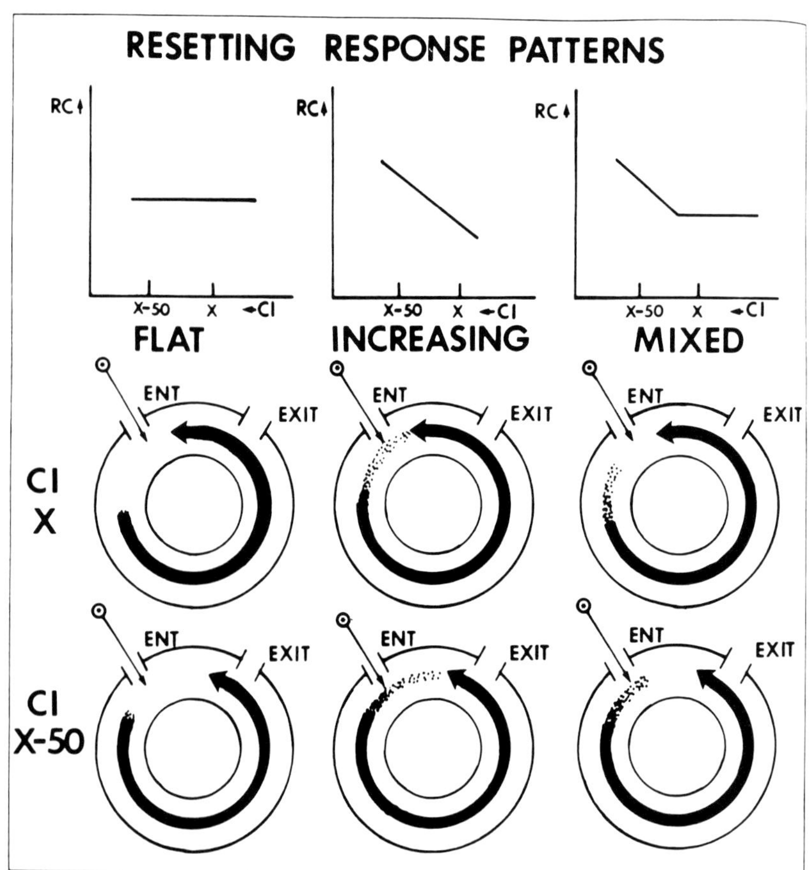

FIGURE 21.2 Resetting response patterns caused by single extrastimulus introduced during tachycardia. Three different reentrant circuits are shown. A fully excitable gap containing no partially refractory tissue is shown on the left. In the middle circuit, all of the tissue in the excitable gap is partially refractory. The example on the right shows an excitable gap containing both fully and partially recovered tissue. Late coupled premature extrastimuli are shown entering the circuit with coupling interval (CI) X. Early coupled extrastimuli (CI X-50) enter the excitable gap at an earlier time. The graphs above show the relationship between the prematurity of the extra stimulus on the abscissa and the resultant return cycle length (RC) on the ordinate for each of the three conditions. See text for discussion. *Reprinted from Stamato et al.,*[60] *with permission.*

without disturbance of the ventricular activation sequence and resultant QRS complex suggests that pacing is being performed from within the tachycardia circuit. However, a change in ventricular activation during resetting does not necessarily exclude stimulation from within this site.[61]

Entrainment

Understanding resetting aids in comprehension of the concept of entrainment. Entrainment is continuous orthodromic resetting of a tachycardia. Although resetting can occur in arrhythmias other than reentrant, entrainment is unique to reentry.

The term "entrainment" will refer to one that is currently used in the electrophysiology literature to describe a phenomenon whereby overdrive pacing during a reentrant arrhythmia creates impulses that enter the tachycardia circuit and accelerate the rhythm to the pacing rate without interrupting the tachycardia. That is, impulses enter the circuit during the excitable gap between trailing and leading edges of the wave front (Fig. 21.3). However, as noted by Jalife et al.[62] entrainment has been used elsewhere to describe the behavior of certain nonreentrant biologic and biochemical rhythms.[63–67] Its use here only pertains to the electrophysiologic phenomenon associated with reentrant arrhythmias.

Four criteria have been put forth to recognize entrainment[68] (Table 21.2). Each of the four criteria of entrainment proves that the rhythm has been entrained, but their lack does not exclude entrainment of the tachycardia.

The ability to entrain a rhythm strongly suggests reentry as the mechanism for the arrhythmia. It has been reported that an automatic focus may, at times, accelerate with pacing at a slightly greater pacing cycle length[69] and thus appears to be entrained. However, an automatic rhythm cannot be terminated by pacing techniques. An example of this is a parasystolic rhythm, which may sometimes be accelerated by rapid pacing, but is not terminable.[67] Triggered rhythms may also be accelerated by pacing. It is important to emphasize that in none of the mechanisms, other than reentry, does progressive fusion occur at different pacing cycle lengths.

The mechanism of atrioventricular (AV) reciprocating tachycardia associated with an accessory pathway is well accepted as being due to macroreentry, the circuit of which incorporates the atrium, AV node/His-Purkinje system, ventricle, and accessory pathway. During orthodromic reciprocating tachycardia (ORT), the AV node and His-Purkinje system are the anterograde limb and the accessory pathway is the retrograde limb. During antidromic reciprocating tachycardia (ART) the direction of the propagation is reversed with anterograde conduction over the accessory pathway and retrograde conduction via the His-Purkinje system and AV node. During the discussion that follows, ORT and ART will be used to designate the two arrhythmias associated with an accessory pathway, as described above. However, when discussing the direction of wave front propagation in the circuit of other reentrant arrhythmia, the term "orthodromic" is used to denote entrance of a wave front into the circuit in the same direction as the spontaneous tachycardia impulse propagation and the term

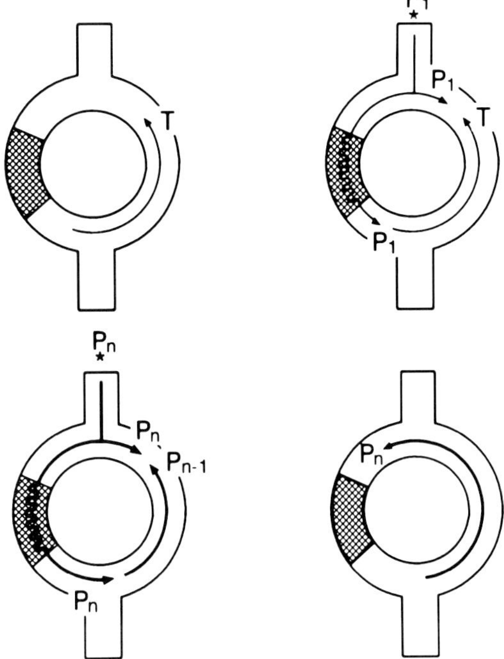

FIGURE 21.3 Entrainment of tachycardia by pacing. The tachycardia impulse (T) travels around the reentrant circuit in the orthodromic direction as illustrated in the upper left panel. The first pacing impulse (P_1) enters the reentrant circuit in the antidromic direction and collides with the last impulse of tachycardia, but also travels in the orthodromic direction so as to continue the tachycardia (**upper right panel**). Each subsequent paced impulse produces the same effect as the first one until the last paced impulse (P_n) collides in the antidromic direction with the next to the last paced impulse (P_{n-1}) (**bottom left panel**) and also in the orthodromic direction where, finding no subsequent paced impulse with which to collide, it continues around the circuit as the next tachycardia impulse (**lower right panel**).

TABLE 21.2 Criteria for Entrainment

1. Pacing during tachycardia results in constant fusion except for the last entrained complex, which, after pacing is discontinued, is not fused and occurs at the pacing cycle length.
2. Pacing at different cycle lengths during tachycardia yields different degrees of fusion.
3. When pacing at a rate that can terminate tachycardia, localized conduction block for one complex occurs and that site is then activated from a different direction and with a shorter conduction time by the next paced impulse.
4. When pacing during tachycardia at two constant cycle lengths that are faster than the tachycardia but that do not terminate it, changes in conduction time to and morphology of an intracardiac electrogram occur.

Modified from Henthorn et al.[68]

"antidromic" denotes entrance of a wave front into a circuit in the opposite direction of the spontaneous impulse propagation.

Waldo et al.,[70] in a study of atrial flutter in patients after cardiac surgery, first described entrainment of a cardiac arrhythmia in man. While attempting to terminate atrial flutter by overdrive pacing, they interpreted the ability to pace the atria at a faster rate than the flutter rate without significantly changing the morphology of the flutter waves as evidence for entrainment. In their initial paper, they did not think that this demonstration of entrainment completely excluded automaticity or triggered activity. The important point was made that the arrhythmia could not be terminated unless pacing was carried out at a rate faster than that which could entrain the rhythm. They also noted that the duration of pacing and probably the distance from the arrhythmia circuit were important factors. In this and subsequent papers they established the criteria summarized in Table 21.2.

First Criterion

A reentrant circuit is shown in Figure 21.3. Pacing entrains the rhythm by entering the circuit in both the orthodromic and antidromic directions. The antidromic impulse collides with the wave front of the spontaneous tachycardia and both are extinguished. The orthodromic wave front of the paced impulse gradually traverses the region of slow conduction and finds that tissue distal to this region has already recovered from the previous tachycardia depolarization. Therefore, it continues around the circuit until colliding with the antidromic impulse of the next paced complex. The orthodromic impulse of this next paced complex enters the region of slow conduction as did its predecessor and exits after the tissue distal to this region has already recovered. This sequence continues until pacing is terminated. At that time, the last paced impulse exits the region of slow conduction, continues through the circuit, and, as there is no subsequent pacing impulse to extinguish it, continues around the circuit to maintain tachycardia.

Since all paced impulses enter the circuit during the excitable gap, each continuously resets and advances the tachycardia. Thus, when pacing is terminated, the last paced impulse continues around the circuit to activate it orthodromically at the pacing cycle length. This can be recorded with electrodes placed at appropriate sites in the tachycardia circuit and can be sometimes recognized on the surface electrocardiogram.

As each paced impulse enters the tachycardia circuit, the previous paced impulse (or, in the case of the first paced impulse, the tachycardia impulse) exits the circuit from another site. Activation of the heart simultaneously by the pacing stimulus and the wave front exiting the reentrant circuit may yield fusion on the surface electrocardiogram. The relative contributions of each component depend upon the pacing cycle length, the region of exit from the circuit, and the pacing site.

As pacing is terminated, the last paced impulse in the reentrant circuit exits when no pacing is occurring. Since the excitable gap is the same as prior to pacing, the tachycardia returns to its previous spontaneous rate.

Sites activated orthodromically show a cycle length of the first postpacing electrogram equal to the pacing cycle length. As there is no next pacing impulse traveling in the antidromic direction, this last entrained impulse does not exhibit fusion and has the morphology of spontaneous tachycardia. This demonstrates the first criterion of entrainment (Table 21.2).

Second Criterion

MacLean et al.[71] showed the importance of different degrees of fusion at different pacing cycle lengths in the demonstration of entrainment (Table 21.2). Pacing at progressively faster cycle lengths activates more of the reentrant circuit in the antidromic direction. Also, each paced impulse enters the excitable gap at an earlier time. The previous impulse will not have had time to activate as much tissue outside the circuit as at the slower pacing cycle length. Therefore, the amount of fusion seen on the surface electrocardiogram (ECG) will be changed and the con-

tribution of the paced impulse will increase at the faster pacing cycle length.

Progressive fusion at decreasing pacing cycle lengths during entrainment of tachycardia excludes the possibility of automaticity or a protected focus (for example, entrance block into a microreentrant focus) as mechanisms of the tachycardia. In those cases, overdrive pacing would yield solely the morphology of the pacing site or variable fusion at each pacing rate, respectively.

Third Criterion

Waldo et al.(72) added a third criterion for entrainment. By recording intracardiac electrograms during entrainment of ORT during atrial pacing, they showed that the first pacing impulse enters the circuit in both the orthodromic and antidromic directions (Fig. 21.3). This paced impulse collides in the antidromic direction with the leading edge of the wave front of the spontaneous tachycardia. At the same time, the paced impulse that is traveling in the orthodromic direction continues around the circuit and exits the reentrant circuit earlier than the expected spontaneous tachycardia impulse would have since the paced impulse entered the circuit ahead of the wave front of the spontaneous wave front (that is, in the excitable gap). The second paced impulse then enters the reentrant circuit in both directions, colliding in the antidromic direction with the leading edge of the first paced impulse and continuing in the orthodromic direction as did the previous impulse. Thus, it is evident that during entrainment, each paced impulse both terminates and reinitiates the reentrant wave front, thereby continuously resetting the tachycardia.

Fusion occurs by the collision of the antidromic and orthodromic impulses. For fusion to happen, conduction block cannot occur before the antidromic impulse and the previous orthodromic impulse collide. The site and occurrence of fusion depend upon the conduction time through the various segments of the reentrant circuit, especially through the region of slow conduction. Further, fusion also varies with the pacing site or, more directly, with the conduction time from the pacing site to the various regions of the circuit.

As the antidromic impulse activates more of the reentrant loop at faster pacing rates, less of the circuit is available from which to record orthodromic impulses. If, for example, during ORT rapid atrial pacing causes fusion in the atrium and orthodromic activation of the ventricles, then at a faster pacing cycle length block may occur in the region of slow conduction (that is, AV node in this example) and the next pacing impulse will activate the ventricles antidromically and the stimulus to ventricular activation time will become shorter, as the impulse no longer has to traverse the region of slow conduction. The previous spontaneous tachycardia will terminate by the occurrence of regional block. This demonstrates the third criterion of entrainment (Table 21.2).

Fourth Criterion

As previously mentioned, progressive fusion demonstrated on the surface ECG at two different pacing cycle lengths satisfies one of the criteria of entrainment. It follows that when this is occurring on the surface, the same phenomenon should be demonstrable with intracardiac recordings at critical sites. With progressive fusion, some portions of the heart are activated in one direction (from orthodromic conduction through the reentrant circuit) at the slower pacing cycle length and in a different direction (antidromically from the pacing site) at the faster pacing cycle length. In an apparent paradox, the conduction time to that local electrogram at the faster pacing cycle length is shorter than the conduction time at the slower pacing cycle length (Figs. 21.4 and 21.5). This is because the reentrant circuit, including the region of slow conduction, is not traversed prior to that site's activation at the faster pacing cycle length. This serves as a fourth criterion for entrainment.(68) Concomitant with the change in activation time of that particular site, a change in electrogram morphology occurs. Thus, fusion is demonstrated at an intracardiac level. Note that the third and fourth criteria differ, as tachycardia termination occurs in the former.

In 1988, Henthorn et al.(68) and Almendral et al.(73) published articles within 2 months of each other describing this fourth criterion for entrainment. Almendral et al.(73) illustrated this criterion in VT, and Henthorn et al.(68) in AV reentrant tachycardia, atrial tachycardia, atrial flutter, and VT. This fourth criterion demonstrates fusion using intracardiac electrograms when it cannot be demonstrated by the surface ECG.

To record the change in activation time and electrogram morphology, the recording electrode must be near a critical point in the reentrant circuit to observe activation from one direction at one pacing cycle length and from another direction at a different pacing cycle length. Thus, a microreentrant circuit such as this may be difficult to demonstrate in AV node reentrant tachycardia. Further, problems may also arise if extreme conduction delay is encountered in the region of slow conduction such that sites distal to this region are always activated by the antidromic wave front and a change in conduction time and electrogram morphology never occurs.

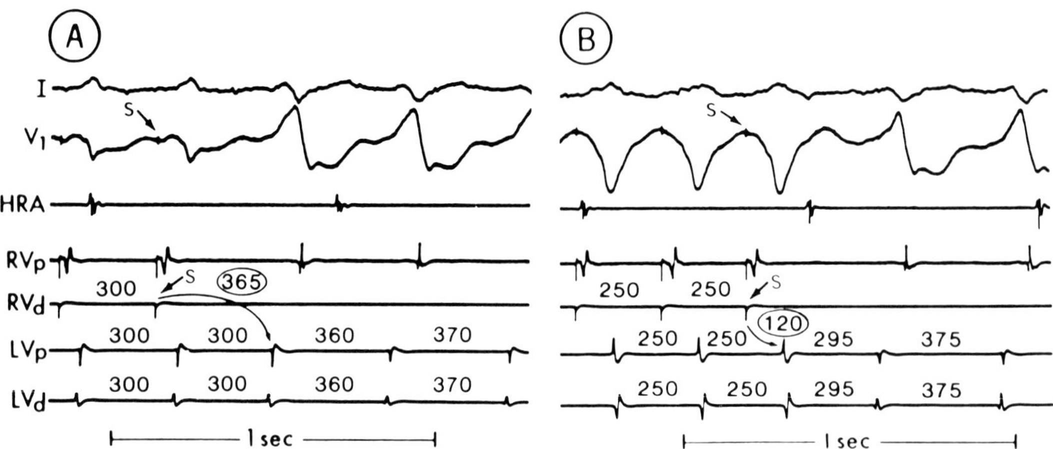

FIGURE 21.4 Progressive fusion at intracardiac sites. During VT, the right ventricle is paced at cycle lengths 300 ms (**A**) and 250 ms (**B**). Progressive fusion is seen on the surface electrocardiographic tracings. In **A**, the left ventricular (LV) sites are activated in the orthodromic direction from the paced impulses with the first nonpaced complex occurring at the pacing cycle length and being nonfused on the surface electrocardiogram. The time from the last pacing artifact (S) to the proximal pair of electrodes of the left ventricular catheter (LV_p) is 365 ms. At a faster pacing cycle length (**B**), the LV sites are activated in the antidromic direction with a paradoxically shorter activation time (120 ms). This occurs because the last paced impulse did not have to traverse the region of slow conduction. In **A**, note the similarity of the LV electrograms of the paced impulses to that of the spontaneous tachycardia electrograms. In **B**, when the LV sites are activated in the antidromic direction, the electrogram morphology has changed. *Reprinted from Henthorn et al.,*[68] *with permission.*

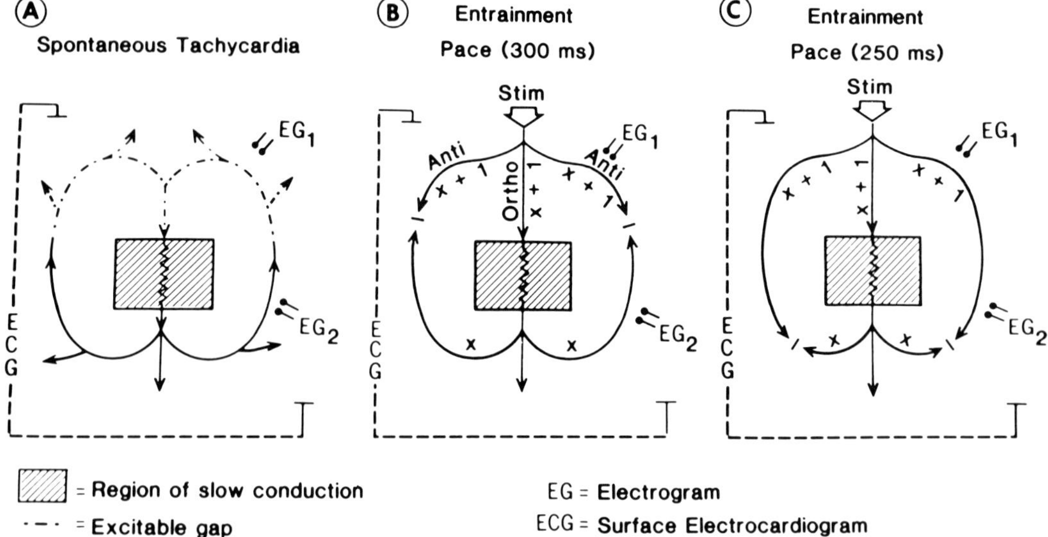

FIGURE 21.5 Illustration of the phenomenon seen in Fig. 21.4. The reentrant circuit is the figure-eight configuration thought to occur in ventricular tachycardia. Intracardiac electrograms are recorded from two sites (EG_1 and EG_2). During spontaneous tachycardia, both sites are activated in the orthodromic direction. With pacing at cycle length 300 ms, EG_1 is activated in the antidromic direction and EG_2 in the orthodromic direction. The electrogram of EG_2 would be the same as that of spontaneous tachycardia and the activation time of EG_2 from the pacing stimulus is relatively long as the region of slow conduction must first be traversed. At the faster pacing cycle length of 250 ms, EG_2 is now activated in the antidromic direction and its electrogram morphology has changed. Likewise, the activation time from the pacing stimulus is shorter than in **B** since it is activated without first conducting over the region of slow conduction. *Reprinted from Henthorn et al.,*[68] *with permission.*

It is obvious that, as with the other three criteria, the site and rate of pacing are critical to being able to demonstrate entrainment. Each criterion reflects the phenomena present in the other three criteria, but each independently demonstrates entrainment.

Transient dissociation of part of the atrium[(74)] and ventricle,[(75)] which has been demonstrated during AV reciprocating tachycardia, must not be confused with entrainment. When a portion of the heart that is not an essential portion of the reentrant circuit is captured and the tachycardia wave front continues uninterrupted along the reentrant pathway, entrainment does not occur.

Site of Pacing

Okumura et al.[(76)] stated that the site of pacing was critical to demonstrating entrainment. Pacing proximal to the region of slow conduction in the circuit was emphasized. Thus, in ORT, pacing would be required from the atrium, and, during ART, pacing should be from the ventricle to demonstrate the criteria for entrainment. During ORT, pacing from the ventricle caused collision in the AV node of the antidromic wave front from the paced impulse with the orthodromic wave front of the previous impulse.[(76)] Therefore, neither atrial nor ventricular fusion would be demonstrable. When pacing is terminated, the final paced wave front entering the circuit orthodromically does not collide in the AV node because its antidromic wave front has already collided with the previous orthodromic impulse. This final paced orthodromic wave front finds the AV node more refractory because of the previous collision. The first postpaced complex cannot occur at the pacing cycle length because of this additional delay within the region of slow conduction (AV node). Electrograms recorded during pacing at different rates were always activated in the same direction and with the same conduction time. The tachycardia was entrained but none of the criteria for entrainment were demonstrated. This was termed "concealed entrainment."

These authors[(76)] concluded from this study that if one can entrain a tachycardia and demonstrate any of the criteria, then the pacing site is proximal to the region of slow conduction in the circuit. Conversely, if one entrains without meeting any of the criteria, then the pacing site is distal to the slow-conducting region. This notion was disproved by Zaman et al.[(77)] who showed that while pacing the ventricles at cycle lengths slightly shorter than the tachycardia cycle length (88 ± 5%), entrainment could be demonstrated by capture of the atrial, ventricular, and His-bundle electrograms at the pacing rate, with the CI of the first postpacing complex occurring at the pacing cycle length in 9 of 18 episodes of ORT. This is explained by the fact that at relatively long pacing cycle lengths, activation of the AV node and His-Purkinje system by the antidromic wave front of the paced impulse may not occur before these sites are activated by the previous orthodromic impulse, and one or more of the previous criteria for entrainment can then be demonstrated (manifest entrainment). Concealed entrainment occurs in this situation when the antidromic wave front of the paced impulse activates the AV node and His-Purkinje system retrogradely, enhancing refractoriness in the AV node and necessarily delaying His activation of the immediate postpacing complex. Pacing bursts of only 5 to 14 pulses should be used. Longer bursts may terminate entirely, or terminate and then reinitiate tachycardia.[(75)] Since the His bundle is not activated antidromically by the paced impulses, the first postpacing His deflection occurs at an HH interval equal to the pacing cycle length.

To entrain a rhythm, the site of pacing must be appropriately situated to enter the tachycardia circuit during the excitable gap. Mann et al.[(78)] reported a case of a patient with two morphologies of VT, each requiring separate pacing sites to entrain. Endocardial mapping was performed and the earliest activation site of the VT with left bundle branch block morphology was on the right ventricular side of the septum. The earliest activation site of the VT with right bundle branch block morphology was on the left ventricular side of the septum. Entrainment could not be accomplished when pacing was performed from the ventricle ipsilateral to the earliest site of activation (that is, could not entrain the right bundle branch block tachycardia with left ventricular pacing, or the left bundle branch block tachycardia with right ventricular pacing). The paced impulse must be able to enter the reentrant circuit during its excitable gap and not collide with the exiting wave front.

The earliest site of activation during VT is at or orthodromically distal to a region of slow conduction in the VT reentrant circuit. During entrainment, the earliest activation site of spontaneous VT is captured after orthodromic activation of the reentrant circuit with a relatively long conduction time from the pacing impulse. When pacing is stopped, this early site is activated last, at the same interval as the pacing cycle length, and the electrogram occurs in late diastole prior to the onset of the first unfused QRS.[(79)] Using the same technique in a later study, Okumura et al.[(80)] demonstrated the effect of verapamil on the reentrant circuit in patients with idiopathic VT of

left ventricular origin (right bundle branch block/left-axis deviation VT in patients without underlying heart disease) in whom the VT could be entrained and was therefore due to reentry. Verapamil dose of 1 mg infused during entrainment of VT caused prolongation of the conduction time from the pacing site to the site of previous earliest activation (distal to the region of slow conduction in the reentrant circuit). This suggests that conduction through the region of slow conduction is, at least partially, Ca^{2+} channel dependent.

To elucidate the mechanism of chronic sustained VT, Kay et al.[81] prospectively studied 19 patients with sustained VT. Seventeen of these patients had coronary artery disease and two had dilated cardiomyopathy. Entrainment could be demonstrated in 79% of these patients. Thus, reentry is the most likely mechanism for VT in this patient population. The importance of pacing from a site from which each stimulation can engage the reentrant circuit was also illustrated. Left bundle branch block morphology VT could be entrained from the left ventricle in 8 of 11 attempts, from the right ventricular apex in 2 of 10 attempts and from the right ventricular outflow tract in 4 of 5 attempts. As expected, the converse was true for right bundle branch block morphology VT where left ventricular pacing was not able to entrain the rhythm in 10 attempts, pacing from right ventricular apex was successful in 9 of 13 attempts, and that from the right ventricular outflow tract in 5 of 9 attempts. When all VT morphologies were considered together, pacing from the right ventricular outflow tract was more successful in achieving entrainment than pacing from either of the other sites.

When recording from the reentrant circuit itself, the conduction time from the last entrained electrogram to the onset of the surface QRS should be equal to that interval during the spontaneous tachycardia. This point is critical in being able to distinguish whether a late diastolic electrogram that closely precedes a surface QRS is from the reentrant circuit itself or from tissue activated after exit from the circuit. In the latter instance, the relationship to the next QRS during both entrainment and spontaneous tachycardia is not likely to be the same.[73]

Termination of Tachycardia

Understanding entrainment facilitates the understanding of termination of reentrant arrhythmias. Termination happens when conduction block occurs at one or more sites in the reentrant circuit. The inability of rapid ventricular pacing to interrupt VT has been previously explained by assuming overdrive suppression of a protected reentrant focus.[82] Demonstrating progressive fusion during pacing at faster cycle lengths would exclude this interpretation. Failure to terminate the tachycardia could be due to inability to attain the critically short cycle lengths to interrupt the tachycardia,[71] or to the inappropriate site of pacing relative to the site of the reentrant circuit.[76,78]

In reciprocating tachycardia involving an accessory pathway, Waldo et al.[72] demonstrated block of the last atrial paced impulse in both directions. That is, the paced impulse blocked in the antidromic direction at the accessory pathway and in the orthodromic direction in the AV node. This occurred in 12 of their 15 patients. The other three patients showed block at the accessory pathway from the antidromic and orthodromic wave fronts of the same paced impulse. The antidromic wave front blocks at the accessory pathway and does not activate the ventricle; therefore, the morphology of the QRS is narrow. The orthodromic impulse activates the ventricles and blocks retrogradely at the accessory pathway. This occurs most likely because of concealed anterograde conduction of the atrial impulse.

Critical duration and rate of pacing are required to terminate tachycardia; and they are related to the time required to develop block, usually orthodromic in the region of slow conduction. Depending upon the timing of pacing offset, the same critical cycle length may interrupt tachycardia on one occasion but not on another. For example, during ventricular pacing in ORT, 2:1 conduction block may occur retrogradely over the accessory pathway. If pacing is terminated at a time when conduction block occurs, tachycardia will be interrupted. Alternatively, if ventriculoatrial conduction occurs in the last paced complex, tachycardia will continue. In the latter circumstance, it cannot be said that entrainment occurred because tachycardia had been interrupted and reinitiated with the entire heart being activated only by every *other* paced impulse. Contrast this with the continuous resetting of the tachycardia by *each* paced impulse during entrainment as described above.

Permanent Form of Junctional Reciprocating Tachycardia

Entrainment has been also demonstrated in the slowly conducting accessory pathway associated with the permanent form of junctional reciprocating tachycardia (PJRT).[83] Compared with the typical rapidly conducting accessory pathway, atrial pacing during ORT in PJRT does not produce atrial fusion because the antidromic atrial impulse from pacing collides with the or-

thodromic impulse in the slowly conducting accessory pathway. In the usual rapidly conducting accessory pathway, the retrograde impulse would have already exited the accessory pathway and fusion would have occurred in the atrium during pacing at the appropriate cycle length.

During ventricular pacing at different cycle lengths, progressive ventricular fusion is seen. Therefore, while entrainment during ventricular pacing fulfills criteria 1 and 2, entrainment during atrial pacing does not. In both atrial and ventricular pacing situations, tachycardia termination occurs when conduction block is produced in both directions. Pacing from either site yields conduction block due to refractoriness in the orthodromic direction and to collision in the antidromic direction.

AV Nodal Reentrant Tachycardia

Entrainment of AV nodal reentrant tachycardia (AVNRT) with atrial pacing was first demonstrated by Portillo et al.[84] The tachycardia was advanced to the pacing rate and, upon termination of pacing, the first ventricular activation occurred at the pacing cycle length before subsequently resuming the original tachycardia cycle length. This is explained by the pacing impulses entering the slow AV nodal pathway during the tachycardia's excitable gap and, at the same time, entering the circuit in an antidromic direction to collide with the previous retrograde impulse in the AV nodal fast pathway.

In another study of entrainment in AVNRT,[85] either atrial or ventricular pacing could produce concealed entrainment. In most cases, none of the four criteria for entrainment could be demonstrated. The atrial and ventricular rates could be increased to the pacing rate and, on termination of pacing, ventricular activation after the last atrial paced complex or atrial activation after the last ventricular paced complex occurred at the pacing cycle length, but no atrial or ventricular fusion occurred. Tachycardia resumed at its previous spontaneous cycle length following the last entrained complex. In only 3 of their 18 patients could they demonstrate any of the four criteria for entrainment.[85] In these patients, atrial pacing during AVNRT entrained the rhythm by activating the slow pathway orthodromically until anterograde block occurred in the slow pathway. The antidromic wave front of that paced impulse (n) collided in the fast pathway with the orthodromic wave front of the previous paced impulse (n − 1). The following paced impulse (n + 1) conducted anterogradely in the antidromic direction of the circuit and activated the ventricle from a different direction and with a shorter conduction time (third criterion).

These studies[84,85] add firm evidence to the hypothesis that AV nodal tachycardia is a reentrant rhythm utilizing two functionally different AV nodal pathways. Further, these two studies are relevant to the debate concerning the question whether the reentrant circuit responsible for AVNRT is entirely intranodal or whether it involves part of the atrium. A discussion on that topic is beyond the scope of this review, and the reader is referred elsewhere.[84–92]

Ventricular Tachycardia

Although earlier reports noted that VT could be terminated by overdrive pacing,[93–95] it was not until 1981 that MacLean et al.[71] noted that failure to interrupt VT was often associated with transient entrainment. In addition to being able to entrain VT from a ventricular site, entrainment of VT by atrial pacing has been demonstrated with[96] and without[97] the criteria for manifest entrainment.

As emphasized in an editorial by Brugada and Wellens,[98] entrainment is dependent upon a number of conditions: (a) Pacing must be performed at a site from which the paced impulse can gain entrance to the arrhythmia circuit. (b) The arrhythmia focus must not be protected by an entrance block. (c) A region of excitable gap must be present in the reentrant loop. (d) The paced impulses must be able to accelerate the tachycardia circuit to the paced rate. Conduction during spontaneous tachycardia cannot be at its upper rate limit.

At present, the ability to entrain a tachycardia remains the most useful criterion to demonstrate reentry and must be sought when attempting to ascribe a specific mechanism to an arrhythmia. However, the inability to reset or entrain does not exclude reentry as the mechanism. In the leading circle model of reentry,[16] the wave front is traveling at its maximum velocity with the leading edge of the impulse being right at its own tail of refractoriness. The impulse circles a functional region of nonconduction, which can be normal tissue that is continuously activated by the encircling wave front. In this scenario, no excitable gap is present, and the rhythm cannot be entrained.

AUTOMATICITY

Table 21.1 shows that automatic rhythms cannot be initiated or terminated by premature stimulation.[52,99,100] Thus, their behavior can only be assessed if they begin spontaneously in the electrophysiology laboratory. Automatic rhythms demonstrate warm-up whereby the first few cycle lengths gradually shorten until attaining the stable cycle length of the tachycardia.[52]

Both normal and abnormal automatic rhythms are temporarily suppressed, but not terminated, by overdrive pacing. The duration of this overdrive suppression of automaticity is usually related to the rate and duration of overdrive pacing and is dependent on maximum diastolic potential.[101] However, when overdrive pacing is performed at rates only slightly faster (10 to 15%) than the underlying rhythm, automatic rhythms may accelerate.[102] It is critical not to confuse this overdrive acceleration with entrainment.

As mentioned in the discussion of reentry, automatic rhythms may also be reset by premature extrastimuli.[51,52,99,100] This indicates absence of entrance block that would result in fully compensatory pauses. Parasystole is an automatic rhythm thought to be protected by entrance block but amendable to electrotonic modulation.[61,103,104]

TRIGGERED ACTIVITY

Like automatic and reentrant rhythms, triggered rhythms may undergo gradual acceleration shortly after initiation (warm-up). The response to programmed electrical stimulation is similar to that of reentrant arrhythmias[105] (Table 21.1). Triggered rhythms may mimic reentrant rhythms to some extent; they may be initiated and terminated by programmed stimulation and may accelerate to a faster rhythm during overdrive pacing.[40,42,43,106] However, it is important to repeat that the ability to entrain an arrhythmia can distinguish reentry from triggered activity. Also, it has been noted that the CI of triggered complexes is directly proportional to the prematurity of the extrastimulus.[48] The converse usually occurs in reentrant rhythms.[49] However, this behavior does not always separate the two mechanisms because the study of Moak and Rosen[48] showed that the typical behavior of DAD-induced triggered activity occurred in most but not in all experiments (83%).

Sung et al.[106–109] have shown that in patients with VT having characteristics consistent with triggered activity verapamil either terminated or prevented electrical induction of arrhythmia. Other patients in whom VT was more consistent with reentry and catecholamine-sensitive automaticity did not respond to verapamil. This pharmacologic manipulation may be useful in the electrophysiology laboratory when trying to determine the mechanism of an arrhythmia.

The behavior of triggered rhythms in humans awaits further investigation. The advent of a safe and reliable method for recording monophasic APs may assist in documenting the occurrence of afterpotentials in clinical situations.[110,111]

ACKNOWLEDGMENTS

The author wishes to express his gratitude to Douglas L. Packer, MD, for his critical review of the manuscript and to Renee Benson for her tireless, expert secretarial skills.

REFERENCES

1. Waller AD: A demonstration on man of electromotive changes accompanying the heart's beat. *J Physiol* 1887;8:229–234.
2. Einthoven W: Le telecardiogramme. *Arch Int Physiol* 1906;4:132–164.
3. Lewis T: *The Mechanism and Graphic Registration of the Heart Beat*, 3d Ed. London, Shaw and Sons, 1925.
4. Katz LN, Pick A: *Clinical Electrocardiography. I. The Arrhythmias*. Philadelphia, Lea & Febiger, 1956.
5. Pick A, Langendorf R: *Interpretation of Complex Arrhythmias*. Philadelphia, Lea & Febiger, 1979.
6. Fisch C: Evolution of the clinical electrocardiogram. *J Am Coll Cardiol* 1989;14:1127–1138.
7. Cooper JK: Electrocardiography 100 years ago. Origins, pioneers, and contributors. *N Engl J Med* 1986;315:461–464.
8. Durrer D, Schoo L, Schuilenburg RM, Wellens HJJ: The role of premature beats in the initiation and the termination of supraventricular tachycardia in the Wolff-Parkinson-White syndrome. *Circulation* 1967;36:644–662.
9. Schuilenburg RM, Durrer D: Atrial echo beats in the human heart elicited by induced atrial premature beats. *Circulation* 1968;37:680–693.
10. Wellens HJJ, Schuilenburg RM, Durrer D: Electrical stimulation of the heart in patients with Wolff-Parkinson-White syndrome, Type A. *Circulation* 1971;43:99–114.
11. Langendorf R, Pick A: Artificial pacing of the human heart. Its contribution to the understanding of the arrhythmias. *Am Heart J/Am J Cardiol* 1971;28:516–525.
12. Scherlag BJ, Helfant RH, Damato AN: A catheterization technique for His bundle stimulation and recording in the intact dog. *J Appl Physiol* 1968; 25:425–428.
13. Scherlag BJ, Lau SH, Helfant RH, Berkowitz WD, Stein E, Damato AN: Catheter technique for recording His bundle activity in man. *Circulation* 1969; 39:13–18.
14. Durrer D, Lie KI, Janse MJ, Schuilenburg RM: Mechanisms of tachyarrhythmias, past and present. *Eur J Cardiol* 1978;8:281–297.
15. Mines GR: On circulating excitations in heart muscles and their possible relation to tachycardia and fibrillation. *Trans R Soc Can* (ser 3, sect IV) 1914;8:43–52.
16. Allessie MA, Bonke FIM, Schopman FJG: Circus movement in rabbit atrial muscle as the mechanism of tachycardia. III. The "leading circle" concept: A new model of circus movement in cardiac tissue without involvement of an anatomical obstacle. *Circ Res* 1977;41:9–18.
17. Akhtar M, Damato AN, Batsford WP, Ruskin JN, Ogunkelu JB, Vargas G: Demonstration of re-entry within the His-Purkinje system in man. *Circulation* 1974;50:1150–1162.

18. Akhtar M, Gilbert C, Wolf FG, Schmidt DH: Reentry within the His-Purkinje system. Elucidation of reentrant circuit using right bundle branch and His bundle recordings. *Circulation* 1978;58:295–304.
19. Guerot C, Valere PE, Castillo-Fenoy A, Tricot R: Tachycardie par reentrée de branche à branche. *Arch Mal Coeur* 1974;67:1–11.
20. Reddy CP, Slack JD: Recurrent sustained ventricular tachycardia. Report of a case with His-bundle branch's reentry as the mechanism. *Eur J Cardiol* 1980;11:23–31.
21. Welch WJ, Strasberg B, Coelho A, et al.: Sustained macroreentrant ventricular tachycardia. *Am Heart J* 1982;104:166–169.
22. Touboul P, Kirkorian G, Atallah G, Moleur P: Bundle branch reentry: A possible mechanism of ventricular tachycardia. *Circulation* 1983;67:674–680.
23. Boineau JP, Cox JL: Slow ventricular activation in acute myocardial infarction. A source of re-entrant premature ventricular contractions. *Circulation* 1973;48:702–713.
24. Downar E, Harris L, Mickleborough LL, Shaikh N, Parson ID: Endocardial mapping of ventricular tachycardia in the intact human ventricle: Evidence for reentrant mechanisms. *J Am Coll Cardiol* 1988;11:783–791.
25. Josephson ME, Horowitz LN, Farshidi A: Continuous local electrical activity. A mechanism of recurrent ventricular tachycardia. *Circulation* 1978;57:659–665.
26. El-Sherif N, Mehra R, Gough WB, Zeiler RH: Reentrant ventricular arrhythmias in the late myocardial infarction period. Interruption of reentrant circuits by cryothermal techniques. *Circulation* 1983;68:644–656.
27. Pag PL, Cardinal R, Shenasa M, Kaltenbrunner W, Cossette R, Nadeau R: Surgical treatment of ventricular tachycardia. Regional cryoablation guided by computerized epicardial and endocardial mapping. *Circulation* 1989;80(suppl I):I-124–I-134.
28. Wit AL, Rosen MR, Hoffman BF: Electrophysiology and pharmacology of cardiac arrhythmias. II. Relationship of normal and abnormal electrical activity of cardiac fibers to the genesis of arrhythmias. *Am Heart J* 1974;88:515–524.
29. Vassalle M: The relationship among cardiac pacemakers. Overdrive suppression. *Circ Res* 1977;41:269–277.
30. Gilmour RF Jr, Zipes DP: Abnormal automaticity and related phenomenon, in Fozzard HA, Haber E, Jennings RB, Katz AM, Morgan HE (eds): *The Heart and Cardiovascular System*. New York, Raven Press, 1986, pp 1239–1257.
31. Rosen MR, Gelband H, Merker C, Hoffman BF: Mechanisms of digitalis toxicity. Effects of ouabain on phase four of canine Purkinje fiber transmembrane potentials. *Circulation* 1973;47:681–689.
32. Zipes DP, Arbel E, Knope RF, Moe GK: Accelerated cardiac escape rhythms caused by ouabain intoxication. *Am J Cardiol* 1974;33:248–253.
33. Kieval RS, Johnson NJ, Rosen MR: Triggered activity as a cause of bigeminy. *J Am Coll Cardiol* 1986;8:644–647.
34. Jackman WM, Friday KJ, Anderson JL, Aliot EM, Clark M, Lazzara R: The long QT syndromes: A critical review, new clinical observations and a unifying hypothesis. *Prog Cardiovasc Dis* 1988;31:115–172.
35. Le Marec H, Dangman KH, Danilo P Jr, Rosen MR: An evaluation of automaticity and triggered activity in the canine heart one to four days after myocardial infarction. *Circulation* 1985;71:1224–1236.
36. Damiano BP, Rosen MR: Effects of pacing on triggered activity induced by early afterdepolarizations. *Circulation* 1984;69:1013–1025.
37. Levine JH, Spear JF, Guarnieri T, et al.: Cesium chloride-induced long QT syndrome: Demonstration of afterdepolarizations and triggered activity in vivo. *Circulation* 1985;72:1092–1103.
38. Bailie DS, Inoue H, Kaseda S, Ben-David J, Zipes DP: Magnesium suppression of early afterdepolarizations and ventricular tachyarrhythmias induced by cesium in dogs. *Circulation* 1988;77:1395–1402.
39. Bonatti V, Rolli A, Botti G: Monophasic action potential studies in human subjects with prolonged ventricular repolarization and long QT syndromes. *Eur Heart J* 1985;6(suppl D):133–143.
40. Cranefield PF, Aronson RS: *Cardiac Arrhythmias: The Role of Triggered Activity and Other Mechanisms.* Mount Kisco, NY, Futura Publishing, 1988.
41. Cranefield PF: Action potentials, afterpotentials, and arrhythmias. *Circ Res* 1977;41:415–423.
42. Wit AL, Rosen MR: Afterdepolarizations and triggered activity, in Fozzard HA, Haber E, Jennings RB, Katz AM, Morgan ME (eds): *The Heart and Cardiovascular System.* New York, Raven Press, 1986, pp 1449–1490.
43. Wit AL, Cranefield PF: Triggered and automatic activity in the canine coronary sinus. *Circ Res* 1977;41:435–445.
44. Zipes DP, Foster PR, Troup PJ, Pedersen DH: Atrial induction of ventricular tachycardia: Reentry versus triggered automaticity. *Am J Cardiol* 1979;44:1–8.
45. Rosen MR, Gelband H, Marker C, Hoffman BF: Mechanisms of digitalis toxicity: Effects of ouabain on phase four of Purkinje fiber transmembrane potentials. *Circulation* 1973;47:681–689.
46. Cranefield PF, Aaronson RS: Initiation of sustained rhythmic activity by simple propagated action potentials in canine cardiac Purkinje fibers exposed to sodium-free solution of ouabain. *Circ Res* 1974;34:477–481.
47. Wit AL, Cranefield PF: Triggered activity in cardiac muscle fibers of the simian mitral valve. *Circ Res* 1976;38:85–98.
48. Moak JP, Rosen MR: Induction and termination of triggered activity by pacing in isolated canine Purkinje fibers. *Circulation* 1984;69:149–162.
49. Wellens HJJ, Duren DR, Lie KI: Observations on mechanisms of ventricular tachycardia in man. *Circulation* 1976;54:237–244.
50. Wellens HJJ, Schuilenburg RM, Durrer D: Electrical stimulation of the heart in patients with ventricular tachycardia. *Circulation* 1972;46:216–226.
51. Strauss HC, Saroff AL, Bigger JT, Giardina EGV: Premature atrial stimulation as a key to the understanding of sinoatrial conduction in man. Presentation of data and critical review of the literature. *Circulation* 1973;47:86–93.
52. Goldreyer BN, Gallagher JJ, Damato AN: The electrophysiologic demonstration of atrial ectopic tachycardia in man. *Am Heart J* 1973;85:205–215.
53. Inoue H, Inoue K, Matsuo H, Kuwaki K, Shirai T, Murao S: Resetting of tachycardia cycle by single and double ventricular extrastimuli in recurrent sustained ventricular tachycardia. *PACE* 1984;7:3–9.
54. Almendral JM, Rosenthal ME, Stamato NJ, et al.:

Analysis of the resetting phenomenon in sustained uniform ventricular tachycardia: Incidence and relation to termination. *J Am Coll Cardiol* 1986;8:294–300.
55. Almendral JM, Stamato NJ, Rosenthal ME, Marchlinski FE, Miller JM, Josephson ME: Resetting response patterns during sustained ventricular tachycardia: Relationship to the excitable gap. *Circulation* 1986;74:722–730.
56. Rosenthal ME, Stamato NJ, Almendral JM, Gottlieb CD, Josephson ME: Resetting of ventricular tachycardia with electrocardiographic fusion: Incidence and significance. *Circulation* 1988;77:581–588.
57. Stevenson WG, Weiss JN, Wiener I, et al.: Resetting of ventricular tachycardia: Implications for localizing the area of slow conduction. *J Am Coll Cardiol* 1988;11:522–529.
58. Kay GN, Epstein AE, Plumb VJ: Resetting of ventricular tachycardia by single extrastimuli. Relation to slow conduction within the reentrant circuit. *Circulation* 1990;81:1507–1519.
59. Kay GN, Epstein AE, Plumb VJ: Region of slow conduction in sustained ventricular tachycardia: Direct endocardial recordings and functional characterization in humans. *J Am Coll Cardiol* 1988;11:109–116.
60. Stamato NJ, Frame LH, Rosenthal ME, Almendral JM, Gottlieb CD, Josephson ME: Procainamide-induced slowing of ventricular tachycardia with insights from analysis of resetting response patterns. *Am J Cardiol* 1989;63:1455–1461.
61. Stevenson WG, Nademanee K, Weiss JN, et al.: Programmed electrical stimulation at potential ventricular reentry circuit sites. Comparison of observations in humans with predictions from computer simulations. *Circulation* 1989;80:793–806.
62. Jalife J, Michaels DC: Phase-dependent interactions of cardiac pacemakers as mechanisms of control and synchronization in the heart, in Zipes DP, Jalife J (eds): *Cardiac Electrophysiology and Arrhythmias.* Orlando, FL, Grune & Stratton, 1985, pp 109–119.
63. Perkel DH, Schulman JH, Bullock TH, Moore GP, Segundo JP: Pacemaker neurons: Effects of regularly spaced synaptic input. *Science* 1964;145:61–63.
64. Winfree AT: Oscillatory glycolysis in yeast. The pattern of phase resetting by oxygen. *Arch Biochem Biophys* 1972;149:388–401.
65. Guevara MR, Glass L: Phase locking, period doubling bifurcations and chaos in a mathematical model of a periodically driven oscillator: A theory for the entrainment of biological oscillators and the generation of cardiac dysrhythmias. *J Math Biology* 1982;14:1–23.
66. Castellanos A, Luceri RM, Moleiro F, et al.: Annihilation, entrainment and modulation of ventricular parasystolic rhythms. *Am J Cardiol* 1984;54:317–322.
67. Jalife J, Moe GK: A biologic model of parasystole. *Am J Cardiol* 1979;43:761–772.
68. Henthorn RW, Okumura K, Olshansky B, Plumb VJ, Hess PG, Waldo AL: A fourth criterion for transient entrainment: The electrogram equivalent of progressive fusion. *Circulation* 1988;77:1003–1012.
69. Lange G: Action of driving stimuli from intrinsic and extrinsic sources on in situ cardiac pacemaker tissues. *Circ Res* 1965;17:449–459.
70. Waldo AL, MacLean WAH, Karp RB, Kouchoukos NT, James TN: Entrainment and interruption of atrial flutter with atrial pacing. Studies in man following open heart surgery. *Circulation* 1977;56:737–745.
71. MacLean WAH, Plumb VJ, Waldo AL: Transient entrainment and interruption of ventricular tachycardia. *PACE* 1981;4:358–366.
72. Waldo AL, Plumb VJ, Arciniegas JG, et al.: Transient entrainment and interruption of atrioventricular bypass pathway type of paroxysmal atrial tachycardia. A model for understanding and identifying reentrant arrhythmias. *Circulation* 1983;67:73–83.
73. Almendral JM, Gottlieb CD, Rosenthal ME, et al.: Entrainment of ventricular tachycardia: Explanation for surface electrocardiographic phenomenon by analysis of electrograms recorded within the tachycardia circuit. *Circulation* 1988;77:569–580.
74. Portillo B, Castellanos A, Mejias J, Zaman L, Leon-Portillo N, Myerburg RJ: Right atrial-ventricular dissociation and entrainment while pacing from high right atrium and coronary sinus during circus movement tachycardias. *PACE* 1984;7:710–719.
75. Saoudi NC, Castellanos A, Zaman L, Portillo B, Schwartz A, Myerburg RJ: Attempted entrainment of circus movement tachycardias by ventricular stimulation. *PACE* 1986;9:78–90.
76. Okumura K, Henthorn RW, Epstein AE, Plumb VJ, Waldo AL: Further observations on transient entrainment: Importance of pacing site and properties of the components of the reentry circuit. *Circulation* 1985;72:1293–1307.
77. Zaman L, Castellanos A, Saoudi NC, et al.: Significance of pacing cycle lengths in manifest entrainment of orthodromic circus movement tachycardia by ventricular pacing. *Am J Cardiol* 1987;59:1325–1331.
78. Mann DE, Lawrie GM, Luck JC, Griffin JC, Magro SA, Wyndham CRC: Importance of pacing site in entrainment of ventricular tachycardia. *J Am Coll Cardiol* 1985;5:781–787.
79. Okumura K, Olshansky B, Henthorn RW, Epstein AE, Plumb VJ, Waldo AL: Demonstration of the presence of slow conduction during sustained ventricular tachycardia in man: Use of transient entrainment of the tachycardia. *Circulation* 1987;75:369–378.
80. Okumura K, Matsuyama K, Miyagi H, Tsuchiya T, Yasue H: Entrainment of idiopathic ventricular tachycardia of left ventricular origin with evidence for reentry with an area of slow conduction and effect of verapamil. *Am J Cardiol* 1988;62:727–732.
81. Kay GN, Epstein AE, Plumb VJ: Incidence of reentry with an excitable gap in ventricular tachycardia: A perspective evaluation utilizing transient entrainment. *J Am Coll Cardiol* 1988;11:530–538.
82. Josephson ME, Horowitz LN, Farshidi A, Spielman SR, Michelson EL, Greenspan AM: Sustained ventricular tachycardia: Evidence for protected localized reentry. *Am J Cardiol* 1978;42:416–424.
83. Castellanos A, Portillo B, Mejias J, Leon-Portillo N, Saoudi NC, Zaman L: Entrainment of circus movement tachycardia utilizing an accessory pathway with long retrograde conduction times during ventricular and atrial stimulation. *J Am Coll Cardiol* 1985;6:1431–1437.
84. Portillo B, Mejias J, Leon-Portillo N, Zaman L, Myerburg RJ, Castellanos A: Entrainment of atrioventricular nodal reentrant tachycardias during overdrive pacing from high right atrium and coronary sinus. With special reference to atrioventricular dissociation and 2:1 retrograde block during tachycardias. *Am J Cardiol* 1984;53:1570–1576.

85. Brugada P, Waldo AL, Wellens HJJ: Transient entrainment and interruption of atrioventricular node tachycardia. *J Am Coll Cardiol* 1987;9:769–775.
86. Josephson ME, Kastor JA: Paroxysmal supraventricular tachycardia: Is the atrium a necessary link? *Circulation* 1976;54:430–435.
87. Bauernfeind RA, Wu D, Denes P, Rosen KM: Retrograde block during dual pathway atrioventricular nodal reentrant paroxysmal tachycardia. *Am J Cardiol* 1978;42:499–505.
88. Scheinman MM, Gonzalez R, Thomas A, Ullyot D, Bharati S, Lev M: Reentry confined to the atrioventricular node: Electrophysiologic and anatomic findings. *Am J Cardiol* 1982;49:1814–1818.
89. Hariman RJ, Chen CM, Caracta AR, Damato AN: Evidence that AV nodal reentrant tachycardia does not require participation of the entire AV node. *PACE* 1983;6:1252–1257.
90. Brugada P, Wellens HJJ: Electrophysiology, mechanisms, diagnosis and treatment of paroxysmal recurrent atrioventricular nodal reentrant tachycardia, in Surawicz B, Reddy CP, Prystowsky EN (eds): *The Tachycardias.* Boston, Martinus Nijhoff, 1984, pp 131–158.
91. Ross DL, Johnson DC, Denniss AR, et al.: Curative surgery for atrioventricular junctional ("AV nodal") reentrant tachycardia. *J Am Coll Cardiol* 1985; 6:1383–1392.
92. Haissaguerre M, Warin JF, Lemetayer P, et al.: Closed-chest ablation of retrograde conduction in patients with atrioventricular nodal reentrant tachycardia. *N Engl J Med* 1989;320:426–433.
93. Bennett MA, Pentecost BL: Reversion of ventricular tachycardia by pacemaker stimulation. *Br Heart J* 1971;33:922–927.
94. Josephson ME, Horowitz LN, Farshidi A, Kastor JA: Recurrent sustained ventricular tachycardia. 1. Mechanisms. *Circulation* 1978;57:431–440.
95. Denes P, Wu D, Dhingra RC, et al.: Electrophysiological studies in patients with chronically recurrent ventricular tachycardia. *Circulation* 1976;54:229–236.
96. Waldo AL, Henthorn RW, Plumb VJ, MacLean WAH: Demonstration of the mechanism of transient entrainment and interruption of ventricular tachycardia with rapid atrial pacing. *J Am Coll Cardiol* 1984;3:422–430.
97. Almendral JM, Gottlieb C, Marchlinski FE, Buxton AE, Doherty JU, Josephson ME: Entrainment of ventricular tachycardia by atrial depolarizations. *Am J Cardiol* 1985;56:298–304.
98. Brugada P, Wellens HJJ: Entrainment as an electrophysiologic phenomenon. *J Am Coll Cardiol* 1984;3:451–454.
99. Scheinman MM, Basu D, Hollenberg M: Electrophysiologic studies in patients with persistent atrial tachycardia. *Circulation* 1974;50:266–273.
100. Gillette PC, Garson A Jr: Electrophysiologic and pharmacologic characteristics of automatic ectopic atrial tachycardia. *Circulation* 1977;56:571–575.
101. Dangman KH, Hoffman BF: Studies on overdrive stimulation of canine cardiac Purkinje fibers: Maximal diastolic potential as a determinant of the response. *J Am Coll Cardiol* 1983;2:1183–1190.
102. Lange G: Action of driving stimuli from intrinsic and extrinsic sources on in situ cardiac pacemaker tissues. *Circ Res* 1965;17:449–459.
103. Jalife J, Moe GK: A biologic model of parasystole. *Am J Cardiol* 1979;43:761–772.
104. Jalife J, Moe GK: Effect of electrotonic potentials on pacemaker activity of canine Purkinje fibers in relation to parasystole. *Circ Res* 1976;39:801–808.
105. Brugada P, Wellens HJJ: The role of triggered activity in clinical ventricular arrhythmias. *PACE* 1984;7:260–271.
106. Sung RJ, Shapiro WA, Shen EN, Morady F, Davis J: Effects of verapamil on ventricular tachycardias possibly caused by reentry, automaticity, and triggered activity. *J Clin Invest* 1983;72:350–360.
107. Sung RJ, Shen EN, Morady F, Scheinman MM, Hess D, Botvinick EH: Electrophysiologic mechanism of exercise-induced sustained ventricular tachycardia. *Am J Cardiol* 1983;51:525.
108. Sung RJ, Keung EC, Nguyen NX, Huycke EC: Effects of beta-adrenergic blockade on verapamil-responsive and verapamil-irresponsive sustained ventricular tachycardias. *J Clin Invest* 1988;81:688–699.
109. Sung RJ, Huycke EC, Lai WT, Tseng CD, Chu H, Keung EC: Clinical and electrophysiologic mechanisms of exercise-induced ventricular tachyarrhythmias. *PACE* 1988;11:1347–1357.
110. Franz MR: Long-term recording of monophasic action potentials from human endocardium. *Am J Cardiol* 1983;51:1629–1634.
111. Franz MR, Burkhoff D, Spurgeon H, Weisfeldt ML, Lakatta EG: In vitro validation of a new cardiac catheter technique for recording monophasic action potentials. *Eur Heart J* 1986;7:34–41.

PART V

Pharmacologic Treatment

Chapter **22**

Effect of Antiarrhythmic Drugs on Membrane Channels in Cardiac Muscle

Andras Varro, MD, PhD, and Borys Surawicz, MD

Several recent reviews have discussed the theories of antiarrhythmic drug actions. The purpose of this chapter is to present this material in a manner that may be helpful to understand the classification of antiarrhythmic drugs based on their effects on cardiac ionic membrane currents, action potential duration (APD), refractoriness (effective refractory period, ERP), conduction, and automaticity. We will focus the discussion on the knowledge accumulated mostly during the last decade. Readers interested in a more detailed treatment of this subject are referred to a number of recent original publications and review articles (see below).

As described in greater detail in other chapters of this book, cardiac arrhythmias may be caused by abnormal impulse formation, abnormal impulse propagation, or both. The abnormal impulse formation, such as early afterdepolarization (EAD) and delayed afterdepolarization (DAD), were reviewed by Hoffman and Rosen,[1] and more recently by Gintant and Cohen.[2] The EAD is implicated in arrhythmias associated with long QT, bradycardia, or toxic effects of drugs that prolong repolarization. The occurrence of DAD is associated with digitalis toxicity, and perhaps other states associated with calcium (Ca^{2+}) overload, for example, myocardial infarction.

Arrhythmias due to abnormal impulse propagation are explained by the reentry phenomenon, which depends critically on the relations between refractoriness and conduction velocity and requires the presence of a unidirectional block in one of the pathways. Factors controlling refractoriness and conduction include: APD, excitability, membrane passive properties, and cell-to-cell coupling. Antiarrhythmic drugs can stop propagation of reentrant impulse, or prevent reentry by either improving or depressing conduction. Improved conduction can either eliminate the unidirectional block or accelerate the returning wave front, which will extinguish reentry by encroaching on the still refractory fibers. Drugs that depress conduction can terminate or prevent reentry by transforming a unidirectional block into a bidirectional block. Drugs that prolong repolarization and refractoriness may prevent or terminate reentry if the reentering impulse finds the fiber inexcitable. Also, reentry may be prevented when the duration of repolarization is unaltered, but the refractoriness is prolonged, that is, the ratio of ERP to APD (postrepolarization refractoriness). To sustain the reentrant arrhythmia, the wave length (refractory period × conduction velocity) in the reentrant circuit must be shorter than the length of the pathway. Therefore, changes in the relation between refractory period and conduction velocity are critical to perpetuation or termination of reentry. The site of unidirectional block is usually attributed to a region of slow conduction, anatomic obstacle, or prolonged refractoriness. Reentry without an anatomical obstacle was shown in the model of leading circle in atrial tissue.[3] In this model, the initiation of reentry is caused by differences in the refractory period in close proximity when premature impulses block at the sites with a long refractory period. Therefore, drugs that change the refractory period may be particularly important in the treatment of arrhythmias due to this mechanism.

Automaticity, conduction, and refractoriness are controlled by the ionic currents flowing through different sarcolemmal channels. There-

655 Avenue of the Americas, New York, NY 10010
Current Topics in Cardiology

fore, the effects of antiarrhythmic drugs on these ionic channels must be considered as important determinants of their therapeutic effects.

Several attempts of classifications have been made to group the antiarrhythmic drugs with similar actions and to provide guidelines for their therapeutic use. The most popular classification (Tables 22.1 and 22.2) was originally proposed by Vaughan Williams in 1970 and 1975[4,5] and modified by others.[6,7] These classifications separate the antiarrhythmic drugs and/or antiarrhythmic actions into four different groups. The class I drugs decrease conduction by depressing the fast sodium current (i_{Na}); class II drugs exhibit antiadrenergic action; class III drugs lengthen repolarization and consequently refractoriness; and class IV drugs inhibit cardiac Ca^{2+} channels. Additionally, a fifth class of antiarrhythmic action was also proposed[8] for drugs that selectively induce bradycardia.

None of the drug classifications is free of inadequacies. Drugs belonging to the same class differ in their interaction with the given channel, an effect that has led to subclassifications.[6] More importantly, most antiarrhythmic drugs fall into more than one class. Also, classifications are based predominantly on studies in healthy cardiac tissues, paced at physiological frequencies. Drugs may produce different changes in the diseased tissues. Also, the heart rate, pH, and level of membrane potential (V_m) influence the effects of antiarrhythmic drugs on both depolarization and repolarization. Also, the classifications do not take into account differences in the effects of antiarrhythmic drugs on different types of cardiac tissues (sinus node, atria, AV node, ventricular, and Purkinje fibers) and on the same cardiac tissues from different species. In addition, the classifications fail to take into account the effects of drugs on passive membrane properties.[9,10] However, in spite of all these shortcomings and ambiguities, the classification has some merits, particularly in relation to the drug action on the membrane currents.

TABLE 22.1 Effect of Antiarrhythmic Drugs on Major Ionic Currents in the Heart

Drugs	i_{Na}	L-type i_{Ca}	i_{K1}	i_K	i_{to}	i_f
Class I						
A. Quinidine	—	—	0 —	—	—	—
Procainamide	—	—	0	—		0 —
Disopyramide	—	—	0 —	—	—	—
B. Lidocaine	—	0 —	0 —	0 —	0	—
Mexiletine	—	—	0	0		
Diphenyl hidantoin	—					
C. Flecainide	—	—	—	—		
Encainide	—					
Ethmozine	—	—				
Lorcainamide	—					
Propafenone	—	—				
Class II						
Propranolol	—					
Metoprolol	0	—	—			
Class III						
Amiodarone	—	—	—	—		
Sotalol			—	—	—	0
Bretylium				—		
Clofilium			0	—	—	
Class IV						
Verapamil	0	—		—		
Diltiazem		—		—		
Nisoldipine		—		—	—	

i_{Na} = inward sodium current, i_{Ca} = inward calcium current, i_{K1} = inward rectifier outward potassium current, i_K = delayed rectifier outward potassium current, i_{to} = transient outward potassium current, i_f = hyperpolarization activated inward pacemaker current (funny current).

TABLE 22.2 Effect of Antiarrhythmic Drugs on Cardiac Action Potential, Conduction, and Automaticity

	Conduct. or $\dot{V}_{max}$			APD			Normal automaticity			
Drugs	VM Pf	AV	Rec. kin. of $\dot{V}_{max}$	VM	Pf	ERP/ APD	SA	Pf	DAD	EAD
Class I										
A. Quinidine	—	—+*	Slow	+	‡	+	0	—	—	+
Procainamide	—		Slow	+	0+	+		—		
Disopyramide	—	—+*	Slow	+	+	0+	—	—		
B. Lidocaine	—	0—	Fast	—	—	+		—	—0	—
Mexiletine	—		Fast	—	—	+			—	—
Diphenylhidantoin	—		Fast	—	—	+			—	
C. Flecainide	—	—	Slow	+0	—	+	—			
Encainide	—	—	Slow	0	—		—	—		
Ethmozine	—	—	Slow	—			—	—		—
Lorcainamide	—	0—	Slow	—	—	+	—	—		
Propafenone	—	—	Slow	0	—	+	0	—		
Class II										
Propranolol	0—	—	Slow	0+†	0—	0				
Metoprolol	0	—		0+†	0					
Class III										
Amiodarone	—	—	Fast	0+†	0—	0	—	—		
Sotalol	0			+	+	0	—			+
Bretylium	0			+	+					
Clofilium	0			+	+	0		—		
Class IV										
Verapamil	0—	—		—	+	0	—	—	—	—
Diltiazem	0—	—		—	+	0	—	—0	—	—
Nisoldipine	0	—		—	+	0	—		—	—

VM = ventricular muscle, Pf = Purkinje fiber, AV = atrioventricular node, SA = sinoatrial node, ERP = effective refractory period, APD = action potential duration, EAD = early afterdepolarization, DAD = delayed afterdepolarization, slow = time constant > 1 s, fast = time constant < 1 s, * = anticholinergic effect, † = after chronic application, ‡ = variable.

EFFECTS MEDIATED THROUGH FAST Na^+ CHANNELS (CLASS I ACTION)

Most of the purely antiarrhythmic drugs depress conduction due to inhibition of the fast i_{Na}. It was recognized long ago[11] that antiarrhythmic drugs depress maximal rate of depolarization in a use- and frequency-dependent manner. To explain the frequency-dependent effect of local anesthetic-type antiarrhythmic drugs, Hille[12] and Hondeghem and Katzung[13,14] proposed the modulated receptor hypothesis. According to this hypothesis (Fig. 22.1), drugs can bind to and dissociate from a receptor postulated to be intimately related to the Na^+ channel. The binding and unbinding are believed to depend on the channel state, with differences in the rate constants of association and dissociation for rested, activated, and inactivated channel states. The drug interaction is assumed to make the channel nonconductive and to alter the channel gating by shifting inactivation into the hyperpolarizing direction. Local anesthetic drugs were shown to preferentially block either activated[15,16] and/or inactivated[17,18] Na^+ channels, while recovery from block occurs during the rested state[14,19,20] (Fig. 22.2).

According to an alternative guarded receptor hypothesis,[21] the frequency-dependent blocking effects of local anesthetic antiarrhythmic drugs on Na^+ channels are attributed to the lack of the blocking agent's access to the channel receptor regulated by the channel gating apparatus. This model postulates the ion channel blockade as a two-stage process: diffusion of drug to a region near the channel binding site and coupling of drug to the binding site resulting in

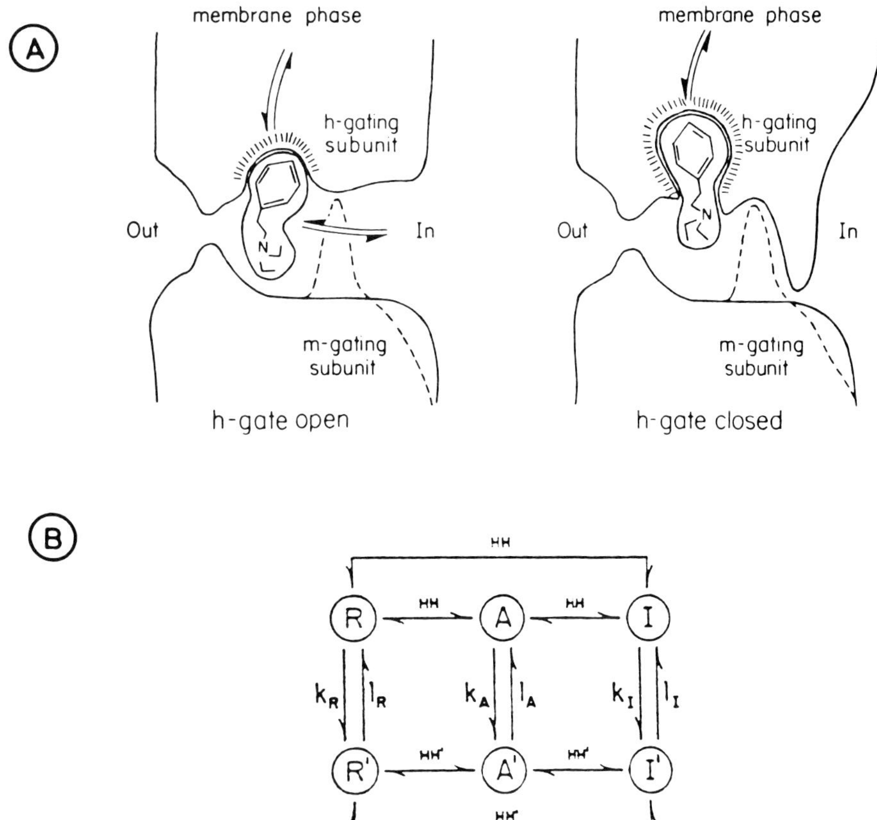

FIGURE 22.1 **A.** Diagram of a local anesthetic molecule binding in the pore of a Na^+ channel in a manner that promotes Na^+ inactivation. The molecule can reach its binding site from the intracellular solution if activation and inactivation gates are both open. It can also reach the site from the membrane phase even if one or both of the gates are closed. The binding site has an important hydrophobic component (shading) and closure of the inactivation gate enhances the hydrophobic interaction. **B.** Schematic diagram illustrating the state-dependent drug binding and unbinding according to the modulated receptor hypothesis. *Reprinted from Hille,*[12] *with permission.*

a nonconducting drug-modified channel. The binding site access through the diffusion path was postulated to be regulated by channel gating. Therefore, relationships for equilibrium properties of guarded receptors differ from those of traditional unguarded receptors.

Antiarrhythmic drugs are weak bases, and, therefore, exist in the body in both neutral and cationic forms. Charged cationic drugs get access to the receptor through the hydrophilic route (Fig. 22.1) only when the channel is open from the inside of the sarcolemma, whereas neutral drugs can get access through the hydrophobic pathways (Fig. 22.1) when channels are closed, that is, during rested and inactivated state.[12] Of importance is that the ratio of the charged to uncharged form of the drug may change with pH. Therefore, the effects of the antiarrhythmic drugs on the Na^+ channel can be influenced by pH changes,[22,23] which may be of importance during ischemia. It should be also emphasized that the antiarrhythmic drugs that bind preferentially to inactivated Na^+ channels exert a greatly enhanced effect in diseased depolarized tissues.

Although all antiarrhythmic drugs with local anesthetic action studied so far have been shown to depress the fast i_{Na} and consequently maximal rate of depolarization ($\dot{V}_{max}$) and conduction velocity (see later) in a frequency-dependent manner, the frequency dependence may vary, which gives rise to a subclassification.[6] Some drugs attain steady-state depression of i_{Na} or $\dot{V}_{max}$ after a few beats (fast-onset kinetics) and recover rapidly from block after cessation of stimulation (fast-offset kinetics). Other drugs have both slow onset and offset kinetics of Na^+ channel block. The onset and offset kinetics of the Na^+ channel block of a given drug can be predicted from their

physico-chemical properties, such as lipid solubility and molecular weight.[24,26] Drugs with slow-offset kinetics of Na^+ channel block depress fast i_{Na} within a wide range of frequencies, while the effects of drugs with fast-offset kinetics are limited only to high frequencies or early premature intervals. Therefore, drugs with slow-recovery kinetics would depress conduction within a wide range of heart rates and would have similar blocking effects on premature and basic impulses.

The frequency-dependent effects of local anesthetic-type antiarrhythmic drugs on i_{Na} and maximum rate of depolarization $\dot{V}_{max}$ can be extended to the effect on conduction even though the relationship between the fast i_{Na}, $\dot{V}_{max}$, and conduction velocity are not linear.[26–28] The maximal rate of depolarization of the action potential (AP) is related to the square root of conduction velocity,[28,29] both in the presence and in the absence of Na^+ channel-blocking drugs. The frequency-dependent effect of antiarrhythmic drugs on conduction has been confirmed both in vitro,[29–31] and in vivo in animals[32–34] and in patients (frequency-dependent widening of the QRS).[35,36]

It has been emphasized that impulse conduction is anisotropic,[37,38] that is, conduction velocity is different in the longitudinal and the transverse direction due to the different arrangement of intercalated discs along the conduction pathways. Preliminary studies suggest that antiarrhythmic drugs differ in their effect on anisotropic conduction.[39,40]

Of additional interest are recent observations[41,42] that local anesthetic drugs may competitively displace each other from the same binding site. This is a basis of a yet clinically untested hypothesis that drug toxicity caused by drugs with slow-recovery kinetics may be effectively treated using drugs with fast-recovery kinetics.

Another effect inherent in the action of local anesthetic-type antiarrhythmic drugs on the fast inward i_{Na} is an increased ratio of refractoriness to APD, that is, the ERP/APD. Campbell[6] postulated that drugs with fast-onset and offset kinetics will increase ERP/APD more than those with slow-onset and offset kinetics, and concluded that class IB drugs have the largest ERP/APD ratio, class IC the smallest, and class IA an intermediate ratio. However, the effect of drug on kinetics of $\dot{V}_{max}$ appears to be not the only factor controlling the ERP/APD ratio because contrary to Campbell's postulate,[6] amiodarone, which exhibits both fast-onset and offset kinetics of $\dot{V}_{max}$ block,[43] did not increase the ERP/APD ratio.[44]

Depression of the fast inward i_{Na} can not only influence the conduction but can also alter repolarization and refractoriness in cardiac cells by the effect on APD. Voltage-clamp experiments in cardiac Purkinje fibers have shown that tetrodotoxin (TTX), a specific blocker of the Na^+ channel, shifted the steady-state current voltage relation in the outward direction with a resulting overlap of the steady-state i_{Na} activation and inactivation curves.[45] This process creates a window for small but steady-state inward depolarizing i_{Na} flowing during the plateau phase of the AP. The so-called window current has been shown to be depressed by quinidine,[46] lidocaine,[46,47] and procainamide,[47] an effect believed to account for the APD shortening produced by these drugs in the Purkinje fibers. In addition, a very slowly inactivating component of the fast inward i_{Na} has been described in Pur-

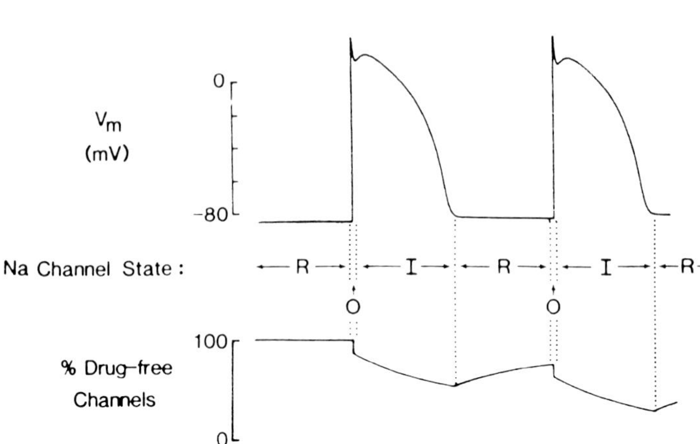

FIGURE 22.2 Schematic illustration of the time-dependent changes in Na^+ channel states, rested (R), open (O), and inactivated (I), associated with cardiac action potentials (**top**) and the resulting changing level of Na^+ channel block (**bottom**) by a local anesthetic-type antiarrhythmic drug. Note that this drug causes some block during the open channel state (upstroke of the AP) and additional block during the inactivated state (plateau). During diastole, partial recovery from block occurs because the affinity of the drug for the channel is low. *Reprinted from Clarkson and Hondeghem,[20] with permission.*

kinje fiber.[48,49] This current component may help to maintain the plateau phase of the cardiac AP, and if decreased by the local anesthetic-type antiarrhythmic drugs will also shorten APD.

In a recent study in guinea pig ventricular myocytes, two types of late single i_{Na} currents were described in single-channel recordings.[50] One of these was called a "background type," independent of the holding potential and consisting of brief random channel openings throughout the depolarization. The other, called "burst type," showed very rare, but sustained, channel openings with rapid interruptions. In the same experiments, whole-cell voltage-clamp measurements yielded a late i_{Na}, which was abolished by TTX with a concomitant shortening of APD. These results confirm the importance of the i_{Na}'s participation in the repolarization of cardiac cells, but further studies are needed to understand the exact mechanisms.

ANTIADRENERGIC MECHANISM (CLASS II ACTION)

The antiarrhythmic drugs in this category depress sinoatrial (SA) automaticity and slow impulse conduction through the atrioventricular (AV) node. They may suppress arrhythmogenic mechanism of adrenergic stimulation resulting from increased automaticity and contractile force, that is, factors that increase the energy demand/supply ratio in the heart. In addition, the effects of β-adrenergic blockers can be caused by the effect on Ca^{2+} currents (i_{Ca}). The enhanced i_{Ca} facilitates both normal[51,52] and abnormal automaticity[53] because β-1 receptor stimulation increases intracellular cyclic AMP levels via the adenyl cyclase system.[52] Elevated cyclic AMP level increases the inward L-type i_{Ca} by increasing the probability of channel openings and by increasing the number of functioning channels.[52] The effects of β-1 receptor stimulation on T-type Ca^{2+} channels have been conflicting. In some studies, the current was increased,[54] but several investigators[55–57] failed to observe changes in this current in the presence of β-agonists.

β-2 receptor stimulation, which is also coupled to the adenyl cyclase system,[58] was reported to increase L-type i_{Ca} in guinea pig atrial, but not in the ventricular, myocytes,[59] probably because of the paucity of the β-2 receptors in this preparation. β-Adrenergic stimulation also increases the transient inward current, responsible for DAD, by cyclic AMP-dependent direct modulation of the sarcoplasmic Ca^{2+} release.[60]

The effect of β-agonists on APD is complex, because the APD lengthening effect of enhanced i_{Ca} resulting from β-adrenergic stimulation is counteracted by several shortening effects of β-receptor stimulation on one or more potassium (K^+) currents (i_K)[61,62] and perhaps chloride (Cl^-) current (i_{Cl}).[63,64] Also, it has been reported that chronic β-adrenergic blockade lengthens repolarization in cardiac muscle.[65]

In conclusion, it is well established that antiadrenergic drugs are useful in controlling different types of supraventricular and ventricular arrhythmias[66,67] not exclusively by opposing the effect of sympathetic stimulation. However, our understanding of the mechanism by which modulation of cardiac ionic channels by antiadrenergic drugs contributes to the antiarrhythmic action is incomplete.

LENGTHENING REPOLARIZATION (CLASS III ACTION)

The duration of cardiac AP is one of the most important factors that determine the duration of the effective refractory period. APD is determined by complex interactions of several depolarizing and repolarizing currents, all of which can be affected by the drugs. Repolarization in cardiac cells can be prolonged by changes in the magnitude or kinetics of several different current systems. One of the mechanisms by which antiarrhythmic drugs increase APD is decreasing the magnitude or changing the kinetics of the delayed rectifier K^+ current (i_K). This outward repolarizing current activates slowly at potentials more positive to -20 mV and deactivates at negative potentials. Another important current that determines resting potential and terminates repolarization in cardiac cells is the time-independent inward rectifying K^+ current (i_{K1}). A decrease of this current would lengthen cardiac APD and refractoriness, an action postulated for several antiarrhythmic drugs.[68–71]

APD can also be lengthened by increasing inward current. This is believed to be the effect of some toxins and alkaloids, like *Anemonia sulcata* toxin (ATX II) veratrine, and DPI 201-106, a newly developed positive inotropic agent. These compounds lengthen APD in cardiac Purkinje[72] and ventricular muscle fibers[73,74] by slowing the inactivation of the fast inward i_{Na}.[74] This type of action reveals a new approach for developing antiarrhythmic drugs capable of increasing APD by mechanisms other than inhibiting outward i_K.

Drugs that directly increase the inward i_{Ca}, like dihydropyridine Ca^{2+} channel agonists Bay K 8644,[52] also increase ventricular APD, but may not be useful in practice because of the reported induction of EAD,[53] and coronary and vascular spasms.[75]

The electrogenic Na/Ca exchange system generates both outward and inward membrane currents during the AP depending on the membrane potential and intracellular Ca^{2+} activ-

ity.[76] It is believed[77] that at resting potential the Na/Ca exchange system mediates the movement of 3 Na^+ ions into the cell for every Ca^{2+} ion transported outward, resulting in a net inward current. At the plateau potential range, this current is reversed and mediates the transfer of 3 Na^+ ions outward for every Ca^{2+} ion inward, resulting in a net outward current. Accordingly, the inhibition of the Na/Ca exchange system, reported to be induced by quinidine and procainamide,[78] may be expected to lengthen the AP plateau and decrease the rate of spontaneous diastolic depolarization near the resting potential.[78]

It has to be emphasized that all drugs that lengthen repolarization regardless of their ionic mechanism may induce EADs in Purkinje fibers and/or increase dispersion of repolarization,[79] that is, processes that may aggravate arrhythmias. It has been suggested that the proarrhythmic complication may be lessened by combination of these drugs with other types of antiarrhythmic drugs.[72,80]

Acute hypoxia shortens repolarization[81,82] and favors arrhythmogenesis.[81] The hypoxia-induced APD shortening has been attributed to a decrease of inward i_{Ca}[81,82] or/and increase of ATP-dependent i_K.[81,83] The latter current is normally suppressed at physiological intracellular ATP levels[84] and activated when the intracellular ATP concentration decreases drastically.[84] Therefore, drugs that selectively inhibit this ATP-dependent i_K such as the antidiabetic drug glibenclamide[85] have been shown to prevent or lessen the hypoxia-induced APD shortening without affecting the repolarization of the normal cardiac cells, an action which may decrease ischemia-induced dispersion of repolarization. Indeed, in experimental arrhythmia models, ATP-dependent K^+ channel blockers protected from arrhythmias induced by acute coronary ligation.[86,87] Recent study showed also that in therapeutically relevant concentrations, quinidine and amiodarone inhibited the equivalent of ATP-sensitive i_K measured by the rate of rubidium uptake, and suggested that this effect might contribute to the antiarrhythmic action of these drugs.[88] Further research is needed to clarify the possible antiarrhythmic effects of the ATP-sensitive i_K inhibition in humans.

INHIBITION OF INWARD i_{Ca} (CLASS IV ACTION)

The inhibition of the inward i_{Ca} may have both a direct antiarrhythmic effect and an indirect effect due to the effects on hemodynamics and myocardiac ischemia.

Two types (T- and L-type) of i_{Ca} have been described in cardiac tissues.[54,57] These two i_{Ca} differ in voltage range of activation and inactivation, kinetics, single-channel conductance, and sensitivity to pharmacological agents.

The T-type i_{Ca} is believed to play a role in the pacemaker function of the SA[89] and Purkinje fibers.[56,90] Agents that block T-type i_{Ca} selectively do not exist yet, but if developed, they could be potentially useful in the treatment of disturbed automaticity.

The L-type i_{Ca} is the major current involved in the depolarization of the SA[89] and AV nodal cells.[91] Therefore, drugs that decrease the magnitude of the L-type i_{Ca} slow sinus node automaticity and depress AV conduction. They may also affect automaticity and conduction in other cardiac tissues depolarized to the level at which inward i_{Ca} becomes activated.

In analogy to the use-dependent drug effect on the Na^+ channel, the effect of drugs on the i_{Ca} is also voltage and frequency dependent and has been interpreted within the framework of the modulated and guarded receptor hypotheses.

In Purkinje fibers, EADs, induced by Bay K 8644, a specific activator of the L-type calcium channel,[53] were effectively suppressed by Ca^{2+} channel inhibitors.[53] These and other findings suggest that blockade of Ca^{2+} may eliminate or prevent arrhythmias caused by EAD. Also, DAD and triggered automatic rhythms dependent on both EADs and DADs will be suppressed.

Intracellular Ca^{2+} concentration ($[Ca^{2+}]_i$) increases during ischemia.[81] This may slow impulse conduction by electrical uncoupling of the cardiac cells.[92] Drugs that decrease inward i_{Ca} and consequently prevent Ca^{2+} overload may reduce or abolish the ischemia-induced conduction delay.[93–95]

BRADYCARDIC ACTION (CLASS V ACTION)

The clonidine derivative, alinidine, and the verapamil derivative, falipamil, have been termed "specific bradycardic agents,"[96] because they decrease heart rate by a direct effect on the sinus node. Recently, voltage-clamp studies have shown that both alinidine[97,98] and falipamil[99,100] reduced the inward pacemaker current, i_f. However, bradycardiac effect was present when i_f was inhibited by cesium.[101] Both alinidine[97] and falipamil[102] influence ionic currents other than i_f, and at this time the exact ionic mechanism by which these drugs slow the sinus rate is unsettled.

SUMMARY OF THE MOST IMPORTANT DRUG EFFECTS

Quinidine

The effects of quinidine on different cardiac ionic channels are shown in Table 22.1. Quinidine, as

expected from its effect on $\dot{V}_{max}$, reduced in a use-dependent manner the magnitude of the fast inward i_{Na} in rat ventricular myocytes,[103] in rabbit Purkinje fibers,[46] and in guinea pig ventricular myocytes.[104] Quinidine preferentially blocks the open or activated Na^+ channels.[15] Recent study has shown that the recovery from quinidine-induced Na^+ channel block occurred not only during the rested state, but also during the activated state.[104] Quinidine also reduces the Na^+ window inward current[46] or the slowly inactivating component of the fast inward i_{Na}.[70]

Quinidine depresses inward i_{Ca} in frog atrial[105] and in cat papillary muscle,[106] in dog Purkinje cells,[107] in dog ventricular myocytes,[70] in guinea pig ventricular myocytes,[69] and in rabbit AV node.[108] Quinidine delays the recovery of the i_{Ca} from inactivation.[105] The depressing effect of quinidine on the L-type i_{Ca} was use-dependent both in dog ventricular myocytes[70] and in rabbit AV node.[108]

The effect of quinidine on the time-independent i_{K1} has varied in different studies depending on different experimental conditions. Some investigators reported moderate decrease,[69,70] and others no significant change.[46,47,107,109] In contrast, there is an agreement that quinidine, independently of the species, preparation, and experimental conditions, markedly inhibits the time-dependent delayed rectifier outward K^+ current (i_K).[46,69,105,108,110,111] In guinea pig ventricular myocytes, Roden et al.[110] found that quinidine not only depresses the magnitude of the i_K, but also delays the activation of this current without significantly changing its activation and deactivation kinetics. In rabbit SA and AV node,[111] quinidine slowed the deactivation of i_K, an effect linked to its modulation of the pacemaker function.

Quinidine also reduces the magnitude of the transient outward potassium current (i_{to}), and delays its recovery from inactivation.[109] The depressing effect of quinidine on i_{to} may play a role in the modulation of the frequency-dependent changes in repolarization in Purkinje fibers and in ventricular muscle fiber, but it is probably of greater importance in atrial tissue, where this current is believed to play a major role in controlling repolarization.[112]

Quinidine decreases the magnitude of the pacemaker current (i_f) in sheep Purkinje fibers[47] and the rabbit AV node.[108] This effect, however, appears to play a small role in reducing the rate of diastolic depolarization in vivo, probably because of the counteracting anticholinergic action of quinidine.

In sheep Purkinje fibers, quinidine reduced the magnitude of the transient inward current.[113] As mentioned earlier, quinidine has been reported to depress the Na/Ca exchange system[78] and the ATP-sensitive i_K,[88] but further studies are needed to clarify the clinical significance of these effects.

Studies using microelectrodes (refs. in 66 and 67) in isolated superfused cardiac preparations (Table 22.2) have shown that quinidine at therapeutically relevant concentrations decreases $\dot{V}_{max}$ and conduction velocity in different cardiac preparations. The effect of quinidine on $\dot{V}_{max}$ is use dependent with slow-onset and slow-offset kinetics[6] and the reported recovery of $\dot{V}_{max}$ time constants ranging from 4 to 8 s[23,114] in cardiac ventricular and Purkinje fibers. This indicates that the depressing effect of the drug on $\dot{V}_{max}$ and conduction extends over a wide range of frequencies.

Quinidine lengthens repolarization of atrial and ventricular muscle, but the effect on repolarization in Purkinje fibers varies depending on concentrations and experimental conditions, such as the stimulation frequency[115] and prematurity interval.[116] Quinidine also decreases the range of premature APD in dog Purkinje fibers,[116] an effect that may be expected to diminish the dispersion of repolarization in premature impulses. Quinidine increases ERP/APD and slows spontaneous Purkinje fiber frequency without significantly changing the SA node rate. However, in vivo the sinus node rate is increased usually and AV conduction time is decreased because of quinidine's anticholinergic effect. Quinidine was reported to decrease DAD,[117] but it can elicit EAD.[118–120]

Procainamide and Disopyramide

The effects of procainamide and disopyramide on ionic currents are similar to those of quinidine, but have been studied less frequently. Both procainamide and disopyramide depress the inward i_{Na}[47] in a use-dependent manner.[121] Disopyramide preferentially binds to the open Na^+ channel.[121]

Both drugs depress the inward i_{Ca}[122,123] and the i_K (Fig. 22.3)[122,123] without significantly altering the time-independent i_{K1}.[47,123] Disopyramide was reported to decrease the transient outward i_K in sheep Purkinje fiber.[124] i_f was reported to be decreased by disopyramide in the rabbit sinus node[125] and in sheep Purkinje fibers.[124]

The effects of procainamide and disopyramide on APs and automaticity are very similar to those of quinidine and are summarized in Table 22.2 (refs. in 66 and 67).

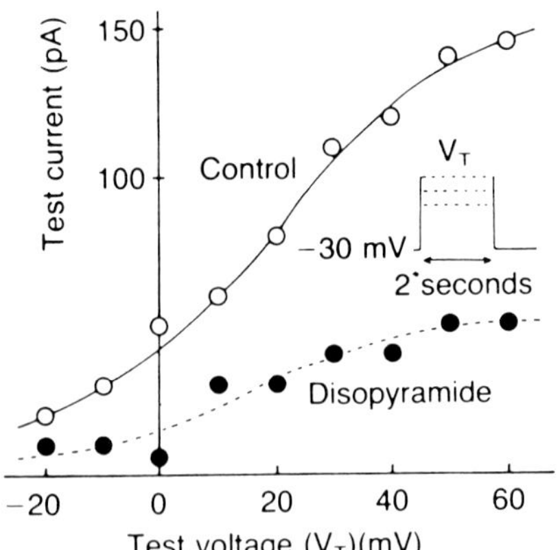

FIGURE 22.3 Effects of 11 μM disopyramide on the outward tail current in ventricular myocyte. The voltage protocol is shown in the inset. The vertical axis is the tail current amplitude (that is, the delayed outward K^+ current, i_K) on the repolarization from test voltages to −30 mV. *Reprinted from Hiraoka et al.,*[123] *with permission.*

Lidocaine

The effects of lidocaine on cardiac ionic channels and AP characteristics are summarized in Tables 22.1 and 22.2. Lidocaine reduced the magnitude of the fast inward i_{Na} in various species and in various types of cardiac tissues, including rat,[18,103] guinea pig[126] and chicken ventricle,[127] rabbit Purkinje fibers,[17,46] and sheep Purkinje fibers.[47] This effect was use dependent.[17,18] Lidocaine depresses the Na^+ window current[46] and/or the slowly inactivating component of the fast i_{Na}.[124,127,128] Some studies indicated that lidocaine preferentially binds to the inactivated Na^+ channel,[17,18] but others found that lidocaine binds to both open and inactivated Na^+ channels.[129]

Nilius et al.[130] showed that 5 μM lidocaine decreased both the opening probability and the mean open time of the single Na^+ channel without changing the unitary current through the channel (Fig. 22.4). They interpreted this as a direct action of lidocaine on the open formation of the channel protein.[130] Although the peak inward i_{Na} was depressed only by 40%, the late burst-type openings of the single Na^+ channels were completely abolished by 5 μM lidocaine. This is consistent with the observation that in cardiac muscle APD shortening occurs at lower lidocaine concentration than the decrease of $\dot{V}_{max}$. The results of Nilius et al.[130] in the single Na^+ channel were confirmed by McDonald et al.[131] However, Grant et al.[132] found that 80 μM lidocaine decreased in a use-dependent manner the opening probability of the single Na^+ channels without changing the mean opening time of the channels. In this study,[132] the late burst-type openings decreased, but did not disappear in the presence of even high concentrations of lidocaine. It appears that further studies are needed to understand fully the mechanism of the lidocaine-induced Na^+ channel block at the single-channel level.

The effect of lidocaine on other cardiac ionic currents is still controversial. In therapeutically relevant concentrations, lidocaine did not affect the inward i_{Ca} in dog[128] and guinea pig[133] ventricular myocytes. In other studies, however, moderate decrease of the amplitude of the inward i_{Ca} was reported.[124,126,127,134] Early voltage-clamp experiments attributed the lidocaine-induced APD shortening to the increase of the inward rectifier outward i_K.[135] Recent studies, however, showed that lidocaine either did not affect this current[124,126,128] or even decreased it.[127,136] Unlike quinidine and disopyramide, lidocaine did not affect the transient outward i_K current.[124] At therapeutic concentrations, lidocaine either did not affect the i_K[126] or only slightly depressed it.[46] Therefore, the lidocaine-induced APD shortening is best explained by the depression of the i_{Na} system.

Lidocaine was reported to decrease the hyperpolarization-induced inward i_f,[47,124] an action that decreases the spontaneous rate of depolarizations in Purkinje fibers.

The transient inward current was depressed by lidocaine in Purkinje fibers, but not in guinea pig ventricular muscle cells.[133] From this, differences between the effects of the drug on triggered automaticity in Purkinje fiber and ventricular muscle have been postulated.

Conventional microelectrode studies (refs. in 66 and 67; Table 22.2) have shown that lidocaine decreases conduction velocity and $\dot{V}_{max}$ in different types of cardiac tissues. This effect of lidocaine is use dependent with relatively fast onset and offset kinetics. The reported recovery of $\dot{V}_{max}$ time constant values at 35 to 37°C ranged from 100 to 200 ms,[22,114] indicating that lidocaine depresses $\dot{V}_{max}$ and conduction only at fast heart rate or at short premature coupling intervals. Similar to other Na^+ channel blockers, the effect of lidocaine on $\dot{V}_{max}$ is increased in depolarization tissues.

The APD shortening effect of lidocaine depends on the stimulation frequency[115] and prematurity.[116] Lidocaine slowed the kinetics of restitution of APD in dog Purkinje fibers, an effect that may be expected to decrease the dispersion of repolarization in premature impulses.[116] At therapeutic concentrations, lidocaine increases

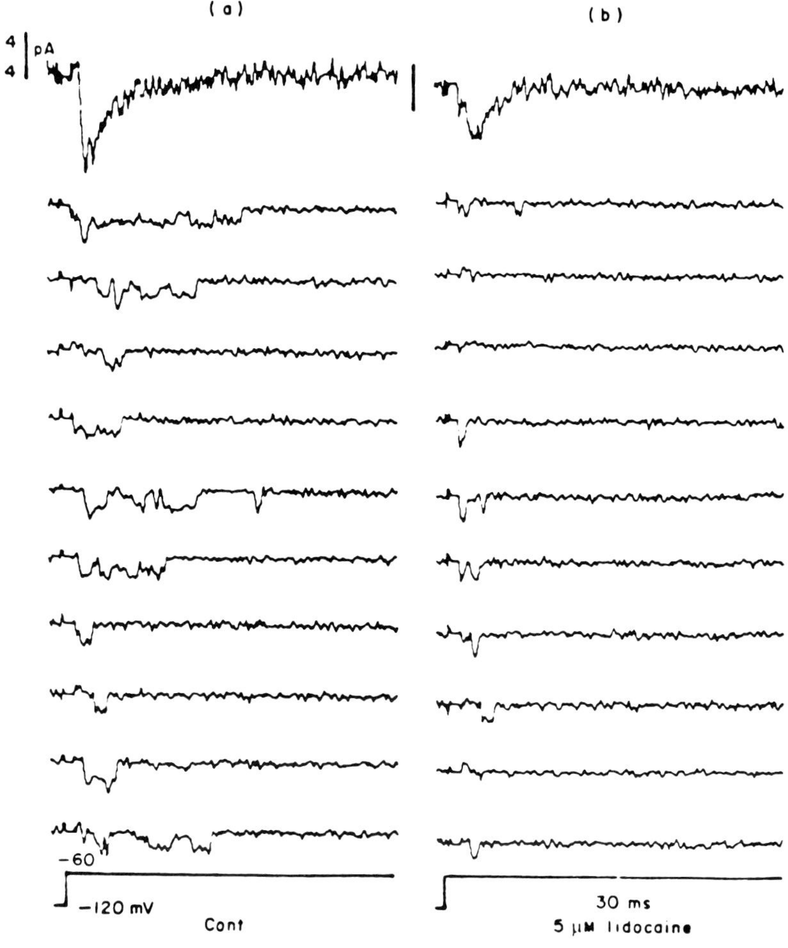

FIGURE 22.4 Effect of 5 μM lidocaine applied from inside to a cell free patch. Downward deflections reflect single Na^+ channel openings. **a.** Control, 30 ms step from −120 to −60 mV. **b.** Channel activity after 5 μM lidocaine. Note the appearance of the less frequent very short openings in the presence of 5 μM lidocaine. The averaged currents were obtained from 76 sweeps. *Reprinted from Nilius et al.,*[130] *with permission*

ERP/APD and often causes postrepolarization refractoriness. Lidocaine does not affect sinus node automaticity but depresses normal automaticity in Purkinje fibers. Both EAD[137] and DAD[138] were reported to be abolished by lidocaine in the Purkinje fibers.

Mexiletine and Phenytoin

Mexiletine and phenytoin were reported to depress the fast i_{Na} in cardiac muscle.[139–141] Mexiletine at therapeutically relevant concentration decreased the inward i_{Ca} in guinea pig ventricular myocytes, an effect believed to assist the shortening of APD induced by the depression of the inward i_{Na}.[134]

In the rabbit SA node cells, 40 to 100 μM mexiletine, in addition to reducing the inward i_{Ca}, also decreased the i_K and the i_f.[142] Table 22.2 shows that the effects of mexiletine and phenytoin on AP characteristics and automaticity are similar to those of lidocaine.

Flecainide, Encainide, Ethmozine, Lorcainide, and Propafenone

All the above drugs (classified as IC) block the fast i_{Na} at relatively low concentrations and in a use-dependent manner.[16,143–145] Ethmozine in rat ventricular myocytes was reported to block predominantly the inactivated Na^+ channels.[144] However, flecainide in guinea pig ventricular myocytes blocked the open Na^+ channels.[16] The unblocking of flecainide from Na^+ channels oc-

curs both in the rested and activated state.[16] Propafenone decreased the mean opening probability of the single Na^+ channel without changing either their mean open time distribution or the unitary current size.[145] Propafenone blocks both the open and the inactivated Na^+ channels.[146]

The inward i_{Ca} was decreased by flecainide,[147] ethmozine[147] in frog ventricular myocytes, and propafenone in the rabbit SA node,[148] an effect that may, in part, contribute to both the antiarrhythmic and the negative inotropic effects of these drugs.[147]

Both flecainide[149] and propafenone[148] decrease the i_K, and propafenone also depresses i_f.[148]

Studies using conventional microelectrodes (refs. in 66 and 67) showed that flecainide, encainide, ethmozine, lorcainide, and propafenone at low concentrations decrease the $\dot{V}_{max}$ in a use-dependent manner in different cardiac preparations. The reported recovery times of $\dot{V}_{max}$ from block ranged from 4.8 to 20.3 s.[114,146,150] This indicates that in analogy to quinidine, disopyramide, and procainamide, these drugs depress $\dot{V}_{max}$ and conduction within a wide range of basic rates and premature coupling intervals. Flecainide, encainide, ethmozine, lorcainide, and propafenone shorten APD in Purkinje fibers but either do not change or slightly lengthen APD in ventricular muscle fibers. Flecainide slowed the kinetics of restitution and decreased the range of premature APD in dog Purkinje fibers.[116] In ventricular muscle, flecainide increased APD more in the premature APs than at similar cycle lengths at normal heart rates.[151]

Flecainide, encainide, and propafenone slow the AV nodal conduction and the sinus node automaticity, decrease Purkinje fiber automaticity, and increase the ERP/APD ratio. Propafenone has additional β-adrenergic blocking properties.

Propranolol and Metoprolol

As discussed earlier, propranolol and metoprolol, as well as other β-receptor antagonists, can indirectly influence the delayed rectifier outward i_K, the inward Ca^{2+} and Cl^- and the hyperpolarization-activated inward pacemaker currents due to their β-adrenergic blocking properties.

At concentrations (10 μg/ml) that are higher than those needed to produce a β-adrenoceptor block, propranolol decreases the fast inward i_{Na} and its rate of reactivation in frog atrial muscle.[152] Propranolol also depresses the Na/Ca exchange system.[78] Metoprolol (10 μg/ml) reduced the inward rectifier outward i_K and inward i_{Ca}[153] in cat ventricular myocytes, but concentrations up to 50 μM did not change $\dot{V}_{max}$ and APD in guinea pig ventricular muscle.[154]

Similar to amiodarone, the effects of propranolol and metoprolol on the AP differ following acute and chronic administration. After acute exposure, propranolol decreased $\dot{V}_{max}$ in a use-dependent manner[114,155] with the reported recovery time constants from the use-dependent block ranging from 1.35 to 14 s.[114,155] Acute treatment with propranolol either did not change or decreased the APD in atrial[156] and ventricular muscle fibers[157] and shortened APD in Purkinje fibers,[157] but after prolonged treatment, both propranolol and metoprolol significantly lengthened repolarization in atrial and ventricular muscle, but not in the Purkinje fibers.[65]

Amiodarone

Amiodarone, which was originally assigned to class III antiarrhythmic drug action, has multiple effects on various cardiac ionic channels and receptors (Table 22.1). At low concentration, amiodarone depresses the fast i_{Na}[145,158,159] in a strongly use-dependent manner[145,159] (Fig. 22.5). Amiodarone is reported to block the rested and inactivated, but not the activated, i_{Na}.[159,160] At the single-channel level, amiodarone decreases the opening probability of the i_{Na} without altering its unitary current amplitude and mean open time[145] (Fig. 22.5).

At therapeutically relevant concentrations, amiodarone blocked the inward i_{Ca} in ferret[158] and in guinea pig ventricle[161] (Fig. 22.6). This effect was use dependent and the drug binding occurred mostly during the inactivated state of the Ca^{2+} channels.[161] Although no detailed analysis is available about the effect of amiodarone on the outward i_K, the published data suggest that the drug blocks both the inward rectifier[162] and the delayed rectifier[158,163] outward i_K.

Conventional microelectrode studies have shown that amiodarone depressed $\dot{V}_{max}$ in different types of cardiac tissues.[43,160,164] This effect was use dependent[43,160] and, similar to lidocaine and mexiletine, was characterized by fast onset and offset kinetics of $\dot{V}_{max}$ block.[43,165] The reported time constants of recovery of $\dot{V}_{max}$ range from 282 to 1,630 ms.[43,44,160,165] The effect of amiodarone on $\dot{V}_{max}$ is greatly enhanced at depolarized potentials.[160] Consequently, amiodarone blocks $\dot{V}_{max}$ and conduction at fast heart rate, at short prematurity interval, and in diseased tissue more than in healthy tissue at normal heart rate. Amiodarone also affects passive membrane properties,[10] and its effect on impulse conduction is anisotropic, that is, it de-

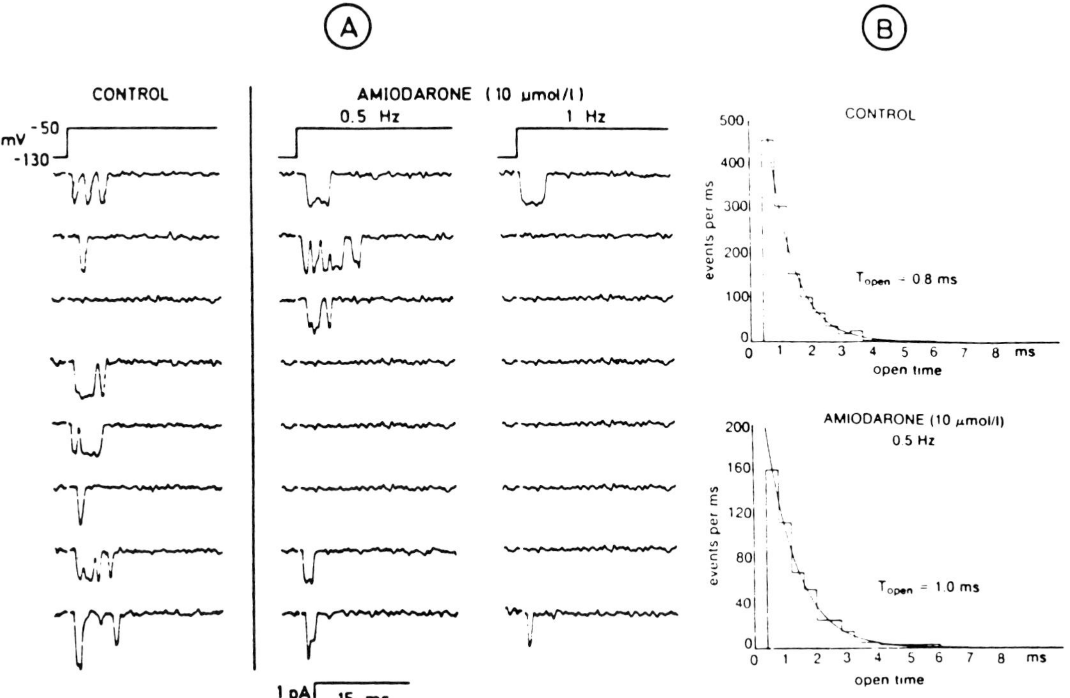

FIGURE 22.5 A. Successive recordings of elementary i_{Na} before (control) and after treating the myocyte with amiodarone on repetitive stimulation at 0.5 and 1 Hz, respectively. Downward deflections represent single Na^+ channel openings. The patch contained two functioning Na^+ channels. Note the less frequent channel openings in the presence of amiodarone, especially at the higher pulse frequency. **B.** Open-time histograms of the same patch before (top) and after (bottom) amiodarone treatment. Note that the open-time distribution did not change in the presence of amiodarone. *Reprinted from Kohlhardt and Fichtner,*[145] *with permission.*

creases conduction more in the longitudinal than in the transverse direction.[166]

Amiodarone slowed the AV nodal conduction in a use-dependent manner[167] and also depressed the slow-response APs.[167,168] The effect of this drug on repolarization varies depending on the type of cardiac tissue and on the mode of administration. After acute application, amiodarone either did not change,[169–172] or only moderately increased[165,173] APD in cardiac atrial or ventricular muscle. However, chronic administration of amiodarone greatly lengthened repolarization in both atrial[174,175] and ventricular muscle.[160,171,174,175] In Purkinje fibers, the acute effect of amiodarone moderately shortened APD,[171,173] but chronic administration did not change significantly repolarization.[171] ERP/APD ratio was not changed either after acute[44,171] or chronic amiodarone application in ventricular and Purkinje fibers.[171] Amiodarone decreased automaticity in the SA node[173,174] and in the Purkinje fibers,[176] abolished depolarization-induced automaticity,[160] DAD, and triggered automaticity.[177] The drug also interacts with the β-,[178] α-,[178] and muscarinic[179] receptors, and alters Na/K pump activity.[180]

Bretylium, Sotalol, and Clofilium

The above drugs depress the i_K.[68,181–183] Sotalol also was reported to decrease the i_{K1} in rabbit[68] and in sheep[71] Purkinje fibers. The i_{to} was decreased by sotalol in sheep Purkinje fiber[71] and by clofilium in rat ventricular myocytes.[184] Sotalol did not change the i_f in Purkinje fibers.[71]

Bretylium, sotalol, and clofilium increased APD in atrial,[185,186] ventricular,[186–188] and Purkinje fibers[185,188] without altering $\dot{V}_{max}$, the ERP/APD ratio, and AV nodal conduction. The APD lengthening effect is greater in Purkinje than in ventricular muscle fibers[185,188] and at long cycle lengths.[115,185,188,189] Sotalol[190] and clofilium[189] slightly increased the spontaneous cycle length of sinus nodal cells and Purkinje fibers due to

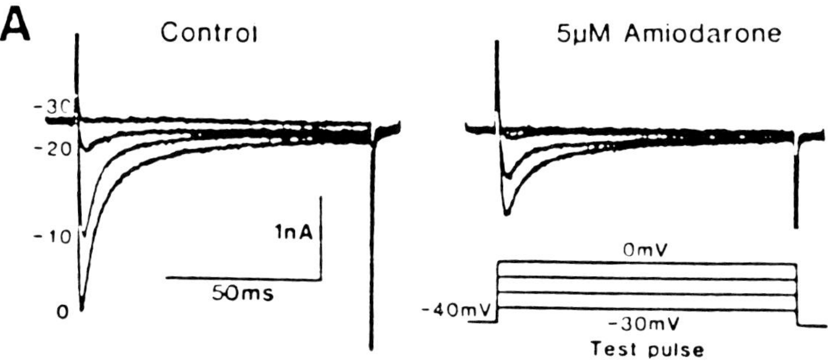

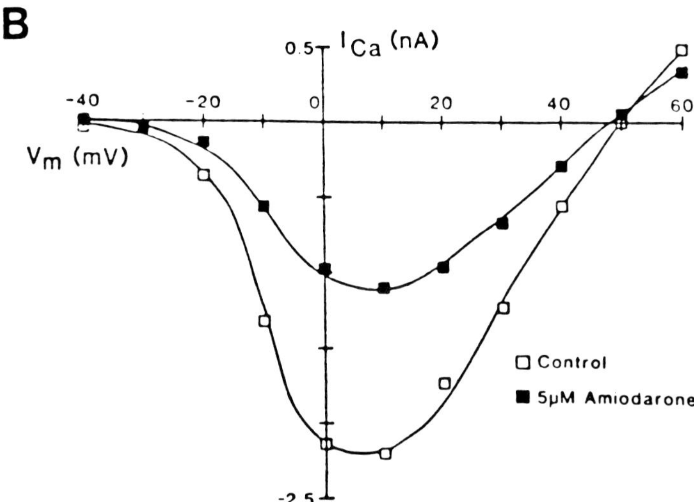

FIGURE 22.6 Amiodarone effects on the current–voltage relationship of i_{Ca}. **A.** A family of i_{Ca} recordings elicited on the step depolarizations from a holding potential of −40 mV to selected test potentials of −30 to 0 mV. Control i_{Ca} tracings are shown on the left side of **A**, whereas the right side shows i_{Ca} recorded during exposure to 5 μM amiodarone. The numbers on the left side of **A** indicate the test potentials. **B.** The current–voltage relationship before and during application of 5 μM amiodarone. Amiodarone reduced i_{Ca} amplitude to one-half of its control value without affecting the shape of the current–voltage relationship. *Reprinted from Nishimura et al.,*[161] *with permission.*

lengthening of repolarization without changing the rate of diastolic depolarization. At high concentrations or at slow frequencies, sotalol can induce EAD in dog[72] and sheep[80] Purkinje fibers.

Verapamil, Diltiazem, and Dihydropyridine Derivatives

These drugs block cardiac L-type Ca^{2+} channels in a use- and voltage-dependent manner[191,192] at micromolar or lower concentrations. Verapamil and diltiazem bind to the inactivated[193] and open[194] Ca^{2+} channels and nisoldipine binds preferentially to the inactivated Ca^{2+} channels.[194] Verapamil and diltiazem inhibit the inward i_{Ca} within a relatively wide range of frequencies, whereas the use-dependent inhibitory effect of nisoldipine is evident only at frequencies higher than 60/min.[194] Each of these drugs slowed the reactivation kinetics of the inward i_{Ca}, but the recovery kinetics in the presence of verapamil and diltiazem were considerably slower than in the presence of nisoldipine.[192]

Verapamil at concentrations as high as 2 mg/liter did not affect significantly the fast inward i_{Na}.[195] However, at higher than the range of therapeutic concentrations, verapamil, diltiazem, and nisoldipine had an inhibiting effect on other current systems. Verapamil[158] and diltiazem[196] depressed i_K, and nisoldipine, at concentrations of 10 μM and 33 μM inhibited i_{to}[197] and i_K[196] in calf Purkinje fibers and frog atrial myocytes, respectively.

In the studies using conventional microelectrodes,[66] verapamil, diltiazem,[198] and nisoldipine either did not change or, at high concentrations, slightly depressed $\dot{V}_{max}$ in normal fast-response cardiac fibers. In contrast, slow-response APs were depressed by these drugs at low concentrations.[199,200] Diltiazem and nisoldipine shorten APD in Purkinje[201] and ventricular muscle,[202] whereas verapamil lengthens APD in Purkinje and shortens APD in ventricular muscle. In vivo verapamil[203] and diltiazem[204] decrease AV conduction in a frequency-dependent manner, decrease sinus node automaticity, and suppress EAD, DAD, and abnormal triggered automaticity.[53,80,138,205]

REFERENCES

1. Hoffman BF, Rosen MR: Cellular mechanisms for cardiac arrhythmias. *Circ Res* 1981;49:1–15.
2. Gintant GA, Cohen IS: Advances in cardiac cellular electrophysiology: Implications for automaticity and therapeutics. *Annu Rev Pharmacol Toxicol* 1988;28:61–81.
3. Allessie MA, Bonke FIM, Schopman FJG: Circus movement in rabbit atrial muscle as a mechanism of tachycardia. III. The "leading circle" concept: A new model of circus movement in cardiac tissue without the involvement of anatomical obstacle. *Circ Res* 1977;41:9–18.
4. Vaughan Williams EM: Classifications of anti-arrhythmic drugs, in Sandoe E, Flensted-Jensen E, Olesen KH (eds): *Symposium on Cardiac Arrhythmias*. Soderstalje, Sweden, AB Astra, 1970, pp 449–472.
5. Vaughan Williams EM: Classification of anti-dysrhythmic drugs. *Pharmac Therap B* 1975;1:115–138.
6. Campbell TJ: Kinetics of onset of rate-dependent effects of class I antiarrhythmic drugs are important in determining their effects on refractoriness in guinea-pig ventricle, and provide a theoretical basis for their subclassification. *Cardiovasc Res* 1983; 17:344–352.
7. Harrison DC: Antiarrhythmic drug classification: New science and practical applications. *Am J Cardiol* 1985;56:185–187.
8. Millar JS, Vaughan Williams EM: Anion antagonism—a fifth class of antiarrhythmic action? *Lancet* 1981;i:1291–1293.
9. Arnsdorf MF, Bigger JT: The effect of lidocaine on components of excitability in long mammalian cardiac Purkinje fibers. *J Pharmacol Exp Ther* 1975; 195:206–215.
10. Levine JH, Moore EN, Kadish AH, Weisman HF, Balke CW, Hanich RF, Spear JF: Mechanisms of depressed conduction from long-term amiodarone therapy in canine myocardium. *Circulation* 1988;87:684–691.
11. Johnson EA, McKinnon MG: The differential effect of quinidine and pyralamine on the myocardial action potential at various rates of stimulation. *J Pharmacol Exp Ther* 1957;120:460–468.
12. Hille B: Local anaesthetics: Hydrophylic and hydrophobic pathways for the drug-receptor reaction. *J Gen Physiol* 1977;69:497–515.
13. Hondeghem LM, Katzung BG: Time- and voltage-dependent interactions of antiarrhythmic drugs with cardiac sodium channels. *Biophys Biochim Acta* 1977;472:373–398.
14. Hondeghem LM, Katzung BG: Antiarrhythmic agents: The modulated receptor mechanism of action of sodium and calcium channel-blocking drugs. *Annu Rev Pharmacol Toxicol* 1984;24:387–423.
15. Colatsky TJ: Quinidine block of cardiac sodium channels is rate- and voltage-dependent. *Biophys J* 1982;37:343a(abstract).
16. Anno T, Hondeghem LM: Interactions of flecainide with guinea pig cardiac sodium channels. Importance of activation unblocking to the voltage dependence of recovery. *Circ Res* 1990;66:789–803.
17. Bean BP, Cohen JC, Tsien RW: Lidocaine block of cardiac sodium channels. *J Gen Physiol* 1983;613–642.
18. Sanchez-Chapula J, Tsuda Y, Josephson IR: Voltage- and use-dependent effects of lidocaine on sodium current in rat single ventricular cells. *Circ Res* 1983; 52:557–565.
19. Courtney KR: Quantifying antiarrhythmic drug blocking action potentials in guinea-pig papillary muscle. *J Mol Cell Cardiol* 1983;15:749–757.
20. Clarkson CW, Hondeghem LM: Mechanism for bupivacaine depression of cardiac conduction: Fast block of sodium channels during the action potential with slow recovery from block during diastole. *Anesthesiology* 1985;47:542–550.
21. Starmer CF, Grant AO, Strauss HC: Mechanisms of use-dependent block of sodium channels in excitable membranes by local anesthetics. *Biophys J* 1984;46:15–27.
22. Grant AO, Strauss LJ, Wallace AG, Strauss HC: The influence of pH on the electrophysiological effects of lidocaine in guinea pig ventricular myocardium. *Circ Res* 1980;47:542–550.
23. Grant AO, Trantham JL, Brown KK, Strauss HS: PH-dependent effects of quinidine on the kinetics of dV/dtmax in guinea pig ventricular myocardium. *Circ Res* 1982;50:210–217.
24. Courtney KR: Interval-dependent effects of small antiarrhythmic drugs on excitability of guinea-pig myocardium. *J Mol Cell Cardiol* 1980;12:1273–1286.
25. Campbell TJ: Importance of physico-chemical properties in determining the kinetics of the effects of class I antiarrhythmic drugs on maximum rate of depolarization in guinea-pig ventricle. *Br J Pharmacol* 1983;80:33–40.
26. Cohen CJ, Bean BP, Tsien RW: Maximal upstroke velocity as an index of available sodium conductance: Comparison of maximal upstroke velocity and voltage clamp measurements of sodium current in rabbit Purkinje fibers. *Circ Res* 1984;54:636–651.
27. Sheets MF, Hanck DA, Fozzard HA: Nonlinear relation between V_{max} and I_{Na} in canine cardiac Purkinje cells. *Circ Res* 1988;63:386–398.
28. Walton MK, Fozzard HA: The conducted action potential: Models and comparison to experiments. *Biophys J* 1983;44:9–26.
29. Buchanan JW, Saito T, Gettes LS: The effects of antiarrhythmic drugs, stimulation frequency, and potassium-induced resting membrane potential changes on conduction velocity and dV/dt_{max} in guinea-pig myocardium. *Circ Res* 1985;56:696–703.
30. Nattel S: Relationship between use-dependent effects of antiarrhythmic drugs on conduction and V_{max} in canine cardiac Purkinje fibers. *J Pharmacol Exp Ther* 1987;241:282–288.
31. Varro A, Bodi I, Jednakovits A, Rabloczky G: Frequency-dependent effects of class III and IV antiarrhythmic drugs on the conduction and excitability in rabbit ventricular muscle. *Arch Int Pharmacodyn* 1988;292:157–165.
32. Davis J, Matsubara T, Scheinman MN, Katzung BG, Hondeghem LH: Use-dependent effects of lidocaine on conduction in canine myocardium: Application of the modulated receptor hypothesis in vivo. *Circulation* 1986;74:205–214.
33. Anderson KP, Walker R, Dustman T, Lux RL, Ershler PR, Kates RE, Urie PM: Rate-dependent electrophysiologic effects of long-term administration of amiodarone on canine ventricular myocardium in vivo. *Circulation* 1989;79:948–958.
34. Nattel S, Jing W: Rate-dependent changes in intraventricular conduction produced by procainamide in anaesthetized dogs. A quantitative analysis based

on the relation between phase 0 inward current and conduction velocity. *Circ Res* 1989;65:1485–1498.

35. Morady F, DiCarlo LA, Bearman JM, Krol RB: Rate-dependent effects of intravenous lidocaine, procainamide and amiodarone on intraventricular conduction. *J Am Coll Cardiol* 1985;6:179–185.
36. Nadamanee K, Tevenson WG, Weiss JN, Frame VB, Antimisiaris MG, Suithichaiyakul T, Pruitt CM: Frequency-dependent effects of quinidine on the ventricular action potential and QRS duration in humans. *Circulation* 1990;81:790–796.
37. Spach MS, Miller WT, Colber PC, Kootsey JM, Sommer JR, Mosher CE Jr: The functional role of structural complexities in the propagation of depolarization in the atrium of the dog: Cardiac conduction disturbances due to discontinuities of effective axial resistivity. *Circ Res* 1982;50:175–191.
38. Spach MS, Kootsey JM: The nature of electrical propagation in cardiac muscle. *Am J Physiol* 1983; 244:H3–H22.
39. Kadish AH, Spear JF, Levine JH, Moore EN: The effects of procainamide on the conduction in anisotropic canine ventricular myocardium. *Circulation* 1986;74:616–625.
40. Quinteiro RA, Biagetti MO, deForteza E: Effects of lidocaine on V_{max} and conduction velocity in uniform anisotropic canine ventricular muscle: Possible role of its binding-rate constants. *J Cardiovasc Pharmacol* 1990;15:29–36.
41. Clarkson CW, Hondeghem LM: Evidence for a specific receptor site for lidocaine, quinidine, and bupivacaine associated with cardiac sodium channels in guinea pig ventricular myocardium. *Circ Res* 1985;56:496–506.
42. Sanchez-Chapula J: Electrophysiological interactions between quinidine-lidocaine and quinidine-phenytoin in guinea pig papillary muscle. *Naunyn-Schmiedeberg's Arch Pharmacol* 1985;331:365–375.
43. Varro A, Nakaya Y, Elharrar V, Surawicz B: Use-dependent effects of amiodarone on V_{max} in cardiac Purkinje and ventricular muscle fibers. *Eur J Pharmacol* 1985;112:419–422.
44. Nakaya Y, Elharrar V, Surawicz B: Effect of mexiletine, amiodarone and disopyridine on the excitability and refractoriness of canine cardiac fibers: Possible relation to antiarrhythmic drug action and classification. *Cardiovascular Drugs and Therapy* 1987;1:147–153.
45. Attwell D, Cohen I, Eisner D, Ohba M, Ojeda C: The steady-state TTX-sensitive ("window") sodium current in cardiac Purkinje fibers. *Pflügers Arch* 1979; 379:137–142.
46. Colatsky TJ: Mechanisms of action of lidocaine and quinidine on action potential duration in rabbit cardiac Purkinje fibers. An effect on steady-state sodium currents? *Circ Res* 1982;50:17–27.
47. Carmeliet E, Saikawa T: Shortening of the action potential and reduction of pacemaker activity by lidocaine, quinidine and procainamide in sheep cardiac Purkinje fibers. An effect on Na or K currents? *Circ Res* 1982;50:257–272.
48. Gintant GA, Datyner NB, Cohen IS: Slow inactivation of a tetrodotoxin-sensitive current in canine cardiac Purkinje fibers. *Biophys J* 1984;45:509–512.
49. Carmeliet E: Slow inactivation of the sodium current in rabbit cardiac Purkinje fibers. *Pflügers Arch* 1987; 408:18–26.
50. Kiyosue T, Arita M: Late sodium current and its contribution to action potential configuration in guinea pig ventricular myocytes. *Circ Res* 1989;64:389–397.
51. Noma A, Kotake H, Irisawa H: Slow inward current and its role mediating the chronotropic effect of epinephrine in the rabbit sinoatrial node. *Pflügers Arch* 1980;388:1–9.
52. Tsien RW, Bean BP, Hess P, Lansman JB, Nilius B, Nowycky MC: Mechanism of calcium channel modulation by beta-adrenergic agents and dihydropyridine calcium agonists. *J Mol Cell Cardiol* 1986; 18:691–710.
53. January CT, Riddle JM, Salata JJ: A model for early afterdepolarizations: Induction with the Ca^{2+} channel agonist Bay K 8644. *Circ Res* 1988;62:563–571.
54. Mitra R, Morad M: Two types of calcium channels in guinea pig ventricular myocytes. *Proc Natl Acad Sci USA* 1986;83:5340–5344.
55. Tytgat J, Nilius B, Vereecke J, Carmeliet E: The T-type Ca channel in guinea pig ventricular myocytes is insensitive to isoproterenol. *Pflügers Arch* 1988; 411:704–706.
56. Tseng G-N, Boyden PA: Multiple types of Ca^{2+} currents in single canine Purkinje cells. *Circ Res* 1989; 65:1735–1750.
57. Bean BP: Two kinds of Ca^{2+} currents in single canine atrial cells. Differences in kinetics, selectivity and pharmacology. *J Gen Physiol* 1985;86:1–30.
58. Brodde OE, O'Hara N, Zerkowski AR, Rohm N: Human cardiac beta-1- and beta-2-adrenoceptors are functionally coupled to the adenylate cyclase in right atrium. *J Cardiovasc Pharmacol* 1984; 6:1184–1191.
59. Iijima T, Taira N: Beta-2-adrenoceptor-mediated increase in the slow inward calcium current in atrial cells. *Eur J Pharmacol* 1989;163:357–360.
60. Boller M, Pott L: Beta-adrenergic modulation of transient inward current in guinea-pig cardiac myocytes. Evidence for regulation of Ca^{2+}-release from sarcoplasmic reticulum by a cyclic AMP dependent mechanism. *Pflügers Arch* 1989;415:276–288.
61. Gadsby DC: Beta-adrenoceptor agonists increase membrane K^+ conductance in cardiac Purkinje fibers. *Nature* 1983;306:691–693.
62. Bennett PB, McKinney L, Begenisich T, Kass RS: Adrenergic modulation of the delayed rectifier potassium channel in calf cardiac Purkinje fibers. *Biophys J* 1986;49:839–848.
63. Bahinski A, Nairn AC, Greengard P, Gadsby DC: Chloride conductance by cyclic AMP-dependent protein kinase in cardiac myocytes. *Nature* 1989; 340:718–721.
64. Harvey RD, Hume JR: Autonomic regulation of a chloride current in heart. *Science* 1989;244:983–985.
65. Raine AEG, Vaughan Williams EM: Adaptation to prolonged beta-blockade of rabbit atrial, Purkinje and ventricular potentials, and of papillary muscle contraction. Time-course of development and recovery from adaptation. *Circ Res* 1981;48:804–812.
66. Zipes DP: Management of cardiac arrhythmias: Pharmacological, electrical, and surgical techniques, in Braunwald E (ed): *Heart Disease. A Textbook of Cardiovascular Medicine.* Philadelphia, WB Saunders, 1988, pp 621–657.
67. Karagueuzian HS, Singh BN, Mandel WJ. Antiarrhythmic drugs: Mode of action, pharmacokinetic properties, and clinical application, in Mandel WJ (ed): *Cardiac Arrhythmias. Their Mechanisms, Di-*

agnosis, and Management. Philadelphia, J.B. Lippincott, 1987, pp 697–737.
68. Carmeliet E: Electrophysiologic and voltage clamp analysis of the effects of sotalol on isolated cardiac muscle and Purkinje fibers. *J Pharmacol Exp Ther* 1985;232:817–825.
69. Hiraoka M, Sawada K, Kawano S: Effects of quinidine on plateau currents of guinea-pig ventricular myocytes. *J Mol Cell Cardiol* 1986;18:1097–1106.
70. Salata JJ, Wasserstrom JA: Effects of quinidine on action potentials and ionic currents in isolated canine ventricular myocytes. *Circ Res* 1988;62:324–337.
71. Berger F, Borchard U, Hafner D: Effects of (+)- and (−)-sotalol on repolarizing outward currents and pacemaker current in sheep cardiac Purkinje fibers. *Naunyn-Schmiedeberg's Arch Pharmacol* 1989; 340:696–704.
72. Lathrop DA, Varro A: Modulation of the effect of sotalol on Purkinje strand electromechanical characteristics. *Can J Physiol Pharmacol* 1989;67:1463–1467.
73. Beress L, Ritter R, Ravens U: The influence of the rate of electrical stimulation on the effects of the Anemonia sulcata toxin ATXII in guinea pig papillary muscle. *Eur J Pharmacol* 1982;79:265–272.
74. Buggish D, Isenberg G, Ravens U, Scholtysik G: The role of sodium channels in the effects of the cardiotonic compound DPI 201-106 on contractility and membrane potentials in isolated mammalian heart preparations. *Eur J Pharmacol* 1985;118:303–311.
75. Schwartz A, Grupp IL, Grupp G, Williams JS, Vaghy PL: Effects of dihydropyridine calcium channel modulators in the heart: Pharmacological and radioligand binding correlations. *Biochem Biophys Res Commun* 1984;125:3871–3894.
76. Beuckelmann DJ, Wier WG: Sodium-calcium exchange in guinea-pig cardiac cells: Exchange current and changes in intracellular Ca^{2+}. *J Physiol* 1989;414:499–520.
77. Eisner DA, Lederer WJ: Na-Ca exchange: Stoichiometry and electrogenicity. *Am J Physiol* 1985; 248:C189–C202.
78. Eyolfson DA, Dhalla NS: Interaction of some antiarrhythmic drugs with the heart sarcolemmal Na^{+}-Ca^{2+} exchange system. *Basic Res Cardiol* 1989;84:414–420.
79. Surawicz B: Electrophysiologic substrate of torsade de pointes: Dispersion of repolarization or early afterdepolarizations? *J Am Coll Cardiol* 1989;14:172–184.
80. Hiromasa S, Coto H, Li Z-Y, Maldonado C, Kupersmith J: Dextrorotatory isomer of sotalol: Electrophysiologic effects and interaction with verapamil. *Am Heart J* 1988;116:1552–1557.
81. Janse MJ, Wit AL: Electrophysiological mechanisms of ventricular arrhythmias resulting from myocardial ischemia and infarction. *Physiol Rev* 1989; 69:1049–1169.
82. Carmeliet E: Cardiac transmembrane potentials and metabolism. *Circ Res* 1978;42:577–587.
83. Noma A, Shibasaki T: Membrane current through adenosine-triphosphate-regulated potassium channels in guinea-pig ventricular cells. *J Physiol* 1985; 363:463–480.
84. Kakei M, Noma A, Shibasaki T: Properties of adenosine-triphosphate-regulated potassium channels in guinea-pig ventricular cells. *J Physiol* 1985; 363:441–462.
85. Escande D, Thuringer D, LeGuern S, Courteix J, Laville M, Caveno I: Potassium channel openers act through an activation of ATP-sensitive K^{+} channels in guinea-pig cardiac myocytes. *Pflügers Arch* 1989;414:669–675.
86. Kantor PF, Coetzee WA, Carmeliet EE, Dennis SC, Opie LH: Reduction of ischemic K^{+} loss and arrhythmias in rat heart. Effect of glibenclamide, a sulfonylurea. *Circ Res* 1990;66:478–485.
87. Wolleben CD, Sanguetti MC, Siegl PKS: Influence of ATP-sensitive potassium channel modulators on ischemia-induced fibrillation in isolated rat hearts. *J Mol Cell Cardiol* 1989;21:783–788.
88. Haworth RA, Goknur AB, Berkoff HA: Inhibition of ATP-sensitive potassium channels of adult rat heart cells by antiarrhythmic drugs. *Circ Res* 1989; 65:1157–1160.
89. Hagiwara N, Irisawa H, Kameyama M: Contribution of two types of calcium currents to the pacemaker potentials of rabbit sino-atrial node cell. *J Physiol* 1988;395:233–253.
90. Hirano Y, Fozzard HA, January CT: Characteristics of L- and T-type Ca^{2+} currents in cardiac Purkinje cells. *Am J Physiol* 1989;256:H1478–H1492.
91. Kokubun S, Nishimura M, Noma A, Irisawa H: Membrane currents in the rabbit atrioventricular node cell. *Pflügers Arch* 1982;393:15–22.
92. DeMello WC: Cell-to-cell coupling assayed by means of electrical measurements. *Experientia* 1987;43:1075–1079.
93. Elharrar V, Gaum WE, Zipes DP: Effects of drugs on conduction delay and incidence of ventricular arrhythmias induced by acute coronary occlusion in dogs. *Am J Cardiol* 1977;39:544–549.
94. Hiramatsu Y, Buchanan JW, Knisley SB, Koch GG, Kropp S, Gettes LS: Influence of rate-dependent cellular uncoupling on conduction change during simulated ischemia in guinea pig papillary muscles. Effect of verapamil. *Circ Res* 1989;65:95–102.
95. Nakaya H, Millard RW, Lathrop DA, Gaum WE, Kaplan S, Schwartz A: Flow-independent improvement by diltiazem of ischemia-induced conduction delay in porcine hearts. *J Am Coll Cardiol* 1983; 2:474–480.
96. Kobinger W, Lille C: Specific bradycardia agents—A novel pharmacological class? *Eur Heart J* 1987; 8(suppl L):7–15.
97. Satoh H, Hashimoto K: Electrophysiological study of alinidine in voltage clamped rabbit sino-atrial node cells. *Eur J Pharmacol* 1986;121:211–219.
98. VanGinneken ACG, Bouman LN, Jongsma HJ, Duivenvoorden JJ, Opthof T, Giles WR: Alinide, as a model of the action of specific bradycardic agents on SA node activity. *Eur Heart J* 1987;8(suppl L):365–375.
99. Snyders DJ, VanBogaert P-P: Alinide modifies the pacemaker current in sheep Purkinje fibers. *Pflügers Arch* 1987;410:83–91.
100. Van Bogaert P-P, Goethals M: Pharmacological influence of specific bradycardiac agents on the pacemaker current of sheep cardiac Purkinje fibres. A comparison between three different molecules. *Eur Heart J* 1987;8(suppl L):35–42.
101. Dennis A, Vaughan Williams EM: Further studies of alinidine induced bradycardia in the presence of caesium. *Cardiovasc Res* 1986;20:375–378.
102. Osterrieder W, Pelzer D, Yang Q-F, Trautwein W:

The electrophysiological basis of the bradycardic action of AQA 39 on the sinoatrial node. *Naunyn-Schmiedeberg's Arch Pharmacol* 1981;317:233–237.
103. Lee KS, Hume JR, Giles W, Brown AM: Sodium current depression by lidocaine and quinidine in isolated ventricular cells. *Nature* 1981;291:325–327.
104. Snyders DJ, Hondeghem LM: Effects of quinidine on the sodium current of guinea pig ventricular myocytes. Evidence for a drug-associated rested state with altered kinetics. *Circ Res* 1990;66:565–579.
105. Ducouret P: The effect of quinidine on membrane electrical activity in frog auricular fibres studied by current and voltage clamp. *Br J Pharmacol* 1976;57:163–184.
106. Nawrath H: Action potential, membrane currents and force of contraction in mammalian heart muscle fibers treated with quinidine. *J Pharmacol Exp Ther* 1980;216:176–182.
107. Nakayama T, Fozzard HA: Quinidine reduces outward current in single canine cardiac Purkinje cells. In Beamish RE, Panagia V, Dhalla NS (eds): *Pharmacological Aspects of Heart Disease*. Boston, Martinus Nijhoff, 1987, pp 33–44.
108. Nishimura M, Huan R-M, Habuchi Y, Tsuji Y, Nakanishi T, Watanabe Y: Membrane actions of quinidine sulfate in the rabbit atrioventricular node studied by voltage clamp method. *J Pharmacol Exp Ther* 1988;244:780–788.
109. Imaizumi Y, Giles WR: Quinidine-induced inhibition of transient outward current in cardiac muscle. *Am J Physiol* 1987;253:H704–H708.
110. Roden DM, Bennett PB, Snyders DJ, Balser JR, Hondeghem LM: Quinidine delays I_k activation in guinea pig ventricular myocytes. *Circ Res* 1988;62:1055–1058.
111. Furukawa T, Tsujimura Y, Kitamura K, Tanaka H, Hanuchi Y: Time- and voltage-dependent block of the delayed K^+ current by quinidine in rabbit sinoatrial and atrioventricular nodes. *J Pharmacol Exp Ther* 1989;251:756–763.
112. Giles WR, Imaizumi Y: Comparison of potassium currents in rabbit atrial and ventricular cells. *J Physiol* 1988;405:123–145.
113. Henning B, Zehender M, Meinertz T, Just H: Effect of tetrodotoxin, lidocaine and quinidine on the transient inward current of sheep Purkinje fibres. *Basic Res Cardiol* 1988;83:176–189.
114. Varro A, Elharrar V, Surawicz B: Frequency-dependent effects of several class I antiarrhythmic drugs on V_{max} of action potential upstroke in canine cardiac Purkinje fibers. *J Cardiovasc Pharmacol* 1985;7:482–492.
115. Varro A, Nakaya Y, Elharrar V, Surawicz B: Effect of antiarrhythmic drugs on the cycle length-dependent action potential duration in dog Purkinje and ventricular muscle fibers. *J Cardiovasc Pharmacol* 1986;8:178–185.
116. Varro A, Elharrar V, Surawicz B: Effect of antiarrhythmic drugs on the premature action potential duration in canine cardiac Purkinje fibers. *J Pharmacol Exp Ther* 1985;233:304–311.
117. Wasserstrom JA, Ferrier GR: Effects of phenytoin and quinidine on digitalis-induced oscillatory afterpotentials, aftercontractions, and inotropy in canine ventricular tissues. *J Mol Cell Cardiol* 1982;14:725–736.
118. Nattel S, Quantz MA: Pharmacological response of quinidine induced early afterdepolarizations in canine cardiac Purkinje fibres: Insights into underlying ionic mechanisms. *Cardiovasc Res* 1988;22:808–817.
119. Kaseda S, Gilmour RF, Zipes DP: Depressant effect of magnesium on early afterdepolarizations and triggered activity by cesium, quinidine, and 4-aminopyridine in canine cardiac Purkinje fibers. *Am Heart J* 1989;118:458–466.
120. Roden DM, Hoffman BF: Action potential prolongation and induction of abnormal automaticity by low quinidine concentrations in canine Purkinje fibers. *Circ Res* 1985;56:857–867.
121. Gruber R, Carmeliet E: The activation gate of the sodium channel controls blockade and deblockade by disopyramide in rabbit Purkinje fibres. *Br J Pharmacol* 1989;97:41–50.
122. Senami M, Irisawa H: Effect of procainamide on the membrane currents of the sino-atrial node cells of rabbits. *Jpn J Physiol* 1981;31:225–231.
123. Hiraoka M, Kuga K, Kawano S, Sunami A, Fan Z: New observations on the mechanisms of antiarrhythmic actions of disopyramide on cardiac membranes. *Am J Cardiol* 1989;64:15J–19J.
124. Coraboeuf E, Deroubaix, Second D, Coulombe A: Comparative effects of three class I antiarrhythmic drugs on plateau and pacemaker currents of sheep cardiac Purkinje fibres. *Cardiovasc Res* 1988;22:375–384.
125. Kotake H, Hasegawa J, Hata T, Mashiba H: Electrophysiological effect of disopyramide on rabbit sinus node cells. *J Electrocardiol* 1985;18:377–384.
126. Ono K, Kiyosue T, Arita M: Effects of AN-132, a novel antiarrhythmic lidocaine analogue, and of lidocaine on membrane ionic currents of guinea pig ventricular myocytes. *Naunyn-Schmiedeberg's Arch Pharmacol* 1989;339:221–229.
127. Josephson IR: Lidocaine blocks Na, Ca and K currents of chick ventricular myocytes. *J Mol Cell Cardiol* 1988;20:593–604.
128. Wasserstrom JA, Salata JJ: Basis for tetrodotoxin and lidocaine effects on action potentials in dog ventricular myocytes. *Am J Physiol* 1988;254:H1157–H1166.
129. Matsubara T, Clarkson C, Hondeghem LM: Lidocaine blocks open and inactivated cardiac sodium channels. *Naunyn-Schmiedeberg's Arch Pharmacol* 1987;336:224–231.
130. Nilius B, Benndorf K, Makwardt F: Effects of lidocaine on single cardiac sodium channels. *J Mol Cell Cardiol* 1987;19:865–874.
131. McDonald TV, Courtney KR, Clusin WT: Use-dependent block of single sodium channels by lidocaine in guinea pig ventricular myocytes. *Biophys J* 1989;55:1261–1266.
132. Grant AO, Dietz MA, Gilliam FR III, Starmer CF: Blockade of sodium channels by lidocaine. Single-channel analysis. *Circ Res* 1989;65:1247–1262.
133. Mitchell MR, Plant S: Effect of lidocaine on action potentials, currents and contractions in the absence and presence of ouabain in guinea-pig ventricular cells. *Q J Exp Physiol* 1988;73:379–390.
134. Ono K, Kiyosue T, Arita M: Comparison of the inhibitory effects of mexiletine and lidocaine on the calcium current of single ventricular cells. *Life Sci* 1986;39:1465–1470.
135. Weld FM, Bigger JT: Effect of lidocaine on the early

inward transient current in sheep cardiac Purkinje fibers. *Circ Res* 1975;37:630–639.
136. Satoh H, Hashimoto K: Effect of lidocaine on membrane currents in rabbit sino-atrial node cells. *Arch Int Pharmacodyn* 1984;270:241–254.
137. Arnsdorf MF: The effect of antiarrhythmic drugs on triggered sustained rhythmic activity in cardiac Purkinje fibers. *J Pharmacol Exp Ther* 1977; 201:689–700.
138. Rosen MR, Danilo P: Effects of tetrodotoxin, lidocaine, verapamil and AHR-2666 on ouabain-induced delayed afterdepolarizations in canine Purkinje fibers. *Circ Res* 1980;46:117–124.
139. Sanchez-Chapula J, Josephson IR: Effect of phentoin on the sodium current in isolated rat ventricular cells. *J Mol Cell Cardiol* 1983;15:515–522.
140. Hering S, Bodewei R, Wollenberger A: Sodium current in freshly isolated and in cultured single rat myocardial cells. Frequency and voltage-dependent block by mexiletine. *J Mol Cell Cardiol* 1983; 15:431–444.
141. Yatani A, Akaike N: Blockage of the sodium current in isolated single cells from rat ventricle with mexiletine and disopyramide. *J Mol Cell Cardiol* 1985; 17:467–476.
142. Igawa O, Kotake H, Mashiba H: Electrophysiological actions of mexiletine on rabbit sinoatrial node cells. *Eur J Pharmacol* 1986;122:11–17.
143. Makielski JC, Undrovinas AI, Hanck DA, Sheets MF, Nesterenko VV, Alpert LA, Rosenshtraukh LV, Fozzard HA: Use-dependent block of sodium current by ethmozin in voltage-clamped internally perfused canine cardiac Purkinje cells. *J Mol Cell Cardiol* 1988;20:255–265.
144. Schubert B, Hering S, Bodewei R, Rosenshtraukh LV, Wollenberger A: Use- and voltage-dependent depression by ethmozine (Moricizine) of the rapid inward sodium current in single rat ventricular muscle cells. *J Cardiovasc Pharmacol* 1986;8:358–366.
145. Kohlhardt M, Fichtner H: Block of single cardiac Na^+ channels by antiarrhythmic drugs: The effects of amiodarone, propafenone and diprafenone. *J Membrane Biol* 1988;102:105–119.
146. Honjo H, Watanabe T, Kamiya K, Kodama I, Toyama J: Effects of propafenone on electrical and mechanical activities of single ventricular myocytes isolated from guinea-pig hearts. *Br J Pharmacol* 1989;97:731–738.
147. Scamps F, Undrovinas A, Vassort G: Inhibition of I_{Ca} in single frog cardiac cells by quinidine, flecainide, ethmozin and ethacizin. *Am J Physiol* 1989;256:C549–C559.
148. Satoh H, Hashimoto K: Effects of propafenone on the membrane currents of rabbit sino-atrial node cells. *Eur J Pharmacol* 1984;99:185–191.
149. Follmer CH, Colatsky TJ: Block of delayed rectifier potassium current, I_k, by flecainide and E-4031 in cat ventricular myocytes. *Circulation* 1990;82:289–293.
150. Campbell TJ: Resting and rate-dependent depression of maximum rate of depolarization (V_{max}) in guinea-pig ventricular action potentials by mexiletine, disopyramide and encainide. *J Cardiovasc Pharmacol* 1983;5:291–296.
151. Varro A, Saitoh H, Surawicz B: Effects of antiarrhythmic drugs on premature action potential duration in canine ventricular muscle fibers. *J Cardiovasc Pharmacol* 1987;10:407–114.
152. Tarr M, Luckstead EF, Jurewicz PA, Haas HG: Effect of propranolol on the fast inward sodium current in frog atrial muscle. *J Pharmacol Exp Ther* 1973; 184:599–610.
153. Barrington PL, Ten Eick RE: Characterization of the electrophysiological effects of metoprolol on isolated feline ventricular myocytes. *J Pharmacol Exp Ther* 1989;252:1034–1052.
154. Manley S, Alexopoulos D, Robinson GJ, Cobbe SM: Subsidiary class III effects of beta blockers? A comparison of atenolol, metoprolol, nadolol, oxprenolol and sotalol. *Cardiovasc Res* 1986;20:705–709.
155. Ban T: A kinetic study of effects of propranolol and N-propylajmaline on the rate of rise of action potential in guinea pig papillary muscles. *Jpn J Pharmacol* 1977;27:865–880.
156. Papp JGY, Vaughan Williams EM: A comparison of the antiarrhythmic actions of ICI 50172 and (−)-propranolol and their effects on intracellular cardiac action potentials. *Br J Pharmacol* 1969; 37:391–399.
157. Davis LD, Temple JV: Effects of propranolol on the transmembrane potentials of ventricular muscle and Purkinje fibers. *Circ Res* 1968;22:661–677.
158. Neliat G, Ducouret P, Gargouil YM: Effects of butoprozine on ionic currents in frog atrial and ferret ventricular fibres. Comparison with amiodarone and verapamil. *Arch Int Pharmacodyn* 1982; 255:237–255.
159. Follmer CH, Aomine M, Yeh JZ, Singer DH: Amiodarone-induced block of sodium current in isolated cardiac cells. *J Pharmacol Exp Ther* 1987; 243:187–194.
160. Mason JW, Hondeghem LM, Katzung BG: Block of inactivated sodium channels and of depolarization-induced automaticity in guinea pig papillary muscle by amiodarone. *Circ Res* 1984;55:277–285.
161. Nishimura M, Follmer CH, Singer DH: Amiodarone blocks calcium current in single guinea pig ventricular myocytes. *J Pharmacol Exp Ther* 1989; 251:650–659.
162. Sato R, Hisatome I, Singer DH: Amiodarone blocks the inward rectifier K^+ channel in guinea pig ventricular myocytes. *Circulation* 1987;76(suppl IV):150(abstract).
163. Balser JR, Hondeghem LM, Roden DM: Amiodarone reduces time dependent I_k activation. *Circulation* 1987;76(suppl IV):151(abstract).
164. Northover BJ: Alterations to the electrical activity of atrial muscle isolated from the rat heart, produced by exposure in vitro to amiodarone. *Br J Pharmacol* 1984;82:191–197.
165. Pallandi RT, Campbell TJ: Resting and rate-dependent depression of V_{max} of guinea pig ventricular action potentials by amiodarone and desethylamiodarone. *Br J Pharmacol* 1987;92:97–103.
166. Biagetti MO, deFroteza E, Quinteiro RA: Differential effects of amiodarone on V_{max} and conduction velocity in anisotropic myocardium. *J Cardiovasc Pharmacol* 1990;15:918–926.
167. Nattel S, Talajic M, Quantz MA, DeRoode M: Frequency-dependent effects of amiodarone on atrioventricular nodal function and slow-channel action potentials: Evidence for calcium channel-blocking activity. *Circulation* 1987;76:442–449.

168. Aomine M: Inhibition by amiodarone of slow response action potentials and contraction in guinea pig ventricular muscle. *Gen Pharmacol* 1988; 19:621–623.
169. Kadoya M, Konishi T, Tamamura T, Ikeguchi S, Hashimoto S, Kawai C: Electrophysiological effects of amiodarone on isolated rabbit heart muscles. *J Cardiovasc Pharmacol* 1985;7:643–648.
170. Dietrich H, Borchard U, Hafner D, Hirth C: Antiarrhythmic and electrophysiological actions of flecainide, bepridil and amiodarone on isolated heart preparations during controlled hypoxia. *Arch Int Pharmacodyn* 1985;274:267–282.
171. Varro A, Nakaya Y, Elharrar V, Surawicz B: The effects of amiodarone on repolarization and refractoriness of cardiac fibers. *Eur J Pharmacol* 1988;154:11–18.
172. Aomine M: Does acute exposure to amiodarone prolong cardiac action potential duration? *Gen Pharmacol* 1988;19:615–619.
173. Yabek SM, Kato R, Singh BN: Effects of amiodarone and its metabolite, desethylamiodarone, on the electrophysiologic properties of isolated cardiac muscle. *J Cardiovasc Pharmacol* 1986;8:197–207.
174. Singh BN, Vaughan Williams EM: The effect of amiodarone, a new anti-anginal drug, on cardiac muscle. *Br J Pharmacol* 1970;39:657–667.
175. Ikeda N, Nademanee K, Kannan R, Singh BN: Electrophysiologic effects of amiodarone: Experimental and clinical observation relative to serum and tissue drug concentrations. *Am Heart J* 1984; 108:890–898.
176. Illes M, Papp GY, Borbola J, Nemeth M, Szekeres L: Electrophysiological characteristics of the antiarrhythmic effects of amiodarone on the ventricular musculature and Purkinje fibre of the dog heart. *Acta Physiol Acad Sci Hung* 1977; 49:295(abstract).
177. Ohta M, Karagueuzian HS, McCullen A: Effect of amiodarone on delayed afterdepolarization and triggered automaticity in isolated rabbit ventricular muscle. *J Am Coll Cardiol* 1985; 5:492(abstract).
178. Polster P, Broekhuysen J: The adrenergic antagonism of amiodarone. *Biochem Pharmacol* 1976; 25:131–134.
179. Cohen-Armon M, Schreiber G, Sokolovsky M: Interaction of the antiarrhythmic drug amiodarone with the muscarinic receptor in the rat heart and brain. *J Cardiovasc Pharmacol* 1984;6:1148–1155.
180. Aomine M: Suggestive evidence for inhibitory action of amiodarone on Na^+, K^+-pump activity in guinea pig heart. *Gen Pharmacol* 1989;20:491–496.
181. Snyders DJ, Katzung BG: Intracellular clofilium reduces I_k in isolated adult cardiac myocytes. *Biophys J* 1986;49:401a(abstract).
182. Bkaily G, Payet MD, Benabderrazik M, Sauve R, Renaud J-F, Bacaner M, Sperelakis N: Intracellular bretylium blocks Na^+ and K^+ currents in heart cells. *Eur J Pharmacol* 1988;151:389–397.
183. Arena JP, Kass RS: Block of heart potassium channels by clofilium and its tertiary analogs: Relationship between drug structure and type of channel blocked. *Mol Pharmacol* 1988;34:60–66.
184. Castle NA: Clofilium and its tertiary homolog are potent inhibitors of the transient outward current in rat ventricular myocytes. *J Mol Cell Cardiol* 1990;22:(suppl I):S7.
185. Strauss HC, Bigger TJ, Hoffman BF: Electrophysiological and beta-receptor blocking effects of MJ 1999 on dog and rabbit cardiac tissue. *Circ Res* 1970;18: 683–689.
186. Papp JGY, Vaughan Williams EM: The effect of bretylium on intracellular cardiac action potential in relation to its anti-arrhythmic and local anaesthetic activity. *Br J Pharmacol* 1969;37:380–390.
187. Singh BN, Vaughan Williams EM: A third class antiarrhythmic action. Effects on atrial and ventricular intracellular potentials, and other pharmacological actions on cardiac muscle, of MJ 1999 and AH 3474. *Br J Pharmacol* 1970;39:675–687.
188. Lathrop DA: Electromechanical characterization of racemic sotalol and its optical isomers on isolated canine ventricular muscles and Purkinje strands. *Can J Physiol Pharmacol* 1985;63:1506–1512.
189. Steinberg MI, Sullivan ME, Wiest SA, Rockhold FW, Molloy BB: Cellular electrophysiology of clofilium, a new antifibrillatory agent, in normal and ischemic canine Purkinje fibers. *J Cardiovasc Pharmacol* 1981;3:881–895.
190. Campbell TJ: Cellular electrophysiological effects of D- and DL-sotalol in guinea-pig sino-atrial node, atrium and ventricle and human atrium: Differential tissue sensitivity. *Br J Pharmacol* 1987; 90:593–599.
191. Ehara T, Daufmann R: The voltage- and time-dependent effects of (−)-verapamil on the slow inward current in isolated cat ventricular myocardium. *J Pharmacol Exp Ther* 1978;207:49–55.
192. Uehara A, Hume JR: Interactions of organic calcium channel antagonists with calcium channels in single frog atrial cells. *J Gen Physiol* 1985; 85:621–647.
193. Kanaya S, Arlock P, Katzung BG, Hondeghem LM: Diltiazem and verapamil preferentially block inactivated cardiac calcium channels. *J Mol Cell Cardiol* 1983;15:145–148.
194. Sanguinetti MC, Kass RS: Voltage-dependent block of calcium channel current in the calf cardiac Purkinje fiber by dihydropyridine calcium channel antagonists. *Circ Res* 1984;55:336–348.
195. Kohlhardt M, Bauer B, Krause H, Fleckenstein A: Differentiation of transmembrane Na and Ca channels in mammalian cardiac fibers by use of specific inhibitors. *Pflügers Arch* 1972;335:309–322.
196. Hume JR: Comparative interactions of organic Ca^{++} channel antagonists with myocardial Ca^{++} and K^+ channels. *J Pharmacol Exp Ther* 1985; 234:134–139.
197. Kass RS: Nisoldipine: A new, more selective calcium current blocker in cardiac Purkinje fibers. *J Pharmacol Exp Ther* 1982;223:446–456.
198. Dangman KH, Ziam S, Miura DS: Local anaesthetic and negative chronotropic effects of diltiazem on canine cardiac Purkinje fibers. *Am Heart J* 1989; 117:1271–1277.
199. Tung L, Morad M: Voltage- and frequency-dependent block of diltiazem on the slow inward current and generation of tension in frog ventricular muscle. *Pflügers Arch* 1983;398:189–198.
200. Kohlhardt M, Haap K: The blockade of V_{max} of the atrioventricular action potential produced by the slow channel inhibitors, verapamil and nisoldipine. *Naunyn-Schmiedeberg's Arch Pharmacol* 1981;316:178–185.

201. Lathrop DA, Valle-Aguilera JR, Millard RW, Gaum WE, Hannon DW, Francis PD, Nakaya H, Schwartz A: Comparative electrophysiologic and coronary hemodynamic effects of diltiazem, nisoldipine and verapamil on myocardial tissue. *Am J Cardiol* 1982;49:613–620.
202. Satoh H, Bailey JC, Surawicz B: Alternans of action potential duration after abrupt shortening of cycle length: Differences between dog Purkinje and ventricular muscle fibers. *Circ Res* 1988;62:1027–1040.
203. Ellenbogen KA, German LD, O'Callaghan WG, Colavita PG, Marchese AC, Gilbert MR, Strauss HC: Frequency-dependent effects of verapamil on atrioventricular nodal conduction in man. *Circulation* 1985;72:344–352.
204. Talajic M, Nayebpour M, Jing W, Nattel S: Frequency-dependent effects of diltiazem on the atrioventricular node during experimental atrial fibrillation. *Circulation* 1989;80:380–389.
205. Amerini S, Giotto A, Mugetti A: Effect of verapamil and diltiazem on calcium-dependent electrical activity in cardiac Purkinje fibres. *Br J Pharmacol* 1985;85:89–96.

Chapter **23**

Effect of Drugs on Arrhythmia Substrate in Vivo: Update

Bruce G. Hook, MD, and Mark E. Josephson, MD

A great deal of information has become available regarding the ionic and cellular bases for antiarrhythmic drug action. In addition, the application of intracardiac recording and stimulation techniques has made it possible to evaluate the electrophysiologic effects and sites of action of the entire spectrum of antiarrhythmic agents. The ability to reproducibly initiate and terminate supraventricular and ventricular tachycardias (VT) has allowed electrophysiologists to determine the mechanism for many of these arrhythmias. By examining the effects of these antiarrhythmic agents on initiation and termination of tachycardias it is now possible to predict which agent may be most successful for the treatment of a particular arrhythmia. In this chapter we will discuss the current classification and actions of antiarrhythmic agents and the value of programmed stimulation in developing optimal therapy for tachyarrhythmias.

CLASSIFICATION OF ANTIARRHYTHMIC AGENTS

Ideally, the choice of antiarrhythmic agent should be based on the mechanism of the arrhythmia in question and the known electrophysiologic effects of the drug on the arrhythmia and its site of origin. For some arrhythmias, for example, reciprocating tachycardias in association with preexcitation syndromes, this is the case. Results of electrophysiologic testing can be used to identify the weak link in the tachycardia circuit, that is, the atrioventricular (AV) node or accessory pathway, and selection of antiarrhythmic therapy can be chosen to affect this particular site. Similarly, the site of origin and mechanism of AV nodal reentrant tachycardia is well known,[1] thereby allowing targeting of therapy to a specific pathway of the circuit. In the case of ventricular tachyarrhythmias, however, this is not the case. While the mechanism of VT associated with coronary artery disease has been fairly well established to be reentrant in nature,[2] the specifics of the tachycardia circuit remain poorly understood. Because of a limited understanding of the mechanism of VT at the microscopic and cellular level, the precise mechanism of action of antiarrhythmic agents in ventricular arrhythmias remains unknown. Consequently, therapy is empiric and efficacy is assessed with repeated electrophysiologic testing. This empirically chosen and electrophysiologically assessed therapy, which is based on in vitro drug classifications of how drugs affect ionic currents and not on how they influence arrhythmias, has been associated with only modest success in the prevention of recurrent ventricular tachyarrhythmias.[3] Thus, a better understanding of how drugs affect arrhythmogenic tissues in vivo is needed to gain a better understanding of why a given antiarrhythmic agent is either antiarrhythmic or proarrhythmic.

IN VITRO CLASSIFICATION OF ANTIARRHYTHMIC AGENTS

The current classification of antiarrhythmic drugs is based on the ability of these agents to block specific ionic currents (for example, sodium, potassium, or calcium), as well as those that block β-adrenergic receptors.[4–6] Class I agents affect primarily the fast sodium (Na^+) channel and have been subdivided into IA, IB, and IC. Type IA agents include quinidine, procainamide, and disopyramide, all of which slow the maximum rate of rise of the action potential ($\dot{V}_{max}$) as well as increasing action potential duration (APD). Class IB drugs have little effect on $\dot{V}_{max}$ at intermediate or slow heart rates and are

655 Avenue of the Americas, New York, NY 10010
Current Topics in Cardiology

reported to decrease APD. This group includes lidocaine, phenytoin, mexiletine, and tocainide. Type IC drugs, including flecainide, encainide, lorcainide, and propafenone, primarily slow conduction by depressing $\dot{V}_{max}$, with little effect on APD. Type II agents include β-adrenergic receptor blocking drugs including propranolol, metoprolol, nadolol, timolol, atenolol, and pindolol. Class III agents are those that block the potassium (K^+) current (i_K), thus resulting in prolongation of APD and refractoriness. This group includes amiodarone, bretylium, clofilium, and sotalol. Class IV agents block calcium (Ca^{2+}) currents (i_{Ca}) and include verapamil, nifedipine, diltiazem, and nicardipine.

While this classification is useful in describing the cellular and ionic effects of these agents, there are several limitations in using this schema in selecting antiarrhythmic therapy for a given arrhythmia.

a. The classification is based on in vitro studies of primarily normal Purkinje fibers, with little attention to effects in abnormal tissue (ischemic and/or chronically infarcted). For example, ischemic fibers have lower resting membrane potentials and depressed fast Na^+ channel activity and, in addition, demonstrate abnormal intercellular coupling.[7,8]

b. Drugs may have multiple mechanisms of action. Drugs may affect multiple channels and/or may both affect the autonomic nervous system as well as have β-adrenergic blocking activity. Moreover, several antiarrhythmic agents can depress myocardial function, resulting in reflex autonomic changes which can affect antiarrhythmic drug action. Finally, several drugs have metabolites with active antiarrhythmic effects, which cannot be assessed during in vitro testing.

c. The effects of antiarrhythmic agents may be different at greater rates of stimulation (use dependence).[9–13] Drugs that bind and dissociate from channels slowly (type IC drugs) show little change in effect as the rate of stimulation is increased. However, type IB agents, which bind and dissociate rapidly, demonstrate a marked increase in effect as the rate of stimulation increases. Class IA drugs show intermediate degrees of binding and dissociating, and consequently their use-dependent effects are intermediate. The modulated receptor hypothesis has been proposed to explain this use-dependent effect.[14–16] This appears to be operative for both Na^+ and Ca^{2+} channels. Effects of rate of stimulation on observed antiarrhythmic drug action may alter conclusions about a particular agent compared to results obtained in a normal isolated Purkinje fiber paced only at a single rate.

d. Changes in extracellular or intracellular ionic concentrations as well as pH can affect membrane currents through ionic channels and may affect drug action.[8,16–19] In addition, increased intracellular Ca^{2+} may decrease intercellular coupling and impair propagation of impulses. The inhibition of the Na/K pump by digitalis and the Na/Ca pump during ischemia may also affect the action of antiarrhythmic drugs.

e. Changes in circulating catecholamine levels and changes in autonomic tone may influence the action of antiarrhythmic agents in vivo. Many of the effects on Na^+, K^+, and Ca^{2+} channels can be reversed by enhanced sympathetic tone or circulating catecholamines.[20] Furthermore, it has been shown that infusions of isoproterenol[21] or epinephrine[22] can reverse the ability of antiarrhythmic agents to suppress induction of sustained VT in man.

f. The influence of drugs on passive membrane properties, in particular, the effects on coupling and propagation of impulses through anisotropic tissue, deserves attention. There is evidence that agents may have disparate effects on longitudinal compared to transverse conduction through myocardial fibers.[23] The role of drug-induced changes in nonuniformly anisotropic tissue remains to be defined.

In summary, the simple classification of antiarrhythmic agents based on effects measured from single-cell microelectrode recordings, whole-cell voltage-clamping, and patch-clamping, provides insufficient information regarding electrophysiologic actions in vivo and even less information relative to a drug's antiarrhythmic effects. To obtain a better understanding of the actions of these drugs in humans, it is important to assess the electrophysiologic properties of these agents in human tissue in vivo.

IN VIVO INVESTIGATIONS

Using standard intracardiac recording and stimulation techniques, the effects of individual drugs on specific cardiac tissue can be determined. A summary of the electrophysiologic properties of various currently available and some promising experimental antiarrhythmic agents in man is shown in Table 23.1.[24–97] These data apply to the effects of these agents on nonischemic, normal tissues measured during sinus rhythm or during pacing from standard right atrial and right ventricular sites. Despite this wealth of data,

TABLE 23.1 In Vivo Electrophysiologic Characteristics of Antiarrhythmic Drugs

	Electrophysiological Effects				
Drug	ERP AVN	ERP HPS	ERP A	ERP V	ERP AP
Quinidine	↓↔↑	↔↑	↑	↑	↑
Procainamide	↔↑	↔↑	↑	↑	↑
Disopyramide	↔↓	↑	↑	↑	↑
Ethmozine	↔	↔	↔↑	↔↑	↑
Lidocaine	↔↓	↔↑	↔	↔	↔
Phenytoin	↔↓	↓	↔	↔	↔
Mexiletine	↔↑	↔↑	↔	↔	↔
Tocainide	↓	↔	↔↓	↔↓	↔↑
Flecainide	↑	↑	↑	↑	↑
Encainide	↑	↑	↑	↑	↑
Lorcainide	↔	↔↑	↔	↔	↑
Propafenone	↔↑	↔	↔↑	↑	↑
Propranolol	↑	↔	↔	↔	↔
Amiodarone	↑	↑	↑	↑	↑
Bretylium	↔↑	↑	↑	↑	↔
Verapamil	↑	↔	↔	↔	↔

	Electrocardiographic Effects					
Drug	Sinus Rate	P-R	QRS	Q-T	A-H	H-V
Quinidine	↔↑	↓↔↑	↑	↑	↓↔↑	↔↑
Procainamide	↔	↔↑	↑	↑	↔↑	↔↑
Disopyramide	↔↑	↔	↔↑	↔↑	↔	↔↑
Ethmozine	↔↓	↔↑	↔↑	↔	↑	↑
Lidocaine	↔	↔	↔	↔	↔↓	↔↑
Phenytoin	↔	↔	↔	↔↓	↔↓	↔
Mexiletine	↔	↔	↔	↔	↔↑	↔↑
Tocainide	↔↓	↔	↔	↔↓	↔↑	↔
Flecainide	↔↓	↑	↑	↑	↑	↑
Encainide	↔	↑	↑	↑	↑	↑
Lorcainide		↔↑	↑	↑	↔	↑
Propafenone	↔↓	↑	↑	↔↑	↑	↑
Propranolol	↓	↔↑	↔	↔↓	↔	↔
Amiodarone	↓	↔↑	↑	↑	↑	↑
Bretylium	↔↓	↔↑	↔	↔↑		
Verapamil	↔↓	↑	↔	↔	↑	↔

Results presented may vary according to tissue type, experimental conditions, and drug concentration.
↑ = increase, ↓ = decrease, ↔ = no change, ↔↑ or ↔↓ = slight inconsistent increase or decrease.
A = atrium, AVN = AV node, HPS = His-Purkinje system, V = ventricle, AP = accessory pathway (WPW),
ERP = effective refractory period—longest S1-S2 interval at which S1 fails to produce a response.

there is still very little known about the influence of antiarrhythmic agents on arrhythmogenic tissues.

In a recent study from our laboratory,[98] the effects of lidocaine and procainamide on intraventricular electrograms from normal right ventricular and abnormal left ventricular regions were recorded during sinus rhythm and pacing at a cycle length of 600 ms. The abnormal left ventricular sites had low amplitude, multicomponent electrograms of long duration, characteristic of sites from which VT arise. Bipolar electrograms were recorded during sinus rhythm with #6 French catheters positioned at an abnormal left ventricular site and normal sites at the right ventricular apex and right ventricular outflow tract. Lidocaine and procainamide had no significant effect on sinus cycle length or elec-

TABLE 23.2 Effects of Lidocaine and Procainamide on Electrogram Amplitude (mV)

	RVA		RVOT		LV	
	C	Drug[a]	C	Drug[a]	C	Drug[a]
Procainamide						
NSR	6.9 ± 3.1	6.7 ± 4.8	5.0 ± 2.6	4.6 ± 2.2	1.6 ± 0.9	1.4 ± 0.8
600	8.5 ± 7.5	8.0 ± 8.8	8.8 ± 4.2	8.7 ± 8.4	1.6 ± 0.8	1.7 ± 1.8
Lidocaine						
NSR	4.8 ± 1.5	4.4 ± 0.8	6.7 ± 3.0	6.2 ± 3.2	1.2 ± 0.6	1.3 ± 0.8
600	11.3 ± 2.8	10.3 ± 1.9	6.8 ± 3.2	6.7 ± 3.2	2.1 ± 1.7	2.0 ± 1.5

C = control.
Reprinted from Schmitt et al.,[98] *with permission.*

trogram amplitude (Table 23.2). Following lidocaine there was no significant change in QRS duration, left ventricular electrogram duration, or right ventricular electrogram duration (Table 23.3). Similarly, at a paced cycle length of 600 ms, there was no change following lidocaine in the electrogram duration or QRS duration (Table 23.3, Fig. 23.1A). In contrast, following procainamide administration (Table 23.4, Fig. 23.1B), the QRS width prolonged from 110 ± 31 to 127 ± 33 ms and the electrogram duration increased significantly at the left ventricular site, the right ventricular apex, and the right ventricular outflow tract during sinus rhythm. Similar effects were seen with pacing at 600 ms.

The observations that lidocaine had no effect on the local electrogram duration and procainamide prolonged indices of conduction in both normal and chronically infarcted myocardium to the same degrees suggest that the tissues responsible for arrhythmogenesis are, in fact, comprised of normal myocardial cells, and that the slowing in conduction reflects cellular uncoupling by intervening fibrosis. If the arrhythmogenic myocardium was composed of fibers exhibiting either depressed Na^+ channels or slow (Ca^{2+}) responses, lidocaine would have affected them in the former case and procainamide would not have affected them in the latter case. In fact, in a few patients we were unable to demonstrate any effect of verapamil on such electrograms, further suggesting that they were not due to i_{Ca}. Direct recordings of action potentials from explanted human hearts in regions of infarction have shown these to be normal.[99,100]

Further studies from our laboratory[101] examining the effect of pacing cycle length on the duration of ventricular electrograms in normal and abnormal areas demonstrated a greater effect of procainamide on conduction through the abnormal arrhythmogenic substrate than on normal tissues at faster paced rates, that is, cycle lengths of 300 ms (Figs. 23.2 and 23.3). Whereas pacing at 600 and 400 ms produced no significant change in electrogram duration at normal and abnormal sites, when the paced cycle length was reduced to 300 ms, procainamide increased significantly electrogram duration at abnormal sites. Procainamide produced a small increase in electrogram duration at normal sites which was comparable to the increase in electrogram duration at abnormal sites at 600- and 400-ms. However, at paced cycle lengths of 300 ms, procainamide-induced increase in electrogram duration was significantly greater at abnormal sites. These findings suggest that in arrhythmogenic ventricular tissue procainamide produces a cycle length-dependent effect on saltatory conduction

TABLE 23.3 Effects of Lidocaine on QRS Width and Electrogram Duration (ms)

	QRS		RVA		RVOT		LV	
	C	L	C	L	C	L	C	L
NSR	122 ± 48	124 ± 48	37 ± 16	40 ± 17	54 ± 24	55 ± 23	94 ± 26	97 ± 24
600	197 ± 19	198 ± 23	30 ± 7	31 ± 5	49 ± 29	52 ± 31	102 ± 30	108 ± 30

C = control, L = lidocaine, RVA = right ventricular apex, RVOT = right ventricular outflow tract, LV = left ventricle, NSR = normal sinus rhythm, 600 = pacing at a cycle length of 600 ms from the RVA.
Reprinted from Schmitt et al.,[98] *with permission.*

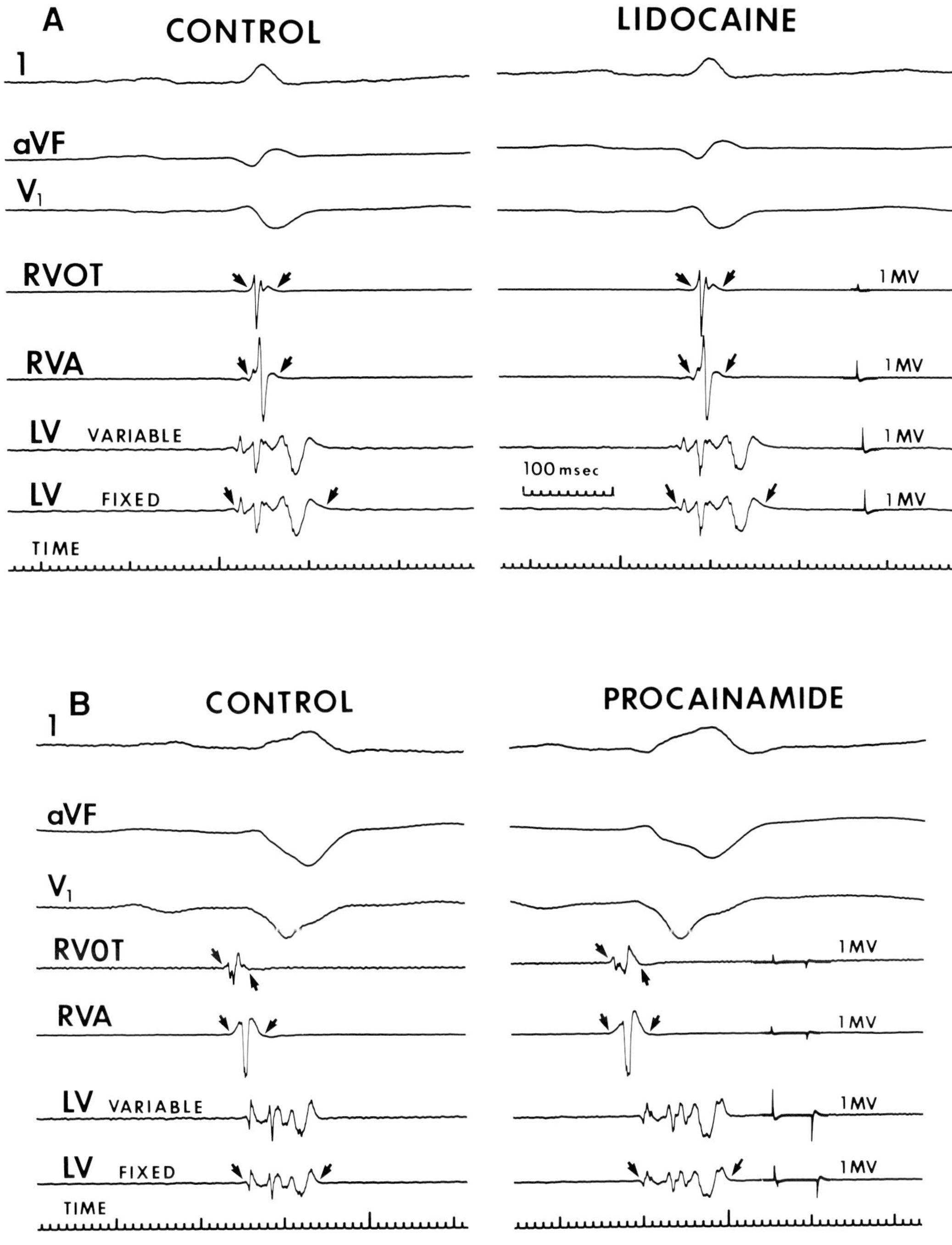

FIGURE 23.1 A. Example of effects of lidocaine on electrograms of surface leads 1, aVF, and V1; intraventricular recordings with variable gain of the right ventricular outflow tract (RVOT) and right ventricular apex (RVA); and recordings of an abnormal site of the left ventricle (LV) with variable and fixed gains. The arrows indicate onset and offset of electrogram duration. *Reprinted from Schmitt et al.,*[98] *with permission.* **B.** Example of effects of procainamide on electrograms of surface leads 1, aVF, and V1; intraventricular recordings with variable gain of the right ventricular outflow tract (RVOT) and right ventricular apex (RVA); and recordings of an abnormal site of the left ventricle (LV) with variable and fixed gains. The arrows indicate onset and offset of electrogram duration. *Reprinted from Schmitt et al.,*[98] *with permission.*

TABLE 23.4 Effects of Procainamide on QRS Width and Electrogram Duration (ms)

	QRS		RVA		RVOT		LV	
	C	P	C	P	C	P	C	P
NSR								
Mean ± SD	110 ± 31	127 ± 33	32 ± 11	38 ± 13	37 ± 12	44 ± 15	84 ± 17	101 ± 23
p value (paired t test)	<.0001		<.0001		<.001		<.001	
% change	+16 ± 6		+19 ± 8		+19 ± 11		+20 ± 9	
600								
Mean ± SD	194 ± 28	226 ± 45	32 ± 9	36 ± 14	34 ± 11	39 ± 13	86 ± 26	108 ± 31
p value (paired t test)	<.001		<.05		<.01		<.001	
% change	+16 ± 8		+13 ± 20		+18 ± 16		+27 ± 22	

C = control, P = procainamide, RVA = right ventricular apex, RVOT = right ventricular outflow tract, LV = left ventricle, NSR = normal sinus rhythm.
Reprinted from Schmitt et al.,[98] with permission.

which is beyond simple use-dependent effects. Kadish et al.[23] have demonstrated that the effects of procainamide on isolated strips of canine epicardium are more pronounced in the longitudinal than the transverse direction when paced at rates of 1,000 to 400 ms. Whether these effects are similar or more pronounced at faster paced rates has not been studied.

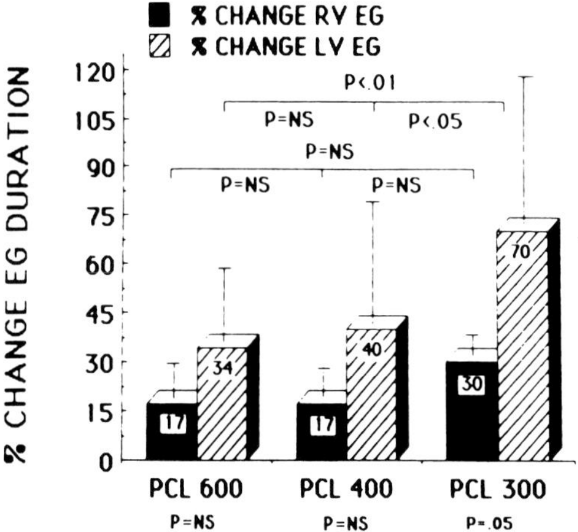

FIGURE 23.2 The percent increase in the electrogram (EG) duration after procainamide administration at normal right ventricular (RV) versus abnormal left ventricular (LV) sites at each paced cycle length (PCL). At a cycle length of 300 ms, a greater increase was observed in the electrogram duration recorded from the abnormal site (p = 0.05). One-way analysis of variance demonstrated a cycle length-dependent difference from 600 or 400 ms to 300 ms in the degree of change in electrogram duration at abnormal but not normal sites (significance bars above bar graphs). *Reprinted from Schmitt et al.,[101] with permission.*

ANALYSIS OF DRUG EFFECTS USING RESETTING AND ENTRAINMENT

Ventricular extrastimuli delivered during a tachycardia may interact with the circuit and provide information relative to the mechanism of the tachycardia and the characteristics of the tachycardia circuit. It has been shown that approximately 60 to 70% of hemodynamically tolerated VT may be reset using single or double ventricular extrastimuli.[102–104] Response patterns have been defined as flat, mixed (flat + increasing), and increasing. The flat portion of the curve, where progressively more premature ventricular extrastimuli are associated with return cycles that are constant, is felt to represent a fully excitable gap within the VT circuit. Increasing curves are thought to represent interaction of the ventricular extrastimulus with tissue that is partially refractory, resulting in delay in conduction of the impulse through the VT circuit. As a result, progressively more premature ventricular extrastimuli encounter more and more delay, producing an increase in the return cycle. Mixed curves represent a fully excitable gap during the flat portion of the curve, and then a portion of relative refractoriness following this. Approximately two-thirds of tachycardias that are able to be reset will demonstrate some flat portion of the curve,[103] consistent with a fully excitable gap within the circuit. By analyzing the influence of drugs on the resetting response curves in the baseline state and then following administration of an antiarrhythmic agent, it is possible to study the effects of the drug on propagation of the impulse through the circuit. Specifically, information regarding changes in the fully excitable gap and on tissue that is relatively refractory can be

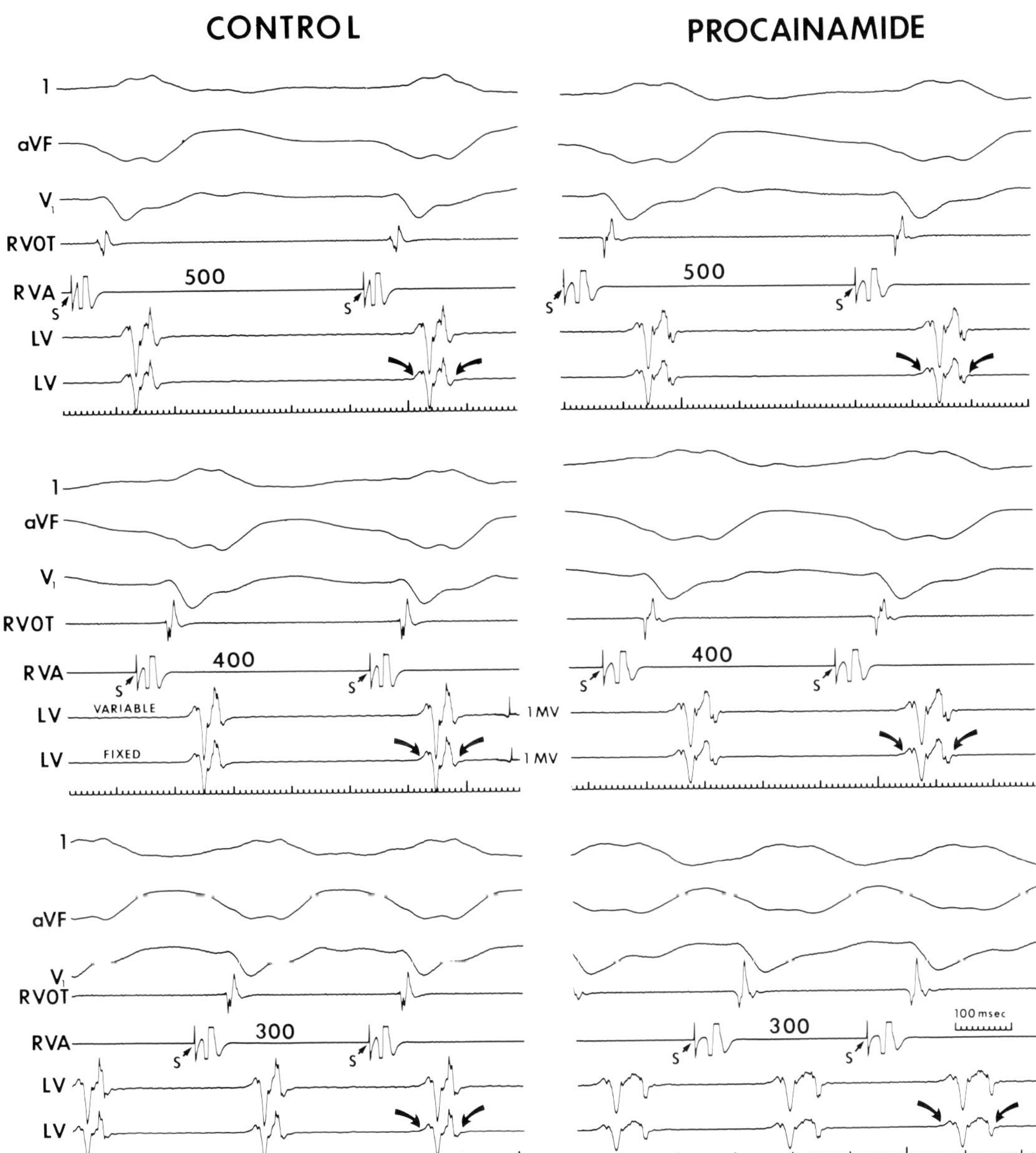

FIGURE 23.3 Example of cycle length-dependent effects on electrograms before (**left**) and after (**right**) administration of procainamide. The duration of the normal right ventricular (RV) and abnormal left ventricular (LV) electrograms did not change with faster paced cycle lengths in the control state. After procainamide, a more pronounced prolongation of the abnormal left ventricular electrogram compared with right ventricular electrogram occurs with shorter cycle lengths. RVA = right ventricular apex, RVOT = right ventricular outflow tract, S = stimulus, curved arrows indicate onset and end of left ventricular electrogram. *Reprinted from Schmitt et al.,*[101] *with permission.*

obtained. To perform this type of analysis requires the presence of unchanged morphology of VT before and after drug, and that the tachycardia is hemodynamically well tolerated in both instances.

In theory, an antiarrhythmic agent may slow VT by decreasing the conduction velocity through the circuit, by prolonging the refractory periods of the tissue within the circuit, or by changing the pathway prior to exit of the impulse from the circuit. In one study from our laboratory examining the effect of procainamide on resetting response patterns during VT,[105] it was found that 12 of 17 VT that could be reset demonstrated a flat portion of the resetting curve after procainamide (Table 23.5). The drug increased VT cycle length from 292 ± 61 to 374 ± 61 ms. The mean effective refractory period, measured at the right ventricle, increased only from 241 ± 21 to 261 ± 24 ms. The finding of some flat portion of the curve in the majority of tachycardias able to be reset suggests that procainamide slowed these VT by decreasing conduction velocity in fully recovered tissue or by changing the pathway of the impulse, and not by prolongation of refractoriness within the circuit. In a preliminary report utilizing resetting response patterns after quinidine and amiodarone,[106] it was found that amiodarone had a greater effect on VT cycle length than quinidine and that four of four VT reset during amiodarone therapy and two of four VT reset on quinidine demonstrated a flat portion of the curve. Further studies from our laboratory with amiodarone have shown that 13 of 15 VT that were reset during amiodarone therapy demonstrated a flat portion of the curve, consistent with slowing of conduction within the circuit or an alteration in the pathway prior to exit from the VT circuit (Table 23.6). Taken together, these data suggest that a fixed increase in refractoriness within the VT circuit is unlikely to account for the increase in VT cycle length seen with procainamide, quinidine, or amiodarone.

TABLE 23.5 Findings in the 16 Tachycardias in Which Reset Response Patterns Were Characterized on Procainamide

VT No.	Reset Response Pattern	Reset Zone	
		Total	Flat
2	F	85	85
5	F + I	105	40
6	I	100	0
7	F + I	160	90
11	I	65	0
12	I	60	0
14	F + I	120	70
15	F	160	160
16	I	90	0
17	F	40	40
18	F	120	120
19	F + I	95	65
20	F + I	80	40
21	F	170	170
22	F	70	70
23	F	30	30

F = flat, I = increasing.
Reprinted from Stamato et al.,[105] with permission.

TABLE 23.6 Findings in the 12 Tachycardias in Which Reset Response Patterns Were Characterized on Amiodarone

VT No.	Reset Response Pattern	Flat Zone (ms)
1	F + I	145
2	F	20
3	F + I	25
4	F + I	80
5	F + I	90
6	F + I	40
7	F	85
8	F + I	110
9	I	0
10	F	95
11	I	0
12	F + I	170

F = flat, I = increasing.

The use of entrainment (continuous resetting of the tachycardia) may also be used in the evaluation of the effects of antiarrhythmic agents on the VT circuit. Kay et al.[107] used this technique to examine the mechanism of procainamide-induced prolongation of VT cycle length in 5 patients with the identical morphology of inducible tachycardia before and after drug. Orthodromic activation time from the right ventricle to the left ventricular local electrogram during VT as well as antidromic conduction time (defined as the interval from a right ventricular paced beat during sinus rhythm to the onset of the left ventricular local electrogram) were measured before and after procainamide. Tachycardia cycle length increased by 27% following procainamide, from 324 ± 23 to 410 ± 42 ms. The increase in orthodromic activation time was 28% following procainamide, compared to an increase of only 9% in antidromic activation time. From these data it was concluded that the effects of procainamide were specific to the tachycardia circuit. A limitation of this study was that the number of stimuli employed at each cycle length that resulted in entrainment was not specified. During entrainment, the first stimulus that resets the VT circuit (n^{th} stimulus) is followed by stimuli that reset the *reset* circuit. Using this technique it is possible for prolongation of orthodromic activation time to be manifested, simply as a result of impingement of successive extrastimuli upon the refractory period of the reset circuit. This can be unrelated to a drug effect on conduction velocity in the tachycardia circuit, and may be due to impingement of the extrastimuli on the refractory period of the preceding reset circuit. It is possible to avoid this confounding effect by analyzing the results of the interaction of a single ventricular extrastimulus with the VT circuit.

In addition to the effects of antiarrhythmic agents on the VT circuit, whether these effects are related to changes in conduction velocity or changes in the functional barriers leading to lengthening of the pathway within the circuit,[(108)] antiarrhythmic agents prolong local refractory periods at the site of ventricular stimulation and/or prolong conduction time from the site of ventricular extrastimuli to the left ventricular site of origin. These effects can influence the ability of extrastimuli to induce VT or reach the VT circuit when it is excitable. We have recently shown in 14 patients in whom VT was rendered noninducible with standard antiarrhythmic agents that repeated ventricular programmed stimulation using a pacing current strength of 10 mA can result in the induction of sustained uniform VT in over 50% of patients. Eight patients demonstrated unimorphic sustained VT with a cycle length of 320 ms with the use of increased pacing current strength. In four of these eight cases, the VT morphology was identical to that seen in the baseline study. These findings suggest that in some patients prevention of VT induction by antiarrhythmic agents is not related to effects of the drugs on the circuit, since tachycardias may still be inducible using higher current strength. Long-term follow-up will be needed to determine whether this subgroup of patients is at higher risk for recurrence of their tachyarrhythmia. Thus, it appears that prevention of initiation of VT by antiarrhythmic agents may be related not only to effects on the VT circuit, but also to changes in refractoriness at the site of ventricular stimulation or prolongation of conduction time of the extrastimuli to the tachycardia site of origin.

EFFICACY OF ANTIARRHYTHMIC AGENTS IN SUSTAINED VENTRICULAR TACHYCARDIA

It is well recognized now that suppression of inducible VT by antiarrhythmic agents improves prognosis compared to patients in whom the arrhythmia cannot be suppressed.[(109–123)] Unfortunately, suppression of inducible VT by antiarrhythmic agents is possible in only 35 to 40% of patients. The criteria by which drug efficacy is established for programmed stimulation have varied. Some have suggested that narrowing of the tachycardia induction zone (that is, the number of extrastimuli required to initiate the tachycardia) may be useful in assessing drug efficacy. In our opinion, prevention of sustained VT is the most appropriate endpoint.[(124)] Although Swerdlow et al.[(125)] suggested that the induction of 15 or less beats was predictive of a good outcome, others have not found this to be true.[(126)] While 15 complexes should not be considered a magical number, it is uncommon for an arrhythmia to last more than 15 or 20 complexes without being sustained. Therefore, the use of 15 complexes merely represent a statistical likelihood that any spontaneous nonsustained arrhythmia would not exceed 15 beats. We do not believe that either the absolute number of extrastimuli used or the relative ease of induction of sustained VT can be used to predict outcome of type I agents or amiodarone. This is shown in Figures 23.4 to 23.6, which represent data derived from 211 patients on type I agents and 106 patients on amiodarone in our laboratory. The relative efficacy of antiarrhythmic drugs is not known for certain. For example, amiodarone only prevents initiation of VT in 10 to 15% of our patients, whether studied at 2 weeks or 3 months. However, amiodarone is only used after several drugs have failed. The same is true for other drugs that have recently

INDUCIBILITY AND OUTCOME
211 PTS

Recurrence	NI	I
No	89	78
Yes	9	35

FIGURE 23.4 Incidence of VT recurrence in patients on type I antiarrhythmic drugs relative to inducibility during electrophysiologic testing. Positive predictive value was 31% and negative predictive value 91%.

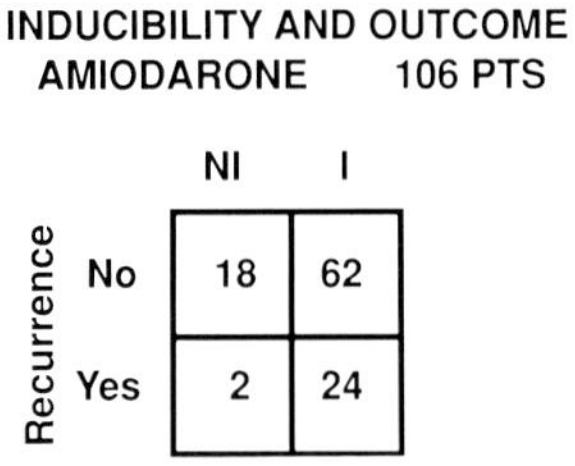

INDUCIBILITY AND OUTCOME
AMIODARONE 106 PTS

Recurrence	NI	I
No	18	62
Yes	2	24

FIGURE 23.5 Incidence of VT recurrence in patients on amiodarone relative to inducibility during electrophysiologic testing. Positive predictive value was 28% and negative predictive value 90%.

been added to our pharmacologic armamentarium (that is, the IC agents). If each of these drugs had been tried as a first-line agent, the relative efficacy might have been different. It can be seen from Figure 23.6 that type IA agents are most likely to prevent induction of sustained VT, with a success rate ranging between 26 and 38%. Type IC agents have a frequency of rendering tachycardias noninducible between 18 and 22%.

Since type IA drugs individually had comparable efficacy, we analyzed whether the response to procainamide, which could be given intravenously in the laboratory, could predict the effect of other type I agents (Fig. 23.7). In a study of 129 patients, Waxman et al.[127] found that if a tachycardia remained inducible on procainamide, it was likely to be inducible on other type I drugs, singly and/or in combination. Conversely, if a drug rendered a tachycardia noninducible, it was likely to be rendered noninducible by another type I agent, although one could not predict what agent that might be. Chi-square analysis of these data yielded a negative predictive value of 83% and a positive predictive value of 87%.

A recent study[128] of 40 patients with uniform monomorphic VT at baseline and after procainamide and oral amiodarone therapy found a significant correlation between VT cycle lengths on the two agents as well as the percent change in cycle length from baseline. More importantly, it was observed that only 15% of the patients had greater slowing of the VT rate during amiodarone therapy as compared to procainamide.

Combination drug therapy does not appear more useful in preventing initiation of ventricular tachyarrhythmias than do single drugs, but may decrease the side effects of high doses of an individual agent. One drug combination, however, that has been successful is quinidine and mexiletine.[129] The addition of procainamide to amiodarone has likewise not increased the incidence of noninducibility, but has led to slowing of the induced tachycardia.[130]

β-Blockers and Ca^{2+} channel blockers are rarely effective in preventing sustained uniform VT. There is, however, one group of patients in whom Ca^{2+} blockers appear to be useful. This is in a group of patients with apparently normal hearts, usually male, who have sustained uniform VT characterized by a right bundle branch block, left superior axis morphology.[131] These patients

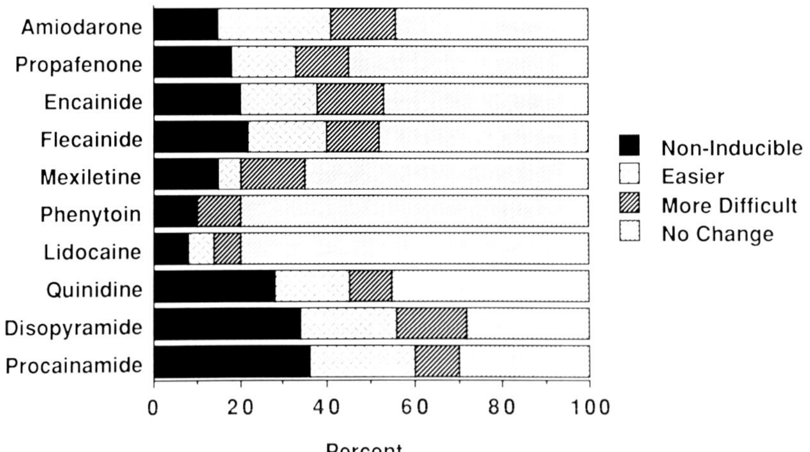

FIGURE 23.6 Effect of drugs on VT induction. Note that type IA agents were most likely to prevent VT induction, while type IB agents were least likely. Amiodarone prevented VT induction in only about 17% of patients.

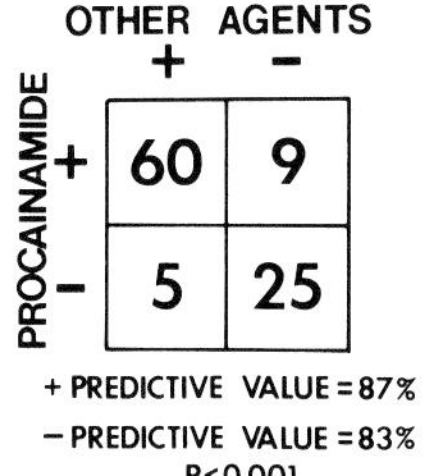

FIGURE 23.7 Patient response during serial electrophysiologic testing compared with the response to procainamide. The plus sign refers to inducibility and the minus sign to noninducibility of VT. *Reprinted from Waxman et al.,*[127] *with permission.*

have tachycardias that can frequently be initiated by atrial pacing as well as with ventricular extrastimuli. In this group of patients, verapamil frequently produces a dose-dependent slowing and/or termination of the tachycardia. In our experience, these patients frequently respond to class IA and IC antiarrhythmic agents. Similarly, VT originating from the right ventricular outflow tract,[133,134] demonstrating a left bundle branch block, left inferior axis morphology appears to be very sensitive to β-blockers. These tachycardias are frequently exercise induced when occurring spontaneously and may be reproduced in the electrophysiology laboratory by ventricular pacing or with infusion of isoproterenol.

Isoproterenol may be of additional benefit in patients with coronary artery disease and sustained VT in whom conventional antiarrhythmic drugs suppress induction of the tachycardia. In a study of 17 patients[21] in whom VT was suppressed with class IA agents, isoproterenol infusion resulted in initiation of the original ventricular tachyarrhythmia in 10 patients while 7 patients remained noninducible. During a mean follow-up period of 13 ± 9 months, 3 of the 10 patients in the first group experienced clinical recurrence of the arrhythmia while there were no recurrences in the second group. These results suggested the use of a β-blocker to counteract the effects of enhanced sympathetic tone may be a useful adjunct to standard antiarrhythmic agents. The beneficial effect of β-blockers on myocardial ischemia may be of added benefit.

Under certain circumstances, persistence of inducible VT may be an acceptable clinical alternative if the rate of VT is slowed to a degree that is not associated with hemodynamic compromise. In a study by Waller et al.[134] of 258 patients with inducible sustained VT undergoing

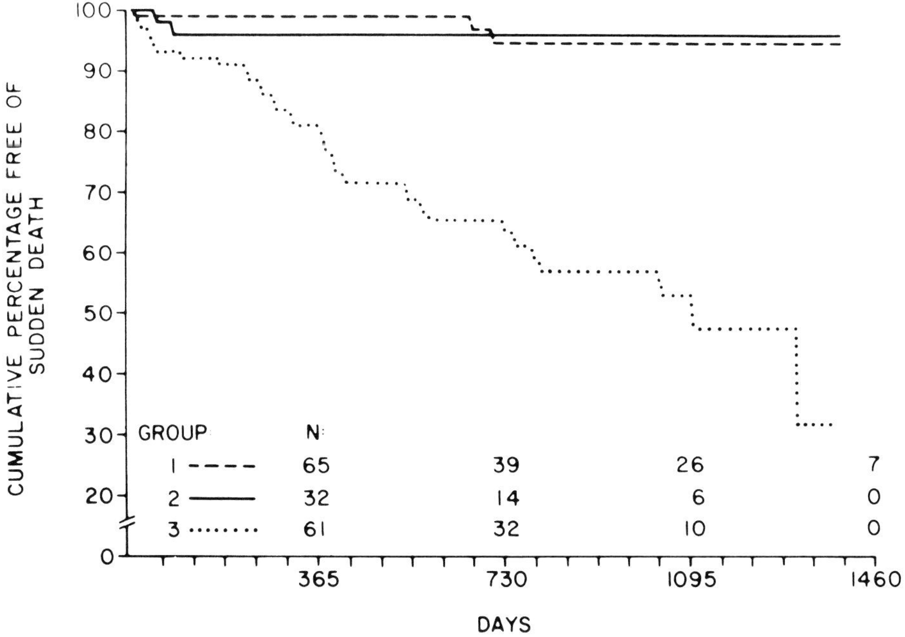

FIGURE 23.8 Life-table analysis of sudden-death events during clinical follow-up. Groups 1 and 2 were not significantly different from each other, but group 3 differed significantly from group 2 ($p < 0.0003$) and group 1 ($p < 0.0001$). The numbers at the bottom of the graph indicate the number of patients who remained available for analysis at years 1, 2, 3, and 4. *Reprinted from Waller et al.,*[134] *with permission.*

electrophysiologic testing, three subgroups were identified based on the response to antiarrhythmic agents. Group 1 included patients in whom the initiation of VT was prevented by the drug. Group 2 were those patients in whom the tachycardia remained inducible, but in whom the tachycardia cycle length increased by greater than 100 ms and the tachycardia did not produce severe symptoms. In group 3 the tachycardia was inducible and either was not significantly slowed or remained poorly tolerated. During follow-up, group 1 showed a significant reduction in the incidence of recurrent VT compared to groups 2 and 3. However, the total mortality and sudden-death mortality was significantly reduced in group 2 compared with group 3 (Fig. 23.8).

ROLE OF ANTIARRHYTHMIC DRUGS IN ATRIAL FLUTTER

The observation that atrial flutter can be reproducibly initiated and terminated in man by programmed stimulation and that it also demonstrates the phenomenon of resetting and entrainment have suggested reentry as the mechanism.[135] Recent studies employing high-density electrode mapping experimentally[136–139] have also confirmed this as the mechanism of atrial flutter in several animal models.

We have recently compared the ability of single and double extrastimuli to reset atrial flutter in a selected population of 18 patients.[140] Resetting with single extrastimuli occurred more often from the HRA than from the coronary sinus and the coupling intervals resetting the flutter were longer from the HRA, indicating a closer proximity to the circuit. Furthermore, regions of the left atrium and the low right atrium could be captured with extrastimuli without affecting the tachycardia. When double extrastimuli were delivered, resetting was observed in all but 1 patient. When the patterns of resetting curves were analyzed, a flat (n = 5) or mixed (flat plus increasing, n = 5) curve was seen in 10 patients, indicating the presence of a fully excitable gap.

It has been known for some time that typical type IA agents, which prolong both atrial refractoriness and conduction, slow the rate of atrial flutter.[5,28,30,32,141] Less data are available concerning the type IC agents encainide, flecainide, and propafenone which also demonstrate slowing of flutter rates. Both classes of drugs appear to slow the flutter cycle length by prolongation of intraatrial conduction. The prolongation of flutter cycle length does not appear to correlate with drug-induced changes in refractoriness. This was shown recently for procainamide by Wu and Hoffman,[47] who demonstrated that slowing of the atrial flutter was due to prolongation of conduction time in an experimental model. Schoels et al.[142] in a preliminary report demonstrated a selective depression of conduction by procainamide on the zone of slow conduction in 6 dogs with sterile pericarditis and atrial flutter. The zone of slow conduction accounted for 57 ± 13% of the revolution time within a region comprising 33 ± 14% of the total length of the reentrant pathway. Procainamide prolonged the cycle length of atrial flutter from 143 ± 26 to 173 ± 38 ms. The increase in conduction time in the region of slow conduction accounted for 91 ± 14% of the total cycle length increase.

The response to extrastimuli delivered during atrial flutter in man before and after disopyramide, flecainide, and amiodarone has recently been examined by Della Bella et al.[143] The increase in cycle length of atrial flutter with disopyramide (25%) and flecainide (37%) was associated with an increase in the duration of the excitable gap of 55% and 105%, respectively. Amiodarone, on the other hand, increased the cycle length by only 4% and decreased the excitable gap by 30%, suggesting a primary effect of the drug on refractoriness, which in turn had little influence on the flutter cycle length. Specific effects of the drugs on the flat and increasing portions of the resetting curves were not reported in this study, however. Furthermore, we have shown a much greater prolongation of flutter cycle length with amiodarone,[144] from 223 ± 32 ms to 282 ± 50 ms. Consequently, the actions of amiodarone on the duration of the excitable gap reported by Della Bella et al. may represent only a partial drug effect.

Ca^{2+} blockers and β-blockers, which have been shown to abolish triggered activity in animal tissues and tissues obtained from human atria at surgery,[141,145] generally have no effect on atrial flutter. Occasionally, Ca^{2+} blockers and β-blockers may terminate new onset atrial flutter, particularly in the postoperative setting in which prior treatment with Ca^{2+} blockers and β-blockers have upregulated the β-receptors.[146] In such instances, the addition of Ca^{2+} blockers and β-blockers may actually prolong atrial refractoriness and terminate the arrhythmia.

ACKNOWLEDGMENTS

Supported in part by grants from the American Heart Association, Southeastern Pennsylvania Chapter, Philadelphia, PA; and grant #HL07346 from the National Heart, Lung, and Blood Institutes, Bethesda, MD.

REFERENCES

1. Josephson ME: *Clinical Cardiac Electrophysiology: Techniques and Interpretations*, 2d ed. Malvern, PA, Lea & Febiger, Chapter 9 (in press).

2. Josephson ME: *Clinical Cardiac Electrophysiology: Techniques and Interpretations*, 2d ed. Malvern, PA, Lea & Febiger, Chapter 12 (in press).
3. Josephson ME: *Clinical Cardiac Electrophysiology: Techniques and Interpretations*, 2d ed. Malvern, PA, Lea & Febiger, Chapter 14 (in press).
4. Vaughn Williams EM: A classification of antiarrhythmic actions reassessed after a decade of new drugs. *J Clin Pharmacol* 1984;24:129–147.
5. Harrison DC: Antiarrhythmic drug classification: New science and practice applications. *Am J Cardiol* 1985;56:185–187.
6. Zipes DP: A consideration of antiarrhythmic therapy. *Circulation* 1985;72:949–956.
7. Fozzard HA, Makielski JC: The electrophysiology of acute myocardial ischemia. *Annu Rev Med* 1985; 36:275–284.
8. Lazzara R, Scherlag BJ: Generation of arrhythmias in myocardial ischemia and infarction. *Am J Cardiol* 1988;61:20A–6A.
9. Johnson EA, McKinnon MG: The differential effect of quinidine and pyrilamine on the myocardial action potential at various rates of stimulation. *J Pharmacol Exp Ther* 1957;120:460–468.
10. Chen CM, Gettes LS, Katzung BG: Effect of lidocaine and quinidine on steady-state characteristics and recovery kinetics of (dV/dt) max in guinea pig ventricular myocardium. *Circ Res* 1975;37:20–29.
11. Courtney KR: Interval-dependent effects of small antiarrhythmic drugs on excitability of guinea pig myocardium. *J Mol Cell Cardiol* 1979;12:1581–1586.
12. Campbell TJ: Importance of physico-chemical properties in determining the kinetics of the effects of class I antiarrhythmic drugs on maximum rate of depolarization in guinea-pig ventricle. *Br J Pharmacol* 1983;80:33–40.
12. Campbell TJ: Kinetics of onset of rate-dependent effects of class I antiarrhythmic drugs are important in determining their effects on refractoriness in guinea pig ventricle and provide a theoretical basis for their subclassification. *Cardiovasc Res* 1983; 17:344–352.
13. Hondeghem LM, Katzung BG: Time- and voltage-dependent interactions of antiarrhythmic drugs with cardiac sodium channels. *Biochim Biophys Acta* 1977;472:373–398.
14. Hondeghem L, Katzung BG: Test of a model of antiarrhythmic drug action. Effects of quinidine and lidocaine on myocardial conduction. *Circulation* 1980;61:1217–1226.
15. Hondeghem LM, Katzung BG: Antiarrhythmic agents: The modulated receptor mechanism of action of sodium and calcium channel-blocking drugs. *Annu Rev Pharmacol Toxicol* 1984;24:387–423.
16. Buchanan JW Jr, Saito T, Gettes LS: The effects of antiarrhythmic drugs, stimulation frequency, and potassium-induced resting membrane potential changes on conduction velocity and dV/dt_{max} in guinea pig myocardium. *Circ Res* 1985;56:696–703.
17. Sada H, Kojima M, Ban T: Effect of procainamide on transmembrane action potentials in guinea pig papillary muscles as affected by external potassium concentration. *Naunyn-Schmiedeberg's Arch Pharmacol* 1979;309:179–190.
18. Grant AO, Strauss LJ, Wallace AC, Strauss HC: The influence of pH on the electrophysiological effects of lidocaine in guinea pig ventricular myocardium. *Circ Res* 1980;47:542–550.
19. Grant AO, Trantham JL, Brown KK, Strauss HC: pH-dependent effects of quinidine on the kinetics of dV/dt in guinea pig ventricular myocardium. *Circ Res* 1982;50:210–217.
20. Gilmour RF Jr, Morrical DG, Ertel PJ, Maesaka JF, Zipes DP: Depressed effects of fast channel blockade on the electrical activation ischemic canine ventricle: Mediation by the sympathetic nervous system. *Cardiovasc Res* 1984;18:405–413.
21. Jazayeri MR, Van Wyhe G, Avitall B, McKinnie J, Tchou P, Akhtar M: Isoproterenol reversal of antiarrhythmic effects in patients with inducible sustained ventricular tachyarrhythmias. *J Am Coll Cardiol* 1989;14:705–711.
22. Morady F, Kou WH, Kadish SH, et al.: Antagonism of quinidine's electrophysiologic effects by epinephrine in patients with ventricular tachycardia. *J Am Coll Cardiol* 1988;12:388–394.
23. Kadish AH, Spear JF, Levine JH, Moore EN: The effects of procainamide on conduction in anisotropic canine ventricular myocardium. *Circulation* 1986; 74:616–625.
24. Prystowsky EN, Lloyd EA, Fineberg N, Zipes DP, Darling E, Heger JJ: A comparison of electrophysiologic effects of antiarrhythmic agents in humans, in Brugada P, Wellens HJJ (eds): *Cardiac Arrhythmias: Where To Go From Here?* Mount Kisco, NY, Futura Publishing, 1987, pp 495–540.
25. Gomes JAC, Kang PS, El-Sherif N: Effect of digitalis on the human sick sinus node after pharmacologic autonomic blockade. *Am J Cardiol* 1981;48:783–788.
26. Alboni P, Shantha N, Filippi L, et al.: Clinical effects of digoxin on sinus node and atrioventricular node function after pharmacologic autonomic blockade. *Am Heart J* 1984;108:1255–1261.
27. Bexton RS, Hellestrand KJ, Cory-Pearce R, Spurrell RAJ, English TAH, Camm AJ: The direct electrophysiologic effects of disopyramide phosphate in the transplanted human heart. *Circulation* 1983; 67:38–45.
28. Shenasa M, Gilbert CJ, Schmidt DH, Akhtar M: Procainamide and retrograde atrioventricular nodal conduction in man. *Circulation* 1982;65:355–362.
29. Josephson ME, Caracta AR, Ricciutti MA, Lau SH, Damato AN: Electrophysiological properties of procainamide in man. *Am J Cardiol* 1974;33:596–603.
30. Schechter JA, Caine R, Friehling T, Kowey PR, Engel TR: Effect of procainamide on dispersion of ventricular refractoriness. *Am J Cardiol* 1983;52:279–291.
31. Josephson ME, Seides SF, Batsford WP, et al.: Electrophysiological effects of intramuscular quinidine on the atrioventricular conducting system in man. *Am Heart J* 1974;85:55–64.
32. Josephson ME, Caracta AR, Gallagher JJ, Lau SH, Damato AN: Effects of lidocaine on refractory periods in man. *Am Heart J* 1972;84:778–786.
33. Josephson ME, Caracta AN, Lau SH, Gallagher JJ, Damato AN: Electrophysiological evaluation of disopyramide in man. *Am Heart J* 1973;86:771–780.
34. Breithardt G, Seipel L, Abendroth RR: Comparison of the antiarrhythmic efficacy of disopyramide and mexiletine against stimulus induced ventricular tachycardia. *J Cardiovasc Pharmacol* 1981;3:1026–1037.
35. Brugada P, Wellens HJJ: Effects of intravenous and oral disopyramide on paroxysmal atrioventricular nodal tachycardia. *Am J Cardiol* 1984;53:88–92.

36. DiMarco JP, Garan H, Ruskin JN: Quinidine for ventricular arrhythmias: Value of electrophysiologic testing. *Am J Cardiol* 1983;51:90–95.
37. Denniss AR, Ross DL, Waywood JA, Cooper MJ, Uther JB: Effect of procainamide, mexiletine, and propranolol on ventricular activation time recorded at cardiac mapping in chronic canine myocardial infarction. *J Electrophysiol* 1988;2:3–9.
38. Goldberg D, Reiffel JA, Davis JC, Gang E, Livelli F, Bigger JT Jr: Electrophysiologic effects of procainamide on sinus node function in patients with and without sinus node disease. *Am Heart J* 1982; 103:75–79.
39. Greenspan AM, Horowitz LN, Spielman SR, Josephson ME: Large dose procainamide therapy for ventricular tachycardia. *Am J Cardiol* 1980;46:453–462.
40. Heger JJ, Nattel S, Rinkenberger RL, Zipes DP: Mexiletine therapy in 15 patients with a drug-resistant ventricular tachycardia. *Am J Cardiol* 1980;45:627–632.
41. Kerr CR, Prystowsky EN, Smith WM, Cook L, Gallagher JJ: Electrophysiological effects of disopyramide phosphate in patients with Wolff-Parkinson-White syndrome. *Circulation* 1982;65:869–878.
42. Marchlinski FE, Buxton AE, Vassallo JA, et al.: Comparative electrophysiologic effects of intravenous and oral procainamide in patients with sustained ventricular arrhythmias. *J Am Coll Cardiol* 1984; 4:1247–1254.
43. Mason JW, Hondeghem LM: Quinidine. *Ann NY Acad Sci* 1984;432:162–176.
44. Morady F, Scheinman MM, Desai J: Disopyramide. *Ann Intern Med* 1982;96:337–343.
45. Nattel S, Zipes DP: Clinical pharmacology of old and new antiarrhythmic drugs. *Cardiovasc Res* 1980; 11:221–248.
46. Wu D, Hung JS, Kuo CT, Hsu KS, Shieh WB: Effects of quinidine on atrioventricular nodal reentrant paroxysmal tachycardia. *Circulation* 1981;64:823–831.
47. Wu KM, Hoffman BF: Effect of procainamide and N-acetyl-procainamide on atrial flutter: Studies in vivo and in vitro. *Circulation* 1987;76:1397–1408.
48. Breithardt G, Borggrefe M, Wiebringhaus E, Seipel L: Effect of propafenone in the Wolff-Parkinson-White syndrome: Electrophysiologic findings and long term followup. *Am J Cardiol* 1984;54:29D–39D.
49. Brugada P, Abdollah H, Wellens HJJ: Suppression of incessant supraventricular tachycardia by intravenous and oral encainide. *J Am Coll Cardiol* 1984; 4:1255–1260.
50. Camm AJ, Hellestrand KJ, Nathan AW, Bexton RS: Clinical usefulness of flecainide acetate in the treatment of paroxysmal supraventricular arrhythmias. *Drugs* 1985;29:7–14.
51. Carey EL Jr, Duff HJ, Roden DM, et al.: Encainide and its metabolites. Comparative effects in man on ventricular arrhythmias and electrocardiographic intervals. *J Clin Invest* 1984;73:539–547.
52. Chilson DA, Heger JJ, Zipes DP, Browne KF, Prystowsky EN: Electrophysiologic effects and clinical efficacy of oral propafenone therapy in patients with ventricular tachycardia. *J Am Coll Cardiol* 1985; 5:1407–1413.
53. Duff HJ, Dawson AK, Roden DM, Oates JA, Smith RF, Woosley RL: Electrophysiologic actions of *O*-demethyl encainide: An active metabolite. *Circulation* 1983;68:385–391.
54. Echt D, Shapiro M, Trusso J, Mason JW, Winkle RA: Treatment with oral lorcainide in patients with sustained ventricular tachycardia and fibrillation. *Am Heart J* 1985;109:28–32.
55. Guehler J, Gornick CC, Tobler HG, et al.: Electrophysiologic effects of flecainide acetate and its major metabolites in the canine heart. *Am J Cardiol* 1985;55:807–812.
56. Hellestrand KJ, Bexton RS, Nathan AW, Spurrell RAJ, Camm AJ: Acute, electrophysiological effects of flecainide acetate on cardiac conduction and refractoriness in man. *Br Heart J* 1982;48:140–148.
57. Horowitz LN, Spielman SR, Webb CR, Morganroth J, Greenspan AM: The clinical electrophysiology of intravenous indecainide. *Am Heart J* 1985;110:784–788.
58. Jackman WM, Zipes DP, Naccarelli GV, Rinkenberger RL, Heger JJ, Prystowsky EN: Electrophysiology of oral encainide. *Am J Cardiol* 1982;49:1270–1278.
59. Prystowsky EN, Klein GJ, Rinkenberger RL, Heger JJ, Naccarelli GV, Zipes DP: Clinical efficacy and electrophysiologic effects of encainide in patients with the Wolff-Parkinson-White syndrome. *Circulation* 1984;69:278–287.
60. Kunze K-P, Kuck K-H, Schluter M, Kuch B, Bleifeld W: Electrophysiologic and clinical effects of intravenous and oral encainide in accessory atrioventricular pathway. *Am J Cardiol* 1984;54:323–329.
61. Mann DE, Luck JC, Herre JM, et al.: Electrophysiologic effects of ethmozine in patients with ventricular tachycardia. *Am Heart J* 1984;107:674–679.
62. Naccarella F, Bracchetti D, Palmieri M, Marchesini B, Ambrosioni E: Propafenone for refractory ventricular arrhythmias: Correlation with drug plasma levels during long-term treatment. *Am J Cardiol* 1984;54:1008–1014.
63. Naccarelli GV, Rinkenberger RL, Dougherty AH, Giebel RA: Encainide: A review of its electrophysiology, pharmacology and clinical efficacy. *Clin Prog Electrophysiol Pacing* 1985;3:268–291.
64. Prystowsky EN, Klein G, Rinkenberger RL, Heger JJ, Nacarelli GV, Zipes DP: Clinical efficacy and electrophysiologic effects of encainide in patients with Wolff-Parkinson-White syndrome. *Circulation* 1984; 69:278–287.
65. Singh BN: Mechanism of action of antiarrhythmic agents: Focus on propafenone. *J Electrophysiol* 1987;1:503–516.
66. Touboul P, Atallah G, Kirkorian G, et al.: Electrophysiologic effects of cibenzoline in humans related to dose and plasma concentrations. *Am Heart J* 1986;112:333–339.
67. Seides SF, Josephson ME, Batsford WP, Weisfogel GM, Lau SH, Damato AN: The electrophysiology of propranolol in man. *Am Heart J* 1974;88:733–741.
68. Marchlinski FE, Buxton AE, Waxman HL, Josephson ME: Electrophysiologic effects of intravenous metoprolol. *Am Heart J* 1984;107:1125–2231.
69. Prystowsky EN, Jackman WM, Rinkenberger RL, Heger JJ, Zipes DP: Effect of autonomic blockade on ventricular refractoriness and atrioventricular conduction in humans: Evidence supporting a direct cholinergic action on ventricular muscle refractoriness. *Circ Res* 1981;49:511–518.
70. Zipes DP, Prystowsky EN, Heger JJ: Amiodarone: Electrophysiologic actions, pharmacokinetics and clinical effects. *J Am Coll Cardiol* 1984;3:1059–1071.
71. Waxman HL, Groh WC, Marchlinski FE, et al.: Amio-

darone for control of sustained ventricular tachyarrhythmia: Clinical and electrophysiologic effects in 51 patients. *Am J Cardiol* 1982;50:1066–1074.
72. Singh BN, Nademanee K, Josephson MA, Ikeda N, Venkatesh N, Kannan R: The electrophysiology and pharmacology of verapamil, flecainide, and amiodarone. Correlations with clinical effects and antiarrhythmic actions. *Ann NY Acad Sci* 1984; 432:210–235.
73. Shenasa M, Denker S, Mahmud R, Lehmann M, Estrada A, Akhtar M: Effect of amiodarone on conduction and refractoriness of the His-Purkinje system in the human heart. *J Am Coll Cardiol* 1984; 4:105–110.
74. Morady F, DiCarlo LA Jr, Krol RB, Baerman JM, deBuitleir M: Acute and chronic effects of amiodarone on ventricular refractoriness, intraventricular conduction and ventricular tachycardia induction. *J Am Coll Cardiol* 1986;7:148–157.
75. Kappenberger LJ, Fromer MA, Steinbrunn W, Shenasa M: Efficacy of amiodarone in the Wolff-Parkinson-White syndrome with rapid ventricular response via accessory pathway during atrial fibrillation. *Am J Cardiol* 1984;54:330–335.
76. Kadish AH, Buxton AE, Waxman HL, Flores B, Josephson ME, Marchlinski FE: Usefulness of electrophysiologic study to determine the clinical tolerance of arrhythmia recurrences during amiodarone therapy. *J Am Coll Cardiol* 1987;10:90–96.
77. Ikeda N, Nademanee K, Kannan R, Singh BN: Electrophysiologic effects of amiodarone: Experimental and clinical observation relative to serum and tissue drug concentrations. *Am Heart J* 1984;108:890–889.
78. Horowitz LN, Spielman SR, Greenspan AM, et al: Use of amiodarone in the treatment of persistent and paroxysmal atrial fibrillation resistant to quinidine therapy. *J Am Coll Cardiol* 1985;6:1402–1407.
79. Heger JJ, Prystowsky EN, Jackman WM, et al: Amiodarone. Clinical efficacy and electrophysiology during long-term therapy for recurrent ventricular tachycardia or ventricular fibrillation. *N Engl J Med* 1981;305:539–544.
80. Ezri MD, Shima MA, Denes P: Amiodarone: A review of its clinical and electrophysiologic effects. *Clin Prog Pacing Electrophysiol* 1983;1:20–9.
81. Brugada P, Wellens HJJ: Effects of oral amiodarone on rate-dependent changes in refractoriness in patients with Wolff-Parkinson-White syndrome. *Am J Cardiol* 1985;56:863–866.
82. Brugada P, Facchini M, Wellens HJJ: Effects of isoproterenol and amiodarone and the role of exercise in initiation of circus movement tachycardia in the accessory atrioventricular pathway. *Am J Cardiol* 1986;57:146–149.
83. Feld GK, Nademanee K, Weiss J, Stevenson W, Singh BN: Electrophysiologic basis for the suppression by amiodarone of orthodromic supraventricular tachycardias complicating pre-excitation syndromes. *J Am Coll Cardiol* 1984;3:1298–1307.
84. Mitchell LB, Wyse DG, Duff HJ: Electropharmacology of sotalol in patients with Wolff-Parkinson-White syndrome. *Circulation* 1987;76:810–818.
85. Sung RJ, Juma Z, Saksena S: Electrophysiologic properties and antiarrhythmic mechanisms of intravenous N-acetyl-procainamide in patients with ventricular dysrhythmias. *Am Heart J* 1983;105:811–819.
86. Touboul P, Atallah G, Kirkorian G, Lamaud M, Moleur P: Clinical electrophysiology of intravenous sotalol, a beta-blocking drug with class III antiarrhythmic properties. *Am Heart J* 1984;107:888–895.
87. Wellens HJJ, Brugada P, Abdollah H, Dassen WR: A comparison of the electrophysiologic effects of intravenous and oral amiodarone in the same patient. *Circulation* 1984;69:120–124.
88. Wynn J, Miura DS, Torres V, et al: Electrophysiologic evaluation of the antiarrhythmic effects of N-acetylprocainamide for ventricular tachycardia secondary to coronary artery disease. *Am J Cardiol* 1985;56:877–881.
89. Wu D, Kou HC, Yeh SJ, Lin FC, Hung JS: Effects of oral verapamil in patients with atrioventricular reentrant tachycardia incorporating an accessory pathway. *Circulation* 1983;67:426–433.
90. Waxman HL, Myerburg RJ, Appel R, Sung RJ: Verapamil for control of ventricular rate in paroxysmal supraventricular tachycardia and atrial fibrillation or flutter: A double-blind randomized crossover study. *Ann Intern Med* 1981;94:1–6.
91. Sung RJ, Shapiro WA, Shen EN, Morady F, Davis J: Effects of verapamil on ventricular tachycardias possibly caused by reentry, automaticity, and triggered activity. *J Clin Invest* 1983;72:350–360.
92. Prystowsky EN: Electrophysiologic and antiarrhythmic properties of bepridil. *Am J Cardiol* 1985; 55:59C–62C.
93. Klein GJ, Gulamhusein S, Prystowsky EN, Carruthers SG, Donner AP, Ko PT: Comparison of the electrophysiologic effects of intravenous and oral verapamil in patients with paroxysmal supraventricular tachycardia. *Am J Cardiol* 1982;49:117–124.
94. Harper RW, Whitford E, Middlebrook K, Federman J, Anderson S, Pitt A: Effects of verapamil on the electrophysiologic properties of the accessory pathway in patients with the Wolff-Parkinson-White syndrome. *Am J Cardiol* 1982;50:1323–1330.
95. Gulamhusein S, Ko P, Carruthers SG, Klein GJ: Acceleration of the ventricular response during atrial fibrillation in the Wolff-Parkinson-White syndrome after verapamil. *Circulation* 1982;65:348–354.
96. Ellenbogen KA, German LD, O'Callaghan WG, et al: Frequency-dependent effects of verapamil on atrioventricular nodal conduction in man. *Circulation* 1985;72:344–352.
97. DiMarco JP, Sellers TD, Berne RM, West GA, Belardinelli L: Adenosine: electrophysiologic effects and therapeutic use for terminating paroxysmal supraventricular tachycardia. *Circulation* 1983; 68:1254–1263.
98. Schmitt CG, Kadish AH, Marchlinski FE, Miller JM, Buxton AE, Josephson ME: Effects of lidocaine and procainamide on normal and abnormal intraventricular electrograms during sinus rhythm. *Circulation* 1988;77:1030–1037.
99. deBakker JMT, vanCapelle FJL, Janse MJ, et al: Reentry as a cause of ventricular tachycardia in patients with chronic ischemic heart disease: Electrophysiologic and anatomic correlation. *Circulation* 1988;77:589–606.
100. deBakker JMT, Coronel R, Tasseron S, et al: Ventricular tachycardia in the infarcted, Langendorff-perfused human heart: Role of the arrangement of surviving cardiac fibers. *J Am Coll Cardiol* 1990; 15:1594–1607.
101. Schmitt C, Kadish AH, Balke WC, et al: Cycle

length-dependent effects on normal and abnormal intraventricular electrograms: Effect of procainamide. *J Am Coll Cardiol* 1988;12:395–403.
102. Almendral JM, Stamato NJ, Rosenthal ME, Marchlinski FE, Miller JM, Josephson ME: Resetting response patterns during sustained ventricular tachycardia: Relationship to the excitable gap. *Circulation* 1986;74:722–730.
103. Rosenthal ME, Stamato NJ, Almendral JM, Gottlieb CD, Josephson ME: Resetting of ventricular tachycardia with electrocardiographic fusion: Incidence and significance. *Circulation* 1988;77:581–588.
104. Almendral JM, Rosenthal ME, Stamato NJ, et al.: Analysis of the resetting phenomenon in sustained uniform ventricular tachycardia: Incidence and relation to termination. *J Am Coll Cardiol* 1986; 8:294–300.
105. Stamato NJ, Frame LH, Rosenthal ME, Almendral JM, Gottlieb CD, Josephson ME: Procainamide-induced slowing of ventricular tachycardia with insights from analysis of resetting response patterns. *Am J Cardiol* 63:1455–1461.
106. Lorca M, Almendral J, Pastor A, et al.: Comparative effects of quinidine and amiodarone in monomorphic ventricular tachycardia. *Circulation* 1989; 80:II-650(abstract).
107. Kay GN, Epstein AE, Plumb VJ: Preferential effect of procainamide on the reentrant circuit of ventricular tachycardia. *J Am Coll Cardiol* 1989; 14:382–390.
108. Janse MJ, Wit AL: Electrophysiological mechanisms of ventricular arrhythmias resulting from myocardial ischemia and infarction. *Physiol Rev* 1989;69:1049–1169.
109. Fisher JD, Cohen HL, Mehra R, Altschuler H, Escher DJW, Furman S: Cardiac pacing and pacemakers II. Serial electrophysiologic-pharmacologic testing for control of recurrent tachyarrhythmias. *Am Heart J* 1977;93:658.
110. Mason JW, Winkle RA: Electrode-catheter arrhythmia induction in the selection and assessment of antiarrhythmic drug therapy for recurrent ventricular tachycardia. *Circulation* 1978;89:971–985.
111. Horowitz LN, Josephson ME, Kastor JA: Intracardiac electrophysiologic studies as a method for the optimization of drug therapy in chronic ventricular arrhythmia. *Prog Cardiovasc Dis* 1980;23:81–98.
112. Doherty JU, Josephson ME: Role of electrophysiologic testing in the therapy of ventricular arrhythmias. *PACE* 1983;6:1070–1083.
113. Swerdlow CD, Winkle RA, Mason JW: Determinants of survival in patients with ventricular tachyarrhythmias. *N Engl J Med* 1983;308:1436–1442.
114. Breithardt G, Borggrefe M, Seipel L: Selection of optimal drug treatment of ventricular tachycardia by programmed electrical stimulation of the heart. *Ann NY Acad Sci* 1984;427:49–66.
115. Rae AP, Greenspan AM, Spielman SR, et al.: Antiarrhythmic drug efficacy for ventricular tachyarrhythmias associated with coronary artery disease as assessed by electrophysiologic studies. *Am J Cardiol* 1985;55:1494–1499.
116. Reddy CP, Chen TJ, Guillory WR: Electrophysiologic studies in selection of antiarrhythmic agents: Use with ventricular tachycardia. *PACE* 1986; 9:756–763.
117. Gottlieb C, Josephson ME: The preference of programmed stimulation-guided therapy for sustained ventricular arrhythmias, in Brugada P, Wellens HJJ (eds): *Cardiac Arrhythmias: Where To Go From Here?* Mount Kisco, NY, Futura Publishing, 1987, pp 421–434.
118. Gottlieb C, Josephson ME: Programmed stimulation in the evaluation of life-threatening or potentially life threatening ventricular arrhythmias (editorial). *Cardiovasc Drugs Therapy* 1987;1:155–159.
119. Swerdlow CD, Peterson J: Prospective comparison of Holter monitoring and electrophysiologic study in patients with coronary artery disease and sustained ventricular tachyarrhythmias. *Am J Cardiol* 1985;56:577–580.
120. Ruskin JN, DiMarco JP, Garan H: Out-of-hospital cardiac arrest. Electrophysiologic observations and selection of long-term antiarrhythmic therapy. *N Engl J Med* 1980;303:607–613.
121. Morady F, Scheinman MM, Hess DS, Sung RJ, Shen E, Shapiro W: Electrophysiologic testing in the management of survivors of out-of-hospital cardiac arrest. *Am J Cardiol* 1983;51:85–89.
122. Skale BT, Miles WM, Heger JJ, Zipes DP, Prystowsky EN: Survivors of cardiac arrest: Prevention of recurrence by drug therapy as predicted by electrophysiologic testing of electrocardiographic monitoring. *Am J Cardiol* 1986;57:113–119.
123. Kim SG, Seiden SW, Matos JA, Waspe LE, Fisher JD: Discordance between ambulatory monitoring and programmed stimulation in assessing efficacy of mexiletine in patients with ventricular tachycardia. *Am Heart J* 1986;112:14–19.
124. Cooper MJ, Hunt LJ, Palmer KJ, et al.: Quantitation of day to day variability in mode of induction of ventricular tachyarrhythmias by programmed stimulation. *J Am Coll Cardiol* 1988;11:101–108.
125. Swerdlow CD, Winkle RA, Mason JW: Prognostic significance of the number of induced ventricular complexes during assessment of therapy for ventricular tachyarrhythmias. *Circulation* 1983; 68:400–405.
126. Platia EV, Reid PR: Nonsustained ventricular tachycardia during programmed ventricular stimulation: Criteria for a positive test. *Am J Cardiol* 1985;56:79–83.
127. Waxman HL, Buxton AE, Sadowski LM, Josephson ME: The response to procainamide during electrophysiologic study for sustained ventricular tachyarrhythmias predicts the response to other medications. *Circulation* 1983;67:30–37.
128. Vaitkus PT, Buxton AE, Josephson ME, Marchlinski FE: Cycle-length response of ventricular tachycardia associated with coronary artery disease to procainamide and amiodarone. *Am J Cardiol* (in press).
129. Duff HJ, Mitchell LB, Manyari D, Wyse DG: Mexiletine-quinidine combination: Electrophysiologic correlates of a favorable antiarrhythmic interaction in humans. *J Am Coll Cardiol* 1987;10:1149–1156.
130. Marchlinski FE, Buxton AE, Miller JM, Vassallo JA, Flores BT, Josephson ME: Amiodarone versus amiodarone and a type IA agent for treatment of patients with rapid ventricular tachycardia. *Circulation* 1986;74:1037–1043.
131. Lin FC, Finley CD, Rahimtoola SH, Wu D: Idiopathic paroxysmal ventricular tachycardia with a QRS pattern of right bundle branch block and left axis

deviation: A unique clinical entity with specific properties. *Am J Cardiol* 1983;52:95–100.
132. Buxton AE, Waxman HL, Marchlinski FE, Simson MB, Cassidy DM, Josephson ME: Right ventricular tachycardia: Clinical and electrophysiologic characteristics. *Circulation* 1983;68:917–927.
133. Buxton AE, Marchlinski FE, Doherty JU, et al.: Repetitive, monomorphic ventricular tachycardia: Clinical and electrophysiologic characteristics in patients with and patients without organic heart disease. *Am J Cardiol* 1984;54:997–1002.
134. Waller TJ, Kay HR, Spielman SR, Kutalek SP, Greenspan AM, Horowitz LN: Reduction in sudden death and total mortality by antiarrhythmic therapy evaluated by electrophysiologic drug testing: Criteria of efficacy in patients with sustained ventricular tachyarrhythmia. *J Am Coll Cardiol* 1987;10:83–89.
135. Josephson ME: *Clinical Cardiac Electrophysiology: Techniques And Interpretations*, 2d ed. Malvern, PA: Lea & Febiger, 1991, Chapter 10 (in press).
136. Allessie MA, Bonke FIM, Schopman JG: Circus movement in rabbit atrial muscle as a mechanism of tachycardia. III. The "leading circle" concept: A new model of circus movement in cardiac tissue without the involvement of an anatomical obstacle. *Circ Res* 1977;41:9–18.
137. Pagé PL, Plumb VJ, Okumura K, Waldo AL: A new animal model of atrial flutter. *J Am Coll Cardiol* 1986;8:872–879.
138. Frame LH, Page RL, Boyden PA, Fenoglio JJ Jr, Hoffman BF: Circus movement in the canine atrium around the tricuspid ring during experimental atrial flutter and during reentry in vitro. *Circulation* 1987;76:1155–1175.
139. Boyden PA, Frame LH, Hoffman BF: Activation mapping of reentry around an anatomic barrier in the canine atrium. Observations during entrainment and termination. *Circulation* 1989;79:406–416.
140. Almendral JM, Arenal A, Abeytus M, SanRoman D, Soriano J, Josephson ME: Incidence and patterns of resetting during atrial flutter: Role in identifying chamber of origin. *J Am Coll Cardiol* 1987;9:153A (abstract).
141. Hordorf AJ, Edie R, Malm JR, Hoffman BF, Rosen MR: Electrophysiologic properties and response to pharmacologic agents of fibers from diseased human atria. *Circulation* 1977;54:774–779.
142. Schoels W, Gough WB, El-Sherif N: Circus movement atrial flutter in the canine sterile pericarditis model: Differential effects of procainamide on the components of the reentrant pathway. *PACE* 1990; 13:513(abstract).
143. Della Bella P, Marenzi G, Tondo C, Guazzi MD: Modification of atrial flutter excitable gap affects arrhythmia termination by overdrive atrial pacing. *Circulation* 1989;80:II-40(abstract).
144. Buxton AE, Marchlinski FE, Doherty JU, Frame LH, Josephson ME: Electrophysiologic effects of amiodarone in patients with atrial flutter and fibrillation (submitted for publication).
145. Wit AL, Cranefield PF: Triggered activity in cardiac muscle fibers of the simian mitral valve. *Circ Res* 1976;38:85–98.
146. Hedberg A, Kempf F Jr, Josephson ME, Molinoff PB: Coexistence of $beta_1$-and $beta_2$-adrenergic receptors in human heart: Effects of treatment with receptor antagonists or calcium entry blockers. *J Pharmacol Exp Ther* 1985;234:561–568.

Chapter **24**

Undesirable Cardiovascular Antiarrhythmic Drug Effects: Incidence, Mechanism, and Practical Considerations

Shmuel Ravid, MD, and Thomas B. Graboys, MD

Antiarrhythmic drugs remain the major therapeutic intervention available for effective prophylactic management of patients with malignant cardiac arrhythmias and out of hospital sudden death.[1,2] Antiarrhythmic drug therapy is frequently associated with a variety of side effects that affect multiple organ systems. Cardiac adverse reactions are the most serious, by virtue of their life-threatening nature; they include aggravation or provocation of tachyarrhythmia, bradyarrhythmia, and hemodynamic depression. By far, the most serious complication of antiarrhythmic drug therapy is worsening of preexisting or induction of new life-threatening sustained ventricular tachyarrhythmia, termed "proarrhythmia."[3] The proarrhythmic effect has been known for many years. Paroxysmal ventricular fibrillation in a patient receiving quinidine sulfate for atrial fibrillation was described almost 70 years ago.[4] In 1964, Seltzer and Wray[5] described the typical occurrence of multifocal ventricular tachycardia with long QT interval in patients receiving quinidine for chronic atrial arrhythmia. However, it was not until recently that the full scale and clinical significance of this complication was appreciated. Velebit and colleagues[6] from our group were the first to attempt a systematic approach to the problem. In a retrospective analysis, they showed that aggravation or provocation of ventricular arrhythmia occurred, with each of the antiarrhythmic drugs being assessed under controlled conditions. Later reports[7–9] concurred with these observations and confirmed that malignant ventricular arrhythmia and sudden cardiac death may result from antiarrhythmic drug therapy. More recently, the disturbing findings of the recent Cardiac Arrhythmia Suppression Trial (CAST) study[10] of substantially higher total mortality and arrhythmic death among patients receiving flecainide or encainide as compared with placebo, most likely due to proarrhythmia, underscore the clinical relevance of the problem. In the following discussion we will review our experience and current knowledge of the mechanisms involved, the magnitude of the problem, and the clinical applications.

MECHANISMS OF PROARRHYTHMIA

The exact mechanism of proarrhythmic response to antiarrhythmic drugs is not clearly understood. However, it is likely to involve the same basic mechanisms associated with arrhythmogenesis, primarily reentry due to abnormal impulse conduction and abnormal impulse initiation triggered by early afterdepolarization.[11–14]

Drug-induced long QT syndrome with torsade de pointes historically has attracted a lot of attention. It was most frequently associated with quinidine therapy,[5,15] but was described also with other class I antiarrhythmic drugs[16,17] and amiodarone.[18,19] Clinical features of quinidine-induced arrhythmia include generally low quinidine plasma concentration, absence of marked QRS prolongation, and almost invariably abrupt heart rate slowing before the initiation of the arrhythmia.[15] The occurrence of this arrhythmia is also enhanced by hypokalemia, hypomagne-

655 Avenue of the Americas, New York, NY 10010
Current Topics in Cardiology

semia, and bradycardia.[15,20] Experimental data reveals that quinidine in clinical concentrations, not only prolongs repolarization, but also may induce afterdepolarization during phases 2 and 3 of the action potential (AP) which, in turn, can initiate triggered arrhythmia manifested as multifocal ventricular tachycardia.[20] Cesium chloride induces long QT syndrome and bradycardia-dependent early afterdepolarizations in vivo.[21,22] Similar findings were reported also in humans with drug-induced polymorphic ventricular tachycardia (VT).[23] The common denominator for these precipitating factors is induction of triggered activity and prolongation of the AP and refractoriness, which results in prolonged repolarization manifested as a longer QT interval on the surface electrocardiogram. On the other hand, torsade de pointes can be terminated by measures that shorten repolarization and abolish triggered activity, for example, pacing[24,25] and magnesium.[26,27]

The fact that drug-induced long QT interval syndrome and triggered activity induced by early afterdepolarizations are provoked and suppressed by similar conditions favors their casual relations and the widely accepted notion that triggered activity is the underlying mechanism for this drug-induced arrhythmia.

The other common form of drug-induced ventricular tachyarrhythmia is incessant monomorphic VT, typically associated with class IC antiarrhythmic drugs, and suggestive of an underlying reentry mechanism.[14] By far, the most important mechanism of ventricular arrhythmogenesis is reentry.[12] The substrate involves scarred myocardium with areas of abnormally slow conduction intercepted by normal myocardium, usually in the margins of previous infarct. An anatomical or functional unidirectional block, through which anterograde conduction cannot proceed, is a prerequisite. Proper timing and balance of slow conduction and refractoriness are required to allow the arrhythmia to be sustained. Potential reentrant circuits may not sustain ventricular tachyarrhythmia because the conduction delay in the slow retrograde limb is too short to allow the normally conducting limb to recover excitability. This may be altered by antiarrhythmic drugs, primarily class IC, which block the fast sodium (Na^+) channels and markedly slow conduction. This conduction slowing may suppress arrhythmia by creating bidirectional block in the slow limb of the reentry circuit. However, in the same manner these antiarrhythmic drugs may paradoxically favor reentry by further depressing conduction in the abnormal myocardium. This allows a potential reentry circuit to manifest and sustain arrhythmia by enabling the normally conducting limb to recover excitability. In this model, drugs, primarily class IC antiarrhythmic agents which produce marked slowing of conduction without affecting significantly repolarization, may potentiate reentry and convert potential reentrant circuits to manifest incessant VT.[14] This correlates with the clinical association between class IC antiarrhythmic drugs and induction of incessant monomorphic VT, especially in patients with significant structural heart disease exhibiting a vulnerable substrate for reentry.

Proarrhythmia may also be related to coexisting conditions that facilitate arrhythmogenesis. Such conditions may include electrolyte abnormalities (for example, hypokalemia or hypomagnesemia), ischemia, or uncontrolled congestive heart failure (CHF).

Concomitant drug administration ("polypharmacy") is an additional, often underappreciated cause for arrhythmogenesis. Drug interaction (for example, quinidine and digitalis[28]) is a well-known mechanism for precipitation of arrhythmia. More often, however, the underlying problem is inappropriate drug dosing or dose escalation,[29] which may be due to: (a) failure to recognize altered renal or hepatic function which may cause slower drug metabolism or excretion resulting in elevated plasma drug levels. This is particularly germane among patients with CHF in whom changes in diuretics, ACE inhibitors, or digitalis drugs may also alter the substrate and provide a nidus for proarrhythmia. (b) Accumulation of active metabolites (for example, *O*-desmethyl encainide or *N*-acetyl procainamide). (c) Genetic determinants of drug metabolism that may result in individually altered rate of metabolism and unpredictable plasma level of active substances. (d) Rapid escalation of the dose before steady state is achieved and active substances continue to accumulate.

It is quite possible that more than one mechanism may be required for the expression of proarrhythmia akin to the combined effects of quinidine, bradycardia, and hypokalemia in the induction of torsade de pointes.

DEFINITION OF PROARRHYTHMIA

The proarrhythmic effect of antiarrhythmic drugs can be classified into two major categories: (a) aggravation of preexisting arrhythmia and (b) induction of new sustained arrhythmia. Both have to be temporarily related to the initiation of drug therapy or to dose escalation, and disappear after the offending agent is discontinued. Both phenomena may occur spontaneously, provoked by exercise stress test or observed solely during electrophysiological studies.

Aggravation of existing arrhythmia, as assessed noninvasively, includes an increase in the frequency of single ventricular premature beats or couplets; the arbitrary definition developed within our group includes a fourfold increase in ventricular premature beats and a tenfold increase in repetitive forms. Increase in the duration of the arrhythmia includes longer duration of a nonsustained VT, incessant VT, and a significant increase (more than 10%) in the rate of preexisting tachyarrhythmia. Development of new arrhythmia includes new onset of VT (sustained or nonsustained), ventricular fibrillation (VF), torsade de pointes, or the conversion of nonsustained to sustained VT.

Definition of proarrhythmic effect during electrophysiological studies is less well established. In the largest study[30] of proarrhythmia during electrophysiological studies, the following criteria were used: (a) initiation of sustained arrhythmia when only nonsustained ventricular arrhythmia was present at baseline; (b) conversion of sustained tachycardia that could be terminated by programmed electrical stimulation at baseline, to one that required cardioversion during drug therapy; (c) initiation of sustained VT by a less aggressive mode of stimulation than was required at baseline. In addition to these criteria, Poser et al.[31] also considered an increase in the rate of tachycardia during electrophysiological studies as a possible proarrhythmic effect.

Although development or induction of new sustained arrhythmia is generally accepted as a valid criterion for proarrhythmic effect, the clinical relevance of increase in the frequency of single and pairs of premature ventricular contractions, and some of the changes observed during electrical stimulation, is less established, despite their statistical validity. Thus far, there is no evidence that these criteria correlate with clinically important spontaneous arrhythmia. In addition, data analysis that is based on frequency is confounded by the significant spontaneous variability in arrhythmia frequency[32,33] and therefore is less reliable.

PREVALENCE OF PROARRHYTHMIA

The prevalence of proarrhythmia by antiarrhythmic drugs is not well established. The data available are derived from retrospective studies involving selected patients with different clinical characteristics, who were studied on a variety of drug protocols. The true incidence of proarrhythmic effect is heavily dependent on several factors: (a) the definition and criteria of proarrhythmia; (b) the specific agent involved; (c) the method of drug evaluation, either noninvasively or by electrophysiological studies, and by the routine use of exercise stress testing for exposure of proarrhythmic response; (d) the clinical characteristic of the patient population. In their pioneering work, Velebit et al.[6] documented a proarrhythmic effect by all seven antiarrhythmics evaluated, in 80 of 772 drug tests (11.1%) ranging from 6% with disopyramide to 15.4% with quinidine. Of the 155 patients evaluated, 53 (34.2%) had worsening or arrhythmia by at least one antiarrhythmic drug. In an expansion of that study,[34] 1,287 drug tests were performed with 11 antiarrhythmic agents in almost 400 patients (Table 24.1). Aggravation of arrhythmia occurred more often with class IC agents compared to class IA agents and the least with class IB agents, mexiletine, and tocainide. Proarrhythmia occurred more frequently when drug selection was guided by electrophysiological studies; in 52 patients who underwent 248 drug trials guided by electrical stimulation, proarrhythmic response was noted in 45 (18%) tests ranging from 5% with disopyramide and tocainide to 37% with encainide.[31] In both studies,[31,34] our group defined proarrhythmia, not only as occurrence of new forms of sustained ventricular arrhythmia, but also as an increase in the frequency of singles and pairs of ventricular premature beats during Holter monitoring. Easier inducibility or faster arrhythmia during electrical stimulation was also

TABLE 24.1 Proarrhythmic Response to Antiarrhythmic Drug Therapy Evaluated Noninvasively

Drug	Patients Tested (n)	Patients with Proarrhythmia (n)	Percent
Disopyramide	102	6	6
Encainide	102	15	15
Ethmozine	82	9	11
Flecainide	26	3	12
Indecainide	16	3	19
Lorcainide	120	9	8
Mexiletine	350	23	7
Procainamide	55	5	9
Propafenone	124	10	8
Quinidine	180	20	15
Tocainide	180	14	8
Overall	1,287	117	9

Reprinted from Podrid et al.,[34] with permission.

considered a proarrhythmic effect. A total of 35% of the proarrhythmic events defined by noninvasive evaluation and 30% by electrophysiological studies accounted for this less established "soft" criterion of proarrhythmia. An interesting observation was that 33% of the proarrhythmic responses were disclosed only during exercise stress testing (Fig. 24.1). Stanton et al.[9] studies 506 patients who underwent 1,268 drug trials for VT or VF. Arrhythmogenic response occurred in 3.4% of the tests and in 6.9% of the patients, ranging from none on lidocaine to 11.8% on encainide. In this study, the criteria for proarrhythmia were stricter and included only occurrence of new sustained ventricular arrhythmia temporarily related to the initiation of an antiarrhythmic

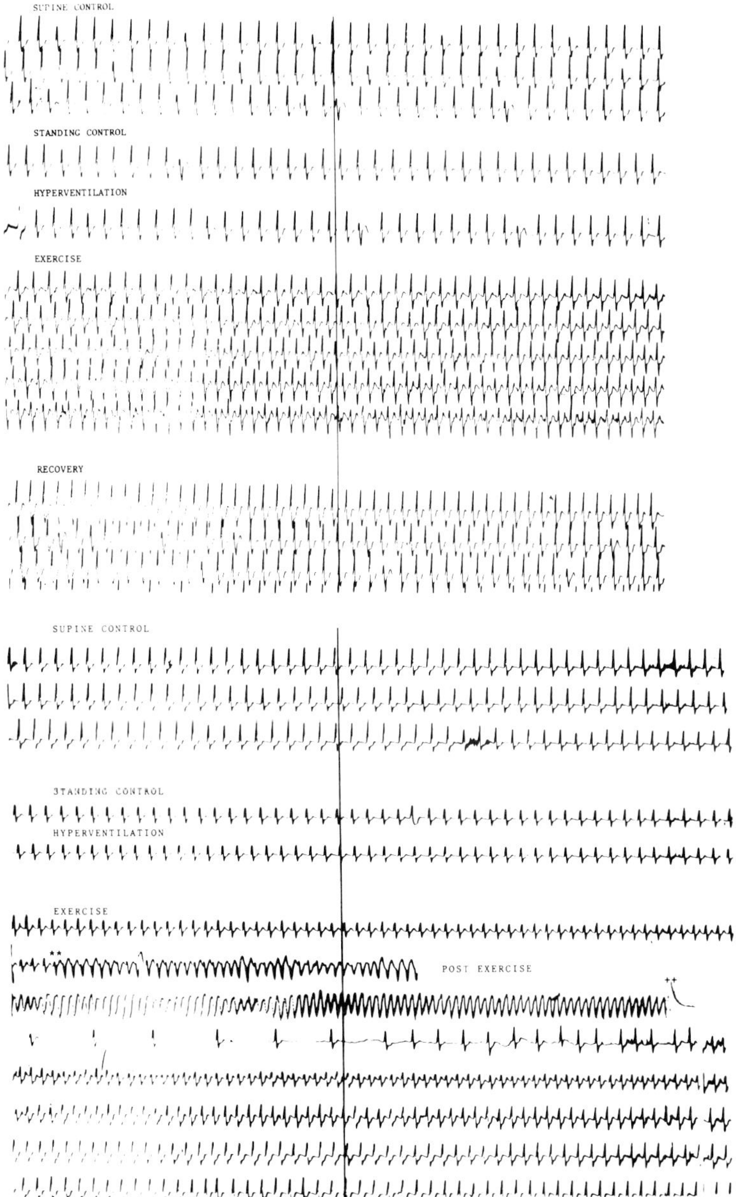

FIGURE 24.1 Continuous electrocardiographic recording of an exercise testing in a patient with nonsustained VT. During control testing on no antiarrhythmics, only single ventricular premature beats were present (**top**). On encainide, VT requiring cardioversion was induced (**bottom**). ** = onset of ventricular tachycardia, + + = direct current electrical cardioversion.

agent, which probably accounts for the lower incidence of proarrhythmic events when compared with the previous study. Several other studies of individual antiarrhythmic agents revealed similar incidence of drug-related arrhythmia worsening.[35-37]

In all the studies, the prevalence of proarrhythmia was consistently higher when evaluated by electrophysiologic studies as compared to noninvasive evaluation. Rae et al.[30] studied the incidence of proarrhythmia in 314 patients who underwent 801 drug tests guided by electrophysiological studies. Conversion of nonsustained to sustained ventricular arrhythmia occurred in 18% of the studies. Conversion of hemodynamically stable sustained VT to an unstable tachycardia or VF requiring cardioversion occurred in 17% of the tests. Induction of the end point arrhythmia by less aggressive mode was noted in 20%. In this study, the overall prevalence of proarrhythmia by the various criteria was 24%. However, it dropped to only 8% when the criterion of arrhythmia induction by one less extrastimulus was excluded, which underscores the importance of proarrhythmia definition for assessment of its accurate prevalence. The drug regimens employed in this study included 14 single drugs and eight combinations, with the highest proarrhythmic response occurring with the class IC drugs flecainide (33%) and indecainide (37%).

CLINICAL PREDICTORS OF PROARRHYTHMIA

Because proarrhythmia is relatively common and potentially life-threatening, prediction of its occurrence and identifying the patients at risk would be desirable. In one study,[38] 47 patients with 51 episodes of proarrhythmia, either on quinidine (IA), mexilitene (IB), or encainide (IC) were matched to 102 "control" tests without aggravation of arrhythmia by the same antiarrhythmic agents. No significant difference between the two groups was found in respect to age, sex, baseline electrocardiographic intervals and drug-induced changes in these intervals, density of arrhythmia on the control monitor and exercise test, dose of the drug or blood level, and the nature of the underlying heart disease. The only variable that was significantly associated with proarrhythmia was the clinical presentation of sustained VT or fibrillation. In these patients, the odds ratio of proarrhythmia was 2.5 times compared to those presenting with nonsustained arrhythmia ($p < 0.01$). Of borderline statistical significance ($p < 0.04$) was left ventricular ejection fractions less than 35% with an odds ratio of 2.2. For the most part, however, proarrhythmia was unpredictable in this study. Proarrhythmic response to one drug has not been shown to predict a similar effect by other drugs.[16] A data base constructed from several studies of flecainide (1,330 patients) and encainide (1,775 patients) revealed that malignant sustained ventricular arrhythmias were associated with significantly higher risk of proarrhythmia.[39,40] In the absence of sustained arrhythmia or structural heart disease, the risk of this complication was very low. High drug dose or rapid escalation of the dose was also associated with significant risk of proarrhythmia. In a study by Stanton and coworkers of 506 patients with VT or VF,[9] similar findings were reported: A higher risk of proarrhythmia was found for patients presenting with sustained compared to nonsustained ventricular arrhythmia. Significant left ventricular systolic dysfunction assessed at the base of the heart by M-mode echocardiography, but not global ejection fraction, also correlated statistically with arrhythmia worsening.

The temporal relationship between the initiation of antiarrhythmic drug therapy or dose escalation and the occurrence of proarrhythmic response is not well defined. It may be an instant event or may occur after many months. According to several reports, primarily involving quinidine,[15] this complication occurred after various intervals ranging from days to months. In a more recent study, an attempt was made to define the clinical characteristics of patients who developed ventricular fibrillation for the first time in association with antiarrhythmic drug therapy.[8] The interval between initiation of therapy and ventricular fibrillation was usually short with a median duration of 3 days. More than 70% of the episodes occurred within 5 days of starting therapy or dose escalation. However, two out of 38 episodes of ventricular fibrillation (6%) occurred after 60 days and one occurred after 21 days. Moreover, in the CAST study[10] the excess in arrhythmic death was evenly distributed throughout the study, suggesting late proarrhythmia.

In conclusion, the prevalence of proarrhythmia is directly related to the patient population, presenting arrhythmias, cardiac diagnosis, ventricular function, and the specific drug. The risk of proarrhythmia, for example, for a patient without structural heart disease presenting with atrial fibrillation is less than 1%. However, for a patient with poor ventricular function, that is, ejection fraction less than 30% whose clinical arrhythmia is recurrent VT, the risk might be as high as 30%. However, the occurrence and timing is unpredictable even in high-risk patients.

HEMODYNAMIC ADVERSE EFFECTS OF ANTIARRHYTHMIC DRUGS

Virtually all the patients with congestive heart failure have ventricular arrhythmia, about 40% have runs of VT, and about half of all deaths are sudden, presumably due to malignant ventricular arrhythmias.[41] Therefore, it is not surprising that many patients with congestive heart failure receive antiarrhythmic therapy, despite unproven efficacy during controlled studies.

The net circulatory effect of antiarrhythmic drugs depends on the balance between their effect on various determinants of pump function, including contractility, peripheral resistance, and autonomic tone. Virtually all antiarrhythmic agents have the potential to suppress cardiac function and to precipitate congestive heart failure.[42] However, excluding calcium (Ca^{2+}) channel blockers and β-adrenergic blockers, the only drugs that are consistently associated with provocation of congestive heart failure are disopyramide[43] and flecainide.[44,45] Both drugs have a significant direct negative inotropic effect, and, in addition, disopyramide also exerts a peripheral vasoconstrictor effect. Even these drugs are safe from a hemodynamic standpoint and are well tolerated by patients with mildly impaired left ventricular function, but without history of congestive heart failure. Gottlieb et al.[46] studied the hemodynamic effects of tocainide, encainide, and procainamide in patients with severe refractory heart failure. They reported a significant decrease in cardiac index and a significant increase in systemic vascular resistance as measured by right heart catheterization with each of the three drugs. However, large-scale clinical reports on the long-term effects of these drugs confirmed that they are well tolerated even by the majority of patients with significantly impaired cardiac function and history of heart failure,[47–49] suggesting that the acute hemodynamic effect does not necessarily predict the long-term clinical outcome.

In a recent study from our group,[49] the incidence of congestive heart failure by six of the new antiarrhythmic drugs including encainide, propafenone, ethmozine, lorcainide, mexiletine, and tocainide was studied. We reported a low incidence of this complication, limited exclusively to patients with a history of congestive heart failure and significant impairment of contractile function. Aggravation of congestive heart failure occurred only in 1.8% of the drug tests, in 3.9% of the patients, and in 9% of the patients with a history of congestive heart failure. Occurrence with one drug did predict the same complication with other agents and it was always reversible upon drug discontinuation or dose reduction.

Thus, the provocation of congestive heart failure by antiarrhythmic drugs, though a potentially serious complication, is infrequent and unpredictable. Therefore, therapy with antiarrhythmic agents, except flecainide and disopyramide, is feasible even in high-risk patients.

ANTIARRHYTHMIC DRUG-INDUCED BRADYARRHYTHMIA

Digitalis, β-adrenergic blockers, and Ca^{2+} channel blockers as well as class I antiarrhythmic drugs can induce sinus bradycardia or sinus arrest. Overdose or a combined effect of these drugs may result in a significant bradyarrhythmia in normals. However a more common clinical phenomenon is aggravation of underlying sinus

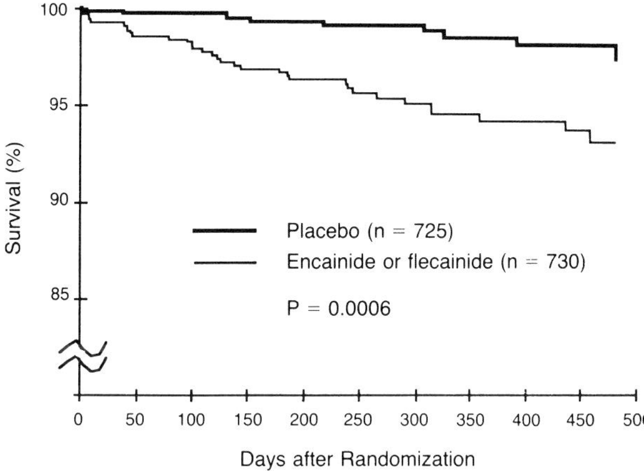

FIGURE 24.2 Survival among 1,455 patients randomly assigned to receive encainide or flecainide or matching placebo. The cause of death was arrhythmia or cardiac arrest. The nominal p value was based on traditional two-sided log rank test adjusted for multiple groups. *Reprinted from Cardiac Arrhythmia Suppression Trial (CAST) Investigators,*[10] *with permission.*

node dysfunction and decreased automaticity by a therapeutic dose of these drugs.[50] In a similar fashion, the same drugs may aggravate or provoke high-grade atrioventricular (AV) block, especially in patients with preexisting AV nodal disease. Patients with prolonged PR interval or second-degree AV block during periods of vagotonia are at higher risk for this complication. Class I antiarrhythmic agents usually prolong HV conduction, with the most potent effect exerted by class IC drugs and to lesser extent by class IA drugs. These antiarrhythmic agents can produce AV block at the level of the His-Purkinje system, primarily in patients with underlying diseased conduction system. Similar, but more prominent, effects on sinus node and AV nodal function were reported also with amiodarone.[51]

Bradyarrhythmias are not a contraindication for antiarrhythmic therapy. However, occasionally implantation of a permanent pacemaker may be required if antiarrhythmic therapy is indicated for life-threatening arrhythmia, and is associated with drug-induced symptomatic bradyarrhythmia.

THE CAST STUDY

Frequent and repetitive forms of premature ventricular beats have been recognized as markers of higher risk for sudden cardiac death after myocardial infarction.[52] CAST[10] is a multicenter, randomized, placebo-controlled, double-blind study designed to test the hypothesis that suppression of these arrhythmias may affect positively the survival of patients with previous myocardial infarction. It followed the successful completion of the Cardiac Arrhythmia Pilot Study (CAPS),[53] which demonstrated the tolerance and efficacy of encainide, flecainide, and moricizine in suppressing asymptomatic or mildly symptomatic ventricular arrhythmias after myocardial infarction. Patients were randomized to blinded therapy with encainide, flecainide, or moricizine if their arrhythmia was suppressed by either drug during an open label phase. The encainide and flecainide arms of the CAST study were terminated prematurely because of definite evidence that the use of these drugs was associated with statistically significant excess mortality and higher incidence of nonfatal cardiac arrest or sudden death (Fig. 24.2).

Thus, efficacious suppression of ventricular arrhythmia after myocardial infarction not only did not improve survival as hypothesized, but on the contrary, was associated with higher incidence of cardiac arrest and sudden death. The harmful effect of flecainide and encainide is presumably due to proarrhythmic effect. In contrast to previous reports,[8] the excess in sudden death and cardiac arrest was equally distributed throughout the 10-month follow-up period of drug treatment and was not limited to the early period of antiarrhythmic therapy. Therefore, it is possible that even a meticulous effort to identify patients with proarrhythmic response may fail because of late expression of the proarrhythmic potential. As already mentioned, the mechanism of proarrhythmia is not completely understood, although in the CAST it is likely to involve reentry that is characteristic for the profound conduction slowing induced by class IC drugs. Ischemia induced by conduction slowing may also play a role, particularly in this patient population with ischemic heart disease and previous myocardial infarction, and may be the reason for the late occurrence of proarrhythmic response.

The main clinical dilemma arising from the CAST results is whether to limit the conclusions to flecainide and encainide or to generalize them to other antiarrhythmic agents. It seems that extrapolation of the CAST findings to other classes of antiarrhythmic drugs and other groups of patients with ventricular arrhythmias should await future randomized trials including the results of the moricizine arm of the CAST study, particularly because this agent possesses both class IA and IB properties. At the present time, however, it is prudent to avoid the use of any antiarrhythmic drugs in patients after myocardial infarction with asymptomatic or only mildly symptomatic ventricular arrhythmia.

CONCLUSION AND CLINICAL IMPLICATIONS

Potentially life-threatening arrhythmias, induced or aggravated by antiarrhythmias, are not uncommon and may occur with all antiarrhythmic drugs. However, despite higher risk for proarrhythmia in patients presenting with sustained ventricular arrhythmia and significant structural heart disease, the occurrence of this complication is largely unpredictable. Therefore, a cautious approach to the use of antiarrhythmic therapy is recommended incorporating the following guidelines:

a. The risk versus benefit of antiarrhythmic therapy should be weighed carefully for each individual. Based on our current knowledge, therapy should be restricted to patients with sustained or symptomatic arrhythmia, primarily sustained VT or VF. Antiarrhythmic therapy is not indicated and should not be used routinely for patients with asymptomatic "potentially malignant" arrhythmia even if associated with significant structural heart disease. β-Adrenergic blockers are the only agents proven to reduce total mortality and sudden cardiac death after myocardial infarction[54] and should be considered the drug of choice for patients with structural heart disease and asymptomatic nonsustained arrhythmias.

b. Exclusion or elimination of underlying factors that may facilitate proarrhythmia (for example, electrolyte imbalance, drug interaction, uncontrolled ischemia, or congestive heart failure) is mandatory prior to initiation of antiarrhythmic drug therapy.

c. Antiarrhythmic therapy for ventricular arrhythmia in patients at risk for proarrhythmic response should be initiated in the hospital while patients are continuously monitored. Once steady state is achieved on a maintenance drug dose, ambulatory electrocardiographic monitoring and an exercise stress test should be repeated to expose potential proarrhythmic effect, even if therapy was guided by electrophysiologic study. For patients in whom the indication for therapy is supraventricular arrhythmia, atrial fibrillation, or preexcitation syndrome, both ambulatory ECG monitoring and exercise testing should also be incorporated in their management to expose potential proarrhythmic effect.

d. Complete familiarity with the pharmacology of each antiarrhythmic agent in use is mandatory. Rapid escalation of antiarrhythmic drug dosage should be avoided specifically for class IC drugs. Whenever feasible, antiarrhythmic drugs with better safety profile should be preferred, for example, class IB agents are safer than class IA agents, which in turn are safer than class IC agents.

REFERENCES

1. Graboys TB, Lown B, Podrid PJ, DeSilva R: Longterm survival of patients with malignant ventricular arrhythmia with antiarrhythmic drugs. *Am J Cardiol* 1982;50:437–443.
2. Ruskin JN, DiMarco JP, Garan H: Out of hospital cardiac arrest: Electrophysiologic observation and selection of long term antiarrhythmic drugs. *N Engl J Med* 1980;303:607–613.
3. Horowitz LN, Zipes DP, Bigger TJ Jr, Campbell RWF, Morganroth J, Podrid PJ, Rosen MR, Woosley RL: Proarrhythmia, arrhythmogenesis or aggravation of arrhythmia—A status report: 1987. *Am J Cardiol* 1987;59:54E–56E.
4. Kern WJ, Bender WL: Paroxysmal ventricular fibrillation with cardiac recovery in a case of auricular fibrillation and complete heart block while under quinidine sulfate therapy. *Heart* 1922;9:269–281.
5. Seltzer A, Wray HW: Quinidine syncope: Paroxysmal ventricular fibrillation occurring during treatment of chronic atrial arrhythmia. *Circulation* 1964; 30:17–26.
6. Velebit V, Podrid PJ, Lown B, Cohen BH, Graboys TB: Aggravation and provocation of ventricular arrhythmia by antiarrhythmic drugs. *Circulation* 1982;65:686–694.
7. Ruskin JN, McGovern BM, Garan H, Minardo JP, Kelly E: Antiarrhythmic drugs: A possible cause of out-of-hospital cardiac arrest. *N Engl J Med* 1983; 309:1302–1306.
8. Minardo JD, Heger JJ, Miles WM, Zipes DD, Prystowsky EN: Clinical characteristics of patients with ventricular fibrillation during antiarrhythmic drug therapy. *N Engl J Med* 1988;319:257–262.
9. Stanton MJ, Prystowsky EJ, Feinberg JS, Miles WM, Zipes DP, Heger JJ: Arrhythmogenic effect of antiarrhythmic drugs: A study of 506 patients treated for ventricular tachycardia or fibrillation. *J Am Coll Cardiol* 1989;14:204–215.
10. The Cardiac Arrhythmia Suppression Trial (CAST) Investigators: Preliminary report: Effect of encainide or flecainide on mortality in a randomized trial of arrhythmia suppression after myocardial infarction. *N Engl J Med* 1989;321:406–412.
11. Cranefield PF: Action potentials, afterpotentials and arrhythmia. *Circ Res* 1977;41:415–423.
12. Josephson ME, Marchlinski FE, Buxton ME, Waxman HL, Doherty JU, Kienzle MG, Falcone R: Electrophysiologic basis for sustained ventricular tachycardia and role of reentry, in Josephson ME, Wellense HJJ (eds): *Tachycardias: Mechanisms, Diagnosis, Treatment.* Philadelphia, Lea and Febiger, 1984, pp 305–323.
13. Rosen MR, Wit AL: Arrhythmogenic effect of antiarrhythmic drugs. *Am J Cardiol* 1987;59:10E–18E.
14. Levine JH, Morganroth J, Kadish AH: Mechanisms and risk factors for proarrhythmia with type Ia compared with Ic antiarrhythmic drug therapy. *Circulation* 1989;80:1063–1069.
15. Roden DM, Thompson KA, Hoffman BF, Woorley RL: Clinical features and basic mechanisms of quinidine induced arrhythmia. *J Am Coll Cardiol* 1986;8:73A–78A.
16. Schweitzer P, Mark H: Torsade de pointes caused by disopyramide and hypokalemia. *Mt Sinai J Med* 1982;49:110–114.
17. Strasberg B, Sclarovsky S, Erdberg A, Duffy CE, Lam W, Swiryn S, Agmon J, Rosen KM: Procainamide induced polymorphous ventricular tachycardia. *Am J Cardiol* 1981;47:1309–1314.
18. McGovern B, Garan H, Ruskin JN: Serious adverse effects of amiodarone. *Clin Cardiol* 1984;7:131–137.
19. Sclarovsky S, Levin RF, Kracoff O, Strasberg B, Arditi A, Agmon J, Tikva P, Aviv T: Amiodarone induced polymorphous ventricular tachycardia. *Am Heart J* 1983;105:6–12.
20. Roden DM, Hoffman BF: Action potential prolongation and induction of abnormal automaticity by low quinidine concentrations in canine Purkinje fibers: Relation to potassium and cycle length. *Circ Res* 1985;56:857–867.
21. Brachmann J, Scharlag BJ, Rosenshtraukh LV, Lazzara R: Bradycardia dependent triggered activity: Relevance to drug induced multiform ventricular tachycardia. *Circulation* 1983;68:846–856.

22. Levine JH, Spear JF, Guarmieri T, Weistfeldt ML, DeLangen CDJ, Becker LC, Moore EN: Cesium chloride induced long Q-T syndrome: Demonstration of afterdepolarization and triggered activity in vivo. *Circulation* 1985;72:1092–1103.
23. Bonatti V, Rolli A, Bott G: Recordings of monomorphic action potentials of the right ventricle in the long Q-T syndrome complicated by severe ventricular arrhythmia. *Eur Heart J* 1983;4:168–172.
24. Karen A, Tzivoni D, Goldman JM, Corcos P, Benhorin J, Stern S: Ventricular pacing in atypical ventricular tachycardia. *J Electrocardiol* 1981;14:201–205.
25. Damiano BP, Rosen M: Effects of pacing on triggered activity induced by early afterdepolarization. *Circulation* 1984;69:1013–1025.
26. Bailie DS, Inoue H, Kaseda S, Ben-David J, Zipes DP: Magnesium suppression of early afterdepolarization and ventricular tachyarrhythmia induced by cesium in dogs. *Circulation* 1988;77:1395–1402.
27. Tzivoni D, Banai S, Schuger C, Benhorin J, Karen A, Gottlieb S, Stern S: Treatment of torsade de pointes with magnesium sulfate. *Circulation* 1988;77:392–397.
28. Leahey EB, Reiffel JA, Drusin RE, Heissenbuttel RH, Lovejoy WP, Bigger JT: Interaction between quinidine and digoxin. *JAMA* 1978;240:533–534.
29. Woosley RL, Roden DM: Pharmacologic causes of arrhythmogenic actions of antiarrhythmic drugs. *Am J Cardiol* 1987;59:19E–25E.
30. Rae AP, Kay HB, Horowitz LN, Spilman SR, Greenspan AM: Proarrhythmic effect of antiarrhythmic drugs in patients with malignant ventricular arrhythmias evaluated by electrophysiologic testing. *J Am Coll Cardiol* 1988;12:131–139.
31. Poser RF, Podrid PJ, Lombardi F, Lown B: Aggravation of arrhythmia induced with antiarrhythmic drugs during electrophysiologic testing. *Am Heart J* 1985;110:9–16.
32. Pratt AM, Delclos G, Wiseman AM: The changing baseline of complex ventricular arrhythmia: A new consideration in assessing long term antiarrhythmic drug therapy. *N Engl J Med* 1985;313:1444–1449.
33. Pratt CM, Slymen DJ, Wierman AM, Young JB, Francis MJ, Seals AA, Quinones MA, Roberts R: Analysis of the spontaneous variability of ventricular arrhythmia: Consecutive ambulatory electrocardiographic recordings of ventricular tachycardia. *Am J Cardiol* 1985;56:67–72.
34. Podrid PJ, Lampert S, Graboys TB, Blatt CM, Lown B: Aggravation of arrhythmia by antiarrhythmic drugs—incidence and predictors. *Am J Cardiol* 1987;59:38E–44E.
35. Morgenroth J, Horowitz LN: Flecainide: Its proarrhythmic effect and expected changes on the surface electrocardiogram. *Am J Cardiol* 1984; 53:(5):89B–94B.
36. Hohnloser S, Lange HW, Raeder EA, Podrid PJ, Lown B: Long term use of tocainide for ventricular arrhythmia. *Circulation* 1986;73:143–149.
37. Tordjman T, Podrid PJ, Raeder E, Lown B: Safety and efficacy of encainide for malignant ventricular arrhythmia. *Am J Cardiol* 1986;58:87C–95C.
38. Slater W, Lampert S, Podrid PJ, Lown B: Clinical predictors of arrhythmia worsening by antiarrhythmic drugs. *Am J Cardiol* 1988;61:349–353.
39. Morganroth J, Anderson JL, Gentzkow GD: Classification by type of ventricular arrhythmia predicts frequency of adverse cardiac events from flecainide. *J Am Coll Cardiol* 1986;8:607–615.
40. Morganroth J: Risk factors for the development of proarrhythmic events. *Am J Cardiol* 1987;59:32E–37E.
41. Francis GS: Development of arrhythmia in the patient with congestive heart failure; pathophysiology, prevalence and prognosis. *Am J Cardiol* 1986; 57:3B–7B.
42. Wilson JR: Use of antiarrhythmic therapy in patients with heart failure: Clinical efficacy, hemodynamic results and relation to survival. *Circulation* 1987; 75(suppl IV):64–73.
43. Podrid PJ, Schoenberger A, Lown B: Congestive heart failure caused by oral disopyramide. *N Engl J Med* 1980;302:614–617.
44. Flecainide Ventricular Tachycardia Study Group: Treatment of resistent ventricular tachycardia with flecainide acetate. *Am J Cardiol* 1986;57:1299–1305.
45. Greene LH, Richardson DW, Hallstrom AP, McBride R, Capon RJ, Baker AH, Roden DM, Echt DS. Congestive heart failure after acute myocardial infarction in patients recieving antiarrhythmic agents for ventricular premature complexes. *Am J Cardiol* 1989; 63:393–398.
46. Gottlieb SS, Kakin ML, Medina N, Yushak M, Packer M: Comparative hemodynamic effects of procainamide, tocainide and encainide in severe chronic heart failure. *Circulation* 1990;81:860–864.
47. Soyka LF: Safety of encainide for the treatment of ventricular arrhythmia. *Am J Cardiol* 1986;58:960–965.
48. Horn HR, Hadidian Z, Johnson JL, Vassulo HG, Williams JH, Young MD: Safety evaluation of tocainide in the American emergency use program. *Am Heart J* 1980;100:1037–1040.
49. Ravid S, Podrid PJ, Lampert S, Lown B: Congestive heart failure induced by six of the new antiarrhythmic drugs. *J Am Coll Cardiol* 1989;14:1326–1330.
50. Bigger TJ Jr, Sahar DI: Clinical types of proarrhythmic response to antiarrhythmic drugs. *Am J Cardiol* 1987;59:2E–9E.
51. Gloor HO, Urthaler F, James TN: Acute effects of amiodarone upon the canine sinus node and atrioventricular region. *J Clin Invest* 1983;71:1457–1466.
52. Bigger JT Jr, Fleiss JL, Kleiger R, Miller JP, Rolnitzky LM: Multicenter Post-Infarction Research Group: The relationships among ventricular arrhythmias, left ventricular dysfunction and mortality in two years after myocardial infarction. *Circulation* 1984;69:250–258.
53. The Cardiac Arrhythmia Pilot Study (CAPS) Investigators: Effects of encainide, flecainide, imipramine and moricizine on ventricular arrhythmia during the year after acute myocardial infarction: The CAPS. *Am J Cardiol* 1988;61:501–509.
54. Six year follow up of the Norwegian Multicenter Study on Timolol after acute myocardial infarction. *N Engl J Med* 1985;313:1055–1058.

Chapter **25**

Pharmacologic Approach to Treatment of Supraventricular Arrhythmias: Update

Philip E. Hill, MD, and John P. DiMarco, MD, PhD

During the last past 20 years, clinical electrophysiologists working in many laboratories have greatly expanded our knowledge of the mechanisms responsible for and the properties of the tissues involved in recurrent supraventricular tachyarrhythmias. These observations have permitted the implementation of new pharmacologic, surgical, and catheter-based therapies for treatment of patients with these arrhythmias. For most patients, drug therapy usually is the first approach attempted, and numerous reports about the usefulness of recently developed drugs in patients with supraventricular arrhythmia have appeared. This chapter will review the current status of antiarrhythmic drug therapy in the management of supraventricular tachyarrhythmias, with particular emphasis on the role of these new agents.

EVALUATION OF EFFICACY

The evaluation of the efficacy of antiarrhythmic drug therapy has always been difficult. There is a general agreement that prevention of arrhythmia-related morbidity and mortality is the goal of all antiarrhythmic therapy. Prevention of morbidity is easy to evaluate when acute arrhythmia termination is the goal of drug treatment. However, a statistically valid evaluation of prophylactic therapy to prevent arrhythmia recurrence is often more difficult. Often, surrogate endpoints such as suppression of either spontaneous asymptomatic arrhythmias during ambulatory electrocardiographic monitoring or arrhythmias induced by programmed electrical stimulation have been employed to evaluate drug therapy. For patients with ventricular arrhythmias, mortality data provides a "gold standard" against which to judge such surrogate endpoints, but mortality is not an appropriate endpoint for the evaluation of therapy for supraventricular arrhythmias since they are rarely life-threatening. It has been widely accepted that for supraventricular arrhythmias the occurrence of symptomatic and/or sustained episodes will be only valid endpoints for drug assessment. However, in most patients these episodes occur sporadically and, even though they are usually not life-threatening, they may require cardioversion or other intervention for termination. Thus, placebo control periods must often be lengthy and entry into a placebo-controlled study may be unattractive to many patients. For these reasons, the design of well-controlled trials to evaluate drugs in supraventricular arrhythmias has proven difficult and, therefore, governmental approval of this indication for many newly developed agents has not been easily attainable.

Recent advancements in the methodology for evaluation of chronic therapy for supraventricular arrhythmias have, however, been made. Pritchett et al. pioneered the use of intermittent transtelephonic ECG monitoring in patients with supraventricular arrhythmias.[1,2] Using this technique, the occurrence of arrhythmias, their mechanisms, and their rates could be confirmed and the duration of any tachycardia-free interval defined.[3] Most patients in these studies displayed a characteristic pattern of arrhythmia occurrence and by use of life-table analysis, the tachycardia-free interval was shown to be an objective quantitative measure of drug effect. An example of this approach is shown in Figure 25.1 from a study by Anderson et al. that reported a placebo-controlled, double-blind cross-over trial of the type IC drug, flecainide, in the prevention

655 Avenue of the Americas, New York, NY 10010
Current Topics in Cardiology

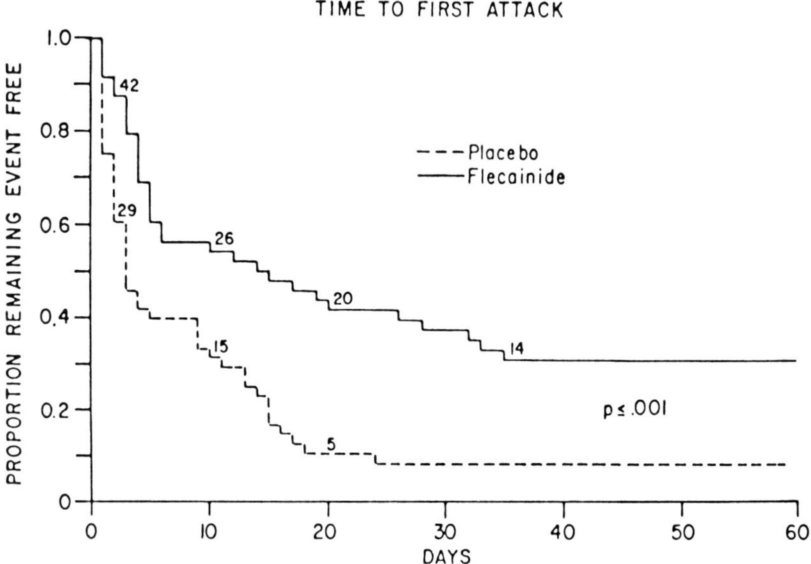

FIGURE 25.1 Life-table analysis of drug efficacy. The data are from a double-blind, placebo-controlled trial[4] of flecainide in patients with paroxysmal atrial fibrillation. Comparison between the two curves allows a drug effect to be established. *Reprinted from Anderson et al.,[4] with permission.*

of atrial fibrillation.[4] Life-table analysis allows the effect of the drug to be demonstrated in a more convincing method than a simple categorization of success or failure.

Although the transtelephonic surveillance approach has many advantages, it also has several significant limitations. Some patients may not exhibit a reproducible pattern of arrhythmias or may have sustained episodes at intervals too long for simple analysis. Patients with asympto-

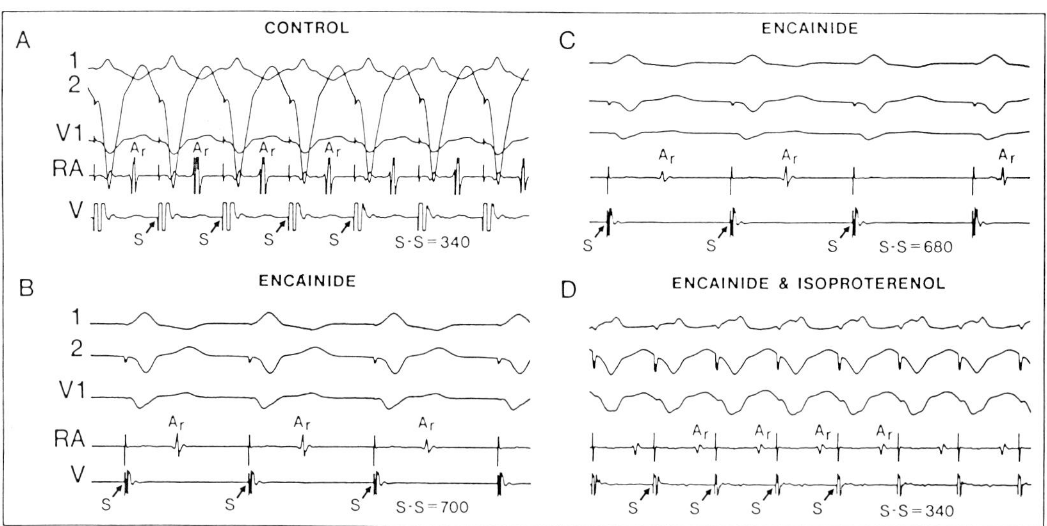

FIGURE 25.2 Reversal of type IC drug effects by isoproterenol. **A.** 1:1 VA conduction at a paced cycle length of 340 ms is present under control conditions. During therapy with encainide, 1:1 VA conduction is lost between 700 ms (**B**) and 680 ms (**C**). When isoproterenol is infused, however (**D**), 1:1 VA conduction down to 340 ms is again observed. A_r = retrograde atrial activation; RA = right atrial electrogram; V = ventricular electrogram. *Reprinted from Akhtar et al.,[5] with permission.*

matic episodes of self-terminating arrhythmias will underestimate their frequency because they will not activate their recorders. Most importantly, patients whose arrhythmias are severely symptomatic or those who require cardioversion to terminate their episodes are often unwilling to participate in the prolonged drug-free period of evaluation needed to establish a baseline for comparison. Despite these limitations, this approach has established a reliable technique for assessing drug efficacy.

Electrophysiologic study is another method commonly used to evaluate drug therapy for supraventricular arrhythmias. It is the only approach that allows a systematic evaluation of a drug's effect on the individual tissues involved in the propagation of a patient's arrhythmia. The effects of a drug on the refractory periods of the atrium, ventricle, atrioventricular (AV) node, and accessory pathways may be measured reproducibly. In patients with Wolff-Parkinson-White syndrome, atrial fibrillation may be initiated with programmed stimulation even if it has not yet occurred clinically, and the potential for life-threatening arrhythmias assessed.

There are limitations to electrophysiologic studies, however. The reentry pathways in paroxysmal supraventricular tachycardia (SVT) involve several cardiac tissues, some or all of which may be responsive to alterations in adrenergic tone. Akhtar et al. demonstrated that isoproterenol infusion may be required for an accurate prediction of future drug efficacy since the effects of many drugs are potentially reversible by catecholamines (Figure 25.2).[5]

NEW ANTIARRHYTHMIC DRUGS

During the last decade, a number of new antiarrhythmic drugs have been developed. Most were originally tested in patients with ventricular arrhythmias and have been released for general use for that indication. However, it has been recognized that these drugs have very significant effects on the tissues involved in supraventricular arrhythmias and, in many cases, they may represent significant advances over previously available agents.

Although there are many limitations to the Vaughn-Williams classification system for antiarrhythmic agents, it remains convenient to use it to describe the general properties of groups of drugs. Flecainide, encainide, and propafenone are type IC drugs, and their principal electrophysiologic effects are due to fast sodium (Na^+) channel blockade during the upstroke of action potential (AP). Sotalol has β-blocking actions (type II) and also prolongs action potential duration (APD) (type III). Amiodarone has primarily type III actions, but also produces some Na^+ channel blockade and is a noncompetitive β-adrenergic antagonist. Verapamil and diltiazem are Ca^{2+} channel blockers (type IV), and their electrophysiologic effects are mostly confined to the AV and, to a lesser degree, the sinus nodes. Adenosine does not fit conveniently into the Vaughn-Williams classification schema. Its value as an antiarrhythmic drug is due to its ability to produce transient AV nodal blockade. The significance of the electrophysiologic properties of these agents in terms of managing patients with supraventricular arrhythmias is shown in Table 25.1.

TABLE 25.1 Effects of New Antiarrhythmic Drugs on Refractory Periods

	Atrium	AV Node	Accessory Pathways
Type IC	0 or sl ↑	↑	↑ ↑ ↑
Flecainide			
Encainide			
Propafenone			
Type III	↑ ↑	↑ ↑ ↑	↑ ↑
Sotalol			
Amiodarone			
Type IV	0 or sl ↓	↑ ↑ ↑	0 or sl ↓
Verapamil			
Diltiazem			
Adenosine	↓ ↓	↑ ↑ ↑	0

TERMINATION OF PAROXYSMAL SUPRAVENTRICULAR TACHYCARDIA

Patients with sustained episodes of SVT may have a variety of presentations. For the patient with severe hemodynamic compromise, urgent cardioversion is the usual procedure of choice.[6] In most cases, however, the patient is, or can be made, relatively comfortable, and a systematic approach to terminating the patient's arrhythmia physiologically or pharmacologically can be made after review of the patient's electrocardiogram to determine the probable mechanism for the arrhythmia.

Vagal maneuvers still have an important role in the management of acute episodes of paroxysmal supraventricular tachycardia (PSVT). Many patients can be instructed in their use and can self-apply them to abort an episode. It is important to stress to the patient that vagal maneuvers should be tried early in the episode before adrenergic reflexes, which will antagonize any vagal stimulus, are fully activated. Several different maneuvers including carotid sinus massage, Valsalva maneuver, dependent head-down body position, and facial immersion in cold water (the diving reflex) may be used to terminate acute episodes of PSVT. Mehta et al. studied the ability of four vagotonic maneuvers for terminating PSVT.[7] The Valsalva maneuver in the supine position was effective for termination in 54% of cases, compared to 17% for right carotid massage and 5% for left carotid massage. The diving reflex was effective in 17%. Waxman et al. also found the Valsalva maneuver to produce a more potent vagal effect than carotid sinus massage.[8] Sreeram and Wren in 1990 retrospectively analyzed the efficacy of digoxin, verapamil, and direct-current cardioversion versus a vagal maneuver, iced water applied to the face (diving reflex), in infants less than 1 year of age.[9] They found the diving reflex effective and safe and suggested that it was the treatment of choice for infants. Overdosing was common with digoxin, recurrence was frequent with DC cardioversion, and one fatality occurred with verapamil. However, despite these optimistic reports, many patients with sustained episodes of tachycardia will not respond to standard vagal maneuvers.

Prior to the introduction of intravenous Ca^{2+} channel blockers, termination of sustained episodes of PSVT was difficult, and many patients required electrical cardioversion. After verapamil became available, this problem became much less common. More recently, other agents have been introduced and they appear to offer alternatives to the use of verapamil (Table 25.2).

The two most common forms of PSVT, AV nodal reentry, and AV reciprocating tachycardia, both require intact AV nodal conduction for their propagation. Ca^{2+} channel blockers increase refractoriness in the AV node and thereby facilitate block within the node. This single occurrence of block is enough to interrupt the reentry circuit and terminate the arrhythmia. Studies on the ability of verapamil to terminate episodes of PSVT in which the mechanism of the arrhythmia is known have reported uniformly high rates of success.[10–12] Although experience with intravenous diltiazem is much more limited, it appears to be similarly effective.[13]

When given for either AV nodal or AV reentry, verapamil is usually well tolerated and safe. However, verapamil is also a potent vasodilator, and if an arrhythmia does not terminate, significant hypotension may result. Neonates seem to be particularly susceptible to adverse hemodynamic reactions, and verapamil should probably not be used in this age group.[9]

TABLE 25.2 Intravenous Antiarrhythmic Therapy for Acute Termination of PSVT

Agent (Ref.)	Site of Action	Dose Range (mg IV)	Estimated Efficacy (%)
Calcium blockers			
Verapamil[10–12]	AV node	5 to 12.5	>85
Diltiazem[13]	AV node	17 to 25	>85
Purinergic agonists			
Adenosine[14–24]	AV node	2.5 to 20	>90
ATP[24–29]	AV node	6 to 30	>90
β-Blockers			
Metoprolol[30]	AV node	2 to 20	50
Flestolol[31]	AV node	7 to 15	40
Types IA, IC, and III			
Flecainide[32–34]	AV node or AP	20 to 140	>80
Propafenone[35]	AV node or AP	100 to 200	>80
Sotalol[36]	AV node or AP	12 to 75	75
Amiodarone[37]	AV node or AP	300	80

AP = accessory pathway, ATP = adenosine triphosphate, AV = atrioventricular, PSVT = paroxysmal supraventricular tachycardia, Ref. = reference(s), IV = intravenous.

Adenosine, and outside the United States, adenosine triphosphate (ATP) are now recognized as extremely effective agents for terminating acute episodes of AV nodal or AV reentrant tachycardia. The electrophysiologic effects of adenosine were first described by Drury and Szent-Györgyi in 1929.[38] Thirty years ago, the intravenous injection of another purinergic compound, ATP, to terminate PSVT was reported in Europe.[26,27] More recently, it has been recognized that adenosine, a degradation product of ATP, acts via a specific cell-surface receptor, designated A_1, to produce a variety of electrophysiologic effects on the heart. Adenosine is also a coronary and systemic vasodilator, but this action is mediated via a different purinergic receptor, designated A_2.[39,40]

Adenosine is rapidly cleared from the circulation by both enzymatic degration and uptake by blood and endothelial cells.[40] The estimated elimination half-life after injection is only a few seconds. When injected as a rapid intravenous bolus, adenosine produces a direct slowing of sinus rate and AV nodal conduction and facilitates AV nodal block. The duration of these direct effects is usually less than 10 s after their onset, and they are then followed by a reflex-mediated period of sinus tachycardia. When given as a continuous infusion, only sinus tachycardia is seen, since the reflex effects predominate.[40]

As mentioned above, the two most common forms of PSVT, AV nodal reentry and AV reentry, both require intact AV nodal conduction for maintenance of the arrhythmia. Production of transient AV nodal block should terminate these arrhythmias, and a number of reports on the use of adenosine for terminating SVT have appeared.[14–24] Most studies have used a protocol of stepwise increases in dosage until termination occurred. Due to the drug's extremely rapid clearance, sequential doses administered at 1- to 2-min intervals are not additive, and each dose constitutes an individual exposure. Virtually all episodes of AV node and AV reentry can be terminated using such a dosing sequence. In one large study, doses of up to 12 mg converted arrhythmias to sinus rhythm in 92% of the patients studied.[18] Conversion usually occurs within 15 to 30 s after injection during the drug's first pass through the cardiac circulation.

Side effects are common after adenosine injection, but they are usually mild and of short duration. The most frequently reported effects are dyspnea, chest tightness, and facial flushing. A few seconds of bradycardia and/or ventricular ectopy are seen in approximately one-third of patients. Since the effects of adenosine are so brief, reinitiation of the tachycardia may occur, and patients should be monitored until stable. Dipyridamole markedly potentiates adenosine's effects, and methylxanthines are A_1 receptor blockers. Adenosine should probably not be used in patients on these medications.[18]

Adenosine and verapamil are now the two best available agents for termination of PSVT. Only limited data directly comparing the agents are available.[20,25] In two studies, the drugs appeared to have approximately equal efficacy. Both may be rarely associated with conversion to atrial fibrillation. Although side effects after adenosine are more common, they are of briefer duration than those associated with verapamil. Adenosine appears to be the agent of choice in infants less than 1 year of age, in patients who are hypotensive, and in cases where the diagnosis of the arrhythmia mechanism is uncertain. In these situations, the lack of adverse hemodynamic effects gives adenosine a distinct advantage over verapamil.

The use of other agents including β-adrenergic blockers, sotalol, type IC agents, and amiodarone to terminate episodes of PSVT has been reported. These agents either act by producing block in an accessory pathway or by slowing antegrade or retrograde AV nodal conduction. These agents appear to offer no advantages over verapamil and adenosine in most patients because their administration is much less convenient and their effects delayed.

CHRONIC PROPHYLAXIS OF PSVT

Patients with infrequent and well-tolerated episodes of PSVT may not require chronic prophylaxis. In other patients, however, long-term suppression is warranted to decrease symptoms during episodes and to eliminate the need for emergency room visits or hospitalizations. For many years, cardiac glycosides were the standard first step in therapy. Cardiac glycosides slow conduction and prolong refractoriness in the AV node. Therapy is inexpensive and convenient, with a single daily dose possible. Unfortunately, there are few controlled data available to confirm the efficacy of therapy with cardiac glycosides, and many patients will have recurrences despite adequate plasma levels of digoxin or digitoxin. In addition, since cardiac glycosides shorten the antegrade refractory period of some accessory pathways, their use is not generally recommended in patients with preexcitation. Ca^{2+} channel blockers such as verapamil or diltiazem may be effective during chronic therapy. These agents have been shown to be effective in 60 to 80% of patients studied in randomized, double-blind trials.[11,12,41,42] β-Adrenergic blockers may also be effective due to their actions on the AV node.[43] A trial of one of these drugs is usually the first step in chronic therapy.

When agents that affect primarily the AV node are ineffective or cannot be tolerated, type IA or type IC drugs may prove effective (Table 25.3). Type IA drugs have modest effects on accessory pathway conduction and on the retrograde "fast" pathway fibers in AV nodal reentry. Type IC drugs, particularly at relatively high doses, produce significantly greater effects on accessory pathways (Fig. 25.3) and also influence the AV node. In many patients with preexcitation, complete accessory pathway block may be achievable, even in pathways with short refractory periods under baseline conditions. As mentioned above, isoproterenol, and by inference, stress or exercise can reverse the effects of all type I drugs on accessory pathways.[5] Therefore, addition of a β-blocking agent to a type IA or IC drug would be expected to increase efficacy. Propafenone, due to its intrinsic β-blocking potential, should probably be used alone.

TABLE 25.3 Chronic Suppression of Paroxysmal Supraventricular Tachycardia

Agent (Ref.)	Site of Action	Dose Range (mg day)	Estimated Efficacy (%)
Calcium blockers			
Verapamil[11,12,41]	AV node	240 to 480	>70
Diltiazem[42]	AV node	90 to 270	60
β-Blockers			
Nadolol[43]	AV node	80 to 160	>60
Type IA agents			
Procainamide[44,45]	AV node or AP	1,000 to 2,000	>60
Type IC agents			
Flecainide[34]	AV node or AP	200 to 300	>65
Encainide[5,46]	AV node or AP	75 to 200	>75
Propafenone[33,47,48]	AV node or AP	450 to 900	>75
Type III agents			
Sotalol[49–52]	AV node or AP	240 to 320	>70
Amiodarone[53,54]	AV node or AP	200 to 600	>75

AP = accessory pathway, AV = atrioventricular, Ref. = reference(s).

Only a few controlled studies concerning the long-term efficacy of antiarrhythmic drugs for suppression of recurrent supraventricular tachycardia are available (Table 25.2). They seem to indicate that type IC agents are more efficacious than type IA drugs, with expected response rates with the former group of 60 to 80%. Further data concerning the value of these drugs are needed to define better their place in management.

Sotalol is a new agent that produces β-adrenergic blockade and also prolongs APD.[49,50] At electrophysiologic study, the refractory periods of atrium, ventricle, AV node, and accessory pathways are all prolonged. Only a few reports concerning its chronic use for suppression of chronic SVT are available,[49–53] but these preliminary data appear quite promising and estimated response rates appear to be similar to those for type IC agents.

Amiodarone is highly effective in many supraventricular arrhythmias.[53,54] It produces noncompetitive β-adrenergic blockade, slows conduction throughout the heart, and prolongs refractory periods in the atrium, ventricle, AV node, and accessory pathways. As a result of these effects, oral amiodarone is often effective in patients with recurrent PSVT, with an expected response rate of 60 to 80% at a daily dose of 100 to 300 mg daily. The major factor limiting the use of amiodarone is its potential for toxicity. Photosensitivity and thyroid dysfunction may occur even at the lowest therapeutic doses. Toxicities in other organ systems including pulmonary fibrosis, hepatic enzyme elevations, skin discol-

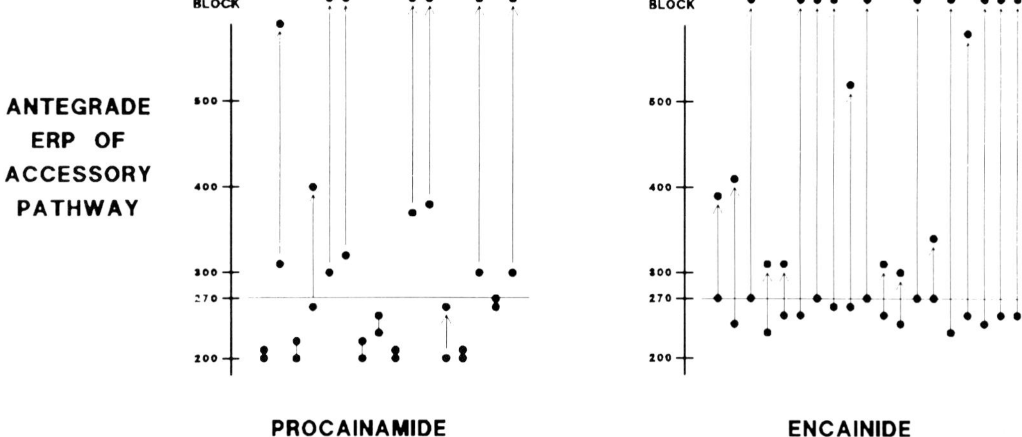

FIGURE 25.3 Procainamide and encainide in Wolff-Parkinson-White syndrome. Data plotted are the antegrade effective refractory periods (in ms) before and after drug in 16 patients treated with procainamide and 18 patients treated with encainide. Note that encainide more frequently produces marked increases in refractoriness or conduction block.

oration, and neuropathies tend to increase in frequency at higher doses and frequently limit effective therapy. In general, amiodarone should be used in patients with non-life-threatening arrhythmias only after other possible forms of treatment have been exhausted.

In summary, the availability of new, more potent agents has greatly improved drug therapy to prevent recurrence of the more common forms of PSVT. The choice between pharmacologic and nonpharmacologic therapy is now more commonly based on the patient's preference to be free of the need for drug treatment rather than an inability to find an effective agent.

TERMINATION OF ATRIAL FIBRILLATION AND FLUTTER

Atrial fibrillation and atrial flutter remain difficult management problems. In contrast to the classic reentrant PSVT in which the electrophysiologic substrates have been well characterized, atrial fibrillation and flutter occur in a wide variety of substrates, ranging from structurally normal to severely diseased atria. It has long been recognized that atrial flutter was an unstable rhythm that required pharmacologic or electrical conversion and few patients can remain in atrial flutter chronically. Recent data have shown the value of maintaining sinus rhythm in patients with recent-onset or paroxysmal atrial fibrillation. Those patients who can be maintained in sinus rhythm will have greater exercise capacity, reduced risk of embolization, and a probably decreased mortality than patients who remain in atrial fibrillation. These observations have made studies on the ability of drugs to convert these arrhythmias more important.

New-onset atrial fibrillation will frequently terminate spontaneously. In one study designed to test the efficacy of digoxin for converting new-onset atrial fibrillation, 44% of the patients who received placebo converted to sinus rhythm.[55] The conversion rate in those who received digoxin was not significantly different. In contrast, when the patient has been in atrial fibrillation for more than 1 week before conversion is attempted, placebo conversion rates are substantially lower but drug convertors are similarly less frequent.[56] Many studies have included patients with varying durations of atrial fibrillation, and this makes an accurate estimation of any drug's efficacy for terminating atrial fibrillation from published data difficult.

As shown in Table 25.4, reported conversion rates in recent-onset atrial fibrillation are higher with type IC drugs such as flecainide or propafenone than have been reported with drugs such

TABLE 25.4 Conversion of Atrial Fibrillation or Atrial Flutter

Agent (Ref.)	Dose Range (mg)	Estimated Efficacy (%)
Cardiac glycosides		
Digoxin[55]	1.4 (O)	50[1]
Type IA agents		
Quinidine[57]	1,200 to 2,400 (O)	40
Type IC agents	100 to 200 (IV)	
Propafenone[58]	200 to 300 (O)	60
Flecainide[33,34,59,60]		60
Type III agents	400 to 800 (O)	
Amiodarone[61,62]	80 to 320 (IV)	30 to 70
Sotalol[56]		70

IV = intravenous, O = orally.

[1] Placebo conversion rate in this study was 44%.

as quinidine. In one trial that directly compares flecainide and quinidine, however, quinidine proved to be equally effective.[60] Amiodarone has also been reported to be highly successful for converting recent onset atrial fibrillation. In the series reported by Santos et al., 86% of patients with recent-onset atrial fibrillation converted to sinus rhythm after treatment with amiodarone.[61] In a second study, reported by Blevins et al., however, only 8 of 25 (32%) patients with chronic atrial fibrillation converted after initiation of amiodarone therapy.[62]

Caution should be used when type IC drugs are used to convert atrial arrhythmias, since both atrial and ventricular proarrhythmias may occur. Falk reported a possible interaction between exercise and new wide complex tachycardias, in 3 patients receiving flecainide for atrial fibrillation.[63] This arrhythmia may have been atrial flutter with aberrant 1:1 conduction. We have observed similar episodes. In view of these observations, the increase in mortality in the treated groups in the Cardiac Arrhythmia Suppression Trial,[64] and the increase in congestive heart failure among patients treated with flecainide in the Cardiac Arrhythmia Pilot Trial,[65] it would be prudent for all cardioversion attempts with type IC drugs to be conducted during in-hospital monitoring.

CHRONIC PROPHYLAXIS OF ATRIAL ARRHYTHMIAS

Prevention of recurrent episodes of either paroxysmal atrial fibrillation or atrial flutter remains one of the most difficult tasks in antiarrhythmic

TABLE 25.5 Suppression of Atrial Fibrillation and Flutter

Agent (Ref.)	Dose Range (mg IV)	Estimated Efficacy (%)	Comments
Type IA agents			
Quinidine[57,67]	1,200 to 1,600	50	
Disopyramide[70]	400 to 1,600	65	Prior drug failures
Type IC agents			
Encainide[67]	75 to 200	70	Prior drug failures
Flecainide[4,34,68]	200 to 300	60 to 80	Prior drug failures
Propafenone[69–71]	300 to 1,200	40 to 75	Prior drug failures
Type III agents			
Sotalol[56,70]	160 to 960	50 to 60	Prior drug failures
Amiodarone[53,72–74]	100 to 600	50 to 90	Prior drug failures

A.Fib. = atrial fibrillation, Ref. = reference(s).

management. In some patients, particularly those with intrinsic sinus node dysfunction and those with markedly dilated atria, maintenance of sinus rhythm is virtually impossible and chronic control of ventricular rate, either by drug therapy or nonpharmacologic modification of AV conduction, becomes the goal of therapy. In most patients, however, sinus rhythm is either intermittently present or can be restored by drug treatment or electrical cardioversion. In these patients, prophylactic antiarrhythmic therapy is usually prescribed in an attempt to maintain sinus rhythm chronically.

Quinidine has traditionally been the treatment of choice to maintain sinus rhythm, but it is often ineffective and is associated with high incidence of side effects. Procainamide also has considerable toxicity when used long-term. Disopyramide has fewer side effects in young, relatively healthy patients, but its anticholinergic actions may be troublesome and its negative inotropic effects make it less attractive in older patients with significant structural heart disease.[66] For these reasons, there has been considerable interest in the use of new antiarrhythmic drugs for this purpose.

A number of studies have appeared which examined the role of type IC and type III drugs for the prevention of recurrent episodes of either paroxysmal or sustained atrial fibrillation and flutter (Table 25.5). In one placebo-controlled trial, flecainide prolonged the median tachycardia-free interval from 3 to 14.5 days in patients with paroxysmal atrial fibrillation.[4] Only one-third of patients, however, remained totally free of recurrent episodes during 2 months of therapy. Another controlled trial,[68] however, showed no difference between survival free of recurrent arrhythmia after cardioversion between patients who received flecainide and those who received no treatment. Antman et al. used a "staged care" approach in 109 patients who had previously failed treatment with a type IA drug.[71] Propafenose was first tried, followed by sotalol, in patients with recurrences on the former drug. Using this approach, 55% of the patients in this study, all of whom had failed one or more prior drugs, could be maintained in sinus rhythm. Amiodarone has also been reported to have an effect in 50 to 80% of patients who had failed other therapies. However, as noted by Brodsky et al., patients with left atrial dimensions greater than 60 mm were unlikely to respond even to this agent.[74] As yet, no simple solution to the problem of preventing atrial fibrillation has been found and the use of serial trials of multiple agents remains the norm.

CONCLUSIONS

The development and introduction of many new antiarrhythmic drugs has greatly enhanced our ability to manage patients with supraventricular arrhythmias. As both pharmacologic and nonpharmacologic options have improved, we are now reaching the point at which most patients can be safely and effectively managed.

REFERENCES

1. Pritchett ELC, Zimmerman JM, Hammill KF, Reiter MJ, Hammill SC: Electrocardiogram recording by telephone in antiarrhythmic drug trials. *Chest* 1982;81:473–476.
2. Sintetos AL, Roark SF, Smith MS, McCarthy EA, Lee KL, Pritchett ELC: Incidence of symptomatic tachycardia in untreated patients with paroxysmal supraventricular tachycardia. *Arch Intern Med* 1986;146:2205–2209.

3. Pritchett ELC, Hammill SC, Reiter MJ, et al.: Life-table methods for evaluating antiarrhythmic drug efficacy in patients with paroxysmal atrial tachycardia. *Am J Cardiol* 1983;52:1007–1012.
4. Anderson JL, Gilbert EM, Alpert BL, et al.: Prevention of symptomatic recurrences of paroxysmal atrial fibrillation in patients initially tolerating antiarrhythmic therapy. *Circulation* 1989;80:1557–1570.
5. Akhtar M, Niazi I, Naccarelli GV, et al.: Role of adrenergic stimulation by isoproterenol in reversal of effects of encainide in supraventricular tachycardia. *Am J Cardiol* 1988;62:45L–49L.
6. Ornato JP: Management of paroxysmal supraventricular tachycardia. *Circulation* 1986;74(suppl IV):IV-108–IV-110.
7. Mehta D, Wafa S, Ward DE, Camm AJ: Relative efficacy of various physical manoeuvres in the termination of junctional tachycardia. *Lancet* 1988;i:1181–1185.
8. Waxman NB, Wald RW, Sharma AD, Huerta F, Cameron DA: Vagal techniques for termination of paroxysmal supraventricular tachycardia. *Am J Cardiol* 1980;46:655–664.
9. Sreeram N, Wren C: Supraventricular tachycardia in infants: Response to treatment. *Arch Dis Child* 1990;65:127–129.
10. Wellens HJJ, Tan LL, Bar FWR, Duren DR, Lie KI, Dohmen HM: Effect of verapamil studied by electrical stimulation of the heart in patients with paroxysmal reentrant supraventricular tachycardia. *Br Heart J* 1977;39:1058–1066.
11. Singh BN, Nademanee K, Baky SH: Calcium antagonists. Clinical use in the treatment of arrhythmias. *Drugs* 1983;25:125–153.
12. Akhtar M, Tchou P, Jazayeri M: Use of calcium channel entry blockers in the treatment of cardiac arrhythmias. *Circulation* 1989;80(suppl IV):IV-31–IV-39.
13. Huycke EC, Sung RJ, Dias VC, Milstein S, Hariman RJ, Platia EV: Intravenous diltiazem for termination of reentrant supraventricular tachycardia: A placebo-controlled, randomized, double-blind, multicenter study. *J Am Coll Cardiol* 1989;13:538–544.
14. Rankin AC, Rae AP, Oldroyd KG, Cobbe SM: Verapamil or adenosine for the immediate treatment of supraventricular tachycardia. *Q J Med* 1990;74:203–208.
15. DiMarco JP, Sellers TC, Lerman BB, Greenberg ML, Berne RM: Diagnostic and therapeutic use of adenosine in patients with supraventricular tachyarrhythmias. *J Am Coll Cardiol* 1985;6:417–425.
16. Rankin AC, Oldroyd KG, Chong E, Rae AP, Cobbe SM: Value and limitations of adenosine in the diagnosis and treatment of narrow and broad complex tachycardia. *Br Heart J* 1989;62:195–203.
17. Overholt ED, Rheuban KS, Gutgesell HP, Lerman BB, DiMarco JP: Usefulness of adenosine for arrhythmias in infants and children. *Am J Cardiol* 1988;61:336–340.
18. DiMarco JP, Miles W, Akhtar M, et al.: Adenosine for paroxysmal supraventricular tachycardia: Dose ranging and comparison with verapamil in placebo-controlled, multicenter trials. *Ann Intern Med* 1990;113:104–110.
19. Watt AH, Bernard MS, Webster J, Passani SL: Intravenous adenosine in the treatment of supraventricular tachycardia: A dose-ranging study and interaction with dipyridamole. *Br J Clin Pharmacol* 1986;21:227–230.
20. Garratt C, Linker N, Griffith M, Ward D, Camm J: Comparison of adenosine and verapamil for termination of paroxysmal junctional tachycardia. *Am J Cardiol* 1989;64:1310–1316.
21. Till J, Shinebourne EA, Rigby ML, Clarke B, Ward DE, Rowland E: Efficacy and safety of adenosine in the treatment of supraventricular tachycardia in infants and children. *Br Heart J* 1989;62:204–211.
22. Munoz A, Leenhardt A, Sassine A, Galley P: Therapeutic use of adenosine for terminating spontaneous paroxysmal supraventricular tachycardia. *Eur Heart J* 1984;5:735–738.
23. Clarke B, Till J, Rowland E, Ward DE: Rapid and safe termination of supraventricular tachycardia in children by adenosine. *Lancet* 1987;i:299–301.
24. Rankin AC, Oldroyd KB, Chong E, Dow JW: Adenosine or adenosine triphosphate for supraventricular tachycardias? Comparative double-blind randomized study in patients with spontaneous or inducible arrhythmias. *Am Heart J* 1990;119:316–323.
25. Belhassen B, Pelleg A, Shoshani D, Geva B, Laniado S: Electrophysiologic effects of adenosine-5′-triphosphate on atrioventricular reentrant tachycardia. *Circulation* 1983;68:827–833.
26. Komor K, Garas Z: Adenosine triphosphate in paroxysmal tachycardia (letter). *Lancet* 1955;ii:93–94.
27. Somlo E: Adenosine triphosphate in paroxysmal tachycardia (letter). *Lancet* 1955;i:1125.
28. Belhassen B, Glick A, Laniado S: Comparative clinical and electrophysiological effects of ATP and verapamil on paroxysmal reciprocating junctional tachycardia. *Circulation* 1988;77:795–803.
29. Greco R, Musto B, Arienzo V, Alborino A, Garofalo S, Marsico F: Treatment of paroxysmal supraventricular tachycardia in infancy with digitalis, adenosine-5′-triphosphate, and verapamil: A comparative study. *Circulation* 1982;66:504–508.
30. Moller B, Ringquist C: Metoprolol in the treatment of supraventricular tachyarrhythmias. *Ann Clin Res* 1979;11:34–41.
31. Swerdlow C, Peterson J, Liem LB, Blake K, Franz MR, Laddu A: Electropharmacology of flestolol for supraventricular tachycardia without associated structural heart disease. *Am J Cardiol* 1987;60:1055–1060.
32. Hoff PI, Tronstad A, Oie B, Ohm O: Electrophysiologic and clinical effects of flecainide for recurrent paroxysmal supraventricular tachycardia. *Am J Cardiol* 1988;62:585–589.
33. Crozier IG, Ikram H, Kenealy M, Levy L: Flecainide acetate for conversion of acute supraventricular tachycardia to sinus rhythm. *Am J Cardiol* 1987;59:607–609.
34. Anderson JL, Jolivette DM, Fredell PA: Summary of efficacy and safety of flecainide for supraventricular arrhythmias. *Am J Cardiol* 1988;62:62D–66D.
35. Hammill SC, McLaran CJ, Wood DL, Osborn MJ, Gersch BJ, Holmes DR Jr: Double-blind study of intravenous propafenone for paroxysmal supraventricular reentrant tachycardia. *J Am Coll Cardiol* 1987;9:1364–1368.
36. Teo KK, Harte M, Morgan JH: Sotalol infusion in the treatment of supraventricular tachyarrhythmias. *Chest* 1985;87:113–118.
37. Waleffe A, Bruninx P, Kulbertus HE: Effects of amiodarone studied by programmed electrical stimula-

tion of the heart in patients with paroxysmal reentrant supraventricular tachycardia. *J Electrocardiol* 1978;11:253–260.
38. Drury AN, Szent-Györgyi A: The physiological activity of adenine compounds with especial reference to their action upon the mammalian heart. *J Physiol* 1929;68:213–237.
39. Clemo HF, Belardinelli L: Effect of adenosine on atrioventricular conduction. I: Site and characterization of adenosine action in the guinea pig atrioventricular node. *Circ Res* 1986;59:427–436.
40. Belardinelli L, Linden J, Berne RM: The cardiac effects of adenosine. *Prog Cardiovasc Dis* 1989;32:73–97.
41. Mauritson DR, Winniford MD, Walker WS, Rude RE, Cary JR, Hillis LD: Oral verapamil for paroxysmal supraventricular tachycardia. *Ann Intern Med* 1982;96:409–412.
42. Yeh SJ, Fu M, Lin FC, Lee YS, Hunmg JS, Wu D: Serial electrophysiology studies of the effects of oral diltiazem on paroxysmal supraventricular tachycardia. *Chest* 1985;87:639–643.
43. Chang MS, Sung RJ, Tai TY, Lin SL, Liu PH, Chiang BN: Nadolol and supraventricular tachycardia: An electrophysiologic study. *J Am Coll Cardiol* 1983;2:894–903.
44. Benson DW, Dunnigah A, Green TP, Benditt DG, Schneider SP: Periodic procainamide for paroxysmal tachycardia. *Circulation* 1985;72:147–152.
45. Wellens HJJ, Bar FW, Dassen WRM, Brugada P, Vanagt EJ, Farre J: Effect of drugs in the Wolff-Parkinson-White syndrome. *Am J Cardiol* 1980;46:665–669.
46. Rinkenberger RL, Naccarelli GV, Miles WM, et al.: Encainide for atrial fibrillation associated with Wolff-Parkinson-White syndrome. *Am J Cardiol* 1988;62:261–301.
47. Breithardt G, Borggrefe M, Wiebringhaus E, Seipel L: Effect of propafenone in Wolff-Parkinson-White syndrome: Electrophysiologic findings and long-term follow-up. *Am J Cardiol* 1984;54:29D–39D.
48. Musto B, D'Onofrio A, Cavallaro C, Musto A: Electrophysiological effects and clinical efficacy of propafenone in children with recurrent paroxysmal supraventricular tachycardia. *Circulation* 1988; 78:863–869.
49. Singh BN, Deedwania P, Nademanee K, Ward A, Sorkin EM: Sotalol. A review of its pharmacodynamic and pharmacokinetic properties, and therapeutic use. *Drugs* 1987;34:311–349.
50. Singh BN, Nademanee K: Sotalol: A beta blocker with unique antiarrhythmic properties. *Am Heart J* 1987;114:121–139.
51. Kunze KP, Schluter M, Kuck KH: Sotalol in patients with Wolff-Parkinson-White syndrome. *Circulation* 1987;75:1050–1057.
52. Sahar DI, Reiffel JA, Bigger JT Jr, Squatrito A, Kidwell GA: Efficacy, safety, and tolerance of d-sotalol in patients with refractory supraventricular tachyarrhythmias. *Am Heart J* 1989;117:562–568.
53. Kopelman HA, Horowitz LN: Efficacy and toxicity of amiodarone for the treatment of supraventricular tachyarrhythmias. *Prog Cardiovasc Dis* 1989; 31:355–366.
54. Mason JW: Drug therapy: Amiodarone. *N Engl J Med* 1987;316:455–466.
55. Falk RH, Knowlton AA, Bernard SA, Gotlieb NE, Battinelli NJ: Digoxin for converting recent-onset atrial fibrillation to sinus rhythm. *Ann Intern Med* 1987;106:503–506.
56. Singh SN, Saini RK, DiMarco JD, Kluger J, Gold R, and the Sotalol Study Group: The efficacy and safety of sotalol in digitalized patients with chronic atrial fibrillation. *J Am Coll Cardiol* 1990;15:27A(abstract).
57. Sodermark T, Edhag O, Sjogren A, et al.: Effect of quinidine on maintaining sinus rhythm after conversion of atrial fibrillation or flutter. *Br Heart J* 1975;37:486–492.
58. Bianconi L, Boccadamo R, Pappalardo A, Gentili C, Pistolese M: Effectiveness of intravenous propafenone for conversion of atrial fibrillation and flutter of recent onset. *Am J Cardiol* 1989;64:335–338.
59. Goy JJ, Kaufmann U, Kappenberger L, Sigwaart U: Restoration of sinus rhythm with flecainide in patients with atrial fibrillation. *Am J Cardiol* 1988;62:38D–40D.
60. Borgeat A, Goy J, Maendly R, Kaufmann U, Grbic M, Sigwart U: Flecainide versus quinidine for conversion of atrial fibrillation to sinus rhythm. *Am J Cardiol* 1986;58:496–498.
61. Santos AL, Alexo AM, Landeiro J: Conversion of atrial fibrillation to sinus rhythm with amiodarone. *Acta Med Port* 1979;1:15–23.
62. Blevins RD, Kerin NZ, Benaderet D, et al.: Amiodarone in the management of refractory atrial fibrillation. *Arch Intern Med* 1987;147:1401–1404.
63. Falk RH: Flecainide-induced ventricular tachycardia and fibrillation in patients treated for atrial fibrillation. *Ann Intern Med* 1989;111:107–111.
64. The Cardiac Arrhythmia Suppression Trial Investigators: Preliminary report: Effect of encainide and flecainide on mortality in a randomized trial of arrhythmia suppression after myocardial infarction. *N Engl J Med* 1989;321:406–412.
65. The Cardiac Arrhythmia Pilot Study Investigators: Effects of encainide, flecainide, imipramine and moricizine on ventricular arrhythmias during the year after acute myocardial infarction: The CAPS. *Am J Cardiol* 1988;61:501–509.
66. Bauman JL, Gallastegui J, Strasberg B, Swiryn S, Hoff J, Welch WJ: Long-term therapy with disopyramide phosphate: Side effects and effectiveness. *Am Heart J* 1986;111:654–660.
67. Makynen PJ, Koskinen PJ, Saaristo TE, Kiisanantti RK: Comparison of encainide and quinidine for supraventricular tachyarrhythmias. *Am J Cardiol* 1988;62:55L–59L.
68. Van Gelder IC, Crijns HJ, Van Gilst WH, Van Wijk LM, Hamer HP, Lie KI: Efficacy and safety of flecainide acetate in the maintenance of sinus rhythm after electrical cardioversion of chronic atrial fibrillation or atrial flutter. *Am J Cardiol* 1989;64:1317–1321.
69. Antman EM, Beamer AD, Cantillon C, McGowan N, Goldman L, Friedman PL: Long-term oral propafenone therapy for suppression of refractory symptomatic atrial fibrillation and atrial futter. *J Am Coll Cardiol* 1988;12:1005–1011.
70. Kerr CR, Klein GJ, Axelson JE, Cooper JC: Propafenone for prevention of recurrent atrial fibrillation. *Am J Cardiol* 1988;61:914–916.
71. Antman EM, Beamer AD, Cantillon C, McGowan N,

Friedman PL: Therapy of refractory symptomatic atrial fibrillation and atrial flutter: A staged care approach with new antiarrhythmic drugs. *J Am Coll Cardiol* 1990;15:698–707.
72. Graboys TB, Podrid PJ, Lown B: Efficacy of amiodarone for refractory supraventricular tachyarrhythmias. *Am Heart J* 1983;106:870–875.
73. Horowitz LN, Spielman SR, Greenspan AM, et al.: Use of amiodarone in the treatment of persistent and paroxysmal atrial fibrillation resistant to quinidine therapy. *J Am Coll Cardiol* 1985;6:1402–1407.
74. Brodsky MA, Allen BJ, Walker CJ 3d, Casey TP, Luckett CB, Henry WI: Amiodarone for maintenance of sinus rhythm after conversion of atrial fibrillation in the setting of a dilated left atrium. *Am J Cardiol* 1987;60:572–575.

Chapter **26**

Approach to Suppression of Ventricular Arrhythmias: Update

Gerald V. Naccarelli, MD, and Anne H. Dougherty, MD

Over the past decade, a major advance in the pharmacologic suppression of ventricular arrhythmias has been the commercial approval of multiple, oral antiarrhythmic drugs for the treatment of ventricular arrhythmias. Prior to 1983, only oral quinidine, procainamide, disopyramide, and propranolol were available for the treatment of ventricular arrhythmias. Since 1983, tocainide, mexiletine, amiodarone, flecainide, encainide, propafenone, and most recently moricizine have become available to the practitioner. Multiple reviews are available for in-depth study of the electrophysiology, pharmacokinetic properties, antiarrhythmic efficacy, and adverse reaction profiles of these drugs.[1–3] Recent findings from the Cardiac Arrhythmia Suppression Trial (CAST) increased the confusion of the role of pharmacologic agents in the treatment of ventricular arrhythmias.[4,5] Although antiarrhythmic drug suppression of inducible sustained ventricular tachycardia (VT-S) and ventricular fibrillation (VF) in the electrophysiology laboratory appears to be associated with a more benign prognosis,[6–8] spontaneous suppression of premature ventricular complexes (PVCs) and complex ventricular ectopy using noninvasive testing has remained controversial.[4,6,9]

CONSIDERATIONS IN CHOOSING ANTIARRHYTHMIC DRUGS

A priori, no antiarrhythmic drug will be effective all of the time. Therefore, there are several factors that may be useful in guiding initial selection of antiarrhythmic drug therapy. Table 26.1 lists some of these considerations. The type of arrhythmia is important. If one is treating symptomatic PVCs, the use of a drug that is an effective PVC suppressor would be a good choice, but one would want to avoid end-organ toxicity and proarrhythmia. The response to prior therapy has some importance. A patient who may have done well on a drug like tocainide, but had some adverse effects, might be considered for a trial with a similar drug such as mexiletine before considering other drug classes. Coexisting disease states need to be kept in consideration. In a patient who has arthritis, procainamide might not be the best first choice. In a patient with significant gastrointestinal problems, quinidine, mexiletine, and tocainide may not be the best initial choice of therapy. Concomitant medications need to be considered because there is always the potential for drug interactions. In patients requiring digoxin therapy, quinidine, propafenone, and amiodarone can be used safely, but only if the digoxin dose and levels are watched more closely due to significant drug interactions. Left ventricular function tends to be one of the most important considerations in choosing an antiarrhythmic drug. In patients with an ejection fraction of 15% and/or New York Heart Association class III to IV congestive heart failure, drugs with significant negative inotropic effect, such as disopyramide, β-blockers, and flecainide should be avoided.[10] Coexisting atrial and ventricular arrhythmias occur in 12% of patients in our practice. Drugs that are effective in both types have advantages over drugs only effective in the ventricular arrhythmias. One should also consider the risk of proarrhythmia.[10] In patients with SVT and left ventricular dysfunction, the risk of proarrhythmia is highest with a type IC drug. These should be avoided if possible in this situation.

655 Avenue of the Americas, New York, NY 10010
Current Topics in Cardiology

TABLE 26.1 Considerations in Choosing Antiarrhythmic Drugs

1. Type of arrhythmia
2. Response to prior therapy
3. Coexisting disease states
4. Concomitant medications
5. Left ventricular function
6. Coexisting atrial and ventricular arrhythmias
7. Risk of proarrhythmia
8. Risk of subjective toxicity
9. Risk of end-organ toxicity
10. Hepatic or renal dysfunction
11. Patient compliance
12. Negative inotropic potential of some antiarrhythmic drugs

Drug selection might be altered in patients with hepatic or renal disease. In patients with hepatic dysfunction, drugs that are predominantly excreted renally might be the best choice. In cases of renal dysfunction, it may be safer to use drugs that are hepatically metabolized. Patient compliance is important. Drugs that are used in the b.i.d. format may be more easily used in patients taking on a q.i.d. format.

COMPARATIVE ANTIARRHYTHMIC EFFICACY

Table 26.2 lists the comparative efficacy of various drugs in treating ventricular arrhythmias. Efficacy for PVC reduction and the treatment of nonsustained VT was obtained from reviewing multiple studies based on Holter monitoring.[11] Amiodarone and type IC agents have the best efficacy of suppressing PVCs. Drugs such as the β-blockers have the lowest efficacy rates. This efficacy in patients with benign symptomatic PVCs or symptomatic nonsustained VT may be useful. However, these efficacy rates are defined by noninvasive criteria and may not have a positive impact on survival. In fact, the recently published CAST trial[4] in postmyocardial infarction patients showed a marked reduction in PVCs with encainide and flecainide, yet this was associated with an increased, not a decreased, mortality. The β-blocker heart attack trials including propranolol, metoprolol, timolol, and acebutalol have shown only mild reductions in PVCs yet significant improvement in survival, suggesting that survival may be improved by a complex interaction of anti-ischemic, antisympathetic, and antiarrhythmic effects.[12–15]

In Table 26.2, efficacy rates defined in patients with VT-S/VF have been obtained from reviewing the literature from programmed electrical stimulation studies.[16] Although amiodarone is only 20% effective, as defined by noninducibility in the electrophysiology lab, clinical studies have shown a consistent efficacy rate of about 60%, even in patients who remain inducible and have failed multiple antiarrhythmic drugs.[17] Early results with sotalol suggest a slightly improved efficacy rate over the type IA and IC agents. Drugs that have the least efficacy by EP criteria for VT-S are the β-blockers. These are predominantly used in VT-S patients that are aggravated by exercise.[18,19] Combination therapy will usually result in slightly improved greater efficacy than each individual drug alone.[20,21]

TABLE 26.2 Comparative Efficacy of Antiarrhythmic Drugs in Ventricular Arrhythmias

	PVC/VT-NS (Holter) (%)	Sustained VT-S/VF (PES) (%)
Quinidine	65	20 to 25
Procainamide	60	20 to 25
Disopyramide	60	20 to 25
Tocainide	50	10 to 15
Mexiletine	50	10 to 20
Moricizine	65	20
Flecainide	75	20 to 25
Encainide	75	20 to 25
Propafenone	70	20 to 25
Propranolol	45	5
Acebutolol	45	5
Amiodarone	75	20 (60)[a]
Sotalol	55	30

[a] Higher spontaneous efficacy rates despite PES inefficacy.

The efficacy rates noted in this table involve averages from multiple studies. With multiple drugs it has been shown in VT-S that the presence of left ventricular dysfunction and ejection fraction lower than 30% is usually associated with lower efficacy rates.[22–24] Thus, in patients with preserved left ventricular function and VT-S, especially those who are hemodynamically stable during tachycardia and who have experienced a beneficial response to intravenous procainamide, serial drug testing using these drugs may be very reasonable in the electrophysiology laboratory.[25–27] In patients who would be predicted to have a low yield of antiarrhythmic efficacy (depressed ejection fraction, continued VT induction after intravenous procainamide), alternative nonpharmacologic approaches may be considered, especially if the patient has hemodynamically unstable VT.

TABLE 26.3 Negative Inotropic Potential of Antiarrhythmics

Most negative inotropy	Disopyramide
	Verapamil
	↓
	Propranolol
	Acebutolol
↓	↓
	Flecainide
	↓
	Propafenone
	Sotalol
	Diltiazem
	↓
	Encainide
↓	
↓	Lidocaine
	Quinidine
	Procainamide
	Tocainide
Least negative inotropy	Mexiletine
	Moricizine
	Amiodarone
	Bretylium

ANTIARRHYTHMIC DRUGS: COMPARATIVE HEMODYNAMICS

Table 26.3 lists the comparative negative inotropic potential of the antiarrhythmic drugs.[28] Notably, disopyramide and verapamil have the most negative inotropic effect. Other drugs with significant negative inotropic effect include propranolol, acebutolol, and flecainide. Propafenone, sotalol, and encainide have lesser negative inotropic effect, but greater than other IA and IB agents and amiodarone.

ANTIARRHYTHMIC DRUGS: SUBJECTIVE TOXICITY

Table 26.4 shows a comparison of subjective toxicity of antiarrhythmic drugs.[2,5] Subjective toxicity is usually a nuisance but is not life-threatening. However, frequently patients will have effective control of their arrhythmias, which may be discontinued because of subjective toxicity. Commonly used oral antiarrhythmics that have the highest incidence of toxicity include quinidine (gastrointestinal adverse effects) and tocainide and mexiletine (neurologic and gastrointestinal adverse effects). Drugs with the least potential for subjective toxicity include moricizine, flecainide, encainide, and the β-blockers.

TABLE 26.4 Comparative Toxicity of Antiarrhythmic Drugs

	Subjective Toxicity	End-Organ Toxicity
Quinidine	+ +	+ +
Procainamide	+	+ +
Disopyramide	+	±
Tocainide	+ +	+
Mexiletine	+ +	±
Propafenone	+	±
Moricizine	+	±
Flecainide	±	±
Encainide	±	±
Phenytoin	+ +	+
Propranolol	+	±
Acebutolol	+	±
Amiodarone	+	+ + + +
Sotalol	+	±

+ + + + = very commonly occurrs, + + = commonly occurs, + = occasionally occurs, ± = rarely occurs.

ANTIARRHYTHMIC DRUGS: END-ORGAN TOXICITY

End-organ toxicity caused by antiarrhythmics is of concern.[2,5] Although amiodarone is an effective antiarrhythmic agent, its potential to cause interstitial fibrosis, drug-induced hepatitis, and thyroid toxicity causes the drug to be used predominantly in patients with life-threatening ventricular arrhythmias in whom the risk/benefit ratio is acceptable.[29] Quinidine and procainamide have significant end-organ toxicity. Quinidine can cause thrombocytopenia and drug-induced hepatitis. Procainamide commonly causes a drug-induced lupus syndrome and has been associated with agranulocytosis. Tocainide has been associated with a rare incidence of agranulocytosis and pulmonary fibrosis. Because of these adverse effects and its low efficacy rates, tocainide is usually used as a second-line agent. All the other oral antiarrhythmic agents have only rare potential for end-organ toxicity and have some advantage in this regard.

ANTIARRHYTHMIC DRUGS: PROARRHYTHMIC POTENTIAL

Of significant concern is the proarrhythmic potential of antiarrhythmic drugs[30–32] (Table 26.5). Clinically significant proarrhythmia predominantly consists of torsade de pointes, which

TABLE 26.5 Proarrhythmic Potential of Antiarrhythmic Drugs

	Proarrhythmia Nonsustained VT Patients	Proarrhythmia Sustained VT Patients	TDP
Quinidine	+	+ +	+ + +
Procainamide	+	+ +	+ +
Disopyramide	+	+ +	+ +
Tocainide	±	+	0
Mexiletine	±	+	0
Moricizine	+	+	0
Flecainide	+[a]	+ + +	±
Encainide	+[a]	+ + +	±
Propafenone	+	+ +	±
β-Blockers	±	+	0
Amiodarone	±	+	+
Sotalol	+	+	+ +

[a] Incidence probably higher in patients postmyocardial infarction as per CAST study.[4]

TDP = torsade de pointes.

occurs predominantly with type IA antiarrhythmic agents and sotolol or incessant VT-S. Torsade de pointes is sometimes dose related, but often is idiosyncratic.[33] Quinidine is the number one cause of torsade de pointes in the United States and worldwide. This may be secondary to quinidine being such a popular antiarrhythmic agent, but it also appears that the incidence is higher for quinidine than procainamide or dysopyramide. The incidence of torsade de pointes with other antiarrhythmic drugs is rare.

In reviewing the incidence of proarrhythmia, it is important to look at the baseline arrhythmia that is being treated. In patients with symptomatic benign, potentially malignant PVCs or nonsustained VT, the incidence of proarrhythmia is reasonably low for all antiarrhythmic drugs. Excluding torsade de pointes, the incidence is less than 3 to 4%. However, drugs such as tocainide, mexiletine, β-blockers, and amiodarone appear to have a lower proarrhythmic profile than the other type IA and IC drugs. In patients with VT-S, the incidence of proarrhythmia is highest in patients with the worst left ventricular function. In sustained VT patients, proarrhythmia is highest with drugs such as flecainide and encainide with an incidence of 10 to 20%. Drugs with the most benign proarrhythmic profile in VT-S include tocainide, mexiletine, β-blockers, and amiodarone. These percentages should be weighed and the proper selection of antiarrhythmic drugs in given patients.

ACUTE VENTRICULAR ARRHYTHMIAS

Treatment Considerations

In patients who have acute ventricular arrhythmias that occur during surgery or acute myocardial infarction, acute treatment may be necessary without the need for chronic oral therapy. Acute ventricular arrhythmias, even ventricular fibrillation, in this setting have very little prognostic significance.[34] The occurrence of these arrhythmias is not predictive of the need for long-term antiarrhythmic therapy. Despite this, it is common for physicians empirically to treat patients with intravenous drugs such as lidocaine (1 to 4 mg/min)[35] after loading, intravenous procainamide (1 to 4 mg/min), bretylium, or β-blockers. Prior to committing any patients to chronic treatment, risk stratification should be performed.

Subacute Ventricular Arrhythmias in the Postmyocardial Infarction Setting

Due to the results of the CAST study,[4] the treatment of asymptomatic ventricular arrhythmias in the postmyocardial infarction setting has become increasingly controversial. In this setting, multiple studies have shown that the presence of frequent PVCs, complex ventricular arrhythmias, especially if associated with left ventricular dysfunction, is associated with an increased incidence of mortality and sudden cardiac death.[36–39] Because of these observations, approaches to risk stratify patients using ejection fractions measured by radionuclide ventriculogram, Holter monitoring to quantitate and qualitate ventricular arrhythmia, stress testing to screen out exercise-induced arrhythmia and latent ischemia, and signal averaging to identify patients with late potentials have been used.[40–42]

Prior to the results of the CAST study, patients were often treated with antiarrhythmic drugs in the hope that the risk and incidence of sudden death could be reduced. However, except for empiric treatment with timolol, propranolol, metoprolol, and, most recently, acebutolol, no study has demonstrated that antiarrhythmic therapy alters prognosis of this group of patients.[43] In the CAST study, patients who were effectively treated with encainide and flecainide with suppression of ventricular arrhythmias had a two to three times increase in mortality and sudden cardiac death as compared to placebo. Although the CAST trial can be constructively criticized in several aspects, including the low placebo mortality, the results of this trial influenced our thinking regarding this group of patients.[5,44]

Given the current state of knowledge, we would make the following recommendations in postmyocardial infarction patients who have asymptomatic ventricular arrhythmias: (a) In patients at moderate to high risk of sudden death, cardio-protective doses of an appropriately studied β-blocker should be the first drug of choice; (b) encainide and flecainide should be avoided in these patients given the results of the CAST findings; and (c) the role of chronic therapy with other role of drugs remains to be determined. In patients who are symptomatic, a decision to treat first with a β-blocker and then with other drugs should be on a case-by-case basis.

Treatment of Benign Ventricular Arrhythmias

Patients with benign ventricular arrhythmias include patients with PVCs, couplets, and nonsustained VT in the setting of normal left ventricular function and minimal organic heart disease. The indication for treating these patients is symptom relief, since their risk for sudden cardiac death is minimal. Multiple studies, including that of Kennedy et al.,[45] demonstrated a benign prognosis in healthy patients with frequent and complex arrhythmias with normal left ventricular function.

In choosing an antiarrhythmic drug for this situation, it is important for the treatment not to be worse than the initial disease process. Since patients do not have a life-threatening arrhythmia, one is treating patients to avoid symptoms of palpitations, dizziness, or presyncope. The best drug in a given patient should relieve symptoms without significant proarrhythmic potential, or subjective or end-organ toxicity. Given this restriction, most antiarrhythmic drugs can be used in this setting, except for amiodarone. Overall, the safest drugs to use in this group of patients are the β-blockers and mexiletine. Although these drugs have the lowest efficacy rates, their incidence of proarrhythmia and end-organ toxicity is low. Therefore, they may be safe to use in the outpatient setting with minimal risk of proarrhythmia. The class IA drugs can be used as initial therapy, but these drugs may be limited due to some of their subjective side effects. Prior to the CAST trial, the IC drugs encainide and flecainide were frequently used in this situation. This was based on the fact that these drugs had high rates of efficacy as defined by Holter monitoring, and low end-organ and subjective toxicity. However, the results of the CAST study make them less desirable as initial therapy, and they should be used only after other agents have failed. Newer drugs, such as propafenone and mexiletine, only have Food and Drug Administration (FDA) approval for life-threatening ventricular arrhythmias. Despite this fact, these drugs have been used frequently in patients with benign symptomatic arrhythmias with excellent results.

Potentially Lethal Ventricular Arrhythmias

Patients with potentially lethal ventricular arrhythmias usually have frequent PVCs, couplets, and nonsustained runs of VT. These patients are separated from those with benign arrhythmias in that they usually have organic heart disease and left ventricular dysfunction. The presence of nonsustained VT in this setting has been associated with increased chances of sudden cardiac and pump death.[46–48] Although angiotensin converting enzyme (ACE) inhibitors have been shown to decrease pump death, sudden cardiac death has not been affected.[49] Because of this, treatment indications have included symptom relief and the hope that treatment may alter an otherwise bad prognosis and decrease the incidence of sudden cardiac death. To date there have been no well-controlled studies to demonstrate that antiarrhythmic drugs in this setting can decrease the incidence of sudden cardiac death.

If a decision is made to treat high risk in this category, most available drugs can be used with certain exceptions: (a) negatively inotropic drugs, such as disopyramide, β-blockers, flecainide, and propafenone should be used with caution; (b) encainide and flecainide should be used with caution given the results of the CAST trial; (c) since the risk of proarrhythmia is higher in this group than in benign patients, high-risk patients should be examined by telemetry if less proarrhythmic drugs, such as β-blockers and mexiletine, are used; (d) in high-risk patients who have failed to have control of their arrhythmia, especially if there is coexisting atrial fibrillation with hemodynamic compromise, low-dose amiodarone might be considered,[50] especially if combination therapy has failed. Ongoing studies with amiodarone in this setting are currently in progress.

Due to the incidence of proarrhythmia in this group of patients, the concern of proarrhythmia is a factor in choosing antiarrhythmic drugs and in considering patients for hospitalization during dosing. In high-risk patients with left ventricular dysfunction, prior antiarrhythmic failure, prior proarrhythmic response, baseline conduction abnormalities, elderly patients, or patients with metabolic abnormalities, in-house therapy may be preferable.

Pharmacologic Treatment of Patients with Sustained Ventricular Tachyarrhythmias

In patients with VT-S and VF, symptom relief remains a treatment indication to avoid near syncope and syncope. Although studies have demonstrated that serial electrophysiologic testing[(6–8,16)] and Holter-guided therapy[(51)] may prevent recurrences of VT-S episodes, no definitive study has shown that treatment of this group of patients prevents sudden death.

The selection of antiarrhythmic therapy in this group of patients differs from that of lower-risk patients because of several factors: (a) patients with VT-S are at high risk for sudden death and therefore the risk of using drugs with adverse effects may be more acceptable; (b) all antiarrhythmic drugs are less effective in VT-S than in less severe ventricular arrhythmias; (c) the higher incidence of left ventricular dysfunction in this group of patients may minimize the use of drugs with negative ionotropic potential. In addition, in patients with left ventricular dysfunction, the efficacy rate of all drugs is lower, (d) the risk of proarrhythmia is higher in this group of patients who need treatment the most. The risk of proarrhythmia is highest in patients with left ventricular dysfunction. IC drugs, such as encainide and flecainide, have the highest proarrhythmic profile.

Because of these considerations, our initial choices of drugs are procainamide, quinidine, and mexiletine. This is based on the fact that these drugs work from 15 to 20% of the time as monotherapy, and they have benign hemodynamic profiles and better proarrhythmic profiles in this group of patients compared to the class IC agents. If these drugs are not effective, we consider a class IA/IB combination. In patients with ejection fractions greater than 30%, we will consider treatment with a classic class IC drug such as encainide or flecainide; in fact, these drugs work in excess of 30% in this situation. If the ejection fraction is less than 30% and classes IA and IB and combination therapy have been ineffective, we frequently consider the use of amiodarone or alternative nonpharmacologic therapies. In patients with hemodynamically stable SVT, we typically consider the above antiarrhythmic approach guided by serial electrophysiologic studies. In patients with hemodynamically unstable ventricular tachyarrhythmias, we typically use drugs as adjunctive therapy along with an automatic implantable defibrillator.

SUMMARY

Antiarrhythmic therapy, despite the above guidelines, is often empiric. Regardless of the drug chosen, proper judgment of efficacy and toxicity needs to be employed. In-depth knowledge of the electrophysiology, pharmacokinetics, efficacy, and adverse effect profile will aid in the proper antiarrhythmic drug selection.

REFERENCES

1. Kreeger RW, Hammill SC: New antiarrhythmic drugs: Tocainide, mexiletine, flecainide, encainide, amiodarone. *Mayo Clin Proc* 1987;62:1033–1050.
2. Naccarelli GV, Rinkenberger RL, Dougherty AH, Berns E, Crandall JR: Pharmacologic therapy of arrhythmias. *Hosp Prac* 1988;10:183–210.
3. Salerno DM: Review: Antiarrhythmic drugs; 1987. Part I–J, *Electrophysiology* 1987;1:217. Part II, 1987;1:300–319. Part III, 1987;1:435–465. Part IV, 1988;2:55–87.
4. The Cardiac Arrhythmia Suppression Trial Investigators: Preliminary report: Effect of encainide and flecainide on mortality in a randomized trial of arrhythmia suppression after myocardial infarction. *N Engl J Med* 1989;321:406–412.
5. Morganroth J, Bigger JT: Pharmacologic management of ventricular arrhythmias after the Cardiac Arrhythmia Suppression Trial. *Am J Cardiol* 1990;65:1497–1503.
6. Mitchell LB, Duff HJ, Manyari DE, Wyse DG: A randomized clinical trial of the noninvasive and invasive approaches to drug therapy of ventricular tachycardia. *N Engl J Med* 1987;317:1681–1687.
7. Wilber DJ, Garan H, Kelly E, McGovern B, Newell J, Ruskin JN: Out-of-hospital cardiac arrest: Role of electrophysiologic testing in prediction of long-term outcome. *N Engl J Med* 1987;318:19–24.
8. Mason JW, Winkle RA: Accuracy of the ventricular tachycardia-induction study for predicting long-term efficacy and inefficacy of antiarrhythmic drugs. *N Engl J Med* 1980;303:1073–1077.
9. Kim SG, Seiden SW, Matos JA, Waspe L, Fisher JD: Discordance between ambulatory monitoring and programmed stimulation in assessing efficacy of class IA antiarrhythmic agents in patients with ventricular tachycardia. *J Am Coll Cardiol* 1985;6:539–544.
10. Podrid PJ, Schoeneberger A, Lown B: Congestive heart failure caused by oral disopyramide. *N Engl J Med* 1980;302:614–617.
11. Salerno DM, Gillingham KJ, Berry DA, Hodges M: A comparison of antiarrhythmic drugs for the suppression of ventricular ectopic depolarizations: A meta-analysis. *Am Heart J* 1990;120:340–353.
12. The Norwegian Multicenter Study Group: Timolol-induced reduction in mortality and reinfarction in patients surivving acute myocardial infarction. *N Engl J Med* 1981;304:801–807.
13. Beta-Blocker Heart Attack Trial Research Group: A randomized trial of propranolol in patients with

acute myocardial infarction. I. Mortality results. JAMA 1982;247:1707–1714.
14. Hjalmarson A, Herlitz J, Malek I, et al.: Effect of mortality of metoprolol in acute myocardial infarction: A double-blind randomized trial. *Lancet* 1981; ii:823–827.
15. Boissel JP, Leizorowitz A, Picolet H, Ducruet T, ASPI Investigators: Efficacy of acebutolol after acute myocardial infarction (the ASPI trial). *Am J Cardiol* 1990;66:24C–31C.
16. Rae AP, Greenspan AM, Spielman SR, et al.: Antiarrhythmic drug efficacy for ventricular tachyarrhythmias associated with coronary artery disease assessed by electrophysiologic studies. *Am J Cardiol* 1985;55:1494–1499.
17. Naccarelli GV, Fineberg NS, Zipes DP, Heger JJ, Duncan G, Prystowsky EN: Amiodarone: Risk factors for recurrence of symptomatic ventricular tachycardia identified at electrophysiologic study. *J Am Coll Cardiol* 1985;6:814–821.
18. Woelfel A, Foster JR, Simpson RJ, Gettes LS: Reproducibility and treatment of exercise-induced ventricular tachycardia. *Am J Cardiol* 1984;53:751–756.
19. Palileo EV, Ashley WW, Swiryn S, Bauernfeind RA, Strasberg B, Petropoulos AT, Rosen KM: Exercise provocable right ventricular outflow tract tachycardia. *Am Heart J* 1982;104:185–193.
20. Duff HJ, Mitchell LB, Manuari D, Wyse DG: Mexiletine-quinidine combination: Electrophysiologic correlates of a favorable antiarrhythmic interaction in humans. *J Am Coll Cardiol* 1987;10:1149–1156.
21. Greenspan AM, Spielman SR, Webb CR, et al.: Efficacy of combination therapy with mexiletine and a type IA agent for inducible ventricular tachyarrhythmias secondary to coronary artery disease. *Am J Cardiol* 1985;56:277–284.
22. Berns E, Naccarelli GV, Dougherty AH, Povia C, Rinkenberger RL: Mexiletine: Lack of predictors of clinical response in patients treated for life-threatening tachyarrhythmias. *J Electrophysiol* 1988;2:201–206.
23. Tardjman T, Podrid PJ, Roeder E, Lown B: Safety and efficacy of encainide for malignant ventricular arrhythmias. *Am J Cardiol* 1986;58:87C–95C.
24. Hession M, Lampert S, Podrid PJ, Lown B: Ethmozine therapy in patients with complex arrhythmia. *Am J Cardiol* 1987;60:59F–66F.
25. Waxman HL, Buxton AE, Sadowski LM, Josephson ME: Response to procainamide during electrophysiologic study for sustained ventricular tachycardia predicts response to other drugs. *Circulation* 1982;67:30–37.
26. Meissner MD, Kay HR, Horowitz LN, et al.: Relation of acute antiarrhythmic drug efficacy to left ventricular function in coronary artery disease. *Am J Cardiol* 1988;61:1050–1055.
27. Spielman SR, Schwartz JG, McCarthy DM, et al.: Predictions of the success or failure of medical therapy in patients with chronic recurrent sustained ventricular tachycardia: A discriminant analysis. *J Am Coll Cardiol* 1983;1:401–408.
28. Ranid S, Podrid PJ, Lambert S, Lown B: Congestive heart failure induced by 6 of the newer antiarrhythmic drugs. *J Am Coll Cardiol* 1989;14:1326–1330.
29. Naccarelli GV, Rinkenberger RL, Dougherty AH, Giebel RA: Amiodarone: Pharmacology, antiarrhythmic and adverse effects. *Pharmacotherapy* 1985; 5(6):298–313.
30. Vellbit V, Podrid P, Lown B, Cohen BH, Graboys TB: Aggravation and provocation of ventricular arrhythmias by antiarrhythmic drugs. *Circulation* 1982; 65:886–894.
31. Stanton MS, Prystowsky EN, Fineberg NS, Miles WM, Zipes DP, Heger JJ: Arrhythmogenic effects of antiarrhythmic drugs: A study of 506 patients treated for ventricular tachycardia or fibrillation. *J Am Coll Cardiol* 1989;14:209–215.
32. Minardo JD, Heger JJ, Miles WH, Zipes DP, Prystowsky EN: Clinical characteristics of patients with ventricular fibrillation during antiarrhythmic drug therapy. *N Engl J Med* 1988;319:257–262.
33. Desertenne F: La tachycardie ventriculaire a deux foyers apposes variables. *Arch Mal Coeur* 1966; 59:263–272.
34. Vismara LA, Amsterdam EA, Mason DT: Relation of ventricular arrhythmias in the late hospital phase of acute myocardial infarction to sudden death after hospital discharge. *Am J Med* 1975;59:6–12.
35. Lie KI, Wellens HJJ, VanCapelle FJ, Durrer D: Lidocaine in the prevention of primary ventricular fibrillation. *N Engl J Med* 1974;291:1324–1326.
36. Mukharji J, Rude RE, Poole WK, et al., and the MILIS Study Group: Risk factors for sudden death after acute myocardial infarction: Two-year followup. *Am J Cardiol* 1984;54:30–36.
37. Bigger JT, Fleiss JL, Kleiger R, et al., and the Multicenter Post-Infarction Research Group: The relationships among ventricular arrhythmias, left ventricular dysfunction and mortality in the 2 years after myocardial infarction. *Circulation* 1984; 69:250–258.
38. Ruberman W, Weinblatt E, Goldberg JK, et al.: Ventricular premature beats and mortality after myocardial infarction. *N Engl J Med* 1977;297:750–757.
39. Schulze R, Rouleau J, Bowers S, Strauss HW, Pitt B: Ventricular arrhythmias in the late hospital phase of acute myocardial infarction: Relation of left ventricular function detected by gated cardiac blood pool scanning. *Circulation* 1975;52:1006–1011.
40. The Multicenter Post-Infarction Research Group: Risk stratification and survival after myocardial infarction. *N Engl J Med* 1983;309:331–336.
41. Gomes JA, Winters SL, Stewart D, et al.: A new noninvasive index to predict sustained ventricular tachycardia and sudden death in the first year after myocardial infarction: Based on signal averaged electrocardiogram, radionuclide ejection fraction and Holter monitoring. *J Am Coll Cardiol* 1987; 10:349–357.
42. Kuchar DL, Thorburn CW, Sammel NL: Prediction of serious arrhythmic events after myocardial infarction: Signal averaged electrocardiogram, Holter monitoring and radionuclide ventriculography. *J Am Coll Cardiol* 1987;9:531–538.
43. Furberg CD: Effect of antiarrhythmic drugs on mortality after myocardial infarction. *Am J Cardiol* 1983;52:32C–36C.
44. Naccarelli GV, Dougherty AH: Risk and benefits of antiarrhythmic therapy. *J Arrhythmia Management* 1990:33–39.
45. Kennedy HJ, Whitlock JA, Sprauge MK, et al.: Long-

term followup of asymptomatic healthy subjects with frequent and complex ventricular ectopy. *N Engl J Med* 1985;312:193–197.
46. Chakko CS, Gheorghiade M: Ventricular arrhythmias in severe heart failure: Incidence, significance and effectiveness of antiarrhythmic therapy. *Am Heart J* 1985;109:497–504.
47. Huang SK, Messer JV, Denes P: Significance of ventricular tachycardia in idiopathic dilated cardiomyopathy: Observations in 35 patients. *Am J Cardiol* 1983;51:507–512.
48. Wilson JR, Schwartz JS, St. John Sutton M, et al.: Prognosis in severe heart failure: Relation to hemodynamic measurements and ventricular ectopic activity. *J Am Coll Cardiol* 1983;2:403–410.
49. The CONSENSUS Trial Study Group: Effects of enalapril on mortality in severe congestive heart failure. *N Engl J Med* 1987;316:1429–1435.
50. Cleland JGF, Dorgie HJ, Findlay IN, Wilson JT: Clinical hemodynamic and antiarrhythmic effects of long-term treatment with amiodarone of patients in heart failure. *Br Heart J* 1987;57:436–445.
51. Graboys TB, Loawn B, Podrid PJ, DeSilva R: Long-term survival of patients with malignant ventricular arrhythmias treated with antiarrhythmic drugs. *Am J Cardiol* 1982;50:437–443.

Chapter **27**

Sudden Cardiac Death

A. Bayes de Luna, J. Guindo, M. Fiol, and J. M. Dominguez de Rozas

Cardiac death is the most frequent cause of mortality in the Western world. Of the different forms of cardiac death,[1] the sudden death stands out due to its dramatic presentation and the socioeconomic implications. The victims of sudden cardiac death may be apparently healthy people, although more often they are patients with heart disease who are enjoying a relatively good quality of life and playing an active role within their families and society. At present, the incidence of sudden death is decreasing due to progressive decline of ischemic heart disease in the United States and other industrialized countries,[2–7] which is believed to be due to changes in diet and life-style. In patients with established ischemic heart disease, this decline can also be attributed to improved medical therapy for patients at risk and more effective systems for resuscitating victims of out-of-hospital cardiac arrest. Nevertheless, sudden cardiac death continues to be responsible for nearly 400,000 cases per year in the United States, and still represents a major challenge confronting contemporary cardiology.[8]

DEFINITION

There is no universally accepted definition of sudden death. The term has been used variously by epidemiologists, clinicians, pathologists, and nonmedical personnel,[1–28] resulting in difficulties in the comparisons of published studies. The main difference concerns the determination of the time interval between initiation of symptoms and cardiac arrest.

In this chapter we do not discuss the sudden cardiac arrest that occurs in the course of acute heart disease, such as acute myocardial infarction, when the patient is already admitted to a coronary care unit. Neither do we discuss the sudden death that frequently occurs in the terminal phase of many illnesses.[29] In this chapter we are concerned with ambulatory sudden death, that is, death that appears suddenly and unexpectedly in an ambulatory patient who may have stable heart disease and whose condition is apparently stable. Within this group we include all patients who die in the hyperacute phase of myocardial infarction before admission to the coronary care unit.

Ambulatory sudden death can be defined as death of natural cause occurring unexpectedly and instantaneously or within the first hour of the onset of premonitory symptoms. The fundamental elements of this definition of sudden death are: (a) natural event, thus excluding accidents, homicide, and suicide; (b) unexpected occurrence in healthy persons or those with known heart disease but in an apparently stable condition before death; and (c) rapid progression to cardiac arrest after the onset of symptoms or acute decompensation.

The most debated point among these three elements is the time interval from the onset of premonitory signs and symptoms to death. As mentioned earlier, we require a 1-h time interval from the onset of premonitory symptoms to cardiac arrest. However, some authors[30–32] accept as sudden any death occurring within 6, 12, and even 24 h of the onset of symptoms after acute decompensation.

Sudden cardiac death only exceptionally appears in patients without apparent heart disease and heart disease is present in 90% of patients who die suddenly. In the remaining, sudden death is usually related to neurological problems, massive bleeding, and so forth.

Sudden cardiac death may be classified into two groups[24]: death associated with coronary disease and death associated with other cardiac causes. Ischemic heart disease is present in more than 80% of cases of ambulatory sudden death

655 Avenue of the Americas, New York, NY 10010
Current Topics in Cardiology

in the Western world.[33,34] Sudden death is the first manifestation of coronary heart disease in 20 to 25% of patients.[35] About 75% of those who die suddenly have an old myocardial infarction, but recent infarction is found only in about 20% of victims of sudden death.[12,36] Less frequently, sudden cardiac death occurs in patients with other heart diseases, such as hypertrophic or dilated cardiomyopathy, valvular heart disease, for example, aortic stenosis, mitral valve prolapse, long QT syndrome, Wolff-Parkinson-White syndrome, congenital heart disease, for example, after correction of Fallot's tetralogy, coronary anomalies, pulmonary embolism, dissecting aneurysm, or some special situations such as unusual effort or extreme obesity.

Regarding the clinical state prior to sudden cardiac death, we differentiate between prodromata and premonitory signs.[9] Prodromata can occur days or weeks before sudden death and include the new cardiac signs or symptoms, or worsening of those already present (chest pain, dyspnea, palpitations, dizziness, syncope, and so forth). Premonitory signs or symptoms of sudden death are those that appear as a result of acute decompensation. They are closely related to the final event and serve as time reference in the definition of sudden death.

Depending on the triggered mechanism, sudden death may be classified into two groups according to Hinkle and Thaler[1]: sudden death due to a primary arrhythmia and sudden death in which the terminal arrhythmia is secondary to a hemodynamic catastrophe (Table 27.1). In the first group are included subjects who collapse suddenly and the pulse stops without previous circulatory failure (arrhythmic death); in the second group, the pulse only stops after peripheral circulatory collapse has occurred (death from circulatory failure). Nevertheless, the study of Hinkle and Thaler included some patients who died suddenly, but with more than 1 h between the onset of symptoms and the final event. The most relevant clinical findings in their study[1] were:

a. Thirty-three percent of arrhythmic deaths and 10% of deaths due to circulatory failure occurred during an episode of acute myocardial ischemia, with a statistically significant difference.

b. Forty-five percent of arrhythmic deaths occurred in patients with a history of congestive heart failure, but without circulatory collapse prior to death.

c. Ninety percent of the terminal events that lasted less than 1 h from the onset of symptoms were arrhythmic deaths, while 74% of those lasting more than 1 h were circulatory deaths.

d. Eighty percent of out-of-hospital deaths were arrhythmic, and 71% of those occurring during hospitalization were due to circulatory collapse.

e. Ninety percent of deaths from heart disease were arrhythmic, while 86% in which the primary cause was other than heart disease were due to circulatory failure.

This study conclusively demonstrates that sudden death occurring in the first hour after the onset of symptoms is generally due to primary arrhythmia, occurs most commonly out of hospital, and is associated with heart disease, usually ischemic heart disease.

TABLE 27.1 Classification of Cardiac Deaths According to the Circulatory State Immediately Before Death

Categories	Total Deaths n = 142 (%)
I. Arrhythmic deaths (abrupt loss of consciousness and loss of pulse in the absence of previous circulatory collapse)	57.7
II. Death in circulatory failure (progressive circulatory failure and circulatory collapse before loss of pulse)	41.5
III. Unclassified deaths	0.7

Adapted from Hinkle and Thaler.[1]

SUDDEN CARDIAC DEATH: A MULTIFACTORIAL PROBLEM

Sudden cardiac death is usually induced by the cumulative effect of several factors. Some of these are better known than others, although their exact interaction is not clear.[37–40] From a clinical point of view it is obvious that the final cause of sudden death is malignant arrhythmia, more often a ventricular tachyarrhythmia than bradyarrhythmia. This malignant arrhythmia is induced by a series of modulating or triggering mechanisms acting upon vulnerable myocardium. In the subsequent text, we will consider the terminal arrhythmias responsible for sudden death, the factors that make the myocardium vulnerable, and the modulating or triggering mechanisms acting on the vulnerable myocardium that induce the malignant arrhythmia responsible for sudden death.

Terminal Arrhythmias

Ventricular fibrillation (VF) is most frequently responsible for sudden ambulatory death. This can be seen not only in patients with cardiac arrest in the coronary unit,[41,42] but in community-based emergency rescue systems and mobile coronary units,[36,43–49] and in patients who die suddenly while wearing Holter devices.[33,34,50–59] In the Miami study of 352 patients who suffered an out-of-hospital sudden death,[44,45] 62% had VF on the arrival of the mobile coronary care team which identified the rhythm within 4 min upon arrival. Sustained ventricular tachycardia (VT) with hypotension was present in 7%, and various bradyarrhythmias or asystole in 31% of cases. It should be emphasized that the prognosis of patients who suffered out-of-hospital cardiac arrest differs depending on whether it was associated with an acute myocardial infarction. In the 20% of patients in whom sudden death was associated with acute myocardial infarction, the prognosis is probably similar to other myocardial postinfarction patients which in the Seattle study[12,36,43] was associated with a sudden death risk of only 2% per year. However, in 80% of patients resuscitated from an out-of-hospital cardiac arrest without demonstration of acute myocardial infarction, the risk of recurrence in the first year was 22%.

To study the factors that may predispose to VF in acute myocardial infarction, Adgey et al.[49] examined 48 consecutive patients who developed VF outside the hospital after the arrival of a mobile coronary unit. An R on T extrasystole was the most important factor in the initiation of VF occurring in 33 patients (69%). In 6 cases (12%), VF was initiated by a late cycle extrasystole or idioventricular rhythm in only 9 cases (19%) by a classical VT. This clearly differs from our results in patients who experienced sudden ambulatory death while wearing Holter devices.[33,34] This difference may be due to the fact that most of Adgey's patients died due to VF secondary to acute myocardial infarction whereas few patients in our group had evidence of acute ischemia.

In the past years, isolated cases and small series of sudden cardiac death recorded in patients with ambulatory Holter devices have been published.[50–59] They have been particularly useful in providing information regarding the final events and the electrical triggering mechanisms of ambulatory sudden cardiac death. In a recent survey reviewing the Holter tapes of 233 patients

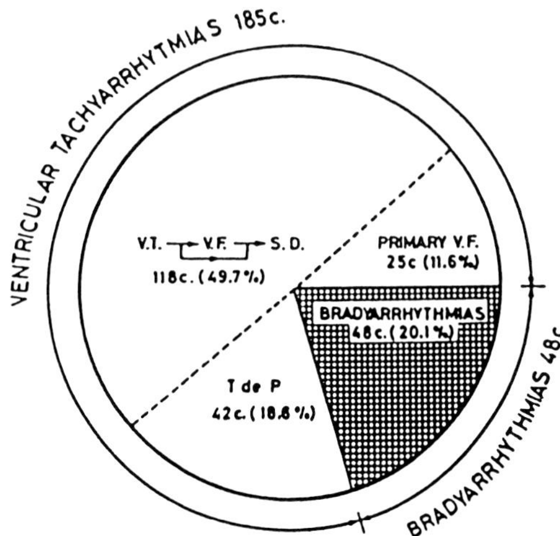

FIGURE 27.1 Causes of ambulatory sudden death recorded by Holter electrocardiography. VT = ventricular tachycardia, VF = ventricular fibrillation, Primary VF = primitive ventricular fibrillation, T de P = torsade de pointes.

who died suddenly[33,34] (Fig. 27.1), we observed that approximately 80% of deaths were due to ventricular tachyarrhythmia (10% primary VF, 50% classical VT that degenerated into VF, and somewhat less than 20% torsade de pointes). Patients who died from ventricular tachyarrhythmia could be divided into two well-defined subgroups. One was composed of individuals who died from torsade de pointes degenerating into VF (Fig. 27.2), and the other consisted of deaths due to VF which sometimes appeared abruptly (Fig. 27.3) but more often was preceded by classical VT (Fig. 27.4). Patients who died from some type of bradyarrhythmia, generally depression of sinoatrial (SA) node and subsidiary automaticity (Fig. 27.5), and less frequently advanced AV block, constituted the remaining 20% of deaths. These groups of patients were clearly differentiated and had specific clinical features (Table 27.2).

In the group with torsade de pointes, only 50% of patients had ischemic heart disease, more than 20% had no heart disease, and an iatrogenic cause was suspected in at least two-thirds of the cases. However, in the group with VF, proarrhythmia was suspected in approximately 10% of patients. In the Holter recordings of patients with torsade de pointes, slowing of the baseline heart

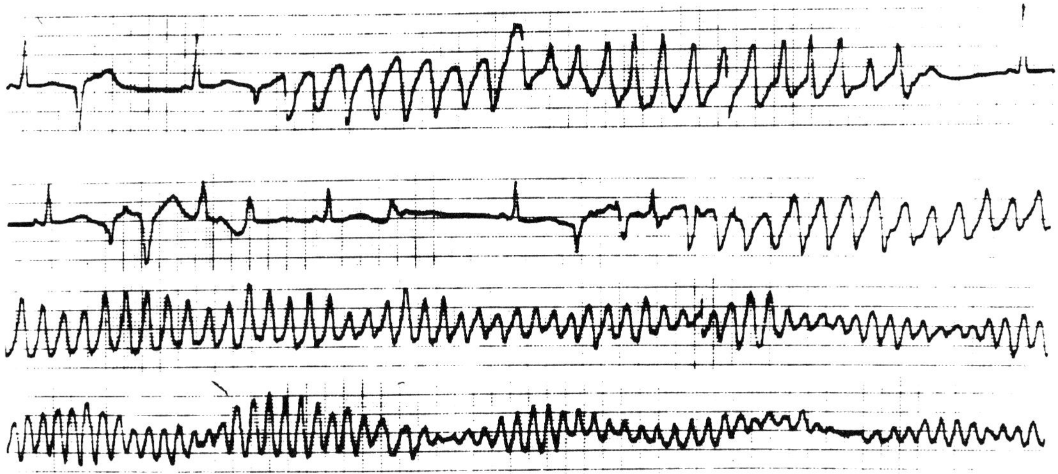

FIGURE 27.2 Typical pattern of torsade de pointes in a woman treated with quinidine for nonsustained VT in the absence of heart disease. A long sequence of torsade de pointes induces VF. *Reprinted from Bayés de Luna et al.,*[34] *with permission.*

rate was often detected in the hours preceding death, which may explain the fact that most deaths occurred at night.

The group with VF frequently had known heart disease (more than 80% ischemic heart disease) and poor left ventricular function. The Holter electrocardiograms (ECGs) showed that when the basal rhythm was sinus, as occurred in two-thirds of cases, it accelerated in the hours prior to death. Of the remaining cases, supraventricular tachycardia (SVT) appeared in some, while in others with preexisting atrial fibrillation, ventricular rate accelerated. Although most of these patients had known ischemic heart disease, only 12.7% had ST segment changes in the baseline ECG prior to sudden death. In contrast, in the group of bradyarrhythmias, ST segment changes prior to sudden death occurred in almost 90% of cases. The frequency of ventricular premature complexes that appeared in the Holter recordings prior to sudden death was highly variable, but cases with frequent or complex ventricular premature complexes prevailed over those in which they were isolated or absent. Moreover, their number increased significantly in the hour prior to the fatal event. The ventricular premature complex that triggered the fatal arrhythmia often followed a relatively long pause (Fig. 27.3). Both in this group and that of torsade de pointes, the patient would generally have survived if the arrhythmia could have been terminated.

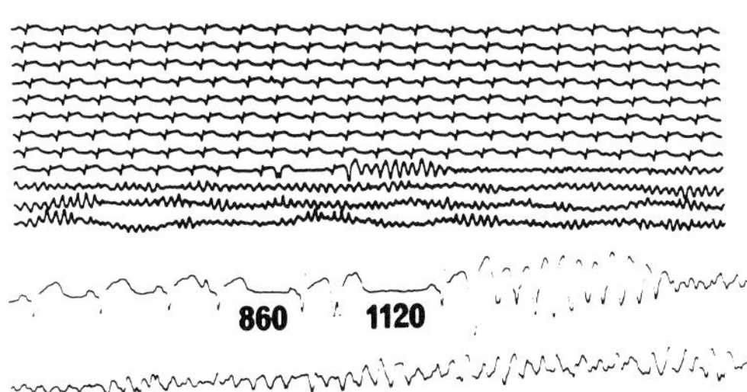

FIGURE 27.3 Ambulatory sudden death due to primary VF initiated by a ventricular extrasystole with short coupling interval occurring after a postextrasystolic pause (1120 ms). *Reprinted from Bayés de Luna et al.,*[34] *with permission.*

Patients in whom the final event was bradyarrhythmia usually had serious advanced heart disease with high incidence of congestive heart failure. Also, acute coronary insufficiency was common, often manifested by ST segment alterations, although in a few cases the bradyarrhythmia was reversible. When bradyarrhythmia was secondary to a hemodynamic catastrophe that was most often an event such as electromechanical dissociation, cardiac rupture, and so forth, patients did not survive. These cases corresponded to death due to circulatory failure of Hinkle and Thaler classification.[1]

Markers of Sudden Cardiac Death Caused by Malignant Ventricular Arrhythmias

Sustained VT or VF in a healthy individual without heart disease, and not using potentially ar-

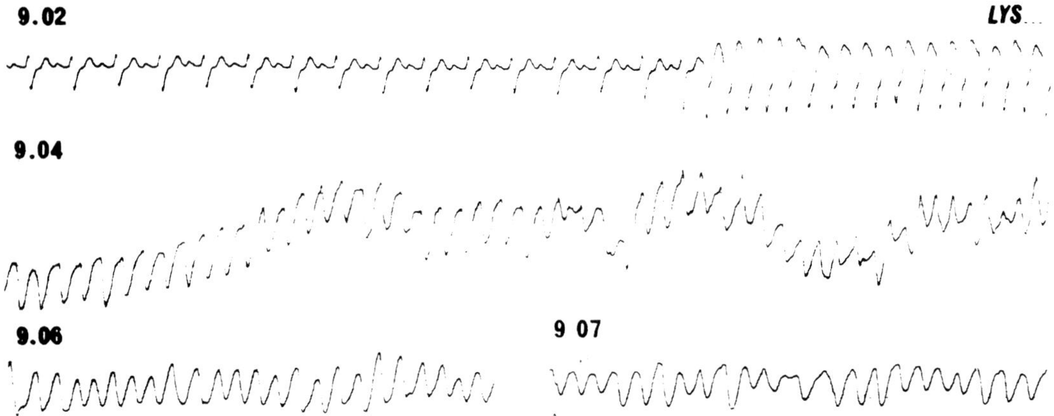

FIGURE 27.4 Ambulatory sudden death due to VF in a coronary patient treated with amiodarone for complex premature ventricular contraction. A monomorphic sustained VT occurred at 9:02 AM and VF ensued at 9:04 AM after an increase of VT rate and QRS duration. *Reprinted from Bayés de Luna et al.,*[34] *with permission.*

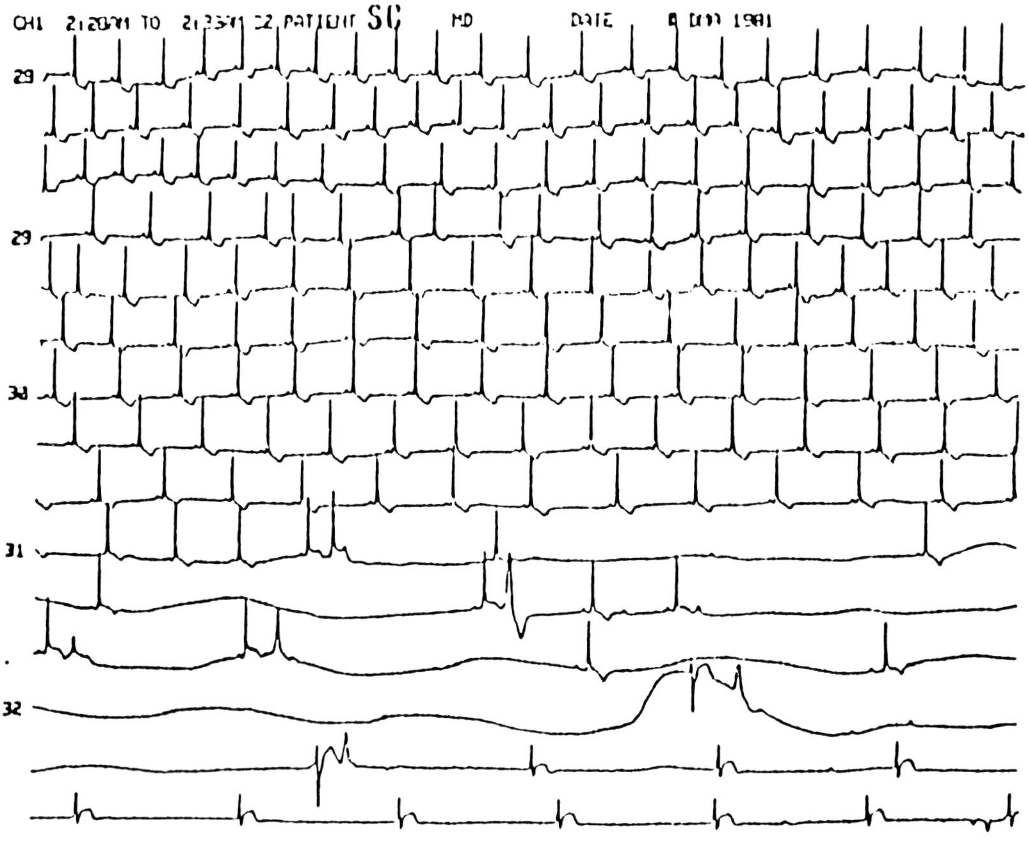

FIGURE 27.5 Ambulatory sudden death due to a progressive depression of sinus and subsidiary automaticity in a postmyocardial infarction patient in subacute phase. The primary cause of death was cardiac rupture with electromechanical dissociation.

TABLE 27.2 Clinical Characteristics of Different Groups of Ambulatory Sudden Death

	Ventricular Tachyarrhythmias		Bradyarrhythmias
	Group I VT-VF (%)	Group II T de P (%)	Group III (%)
Number of cases	143 (61.3)	42 (18.6)	48 (20.1)
Mean age (years)	70	60	70
Sex (males)	76	40	39
Coronary artery disease	84	50	70
No known heart disease	4	22	10
Antiarrhythmic treatment	63.8	76	?
Proarrhythmia (probable)	10	66	?

VT-VF = Classical ventricular tachycardia which triggers ventricular fibrillation or primary ventricular fibrillation, T de P = torsade de pointes.

rhythmogenic drugs, has been very rarely observed. In such cases, exhaustive studies have not been always performed to rule out possible alterations of the autonomic nervous system or other factors that could explain the final event, for example, concealed form of the long QT syndrome. Thus, a vulnerable myocardium is almost always present in victims of malignant ventricular arrhythmias and ambulatory sudden cardiac death. We will now consider the characteristics of vulnerable myocardium, first in postmyocardial infarction patients and then in another group of patients.

Postinfarction patients. Figure 27.6 shows different markers of vulnerable myocardium and the triggering factors that may induce malignant ventricular arrhythmias in the postmyocardial infarction patients. Following are the most important characteristics of these different markers:

a. Anatomic. The scar of myocardial infarction and ventricular aneurysm facilitate the appearance of sustained reentrant ventricular arrhythmias.[60,61] The mechanism is supported by signal-averaged electrocardiogram[62,63] and electrophysiological studies.[64] Also, the severity of coronary artery disease parallels the risk of sudden death.[65] Patients with three-vessel disease or left main coronary artery disease are at greatest risk of sudden death. Left ventricular hypertrophy is another risk factor for sudden death. The Framingham study has shown that left ventricular hypertrophy is frequently associated with complex ventricular arrhythmias and is a marker of sudden cardiac death in patients with ischemic heart disease, independent of blood pressure.[66,67] In the study of Cooper et al.,[68] left ventricular hypertrophy was associated with worse survival, regardless of ventricular function and the number of severely narrowed coronary arteries, but the data were insufficient to distinguish sudden from nonsudden cardiac death.

b. Functional. Depressed left ventricular function contributes to increased vulnerability and has important prognostic implications.[69,70] Recent multicenter studies have established that left ventricular ejection fraction below 40 to 45% is an independent marker of total and cardiac sudden death.[71,72] In patients with very low ejection fraction and with advanced heart failure (NYHA class III or IV), the risk of sudden cardiac death is very high and the annual mortality generally ranges from 10 to 50%, depending on the NYHA functional class (30 to 50% in NYHA classes III to IV)[73,74]; approximately half of these deaths are sudden.

c. Ischemic. More than 80% of ambulatory sudden cardiac deaths occur in patients with ischemic heart disease.[33,34] In postmyocardial infarction patients residual ischemia is a distinct risk marker of new ischemic events and cardiac death. Postinfarction angina is associated with greater cardiac mortality, although no increase in sudden death has been demonstrated due to the small sample size in the reported studies.[75–78] However, in postmyocardial infarction patients, both in the CASS study[79] and in Theroux et al.'s study,[80] the ST segment depression during the exercise test was associated with increased risk, independent from the presence of angina.

Both the anatomic and the functional data reviewed above make it evident that sud-

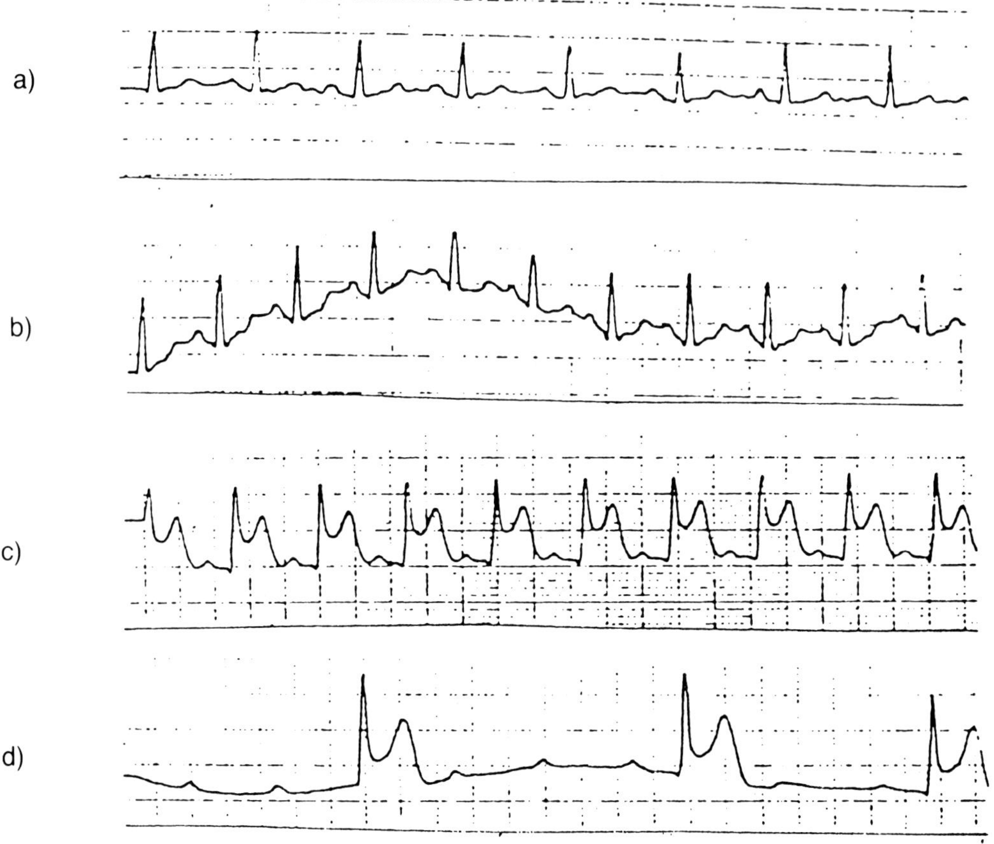

FIGURE 27.6 a to d. Paroxysmal AV block during a crisis of Prinzmetal angina induced by exercise testing. *Reprinted from Bayés de Luna A. Textbook of clinical electrocardiography. Martinus Nijhoff Publishing, Dordrecht 1988, with permission.*

den cardiac death appears in presence of silent or symptomatic "ischemic milieu." It has been shown that cardiac arrest,[81] or sudden cardiac death[82] often appear in patients with silent myocardial ischemia, but this does not mean that silent ischemia is the trigger of sudden death (see later).

d. Electrical. Frequent or complex premature ventricular complexes[71,72,83–88] are reported to be an independent marker of poor prognosis. Ruberman et al.[84] found that they were associated with higher mortality, regardless of presence or absence of congestive heart failure, but the mortality rate was still higher when both were present. Bigger[71,85] has shown that the number of premature ventricular complexes per hour is an independent marker of sudden cardiac death. When their number rises from 1/h to 10/h, mortality increases appreciably but tends to stabilize from then on. The importance of premature ventricular contractions as markers of sudden cardiac death was also found in the MILIS study.[72] Both in the MILIS and MPIP studies,[72,85] the risk was greater in the presence of depressed left ventricular function. In patients with acute myocardial infarction, the appearance of intraventricular conduction disturbance, particularly right bundle branch block or bifascicular block, represents a risk factor for sudden cardiac death in the following weeks. Contrary to initial belief, sudden death in these patients is not usually caused by bradyarrhythmias, but by ventricular tachyarrhythmias.[89–93]

e. Autonomic nervous system. Various alterations of the autonomic nervous system,[94–106] for example, lack of R-R interval variability,[95,96] long QT interval,[97–102] and baroreflex sensitivity,[102–106] have been implicated as markers of malignant ventricular arrhythmias. Kleiger et al.[95,96] observed that in patients with absent or less than 50 ms R-R interval variability in the surface ECG mortality was 50%, but with more than 100 ms variability was 15%. Long QT interval in the surface ECG is also a possible marker of sudden cardiac death.[97,98] In a recent pilot study in postmyocardial infarction patients, we have shown that the dynamic behavior of the QT interval in Holter tapes may stratify risk of malignant ventricular arrhythmias. Patients with frequent peaks of QTc interval of more than 500 ms are much more likely to develop malignant ventricular arrhythmias.[99,100] A recent study[101] reports an abnormal repolarization response to exercise, that is, a paradoxical increase in the QTc intervals in patients who experience a proarrhythmic effect using type IC antiarrhythmic agents. In addition, experimental and clinical studies have shown that reduced baroreflex sensitivity after myocardial infarction is a strong risk predictor of sudden cardiac death.[102–106]

f. Coronary risks factors. Multivariate analysis in the Framingham study[107,108] has documented that the classical risk factors of ischemic heart disease represent a marker of sudden death in general population. In postinfarction patients, the predictive value of blood pressure and lipid alterations is considerably less for two reasons: decreased blood pressure may be an expression of poor contractility, and different drugs may interfere with lipid metabolism. In some studies, however, it has been demonstrated that serum lipid levels,[109,110] high blood pressure,[111] and cigarette smoking[112] influence morbidity and mortality. From all such studies in postinfarction patients, only hypercholesterolemia was proved to be associated with an increase in both total and sudden cardiac death.[109]

Patients with ischemic heart disease other than those with myocardial infarction have the same risk markers as the postinfarction patients. However, these patients are at a lesser risk of "malignant" ventricular arrhythmia, particularly classical sustained VT, because they have no myocardial scars or ventricular aneurysms. In patients with stable angina pectoris, the risk of myocardial infarction parallels the risk of sudden cardiac death. Patients with vasospastic angina also are at risk of sudden death, although to lesser extent, and the risk has been reduced considerably by the use of calcium (Ca^{2+}) channel blockers.[113,114] Isolated cases of sudden cardiac death have been described in patients with coronary spasm without associated coronary lesions.[115] It has been also suggested that patients with normal coronary vessels and coronary spasm may have more reperfusion arrhythmias.[116] However, the clinical importance of reperfusion arrhythmias is less certain even though isolated cases of VF after reperfusion have been described.[117]

In patients with hypertension, left ventricular enlargement contributes to myocardial vulnerability and has been associated with ventricular arrhythmias and sudden cardiac death.[118–120] This also partly explains myocardic vulnerability in patients with ischemic heart disease.

In patients with hypertrophic cardiomyopathy, myocardial vulnerability, besides left ventricular enlargement, can be also related to the special arrangement of the myocardial fibers.[121–123a) It has been shown recently that fiber disarray without myocardial hypertrophy may be associated with sudden death.[123a] Cripps et al.[123b] recently showed that presence of late potentials in signal-averaged ECG may be useful in the prediction of sudden death, particularly in the young patients with hypertrophic cardiomyopathy.

In patients with dilated cardiomyopathy, the vulnerability of the myocardium is mostly related to depressed left ventricular function and ventricular arrhythmias, although the latter is subject to much controversy.[124–130] Some investigators consider ventricular arrhythmias as independent markers of sudden cardiac death,[124–127] but others do not.[128–130] We believe that although the risk of sudden death fundamentally depends on the degree of left ventricular dysfunction, the presence of ventricular arrhythmias is an additional independent risk marker.

Arrhythmogenic right ventricular dysplasia is a special type of dilated cardiomyopathy of the right ventricle with risk of sustained ventricular tachycardias and sudden death.[131–133] This cardiomyopathy needs to be considered in all patients with frequent premature ventricular complexes with left bundle branch block morphology.

Sudden cardiac death may also occur in patients with aortic valve stenosis[134–136] or mitral valve prolapse.[137,138] Nevertheless, considering the high prevalence of mitral valve prolapse in the general population, as demonstrated by echocardiography and frequent presence of complex ventricular arrhythmias, the incidence of sudden death is very low. Patients at increased risk of sudden death are those with significant mitral regurgitation, myxomatous valve degeneration, repolarization alterations in the inferior leads, complex ventricular arrhythmias, history of syncope, or family history of sudden cardiac death.[137,138]

Sudden cardiac death can occur occasionally in patients with certain congenital heart diseases, for example, coronary artery anomalies,

Ebstein's disease, primary pulmonary hypertension, and more often in postoperative patients with tetralogy of Fallot, transposition of the great vessels, complete AV canal, double-outlet right ventricle, interatrial septal defect, and so forth.[139–144]

In patients with Wolff-Parkinson-White syndrome,[145–150] the myocardium is vulnerable due to the presence of anomalous conduction pathways with short refractory periods. In these cases atrial fibrillation with a very rapid ventricular response can induce malignant ventricular arrhythmias, particularly in the presence of multiple accessory pathways and/or heart disease. Table 27.3 shows clinical and electrophysiological features enhancing risk of VF in patients with Wolff-Parkinson-White syndrome, as determined by the results of the European Registry of Sudden Death in the Wolff-Parkinson-White syndrome.[145]

In the case of a long QT syndrome, prolonged and nonhomogenous repolarization (vulnerable myocardium) facilities the appearance of malignant ventricular arrhythmias,[151–155] which may be precipitated by physical or physiological stress (modulating factors).

Dissection of the aorta and massive pulmonary embolism can result in sudden death.[156–161] Dissection of the aorta in young patients is often associated with Marfan's syndrome,[156,157] and in the elderly with arterial hypertension of long duration.[158] Massive pulmonary embolism is usually thromboembolic,[159] but may be caused by fat, for example, after orthopedic surgery, gas, or amniotic fluid (for example, during childbirth).[160,161]

Finally, in a very small group of patients, there are no cardiac or other alterations detectable by the present diagnostic methods. Possible causes include subclinical myocarditis,[162–164] sudden onset of severe ionic or metabolic alteration, and/or an abnormality of the autonomous nervous system, such as acute lengthening of the QT interval. In these cases, an apparently normal myocardium may become vulnerable and an appropriate trigger such as physical or mental stress could cause sudden cardiac death.

Triggers of Malignant Ventricular Arrhythmias

One or more markers of vulnerable myocardium may exist for a long time without an appearance of a malignant arrhythmia as long as there is no triggering mechanism (Fig. 27.6).

Following are the most important triggering mechanisms of malignant ventricular arrhythmias:

a. Changing function of the autonomic nervous system. Although different parameters modulated by the autonomous nervous system may be considered as markers of malignant ventricular arrhythmias,[94–106] the increase of these markers or the changes in the function of the autonomic nervous system may trigger the final event. Studies in patients wearing a Holter device have shown[33,34] that heart rate increased before the onset of ventricular tachyarrhythmias resulting in VF, which suggests increased sympathetic activity. Other authors[165] have not reported this finding perhaps because they have not compared the mean heart rate within the hour prior to the event with that of earlier hours as we have done. The presence of faster heart rates detected by Holter recording during the hour before sudden death suggests that β-blockers could prevent sudden cardiac death by lowering the basal heart rate in postmyocardial infarction patients.

b. Physical and mental stress. Sudden cardiac death can be triggered by mental stress,[166–171a] but this happens only occasionally. Also,

TABLES 27.3 Patients with Wolff-Parkinson-White Syndrome and Risk of Ventricular Fibrillation

	p	Specificity (%)	Sensitivity (%)
Clinical features			
1. Documentation of more than one type of spontaneous SVT	<0.01	96	26
2. Documentation of spontaneous AFl	<0.01	98	22
3. Documentation of AFl + spontaneous AF	<0.05	98	13
Electrophysiological features			
1. Shorter R-R interval during induced AF, <180 ms	<0.001	96	60
2. More than one anomalous pathway	<0.01	96	27
3. Shorter CL during 1:1 VA conduction by the anomalous pathway, <240 ms	<0.01	87	66

SVT = supraventricular tachycardia, AFl = atrial flutter, AF = atrial fibrillation, CL = cycle length, VA = ventriculoatrial, ms = milliseconds.

exercise can trigger sudden cardiac death both in patients with ischemic heart disease and in those with other diseases, although the incidence of malignant ventricular arrhythmias during exercise testing is low.[172–175] Both physical and mental stress are known triggers of sudden cardiac death in patients with a long QT syndrome.

c. The appearance or increased severity of ischemia or premature ventricular complexes. Both Lewis [56] and we[33,34] found a frequent increase in the number of premature ventricular complexes before sudden cardiac death.

Both symptomatic and silent acute ischemia are believed to be responsible for the majority of sudden deaths in patients with acute myocardial infarction (4 to 18% of patients admitted to a coronary unit during the first hour of infarction).[176–178] However, acute ischemia is not often responsible for ambulatory sudden death in other settings. Direct evidence for association between an acute ischemic crisis and sudden death is limited to case reports.[179–181] In our experience,[33,34] the incidence of ST segment alterations before ambulatory sudden death is low, although it varies depending on the type of lethal arrhythmia. Only 12.5% of patients who died due to VF showed previous repolarization changes. In contrast, about 90% of patients who died due to bradyarrhythmias showed ST segment alterations that suggested acute ischemia before the final arrhythmia. Therefore, only 30% of patients who died suddenly wearing a Holter device had evidence of ischemia as a triggering mechanism. We assume that ST segment alterations represent true ischemia because the studies of Deanfield[182,183] identified the ST depression during Holter monitoring (symptomatic or silent) as evidence of true myocardial ischemia. Also, each of the four patients who had acute myocardial infarction while wearing a Holter device showed marked ST elevation.[181] This was also observed in many of our patients who died suddenly due to bradyarrhythmias. Admittedly, acute myocardial infarction may be silent,[181] but absence of electrocardiographic changes is extremely rare.

The low probability of new ischemic crises being responsible for the majority of cases of ambulatory sudden death is also supported by the very low incidence of malignant ventricu lar arrhythmias during positive exercise tests,[184] percutaneous coronary angioplasty (PTCA),[185] and Prinzmetal's angina,[113] the low incidence of cardiac symptoms before sudden death,[186] the low incidence of myocardial infarction in out-of-hospital sudden death,[12] the relatively low incidence of sudden cardiac death during acute ischemia in the Hinkle and Thaler study,[1] and the low incidence of ventricular arrhythmias during subendocardial ischemia.[187–189]

In patients who die due to VF preceded by a monomorphic sustained ventricular tachycardia, which includes the majority of ambulatory sudden deaths,[33,34] ischemia probably does not play an important role because it is not a requirement for induction of sustained ventricular tachycardias, nor has it been shown that ischemia induces late potentials.[190] More likely, ischemia plays a role in malignant ventricular arrhythmias induced by exercise and in VF of abrupt onset. Also it is possible that mild ischemia undetectable by current methods may induce malignant arrhythmias when associated with other triggers. This hypothesis merits further studies.

The appearance of paroxysmal supraventricular arrhythmias or pauses following premature ventricular or supraventricular complexes may also trigger malignant ventricular arrhythmias. This has been shown by other authors and by us.[33,34,191] Furthermore, as previously mentioned, in patients with Wolff-Parkinson-White syndrome, sudden death is almost always secondary to rapid atrial fibrillation which degenerates into VF.[145–150]

Electrolyte imbalance or metabolic derangements can act as triggers precipitating malignant ventricular arrhythmias, especially when premature ventricular impulses and a vulnerable myocardium are present.[192,193] Recent investigations suggest that electrolyte imbalance may be an important immediate precipitating cause of fatal ventricular tachyarrhythmias in patients with severe left ventricular dysfunction and preexisting structural, hemodynamic, or neurohormonal factors.[192]

d. Drugs. The administration of drugs, particularly antiarrhythmic agents, may often trigger ambulatory sudden death, usually in the presence of electrolyte imbalance. As stated previously, proarrhythmia is suspected in approximately 66% of cases of ambulatory sudden death manifested by torsade de pointes but only in 10% of cases of VF.[33,34] The unfavorable results of the CAST study[194] and the repeated evidence of the danger of proarrhythmia resulting from administration of several drugs,[195–204] may be expected to decrease its incidence in the future.

Markers and Triggering Mechanisms of Sudden Death Due to Bradyarrhythmias

Patients who die suddenly due to bradyarrhythmia as the final event have advanced heart disease. Survival is very unlikely, even if the bradyarrhythmia is corrected by implanting a pacemaker, because such patients often have severe congestive heart failure, massive myocardial infarction, or mechanical complication of acute infarction which facilitates electromechanical dissociation. As stated earlier, these patients usually belong to type II of Hinkle and Thaler's classification.[1]

In our series of patients with ambulatory sudden death,[33,34] AV block was present in only 20% of cases who died due to bradyarrhythmia (almost 4% of all deaths). In some of these cases, and in a few patients with sick-sinus syndrome,[205] bradyarrhythmias were reversible, but in the majority bradyarrhythmia was caused by irreversible depression of the sinus rhythm and subsidiary automaticity as a result of an irreversible catastrophe.

Patients with cardiac arrest due to reversible bradyarrhythmia account for very few cases of ambulatory sudden cardiac death, mainly because pacemakers are implanted in time. However, vulnerable myocardium is also believed to exist in some cases of sudden cardiac death due to reversible bradyarrhythmias. In these, guidelines for pacemaker implantation indications have been published elsewhere.[206] In the absence of clinical findings, the following ECG findings may indicate implantation of pacemakers:

a. True trifascicular block,[207] that is, alteration of right bundle branch block plus hemiblock of the anterior division, with right branch block plus hemiblock of the posterior division.

b. Paroxysmal AV block[208] with several blocked P waves or long pauses, due to depressed sinus automaticity, particularly when it occurs during daily activities. At rest it may be asymptomatic, particularly in the elderly.

c. Some types of bifascicular block have an elevated risk of producing advanced atrioventricular block: masked bifascicular block, that is bifascicular block without S wave in V1 and V2[209] and bifascicular block with intrahisian block.[210]

Risk markers of ambulatory sudden death due to bradyarrhythmia differ from patient to patient, but sudden death does not usually occur until some triggering mechanisms or modulating factors appear. In our opinion, ischemic crises, extreme vagal tone, and the administration of different drugs are particularly important as modulating and/or triggering mechanisms of reversible malignant bradyarrhythmias. These triggering mechanisms sometimes induce potentially lethal bradyarrhythmia in the absence of evident markers.

There is no doubt that acute myocardial ischemia, particularly if severe and of long duration as occurs in acute inferior infarction, may occasionally trigger bradyarrhythmia, especially paroxysmal AV block which is responsible for a few cases of sudden death. Transient acute ischemia, particularly that related to coronary arterial spasm, may also trigger paroxysmal AV block and even sudden death (Fig. 27.7).

Vagal tone has a depressing effect on automaticity and/or conduction, and if depression is extreme, either spontaneous or induced by carotid sinus hypersensitivity, it may cause syncope,[211] although it rarely produces sudden death.

All antiarrhythmic agents and some other cardioactive drugs can induce severe reversible

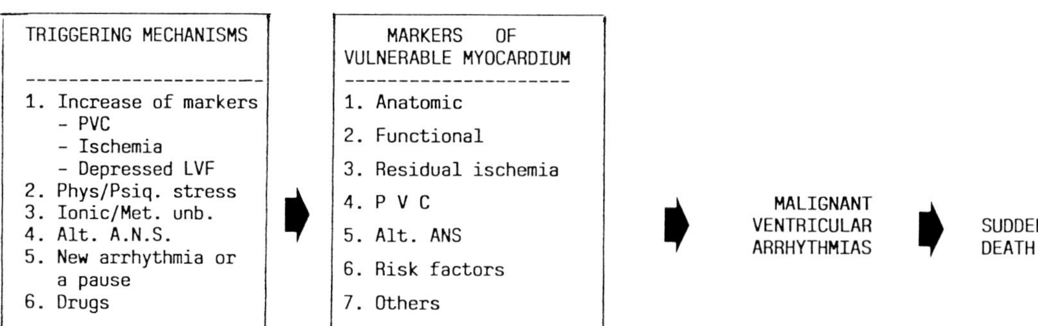

FIGURE 27.7 The multifactorial problem of sudden cardiac death due to malignant ventricular arrhythmias in postmyocardial infarction patients. In the presence of a vulnerable myocardium, the triggering mechanism leads to malignant ventricular arrhythmias and sudden cardiac death.

bradyarrhythmia. We think that malignant bradyarrhythmias may explain at least some cases of late proarrhythmia in the CAST trial.[203] It is well known that the high antiarrhythmic efficacy of type IC antiarrhythmic drugs is accompanied by marked depression of automaticity and cardiac conduction. However, there are no references in any of the numerous comments and editorials which have appeared since the publication of CAST[195–202] to the possibility that severe bradyarrhythmia may explain late proarrhythmia in some cases. The following findings could be of interest to test this hypothesis.

a. To determine whether history of sustained VT was similar in the group treated with flecainide and encainide and in the placebo group. If all deaths in the CAST study were due to malignant ventricular arrhythmias, we would expect to find greater incidence of history of sustained VT in the treated group.

b. To determine whether bradycardia or AV block or long pauses in the Holter ECG were present in the last control ECG before death.

c. Occurrence of sudden death during the night would also suggest that the primary bradyarrhythmia or one leading to torsade de pointes was the cause of sudden death.

We know that not many CAST patients had sustained VT,[212] but the remaining questions are yet to be answered.

HOW TO DETECT PATIENTS AT RISK OF SUDDEN CARDIAC DEATH

Many methods and techniques are available to detect patients at risk. Their value varies depending on the underlying heart disease. These methods and procedures are summarized in Table 27.4.[213–262]

It is important to emphasize that since sudden cardiac death is a multifactorial problem, a multifactorial approach for risk stratification is mandatory. Several studies have shown that the combined use of different parameters increases sensitivity, specifity, and positive predictive accuracy in detection of patients at risk.[249–251] In the study of Gomes et al.,[249] patients with abnormal signal averaging, ejection fraction below 40%, and complex ventricular arrhythmias had 50% incidence of sustained VT or sudden death at 1 year, whereas no arrhythmic events occurred in patients in whom all three variables were normal. Kuchar et al.[250] showed that an abnormal signal-averaged ECG in the presence of an ejection fraction less than 40% identified patients with a 35% probability of sudden death or sustained symptomatic VT during a mean follow-up of 14 months. By contrast, in patients with left ventricular dysfunction but normal signal-averaged tracing, the risk of arrhythmic events was 4%. Cripps et al.[251] showed that in postmyocardial infarction patients, the combination of Killip class >II, presence of late potentials, and positive effort test had 100% positive predictive accuracy for sudden death or symptomatic VT, whereas presence of late potentials alone had a positive predictive value of only 22%.

In future years, our aim should be to develop risk indices to identify more accurately high-risk individuals to apply the most effective preventative procedures to prevent sudden cardiac death.

TABLE 27.4 Methods to Detect Patients at Risk of Sudden Cardiac Death

Clinical history and physical examination.[213–221] Age, coronary risk factors, dyspnea, rales, syncope, angina.

Surface Electrocardiogram.[97,98,222–230] Left ventricular hypertrophy, ST depression, sinus-atrial, atrioventricular, and intraventricular block, localization and number of Q waves, long QT interval, sinus bradycardia, sinus tachycardia, premature ventricular contractions.

Exercise testing.[79,80,101,184,231–240] ST segment depression, blood pressure response (exercise hypotension), poor exercise capacity, induced serious ventricular arrhythmias, induced atrioventricular or ventricular block, paradoxical increase in QT interval.

Holter ECG.[33,34,50–59,71,72,95,96,99,100,113,187–189,241] Heart rate, ST segment alterations, RR variability, premature ventricular contractions, dynamic QT.

Electrophysiological studies.[64,242–247] Induction of sustained ventricular tachycardia, fragmented activity, sinus node dysfunction, conduction abnormalities, presence of rapid and multiple accessory pathways.

Signal-Averaging ECG.[62,63,190,248–251] Late potentials.

Radionuclide ventriculography.[70–72,252–257] Depressed left ventricular ejection fraction (basal and/or lack of increase during exercise), wall motion abnormalities, ventricular aneurysm.

Echocardiography-Doppler.[258,259] Depressed ventricular ejection fraction, wall motion abnormalities, ventricular aneurysm, valvular dysfunction.

Cardiac catheterization.[69,260–262] Extent of coronary artery obstruction, depressed left ventricular ejection fraction, wall motion abnormalities, ventricular aneurysm, valvular dysfunction.

REFERENCES

1. Hinkle LE, Thaler HT: Clinical classification of cardiac deaths. *Circulation* 1982;3:457.
2. Goldberg RJ: Declining out-of-hospital sudden cor-

onary death rates. Additional pieces of the epidemiologic puzzle. *Circulation* 1989;79:1369.
3. Stern MP: The recent decline in ischemic heart disease mortality. *Ann Intern Med* 1979;91:630.
4. Kuller LH, Peper JA, Dai WS, et al.: Sudden death and the decline in coronary heart disease mortality. *J Chronic Dis* 1986;39:1001.
5. Gillum RF: Sudden coronary death in the United States 1980–1985. *Circulation* 1989;79:756.
6. Gillum RF: Geographic variation in sudden coronary death. *Am Heart J* 1990;119:380.
7. Goldstein S: Toward a new understanding of the mechanisms and prevention of sudden death in coronary heart disease. *Circulation* 1990;82: (in press).
8. Lown B: Sudden cardiac death: The major challenge confronting contemporary cardiology. *Am J Cardiol* 1979;43:313.
9. Myerburg RJ, Castellanos A: Cardiac arrest and sudden cardiac death, in Braunwald E (ed): *Heart Disease*. Philadelphia, WB Saunders, 1987, p 742.
10. Josephson ME: *Sudden Death. Cardiovascular Clinics.* Philadelphia, FA Davis, 1985.
11. James TN: 15th Bethesda Conference Report: Sudden Cardiac Death. *J Am Coll Cardiol* 1985;5:1B.
12. Cobb LA, Werner JA: Mechanisms, predictors and prevention of sudden cardiac death, in Hurst JW (ed): *The Heart.* New York, McGraw-Hill, 1990, p 604.
13. Iturralde Torres P: Muerte súbita. *Arch Inst Cardiol Mex* 1986;56:539.
14. Morganroth J, Horowitz LN: *Sudden Cardiac Death.* London, Grune & Stratton, 1985.
15. Goldstein S: The necessity of a uniform definition of sudden coronary death: Witnessed death within 1 hour of the onset of acute symptoms. *Am Heart J* 1982;103:156.
16. Prineas RJ, Blackburn: Sudden coronary death outside hospital. *Circulation* 1975;52(suppl III):1.
17. Corday E, Dodge HT: Symposium on identification and management of the candidate for sudden cardiac death. *Am J Cardiol* 1975;39:813.
18. Goldstein S: Sudden death and coronary heart disease. Mount Kisco, NY, Futura Publishing, 1974.
19. Sonneblick EH, Lesch M: *Sudden Cardiac Death.* New York, Grune & Stratton, 1981.
20. Zipes DP: Sudden cardiac death. *Am J Med* 1981;70:1151.
21. Greenberg HM, Dwyer EM: Sudden coronary death. *Ann NY Acad Sci* 1982;382:1.
22. Lown B: Sudden cardiac death 1978. *Circulation* 1979;60:1593.
23. Taussig AS, Whitworth HB: Sudden cardiac death, in Hurst JW (ed): *Clinical Essays on the Heart.* New York, McGraw-Hill, 1985, vol. 5, p 159.
24. Roberts WC: Sudden cardiac death: Definition and causes. *Am J Cardiol* 1986;57:1
25. Guarnieri T, Levine JH, Griffith LSC, Velhi EP: When "sudden cardiac death" is not so sudden: Lessons learned from the automatic implantable defibrillator. *Am Heart J* 1988;115:205.
26. James TN, Viklert AM: The Fourth USA/URSS Symposium on Sudden Cardiac Death. *J Am Coll Cardiol* 1986;8(suppl A):1.
27. Dimarco JP, Haines DE: Sudden cardiac death. *Curr Probl Cardiol* 1990;15:183.
28. Guindo J, Bayés de Luna A: Muerte súbita en la cardiopatia isquémica. *Rev Esp Cardiol* 1989;42:116.
29. Wang F, Lien W, Fong T, et al.: Terminal cardiac electrical activity in adults who die without apparent cardiac disease. *Am J Cardiol* 1986;58:491.
30. Lie JT, Fitus JL: Pathology of the myocardium and the conduction system in sudden cardiac death. *Circulation* 1975;51(suppl III):41.
31. Reichenbach DD, Moss NS, Meyer E: Pathology of the heart in sudden cardiac death. *Am J Cardiol* 1987;39:865.
32. Davies MJ, Thomas A: Thrombosis and acute coronary artery lesions in sudden cardiac ischemic death. *N Engl J Med* 1984;310:1137.
33. Bayés de Luna A, Coumel Ph, Leclercq JF: Ambulatory sudden death: Mechanisms of production of fatal arrhythmia on the basis of data from 157 cases. *Am Heart J* 1989;117:151–159.
34. Bayés de Luna A, Guindo J, Rivera J: Ambulatory sudden death in patients wearing Holter devices. *J Ambulat Monitoring* 1989;2:3.
35. Kannel WB, Doyle JT, McNamara PM, Quickenton P, Gordon T: Precursors of sudden coronary death: Factors related to the incidence of sudden death. *Circulation* 1975;51:606.
36. Cobb LA, Baum RS, Alvarez H, Schaffer WA: Resuscitation from out-of-hospital ventricular fibrillation: 4 years follow-up. *Circulation* 1975;51(suppl III):223.
37. Coumel P: Factors responsible for sudden death: Triangle or polygon? in Piccolo E, Raviele A, Alboni P (eds): *Aritmie cardiache.* Venecia, Centro Scientifico Editore, 1989, p 279.
38. Wellens HJJ, Brugada P: Sudden cardiac death: A multifactorial problem, in Brugada P, Wellens HJJ (eds): *Cardiac Arrhythmias: Where to Go from Here?* New York, Futura Publishing, 1987, p 391.
39. Bayés de Luna A, Guindo J: *Sudden Cardiac Death.* Barcelona, MCR Publishers, 1989.
40. Surawicz B: Prognosis of ventricular arrhythmias in relation to sudden cardiac death: Therapeutic implications. *J Am Coll Cardiol* 1987;10:435.
41. Oliver MF, Julian DG, Donald KW: Problems in evaluating coronary care units. Their responsibility and their relation to the community. *Am J Cardiol* 1967;20:465.
42. Campbell RWF, Murray A, Julian DG: Ventricular arrhythmias in first 12 hours of acute myocardial infarction. Natural history study. *Br Heart J* 1981;46:351.
43. Greene HL: Sudden arrhythmic cardiac death—Mechanisms, resuscitation and classification: The Seattle perspective. *Am J Cardiol* 1990;65:4B.
44. Myerburg RJ, Zaman L, Luceri RM, Kessler KM, Estes D, Trohman R, Horgan J, Castellanos A: Clinical characteristics of sudden death. Implications on survival, in Josephson ME (ed): *Sudden Cardiac Death.* Philadelphia, FA Davis, 1985, p 107.
45. Myerburg RJ, Kessler KM, Zaman L, et al.: Survivors of prehospital cardiac arrest. *JAMA* 1982;247:1485.
46. Geddes JS: *The Management of Acute Coronary Attack.* London, Academic Press, 1986.
47. Chamberlain DA, White NM, Binning R, Parker WS, Kimber ER: Mobile coronary care provided by ambulance personnel. *Br Heart J* 1973;35: 550(abstract).
48. Baskett PJF, Diamond AW, Cochrane DF: Urban mobile resuscitation: Training and service. *Br J Anesth* 1976;48:377.
49. Adgey AA, Devlin JE, Webb SW, Mulholland HC: Initiation of ventricular fibrillation outside hospital in patients with acute ischemic heart disease. *Br Heart J* 1982;47:55.

50. Leclercq JF, Coumel P, Maison-Blanche P, et al: Mise en évidence des mécanismes déterminants de la mort subite. Enquete coopérative portanat sur 69 cas enregistrés para la méthode do Holter. *Arch Mal Coeur* 1986;79:1024.
51. Kempf FC, Josephson ME: Cardiac arrest recorded on ambulatory electrocardiograms. *Am J Cardiol* 1984;53:1577.
52. Panidis IP, Morganroth J: Sudden death in hospitalized patients: Cardiac rhythm disturbances detected by ambulatory electrocardiographic monitoring. *J Am Coll Cardiol* 1983;2:798.
53. Pratt C, Francis MJ, Luck JC, et al.: Analysis of ambulatory electrocardiograms in 15 patients during spontaneous ventricular fibrillation with special reference to preceding arrhythmic events. *J Am Coll Cardiol* 1983;2:789.
54. Milner PJ, Platia EV, Reid PR, Griffith LSC: Ambulatory electrocardiographic recordings at the time of fatal cardiac arrest. *Am J Cardiol* 1985;56:588.
55. Olshausen KV, Treese N, Pop T, Meyer J, et al.: Sudden death during Holter monitoring. *Dtsch Med Wochenchr* 1985;110:1195.
56. Lewis BH, Antman EM, Graboys TB: Detailed analysis of 24 hour ambulatory ECG recordings during ventricular fibrillation or torsades de pointes. *J Am Coll Cardiol* 1983;2:426.
57. Roelandt J, Klootwijk P, Lubsen J, Janse J: Sudden death during longterm ambulatory monitoring. *Eur Heart J* 1984;5:7.
58. Klein RC, Vera Z, Mason DT, DeMaria AN, Awan NA, Amsterdam EA: Ambulatory Holter monitor documentation of ventricular tachyarrhythmias as mechanism of sudden death in patients with coronary artery disease. *Clin Res* 1979;27:7A.
59. Denes P, Gabster A, Huang SK: Clinical, electrocardiographic and follow-up observations in patients having VF during Holter. *Am J Cardiol* 1981;48:9.
60. Miller JM, Vassallo JA, Kussmaul WG, Cassidy DM, Hargrove WC, Josephson ME: Anterior left ventricular aneurism: Factors associated with the development of sustained ventricular tachycardia. *J Am Coll Cardiol* 1988;12:375.
61. Weiner I, Mindich B, Pitchon R: Determinants of ventricular tachycardia in patients with ventricular aneurisms: Results of intraoperative epicardial and endocardial mapping. *Circulation* 1982;65:856.
62. Breithardt G, Schartzmaier J, Borggrefe M, Haerten K, Seipel L: Prognostic significance of late ventricular potentials after acute myocardial infarction. *Eur Heart J* 1983;4:487.
63. Gomes JA, Mehra R, Barreca P, El-Sherif N, Hariman R, Holtzman R: Quantitative analysis of of the high-frequency components of the signal-averaged QRS complex in patients with acute myocardial infarction: A prospective study. *Circulation* 1985;72:105.
64. Roy D, Marchand D, Theroux P, Watres DD, Pelletier GB, Cartier R, Bourassa MG: Long-term reproducibility and significance of provocable ventricular arrhythmias after myocardial infarction. *J Am Coll Cardiol* 1986;8:32.
65. Roubin GS, Harris PJ, Bernstein R, Kelly DT: Coronary anatomy and prognosis after myocardial infarction in patients 60 years of age and younger. *Circulation* 1983;67:743.
66. Kannel WB, McGee DL, Scharzkin A: An epidemiological perspective of sudden death: 26-year follow-up in the Framingham study. *Drugs* 1984;28(suppl I):1.
67. Kannel WB: Contribution of the Framingham study to preventive cardiology. *J Am Coll Cardiol* 1990;15:206.
68. Cooper RS, Simmons BE, Castañer A, Santhanam V, Ghali J, Mar M: Left ventricular hypertrophy is associated with worse survival independent of ventricular function and number of coronary arteries severely narrowed. *Am J Cardiol* 1990;65:441.
69. Sanz G, Castañer A, Betriu A, et al.: Determinants of prognosis in survivors of myocardial infarction. A prospective clinical angiographic study. *N Engl J Med* 1982;306:1065.
70. Ahnve S, Gilpin E, Henninng H, Curtis G, Ross J Jr: Limitations and advantages of the ejection fraction for defining high risk after acute myocardial infarction. *Am J Cardiol* 1986;58:872.
71. Bigger JT, Fleiss JL, Kleiger R, Miller P, Rolnitzky LM: The relationships among ventricular arrhythmias, left ventricular dysfunction, and mortality in the 2 years after myocardial infarction. *Circulation* 1984;69:250.
72. Mukharji J, Rude RE, Poole WK, et al.: The MILIS Study Group. Risk factors of sudden death after acute myocardial infarction: Two year follow-up. *Am J Cardiol* 1984;54:31.
73. Likoff MJ, Chandler SL, Kay HR: Clinical determinants of mortality in chronic congestive heart failure secondary to idiopathic dilated or to ischemic cardiomyopathy. *Am J Cardiol* 1987;59:634.
74. Packer M: Sudden unexpected death in patients with congestive heart failure. A second frontier. *Circulation* 1985;72:681.
75. Figueras J, Cinca J, Valle V, Rius J: Prognostic implications of early spontaneous angina after acute transmural myocardial infarction. Intern *J Cardiol* 1983;4:261.
76. Shuster EH, Bulkley BH: Early postinfarction angina. Ischemia at distance and ischemia in the infarct zone. *N Engl J Med* 1981;305:1101.
77. Fioretti P, Brower RW, Balakumaran K: Early postinfarction angina. Incidence and prognostic relevance. *Eur Heart J* 1986;7(suppl C):73.
78. Bosch X, Theroux P, Waters DD, Pelletier GB, Roy D: Early postinfarction ischemia: Clinical, angiographic and prognostic significance. *Circulation* 1987;75:988.
79. Weiner DA, Ryan TJ, McCabe CH, Luk S, Chaitman BR, Schefield LT, Tristanu F, Fisher LD: Significance of silent myocardial ischemia during exercise testing in patients with coronary artery disease. *Am J Cardiol* 1987;59:725.
80. Theroux P, Waters D, Halphen C, Debaisieux JC, Mizgala H: Prognostic value of exercise testing soon after myocardial infarction. *N Engl J Med* 1979;301:341.
81. Sharma B, Asinger R, Francis G, Hodges M, Wyeth R: Demonstration of exercise-induced painless myocardial ischemia in survivors of out-of-hospital ventricular fibrillation. *Am J Cardiol* 1987;59:740.
82. Warnes C, Roberts W: Sudden coronary death: Relation of amount and distribution of coronary narrowing at necropsy to previous symptoms of myocardial ischemia, left ventricular scarring and heart weight. *Am J Cardiol* 1984;54:65.
83. Lown B, Wolf M: Approaches to sudden death from coronary heart disease. *Circulation* 1981;64:297.

84. Ruberman W, Wienblatt E, Golberg J: Ventricular premature complexes and sudden death after myocardial infarction. *Circulation* 1981;64:297.
85. The Multicenter Postinfarction Research Group: Risk stratification and survival after myocardial infarction. *N Engl J Med* 1983;309:331.
86. Moss AJ, Davis HT, DeCamilla J, Bayer LW: Ventricular ectopic beats and their relation to sudden and nonsudden cardiac death after myocardial infarction. *Circulation* 1979;60:998.
87. Anderson KP, De Camila J, Moss AJ: Clinical significance of ventricular tachycardia (3 beats longer) detected during ambulatory monitoring after myocardial infarction. *Circulation* 1977;57:890.
88. Kotler M, Talvatznik B, Mowen N, Tominage S: Prognostic significance of ventricular ectopic beats with respect to sudden death in the late post-infarction period. *Circulation* 1973;47:954.
89. Lie KI, Liem KL, Schrlenburgh RM, David GK, Durrer D: Early identification of patients developing late in-hospital ventricular fibrillation after discharge from the coronary unit. A 5.5 year retrospective and prospective study of 1987 patients. *Am J Cardiol* 1978;41:674.
90. Hindman MC, Wagner GS, Jaro M, et al.: The clinical significance of bundle branch block complicating acute myocardial infarction. Two indications for temporary and permanent pacemaker insertion. *Circulation* 1978;58:689.
91. Hauer RN, Lie KI, Liem KL, Durre C: Long-term prognosis in patients with bundle branch block complicating acute anteroseptal myocardial infarction. *Am J Cardiol* 1982;49:1581.
92. Hollander G, Nadiminti V, Lichstein E, Greengart A, Sanders M: Bundke branch block in acute myocardial infarction. *Am Heart J* 1983;105:738.
93. Bayés de Luna A, Dominguez de Rozas JM, Garcia Picart J, Oter R, Songa V: Management of patients with ventricular block in acute myocardial infarction, in Caturelli G (ed): *Cura Intensiva Cardiologica 1987*. Milano, Librex Publishers, 1987, p 66.
94. Butrous GS: Autonomic tone modulation of cardiac arrhythmias, in Camm AJ, Ward DE (eds): *Clinical Aspects of Cardiac Arrhythmias*. Dordrecht, Kluwer Academic Publishers, 1988.
95. Kleiger RE, Miller J Ph, Bigger JT, Moss AJ: Decreased heart rate variability and its association with increased mortality after acute myocardial infarction. *Am J Cardiol* 1987;59:256.
96. Kleiger RE, Miller JP, Krone RJ, Bigger JT, and the Multicenter Postinfarction Research Group: The independence of cycle length variability and exercise testing on predicting mortality of patients surviving acute myocardial infarction. *Am J Cardiol* 1990; 65:408.
97. Schwartz PJ, Wolf SW: QT interval prolongation as predictor of sudden death with myocardial infarction. *Circulation* 1978;57:1074.
98. Anhve S, Helmers C, Lundman T: QTc intervals at discharge after acute myocardial infarction and long-term prognosis. *Acta Med Scand* 1980;208:55.
99. Marti V, Bayés de Luna A, Arriola J, Songa V, Guindo J, Dominguez de Rozas J, et al.: Value of dynamic QTc in arrhythmology. *N Trends Arrhyth* 1988;4:683.
100. Martí V, Bayés de Luna A, Arriola J, Songa V, Guindo J, Dominguez de Rozas JM, Thakor N, Caminal P, Laguna P: Value of dynamic QTc as a marker of malignant ventricular arrhythmias. *Eur Heart J* 1898;9(suppl 1):42.
101. Kadish AH, Weisman HF, Veltri EP, Epstein AE, Slepian MJ, Levine JH: Paradoxical effects of exercise on the QT interval in patients with polymorphic ventricular tachycardia receiving type Ia antiarrhythmic agents. *Circulation* 1990;81:14.
102. Stramba-Badiale M, Vanoli E, De Ferrari GM, Cerati D, Foreman RD, Schwartz PJ: QT interval and baroreflex sensitivity in concious dogs before and after myocardial infarction, in Butrous GS, Schwartz PJ (eds): *Clinical Aspects of Ventricular Repolarization*. London, Farrand Press, 1989, p 395.
103. Billman GE, Schwartz PJ, Stone HL: Baroreceptor reflex control of heart rate: A predictor of sudden death. *Circulation* 1982;66:874.
104. Bigger JT, La Rovere MT, Steinman RC, Fleiss JL, Rottman JN, Rolnitzky LM, Schwartz PJ: Comparison of baroreflex sensitivity and heart period variability after myocardial infarction. *J Am Coll Cardiol* 1989;14:1511.
105. La Rovere MT, Specchia G, Mortara A, Schwartz PJ: Baroreflex sensitivity, clinical correlates, and cardiovascular mortality among patients with a first myocardial infarction. A prospective study. *Circulation* 1988;78:816.
106. Schwartz JT, Vanoli E, Stramba-Badiale M, De Ferrari GM, Billman GE, Foreman RD: Autonomic mechanisms and sudden cardiac death. New insights from analysis of baroreceptor reflex in conscious dogs with and without a myocardial infarction. *Circulation* 1988;78:969.
107. Kannel WB, Thomas HE: Sudden coronary death: the Framingham study. *Ann NY Acad Sci* 1982;382:3.
108. Kannel WB, et al.: Sudden death risk in overt coronary heart disease: The Framingham study. *Am Heart J* 1987;113:799.
109. Coronary Drug Project Research Group: Natural history of myocardial infarction in the Coronary Drug Project: Long-term prognostic importance of serum lipid levels. *Am J Cardiol* 1978;42:489.
110. Phillips AN, Shaper AG, Pocock SJ, Walter M, MacFarlane PW: The role of risk factors in heart attacks occurring in men with pre-existing ischaemic heart disease. *Br Heart J* 1988;60:404.
111. Kannel WB, Sorlie P, Castelli WP, McGee D: Blood pressure and survival after myocardial infarction: The Framingham study. *Am J Cardiol* 1980;45:326.
112. Mulcahy R: Influence of cigarette smoking on morbidity and mortality after myocardial infarction. *Br Heart J* 1983;49:410.
113. Bayés de Luna A, Carreras F, Cladellas M, Oca F, Sagues F, Garcia Moll M: Holter ECG study of the electrocardiographyc phenomena in Prinzmetal angina attacks with emphasis on the study of ventricular arrhythmias. *J Electrocardiol* 1985;18:267.
114. Waters D, Miller D, Szlachcic J, et al.: Factors influencing the long-term prognosis of treated patients with variant angina. *Circulation* 1983;68:258.
115. Puddu PE, Bourassa M, Waters DD, Lasperance J: Sudden death in two patients with variant angina and apparently minimal fixed coronary stenoses. *J Electrocardiol* 1983;16:213.
116. Previtali M, Klersy C, Salerno J: Ventricular arrhythmias in Prinzmetal's variant angina. Clinical

significance and relation to the degree and time course of ST segment elevation. *Am J Cardiol* 1983;52:19.

117. Tzivoni D, Keren A, Granot H, Gottlieb S, Benhorin J, Stern S: Ventricular fibrillation caused by myocardial reperfusion in Prinzmetal's angina. *Am Heart J* 1983;105:323.
118. McLenachan JM, Henderson E, Karen I, Morris, Dargie HJ: Ventricular arrhythmias with hypertensive left ventricular hypertrophy. *N Engl J Med* 1987;317:787.
119. Messerli FH, Ventura HO, Elizari DJ, Dunn FG, Frohlich ED: Hypertension and sudden death, increased ventricular ectopic activity in left ventricular hypertrophy. *Am J Med* 1984;77:18.
120. Levy D, Anderson KM, Savage DD, Balkus SA, Kannel WB, Castelli WP: Risk of ventricular arrhythmias in left ventricular hypertrophy: the Framingham heart study. *Am J Cardiol* 1987;60:560.
121. Nicod P, Polikar R, Peterson KL: Hypertrophic cardiomyopathy and sudden death. *N Engl J Med* 1987;316:780.
122. Maron BJ, Roberts WC, Epstein SE: Sudden death in hypertrophic cardiomyopathy: A profile of 78 patients. *Circulation* 1982;65:1388.
123. Spirito P, Maron BJ: Relation between extent of left ventricular hypertrophy and occurrence of sudden cardiac death in hypertrophyc cardiomyopathy. *J Am Coll Cardiol* 1990;15:1521.

123a. Maron BJ, Kragel AH, Roberts, WC: Sudden death in hypertrophyc cardiomyopathy with normal left ventricular mass. *Br Heart J* 1990;63:308.

123b. Cripps TR, Couniham PJ, Frenneaux MP, Ward DE, Camm AJ, McKenna WJ: Signal-averaged electrocardiography in hypertrophyc cardiomyopathy. *J Am Coll Cardiol* 1990;15:956.

124. Wilson JR, Schwartz JS, Sutton MJ, Ferraro N, Horowitz LN, Reichek N, Josephson ME: Prognosis in severe heart failure: Relation to hemodynamic measurements and ventricular ectopic activity. *J Am Coll Cardiol* 1983;2:403.
125. Von Olshausen K, Schafer A, Mehmel HC, Schwartz F, Senges J, Kubler W: Ventricular arrhythmias in idiopathic dilated cardiomyopathy. *Br Heart J* 1984;51:195.
126. Huang SK, Messer JV, Denes P: Significance of ventricular tachycardia in idiopathic dilated cardiomyopathy: Observations in 35 patients. *Am J Cardiol* 1983;51:507.
127. Chakko CS, Gheorghiade M: Ventricular arrhythmias in severe heart failure: Incidence, significance, and effectiveness of antiarrhythmic therapy. *Am Heart J* 1985;109:497.
128. Meinertz T, Hofman T, Kasper W, Treese M, Bechtold H, Stienen U, Pop T, Leitner E-RV, Andersen D, Meyer J: Significance of ventricular arrhythmias in idiopathic dilated cardiomyopathy. *Am J Cardiol* 1984;53:902.
129. Holmes J, Kubo SH, Cody RJ, Kligfield P: Arrhythmias in ischemic and non-ischemic dilated cardiomyopathy: Prediction of mortality by ambulatory electrocardiography. *Am J Cardiol* 1985; 55:146.
130. Unverferth DV, Magorien RD, Moechsberger ML, et al.: Factors influencing one year mortality of dilated cardiomyopathy. *Am J Cardiol* 1984;54:147.
131. Marcus FI, Fontaine GH, Guiraudon G, Frank R, Laurenceau JL, Malergue CH, Grosgogeat Y: Right ventricular dysplasia: A report of 24 Adult Cases. *Circulation* 1982;65:384.
132. Thiene G, Nava A, Corrado D, Rossi L, Pennelli N: Right ventricular cardiomyopathy and sudden death in young people. *N Engl J Med* 1988;318: 129.
133. Rizzon P, Breithardt G, Chiddo A: Arrhythmogenic right ventricle. *Eur Heart J* 1989;10(suppl D):1.
134. Chizner MA, Pearled DL, DeLeon AC: The natural history of aortic stenosis in adults. *Am Heart J* 1980;99:419.
135. Klein RC: Ventricular arrhythmias in aortic valve disease: Analysis of 102 patients. *Am J Cardiol* 1984;53:1079.
136. Schwartz LS, Goldfischer J, Sprague GJ, Schwartz SP: Syncope and sudden death in aortic stenosis. *Am J Cardiol* 1969;23:647.
137. Kligfield P, Levy D, Devereux RB, Savage DD: Arrhythmias and sudden death in mitral valve prolapse. *Am Heart J* 1987;113:1298.
138. Shappell SD, Marshall CE, Brown RE, Bruce TA: Sudden death and the familial occurrence of midsystolic click, late systolic murmur syndrome. *Circulation* 1973;48:1128.
139. Lambert EC, Menon VA, Wagner HR, et al.: Sudden unexpected death from cardiovascular disease in children: A cooperative international study. *Am J Cardiol* 1974;34:89.
140. Cheitlin MD, DeCastro CM, McAllister HA: Sudden death as a complication of anomalous left coronary origin from the anterior sinus of Valsalva: A not so minor congenital anomaly. *Circulation* 1974;50:780.
141. Quattlebaum TG, Varghese PJ, Neill CA, et al.: Sudden death among postoperative patients with Tetralogy of Fallot. *Circulation* 1976;54:289.
142. Gillette PC, Yeoman MA, Mullins CE, et al.: Sudden death after repair of Tetralogy of Fallot. *Circulation* 1977;56:566.
143. Gillette PC, El-Said GM, Silvarjan N, et al.: Electrophysiological abnormalities after Mustard operation for transposition of the great arteries. *Br Heart J* 1974;36:186.
144. Shen W, Holmes DR, Porter CJ, McGoon DC, Ilstrup DM: Sudden death after repair of double-outlet right ventricle. *Circulation* 1990;81:128.
145. Torner P and the European Registry on Sudden Death in the Wolff-Parkinson-White Syndrome: Ventricular fibrillation in the Wolff-Parkinson-White syndrome. *Circulation* 1988;78(suppl II):23.
146. Klein GJ, Bashore TM, Sellers TD, Pritchett ELC, Smith WM, Gallagher JJ: Ventricular fibrillation in the Wolff-Parkinson-White syndrome. *N Engl J Med* 1979;301:1080.
147. Kaplan MA, Cohen KL: Ventricular fibrillation in the Wolff-Parkinson-White syndrome. *Am J Cardiol* 1969;24:259.
148. Dreifus LS, Haiat R, Watanabe Y, et al.: Ventricular fibrillation. A possible mechanism of sudden death in patients with Wolff-Parkinson-White syndrome. *Circulation* 1971;43:520.
149. Duvernoy WFC: Sudden death in Wolff-Parkinson-White syndrome. *Am J Cardiol* 1977;39:472.
150. Lipsitt LP, Sturner WQ, Oh W, et al.: Wolff-Parkinson-White and sudden-infant-death syndromes. *N Engl J Med* 1979;300:1111.

151. Surawicz B, Knoebel SB: Long QT: Good, bad or indifferent? *J Am Cardiol Coll* 1984;4:398.
152. Jervell A, Lange-Nielson F: Congenital deaf-mutism, functional heart disease with prolongation of QT interval and sudden death. *Am Heart J* 1957;54:59.
153. Romano C: Congenital cardiac arrhythmia. *Lancet* 1965;i:658.
154. Ward OC: A new familial cardiac syndrome in children. *J Ir Med Assoc* 1964;54:103.
155. Schwartz PJ, Locati E: The idiopathic long QT syndrome: Pathogenetic mechanisms and therapy. *Eur Heart J* 1985;6:103.
156. Eagle KA, De Sanctis RW: Diseases of the aorta, in Braunwald E (ed): *Heart Disease: A Textbook of Cardiovascular Medicine*. Philadelphia, WB Saunders, 1988, p 1546.
157. Roberts WC, Honing HS: The spectrum of cardiovascular disease in the Marfan syndrome. *Am Heart J* 1982;104:115.
158. Dreslinski GR: Diseases of the aorta and arterial tree, in Messerli FH (ed): *Cardiovascular Disease in the Elderly*. Boston, Martinus Nijhoff Publishing, 1984, p 205.
159. Goldhaber SZ, Braunwald E: Pulmonary embolism, in Braunwald E (ed): *Heart Disease: A Textbook of Cardiovascular Medicine*. Philadelphia, WB Saunders, 1988, p 1577.
160. Morgan M: Amniotic fluid embolism. *Anaesthesia* 1979;34:20.
161. Aronson ME, Nelson PK: Fatal air embolism in pregnancy resulting from an unusual sexual act. *Obstet Gynecol* 1967;30:127.
162. Strain JE, Grose RM, Factor SM, Fisher JD: Results of endomyocardial biopsy in patients with spontaneous ventricular tachycardia but without apparent structural heart disease. *Circulation* 1983;68:1171.
163. Vignola PA, Kazutaka A, Swayne PS, et al.: Lymphocytic myocarditis presenting as unexplained ventricular arrhythmias. *J Am Coll Cardiol* 1984;4:812.
164. Sugrue DD, Holmes DR, Gersh BJ, et al.: Cardiac histologic findings in patients with life-threatening ventricular arrhythmias of unknown origin. *J Am Coll Cardiol* 1984;4:952.
165. Gomes JA, Alexopoulos D, Winters SL, Deshmukh P, Fuster V, Suh K: The role of silent ischemia, the arrhythmic substrate and the short-long sequence in the genesis of sudden cardiac death. *J Am Coll Cardiol* 1989;14:1618.
166. Lown B, Dasilva RA, Reich P, Murawski B: Psychological factors in sudden cardiac death. *Am J Psychiatry* 1980;137:1325.
167. Lown B: Mental stress, arrhythmias and sudden death. *Am J Med* 1982;72:177.
168. Malliani A, Schwartz PJ, Zanchetti A: Neural mechanisms in life-threatening arrhythmias. *Am Heart J* 1980;100:705.
169. Reich P, DaSilva RA, Lown B, Murawski BJ: Acute psychological disturbances preceding life-threatening ventricular arrhythmias. *JAMA* 1981;286:233.
170. Greene WA, Goldstein S, Moss AJ, Rochester: Psychosocial aspects of sudden death. *Arch Intern Med* 1972;129:725.
171. Brackett CD, Powell LH: Psychosocial and physiological predictors of sudden cardiac death after healing of acute myocardial infarction. *Am J Cardiol* 1988;61:979.
171a. Garcia Sanchez S, Guindo J, Bayés de Luna A: Psychological stress and sudden cardiac death, in Bayés de Luna A, Brugada P, Cosin J, Navarro Lopez F (eds): *Sudden Cardiac Death*. Dordrecht, Kluwer Academic Publishers, 1990, p 75.
172. Siscovick DS, Weiss NS, Fletcher RH, Lasky T: The incidence of primary cardiac arrest during vigorous exercise. *N Engl J Med* 1984;311:874.
173. Kaplan JP: Cardiovascular deaths while running. *JAMA* 1979;242:2578.
174. Northcote RJ, Ballantyne D: Sudden cardiac death in sport. *Br Med J* 1983;287:1357.
175. Maron BJ, Epstein SE, Roberts WC: Causes of sudden death in competitive athletes. *J Am Coll Cardiol* 1986;7:204.
176. Lie KJ, Wellens HJJ, Dorsnar E, Durrer D: Observations on patients with primary ventricular fibrillation complicating acute myocardial infarction. *Circulation* 1975;52:755.
177. El-Sherif N, Myerburg RJ, Scherlang BJ, Befeler B, Aranda JM, Castellanos A, Lazzarra R: Electrocardiographic antecedents of primary ventricular fibrillation. Value of the R-on-T phenomenon in myocardial infarction. *Br Heart J* 1976;38:415.
178. Lawrie DM, Higgins MR, Godman MJ, Julian DG, Donald KM: Ventricular fibrillation complicating acute myocardial infarction. *Lancet* 1968;ii:523.
179. Gradman AH, Bell PA, DeBusk RF: Sudden death during ambulatory monitoring: Clinical and electrocardiographic correlations. Report of a case. *Circulation* 1977;55:210.
180. Hong R, Bhandari A, McKay C, Au P, Rahimtoola S: Lifethreatening ventricular tachycardial and fibrillation induced by painless myocardial ischemia during exercise test. *JAMA* 1987;257:1937.
181. Turitto G, Zanchi E, Prati PL: Acute myocardial infarction during Holter recording. *Am J Cardiol* 1990;63:364.
182. Deanfield JE, Shea MJ, Ribeiro P, et al.: Transient ST segment depression is a marker of myocardial ischemia during daily life. *Am J Cardiol* 1984;54:1195.
183. Deanfield JE, Ribeiro P, Oakley K, et al.: Critical analysis of ST segment changes in normal subjects: Implications for ambulatory monitoring of myocardial ischemia in patients with angina. *Am J Cardiol* 1984;54:1321.
184. Fintel DJ, Platia EV: Exercise testing and cardiac arrhythmias, in Platia EV (ed): *Management of Cardiac Arrhythmias*. Philadelphia, JB Lippincott, 1987, p 28.
185. Meinertz T, Zeheder M, Hohnloser S, Just H: Prevalence of ventricular arrhythmias during silent myocardial ischemia. *Cardiovasc Rev Rep* 1988(suppl):34.
186. Goldstein S, Medendrop SV, Landis JR, Wolfe RA, Leighton R, Ritter G, Vasu M, Acheson A: Analysis of cardiac symptoms preceding cardiac arrest. *Am J Cardiol* 1986;58:1195.
187. Graboys TB, Stein IM, Cueni L, Lampert S, Lown B: Is the presence of silent ischemia associated with the provocation of ventricular arrhythmias? *Circulation* 1987;76(suppl IV):365.
188. Bayés de Luna A, Camacho AM, Guindo J, Oca F, Martí V, Torner P, Martinez Dunker D: Is there a

relationship between crises of silent ischemia and appearance of ventricular arrhythmias? *Eur Heart J* 1989;10:29.
189. Banai S, Stern S, Keren A, Tzivoni D: Increased ventricular ectopic activity during ischemia episodes in ambulatory patients. *J Am Coll Cardiol* 1989;13:33.
190. Turitto G, Zanchi E, Risa AL, Maddaluna A, Saltarocchi ML, Vajola SF, Prati PL: Lack of correlation between transient myocardial ischemia and late potentials on the signal-averaged electrocardiogram. *Am J Cardiol* 1990;65:290.
191. Leclerq JF, Maisonblanche P, Cuchemezand B, Coumel P: Respective role of sympathetic tone and of cardiac pauses in the genesis of 62 cases of ventricular fibrillation recorded during Holter monitoring. *Eur Heart J* 1988;9:1276.
192. Hollenberg NK, Hollifield JW: Potassium/magnesium depletion: Is your patient at risk of sudden death? *Am J Med* 1987;82:1.
193. Birkenhäger WH, Solomon RJ, Wills MR: Electrolyte disturbances and cardiac risks. *Drugs* 1984; 28(suppl 1):1.
194. CAST Investigators: Preliminary Report: Effect of encainide and flecainide on mortality in a randomised trial of arrhythmia suppression after myocardial infarction. *N Engl J Med* 1989; 321:406.
195. Ruskin JN: The cardiac arrhythmia suppression trial (CAST). *N Engl J Med* 1989;321:386.
196. Podrid PJ, Marcus FI: Lessons to be learned from the cardiac arrhythmia suppression trial. *Am J Cardiol* 1989;64:1189.
197. Garratt C, Ward DE, Camm AJ: Lessons from the cardiac arrhythmias suppression trial. *Br Med J* 1989;299:805.
198. Farré J, Brugada P. Extrasistoles ventriculares postinfarto y el uso de fármacos antiarrítmicos: Un peligro anticipado por los electrofisiólogos. *Rev Esp Cardiol* 1989;42:498.
199. Gottlieb SS: The use of antiarrhythmic agents in heart failure: Implications of CAST. *Am Heart J* 1989;118:1074.
200. Akhtar M, Breithardt G, Camm AJ, Coumel P, Janse J, Lazzara R, Myerburg RJ, Schwartz PJ, Waldo AL, Wellens HJJ, Zipes DP: CAST and beyond. Implications of the Cardiac Arrhythmia Suppression Trial. Task Force of the Working Group on Arrhythmias of the European Society of Cardiology. *Eur Heart J* 1990;11:194.
201. Partt CM: The Cardiac Arrhythmia Suppression Trial—Does it alter our concepts of and approaches to ventricular arrhythmias? *Am J Cardiol* 1990;65(suppl B):1B.
202. Vlay SC: Lessons from the past and reflections on the Cardiac Arrhythmia Suppression Trial. *Am J Cardiol* 1990;65:112.
203. Bayés de Luna A, Guindo J, Borja J, Roman M, Madoery C: Recasting the approach of treatment of potentially malignant ventricular arrhythmias after CAST study. *Cardiovasc Drugs Ther* 1990; (in press).
204. Velebit V, Podrid P, Lown B, Cohen BH, Graboys TB: Aggravation and provocation of ventricular arrhythmias by antiarrhythmic drugs. *Circulation* 1982;65:886.
205. Simonsen E, Nielsen JS, Nielsen BL: Sinus node disfunction in 128 patients. A retrospective study with follow-up. *Acta Med Scand* 1980;208:343.
206. Frye RJ: Report of the joint ACC/AHA. Guidelines for permanent cardiac pacemaker implantation. *Circulation* 1984;70:331A.
207. Medrano G, Brenes P, De Michelo A, Sodi D: El bloqueo simultaneo de las divisiones anterior y posterior de la rama izquierda del haz de His (bloqueo bifascicular) y su associación con bloqueos de rama derecha (bloqueo trifascicular). *Arch Inst Card México* 1970;40:752.
208. Tavazzi L, Salerno JA, Cimiente M, Ray M, Bobba P: Electrophysiological mechanisms of paroxysmal AV block, in Bayés de Luna A, Cosin J (eds): *Diagnosis and Treatment of Cardiac Arrhythmias*. Oxford, Pergamon Press, 1979, p 415.
209. Bayés de Luna A, Torner P, Oter R, Oca F, Guindo J, Rivera I, Fort de Ribot R: Study of evolution of masked bifascicular block. *PACE* 1988;11:1517.
210. Puech P, Grollead R, Latour H, Cabasson J, Robin JM, Baissus C, Gilbert M: Diagnostic des blocs tronculaires hissiens par l'enregistrement endocavitaire et la stimulationde fasciceau de His. *Arch Mal Coeur* 1972;65:315.
211. Milstein S, Buetikofer J, Lesser J, Goldenberg IF, Benditt DG, Gornick C, Reyes WJ: Cardiac asystole: A manifestation of neurally mediated hypotension-bradycardia, *J Am Coll Cardiol* 1989;14:1626.
212. Bigger JT: Implications of the Cardiac Arrhythmia Supression Trial for antiarrhythmic drug treatment. *Am J Cardiol* 1990;65:3D.
213. DeBusk RF, Kraemer HC, Nash E: Stepwise risk stratification soon after acute myocardial infarction. *Am J Cardiol* 1983;52:1161.
214. Merrilees MA, Scott PJ, Norris MN: Prognostic after myocardial infarction. Results of 15-year follow-up. *Br Med J* 1984;288:356.
215. Smith JW, Marcus FI, Serokman R, with the Multicenter Postinfarction Research Group: Prognostic of patients with diabetes mellitus after acute myocardial infarction. *Am J Cardiol* 1984;54:718.
216. Tofler GH, Stone PH, Muller JE, Willich SN, Davis VG, Poole WK, Strauss HW, Willerson JT, Jaffe AS, Roberston T, Passamani E, Braunwald E, and the MILIS Study Group: Effects of gender and race on prognosis after myocardial infarction: adverse prognosis for women, particularly black women. *J Am Coll Cardiol* 1987;9:473.
217. The Coronary Drug Project Research Group: Blood pressure in survivors of myocardial infarction. *J Am Coll Cardiol* 1984;4:1135.
218. Killip T, Kimball JI: Treatment of myocardial infarction in a coronary care unit. A two-year experience with 250 patients. *Am J Cardiol* 1967;20:457.
219. Bosch X, Theroux P, Watesrs DD, Pelletier GB, Roy D: Early postinfarction ischemia: Clinical, angiographic and prognostic significance. *Circulation* 1987;75:988.
220. Brugada P, Talajic M, Smeets J, Mulleneers R, Wellens HJJ: The value of the clinical history to assess prognosis of patients with ventricular tachycardia or ventricular fibrillation after myocardial infarction. *Eur Heart J* 1989;10:747.
221. Maisel AS, Gilpin E, Holt B, LeWinter M, Ahnve S, Henning H, Collins D, Ross J: Survival after hospital discharge in matched populations with inferior or

anterior myocardial infarction. *J Am Coll Cardiol* 1985;6:731.
222. Fish C: Role of the electrocardiogram in identifying the patient at increased risk for sudden death. *J Am Coll Cardiol* 1985;5(suppl B):6.
223. Kreger BD, Cupples LA, Kannel WB: The electrocardiogram in prediction of sudden death: Framingham study experience. *Am Heart J* 1987; 113:377.
224. Hands ME, Lloyd BL, Robinson JS, Deklerk N, Thompson PL: Prognostic significance of electrocardiographic site of infarction after correction for enzymatic size of infarction. *Circulation* 1986; 73:885.
225. Marmor A, Geltman EM, Schechtman K, Sobel BE, Roberts R: Recurrent myocardial infarction: Clinical predictors and prognostic implications. *Circulation* 1982;66:415.
226. Hutter AM, DeSanctis RW, Flynn T, Yeatman LA: Nontransmural myocardial infarction: A comparison of hospital and late clinical course of patients with that of matched patients with transmural anterior or transmural inferior myocardial infarction. *Am J Cardiol* 1981;48:595.
227. Nicholson MR, Roubin GS, Bernstein L, Harris PJ, Kelly DT: Prognosis after an initial non-Q-wave myocardial infarction related to coronary artery anatomy. *Am J Cardiol* 1983;52:462.
228. Maisel AS, Ahnve S, Gilpun E, Henning H, Goldberger AL, Collins D, LeWinter M, Ross J: Prognosis after extension of myocardial infarct: The role of Q-wave and non-Q-wave infarction. *Circulation* 1985;71:211.
229. Mullins CB, Atkins JM: Prognoses and management of ventricular conduction blocks in acute myocardial infarction. *Mod Concepts Cardiovasc Dis* 1976;45:129.
230. Bayés de Luna A, Dominguez de Rozas JM, Garcia Picart J, Oter R, Songa V: Management of patients with ventricular block in acute myocardial infarction, in Caturelli G: Cura intensiva cardiologica 1987. Milan, Librex, 1987, p 66.
231. McHenry PL: Role of exercise testing in predicting sudden death. *J Am Coll Cardiol* 1985;5(suppl B):9.
232. Epstein S, Palmeri S, Patterson R: Evaluation of patients after acute myocardial infarction: Indications for cardiac catheterization and surgical intervention. *N Engl J Med* 1982;307:1487.
233. Ericsson M, Granath A, Ohlsen P, et al.: Arrhythmias and symptoms during treadmill testing three weeks after myocardial infarction in 100 patients. *Br Heart J* 1973;35:787.
234. DeBusk RF, Blomqvist CG, Kouchoukos NT, et al.: Identification and treatment of low risk patients after acute myocardial infarction and coronary-artery bypass graft surgery. *N Engl J Med* 1986; 314:161.
235. Sami M, Kraemer H, DeBusk RF: The prognostic significance of serial exercise testing after myocardial infarction. *Circulation* 1979;60:1238.
236. Davidson DM, DeBusk RF: Prognostic significance of a single exercise test 3 weeks after acute myocardial infarction. *Circulation* 1980;61:236.
237. Weld FM, Chu KL, Bigger JT, Rolnitzky LM: Risk stratification with low-level exercise testing 2-weeks after acute myocardial infarction. *Circulation* 1981;64:306.
238. Waters DD, Bosch X, Bouchard A, Moise A, Roy D, Pelletier G, Theroux P: Comparison of clinical variables derived from a limited predischarge exercise testing as predictors of early and late mortality after myocardial infarction. *J Am Coll Cardiol* 1985;5:1.
239. Williams WL, Nair RC, Higginson LA, Baird MG, Allan K, Beanlands DS: Comparison of clinical and treadmill variables for the prediction of outcome after myocardial infarction. *J Am Coll Cardiol* 1984;4:477.
240. DeBusck RF: Specialized testing after recent acute myocardial infarction. *Ann Inter Med* 1989;110:470.
241. Amstrong WF, Mchenry PL: Ambulatory electrocardiographic monitoring: Can we predict sudden death? *J Am Coll Cardiol* 1985;5(suppl B):13.
242. Santarelli P, Bellocci F, Loperfido F, et al.: Ventricular arrhythmia induced by programmed electrical stimulation after acute myocardial infarction. *Am J Cardiol* 1985;55:391.
243. Marchlinski RE, Buxton AE, Waxman HL, et al.: Identifying patients at risk of sudden death after myocardial infarction: Value of the response to programmed stimulation, degree of ventricular ectopic activity and severity of left ventricular disfunction. *Am J Cardiol* 1983;52:1190.
244. Roy D, Marchand E, Theroux P, Waters DD, Pelletier GB, Bourassa MG: Programmed ventricular stimulation in predicting one-year mortality after acute myocardial infarction. *Circulation* 1985; 72:487.
245. Livelli FD, Bigger JT, Reiffel JA, Gang ES, Patton JN, Noethling PM, Rolnitzky LM, Gliklich JI: Response to programmed ventricular stimulation: Sensitivity, specifity and relation to heart disease. *Am J Cardiol* 1982;50:452.
246. Brugada P, Green M, Abdollah H, Wellens HJJ: Significance of ventricular arrhythmias initiated by programmed ventricular stimulation: The role importance of the type of ventricular arrhythmia induced and number of premature stimuli required. *Circulation* 1984;69:87.
247. Hamer A, Vohra J, Hunt D, Sloman G: Prediction of sudden death by electrophysiological studies in high risk patients surviving acute myocardial infarction. *Am J Cardiol* 1982;50:223.
248. Breithardt G, Borggrefe M: Pathophysiological mechanisms and clinical significance of ventricular late potentials. *Eur Heart J* 1986;7:364.
249. Gomes JA, Winters SL, Stewart D, Horowitz S, Milner M, Barreca P: A new noninvasive index to predict sustained ventricular tachycardia and sudden death in the first year after myocardial infarction: Based on signal-averaged electrocardiogram, radionuclide ejection fraction and Holter monitoring. *J Am Coll Cardiol* 1987;10:349.
250. Kuchar DL, Thorburn CW, Sammel NL: Prediction of serious arrhythmic events after myocardial infarction: Signal-averaged electrocardiogram, Holter monitoring and radionuclide ventriculography. *J Am Coll Cardiol* 1987;9:531.
251. Cripps T, Bennet D, Camm J, Ward D: Prospective evaluation of clinical assessment, exercise testing and signal-averaged electrocardiogram in predicting outcome after acute myocardial infarction. *Am J Cardiol* 1988;62:995.
252. Sanford CF, Corbett J, Nicod P, et al.: Value of ra-

dionuclide ventriculography in the immediate characterization of patients with acute myocardial infarction. *Am J Cardiol* 1982;49:637.

253. Morris KJ, Palmeri ST, Califf RM, et al.: Value of radionuclide angiography for predicting specific cardiac events after acute myocardial infarction. *Am J Cardiol* 1985;55:318.
254. Ong L, Green S, Reiser S, Morrison J: Early prediction of mortality in patients with acute myocardial infarction: A prospective study of clinical and radionuclide risk factors. *Am J Cardiol* 1986;57:33.
255. The Multicenter Postinfarction Research Group: Risk stratification and survival after myocardial infarction. *N Engl J Med* 1983;309:331.
256. Corbett JR, Dehmer GJ, Lewis, et al.: The prognostic value of submaximal exercise testing with radionuclide ventriculography before hospital discharge in patients with recent myocardial infarction. *Circulation* 1981;64:535.
257. Nicod P, Corbet JR, Firth BG, et al.: Prognostic value of resting and submaximal exercise radionuclide ventriculography after acute myocardial infarction in high-risk patients with single and multivessel disease. *Am J Cardiol* 1983;52:30.
258. Nishimura RA, Reeder GS, Miller FA, Ilstrup DM, Shub C, Seward JB, Tajik AJ: Prognostic value of predischarge 2-dimensional echocardiogram after acute myocardial infarction. *Am J Cardiol* 1984;53:429.
259. Bhatnagar SK, Moussa MAA, Al-Ysuf AR: The role of prehospital discharge two-dimensional echocardiography in determining the prognosis of survivors of first myocardial infarction. *Am Heart J* 1985;109:472.
260. Burggraf JW, Parker JO: Prognosis in coronary artery disease-angiographic, hemodynamic and clinical factors. *Circulation* 1975;51:146.
261. Humphries JO, Kuller L, Ross RS, Friesinger GC, Page EE: Natural hystory of ischemic heart disease in relation to angiographic findings. *Circulation* 1974;49:489.
262. Trappe HJ, Klein H, Wenzlaff P, Hartwig CA: Natural history of single vessel disease. Risk of sudden coronary death in relation to coronary anatomy and arrhythmia profile. *Eur Heart J* 1989;10:514.

PART VI

Nonpharmacologic Treatment

Chapter **28**

Current Status of Cardiac Pacing for Bradyarrhythmias: Developments of the 1980s

Victor Parsonnet, MD, and Alan D. Bernstein, EngScD

As the twentieth century enters its final decade, the implanted pacemaker is well on its way to becoming a self-contained and largely autoregulating rhythm-management system, whose capabilities for dealing with a great variety of bradyarrhythmias and tachyarrhythmias may eventually include even automatic drug administration. The early history of pacing is well known[1,2]; in this chapter we will describe some of the innovations of the most recent decade, and express personal opinions where appropriate.

Antibradyarrhythmia pacing began 31 years ago with a three-transistor, fixed-rate device powered by a mercury–zinc battery. It was not adequately sealed to prevent the pacemaker from being short-circuited almost immediately by body fluids.[3,4] By 1980 all implantable pacemaker pulse generators were hermetically sealed and thus permanently impervious to fluid invasion. Integrated-circuit technology decreased their power consumption while allowing the incorporation of new capabilities, such as some degree of noninvasive adjustment ("programming") and more sophisticated responses to sensed electrical signals. At that time a typical pulse generator weighed about 120 gm, was the size of a small paperweight, and cost about $3,500. It contained a lithium-iodine battery capable of sustaining its operation for about 5–8 years. Most implantations were performed by surgeons. Introducer techniques for transvenous lead implantation had just been described.[5] Pacemaker programming was largely limited to the most basic functions, primarily pacing rate and stimulus strength.

There were about 100,000 new implants and 19,000 pulse-generator replacements each year in the United States—500 new implants annually per million population, which was considerably more than in any other country. Dual-chamber pacing modes, chiefly DVI and, to a lesser extent, DDD, had been accepted as preferable for patients with stable atrial rhythms, but were actually used in fewer than 10% of patients.[6]* Experiments had already begun in adaptive-rate pacing (ARP), in which auxiliary physiologic variables (respiration rate and blood pH) controlled the pacing rate according to need when a normal rate response was absent.[7–9]**

Formal guidelines for permanent pacing did not exist. There were no clearly defined indications and no set methods for device selection; nor was there any peer-review process by which pacing practices might be evaluated. Antibradyarrhythmia pacing was in a state of experimentation and explosive expansion.

During the 1980s it became the objective of the pacemaker industry to develop physiologic pacing: to design implantable electronic devices that mimicked closely the behavior of the normal cardiac conduction system. Of equal interest was the development of bidirectional communication

* Pacing modes are represented throughout this chapter by the NBG Code (see Table 28.1),[10] which was designed on the basis of an earlier code developed under the auspices of the Intersociety Commission for Heart Disease Resources (ICHD).[11]

** The misleading term "rate-responsive" is sometimes used to describe a pacemaker in which the lower rate limit is modulated according to changes in an auxiliary variable such as mechanical vibration or blood temperature. These devices do not respond to rate, as the term implies, but to changes in the auxiliary variable. Therefore, we prefer the term "adaptive-rate" for modes and devices that incorporate rate modulation of this type.

655 Avenue of the Americas, New York, NY 10010
Current Topics in Cardiology

TABLE 28.1 The NBG Pacemaker Code

Position	I	II	III	IV	V
Category	Chamber(s) paced **O** = None **A** = Atrium **V** = Ventricle **D** = Dual (A + V)	Chamber(s) sensed **O** = None **A** = Atrium **V** = Ventricle **D** = Dual (A + V)	Response to sensing **O** = None **T** = Triggered **I** = Inhibited **D** = Dual (T + I)	Programmability, rate modulation **O** = None **P** = Simple Programmable **M** = Multiprogrammable **C** = Communicating **R** = Rate modulation	Antitachyarrhythmia function(s) **O** = None **P** = Pacing (antitachyarrhythmia) **S** = Shock **D** = Dual (P + S)
Manufacturers' designation only	**S** = single (A or V)	**S** = single (A or V)			

Note: Positions I through III are used exclusively for antibradyarrhythmia function.

Reprinted from Bernstein et al.,[10] *with permission.*

with implanted devices, which would permit extensive telemetry of ongoing measurement results and stored data and noninvasive control of many aspects of operation. Progress in device development has been accompanied by profound changes in clinical perceptions and practice patterns, as we approach the twenty-first century with the amalgam of clinical medicine and implanted-device technology that is cardiac pacing.

PACING DEVICES

Pulse Generators

Pacemaker manufacturers have been persistent in their effort to reduce pulse-generator size and, concomitantly, the size of connectors and leads. The typical, ovoid pulse generator weighs 25–45 gm and is 0.5 to 0.7 cm thick, but experimental prototype pulse generators the size of a 50-cent piece have been implanted in animals. Actually, miniaturization sacrifices ease of handling, connector and lead strength, and, if smaller batteries are used, device longevity, for cosmetic and marketing advantages (Table 28.2); there is no evidence that these trade-offs are worthwhile in terms of patient satisfaction. Almost without exception, today's implanted pulse generators are powered by lithium-iodine cells whose capacity (and, therefore, longevity in an implanted device) is related to their size. Other lithium compounds have been largely abandoned, although new formulations, such as lithium-silver vanadium pentoxide for implantable cardioverter/defibrillators (ICDs), are being evaluated.[12,13] Plutonium as a power source enjoyed brief popularity in the 1970s but has disappeared from use except at a few die-hard centers. It is nonetheless potentially useful, especially for relatively power-hungry adaptive-rate pacemakers (one new device has a 24-microampere idling current, exclusive of pacing, about 10 times that of a conventional VVI pacemaker); it would also offer the advantages of less frequent surgical replacement and reduced overall pacing cost.[14]

Leads and Connectors

Transvenous leads have become thinner, sacrificing fluoroscopic visibility, maneuverability, and ease and safety of extraction for a presumed but undocumented advantage of minimizing venous thrombosis (Table 28.3).

Epimyocardial leads are used very rarely, almost exclusively in conjunction with open-heart surgery and in cases where access veins are no longer available. They are constructed differently from transvenous leads, because they need not accommodate stylets. Their long-term effectiveness is not as well documented as is that of transvenous leads, primarily because of limited data concerning initial thresholds. It is known, how-

TABLE 28.2 Consequences of Pulse-Generator Miniaturization

Attribute	Large	Small
Longer battery life	X	
Accommodate more components	X	
Ease of handling	X	
Avoiding pulse-generator "twiddling"	X	
Connector simplicity	X	
Ease and economy of manufacture	X	
Cosmesis		?
Use in infants		X
Marketing advantage		?

TABLE 28.3 Consequences of Pacing-Lead Miniaturization

Attribute	Thick	Thin
Accommodate multiple conductors	X	
Better fluoroscopic visibility	X	
Maneuverability	X	
Ease of extraction (isodiametric)	X	
Toughness	X	
Connector simplicity	X	
Advoidance of tissue fibrosis at electrode		?
Lower thrombogenicity		?
Cosmesis		?
Martketability		?
Use in infants		X

ever, that the thresholds frequently rise to unacceptably high levels.[15] Moreover, explanting an infected epimyocardial lead entails all of the risks associated with another thoracotomy.[16] Fortunately, transvenous-lead implantation after open-heart surgery is as easily accomplished as that of a primary transvenous lead, and with equally good results.[17]

Apart from electrode characteristics (see below), several aspects of lead design have been shown to be of clinical importance. Two types of insulation are used. Polyurethane is slippery, stiff, and tough enough to resist damage during surgery, but it is more difficult to manufacture. Chemical degradation of this material, at one time of great concern, is no longer a problem. Silicone rubber is more flexible than polyurethane but less slippery; additives are now being used to make it more slippery and thus functionally equivalent to polyurethane in most respects.

Despite intensive competition among manufacturers, some standardization of lead and connector design has been achieved, specifically for the end of the lead that plugs into the pulse-generator header, that is, the terminal, and the connector. Standards have been established for the diameter and length of terminal pins, the presence of sealing rings that are either integral with the header or attached to the lead, and the spacing between contacts in bipolar systems. The standards, known as Voluntary (VS1) and International (IS1), are broadly accepted, but even now not all manufacturers have settled upon a favorite. [18,19] Some combinations of leads and pulse generators are actually incompatible, necessitating careful component choice. This can be facilitated by an up-to-date table (Table 28.4) posted in the implantation room.

There are no standards governing lead material, diameter, or length, although special lengths are available upon request. Allowing for 10 cm of selvage, actual required lead lengths are about 36 cm for right atrial leads and 45 cm for right-ventricular leads when the pulse generator is implanted on the left side. When custom-length leads are not used, the excessive selvage is wrapped around the pulse generator or coiled behind it, increasing the bulk of the implant and the risk of lead damage during lead explantation or pulse-generator replacement. Methods for selecting appropriate lead lengths and shortening excessively long unipolar leads have been suggested.[20]

Electrodes

Pacing electrodes have been developed in immense variety. It has been the goal to produce a surface of the proper material, area, shape, and

TABLE 28.4 Compatibility of Pulse Generators and Leads

	Pulse-Generator Headers		
Pacing Leads	VS1A No Sealing Rings; Long Receptacle	VS1B Sealing Rings; Long Receptacle	VS1/IS1 No Sealing Rings; Short Receptacle
VS1/IS1 Sealing rings; short pin	Compatible	Compatible	Compatible
Cordis Sealing rings; long pin	Compatible	Compatible, but tight fit	Incompatible (will not fit)
Medtronic No sealing rings; long pin	Incompatible (no seal)	Compatible	Incompatible (will not fit)

This table applies primarily to older leads; newer Cordis and Medtronic leads conform to the IS1 standard. Because device changes are sometimes made without changing the model number, careful matching on a model-by-model basis is the safest way to avoid compatibility problems.

texture to provide the lowest possible stimulation threshold together with the highest possible sensing-signal amplitude, while ensuring the physical and functional stability of the electrode. Minimizing the stimulation threshold prolongs battery life and may allow the use of a smaller battery and pulse generator.

Two new lead-design elements have become popular: "active-fixation" leads with grasping elements that maximize physical stability and steroid-eluting leads that minimize the acute stimulation-threshold rise associated with maturation of the electrode/tissue interface.(21) We commonly use active-fixation leads without steroid elution in both atrium and ventricle. The long-term reliability of pacing leads is shown in Figure 28.1.

Unipolar leads were once preferred, mainly because of the larger stimulation artifact seen on the surface-lead electrocardiogram (ECG), but they are being supplanted by bipolar versions, largely because of the growing popularity of dual-chamber pacing. The bipolar electrode configuration reduces far-field sensing and interchannel crosstalk, and provides greater immunity to electromagnetic interference (EMI).(22)

Adaptive-rate pacemakers (ARPs) sometimes require special leads, or auxiliary leads, for measurement of the auxiliary physiologic variables used to modulate the pacing rate. Therefore, only those ARPs that use standard leads can be regarded as forward- and backward-compatible. This includes devices controlled by mechanical vibration, the stimulus-to-T-wave interval, the ventricular depolarization gradient, and some impedance-related variables.

It should be noted that most available leads are slightly thicker at the tip than along the shaft, a design flaw that makes it impossible to remove them without excessive traction or major surgery.(16)

Pacemaker "Programming" Devices

As pacemaker designs increased in complexity during the 1980s, so did the "programming" devices used for noninvasive adjustment of operating parameters such as those listed in Table 28.5. Most significant was the development of microcomputer-based pacemaker programmers, which can send and receive pulse-coded radiofrequency transmissions over an electromagnetic link between a coil contained in the pulse generator and another coil in an external programming "head" placed over the pulse generator and connected to the programming device. Microcomputer-based programmers, many of which are capable of data or electrogram-signal telemetry, are assuming an increasingly active role in follow-up monitoring, clinical data management, and problem solving as well as in adjusting pacemaker operation.

MODES OF PACING

Cardiac pacing came into being as a means of preventing death from complete heart block. The first implantable pacemakers were fixed-rate, nonsensing VOO devices that ensured a safe lower rate limit. Subsequently VVI (ventricular demand) pacing was developed as an effort to avoid competition with spontaneous rhythms. With time and the development of new pacing

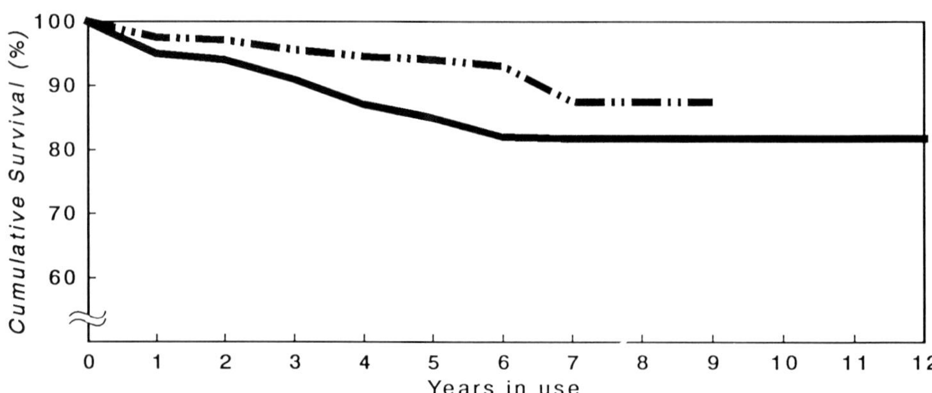

FIGURE 28.1 Actuarial survival of 1,094 atrial (solid curve) and 1,319 ventricular (dashed curve) pacing leads. Survival was 81.8% at 155 months and 87.4% at 110 months for atrial and ventricular leads, respectively. Electrode dislodgment and malposition were the most frequent causes of failure in the entire cohort of atrial (4.7%) and ventricular (2.0%) leads. Insulation degradation caused 20 failures in the 101 polyurethane atrial leads (19.8%) and 11 failures in the 116 polyurethane ventricular leads (9.5%).

TABLE 28.5 Programmable Pulse-Generator Operating Parameters for Dual-Chamber Antibradyarrhythmia Pacing

Timing functions
- Pacing mode
- Lower rate limit
- Maximum atrial-tracking rate
- Response to fast atrial rates (rate fallback, pacemaker "Wenckebach," n-to-one block, etc.)
- Fallback rate
- Fallback deceleration rate
- Response to pacemaker-reentrant tachycardia
- Atrioventricular (AV) interval
- Crosstalk protection: AV interval after early ventricular sensing
- Atrial-channel refractory period
- Ventricular-channel refractory period
- Ventricular-channel blanking period

Adaptive pacing
- Hysteresis-pacing escape interval
  - After sensing
  - After pacing
- AV interval
  - After sensing
  - After pacing
- Response to premature ventricular depolarizations (automatic extension of atrial-channel refractory period)
- Rate-smoothing parameters
- Automatic threshold-adjustment settings
- Rate-modulation parameters
  - Sensitivity
  - Upper rate limit for adaptive-rate pacing

Input and output
- Sensing threshold, atrial and ventricular
- Stimulus amplitude, atrial and ventricular
- Stimulus duration, atrial and ventricular
- Electrode configuration (unipolar or bipolar for sensing and stimulation)
- Telemetry functions
  - Status reports
  - Measurement results
  - Event logging
  - Intracardiac electrograms
  - Timing diagrams
  - Trend analysis
  - "Holter" reports

Special functions
- Stimulation-threshold tracking
- End-of-life indicators
- Response to magnet application
- Emergency-mode parameter values
- External triggering for noninvasive electrophysiologic studies

modes, the indications for pacing expanded to include almost every form of fixed, intermittent, or even suspected heart block. At the same time, atrial modes such as AAI created a more "physiologic" alternative for managing sinus bradycardia.

Dual-Chamber Pacing

Atrioventricular (AV) synchrony, whose importance became increasingly evident, could be maintained either by pacing the atrium and ventricle in sequence or by pacing the ventricle at a suitable interval after a native P wave. In the United States, dual-chamber pacing modes (seven of them: VAT, VDD, DOO, DVI, DDD, DDI, and DDT) are now used in more than 30% of pacemaker patients.[23]

Adaptive Pacing

Continuing efforts to make pacemakers emulate normal physiology have resulted in still more complex devices and functions. Some pulse generators now incorporate adaptive (autoregulating) functions to compensate for chronotropic incompetence by automatically adjusting the pacing rate (see below), shortening the AV interval as the rate increases, adjusting the stimulus strength to maintain an adequate, but not wasteful, capture-safety margin, and recognizing and adapting appropriately to supraventricular tachyarrhythmias. These developments have improved the capability of implanted pacemakers to restore normal cardiac rhythms, but simultaneously have increased the difficulty of choosing the most appropriate pacing mode and operating-parameter values for each patient. Not surprisingly, the need for extensive programmability, telemetry capabilities, and information storage within adaptive pulse generators has become increasingly evident.

Single-Chamber Adaptive-Rate Pacing

During the last 4 years adaptive-rate ventricular demand pacing (VVIR) has become the most widely used pacing mode in the United States.[24] The impetus for this development was the gradual recognition that dual-chamber pacing may not achieve optimum hemodynamic benefit when the spontaneous atrial rate fails to rise high enough, or quickly enough, in response to exercise, emotion, or other stimuli that cause rate acceleration in the normal heart. This inadequacy of atrial rate response was first described in 1975 by Ellestad and Wan, who applied the term "chronotropic incompetence" to its various manifestations.[25]

Furthermore, in analyzing the role of rate augmentation in patients with ventricular pacemakers, it became evident that, in the acute setting, the measurable indices of cardiac performance during exercise, such as filling pressure, blood pressure, cardiac output, and exercise tolerance, behaved similarly in the VVIR (adaptive-rate ventricular pacing) and VAT (atrial-synchronous ventricular pacing) modes. Similar results were found in patients whose pacemakers could be programmed alternately to single- and dual-chamber modes. Not illogically, most clinicians came to prefer adaptive-rate ventricular pacemakers to dual-chamber devices, even for patients with normal atrial rhythms, because they were easier to implant and did not require an atrial lead.[26]

In terms of clinical function, not everyone agrees that adaptive-rate ventricular pacing is equivalent to dual-chamber pacing. A major difference, of course, is that single-chamber ventricular pacemakers (with or without rate modulation) cannot maintain AV synchrony. The most obvious disadvantage of this limitation is seen in patients with retrograde (VA) conduction, in whom the random phase relation between atrial and ventricular contractions produces a severe pacemaker syndrome.[27]* In some patients with angina or impaired left-ventricular function an increase in heart rate may have unpredictable countereffects. (Conversely, in some situations these problems may be ameliorated by rate augmentation.)[28] Furthermore, in comparison with modes that maintain AV synchrony, long-term ventricular pacing without synchronous atrial pacing or synchronization to native P waves tends to encourage the eventual development of atrial fibrillation, decrease life expectancy, and increase the risk of thromboembolism. These issues have been summarized succinctly by Camm and Katritsis.[26]

Dual-Chamber Adaptive-Rate Pacing

An obvious response to these problems has been the development of ARPs that pace *both* chambers. These devices hold great promise for the future because, when combined with other adaptive functions, they approach the fully physiologic pacemaker that for years has been everyone's goal.

Adaptive-rate pacing based on respiration rate was proposed as early as 1966,[7] but actually the first practical control variable for rate modulation was physical motion. A piezoelectric crystal, cemented to the interior surface of the pulse-generator housing, produces a transient voltage signal with mechanical flexure, responding to pressure and vibration, and thus to body motion. This sensor is conceptually simple, mechanically and electrically reliable, and requires almost no additional current from the battery for the processing of the signals it generates. Moreover, because the vibration transducer is contained within the pulse generator, this method is completely compatible with existing leads and electrodes. However, the sensor may also respond to nonphysiologic phenomena such as ambient vibration (as in a moving vehicle), pressure on the pulse generator (as during programming), and cold temperatures. Of course it does not respond to physiologic stimuli other than motion, such as catecholamine level, isometric exercise, and metabolic derangements. Thus, it is not a "perfect" sensor for rate modulation.

Auxiliary-variable sensors of many sorts have been tested and a few have reached the marketplace (Table 28.6). Some measure electrical phenomena (such as the stimulus-to-T-wave interval in the ventricle) or the evoked ventricular response (Wilson's ventricular-depolarization gradient). Some detect changes in blood temperature in response to exercise, others use bioimpedance measurements to detect changes in respiratory rate, minute ventilation, stroke volume, or preejection period, while still others measure variations in intracardiac pressure (dp/dt) or oxygen saturation.[26] Sensing systems can be classified in a variety of ways.

Recently, considerable effort has been dedicated to developing dual-chamber pacemakers that employ one or more of these rate-modulation schemes. Several have reached the clinical-evaluation stage while others are still in development or premarket-approval phases (Table 28.6). Each scheme has unique advantages and disadvantages, some of which have yet to be discovered. How the clinician chooses among this enormous and potentially bewildering array of options is discussed below.

The goal is to provide a device that will respond to a full range of stimuli, while maintaining AV synchrony whether or not the native atrial rate is adequate. A single pacemaker may combine two or more auxiliary-variable sensors, for example, using vibration sensing to respond to physical exercise and the stimulus-to-T-wave interval to respond to emotional stress. Thus, both physical activity and catecholamine-level

* The pacemaker syndrome, first described in 1969 by Mitsui et al., is a symptom complex reflecting the objective and subjective consequences of lack of AV synchrony, or inappropriate AV timing resulting from fixed VA conduction. Symptoms range from simple awareness of peculiar or unusual cardiac contractions to congestive heart failure.[29,30]

TABLE 28.6 Pulse Generators for Adaptive-Rate Pacing

Manufacturer	Model	Single- or Dual-Chamber	Pacing Modes	Sensor Type	Variable Sensed	Special Elec-trode	Sensing Con-figuration	Pacing Con-figuration
CPI	Precept VR 1100	S	VOO, VVI(R)	Impedance	Stroke volume, pre-ejection interval	No	Bipolar	Unipolar
CPI	Precept DR 1200	D	DOO, DDD(R), DDI(R), DVI(R), VVI(R), SOO, AAI	Impedance	Stroke volume, pre-ejection interval	No	Bipolar	Unipolar
Cook	Kelvin 500 Series	S	SOO(R), SSI(R)	Thermistor	Central venous blood temperature	Yes	Both available	Both available
Cook	Kelvin 510 Series	S	SOO(R), SSI(R)	Thermistor	Central venous blood temperature	Yes	Both available	Both available
Intermedics	Circadia	D	?	Thermistor	Central venous blood temperature	Yes	Programmable	Programmable
Intermedics	Relay	D	?	Accelerometer	Motion	No	Programmable	Programmable
Medtronic	Activitrax 8400, 8402, 8403	S	VOO(R), VVI(R)	Piezoelectric crystal	Vibration	No	Both available	Both available
Medtronic	Activitrax II 8412, 8413, 8414	S	SOO(R), SSI(R), SST	Piezoelectric crystal	Vibration	No	Both available	Both available
Medtronic	Elite 7074, 7075, 7076, 7077	D	DDD(R), DDI(R), DVI(R), SSI(R), SST, DOO(R), SOO(R)	Piezoelectric crystal	Vibration	No	Programmable	Programmable
Medtronic	Legend 8416, 8417, 8418	S	SSI(R), SOO(R)	Piezoelectric crystal	Vibration	No	Programmable	Programmable
Medtronic	Synergyst 7026, 7027	D	DDD, DVI, VVI(R), DOO, SOO	Piezoelectric crystal	Virbration	No	Both available	Both available
Medtronic	Synergyst II 7070, 7071	D	DDD(R), DVI(R), VVI(R), DOO, SOO	Piezoelectric crystal	Vibration	No	Both available	Both available
Pacesetter	Sensolog I 703	S	SOO(R), SSI(R), SST	Piezoelectric crystal	Vibration	No	Both available	Both available
Pacesetter	Sensolog III 2034T	S	SOO(R), SSI(R), SST(R)		Vibration	No	Programmable	Programmable

(continued)

TABLE 28.6 *(continued)*

Manufacturer	Model	Event Counters	Rate Histograms	Electrogram Telemetry	Measurement Telemetry	Demographic Data	Special Features	Market Status
CPI	Precept VR 1100	No	No	Yes	Battery status	Yes	Sensor select, sensor telemetry	Investigational
CPI	Precept DR 1200	No	No	Yes	Battery status	Yes	Sensor select, sensor telemetry, rate smoothing	Investigational
Cook	Kelvin 500 Series	Yes	No	No	Battery status, thermistor data	No	Interim rate, time; sensor telemetry	Marketed
Cook	Kelvin 510 Series	Yes	Yes	No	Battery status, thermistor data	No	Interim rate, time; sensor telemetry	Marketed
Intermedics	Circadia	Yes	Yes	Yes	Battery status, lead "impedance," output energy	Yes	?	Developmental
Intermedics	Relay	Yes	Yes	Yes	Battery status, lead "impedance," output energy	Yes	?	Developmental
Medtronic	Activitrax 8400, 8402, 8403	No	No	No	No	No		Marketed
Medtronic	Activitrax II 8412, 8413, 8414	No	No	No	Battery status, lead "impedance"	No		Marketed
Medtronic	Elite 7074, 7075, 7076, 7077	Yes	Yes	Yes	Battery status, lead "impedance"	No	Low output (0.8 V)	Investigational
Medtronic	Legend 8416, 8417, 8418	No	Yes	No	Battery status, lead "impedance"	No	Low output (0.8 V)	Marketed
Medtronic	Synergyst 7026, 7027	No	No	No	No	No		Marketed
Medtronic	Synergyst II 7070, 7071	No	No	No	No	No		Marketed
Pacesetter	Sensolog I 703	No	Yes	Yes	Battery status, lead "impedance"	Yes	"Vario"	Marketed
Pacesetter	Sensolog III 2034T	No	Yes	Yes	Battery status, lead "impedance"	Yes	"Vario," "Autoset"	Marketed

Manufacturer	Model	Single- or Dual-Chamber	Pacing Modes	Sensor Type	Variable Sensed	Special Electrode	Sensing Configuration	Pacing Configuration
Pacesetter	Synchrony 2020T	D	DDD(R), DDI(R), DVI(R), DOO(R), SSI(R), SOO(R)	Piezoelectric crystal	Vibration	No	Programmable	Programmable
Pacesetter	Synchrony II	D	DDD(R), DDI(R), DVI(R), DOO(R), SSI, SOO(R)	Piezoelectric crystal	Vibration	No	Programmable	Programmable
Telectronics	Meta MV 1202	S	SOO, SSI(R)	Impedance	Minute volume	No	Bipolar	Bipolar
Telectronics	Meta DDDR 1250	D	DDD(R), DDI(R), SSI(R), DOO, SOO, VVT	Impedance	Minute volume	No	Programmable	Programmable

Telectronics	Prism-CL 450A	S	VVO, VVI(R), VVT	Intracardiac electrogram	Ventricular-depolarization gradient	No	Programmable	Unipolar
Vitatron	Quintech TX 911, 915	S	VOO, VVI(R)	Intracardiac electrogram	Stim-T interval	No	Unipolar	Unipolar
Vitatron	Rhythmyx 501	S	VOO, VVI(R), VVT(R)	Intracardiac electrogram	Stim-T interval	No	Unipolar	Unipolar
Biotronik	Ergos-01	S	SOO(R) SSI(R)	Piezoelectric crystal	Vibration	No	Unipolar	Unipolar
Biotronik	Ergos-02	D	DDD(R) SSI(R), DVI(R), DOO(R), SOO(R), VDD	Piezoelectric crystal	Vibration	No	Unipolar	Unipolar
Biotronik	Thermos-01	S	SOO, SSI(R)	Thermistor	Central venous blood temperature	Yes	Unipolar	Unipolar
Biotronik	Thermos-02	S	SOO, SSI(R)	Thermistor	Central venous blood temperature	Yes	Unipolar	Unipolar

Manufacturer	Model	Event Counters	Rate Histograms	Electrogram Telemetry	Measurement Telemetry	Demo-graphic Data	Special Features	Market Status
Pacesetter	Synchrony 2020T	No	Yes	Yes	Battery status, lead "impedance"	Yes	Passive sensor mode	Marketed
Pacesetter	Synchrony II	No	Yes	Yes	Battery status, lead "impedance"	Yes	Passive sensor mode	Developmental
Telectronics	Meta MV 1202	Yes	No	Yes	Battery status, lead "impedance"	No	ADPT function	Marketed
Telectronics	Meta DDDR 1250	Yes	No	Yes	Battery status, lead "impedance"	Yes	Reversion to VVIR at high atrial rate	Investigational
Telectronics	Prism-CL 450A	Yes	No	No	Battery status, lead "impedance"	No	Closed-loop rate control, auto threshold set	Investigational
Vitatron	Quintech TX 911, 915	No	No					Marketed, Europe
Vitatron	Rhythmyx 501	Yes	No				Rate drop at night, nonlinear response	Marketed, Europe
Biotronik	Ergos-01	No	No	No	No	Yes		Investigational
Biotronik	Ergos-02	No	No	No	No	Yes		Investigational
Biotronik	Thermos-01	No	Yes (temp/rate plot)	No	No	Yes		Investigational
Biotronik	Thermos-02	No	Yes (temp/rate plot)	No	No	Yes		Investigational

Pacing modes shown with (R) for conciseness are available both with and without rate modulation; for example, VVI(R) = VVI, VVIR. Stim-T = stimulus-to-T-wave.

After Benedek et al.[52]

changes will be detected. The pacemaker will arbitrate among the sensing-system outputs to decide which sensor to follow (or which sensor should not be followed because it appears to be responding to far-field or spurious phenomena). There are some major problems with these concepts: differentiating physiologic from pathologic atrial rates, prioritizing between different rates derived from different sensors, establishing appropriate upper rate limits for P-wave tracking and sensor-derived rates, and responding appropriately to ventricular arrhythmias. Particularly attractive are pulse generators that combine two or more rate-modulation schemes that use the same transducer or set of leads.

Other Adaptive Functions

Some dual-chamber pulse generators mimic the normal phenomenon of P-R interval shortening as the heart rate increases. Without this shortening, atrial and ventricular contractions could be effectively out of phase at high rates. Some devices provide a slightly longer AV interval after pacing the atrium than after sensing a P wave, to allow time for atrial activation, thus stabilizing the mechanical AV interval between atrial and ventricular contractions. Some rate-modulation control variables, such as the ventricular depolarization gradient, also allow automatic adjustment of the output to ensure an adequate, but not wasteful, capture-safety margin as the stimulation threshold changes with time.

PROGRAMMING AND TELEMETRY

To optimize pacing therapy for each patient, the pacemaker must be "programmed" to select appropriate values for each of about 24 different operating parameters (see Table 28.5). (Adaptive-rate dual-chamber pacemakers may have more than 30 trillion different possible combinations of programmable-parameter choices!) Some programmable parameters are of considerable practical importance, particularly output, rate, and sensitivity, all of which were programmable 10 years ago.

The Need for Programming

When the output parameters (stimulus amplitude and duration) are left at the factory settings, battery depletion may occur in as little as 3 years, while programming appropriate to the actual stimulation thresholds can double the battery life. In some dual-chamber modes, maintaining the output amplitude at 1.5 to 2 times its threshold value also reduces the likelihood of interchannel crosstalk (where one channel senses the afterpotential that follows a stimulus in the other channel) and the inadvertent induction of an arrhythmia or of cardiac standstill.

In the same vein, appropriate choice of the lower rate limit can extend battery life by reducing the number of stimuli per unit time and thus the current drain (conventional sensing requires very little current); it also allows normal physiologic cardiac function to supervene more often. Dual-chamber pacemakers also have upper rate limits above which the pacemaker will not track fast atrial depolarizations or other signals.

Programming for proper sensing of spontaneous atrial activity is especially important because P-wave signals often are of much lower amplitude than R-wave signals, which are more likely to be sensed appropriately without reprogramming.

Of what value are the other programmable parameters listed in Table 28.5? In standard single- and dual-chamber pacing, many of them are of little clinical importance, adding little to the optimization of therapy and the quality and duration of life. A notable exception is the set of parameters that govern the behavior of an adaptive-rate pacemaker in response to changes in its specific sensed variable; the adjustment of sensitivity, "slope," and rate limits may be quite important. Most ARPs require careful clinical testing and adjustment, although the rate-modulation scheme based on measurement of the ventricular depolarization gradient is completely self-calibrating.

The Need for Telemetry

Two-way communication with the implanted device has become increasingly important, and in the more complex modes it is clearly essential to the clinician. A disadvantage of dual-chamber and adaptive-rate pacemakers has been the great difficulty of interpreting the paced ECG. The basic observation, as in the past, of the presence or absence of a pacemaker stimulus, the measurement of interstimulus (AV and VA) intervals, and the verification of rate stability are now often inadequate or even impossible. Effective pacemaker surveillance depends to an ever-increasing degree upon telemetered measurements and electrogram signals, stored information, and analytic timing diagrams generated by the programming device from telemetered data.

ANTITACHYARRHYTHMIA DEVICES

Although this communication is intended primarily as an overview of antibradyarrhythmia pacing, a few words about antitachyarrhythmia pacing are in order because both of these functions are often contained in a single device. Some pacemakers are designed to manage tachyarrhythmias through a variety of interventions ranging from simple underdrive or overdrive pacing to the production of groups of stimuli programmed to fire at preset points in a tachycardia cycle, in the form of pairs or bursts, with fixed or automatically adjusted intervals.[31,32] These devices are programmed with special programming devices, and may be activated either manually (by the patient or the physician) or by automatic tachycardia-detection mechanisms selected by the electrophysiologist. Incorporating such capabilities adds immensely to the complexity of antitachyarrhythmia devices, and frequent telemetry-aided monitoring is necessary to ensure continuing therapeutic effectiveness. Such devices exemplify the complex nature of modern pacing technology. Especially when combined with cardioversion and defibrillation capabilities, it seems clear that the pacemaker devices of the future indeed will be "rhythm-management systems," a term suggested by the late Michael Bilitch.

DEVELOPMENTS IN PACEMAKER-IMPLANTATION TECHNIQUES

The transvenous introducer almost revolutionized pacemaker implantation, bringing it from the fairly exclusive realm of the surgeon to that of the the internist and cardiologist.[33] The method, in fact, proved to be the forerunner of many other invasive procedures in medicine and has expanded immensely the scope of technologies available to nonsurgeons.

During the last 10 years implantation techniques have changed little, transvenous lead implantation being preferred by 95% of implanters. There has been a growing awareness, however, of some potentially lethal complications of the introducer method, among them air embolization, tension pneumothorax, and hemothorax; these do not occur with the cephalic-vein cutdown technique. Nevertheless the introducer is an almost indispensable tool, especially in reoperations when leads must be added or replaced, or when the cephalic or other superficial veins are no longer available. Many implanters still advocate the use of the introducer in all procedures.

The preferred location of the pulse generator is still infraclavicular, superficial to the pectoral fascia. About 10% of implanters wrap the pulse generator in a stretch-Dacron pouch. Two leads are used in 30% of cases, with a strong preference for active-fixation (screw-in) electrodes.[24] Special techniques are needed in infants and children, but even in tiny patients the transvenous method has become popular, with epimyocardial-electrode placement avoided whenever possible. With the development of small pediatric versions, pulse-generator implantation in such patients has become more acceptable.

The nature and frequency of device failures and surgery-related complications has not changed markedly. Polyurethane degradation, once a serious problem, appears to have been associated with a single polyurethane formulation that is no longer in use. Lead-conductor fracture has dropped to a seemingly irreducible level, accounting for about 5% of reoperations.[34] These fractures were once attributed to subtle microscopic defects in the alloys or mishandling by the implanter, but now it appears that the introducer may play a role. The entrance point of the needle into the subclavian vein directly behind the clavicle apparently produces a point of intermittent forceful compression of the conductor coil, which ultimately leads to fracture. (About 90% of fractures occur at that location.)[35]

Lead entrapment is a growing problem, and a potentially lethal one. While it has been proved that noninfected or noncontaminated leads need not be removed, once infection occurs removal is mandatory.[36] A countertraction device for removing these offending foreign bodies has been developed, but as yet it has not been tested widely.[37] The problem could be solved if the tip of the lead and the electrode were not of greater diameter than the shaft, but for reasons known only to themselves most manufacturers have stubbornly ignored this simple solution.

INDICATIONS FOR PACING

The indications for permanent-pacemaker implantation have not changed dramatically during the past decade. What has changed is the degree to which implantation is affected by third parties. In 1983 a report by the Senate Commission on Aging suggested the existence of fraudulent practices among pacemaker manufacturers and medical providers.[38] At almost the same time, there were reports that about one-quarter of all implantations performed in the United States were needless.[39] The implantation rate fell from 509 new implants per million population in 1983 to 385 in 1985.[6,23] Since then the implantation rate has declined further to 359 per million in 1989.[24] As an aftermath of these rather turbulent times, guidelines for pacemaker implantation were published, the most important of which were prepared by a joint committee of the American College of Cardiology (ACC) and the American Heart Association (AHA).[40,41] (An updated edition is in preparation.) They were widely publicized and subsequently adopted in a modified form by local peer-review organizations (PROs). PROs throughout the country initially mandated that all proposed pacemaker implantations in Medicare patients receive approval in advance because these were "high-priced" procedures. Many states have now made this process optional. The entire PRO system has in fact fallen into some disrepute.[42]

According to the ACC/AHA guidelines, indications for implantation are divided into three categories, Class I being those that are generally accepted and Class II those where there may be some disagreement. Class III lists those situations in which implantation is contraindicated. The essence of the guidelines is that there must be a documented electrocardiographic indication, usually accompanied by related symptoms.

SELECTION OF PACING MODE AND IMPLANTED DEVICES

Examples of the most (and least) appropriate indications for use of a selection of the more common pacing modes might be described in simplified fashion as follows.

VVI *Most appropriate:* Symptomatic bradycardia with fixed or frequent atrial tachycardia; rate augmentation contraindicated. *Least appropriate:* Bradycardia in the presence of fixed VA conduction.

AAI *Most appropriate:* Symptomatic sinus pause or arrest; proved normal his-to-ventricle (HV) interval; no likelihood of drug- or vagal-induced AV block; no supraventricular tachyarrhythmias. *Least appropriate:* Sinus bradycardia in the presence of slowed or blocked AV conduction and frequent supraventricular tachyarrhythmias.

DDD *Most appropriate:* Symptomatic bradycardia with normal atrial function and slowed or blocked AV conduction; modest or high level of activity anticipated. *Least appropriate:* Bradycardia with frequent supraventricular tachyarrhythmias.

DDI *Most appropriate:* Symptomatic bradycardia with slowed or blocked AV conduction and frequent but nonsustained supraventricular tachyarrhythmias; modest or high level of activity anticipated. *Least appropriate:* Bradycardia with nonexistent P waves.

Any of these modes (as well as less common modes not listed above) can be combined with rate augmentation. In considering an adaptive-rate mode, one must take into account: (a) the presence of serious morbidity or infirmity that would make the projected activity level low and the use of augmented rates unlikely; (b) the presence of angina or congestive heart failure shown to be aggravated by fast rates; (c) the possible imprudence of adding a device-specific lead if required by the ARP under consideration; (d) any special contraindications to a specific rate-modulation scheme, such as chronic pulmonary disease when a minute-ventilation device is under consideration; and (e) the absence of skilled personnel to perform the required programming and surveillance tasks.

Similarly, in choosing any pacemaker, there are many important considerations: (a) the patient's general medical condition; (b) the projected activity level; (c) the presence of VA conduction; (d) the presence of chronotropic incompetence; (e) the clinical need for rate augmentation; (f) the potential for spontaneous or drug-induced worsening of the conduction abnormalities; (g) the presence of angina or left-ventricular dysfunction, and (h) any technical constraints such as anatomic anomalies or the unavailability of venous access sites.

Computer algorithms that model such decision-making processes have been developed to assist in making choices in a systematic way. An example is shown in Figure 28.2.[43]

PACING COSTS

The already complicated process of choosing appropriately from a bewildering variety of modes, features, and devices is made more difficult still by pressure from hospital administrations and sales representatives and by ever-changing price lists. Some states and hospitals have instituted bulk-purchasing plans and insist that implanters use the cheapest models available through the plans. Most implanters tend to select the models with which they are most familiar, and for which accessories are immediately available (such as connectors, splicing kits, leads, pacemaker systems analyzers, and—most importantly—programming devices).

PACEMAKER FOLLOW-UP

The task of following pacemaker patients is a matter of constant concern. No comprehensive appraisal of the efficacy and costs of existing follow-up methods has been conducted in recent years, and unfortunately as many as 10% of patients in the United States receive no follow-up whatever.[25] In modern pacing, some follow-up is of great importance, especially for managing ARPs and antitachyarrhythmia devices and for

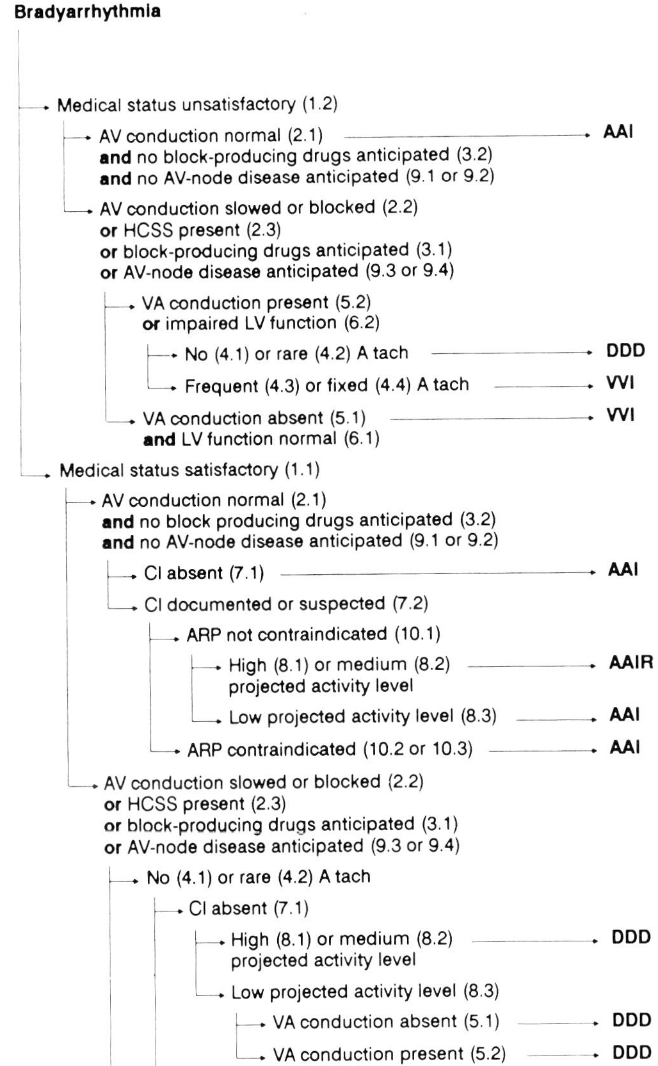

FIGURE 28.2 Tree-structure representation of an algorithm for computer-assisted selection of an appropriate antibradyarrhythmia-pacing mode. ARP = adaptive-rate pacing, AV = atrioventricular, CI = chronotropic incompetence, HCSS = hypersensitive carotid sinus syndrome, LV = left ventricular, VA = ventriculoatrial.

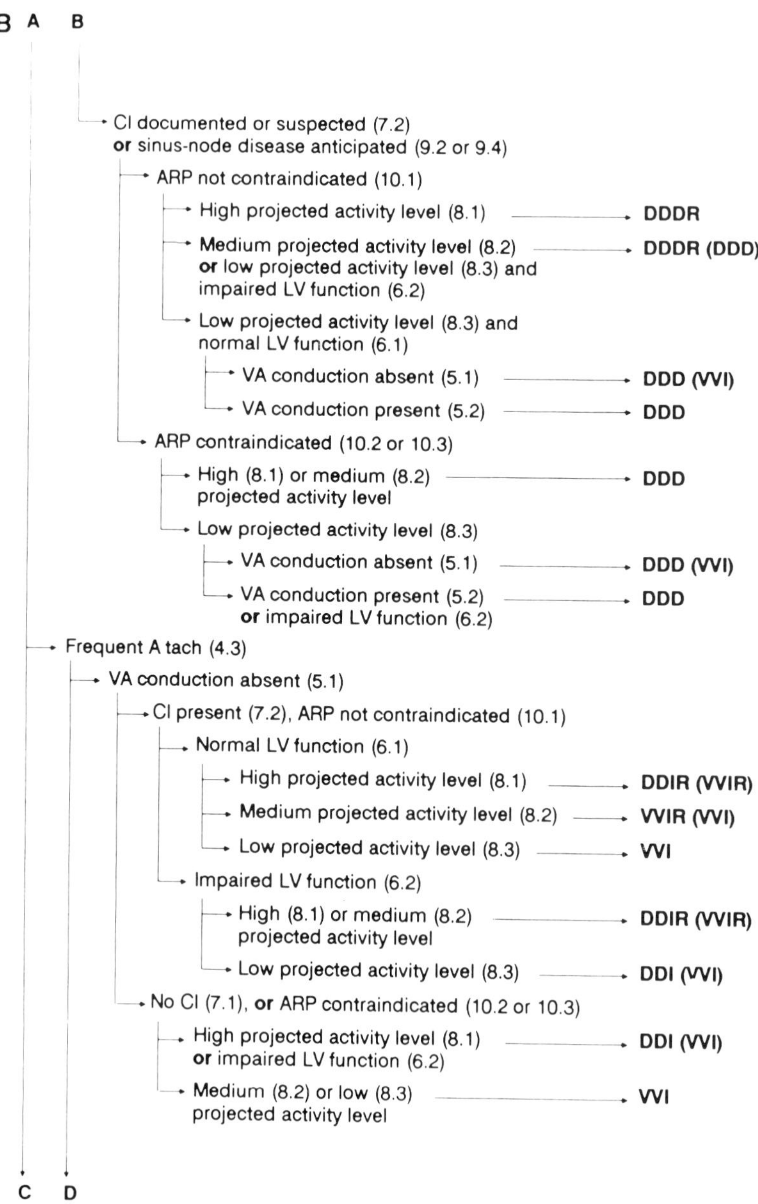

FIGURE 28.2 (*continued*)

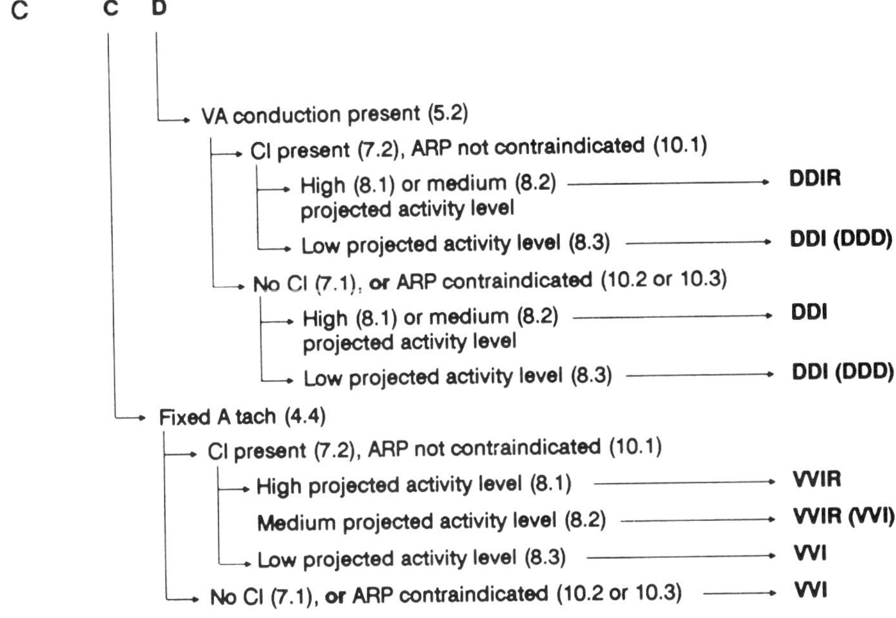

NOTE: It is assumed, for the time being, that available DDDR pulse generators cannot distinguish between normally high atrial rates and atrial tachycardia.

FIGURE 28.2 (*continued*)

achieving most effective and physiologic performance in any pacemaker.

Pacemaker follow-up facilities vary (see Table 28.7), with more elaborate equipment and techniques required as the scope and volume of the follow-up service increase.[44] A follow-up program may include any of the following elements.

a. Patient interviews.
b. Physical examination, at least of the pulse-generator site.
c. Electrocardiographic evaluation of the paced and unpaced rhythm strips (pacing can be inhibited, if the patient is not pacemaker-dependent), and of 12-lead ECGs when appropriate.
d. Waveform analysis: measurement of the amplitude and duration of the electrocardiographic stimulus artifact and observation of changes in artifact morphology; measurement of the magnet-mode and free-running interstimulus intervals.
e. Interrogation of the implant by telemetry.
f. Transtelephonic monitoring (TTM); most TTM systems allow remote recording of a rhythm strip and measurement of the pacing rate and stimulus duration.
g. Programming.
h. Exercise tests, especially to calibrate rate-modulation mechanisms in adaptive-rate pacing.
i. Ambulatory 24-h (Holter) ECG monitoring or transient-event recording.
j. Chest X-rays and echocardiograms.
k. Computer-assisted evaluation and reporting; some systems can acquire and store measurement results automatically and generate reports such as that shown in Figure 28.3.

How follow-up is actually performed in the United States is not clear because what is meant by "clinic evaluation" may vary greatly. The general characteristics enumerated above are derived from our quadrennial national surveys.[6,23,24]

TABLE 28.7 Resources of a Fully Equipped Pacemaker Follow-Up Center for 100 or More Patients

Space
- Examination room(s)
- Office space
- Storage space (for programming devices and other equipment)

Personnel
- Physician(s)
- Nurse/clinician(s)
- ECG technician(s)
- TTM technician(s)
- Business administrator(s)
- Secretarial staff
- Billing clerk
- Electronics technician
- Computer programmer/system manager
- Telephone-answering service

Instrumentation
- TTM transmitters
- TTM receiver(s)
- Crash cart
- External cardioverter/defibrillator
- External pacemaker
- Pacemaker-programming devices (two for each pulse-generator model series)
- Analog (or special-purpose digital) ECG monitor with strip-chart recorder
- 12-Lead electrocardiograph
- Holter monitors
- Transient-event recorders
- Computer facilities (mainframe or local-area network)
- Computer software
  - Follow-up testing
  - Administrative and financial

Office Visits

Periodic visits to a follow-up physician's office is the most common system, but what is done in the office is essentially unknown. It may be a simple rhythm-strip recording and a brief interview and examination, or it may be a full computer-assisted study including 12-lead electrocardiography and programming. The latter method is indistinguishable from the functions of a well-organized pacemaker clinic. TTM is almost always a part of this service.

Pacemaker Clinics

When the number of patients to be followed reaches a "critical mass," perhaps 50, an organized center for pacemaker follow-up provides the most efficient method.[44] There are relatively few of these in the United States, but in Canada and Europe most follow-up programs are run in this fashion. In the United States, patients are seen in the clinic or tested by TTM according to a schedule prescribed by Medicare, which insures 80% of pacemaker patients. Most of the tests described above are performed.

TTM Alone

Some patients are followed only by telephone, either by their private physicians or a proprietary TTM service. At the same time, brief interviews may be held with the patient or an attending person. It is important to realize that this method lacks most of the essential ingredients of basic follow-up, and is inadequate for dual-chamber and adaptive pacing.

Pacemaker Programming

As mentioned earlier, programming is essential to patient care. The pacemaker is programmed with an external device provided by the pacemaker manufacturer, which may or may not be compatible with all of that manufacturer's pulse generators. Small practices tend to use devices produced by a single manufacturer to limit the amount of equipment they must purchase and maintain; large clinics may have dozens of programmers, often two of each model for back-up. Detailed programming is almost always guided by data obtained by telemetry.

When programming an adaptive-rate pacemaker, the rate limits and "slope" (sensitivity of response to changes in the auxiliary rate-control variable) must be established to optimize the rate response. This process may take 30–40 min. Although simple tests, such as arm waving or stair climbing, have been devised to shorten the process, a universally reliable method does not exist.[45] A useful pragmatic test is to ask the patient how he feels after several weeks of living with the latest programmed settings!

PACING SOCIETIES, CREDENTIALS, AND PEER REVIEW

Pacing Societies

In 1979, 20 years after the first pacemaker was implanted, the North American Society of Pacing and Electrophysiology (NASPE) was founded.

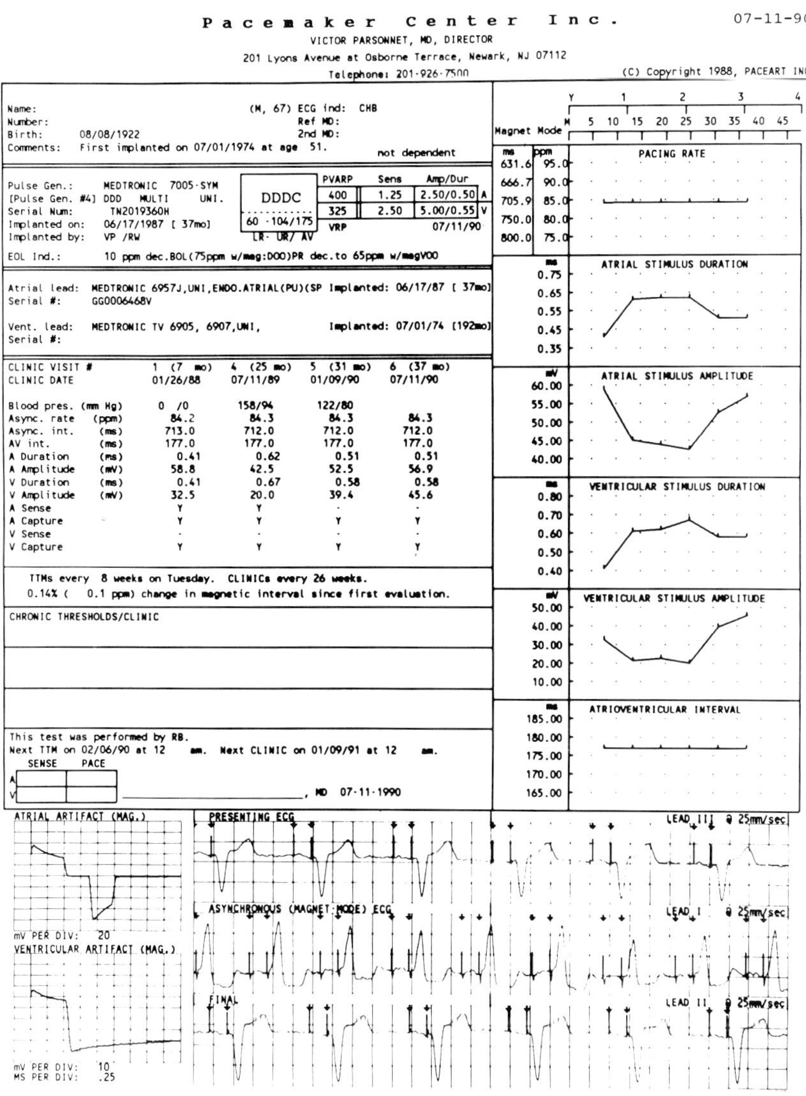

Pacemaker Center Inc. 07-11-90
VICTOR PARSONNET, MD, DIRECTOR
201 Lyons Avenue at Osborne Terrace, Newark, NJ 07112
Telephone: 201-926-7500
(C) Copyright 1988, PACEART INC

Name: (M, 67) ECG ind: CHB
Number: Ref MD:
Birth: 08/08/1922 2nd MD:
Comments: First implanted on 07/01/1974 at age 51. not dependent

Pulse Gen.: MEDTRONIC 7005-SYM
[Pulse Gen. #4] DDD MULTI UNI.
Serial Num: TN2019360H
Implanted on: 06/17/1987 [37mo]
Implanted by: VP /RW

DDDC
60 ·104/175
LR· UR/ AV

PVARP	Sens	Amp/Dur	
400	1.25	2.50/0.50	A
325	2.50	5.00/0.55	V
VRP		07/11/90	

EOL Ind.: 10 ppm dec.BOL(75ppm w/mag:DOO)PR dec.to 65ppm w/magVOO

Atrial lead: MEDTRONIC 6957J,UNI,ENDO.ATRIAL(PU)(SP Implanted: 06/17/87 [37mo]
Serial #: GG0006468V

Vent. lead: MEDTRONIC TV 6905, 6907,UNI, Implanted: 07/01/74 [192mo]
Serial #:

CLINIC VISIT #	1 (7 mo)	4 (25 mo)	5 (31 mo)	6 (37 mo)
CLINIC DATE	01/26/88	07/11/89	01/09/90	07/11/90
Blood pres. (mm Hg)	0 /0	158/94	122/80	
Async. rate (ppm)	84.2	84.3	84.3	84.3
Async. int. (ms)	713.0	712.0	712.0	712.0
AV int. (ms)	177.0	177.0	177.0	177.0
A Duration (ms)	0.41	0.62	0.51	0.51
A Amplitude (mV)	58.8	42.5	52.5	56.9
V Duration (ms)	0.41	0.67	0.58	0.58
V Amplitude (mV)	32.5	20.0	39.4	45.6
A Sense	Y	Y	-	-
A Capture	Y	Y	Y	Y
V Sense	-	-	-	-
V Capture	Y	Y	Y	Y

TTMs every 8 weeks on Tuesday. CLINICs every 26 weeks.
0.14% (0.1 ppm) change in magnetic interval since first evaluation.

CHRONIC THRESHOLDS/CLINIC

This test was performed by RB.
Next TTM on 02/06/90 at 12 am. Next CLINIC on 01/09/91 at 12 am.
SENSE PACE
A
V
____________________, MD 07-11-1990

FIGURE 28.3 A microcomputer-generated pacemaker follow-up report.

Since then its membership has grown to 1,500, and it includes pacemaker clinicians, cardiac electrophysiologists, and associated professionals. The 1990 annual meeting drew 2,000 registrants. NASPE works closely with the International Cardiac Pacing and Electrophysiology Society (ICPES).

Besides promoting pacing and electrophysiology education, NASPE fosters research, sponsors fellowships, and periodically issues policy statements in important areas of current interest. It also supports a biannual examination (the NASPExAM) for special competence in pacing and electrophysiology.[46]

Quality Control

A chronic problem facing hospitals, credentialing bodies, peer-review organizations, and the parent specialty societies has been to find a niche for pacing and electrophysiology within the existing framework of these various bodies. For electrophysiology, at least, there has been a solution in that a board examination has been established in this subspecialty by the American Board of Internal Medicine. The first examination will be given in 1991. Pacing, however, remains an orphan subspecialty, with no systematic guidance from any source other than NASPE. Some degree of quality control has been achieved through the joint publication of guidelines by the ACC and AHA, as described above, but in hospitals structured pacemaker services as subdivisions of cardiology or surgery are almost nonexistent. Thus quality-control issues are left either to chance or sporadic attention by hospital quality-control committees. There is some evidence that limited training and experience and infrequent patient contact have a negative effect on the quality of patient care.[47] Nevertheless, nothing is being done to correct this inadequacy. Soon more automatic cardioverter/defibrillators will be implanted transvenously. This will make the situation even more lamentable and add to the disarray.

Credentials

Despite the lack of formal requirements, efforts have been made to define the necessary training and experience of a pacemaker implanter. The first such definitions were prepared by the ICHD in 1974 and 1983.[48,49] Since then, additional recommendations were made in the ACC/AHA report in 1984,[40] and the 17th Bethesda Conference in 1986.[41] In the second ICHD report, a two-tiered system was suggested, with a modest level of exposure for trainees in general cardiology and a full year of more intensive exposure for those wishing to specialize in pacing. The Boards of Thoracic Surgery and Internal Medicine require "exposure" of an unspecified nature and unspecified amount, and there is little oversight of these requirements. The authors are unaware of formal training requirements imposed by any hospitals or states, but we often receive telephone inquiries about this issue, usually because of local conflicts between surgeons and cardiologists over implantation privileges. No major society appears to recognize the broad range of knowledge and skills that a pacemaker specialist must have, even though these were listed explicitly in the Frye Committee report. As things stand, pacemakers may be implanted by any physician who professes to be able to do so.

Costs

The recent mandated attention to the cost of medical care has had a profound impact on the pacemaker industry and the pacemaker physician. The advent of the system of Diagnosis-Related Groups (DRGs), under which the hospital is paid a single, fixed sum for the entire hospitalization (within certain length-of-stay limits), has made it necessary for hospital administrators to ask (and sometimes demand) that their physicians use the least expensive implant available. Many hospitals have established bulk-purchasing contracts with pacemaker manufacturers, and administrators strike bargains with local sales representatives. At times hospital administrations even ask physicians to limit the number of pacemakers they implant during a given period.

At the same time, physicians' fees have been reassessed by Medicare, through the Health Care Financing Agency (HCFA), by Hsiao et al., in a study of the Harvard Resource-Based Relative Value Scale (RBRVS),[49] and by Congress, through the Omnibus Budget Reconciliation Act (OBRA).[50] Certain surgical operations, including pacemaker implantation, have been identified as overpriced. Consequently, fees from Medicare are being reduced, with projections of as much as 30% over the next few years. The impact these events will have on the frequency of pacemaker procedures and on future professional practices remains to be seen.

OTHER ISSUES

Pacemaker-Industry Developments

Some pacemaker-manufacturing companies have either disappeared or been bought out by others. Market-share patterns have changed. In the United States, Medtronic, Inc. maintains its lead in sales, but others have gained considerably, notable Cardiac Pacemakers, Inc. (now a subsidiary of Eli Lilly), Pacesetter Systems, Inc. (now a subsidiary of Siemens-Elema), Cordis (now merged with Telectronics as Telectronics/Cordis), and Intermedics (now a subsidiary of Sulzer Medica). Three smaller American companies remain in operation (Cardiac Control Systems, Cook, and Biocontrol Technologies, previously known as Coratomic). Several European companies retain footholds here as well (ELA, Biotronik, and Biotec).

Litigation between pacemaker manufacturers is a major embarrassment to everyone.[51] The legal expenses have been gigantic, and certainly have been passed on to consumers.

Another extremely serious issue affects everyone in pacing, including the patient community. Because of stringent requirements for premarket approval (PMA) of medical devices by the U.S. Food and Drug Administration (FDA), and because of the long delays involved in obtaining such approval (perhaps aggravated by underfunding of the FDA itself), innovative clinical research has been forced overseas where controls are less stringent. This has deprived American innovators of the opportunity of testing their own inventions, and has dampened enthusiasm for research and development in the United States.

Pioneers in Pacing

The field of cardiac pacing is now old enough to have identified its pioneers and leaders, some of whom have died prematurely. Among these were Drs. Thalen and Welti in Europe, and Drs. Rosen, Littleford, Bilitch, and Mirowski in the United States. All had made immense and lasting contributions to pacing, and they have been much missed by their colleagues.

FUTURE DEVELOPMENTS

Whether the next decade will be as productive as the last is hard to predict, but many tempting speculations suggest themselves. Research in device technology has been slowed by financial constraints and, in part, by unclear objectives. If past events are any indication, there will surely be a tendency to develop even more complex and exquisitely programmable devices simply because it is possible to do so. One hopes that there also will be an effort to identify important gaps in clinical capabilities. There is evident need for simplification of pacemaker components for the sake of costs and ease of use, for compatibility of devices manufactured by different companies, for a "universal" programmer that can be used with any pacemaker, and for simplified follow-up systems.

The diagnostic features of devices developed for adaptive pacing certainly will be exported to other areas for electronic surveillance of cardiac and other aspects of physiologic function. The treatment of tachyarrhythmias and bradyarrhythmias will merge, blurring further the already unclear distinction among the cardiologist, electrophysiologist, and pacemaker-surgery specialist (if there is such a thing). We may expect to see the emergence of "rhythm-management systems" capable of treating brady- and tachyarrhythmias as well as ventricular tachycardia and fibrillation, incorporating sophisticated autodiagnostic and autoregulating features, and perhaps even the automatic administration of drugs.

On the other hand, the cost of care may become such an insurmountable problem that rationing of services almost certainly will be required. Reconciliation of these opposing forces poses an immense challenge to all those concerned.

ACKNOWLEDGMENTS

The authors are indebted to David Harari and Donna Neglia, RN, for their assistance in compiling some of the material presented in the tables and to Mia Parsonnet, MD, for her counsel in editing the text.

REFERENCES

1. Furman S, Escher DJW: *Principles and Techniques of Cardiac Pacing*. New York, Harper & Row, 1970, pp 1–13.
2. Siddons H, Sowton E: *Cardiac Pacemakers*. Springfield, IL, Charles C. Thomas, 1967.
3. Schechter DC: Background of clinical cardiac electrostimulation. *NY State J Med* 1971;71:2575–2805 and 1972;72:270–1191.
4. Greatbatch W: *Implantable Active Devices*. Clarence, NY, Greatbatch Enterprises, 1983, pp 1–8.
5. Littleford PO, Parsonnet V, Spector SD: Method for the rapid and atraumatic insertion of permanent endocardial pacemaker electrodes through the subclavian vein. *Am J Cardiol* 1979;43:980–982.
6. Parsonnet V, Crawford CC, Bernstein AD: The 1981 United States survey of cardiac pacing practices. *J Am Coll Cardiol* 1984;3:1321–1332.
7. Krasner JL, Noukydis PC, Nardella PC: A physiologically controlled cardiac pacemaker. *J Am Assoc Adv Med Instr* 1966;1:14.
8. Funke HD: Ein Herzschrittmacher mit belastungsabhaengiger Frequenzregulation. *Biomed Technik* 1975;20:225.
9. Cammilli L, Alcidi L, Papeschi G: A new pacemaker autoregulating the rate of pacing in relation to metabolic needs, in Watanabe Y (ed): *Proc Vth International Symposium on Cardiac Pacing*. Amsterdam, Exerpta Medica, 1977.
10. Bernstein AD, Camm AJ, Fletcher RD, et al.: The NASPE/BPEG Generic Pacemaker Code for antibradyarrhythmia and adaptive-rate pacing and antitachyarrhythmia devices. *PACE* 1987;10:794–799.
11. Parsonnet V, Furman S, Smyth NPD, et al.: Implantable cardiac pacemakers: Status report and resource guidelines, 1982. Pacemaker Study Group, Intersociety Commission for Heart Disease Resources. *Circulation* 1983;68:227A.
12. Owens B, ed: *Batteries for Implantable Biomedical Devices*. New York, Plenum Press, 1986.
13. Takeuchi ES, Quattrini PJ, Greatbatch W: Lithium/silver vanadium oxide batteries for implantable defibrillators. *PACE* 1988;11:2035–2039.
14 Napphotz, TA: Personal communication.
15. DeLeon SY, Ilbawi MN, Backer CL, et al.: Exit block in pediatric cardiac pacing: Comparison of the suture-type and fishhook epicardial electrodes. *J Thorac Cardiovasc Surg* 1990;99:905–910.
16. Myers M, Parsonnet V, Bernstein AD: Extraction of implanted transvenous pacing leads: A review of a persistent clinical problem. *Am Heart J* 1991; 121:881–888.
17. Parsonnet V: Stability of permanently implanted endocardial electrodes during open heart surgery. *Am Heart J* 1979;98:812.
18. Calfee RV, Saulson SH: A voluntary standard for 3.2 mm unipolar and bipolar pacemaker leads and connectors. *PACE* 1986;9:1181–1185.
19. Draft International Standard, ISO/DIS 5841–5843: *Cardiac Pacemaker—Part 3: Low Profile Connectors for Implantable Pacemakers*. International Organization for Standardization, UDC 615.817,1989.
20. Parsonnet V, Bernstein AD, Gallagher R, Crawford CC: Preoperative estimation of the proper transvenous electrode lead length. *PACE* 1984;7:475.
21. Sutton R, Guneri S: The impact of steroid eluting

leads on long term pacing in the atrium and ventricle. *J Am Coll Cardiol* 1990;15:68A.
22. Kay GN, Epstein AE, Plumb VJ: Comparison of unipolar and bipolar active fixation atrial pacing leads. *PACE* 1988;11:544–549.
23. Parsonnet V, Bernstein AD, Galasso D: Cardiac pacing practices in the United States in 1985. *Am J Cardiol* 1988;62:71–77.
24. Bernstein AD, Parsonnet V: The 1989 United States survey of cardiac pacing practices. In preparation.
25. Ellestad M, Wan M: Predictive implications of stress testing follow-up of 2700 subjects after maximum treadmill stress testing. *Circulation* 1975;51:363–369.
26. Camm AJ, Katritsis D: Ventricular pacing for sick sinus syndrome—A risky business? *PACE* 1990; 13:695–699.
27. Parsonnet V, Myers M, Perry GY: Paradoxical paroxysmal nocturnal congestive heart failure as a severe manifestation of the pacemaker syndrome. *Am J Cardiol* 1990;65:683–685.
28. Tyers GFO: Current status of sensor-modulated rate-adaptive cardiac pacing. *J Am Coll Cardiol* 1990;15:412–418.
29. Mitsui T, Hori M, Suma K, et al.: The "pacemaking syndrome," in Jacobs E (ed): *Proceedings of the Eighth Annual International Conference on Medical and Biological Engineering*. Chicago, Association for the Advancement of Medical Instrumentation, 1969, pp 29–33.
30. Ausubel K, Furman S: The pacemaker syndrome. *Ann Intern Med* 1985;103:420–429.
31. Barold SS, Falkoff MD, Ong LS, Heinle RA: Cardiac pacing for the treatment of tachycardia, in Barold SS (ed): *Modern Cardiac Pacing*. Mount Kisco, NY, Futura, 1985, pp 693–726.
32. De Belder JA, Malik M, Ward DE, Camm AJ: Pacing modalities for tachycardia termination. *PACE* 1990;13:231–248.
33. Parsonnet V, Bernstein AD: Transvenous pacing: A seminal transition from the research laboratory. *Ann Thorac Surg* 1989;48:738–740.
34. Parsonnet V: Unpublished data.
35. Bonavita G, Perry G, Hesselson AB, Parsonnet V: Incidence of pacemaker lead fracture by location and lead type. *PACE* 1990;13:551.
36. Furman S, Behrens M, Andrews C, Klementowicz P: Retained pacemaker leads. *J Thorac Cardiovasc Surg* 1987;94:770–772.
37. Byrd CL: Use of locking stylets and sheaths as part of an intravascular technique for lead extraction. *PACE* 1990;13:549.
38. Special Committee on Aging, U.S. Senate: Fraud, waste, and abuse in the Medicare pacemaker industry: An information paper. Washington, DC, U.S. Government Printing Office, 1982, document No. 98–116 O.
39. Greenberg A, Kowey PR, Bargmann E, Wolfe SM: Permanent pacemakers in Maryland. Report by Health Research Group, 2000 P Street, NW, Washington, DC 20036, July 1982.
40. Frye RL, Collins JJ, DeSanctis RW, et al.: Guidelines for permanent pacemaker implantation. *J Am Coll Cardiol* 1984;4:434–442.
41. Schlant RC, Adolph RJ, Beller GA, et al.: 17th Bethesda Conference: Adult Cardiology Training. *J Am Coll Cardiol* 1986;7:1192–1218.
42. Falk R: Impact of prospective peer review on pacemaker implantation rates in Massachusetts. *J Am Coll Cardiol* 1990;15:1087–1092.
43. Bernstein AD, Parsonnet V: Computer-assisted mode selection in antibradyarrhythmia pacing. *J Am Coll Cardiol* 1990;15:266A.
44. Parsonnet V, Bernstein AD: Overview of pacemaker follow-up techniques. *Heart House Learning Center Highlights* 1987;3:11–15.
45. Wilkoff B, Corey J, Blackburn G: A mathematical model of the cardiac chronotropic response to exercise. *J Electrophysiol* 1989;3:176–180.
46. Furman S, Bilitch M: NASPExAM. *PACE* 1987;10:278–280.
47. Parsonnet V, Bernstein AD, Lindsay B: Pacemaker-implantation complication rates: An analysis of some complicating factors. *J Am Coll Cardiol* 1989;13:17–21.
48. Parsonnet V, Furman S, Smyth NPD: Implantable cardiac pacemakers: Status report and resource guidelines. Pacemaker Study Group, Inter-Society Commission for Heart Disease Resources (ICHD). *Circulation* 1974;50:A21.
49. Hsiao WC, Braun P, Yntema D, et al.: Estimating physicians' work for a resource-based relative-value scale. *N Engl J Med* 1988;319:835–841.
50. Omnibus Budget Reconciliation Act of 1989, P.L. 101-239.
51. Greatbatch W: Pacemaker patents—Use and misuse. *PACE* 1989;12:115–116.
52. Benedek ZM, Gross J, Furman S: Rate modulated pacemakers. *Newspaper of Cardiology*, February 1990, pp 8–11.

Chapter **29**

Ventricular Tachycardia Ablation: Electrophysiologic Parameters for Selection of the Ablation Site

G. Fontaine, MD, R. Frank, J. Tonet, T. Iwa, Jr., and Y. Grosgogeat

Fulguration (DC catheter endocardial ablation) is a new technique used in the treatment of chronic ventricular tachycardias (VT).[1–5] The experience accumulated by our team over the years has been reported in previous papers.[6–9] Favorable results have been achieved because of multiple factors, including perseverance and special expertise in physics and high-voltage electricity. At present, we think that this procedure is less risky but less effective than surgery either because it must be restricted to a small zone of myocardium and/or because the site considered for ablation has not been precisely determined. The purpose of this chapter is to concentrate on the second subject. The concepts of clinical electrophysiology based on overdrive pacing and recording during VT could provide valid markers that are applicable to any form of ablation, including surgery. Some of these concepts will be reviewed in a simplified schematic fashion.

PACING AWAY FROM A NONPROTECTED TACHYCARDIA FOCUS

When the tachycardia focus (whether automatic or reentrant) is not protected by surrounding scar tissue or by a functional barrier to impulse entry, overdrive pacing (pacing at a rate faster than the VT rate) at a normal site will instantaneously activate the myocardium. However, an isoelectric line between the stimulus artifact and the paced QRS complex could be seen in some leads when the electrical forces resulting from myocardial activation are oriented perpendicularly to the recording lead, which simulates latency between the pacing stimulus and the resultant QRS complex. Inspection of other leads will demonstrate instantaneous activation of myocardium. This phenomenon is of crucial importance when pacing inside the zone of slow conduction is discussed.

Continuous overdrive stimulation will inhibit the tachycardia focus. If a local endocardial electrogram is recorded in the area of normal myocardium, it will reveal a discrete endocardial potential which is inscribed within the QRS complex recorded on the surface ECG (S, Fig. 29.1). At the cessation of pacing, the VT will resume at its previous rate after an interval (X, Fig. 29.1) that is identical to the interval between the two consecutive beats during the underlying tachycardia. Because of the phenomenon of overdrive inhibition, the escape interval after cessation of pacing can be longer than the basic tachycardia cycle. This can occur even when the rate of pacing is increased. Both the local endocardial electrogram (E, Fig. 29.1) and the surface QRS complex will have different morphology due to a change in the activation sequence from the paced beats to the VT beats arising from a different location. This kind of response to overdrive suggests that the pacing site is far from the tachycardia focus and should not be considered for ablation.

655 Avenue of the Americas, New York, NY 10010
Current Topics in Cardiology

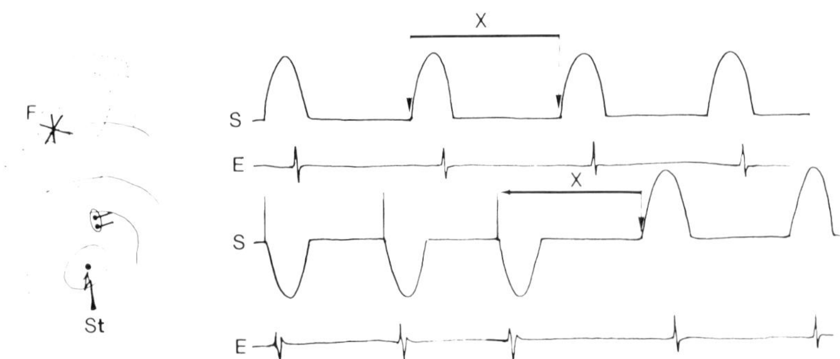

FIGURE 29.1 Effect of overdriving a nonprotected focus of VT by pacing away from the site or origin of tachycardia. The top surface electrogram (S) is schematically diagrammed and the endocardial electrogram (E) recorded by close bipolar electrodes is inscribed at the end of the QRS complex (top tracing). In the lower part of the tracing the tachycardia is overdriven by pacing stimuli (S) and the position of the endocardial signal E is shifted to the beginning of the QRS complex. At cessation of pacing, the escape interval is equal to or longer than the basic VT interval X.

INDUCTION AND TERMINATION OF VENTRICULAR TACHYCARDIA

Wellens and coworkers showed first in 1972 that chronic stable monomorphic ventricular tachycardia can be induced and terminated by programmed pacing,[10] a behavior observed in an appreciable number of cases. The same approach had been previously used in the investigation of the Wolff-Parkinson-White syndrome,[11] and in both cases suggested that ventricular tachycardia was the result of a reentrant phenomenon.[10,12–14] This finding was important because it opened a vast field of investigations. In particular it became possible to initiate VT at the time of surgery and to use the technique of epicardial mapping that had been previously developed for the Wolff-Parkinson-White syndrome.[15]

Simple ventriculotomy, which transsected the site of origin of VT, proved to be effective in some cases.[16] At the same time the technique of epicardial mapping suggested the presence of reentrant pathway and demonstrated differences in conduction speeds in different zones of myocardium.

MAPPING THE ACTIVATION LOOP DURING VENTRICULAR TACHYCARDIA

Direct demonstration of reentry was strongly suggested in 1975 in a patient with Uhl's anomaly.[17,18] In this rare congenital anomaly, the right ventricular wall is so thin (parchment heart) that in zones occupied by myocardium a two-dimensional structure could be investigated without the probability of an epicardial–endocardial reentry loop that can be present in the left ventricular wall.[17–19] Plication of myocardium made for technical reasons in the tissue that was not at the site of the slowest conduction but was an integral part of the reentry pathway resulted in inability to reinduce VT. This confirmed our initial observations in patients in whom a simple incision (simple ventriculotomy) at the site of origin of VT was sufficient to prevent recurrence of sustained VT.[20]

These observations provided the basis for surgical treatment of patients with chronic resistant VT.[21–23] The same concept was confirmed later by showing the reentry wave front in the experimental laboratory.[24]

VENTRICULAR TACHYCARDIA SUBSTRATE

Histological examination shows the presence of tissues that may constitute a possible reentrant pathway, but so far no marker that could have clinical significance has been identified.[25] Electrophysiological data and physiopathological studies have suggested the existence of a discrete structure that has been called the "arrhythmogenic substrate."

Reentry takes place in a particular kind of tissue consisting of a mixture of normal and abnormal fibers that can create a pathway of conduction in which a circus movement may be sustained when certain electrophysiologic conditions are present.

From an electrophysiological standpoint, reentry can be sustained in the presence of two adjacent structures with strongly different conduction properties that are isolated from each other. Arcs of conduction blocks have been shown by El-Sherif et al. and are considered fundamental in the development of reentry in experimental studies.[26] This means that there is a pathological and/or anatomical structure conducting at a reduced speed that can form a narrow isthmus connecting two zones of normal conduction.[27]

THE CRITICAL PATHWAY

Pathological evidence suggests that the area of abnormal conduction in the arrhythmogenic substrate is a complex structure that includes multiple zones of possible abnormal conduction in addition to the critical pathway required to sustain reentry.[28,29] If identification of the fulgurating zone must be precise to be effective, the shock should be directed to an area that is an integral part of the reentrant pathway and is essential for the perpetuation of VT.[27,30]

Slow conduction pathways have been observed in several portions of abnormal myocardium resulting in different VT morphologies. Therefore, it becomes necessary to modify conduction in each relevant part of the reentrant circuit.[24,31–33] The slow pathway consists of only few myocardial cells, which are not sufficient to generate electrical forces that can be recorded on the surface electrocardiogram (ECG). Therefore, the activation of normal myocardium during VT is the only direct marker available from the surface ECG.

The initial QRS forces during VT provide clues to the site where the tachycardia originates. However, this information is not precise and can be misleading even with regard to the site (right versus left ventricle) of VT origin.[34]

VALUE OF ELECTROPHYSIOLOGICAL STUDY

A more selective approach is to use a probe or a catheter in direct contact with the heart.[17,35] Recording a discrete potential at or shortly before the onset of the QRS complex during VT suggests that the "site or origin" of the VT focus has been identified.[34]

However, frequently there are in the area close to the site of VT origin discrete endocardial potentials preceding the onset of QRS complexes, which are outside the critical zone of conduction. This "presystolic" activity can be recorded in a zone that is abnormal but it represents an innocent bystander. Therefore, the presystolic potential associated with the critical zone must be validated prior to attempted ablation.[36,37] It is important to review the electrophysiological parameters in an area that is considered for ablation.

ENTRAINMENT OF VENTRICULAR TACHYCARDIA

The concept of entrainment introduced by Waldo et al. in postoperative atrial flutter has been extended recently to VT and has supported reentry as the mechanism of this arrhythmia.[38–40] Entrainment can be helpful in the selection of the appropriate site for ablation.[41,42]

When reentry occurs around an anatomical obstacle there exists an excitable gap ahead of the activation front (Fig. 29.2). A different activation front arriving at this excitable gap before the circulating reentrant activation front may capture it. Subsequent activation is divided in two limbs. The antidromic activation will meet the upcoming activation front after its entrance into the normal myocardium and will be blocked. The orthodromic activation will activate the excitable gap as well as the myocardium in the orthodromic direction and will preexcite the activation path in the orthodromic direction, thus resetting the tachycardia (Fig. 29.2). This forms the electrophysiological basis for entrainment. During a stable VT, if the heart is paced at a rate faster than the VT rate from an area of normal myocardium located close to the reentry loop, the activation front from each paced beat may reach the excitable gap and capture it before the circulating reentrant wave front. In this manner, the tachycardia is entrained by pacing. At cessation of pacing, the tachycardia resumes at its original rate. There are several consequences of such entrainment:

- **a.** The surface QRS morphology of VT will be altered due to fusion between two sites of activation of the normal myocardium; namely, the exit site from the abnormal conduction zone and the pacing site from the entraining pacing catheter.
- **b.** At any given pacing rate the degree of fusion, that is, the percentage of contribution of activation from the paced site, will be constant from beat to beat (Fig. 29.3). The timing of the endocardial signal will change during fusion if it is located in a zone directly influenced by pacing.
- **c.** The degree of fusion will progressively increase as the pacing rate is increased (Fig. 29.3). The conduction speed through the zone of abnormal conduction remains constant (or actually may decrease) as the pacing rate in-

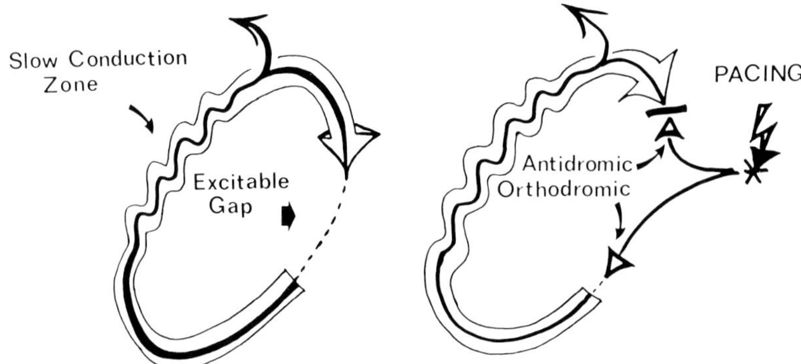

FIGURE 29.2 Main components of a reentrant VT pathway. The zone of slow conduction is indicated by a wavy line as opposed to the zone of normal conduction connecting the exit to the entry site of the zone of slow conduction. The exit site suggests that the activation spreads in several directions and could initiate a figure-eight pattern. Inside the area of normal conduction the left part of the figure represents the excitable gap. On the right side, pacing is performed in normal myocardium close to the reentrant pathway. The excitable gap is activated in two opposite directions: the antidromic direction activation collides with the upcoming wave front; the orthodromic activation preexcites the pathway and resets the tachycardia.

creases. This allows a progressively larger portion of the myocardium to be depolarized by the activation front spreading from the pacing catheter, and a progressively smaller portion by the activation from the exit site of the circus movement.

In addition, the alteration of the morphology of the QRS complex during entrainment depends on the location of the pacing site in relation to the exit site of the zone of abnormal conduction. When pacing is performed close to the exit site of the zone of abnormal conduction, the amount of myocardium that will be activated by the pacing stimulus before it encounters the upcoming activation from the exit site will be smaller than when the pacing site is further away from the exit site (Fig. 29.4). Conversely, there will

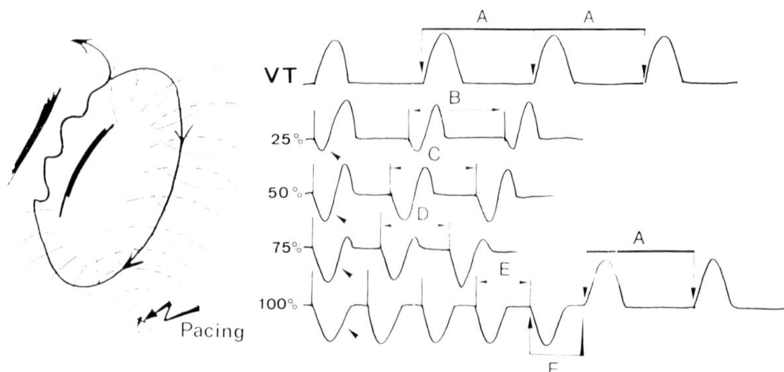

FIGURE 29.3 VT entrainment. The spontaneous VT rate is shown in the upper part of the figure with a cycle length A. Pacing close to the reentrant pathway produces a fusion beat. When the rate of pacing is relatively close to the spontaneous VT rate, a fusion complex incorporates 25% of pacing depolarized ventricle. This indicates that the beginning of the activation is made from the pacing site in a different direction up to the excitable gap. From this point, activation resets the tachycardia with a cycle length B with a minor fusion. When the rate of pacing is increased, a larger percentage of fusion is observed. It is schematically depicted as 50% with a period C. For a period E, 100% of ventricular depolarization is now made by pacing. At cessation of pacing, the first unpaced beat occurs at the rate of pacing. The first beat after the last paced beat has the same morphology as VT morphology and the following cycle has the same tachycardia cycle. Therefore, the first beat after cessation of pacing is entrained but not fused.

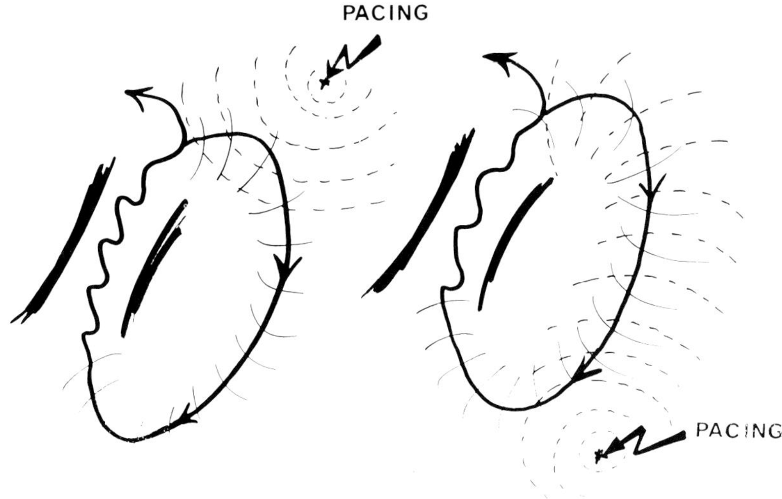

FIGURE 29.4 When pacing is performed in the area of normal conduction, its effect on the ECG and the resulting amount of fusion varies according to the site of pacing in relation to the tachycardial circle. In particular, when pacing is performed close to the exit of the zone of slow conduction (left side of the panel), the maximal amount of fusion is smaller than when pacing close to the zone of slow conduction as shown on the right side of the panel and in which a large amount of fusion could take place.

be a higher degree of fusion when pacing is performed closer to the entry site.[42]

d. The interval between the last paced beat and the first spontaneous beat will be the same as that of the pacing rate, but the morphology of the first spontaneous beat will be the same as that of the underlying VT without any signs of fusion (Fig. 29.3).

ABLATION SITE

A zone located in normal myocardium is obviously not an appropriate site for successful ablation because adjacent normal myocardium will continue to be activated normally and reentry will not be prevented.[43] We have never observed alterations of the QRS complex during sinus rhythm either in the clinical or in the experimental laboratory after standard fulguration of normal ventricular muscle on the right or on the left side of the heart. Hence, no change in the QRS morphology of VT should occur after delivery of the shock to normal myocardium. If there is a change in QRS morphology without change in the tachycardial rate, after shock, the latter has probably modified the exit site of the zone of abnormal conduction. If there is also a change in the tachycardia rate in addition to the change in morphology, the area of abnormal conduction probably has been modified, decreasing the speed of conduction but not impairing the safety factor of propagation sufficiently to prevent reentrant activation. A change in the tachycardia rate without a change in the QRS morphology of VT suggests modification of the zone of abnormal conduction without modification of the exit site.[44]

Finally, entrainment with fusion is a marker of the zone of normal myocardium that is located close to the reentrant pathway and is less appropriate for ablation than the zone of abnormal conduction itself. Ablation of the critical zone of abnormal conduction may be more effective for control of VT than ablation of the classical "site of VT origin."[41,45] Demonstration of entrainment through the zone of abnormal conduction helps to study the behavior of the arrhythmogenic substrate during programmed pacing.

DELAYED POTENTIALS AND THE ZONE OF STRUCTURAL SLOW CONDUCTION

Delayed potentials represented the first evidence consistent with the presence of a zone of slow conduction during sinus rhythm. They were first observed in humans by our group in December 1971, and reported in 1974 after observation of the second case.[46] We proposed that this delayed activity was the result of a particular form of slow conduction that was not generated by membrane currents during action potentials (APs) but more likely caused by slower trans-

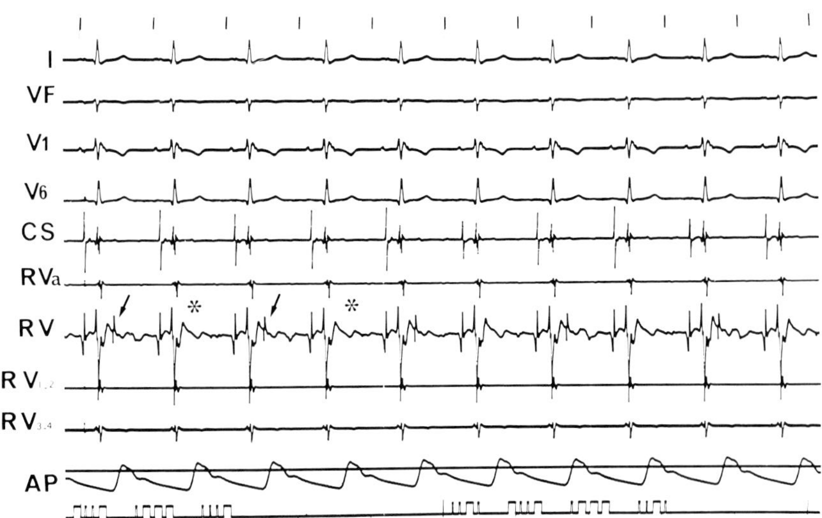

FIGURE 29.5 Typical double potential observed in the right ventricle in a patient with ventricular dysplasia. The catheter has been located under the tricuspid valve, and the right ventricular electrogram (RV) is recording both atrium and ventricle. A typical double-delayed potential is indicated by oblique arrows. However, there is a 2:1 block of this potential as shown by the asterisk. Therefore, it is possible to conclude that this delayed potential plays no role in the VT cycle and that this delayed potential should not be considered for ablation. CS = coronary sinus, RVa = apex of the right ventricle, RV1,2 and RV3,4 = quadripolar catheter in the infundibulum, AP = arterial blood pressure.

mission of activation from fiber to fiber, or perhaps resulting from electrotonic forces. These delayed potentials could explain also the propagation of activation at a relatively slow speed between two consecutive tachycardia beats. The behavior of delayed activity was originally studied by our group during surgery,[47–49] and later by the other groups in the clinical laboratory of electrophysiology.[35] It was also extensively documented in the experimental laboratory.[27,50] We found that:

a. Delayed potentials were of small amplitude and frequently fragmented or doubled (Fig. 29.5).

b. The zone of recording delayed potentials exhibited high pacing threshold.

c. Different forms of spontaneous conduction disturbances were observed (Fig. 29.5).

d. Pacing normal ventricular epicardium near

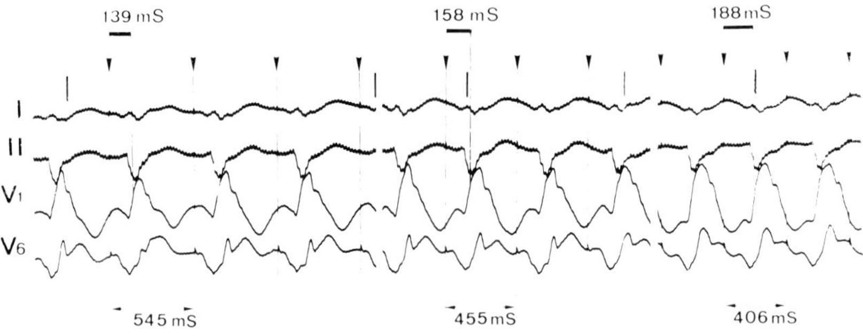

FIGURE 29.6 Time-dependent changes during an increase in pacing rate in the zone of slow conduction. From left to right, when the period of pacing is increased from 545 to 460 ms, the latency interval between the pacing stimulus and the middle of the ascending phase of the ventricular activation is prolonged from 130 to 188 ms. Note that for the high frequencies there is a slight change in the QRS morphology, especially in leads V1 and V6, which demonstrates that there is a real relation between pacing and the ventricular responses.

the sites at which delayed potentials were observed demonstrated that the coupling interval of the delayed potentials was increased with increasing the pacing rate.[48]

e. Premature stimuli delivered inside zones of slow conduction encountered exceedingly long effective refractory periods.

f. Pacing at the site of slow conduction showed that the activation was transmitted to the remaining part of myocardium after a long-lasting isoelectric line[48,51] (Fig. 29.6).

g. The preceding phenomenon exhibited time-dependent properties[48,51] (Fig. 29.6).

At this point, all basic elements constituting a reentry pathway were demonstrated in human myocardium.[52] This reinforced the previously mentioned results obtained at surgery where a possible circus movement was mapped in the case of Uhl's anomaly.[17]

THE MIDDIASTOLIC POTENTIALS

If a zone of slow conduction within the reentry pathway is activated by a slightly premature stimulus, it should produce a premature response after a certain latency but with the same morphology as that of the spontaneous tachycardia. The term perfect "pacemapping" used later to describe this condition is an extension of the original definition of pacemapping. When a more precise language is required, the term "pacemapping" as opposed to the term "pacemapping after delay" or "delayed pacemapping" could be used. However, here again ablation should be directed at the critical pathway. The term "critical" is quite relative. It means that the effect of the ablative procedure involves an area that is modified sufficiently to prevent reentry.

An ablative technique involving a large zone of myocardium will define a "critical" zone which would not be "critical" if a different less extensive ablative technique is employed.[43]

THE ZONE OF SLOW CONDUCTION

During the electrophysiological evaluation of VT in a patient with a circus movement tachycardia, it is possible to obtain information about the extent of the zone of slow conduction. If we consider a two-dimensional structure like the free wall of the right ventricle in a patient with arrhythmogenic right ventricular dysplasia, the zone of slow and normal conduction can be depicted as in Figure 29.7.

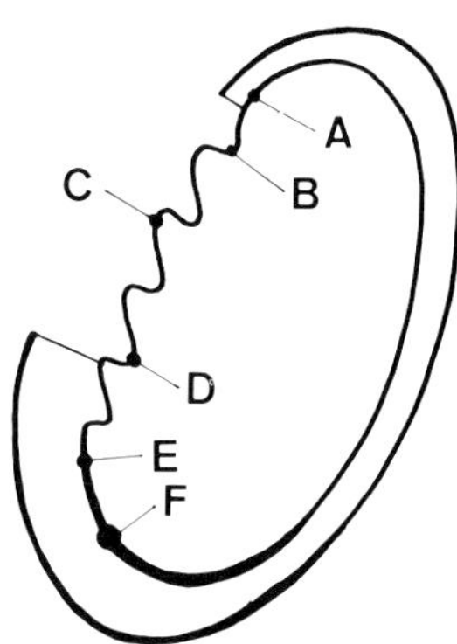

FIGURE 29.7 The reentry loop of VT. The outer solid line indicates the activation of normal myocardium which starts at the exit site of slow conduction. End of normal myocardial activation occurs after its entrance into the slow conduction pathway. Points B, C, and D are located on the zone of slow conduction. Points A, E, and F are located on the normal pathway on the reentry loop just after exit (and before entering the slow-conducting zone).

When the latency interval between the pacing stimulus and the onset of the QRS complex resulting in a perfect delayed pacemapping is identical to the coupling of a presystolic potential during VT, and this interval is short (site B), it can be assumed that the paced site is located close to the exit of the zone of slow conduction. Conversely, when the latency interval is long, the paced site is probably closer to the entry into the zone of slow conduction (site C or D). Therefore, these points define a zone of structural (as opposed to functional) slow conduction.

The three points (B, C, and D) located in the area of slow conduction must be consistent with previously described properties of the circuit. If we consider the adjacent point A, it is also located within the reentrant pathway; it would appear to be appropriate for ablation if it were located close to the exit of the zone of slow conduction. However, it is located within normal myocardium and, therefore, even though it produces a perfect pacemapping, it does not meet the specific slow conduction zone criteria (this point being located after the zone of slow conduction). It is necessary to show that normal electrophysiologic properties create a large amplitude electrogram, which is transmitted to the remaining normal myocardium. Activation at the speed limited by the local Wenckebach point can overdrive the ventricle to its maximal rate.

If we consider point E, it is also located in a zone of normal tissue adjacent to the entry in the zone of slow conduction, and therefore should meet the criteria of the triad but not the specific criteria of the zone of slow conduction. When the excitable gap is small, it depolarizes such a small amount of ventricular muscle that pacing does not modify the QRS complex of VT. Since point E is located before the zone of slow conduction, it cannot transmit the activation at the speed limited by the local Wenckebach point.

FUNCTIONAL BLOCK AND SLOW CONDUCTION

We have mentioned earlier that the reentry pathway is the result of an anatomical or pathological obstacle forming a narrow isthmus, which can be the site of slow conduction. It can be also assumed that part of the reentry circuit is preceded by an isthmus produced by a surrounding functional block, marked as point F. In this case the behavior of transmission will be similar to that at points A and E, that is, showing some of the normal local electrophysiologic properties. However, conduction between this point and the zone of slow conduction can be either normal or abnormal, depending on the electrophysiologic properties producing the functional block. Note that this phenomenon of functional block is possible only before the entry in the area of slow conduction.

It is now possible to understand that the most appropriate area for ablation should be the critical area of slow conduction inside an abnormal structure and not a zone artificially created by a functional block and located too far from the zone of slow conduction to be affected by the ablative procedure. Therefore, we operate within a framework of a concept of a zone of slow conduction located within a zone of abnormal structure. This will be later designated as the zone of "structural slow conduction."

CORRELATION BETWEEN ECG AND THE REENTRY LOOP

The logical starting point of the reentry pathway may be assigned to the beginning of the QRS complex of VT, which is the entry into the conduction path consisting of normal myocardium and representing the fast conduction pathway. This point is, in fact, preceded by a short zone of normal conduction located at the end of the zone of slow conduction, which has not generated sufficient electrical forces to be recorded from the surface.

As a first approximation, it is possible to postulate that the two phenomena, that is, activation of normal myocardium from the site of origin and activation of the structural zone of slow conduction, represent two zones that could be clearly identified. However, at a certain time the reentrant wave front will enter the zone of slow conduction. This is not likely to occur at the end of the QRS complex (Fig. 29.8a) because it would mean that the activation of the normal myocardium is over when reentrant activation invades the zone of slow conduction. Conversely, if the reentrant activation occurs at the beginning of the preceding QRS complex, we can deduce that the activation time spent in the zone of slow conduction is quite long and that a small zone of normal tissue is activated by the reentry loop (Fig. 29.8d). Two other possibilities shown in Figure 29.8, b and c, are more likely.

In most instances activation of the entrance into the area of slow conduction will take place during the inscription of QRS. We have seen that this could be more accurately determined by recording the point that meets the criteria of structural slow conduction in addition to the conditions suggesting the longest activation time in this zone. We and others [53] have seen examples where this delay was so long that the pacing stimulus was observed inside the *previous* QRS complex. This suggests that the exit of the activation originating in the reentry has a small zone in common with the normal myocardium. This may be an interesting area to ablate (Fig. 29.8d).

The same delay cannot be longer than the

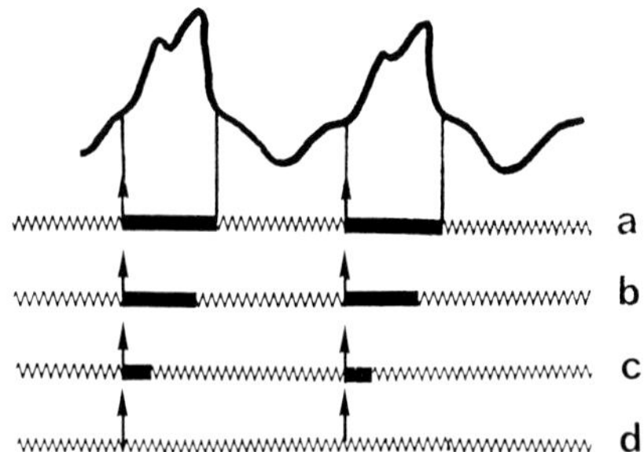

FIGURE 29.8 The connection between the zone of slow conduction and normal myocardium can be deduced from the analysis of the ECG and is assumed to occur at the onset of QRS complex of VT recorded on the surface ECG. However, the entrance into the zone of slow conduction can not be predicted from the surface ECG. This point could be theoretically at the end of the QRS complex but this is an improbable situation which is depicted in a. Most frequently, the connection between the normal myocardium and the reentry loop is located in the middle or at the beginning of the QRS complex as depicted in b and c. In d, a very small zone of normal myocardium is activated by the reentry loop.

basic cycle length of the tachycardia. If this is the case, this suggests that we are dealing with a zone of slow conduction that is in some way *connected* with the reentrant pathway and therefore cannot be "critical" and should not be considered for ablation. This is of real concern because the critical zone of structural slow conduction is surrounded by abnormal myocardium which may introduce a significant error in the electrophysiological results. A zone of slow conduction connected to the critical pathway may be located at any place along the critical zone of slow conduction. We think that it is possible to identify this particular situation by means of electrophysiological measurements.[54]

ENTRAINMENT WITHOUT FUSION

We have postulated that the zone of slow conduction is essential to the tachycardia circuit, and probably plays little or no role in the overall myocardial contraction. Hence, it is logical to assume that the most appropriate site for any ablation is the critical zone of slow conduction itself. In addition, this slow conduction zone may potentially be quite sensitive to ablation energy.

New criteria have been developed (the Sydney criteria)[55] to determine whether the ablating catheter is located in the critical zone of slow conduction. Firstly, during VT, a local endocardial potential of activation (E, Fig. 29.9) is recorded that precedes the surface QRS complex by a certain "presystolic interval." Second, when pacing is performed inside the zone of slow conduction, at a rate slightly faster than the spontaneous VT rate, the pacing artifact is followed by a certain "latency" before the onset of the surface QRS complex. The length of this latency should be equal to the interval between the presystolic activity and the unpaced QRS (PR1, Fig. 29.9). In addition this interval will exhibit time-dependent properties (PR2, Fig. 29.9). Third, pacing in the area of slow conduction during VT exactly reproduces the QRS morphology of the clinical VT, while accelerating the rate of the VT to the overdrive pacing rate.

This new approach is not based on a change in the morphology of the QRS complex secondary to fusion, as seen during classical entrainment. In fact, the QRS morphology remains unchanged. Therefore, this has been called "entrainment without fusion." However, the phenomenon of entrainment is clearly present because the tachycardia rate is increased and the tachycardia resumes at its original rate after cessation of pacing. In this case, the first reentrant beat following the last paced beat occurs after an interval (Fig. 29.9) equal to the period of the initial VT, with the same morphology as the VT (and, of course, the paced beats). This is in contrast to the classical form of entrainment with fusion, in which the first reentrant beat following the cessation of pacing follows the paced beat at the same interval as the pacing rate, and with a different morphology.

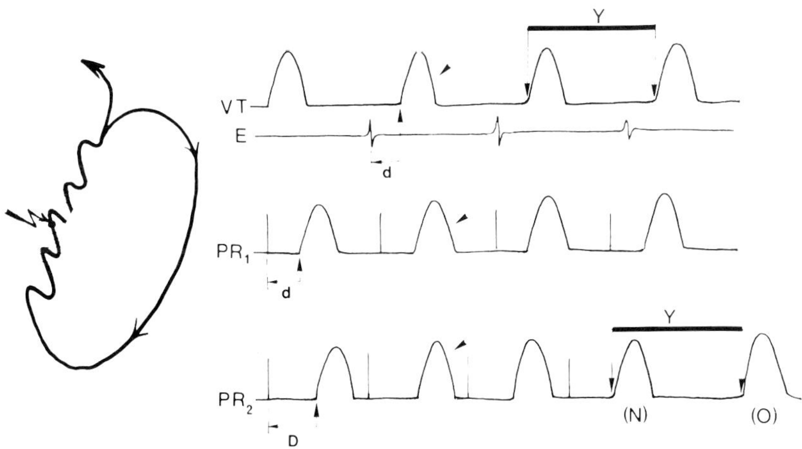

FIGURE 29.9 Criteria of the Sydney triad are: (a) Pacing during spontaneous VT rate (PR1) shows a latency interval (d) before the onset of the QRS complex. (b) This interval is the same as the presystolic interval observed during VT endocardial recording by the same catheter. (c) QRS complex morphology during VT overdrive does not change. When the rate of pacing is increased (PR2), the latency interval (d) is prolonged. The first escape interval (Y) after cessation of pacing is the same as the spontaneous period of the VT cycle.

ENTRAINMENT WITH FUSION THROUGH A ZONE CONNECTED TO THE REENTRANT PATHWAY IN THE AREA OF NORMAL CONDUCTION

This situation involves overdrive pacing in a zone of slow conduction with an exit located on the normal conduction limb of the reentrant pathway. The properties previously described in case of the classical entrainment will be observed, with the exception that pacing will be followed by a certain latency interval. During spontaneous tachycardia, the local potential will appear later due to the connecting zone of slow conduction. At the interruption of pacing, the last paced beat will be entrained and not fused.

ENTRAINMENT WITHOUT FUSION THROUGH A ZONE CONNECTED TO THE REENTRANT PATHWAY IN THE AREA OF SLOW CONDUCTION

Exploration of the zone of slow conduction has shown that many patients with reentrant VT have a pattern during VT in which the presystolic interval is shorter than the latency interval (Fig. 29.10). This could possibly be due to the placement of the pacing catheter in an area of slow conduction that is connected to, but is not part of, the slow conduction zone critical for the reentrant tachycardia. Activation in this area will travel to the critical zone, exit, and depolarize normal myocardium. It will then "normally" reenter the entrance site of the critical zone of slow conduction and advance into the connecting zone of slow conduction. Hence, the connecting zone will be activated twice, and the time spent by the preceding activation front to accomplish its journey will be longer than the tachycardia cycle.

INCORPORATION OF THE TACHYCARDIA CYCLE LENGTH

The approaches outlined for identification of the critical zone have not included the effects of tachycardia cycle length. However, it is well known that tachycardias with the same morphology can have different cycle lengths, suggesting changes in the speed of conduction in the abnormal area or a different input into the area of slow conduction. Therefore, accurate measurements should consider this possibility.

ATYPICAL FORM OF ENTRAINMENT

Frequently, one observes a pattern of short latency while pacing inside the slow conduction zone, but at the same time the endocardial activation occurs at the end of the QRS complex during spontaneous VT. In this case, the time spent by the activation front induced by pacing and the occurrence of the discrete potential is shorter than the reentrant cycle, a situation that is not anticipated logically.

One possible interpretation of this phenomenon is that the pacing stimulus is delivered by the distal electrode inside the zone of slow conduction close to its exit site, but the resulting electrogram is registered by the proximal pole of the bipolar electrode and represents a potential recorded from a different site of slow conduction. Therefore, this site should not be considered for ablation.[56]

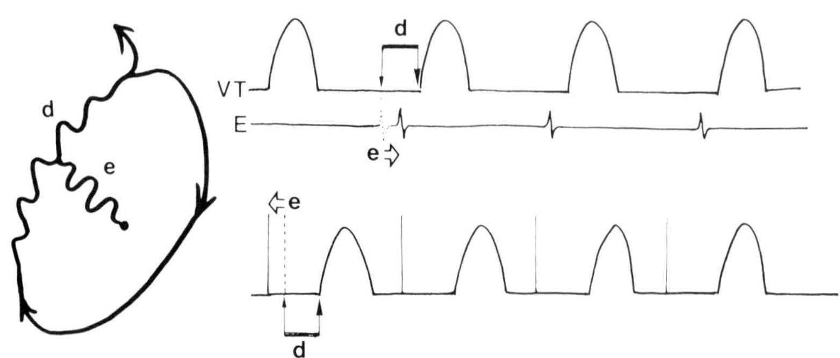

FIGURE 29.10 When the point of interest is located in a zone of slow conduction which is connected to the critical zone of slow conduction, the recording of the local electrogram shows an earlier prematurity but a longer latency interval.

ADDITIONAL PROPERTIES OF THE SLOW-CONDUCTING ZONE

Certain other characteristics of the behavior of the zone of slow conduction have application during the ablation procedure.

a. Induction of VT by increasing the rate of pacing. VT may be sometimes induced by increasing the rate of pacing in the critical zone to the tachycardia rate. This may be the result of antidromic activation of the zone of slow conduction during pacing within the zone of slow conduction. When this activation reaches the myocardium opposite to the previous exit, the resulting QRS complex will have a different configuration. Therefore, a different mechanism of a fusion complex can be observed in this case.

b. Termination of VT without activation. VT may be interrupted by a pacing stimulus delivered in the zone of slow conduction or in a zone of slow conduction without activating the ventricle.[57–59] This is the result of failure of transmission inside the zone of slow conduction which can be caused by several mechanisms. Strictly speaking, it is incorrect to say that the stimulus was delivered during the ventricular refractory period if the latter has not been previously assessed. It could be that the paced area did not transmit the activation to the normal myocardium but modified a partially activated zone. Recording a local electrogram helps to identify this mechanism (Fig. 29.11). Several electrophysiological phenomena at the cellular level could explain this phenomenon.[60]

PRACTICAL METHODOLOGY

It may be convenient to localize the zone of slow conduction by means of perfect or delayed pacemapping. Continuous slight overdrive of the VT is easier than the delivery of premature stimuli during the tachycardia cycle. Interruption of overdrive pacing should cause the tachycardia rate to resume at its original cycle length without change in the QRS complex morphology. Therefore, the risk of changing myocardial properties due to time dependencies is minimal.[61]

Because of identical QRS morphologies of VT during pacing and endocardial recording, it is no longer necessary to measure accurately the beginning of QRS complexes of VT. Any point that is easy to identify on the QRS complex may be chosen. The practical application of this property is exemplified by the use of a Tektronix digital

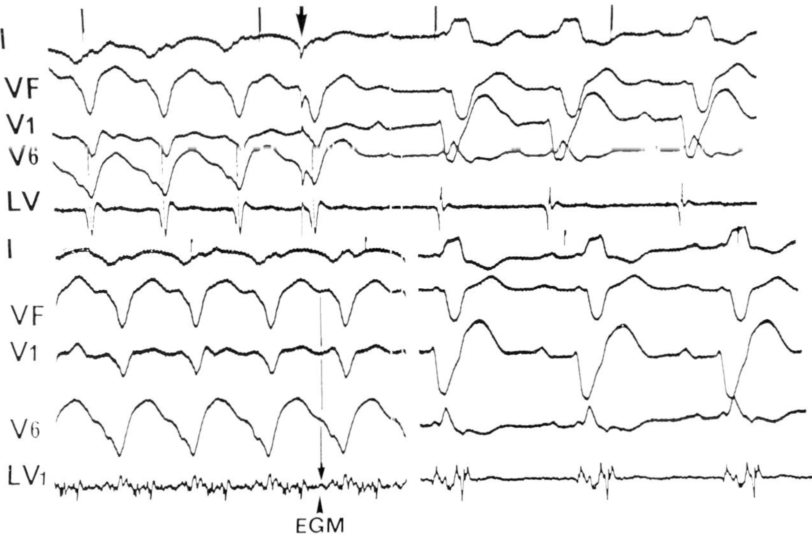

FIGURE 29.11 Interruption of VT by pacing in the zone of slow conduction without ventricular capture. In the upper part of the figure, the pacing stimulus indicated by a vertical arrow, delivered just after the onset of the QRS complex, interrupts VT. The lower part of the figure is taken from a retrospective analysis of tape recording taken just before the previous tracing-pacing when the catheter was recording the local EGM. It shows a consistent pattern of major fragmentation but always exactly the same reproducible pattern from beat to beat. This suggests that the pacing stimuli have made refractory some of the critical fibers of the arrhythmogenic substrate preventing perpetuation of the arrhythmia by modifying synchronization of diseased fibers.

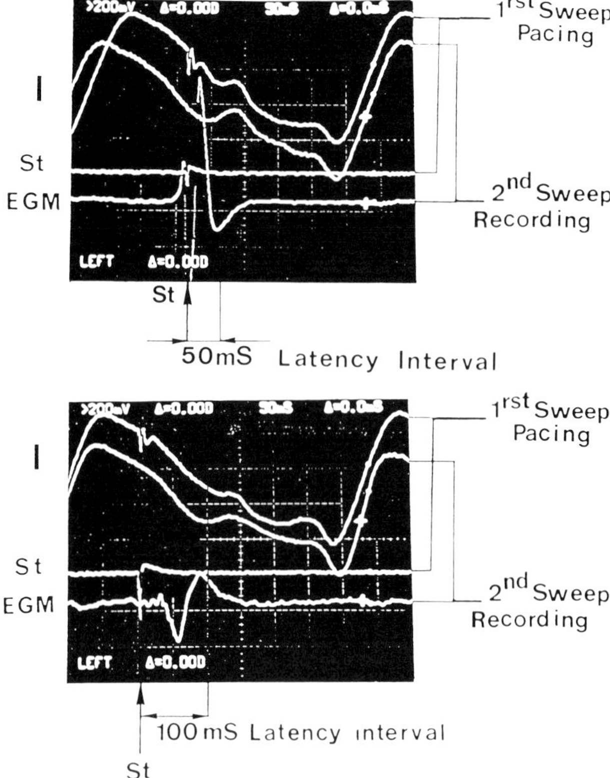

FIGURE 29.12 Use of the Tektronix digital oscilloscope to establish criteria for ablation. During the first sweep, the surface QRS (**top**) is recorded, in addition to the result of pacing in the area of interest. During the second sweep, the spontaneous VT ECG is displayed and the second tracing (lower in each figure) records the electrogram. In the upper figure, the presystolic interval is equal to the latency interval of 50 ms, but the local EGM shows a large amplitude and the intervals are very close to the onset of the QRS complex, suggesting that the zone under consideration pertains to the site of origin of VT. In the lower figure, the latency interval is in the range of 100 ms and the local EGM has a presystolic coupling in the range of 50 ms. This potential is of small amplitude and of low frequency, suggesting that the record is made in the area of structural slow conduction. Despite a difference of 50 ms between the presystolic interval and the latency interval, this point was chosen and ablation will prove to be successful.

storage oscilloscope (Fig. 29.12). The triggering point of the time base may be chosen on the fastest slope of the QRS complex or at its peak, and displayed in the middle (or shifted to the right) of the screen. The presystolic activity as well as the pacing stimulus latency will be displayed on the left of the QRS complexes and stored in two independent memories.[54] This facilitates comparison of the coupling intervals. When the zone under consideration is not met, there is a long interval between the local electrogram recorded during the first sweep and the pacing stimulus and the QRS complex recorded during the second sweep. The goal of the investigator is to reduce this interval to a minimum.

Measurements made in two steps using the oscilloscope procedure are faster than those performed tediously on the recording paper.

CLINICAL RESULTS

The preceding concepts have been based on logical deductions which may be altered by many biological features concerning the inhomogeneity of the zones of slow conduction. Therefore, clinical evaluation of the patient "fulgurated" on the basis of the discussed hypothesis should demonstrate prevention of VT. The most appropriate cases are patients with hemodynamically stable monomorphic VT in whom fibrosis is not extensive and the tachycardia originates in a zone that is not too far from the fulgurating catheter. Patients with arrhythmogenic right ventricular dysplasia (ARVD) and in some cases after myocardial infarction provide appropriate clinical models. Our first case was studied in 1984. This patient had VT of three morphologies, two of which were clinically documented. One VT was fulgurated in the structural area of slow conduction according to the Sydney triad criteria. After stabilization of the myocardium, no tachycardia recurred during the follow-up. This new approach had been used successfully in six other patients in our series.

DISCUSSION

It is obvious that the presented descriptions oversimplify the complex phenomena that occur in diseased myocardium. The zone of slow conduction is not a single, well-circumscribed area but encompasses a larger zone in which there are conceivably several entry and exit sites, and the propagation of activation follows circuitous path-

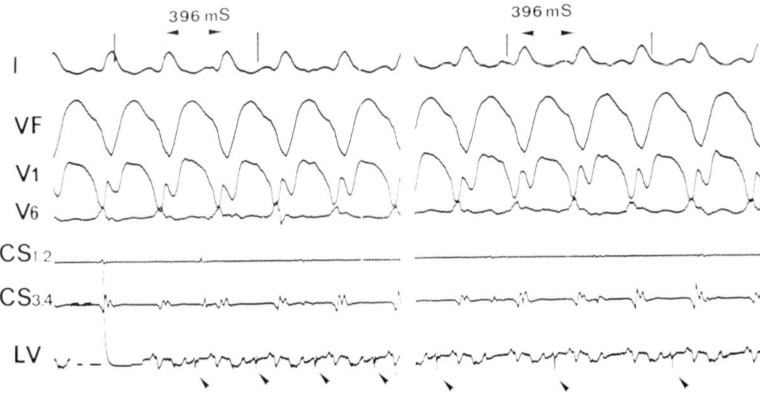

FIGURE 29.13 One of the main limitations of localizing the area appropriate for ablation is the recording of a nonappropriate electrogram. On the left, the oblique arrows show a middiastolic discrete potential that exhibits a 1:1 conduction in relation to the VT cycle. On the right, these potentials now exhibit a 2:1 conduction. It is not part of the reentrant pathway and plays no role on the VT cycle. This site should not be considered for ablation and pacing was not attempted at this site.

ways that have only been partially defined. The velocity of conduction is also practically unknown, probably varying from place to place and unlikely to be the same in both directions.

In such abnormal tissue, it is difficult to record any potential and increased amplification will be necessary. However, the multiphasic nature of fragmented potential could make it difficult to recognize the discrete potential produced by activation wave front. Occurrence of a 2:1 block of discrete potential considered appropriate for ablation is a major difficulty that has been encountered in a few cases (Fig. 29.13).

Pacing in the area of slow conduction can also pose a problem, and a stimulator providing high-amplitude output is mandatory. The interval between the pacing stimulus and the discrete endocardial electrogram cannot be reduced to zero when the critical zone is inaccessible to the exploring catheter, for example, by using endocardial investigation when the site of tachycardia focus is intraseptal or epicardial.

Reasoning alone without careful clinical investigation and without experimental verification of the mechanisms of reentry will probably be of little clinical value. However, the concepts of entrainment with and without fusion, and of the zone of slow conduction with entrance and exit sites, have been helpful in localizing an effectively ablated site. Our results suggest that the zone of slow conduction may be a more appropriate place to deliver ablation energy than the site of earliest activation, as was thought previously. Further studies are necessary to be able to reinterpret our data with these new concepts in mind.

CONCLUSION

Electrophysiology of the zone of slow conduction in VT is complex and we have presented only simplified concepts. The preceding criteria are theoretically applicable to any kind of ablative technique and practical application will further refine our deductive approach.

ACKNOWLEDGMENTS

This work was supported in part by grants from Centre de Recherche sur les Maladies Cardiovasculaires de l'Association Claude Bernard, La Fédération de Cardiologie, and L'Institut National de la Sante et de la Recherche Medicale (INSERM Contrat N° 865005).

REFERENCES

1. Hartzler GO: Electrode catheter ablation of refractory focal ventricular tachycardia. *J Am Coll Cardiol* 1983;2:1107–1113.
2. Puech P, Gallay P, Grolleau R, Koliopoulos N: Traitement par électrofulguration endocavitaire d'une tachycardie ventriculaire récidivante par dysplasie ventriculaire droite. *Arch Mal Coeur* 1984;77:826–835.
3. Fontaine G, Frank R, Tonet JL, Gallais Y, Touzet I, Todorova M, Baraka M, Grosgogeat Y: Treatment of resistant ventricular tachycardia with endocavitary fulguration and antiarrhythmic therapy, compared to antiarrhythmic therapy alone: Experience in 111 consecutive cases with a mean follow-up of 18 months. *Texas Heart Inst* 1986;13:401–418.
4. Morady F, Scheinman MM, DiCarlo LA Jr, Davis JC, Herre JM, Griffin JC, Winston SA, De Buitleir M, Hantler C, Wahr J, Kou WH, Nelson SD: Catheter ablation of ventricular tachycardia with intracardiac shocks: Results in 33 patients. *Circulation* 1987;75:1037–1049.

5. Borggrefe M, Breithardt G, Podczeck A, Rohner D, Budde Th, Martinez-Rubio A: Catheter ablation of ventricular tachycardia using defibrillator pulses: Electrophysiological findings and long-term results. *Eur Heart J* 1989;10:591–601.
6. Fontaine G, Tonet J, Frank R, Gallais Y, Rougier I, Farenq G, Lilamand M, Grosgogeat Y: Catheter ablation for ventricular tachycardia. B. Clinical application, in Saksena S, Goldschlager N (eds): *Electrical Therapy for Cardiac Arrhythmias. Pacing, Antitachycardia Devices, Catheter Ablation.* Philadelphia, WB Saunders, 1990, pp 663–683.
7. Fontaine G, Frank R, Tonet J, Rougier I, Farenq G, Grosgogeat Y: Treatment of rhythm disorders by endocardial fulguration. *Am J Cardiol* 1989;64:83J–86J.
8. Fontaine G, Frank R, Rougier I, Tonet JL, Gallais Y, Farenq G, Lascault G, Lilamand M, Fontaliran F, Chomette G, Grosgogeat Y: Electrode catheter ablation of resistant ventricular tachycardia in arrhythmogenic right ventricular dysplasia. Experience of 13 patients with a mean follow-up of 45 months. *Eur Heart J* 1989;10(suppl D):74–81.
9. Fontaine G, Tonet JL, Frank R, Rougier I: Electrode catheter ablation of ventricular tachycardia by fulguration and antiarrhythmic therapy. Experience of 43 patients with a mean follow-up of 29 months. *Chest* 1989;95:785–797.
10. Wellens HJJ, Schuilenburg RM, Durrer D: Electrical stimulation of the heart in patients with ventricular tachycardia. *Circulation* 1972;46:216.
11. Durrer D, Schoo L, Schuilenburg RM, Wellens HJJ: The role of premature beats in the initiation and termination of supraventricular tachycardia in the WPW syndrome. *Circulation* 1967;36:644.
12. Wellens HJJ, Lie KI, Durrer D: Further observations on ventricular tachycardias as studied by electrical stimulation of the heart. *Circulation* 1974;49:647.
13. Wellens HJJ, Duren DR, Lie KI: Observations on mechanisms of ventricular tachycardia in man. *Circulation* 1976;54:237.
14. Wellens HJJ: Value and limitations of programmed electrical stimulation of the heart in the study and treatment of tachycardias. *Circulation* 1978;57:845–853.
15. Fontaine G, Frank R, Bonnet M, Cabrol C, Guiraudon G: Methode d'etude experimentale et clinique des syndromes de Wolff-Parkinson-White et d'ischemie myocardique par cartographie de la depolarisation ventriculaire epicardique. *Coeur Med Interne* 1973;12:105.
16. Fontaine G, Guiraudon G, Frank R, Coutte R, Dragodanne C: Epicardial mapping and surgical treatment in 6 cases of resistant ventricular tachycardia not related to coronary artery disease, in Wellens HJJ, Lie KI, Janse MJ (eds): *The Conduction System of the Heart.* Leiden, Stenfert Kroese Pub., 1976, p 545.
17. Fontaine G, Guiraudon G, Frank R, Vedel J, Grosgogeat Y, Cabrol C, Facquet J: Stimulation studies and epicardial mapping in ventricular tachycardia: Study of mechanisms and selection for surgery, in Kulbertus HE (ed): *Reentrant Arrhythmias.* Lancaster, MTP Pub., 1977, pp 334–350.
18. Fontaine G, Guiraudon G, Frank R, Vedel J, Grosgogeat Y: Tratamiento quirurgico de las arritmias ventriculares. Valor de los estudios electrofisiologicos y de los potenciales tardios en la taquicardia ventricular de origen isquemico y no isquemico, in Bayes A, Cosin J (eds): *Diagnostico y tratamiento de las arritmias cardiacas.* Barcelona, Ediciones Doyma S.A. Pub., 1978, p 307.
19. Fontaine G, Guiraudon G, Frank R, Vedel J, Grosgogeat Y: Epicardial mapping in 1978, in Bayes A, Cosin J (eds): *Diagnosis and Treatment of Cardiac Arrhythmias.* Oxford, Pergamon Press, 1980, pp 109–125.
20. Fontaine G, Shantha N, Frank R, Tonet JL, Cansell A, Grosgogeat Y: New approaches in the electrophysiological determination of optimal treatment of recurrent tachyarrhythmias, in Greenberg HM, Kulbertus HE, Moss AJ, Schwarz PJ (eds): *Clinical Aspects of Life-Threatening Arrhythmias.* Ann NY Acad Sci 1984, vol 427, pp 67–83.
21. Fontaine G, Guiraudon G, Frank R, Fillette F, Cabrol C, Grosgogeat Y: Value of epicardial mapping in the management of chronic ventricular tachycardia. *Card Rev Rep* 1981;2:766.
22. Waldo AL, Arciniegas JG, Klein HO: Surgical treatment of life-threatening ventricular arrhythmias: The role of intraoperative mapping and consideration of the presently available surgical techniques. *Prog Cardiovasc Dis* 1981;23:247.
23. Harken AH, Horowitz LN, Josephson ME: Comparison of standard aneurysmectomy and aneurysmectomy with direct endocardial resection for the treatment of recurrent sustained ventricular tachycardia. *J Thorac Cardiovasc Surg* 1980;80:527.
24. El-Sherif N, Mehra R, Gough WB, Zeiler RH: Reentrant ventricular arrhythmias in the late myocardial infarction period. Interruption of reentrant circuits by cryothermal techniques. *Circulation* 1983; 68:644–656.
25. Ursell PC, Gardner PI, Albala A, Fenoglio JJ Jr, Wit AL: Structural and electrophysiological changes in the epicardial border zone of canine myocardial infarcts during infarct healing. *Circ Res* 1985;56:436–451.
26. Gough WB, Mehra R, Restivo M, Zeiler RH, El-Sherif N: Reentrant ventricular arrhythmias in the late myocardial infarction period in the dog. 13- Correlation of activation and refractory maps. *Circ Res* 1985;57:432–442.
27. El-Sherif N, Gough WB, Restivo M: Reentrant ventricular arrhythmias in the late myocardial infarction period. 14- Mechanisms of resetting entrainment, acceleration, or termination of reentrant tachycardia by programmed electrical stimulation. *PACE* 1987;10:341.
28. Wit AL: Electrophysiological mechanisms of ventricular tachycardia caused by myocardial ischemia and infarction in experimental animals, in Josephson ME (ed): *Ventricular Tachycardia. Mechanism and Management.* Mount Kisco, NY, Futura Publishing, 1982, pp 33–96.
29. Fenoglio JJ Jr, Pham TD, Harken AH, Horowitz LN, Josephson ME, Wit AL: Recurrent sustained ventricular tachycardia: Structure and ultrastructure of subendocardial regions in which tachycardia originates. *Circulation* 1983;68:518–533.
30. Gessman LJ, Endo T, Egan J, Gallagher JJ, Hastie R, Maroko PR: Dissociation of the site of origin from

the site of cryo-termination of ventricular tachycardia. *PACE* 1983;6:1293–1305.

31. Gessman LJ, Agarwal JB, Endo T, Helfant RH: Localization and mechanism of ventricular tachycardia by ice mapping 1 week after the onset of myocardial infarction in dogs. *Circulation* 1983;68:657–666.
32. Morady F, Frank R, Kou WH, Tonet JL, Nelson SD, Kounde S, De Buitleir M, Fontaine G: Identification and catheter ablation of a zone of slow conduction in the reentry circuit of ventricular tachycardias in humans. *J Am Coll Cardiol* 1988;11:775–782.
33. Frank R, Tonet JL, Kounde S, Farenq G, Fontaine G: Localization of the area of slow conduction during ventricular tachycardia, in Brugada P, Wellens HJJ (eds): *Cardiac Arrhythmias: Where To Go from Here?* Mount Kisco, NY, Futura Publishing, 1987, pp 191–208.
34. Josephson ME, Horowitz LN, Spielman SR, Waxman HL, Greenspan AM, Marchlinski PE, Ezri MD: Sustained ventricular tachycardia. Role of the 12-lead electrocardiogram in localizing site of origin. *Circulation* 1981;64:257–272.
35. Josephson ME, Horowitz LN, Farshidi A, Spear JP, Kastor JA, Moore EN: Recurrent sustained ventricular tachycardia. II- Endocardial mapping. *Circulation* 1978;57:440.
36. Frank R, Fontaine G, Baraka M, Kounde S, Farenq G, Grosgogeat Y: Catheter endocardial mapping in fulguration, in Aliot E, Lazzara R (eds): *Ventricular Tachycardias: From Mechanism to Therapy*. Dordrecht, Martinus Nijhoff, 1987, pp 390–402.
37. Fitzgerald DM, Friday KJ, Wah JA, Lazzara R, Jackman WM: Electrogram patterns predicting successful catheter ablation of ventricular tachycardia. *Circulation* 1988;77:806–814.
38. Okumura K, Olshansky B, Henthorn RW, Epstein AE, Plumb VJ, Waldo AL: Demonstration of the presence of slow conduction during sustained ventricular tachycardia in man: Use of transient entrainment of the tachycardia. *Circulation* 1987;75:369.
39. Kay GN, Epstein AE, Plumb VJ: Incidence of reentry with an excitable gap in ventricular tachycardia: A prospective evaluation utilizing transient entrainment. *J Am Coll Cardiol* 1988;11:530–538.
40. Rosenthal ME, Stamato NJ, Almendral JM, Gottlieb C, Josephson ME: Resetting of ventricular tachycardia with electrocardiographic fusion: Incidence and significance. *Circulation* 1988;77:581–588.
41. Waldo AL, Okumura K, Olshansky B, Henthorn RW: Use of transient entrainment of tachycardia as an aid to application of fulguration, in Fontaine G, Scheinman MM (eds): *Ablation in Cardiac Arrhythmias*. Mount Kisco, NY, Futura Publishing, 1987, pp 277–288.
42. Waldo AL, Henthorn RW: Use of transient entrainment during ventricular tachycardia to localize a critical area in the reentry circuit for ablation. *PACE* 1989;12(part II):231–244.
43. Fontaine G, Frank R, Pavie A, Tonet J, Lascault G, Grosgogeat Y: Ablation of the slow conduction area in chronic ventricular tachycardia. *Cardiology* 1990;77:240–258.
44. Fontaine G, Tonet JL, Frank R, Gallais Y, Touzet I, Kounde S, Farenq G, Baraka M, Grosgogeat Y: Electrode catheter ablation of resistant ventricular tachycardia by endocavitary fulguration associated with anti-arrhythmic therapy. Experience of 38 patients with a mean follow-up of 23 months, in Brugada P, Wellens HJJ (eds): *Cardiac Arrhythmias: Where To Go from Here?* Mount Kisco, NY, Futura Publishing, 1987, pp 539–569.
45. Fontaine G: Du lieu d'origine à la zone à conduction lente. Application au traitement de la tachycardie ventriculaire. *Arch Mal Coeur* 1988;81:145.
46. Fontaine G, Frank R, Vedel J, Vachon JM, Guiraudon G, Grosgogeat Y, Facquet J: La genese de certains troubles du rythme ventriculaire. *Nouv Presse Med* 1974;3:2321.
47. Fontaine G, Guiraudon G, Frank R: Intramyocardial conduction defects in patients prone to ventricular tachycardia. I- The Postexcitation syndrome in sinus rhythm, in Sandoe E, Julian DG, Bell JW (eds): *Management of Ventricular Tachycardia. Role of Mexiletine*. Amsterdam, Excerpta Medica, 1978, pp 39–55.
48. Fontaine G, Guiraudon G, Frank R: Intramyocardial conduction defects in patients prone to ventricular tachycardia. II- A dynamic study of the post-excitation syndrome, in Sandoe E, Julian DG, Bell JW (eds): *Management of Ventricular Tachycardia. Role of Mexiletine*. Amsterdam, Excerpta Medica, 1978, pp 56–66.
49. Fontaine G, Guiraudon G, Frank R: Intramyocardial conduction defects in patients prone to ventricular tachycardia. III- The post-excitation syndrome during ventricular tachycardia, in Sandoe E, Julian DG, Bell JW (eds): *Management of Ventricular Tachycardia. Role of Mexiletine*. Amsterdam, Excerpta Medica, 1978, pp 67–79.
50. Wit AL, Allessie MA, Bonke FIM, Lammers WJEP, Smeets J, Fenoglio JJ Jr: Electrophysiologic mapping to determine the mechanism of experimental ventricular tachycardia initiated by premature impulses. *Am J Cardiol* 1982;49:166–185.
51. Fontaine G, Guiraudon G, Frank R: Mechanism of ventricular tachycardia with and without associated chronic myocardial ischaemia: Surgical management based on epicardial mapping, in Narula OS (ed): *Cardiac Arrhythmias: Electrophysiology, Diagnosis and Management*. Philadelphia, Williams & Wilkins, 1979, pp 516–545.
52. Boineau JP, Cox JL: Slow ventricular activation in acute myocardial infarction. A source of reentrant premature ventricular contractions. *Circulation* 1973;48:702.
53. Stevenson WG, Weiss JN, Wiener I, Nademanee K, Wohlgelernter D, Yeatman L, Josephson ME, Klitzner T: Resetting of ventricular tachycardia: Implications for localizing the area of slow conduction. *J Am Coll Cardiol* 1988;11:522–529.
54. Fontaine G, Frank R, Tonet J, Grosgogeat Y: Identification of a zone of slow conduction appropriate for VT ablation. Theoretical and practical considerations. *PACE* 1989;12(part II):262–267.
55. Fontaine G: Prevention of sudden arrhythmic death. Catheter ablation, in *Proceedings of the 1985 Sydney Opera House Symposium*. Sydney, Telectronics Vectors, October 1986, pp 18–21.
56. Fontaine G, Evans S, Frank R, Tonet J, Iwa T Jr, Lascault G, Grosgogeat Y: Ventricular tachycardia overdrive and entrainment with and without fusion:

Its relevance to the catheter ablation of ventricular tachycardia. *Clin Cardiol* 1990;13:797–803.
57. Prystowsky EN, Zipes DP: Inhibition in the human heart. *Circulation* 1983;68:707–713.
58. Antzelevitch C, Moe GK: Electrotonic inhibition of impulse transmission across inexcitable segments of cardiac tissue. *Circulation* 1982;66(suppl II):358(abstract).
59. Ruffy R, Friday KJ, Southworth WF: Termination of ventricular tachycardia by single extrastimulation during the ventricular effective refractory period. *Circulation* 1983;67:457–459.
60. Antzelevitch C, Moe GK: Electrotonic inhibition and summation of impulse conduction in mammalian Purkinje fibers. *Am J Physiol* 1983;245:H42–H53.
61. Brugada P, Wellens HJJ: Entrainment as an electrophysiologic phenomenon. *J Am Coll Cardiol* 1984;3:451–454.

Chapter **30**

Transcoronary Chemical Ablation of Arrhythmogenic Areas or Pathways

Pedro Brugada, MD, Hans de Swart, MD, and Joep Smeets, MD

Control of cardiac arrhythmias is possible at the present time in a considerable proportion of patients.[1] We have at our disposal a powerful therapeutic armamentarium which includes antiarrhythmic drugs, antitachycardia pacemakers and defibrillators, percutaneous electrical ablation, and surgery. However, only percutaneous electrical ablation and surgery can remove, isolate, or destroy the arrhythmia substrate. When successful they may cure the arrhythmia. Unfortunately, not all patients are candidates for surgical treatment. Particularly, patients with extensive myocardial damage face unacceptably high surgical risk.[2] Percutaneous electrical ablation of ventricular tachycardia has not yet produced the desired results,[3] in part because of the difficulties in localizing the arrhythmogenic area or critical tachycardia pathway, and in part because of the small lesions created by the electric shock in areas of myocardial scarring.

Any arrhythmogenic area or pathway in the heart must have blood supply to maintain electrical activity of the cells involved in the arrhythmia mechanisms. Cells surviving in an area of myocardial infarction play an important role in the mechanisms of ventricular arrhythmias after myocardial infarction.[4] These cells may have normal action potentials (APs), which suggests normal metabolism. The maintenance of normal function requires blood supply, which in part may come from the ventricular cavity but also from anterograde coronary flow through arteries reperfused after a myocardial infarction or from retrograde collateral flow.

Recently, an Indianapolis group has shown that it is possible to terminate aconitine-induced ventricular arrhythmias in the experimental laboratory by injecting alcohols through the "tachycardia-related" coronary artery.[5] On the basis of these concepts, a technique was developed to identify the tachycardia-related coronary artery in patients with ventricular tachycardia (VT) after myocardial infarction.

The atrioventricular (AV) node is supplied by the AV nodal artery, which is readily identifiable in most humans; it originates from the right coronary artery in the majority of patients. The AV node is a critical pathway in the conduction of electrical impulses from atria to ventricles during sinus rhythm and during a large variety of supraventricular arrhythmias. Using the same techniques as in patients with VT, the AV nodal artery was selectively catheterized to chemically destroy the AV node in patients with uncontrollable fast ventricular rates during atrial fibrillation.

PATIENTS WITH INCESSANT VT AFTER MYOCARDIAL INFARCTION

The first transcoronary chemical ablation of incessant VT after myocardial infarction was performed in October 1987. A series of studies in patients with a variety of cardiac arrhythmias were undertaken before this procedure.[6–8] The Medical Ethical Committee at our institution approved the research protocol and all patients gave informed consent. A complete electrophysiologic study and a coronary angiography were performed in 18 patients. Two patients had circus

655 Avenue of the Americas, New York, NY 10010
Current Topics in Cardiology

movement tachycardia using an accessory AV pathway, 1 had incessant left atrial tachycardia, 13 had VT after myocardial infarction, and 2 had idiopathic VT. After correlating the data of the electrophysiologic and angiographic studies, selective catheterization of the tachycardia-related artery was done in these patients. The methodology employed will be described later. We demonstrated that incessant atrial tachycardia could be repeatedly terminated by selective catheterization of a small left atrial branch and administration of iced isotonic saline. We also terminated circus movement tachycardia using retrogradely an accessory AV pathway by selective catheterization of the AV nodal artery. Of 15 patients with VT, reproducible termination of arrhythmia was achieved in 14 when isotonic saline was given through the tachycardia-related coronary artery.[7,8]

The above studies have shown that it is possible to identify the tachycardia-related coronary artery, and that administration of isotonic saline allows identification of the appropriate coronary artery. Encouraged by these results and confronted with patients with incessant VT after myocardial infarction who had no therapeutic alternatives, transcoronary chemical ablation of VT was attempted in 9 patients. In 2 of these, the procedure was not finalized because of inability to identify the tachycardia-related artery in one, and the presence of extensive collateral blood supply in the other. In the latter patient we feared that ethanol could result in extensive myocardial damage via the extensive collateral system. The clinical data of the 7 patients in whom transcoronary chemical ablation was completed are summarized in Table 30.1.

TABLE 30.1 Clinical Data on the 7 Patients Undergoing Transcoronary Chemical Ablation of Ventricular Tachycardia

Patient Number	Age[a]/Sex	Site MI	FUP	Inc. VT	Par. VT
1	61/M	ANT + INF	27	No	No
2	44/M	ANT + INF	23	No	Yes
3	62/M	ANT	18	No	No
4	58/M	ANT	17	No	No
5	59/M	ANT	—	No	—[b]
6	62/M	ANT	16	No	Yes[c]
7	60/M	ANT	3	No	No

[a] Age in years.

[b] Death 15 days after ablation.

[c] Death caused by congestive heart failure.

ANT = anterior, INF = inferior, FUP = duration of follow-up in months after chemical ablation, Inc. VT = incessant VT during follow-up, M = male, MI = myocardial infarction, Par. VT = paroxysmal ventricular tachycardia during follow-up.

All patients suffered from incessant VT uncontrollable with antiarrhythmic drugs after myocardial infarction. No patient was a candidate for arrhythmia surgery either because they already had surgery (4 patients) with failure to control the arrhythmia or because of the extremely poor left ventricular function (all patients). The incessant, continuous nature of VT made it impossible to consider antitachycardia pacing or an implantable defibrillator as a possible therapy. A complete electrophysiologic study was performed in all patients to identify the site of origin of the arrhythmia. Uni- and bipolar endocardial mapping and pacemapping were used for that purpose. Thereafter, a coronary angiography and a left ventriculography were done. A possible tachycardia-related coronary artery was identified in the coronary angiogram by correlating electrical and anatomic data. The tachycardia-related coronary artery had to supply the area where VT originated. The tachycardia-related coronary artery was selectively catheterized using an 8F guiding catheter, a 2.4F to 3F perfusion catheter with a metallic marker on the tip, and a 0.014-in angioplasty guide wire. The guide wire was placed as distal as possible in the tachycardia-related artery. Thereafter, the perfusion catheter was advanced as far as possible in that artery and the guide wire retired. To confirm selective catheterization of the tachycardia-related artery, 10 ml of iced isotonic saline were injected as fast as possible through the perfusion catheter during VT. Repeated termination of VT with iced saline confirmed that critical area or pathway for perpetuating the arrhythmia was rendered ischemic and/or cooled preventing the electric impulse conduction in the area or pathway. An example of termination of VT by saline is shown in Figure 30.1. In the 7 patients undergoing chemical ablation, 2 to 6 ml of 96% ethanol were given during VT through the tachycardia-related coronary artery. Patient 1 was the only one who received 6 ml of ethanol in a total of two sessions. In the other patients, only 2 ml were administered. Ethanol was used based on the reported experiences by Inoue et al.[5] and its previous use for ablation of renal tumors in humans.[9,10]

Ethanol abruptly terminated VT in all 7 patients. After ablation the arrhythmia could not be reinduced by pacing immediately and after 1 week in 4 of 5 patients who underwent a repeated electrophysiologic study. One patient died 15 days after chemical ablation of an asystole caused by accidental procainamide intoxication. Another patient died of congestive heart failure 16 months after successful chemical ablation. The other 5 patients are alive. The functional class for dyspnea improved in all after control of

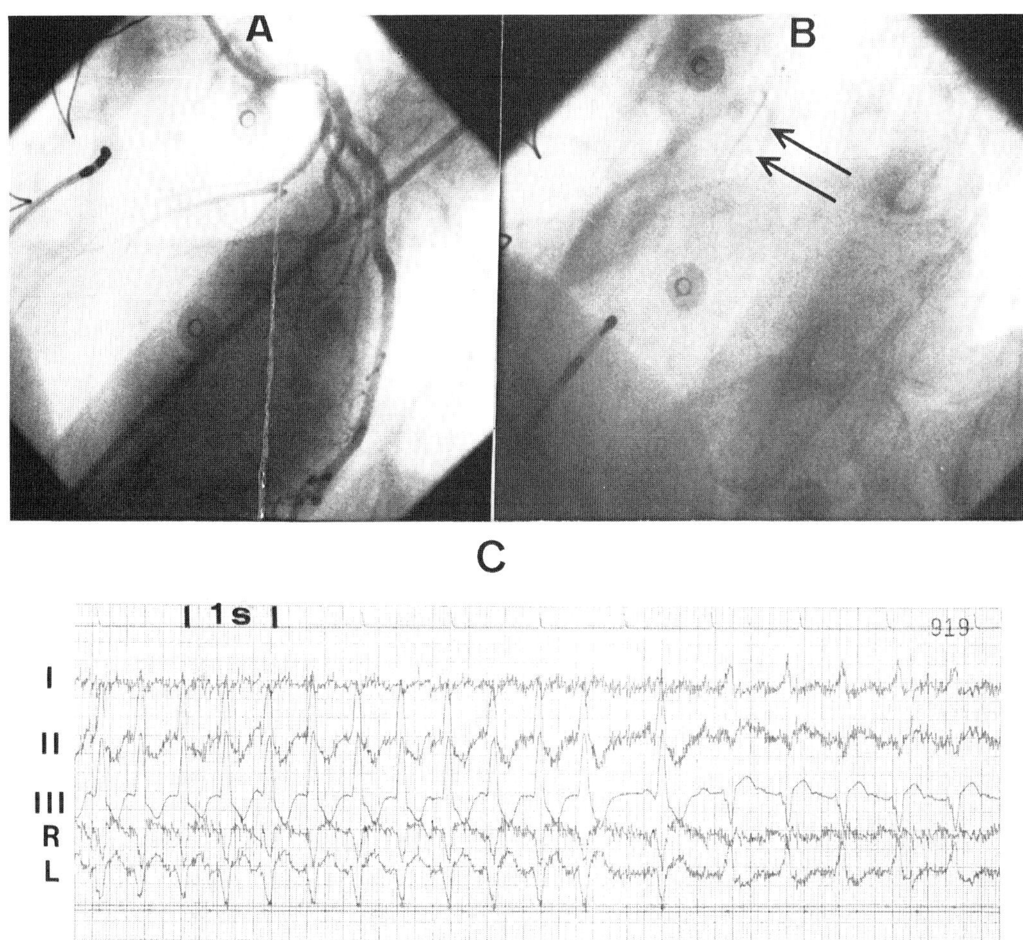

FIGURE 30.1 Selective catheterization of the first septal artery in a patient with incessant VT after myocardial infarction arising high in the intraventricular septum. A. The left coronary angiography in the left anterior oblique projection with angulation. B. Selective catheterization of the first septal artery (arrows). C. Termination of VT when iced saline was given through the first septal branch.

incessant VT. Two patients had a single recurrence of a slow-rate VT 1 year after chemical ablation. The other patients are asymptomatic. The longest follow-up is 27 months for the first patient at this time (Table 30.1).

TRANSCORONARY CHEMICAL ABLATION OF AV CONDUCTION[11]

Transcoronary chemical ablation of the AV node was undertaken in 11 patients with uncontrollable fast ventricular rates during atrial fibrillation (Table 30.2). For this purpose, the AV nodal artery was selectively catheterized using a 2F to 3F catheter with a metallic marker on the tip (Fig. 30.2). Confirmation of selective catheterization of the AV nodal artery was obtained by injecting iced saline or contrast material and observing the occurrence of transient complete AV block. Recording of an electrogram of the bundle of His confirmed in 4 patients that AV block occurred proximal to the bundle of His. The procedure was attempted in a total of 13 patients. In 1 patient, the presence of severe coronary artery disease prevented safe catheterization of the AV nodal artery and the procedure was not attempted. In another patient, transient AV block could not be produced with saline or contrast material in spite of catheterization of several small coronary branches arising from the crux of the right coronary artery. Chemical ablation was not performed in this patient. In the remaining 11 patients, after confirmation of selective catheterization of the AV nodal artery, 1 ml of ethanol was administered with the catheter placed as far as

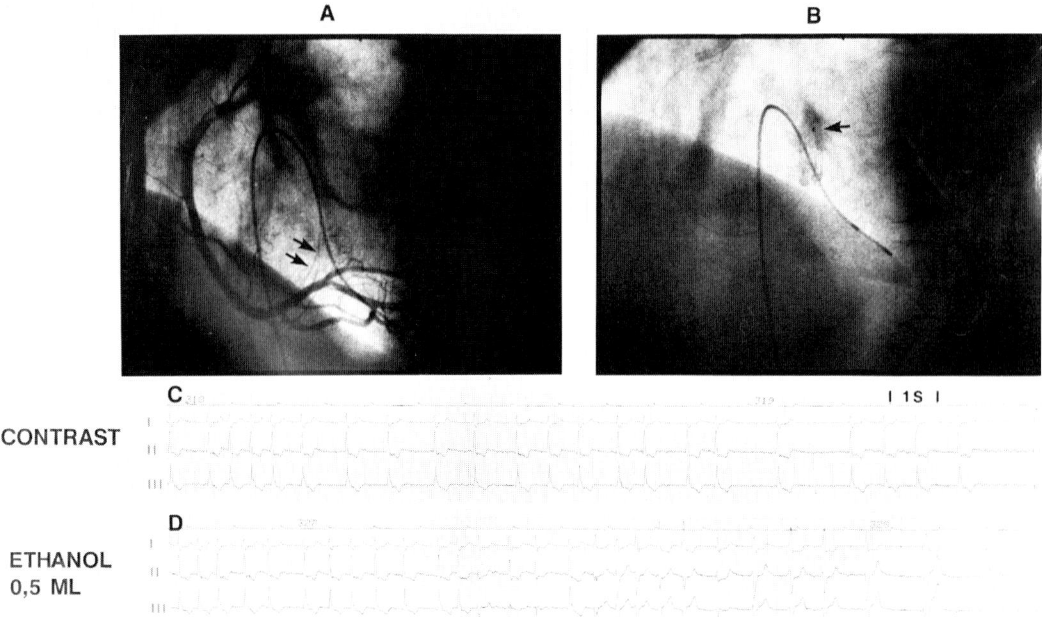

FIGURE 30.2 A. The right coronary angiography in the left anterior oblique projection and the AV nodal artery (arrows). **B.** Selective catheterization of the AV nodal artery (arrow). In the bottom panels, the effect of contrast material and ethanol during atrial fibrillation is shown. Note that contrast resulted in transient AV block (**C**). Several extrasystoles are observed before complete AV block occurs after administration on ethanol (**D**). The spontaneous ventricular rate during atrial fibrillation was much faster as observed in **C.** Upon catheterization of the AV nodal artery, the ventricular rate decreases in all patients because of ischemia of the AV node.

possible in that artery. Patient 4 received 2 ml of ethanol. The reasons for that and the complication that occurred will be described later in detail. Permanent AV block was produced in 8 patients. In the other 3 patients, conduction was sufficiently modified as to control symptoms without using antiarrhythmic drugs. In all patients, a permanent ventricular rate-responsible pacemaker was implanted. In 1 patient with modification of AV conduction and known with

TABLE 30.2 Clinical Data on 11 Patients Undergoing Transcoronary Chemical Ablation of Atrioventricular Conduction

Patient Number	Age/Sex	Reasons for TCA-AVC	QT	FUP	AVC	Complications
1	64/F	Failed EA	96	14	CB	None
2	70/F	Failed EA, PM	85	14	CB	None
3	64/M	Failed EA, PM	65	12	MC	None
4	69/F	PM	338	11	CB	INF MI
5	60/M	Failed EA, PM	37	9	MC	None[a]
6	69/M	Patient preference	29	8	CB	None
7	54/M	PS	220	7	CB	None
8	56/M	Patient preference	12	6	CB	None
9	55/M	Patient preference	40	3	MC	None
10	60/M	Failed EA	29	3	CB	None
11	59/M	Patient preference	42	2	CB	None

[a] Patient undergoing bypass surgery and sectioning of the bundle of His at a later date.

AVC = atrioventricular conduction after ablation, CB = complete block in conduction, EA = electrical ablation, F = female, INF MI = inferior wall myocardial infarction, MC = modification of atrioventricular conduction, QT = maximal oxalacetyc transaminase value after ablation (normal less than 40 units), PM = pacemaker, PS = previous surgery because of an accessory atrioventricular pathway, TCA-AVC = transcoronary chemical ablation of atrioventricular conduction. Other abbreviations as in Table 30.1.

coronary artery disease, angina pectoris worsened during follow-up. Repeated coronary angiography did not show any progression of the disease but it was decided to perform coronary bypass surgery, and the bundle of His was sectioned at the same time.

In patient 4, complete AV block or modification of AV conduction did not occur after a first session where a dose of 1 ml of ethanol was administered. The procedure was repeated 1 day later and 2 ml of ethanol were given. Complete AV block occurred, but the patient developed an inferior wall myocardial infarction. Analysis of the angiographic data suggested that this dose was too large and caused reflow into the distal right coronary artery. The left ventricular ejection fraction was 50% after the infarction. The patient developed ventricular fibrillation (VF) while still in-hospital and was given oral amiodarone. No other complications have occurred during the follow-up.

DISCUSSION

The data presented here show that it is possible to identify and selectively catheterize a small coronary artery providing blood supply to an arrhythmogenic area or critical tachycardial pathway. Administration of cold isotonic saline or contrast material allows confirmation of selective catheterization of the tachycardia-related artery by producing conduction block or termination of the arrhythmia. Administration of a chemical (ethanol in our study) can destroy the arrhythmogenic area or pathway and cure the arrhythmia or control some of its consequences, like fast ventricular rates during atrial fibrillation. It is obvious that refinements of this technique are required to establish its place in the treatment of cardiac arrhythmias. At the present stage, it offers an alternative to be considered in selected patients devoid of other therapeutic options. However, several precautions are necessary when using this method.

Experienced angiographers and electrophysiologists should cooperate closely. Catheterization of small coronary arteries (about 1 to 2 mm in diameter) should be performed by a catheterization team with extensive experience in different angioplasty techniques. An experienced electrophysiologist is needed to make the decision about the artery that should be selectively catheterized. Identification of the AV nodal artery is usually relatively easy, but sometimes may be difficult and, as it happened in one of our patients, even impossible. If administration of saline or contrast does not result in AV block, another branch has to be selected and the procedure restarted. The ethanol should be given only when one is completely sure that the correct artery has been identified.

Greater patience and skills are required when considering chemical ablation of VT or other arrhythmias (a case of incessant atrial tachycardia has been successfully ablated with ethanol in Sao Paulo, Brazil, by Dr. Eduardo Sosa and coworkers) than during AV nodal ablation.

Ethanol destroys myocardium and occlusion of the tachycardia-related artery is common. An enzyme rise, sometimes minimal, was observed in all patients. A destruction of an area different from the target will not control the arrhythmia and cause unnecessary myocardial damage.

It has to be emphasized that the tachycardia-related coronary arteries are located very distally in the coronary artery system with diameters of 1 to 2 mm. Ethanol should never be injected into a proximal branch because it will produce an extensive myocardial infarction. Also, as observed in our fourth patient undergoing chemical ablation of the AV node, reflow of the ethanol has to be prevented by wedging the catheter as distally as possible. Catheters of different diameters or with a distal balloon, when available, may be useful for that purpose.

The technique of transcoronary chemical ablation of VT is too young to determine its definite place in the management of cardiac arrhythmias, and more information is required. At present, the technique should be reserved for patients not having any other therapeutic options. In patients with uncontrollable fast ventricular rates during atrial fibrillation, the technique is simpler than in patients with VT after myocardial infarction. However, unless special reasons are present (like in several of our patients), it should not be used as first choice by teams without experience in lieu of electrical ablation. The limited number of patients in whom we have performed chemical ablation within nearly a 3-year period since the first procedure shows how selective we have been resorting to this technique.

ACKNOWLEDGMENT

The continuous support of H. Wellens is acknowledged.

REFERENCES

1. Brugada P: Op weg naar een gouden tijdperk in de Ritmologie, in: *Cardiologie van de toekomst*. Utrecht, ICIN Publications. Wetenschappelijke uitgeverij Bunge, 1989.
2. Harken AH, Josephson ME: Surgical management of ventricular tachycardia, in Josephson ME, Wellens HJJ (eds): *Tachycardia: Mechanisms, Diagnosis and Treatment*. Philadelphia, Lea & Febiger, 1984, pp 475–487.
3. Hartzler GO: Electrode catheter ablation of refrac-

tory focal ventricular tachycardia. *J Am Coll Cardiol* 1983;2:1107–1113.

4. Friedman PL, Steward JR, Fenoglio Jr JJ, Wit AL: Survival of subendocardial Purkinje fibers after extensive myocardial infarction in dogs: In vitro and in vivo correlations. *Circ Res* 1973;33:597–611.
5. Inoue H, Waller BF, Zipes DP: Intracoronary ethyl alcohol or phenol injection ablates aconitine-induced ventricular tachycardia in dogs. *J Am Coll Cardiol* 1987;10:1342–1349.
6. Brugada P, de Swart H, Smeets JLRM, Wellens HJJ: Termination of tachycardias by interrupting blood supply to the arrhythmogenic area. *Am J Cardiol* 1988;62:387–392.
7. Brugada P, de Swart H, Smeets JLRM, Wellens HJJ: The role of collateral blood supply in ventricular tachycardia after myocardial infarction. *Eur Heart J* 1988;9:1104–1111.
8. Brugada P, de Swart H, Smeets JLRM, Wellens HJJ: Transcoronary chemical ablation of ventricular tachycardia. *Circulation* 1989;79:475–482.
9. Ellman BA, Parkhill BJ, Curry III TS, Marcus PB, Peters PC: Ablation of renal tumors with absolute ethanol: A new technique. *Radiology* 1981;141:619–626.
10. Ellman BA, Parkhill BJ, Marcus PB, Peters PC: Renal ablation with absolute ethanol: Mechanism of action. *Invest Radiol* 1984;19:416–423.
11. Brugada P, de Swart H, Smeets JLRM, Wellens HJJ: Transcoronary chemical ablation of atrioventricular conduction. *Circulation* 1990;81:757–761.

Chapter **31**

Pacing, Cardioversion, and Defibrillation for Ventricular Tachyarrhythmias

John J. Hayes, MD, and Gust H. Bardy, MD

Implantable device therapy for ventricular tachycardia (VT) and ventricular fibrillation (VF) has advanced substantially since the initial clinical use of automatic implantable defibrillators in 1980.[1] From the first fixed-rate nonprogrammable defibrillator, the implantable defibrillator has evolved into a multistaged, multiprogrammable device capable of a variety of sense algorithms, bradycardia and antitachycardia pacing, low-energy and high-energy cardioversion, as well as high-energy defibrillation. It is the purpose of this chapter to examine the basic components of today's multiprogrammable antiarrhythmia devices and to assess the expanding role of device therapy for VT and VF. The mechanics and merits of antitachycardia pacing and synchronized low-energy cardioversion will be discussed together with an examination of the relationship of arrhythmia sensing to VT and VF therapy, the importance of pulse waveform on defibrillation efficacy, and the utility of nonthoracotomy or transvenous lead systems compared to epicardial lead systems for defibrillation.

ANTITACHYCARDIA PACING

Although high-energy shocks are effective in terminating episodes of VT and VF, they do not necessarily represent optimal therapy for the treatment of monomorphic VT. A high-energy shock delivered to an awake patient may be extremely uncomfortable and frightening. In a patient with frequent episodes of VT, the problem of discomfort is compounded by rapid depletion of battery supply. Antitachycardia pacing can be as effective as high-energy shocks yet better tolerated and more energy efficient. Antitachycardia pacing therapy may also be more rapidly delivered because it avoids the 6 to 10 s needed to charge the capacitor to high energies. Unfortunately, antitachycardia pacing has played a limited role in the therapy of monomorphic VT thus far. However, the recent advances in device technology are likely to expand the utility of antitachycardia pacing in the future.

Although antitachycardia pacing has not been part of the standard automatic implantable defibrillator, it has been employed in a limited fashion in pacemakers for termination of recurrent monomorphic VT since 1974.[2,3] Despite scattered reports of the effectiveness of antitachycardia pacemakers in terminating VT, many patients with antitachycardia pacemakers continue to be at risk for sudden cardiac death.[3] The reasons for this are probably threefold. First, patients with VT are also likely to have VF.[4] No pacing intervention has been shown to terminate VF. Second, any attempt to overdrive pace terminate VT carries a significant risk of accelerating the VT.[3,5–8] The incidence of acceleration during antitachycardia pacing has ranged from 18 to 43%.[3,5–8] Third, the natural history of VT is unpredictable and efficacy of antitachycardia pacing at one time does not assure continuing success at another time. For these reasons, stand-alone antitachycardia pacemakers have not been popular unless accompanied by a backup implantable defibrillator. However, potential interactions between an implantable defibrillator and a pacemaker[9] have prevented widespread use of this combination of devices, except in select cases.[10]

Data regarding specific types of antitachycardia pacing maneuvers are scarce. Part of the difficulty inherent in determining the usefulness of any particular pacing intervention is the need

655 Avenue of the Americas, New York, NY 10010
Current Topics in Cardiology

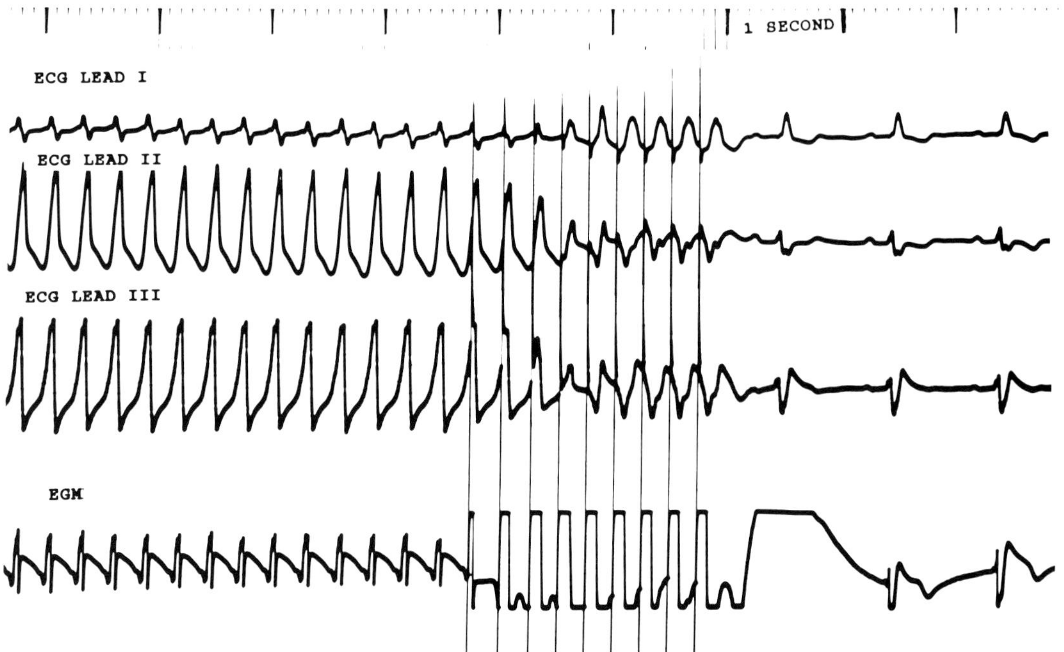

FIGURE 31.1 Termination of VT by autodecremental (ramp) pacing using the parameters listed in Figure 31.3. In this example the device is programmed to deliver the pacing sequence after detecting 16 consecutive cycles below 380 ms. The tachycardia has a cycle length of approximately 300 ms and ramp pacing begins at 290 ms, 97% of the VT cycle length. Each subsequent stimulus decrements by 10 ms until the minimum interval of 260 ms is reached, at which point subsequent stimuli continue at 260 ms intervals. EGM is the bipolar intracardiac electrogram.

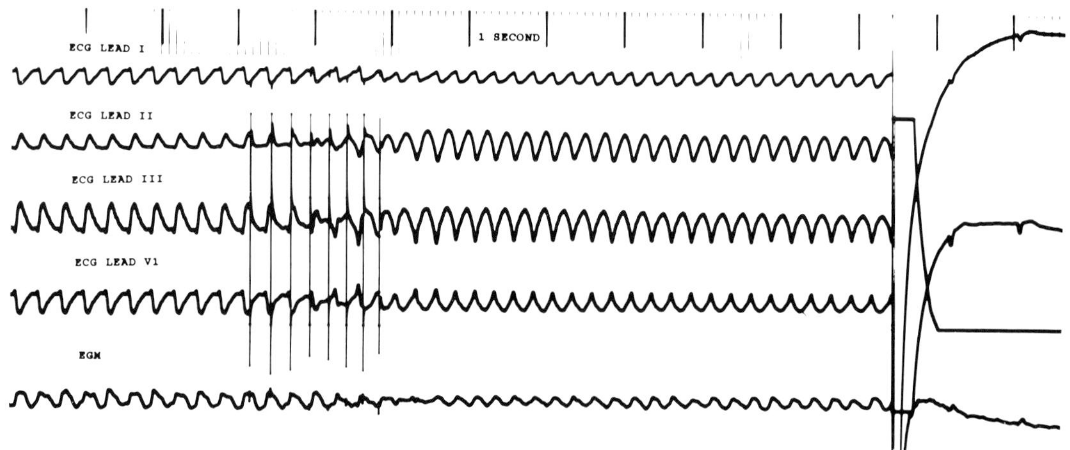

FIGURE 31.2 VT acceleration during autodecremental overdrive pacing. The tachycardia cycle length is initially 300 ms, but accelerates to 260 ms following the eight paced beats. The tachycardia cycle length fell below the fibrillation detection interval, and the patient was subsequently successfully converted to sinus rhythm with a 5 J "defibrillation" pulse. EGM is the bipolar intracardiac electrogram.

for repetitive induction of the same 12-lead electrocardiogram (ECG) documented VT morphology during electrophysiologic testing. This may be difficult because many patients have pleomorphic VT (that is, multiple monomorphic VT morphologies).(11) In our experience, each of the 15 patients with VT undergoing repeated programmed electrical stimulation at the time of multiprogrammable antiarrhythmia device implant and during serial outpatient device testing had pleomorphic VT. Therefore, prospective systematic quantitative analysis of any pacing intervention requires prolonged electrophysiologic testing with clinically unacceptable frequencies of VT inductions.

Antitachycardia pacing methods traditionally have included scanning of the diastolic interval with single and double extrastimuli and with fixed-rate multicycle burst pacing. Scanning diastole with one or two extrastimuli was first shown to be effective by Wellens et al. in 1972.(12) Unfortunately, the success rate for this method is poor except for termination of slow VT.(6) Fixed-rate multistimulus burst or overdrive pacing has been shown to be more effective than single or double extrastimuli but carries a higher incidence of acceleration.(5–7)

Depending upon the study, burst pacing is successful in up to 87% of patients but acceleration occurs in up to 43%.(5–7) An unacceptably high incidence of VT acceleration with burst pacing has led to the exploration of alternative pacing methods. Autodecremental overdrive pacing is one of the recently proposed alternatives to fixed-rate burst pacing where successive stimuli are coupled at progressively shorter intervals limited by some predetermined minimum pacing cycle length (Fig. 31.1). If the first pacing attempt fails to terminate the VT, one more stimulus is added to the train during the subsequent attempt. The only comparison of overdrive burst pacing to autodecremental overdrive pacing has been published by Charos et al.(13) In this study, autodecremental pacing was found to be more efficacious than burst pacing by terminating 92% of VT episodes compared to 56% for burst pacing. Burst pacing accelerated 2% of VT episodes, but no VT acceleration was observed with autodecremental overdrive pacing. With autodecremental overdrive pacing, the number of stimuli required to terminate VT ranged from 5 to 12, with more episodes requiring 7 or 8 stimuli when the starting pacing cycle length was 90 to 97% of the VT cycle length. In this study each successive stimulus was decremental by 3.5% of the VT cycle length, which averaged 8 ms per cycle. The Charos study was limited in its scope, however, because of the failure to control the number of stimuli used with each method, the minimum coupling interval for each method, and the number of iterations before declaring success or failure. Moreover, Charos's study was limited to a small number of patients with hemodynamically well tolerated, repetitively inducible VT with cycle lengths greater than 280 ms. Our experience differs from the results of Charos et al. We found that decremental overdrive pacing can accelerate VT, even at cycle lengths greater than 280 ms as shown in Figure 31.2. Consequently, further investigation is required before a reliable pacing algorithm can be selected for any individual patient.

Current multiprogrammable antiarrhythmia devices require the user to select several parameters to initiate antitachycardia pacing (Figure 31.3). One may choose adaptive burst overdrive

```
VT THERAPY #1:
 THERAPY TYPE                        RAMP
 VT THERAPY ENABLE                 ON/OFF
 INITIAL # OF S PULSES                  9
 FIRST R-S INTERVAL                  97 %
 PER PULSE DECREMENT                 10 MS
 # OF SEQUENCES                         2
 MINIMUM INTERVAL                   260 MS

VT THERAPY #2
 THERAPY TYPE                       BURST
 VT THERAPY ENABLE                 ON/OFF
 # OF S PULSES                          9
 S-S INTERVAL                        97 %
 PER SEQUENCE DECREMENT              10 MS
 # OF SEQUENCES                         2
 MINIMUM INTERVAL                   260 MS

VT THERAPY #3
 THERAPY TYPE                CARDIOVERSION
 VT THERAPY ENABLE                 ON/OFF
 CV PULSE WIDTH                     6.8 MS
 CV ENERGY (JOULES)                 0.2 J
 CV CURRENT PATHWAY                  SEQ
```

FIGURE 31.3 Programmable parameters for three antitachycardia therapies available in a multiprogrammable antiarrhythmia device (Medtronic model 7217B). The user can choose among ramp (autodecremental) pacing, burst pacing, or cardioversion. For pacing therapies, the user must then choose the number of stimuli, the initial pacing cycle length as a percentage of the tachycardial cycle length, and the number of pacing sequences. The cycle length of each successive ramp pacing stimulus is a function of the programmed parameter "per pulse decrement." The number of stimuli will increase by one for each successive sequence. The paced cycle length for burst pacing will decrease by the specified per sequence decrement. The paced cycle length will be limited by the minimum pacing interval programmed. For cardioversion therapy, energy, pulse width, and current pathway must be specified. Pulsing can occur over one current pathway, two pathways simultaneously, or two pathways sequentially.

pacing, adaptive autodecremental overdrive pacing, or sequence of one after the other. Once the pacing mode is selected, one chooses the number of pacing stimuli, usually 5 to 15. Pacing can begin over a broad range of percentages of the VT cycle length depending on user preference and patient response. The user is also responsible for choosing the number of attempts to pace-terminate VT, in the timing of successive attempts, and number of pacing cycle lengths. Proper selection of pacing parameters requires careful testing in the electrophysiology laboratory before hospital discharge. Excessive antitachycardia pacing interventions should be avoided. We believe that repetition of pacing sequences beyond four attempts may lead to prolonged hypotension and delay delivery of definitive therapy.

LOW-ENERGY CARDIOVERSION

Similar to antitachycardia pacing, low-energy cardioversion for termination of VT has not been used in implantable defibrillators until recently. Most of the studies in low-energy cardioversion have involved transvenous lead systems. Consequently, much of what is known of low-energy cardioversion is not transferable to implantable epicardial devices. Furthermore, as with studies examining antitachycardia pacing, controlled examination of energy settings in a prospective fashion repeatedly in the same 12-lead ECG-documented VT has not been done. In spite of limited data, it appears that low-energy cardioversion of VT is no less complicated and no more efficient than the antitachycardia pacing. However, success rates and acceleration rates for the two approaches may differ for different subgroups of patients.

The feasibility of low-energy cardioversion of VT was first shown by Zipes et al.[14] In this transvenous study using the Medtronic 6880 lead, a 6 ms truncated exponential pulse of 0.025 to 2.0 J terminated VT in 5 of 7 patients. Shocks of less than 0.5 J were tolerated with minimal discomfort. However, in one of the 7 patients, VT accelerated to VF following a pulse of 0.075 J, indicating potential harm of this method. In addition, it is unclear from this study how many attempts were made before low-energy cardioversion was deemed successful or not.

Waspe and coworkers studied 13 patients and found a 62% success rate for 50 shocks of 0.01 to 5.0 J amplitude.[15] VT acceleration occurred in 14% of cardioversion attempts. In 4 patients, there appeared to be no difference in efficacy between antitachycardia burst pacing and low-energy cardioversion. This study essentially confirmed both the feasibility and the hazard of low-energy cardioversion. It did not, however, define the range of useful pulsing energies and the circumstances where low-energy cardioversion would be harmful. For example, it is not known whether the occurrence of acceleration at one energy setting signifies that the next increment in energy will terminate or accelerate the same VT.

Ciccone et al.[16] found that low-energy transvenous cardioversion with the Medtronic 6880 lead, defined as 2.2 J or less, was successful in 56% of patients with VT. Similar to antitachycardia pacing, there was an 8% incidence of VT acceleration per shock. Another concern was a 18% incidence of bradyarrhythmias and a 9% incidence of supraventricular arrhythmias following low-energy cardioversion. Although 57% of the patients reported as tolerable shocks up to 0.5 J, only 24% felt that shocks greater than 1.0 J were tolerable. These data suggest that low-energy transvenous cardioversion may be no more effecacious than antitachycardia pacing yet may result in more complications. Furthermore, this study suggests that high-energy cardioversion, defined as pulses up to 10 J, was not necessarily superior to low-energy cardioversion. Using pulses of up to 10 J, only 72% of VT episodes were converted. In addition, there was a higher incidence of transient bradyarrhythmias (31%) and supraventricular tachyarrhythmias (16%). These findings imply that if a patient's VT can be cardioverted with transvenous pulses, the use of low energy may be associated with fewer complications.

Saksena et al.[17] compared transvenous cardioversion with fixed-rate burst ventricular pacing for VT termination, and found that both had similar efficacy (83% versus 80% success rate) and similar incidence of VT acceleration (11% versus 6%), but pacing was better tolerated and less likely to cause postconversion supraventricular arrhythmias (23% versus 3%). As in other studies, the success of either method in this study depended on tachycardia cycle length.[6,7,16] Slower VTs were more likely to convert and less likely to accelerate with either therapy. Cardioversion was performed with a lead system now recognized to be suboptimal. Epicardial leads or one of the more effective transvenous lead systems currently employed may have led to a different outcome.

Using a catheter-chest patch system, Lindsay et al.[18] found that VT could be reproducibly terminated in 96% of patients, with the choice of therapy based on tachycardia cycle length. VTs with cycle lengths less than 300 ms were treated with high-energy cardioversion, while slower VTs were treated with rapid pacing followed by cardioversion if pacing was unsuccessful. While the acceleration rate for antitachycardia pacing for slow VTs was relatively low (5%), the incidence of acceleration for cardioversion was substantially higher (20 to 40%) when 5- to 15-J pulses were used. This was the first study to suggest an approach to tachycardia based upon rate. Once again, however, these findings must be viewed within the context of the lead system and pulsing method employed. The success or failure of cardioversion, as well as antitachycardia pacing, may change for different transvenous systems, epicardial lead systems, or alternative pulsing methods.

These studies demonstrated that although rapid ventricular pacing or low-energy cardioversion could be effective in terminating VT, both methods would be safer and more applicable if back-up bradycardia pacing and high-energy cardioversion/defibrillation were available. Early experience with a permanently implanted low-energy transvenous cardioverter[19] confirmed the need for high-energy defibrillation capability. One of 7 patients with this device was reported to have developed VF as a consequence of an attempted cardioversion of monomorphic VT. The resultant VF could only be terminated with a transthoracic rescue pulse.

In present devices capable of antitachycardia pacing, low-energy cardioversion, and high-energy pulsing, the role of low-energy cardioversion has yet to be better defined. The results of the automatic use of low-energy cardioversion have been mixed. Yee et al. reported 100% success rates for low-energy cardioversion either as initial therapy or as back-up therapy for antitachycardia pacing.[20] In contrast, Winkle et al. found low-energy cardioversion to accelerate VT to VF 50% of the time.[21] Our own experience with low-energy cardioversion suggests that this technique has limited effectiveness and cannot substitute for other therapeutic alternatives. As illustrated in Figures 31.4 and 31.5, low-energy cardioversion can provide a clean termination of VT as well as acceleration to VF. As such its efficacy must be evaluated in each patient.

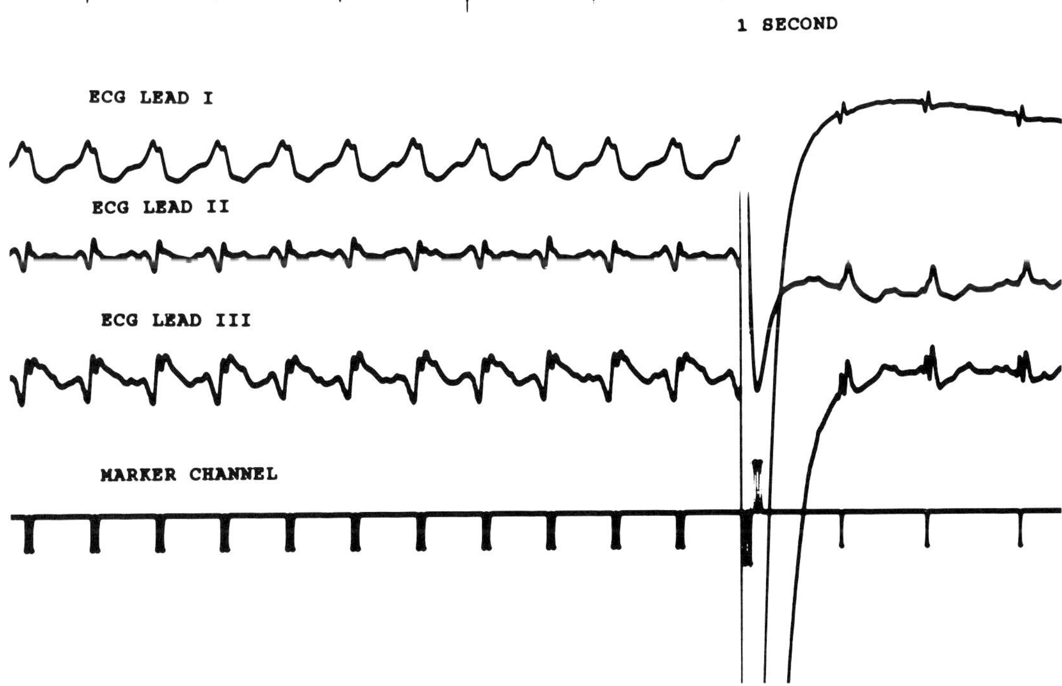

FIGURE 31.4 Low-energy cardioversion terminating VT by a Medtronic 7217B antiarrhythmia device. VT with cycle length of 520 ms was detected and converted to sinus rhythm with a single 0.2-J synchronized pulse through an epicardial patch electrode system. The marker channel from the device outputs varying size and width pulses to indicate what the device is sensing. The double markers during VT indicate sensing of intervals below the VT detection interval. Following cardioversion there are single markers that indicate sensing of intervals above the VT detection interval.

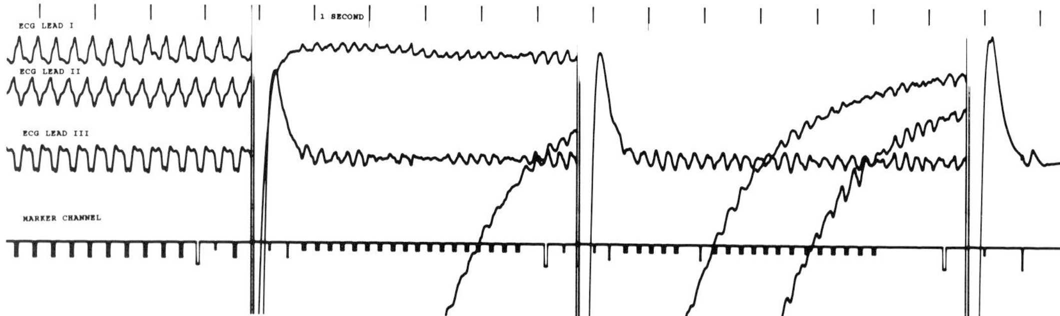

FIGURE 31.5 Low-energy cardioversion accelerating VT to VF. VT with a cycle length of 320 ms was automatically detected and a 2-J synchronized pulse was delivered through epicardial patch electrodes. The 2-J cardioversion pulse accelerated VT to VF which was subsequently detected and converted to sinus rhythm with a 10-J defibrillation pulse. An intervening 4-J defibrillation pulse failed to terminate VF. The marker channel from the device outputs varying pulses to indicate what the device is sensing. In the first part of the tracing, the double pulses indicate sensing of VT. In the second part of the tracing the shorter double pulses indicate sensing of VF. The broad single marker indicates completion of capacitor charging.

VT AND VF DETECTION

Antitachycardia pacing, cardioversion, and defibrillation are dependent upon the accurate automatic detection of VT and VF. Several methods for automatic VT and VF detection have been described.[22,23] Up to now, the predominant method used in antitachycardia pacemakers and the automatic implantable cardiovertor defibrillator (AICD) has been a simple single rate criterion as the major determinant of tachycardia detection. A rate cutoff alone, however, cannot distinguish between monomorphic VT and sinus tachycardia or between monomorphic VT and atrial fibrillation, neither of which requires device intervention. Furthermore, a single rate cutoff doesn't allow for the staging of distinct therapies for VT separate from VF.

The AICD also has an optional "probability density function," the purpose of which is to examine the time during which the intracardiac electrogram strays from the isoelectric baseline. In theory, VF and VT should stray more than supraventricular tachycardia (SVT). However, it has never been documented to be effective in man and has not eliminated device triggering by sinus tachycardia or atrial fibrillation.[24–26]

Other programmable criteria used in antitachycardia pacemakers and recently incorporated into multiprogrammable antiarrhythmia devices include: suddenness of onset, rate stability, and sustained rate. The "onset" criterion is designed to differentiate sinus tachycardia from VT based on the observation that most sinus tachycardias begin gradually and most VTs abruptly.[27] The "stability" criterion is intended to avoid detection of atrial fibrillation by excluding from detection tachycardias in which cycle length varies from beat to beat. "Sustained rate" has the purpose of preventing device intervention during nonsustained VT. Although each of these criteria may improve specificity of VT detection, their ultimate value is uncertain due to longest experience.

These new options for VT detection also have problems with sensitivity of VT and VF detection. For example, VT detection may fail when VT detection criteria are overspecified. The stability function, designed to avoid inappropriate device treatment of atrial fibrillation, relies upon the usual regularity of VT cycle length as opposed to the usual irregularity of atrial fibrillation as a distinguishing feature. Programming a stability value of 50 ms, for example, would reset the VT detection counter for any tachycardia faster than the VT rate cutoff that also varies in cycle length by more than 50 ms. This function can prevent inappropriate therapy of atrial fibrillation as evidenced in Figure 31.6. However, the use of stability may prevent the detection of polymorphic VT, with a variable cycle length, or the recognition of monomorphic constant cycle length VT with superimposed intermittent ventricular premature depolarizations (VPDs).

The advanced features of multiprogrammable antiarrhythmia devices can also lead to VF undersensing. VF may go undetected, for example, when an excessively broad VT detection zone constricts or limits the VF detection zone. This problem occurs when a fast monomorphic VT coexists with relatively slow VF. If the physician wants to treat this patient's fast VT with antitachycardia pacing, for example, the cutoff rate for VT detection must be high, which in turn

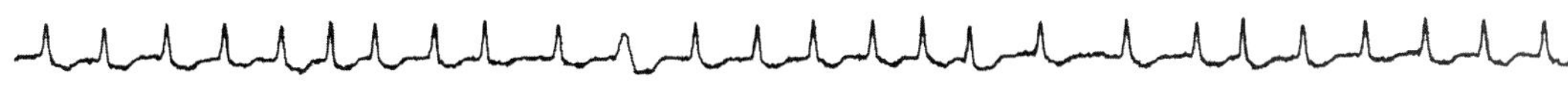

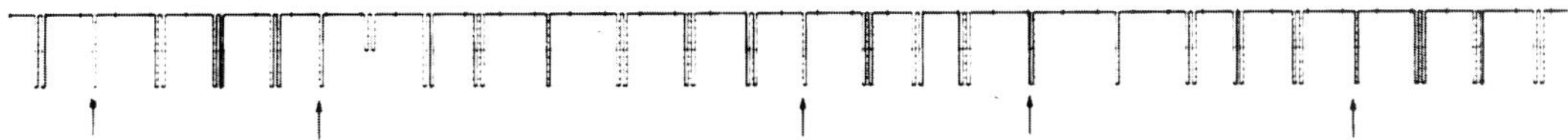

FIGURE 31.6 Atrial fibrillation with rapid ventricular response that fulfills the rate criterion for VT detection, but is not detected by the device because the rate stability criterion is not satisfied. The VT detection interval is 400 ms and the stability criterion is 60 ms. A single detection marker is declared for cycles whose variation from the preceding cycle is greater than 60 ms, even though the cycle length is shorter than the VT detection interval (arrows). VT markers (double outputs) are shown only when successive intervals are coupled at 60 ms or less. Consecutive VT markers must total a programmed "number of intervals to detect" before VT therapy is initiated. Each time a single detection marker is declared, the VT counter is reset. A single VF marker (short double output) declares one interval with a cycle length less than the programmed fibrillation detection interval of 280 ms.

raises the cutoff rate for VF detection (to avoid a high-energy shock for VT). Under these circumstances, if this patient has slow VF, perhaps because of antiarrhythmic drug therapy, the VF cycle length may fall into the VT detection zone rather than the VF detection zone. This would lead to VF being treated inappropriately with antitachycardia pacing.

Prior to the advent of multiprogrammable devices, staged sensing of VT and VF was removed from physicians' purview. The option to stage VT detection and therapy differently from VF detection and therapy complicates sensing, particularly that of VF. Too much emphasis on VT detection and therapy increases the likelihood that VF will be undersensed or treated inappropriately as monomorphic VT with pacing or low-energy cardioversion therapies.

In present multiprogrammable antiarrhythmia devices, VF detection is also dependent upon proper selection of sensitivity settings. Variable intracardiac electrogram amplitude during VF makes detection of each VF signal more difficult than detection of each monomorphic VT signal (Fig. 31.7). The variability in electrogram amplitude during VF can lead to undersensing because of the combined effects of the sensitivity setting floor and the dynamics of an automatic gain control (Fig. 31.8). Because VF electrogram amplitude can be substantially smaller than sinus rhythm or VT electrogram amplitude, sensitivity settings must be higher to detect VF than to detect bradyarrhythmias or monomorphic VT. However, if the setting is too sensitive, oversensing of T waves and ambient muscle and respiratory noise can precipitate inappropriate discharges. An automatic gain control compensates for oversensing somewhat by resetting the threshold for signal detection from the preceding signal (Fig. 31.9). This should avoid undersensing and oversensing in most circumstances (Fig. 31.10). However, the sometimes marked variation in electrogram amplitude during VF can make VF detection precarious unless the VF detection parameters are programmed cautiously with respect to the VT detection parameters. VF detection in difficult settings, for example, in patients treated with amiodarone, may require maximum sensitivity settings together with low fibrillation detection rates and low numbers of cycles used in detection. VF detection in these circumstances requires a balance in the programming of the device, sometimes sacrificing antitachycardia pacing or low-energy cardioversion therapies of monomorphic VT in favor of a more forgiving detection schema for the more serious arrhythmia of VF.

Several other methods for improving specificity of detection have been investigated recently, but not yet included in an implantable device. These include atrial sensing,[28] frequency spectra of electrograms,[29] right ventricular

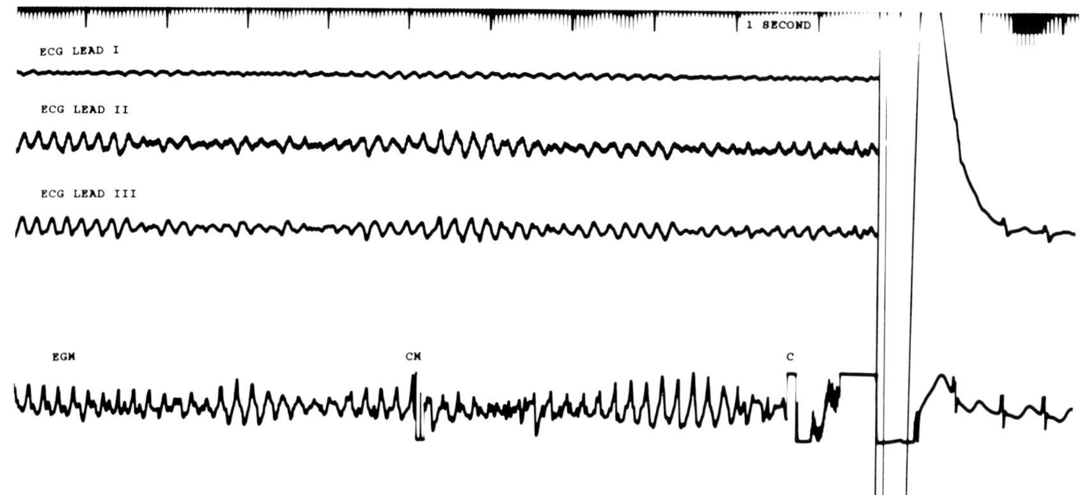

FIGURE 31.7 VF detection by a multiprogrammable antiarrhythmia device. The bottom electrogram (EGM) is the bipolar intracardiac electrogram sensed by the device and transmitted by telemetry. Note the variable amplitude of intracardiac electrograms during VF. In the middle of the VF episode, a cancel magnet marker (CM) can be seen that allows the device to begin detection. Prior to a 5-J defibrillation pulse, another marker (C) indicates that the capacitor is fully charged.

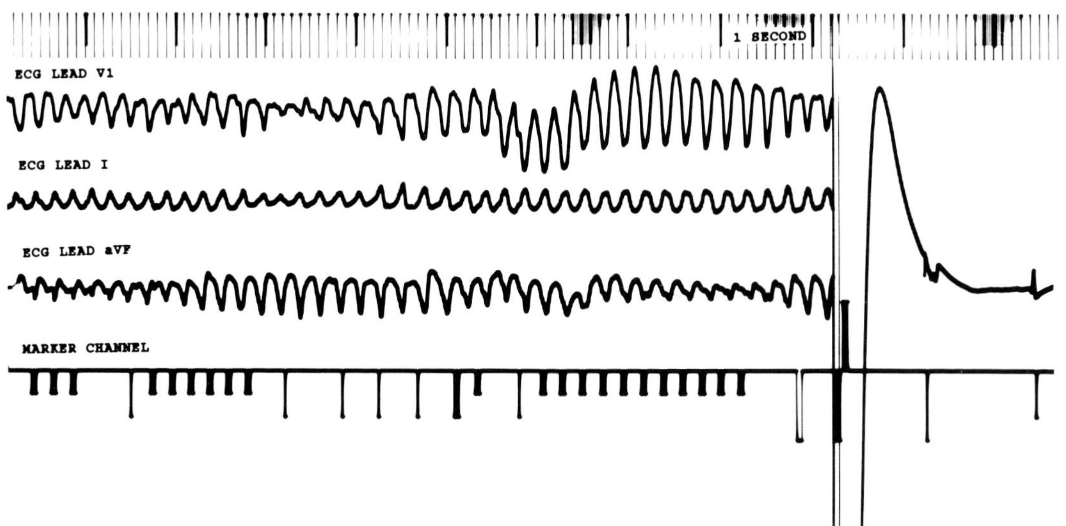

FIGURE 31.8 Impaired sensing of VF due to low sensitivity setting. In this example, VF has a cycle length below the programmed fibrillation detection interval. However, as demonstrated in the telemetry marker channel, many intervals were not sensed. The short, double pulses mark sensed events satisfying the fibrillation detection interval and the tall, double pulse marks one sensed event within the tachycardia detection interval. The tall, single pulses mark sensed beats above the VT detection interval. After 9 s of VF, sufficient cycles were eventually detected to trigger delivery of a 5-J defibrillation pulse. The programmed sensitivity was 2.4 mV. Impaired sensing was corrected by increasing the sensitivity to 0.3 mV.

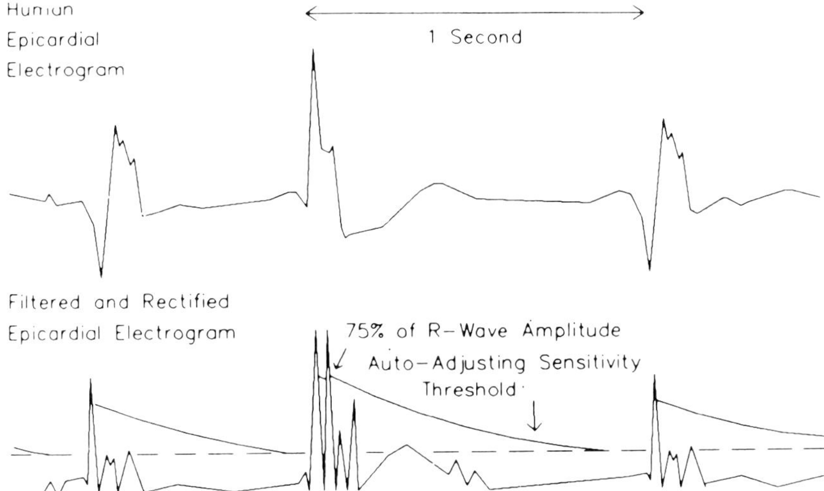

FIGURE 31.9 Autoadjusting sensitivity threshold (automatic gain control) in Medtronic 7217B antiarrhythmia device. To avoid oversensing of T waves, yet allow very low sensing thresholds for VF detection, the sensing threshold is automatically raised to 75% of the sensed electrogram amplitude following a sensed event. The threshold then exponentially decays back to the programmed setting.

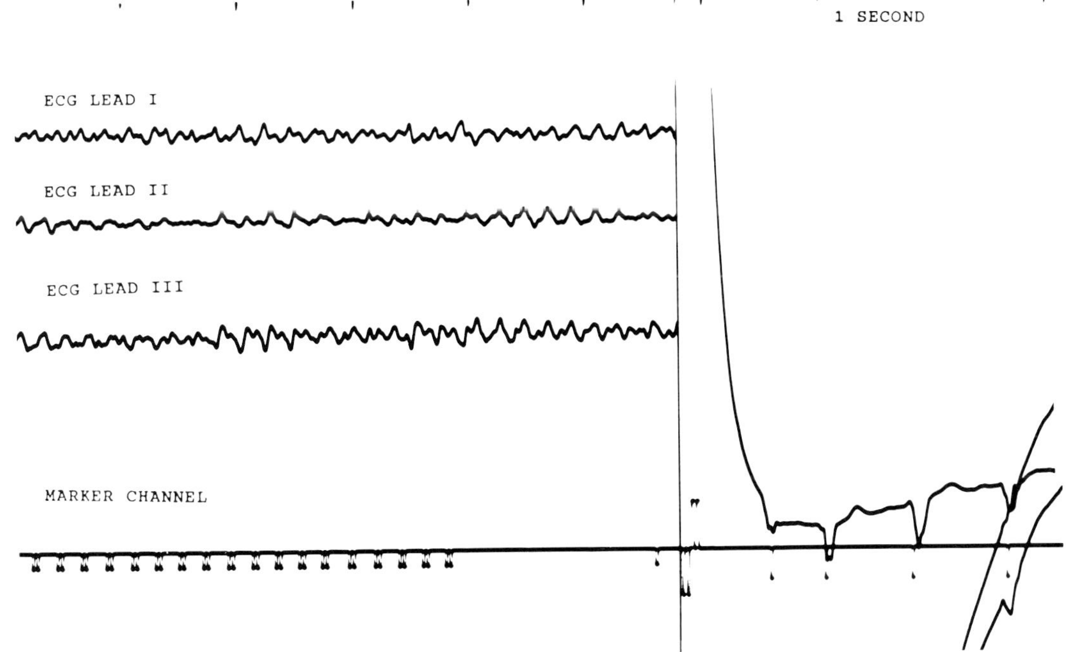

FIGURE 31.10 An example of prompt unequivocal VF detection when each cycle below the programmed fibrillation detection interval is sensed (in this case 18 cycles less than 280 ms). The sensitivity setting was 0.3 mV. A 10-J pulse was delivered terminating VF. The cycles sensed below the fibrillation detection interval are the short double pulses on the telemetry marker channel. Differences in the size of the marker channel pulses from one illustration to another reflect variation in amplifier gain settings in the different examples.

impedance,[30] and right ventricular pressure.[31] The latter two methods are attractive because they might reflect the hemodynamic significance of the arrhythmia. Future considerations notwithstanding, no single detection parameter will likely identify all arrhythmias correctly. A combination of criteria tailored to the individual patient will be more realistic for the foreseeable future.

DEFIBRILLATION WAVEFORMS

The defibrillation waveform used in the AICD has been a monophasic truncated exponentially decaying waveform with fixed tilt of approximately 65% from a 120 μF capacitor (Fig. 31.11a). The origin of this pulse is in part a practical choice and in part guided by early animal studies.[32,33] Today, however, the role of pulsing waveform in defibrillation efficacy can be expanded. Most recently, the focus has been on biphasic waveforms (Fig. 31.11b). Several studies have shown that biphasic waveforms in general defibrillate with less energy than monophasic waveforms.[34–38] However, efficacy for biphasic waveforms varies substantially, depending on shape.[39] Tilt and pulse width affect biphasic waveform performance just as they do monophasic waveforms.

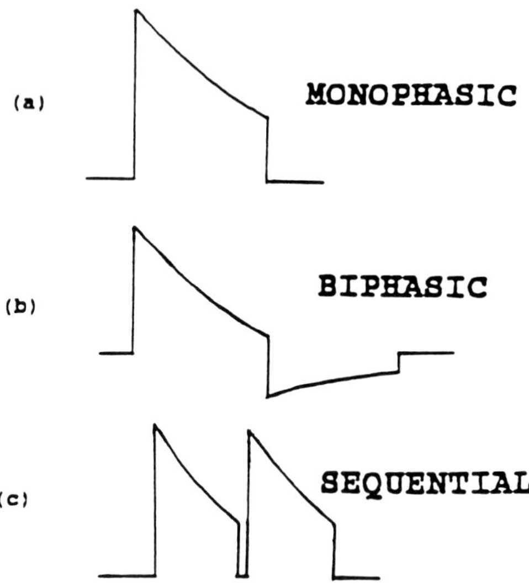

FIGURE 31.11 Example of the different defibrillation waveforms utilized in implantable defibrillators. All the pulses are truncated exponentially decaying. The degree of waveform decay is expressed as % tilt (% TILT = 100 × (1 − V_T/V_L), where V_T = trailing edge voltage and V_L = leading edge voltage. **a.** The monophasic pulse with 65% tilt has been the standard defibrillation waveform. **b.** The biphasic waveform is made by inverting the output polarity so that the trailing edge voltage of the positive phase becomes the leading edge voltage of the negative phase. **c.** A sequential pulse is generated by discharging two capacitors in rapid sequence.

In contrast to a multitude of animal studies in defibrillation, few studies have examined biphasic pulses in man. Using a biphasic waveform with only a small negative phase, Winkle et al. showed a slight improvement in biphasic defibrillation efficacy compared to that observed with monophasic pulse defibrillation.[40] However, biphasic defibrillation was more efficient only at pulsing values less than 6.4 J. This finding essentially removes the advantage of such a waveform. The need for increased defibrillation efficiency is at the higher end of the pulsing range rather than at the lower end.

Our examination of biphasic waveform defibrillation in man used a biphasic pulse with a much larger negative component than that used by Winkle.[40] The waveform was generated with a single 120 μF capacitor by switching the polarity of the trailing edge of the positive phase so that it became the leading edge of the negative phase. This waveform therefore could easily be incorporated into implantable devices without the need for a second capacitor and therefore a larger pulse generator. The results of the study showed that biphasic defibrillation was more effective than monophasic defibrillation, especially in patients with higher defibrillation thresholds. In general, the defibrillation threshold was reduced by 26% with the biphasic pulse. Additionally, despite a general advantage of biphasic pulsing to monophasic pulsing, exceptions were present where monophasic pulses outperformed biphasic pulses. These findings indicate significant individual variation of response to different waveforms.

The other major development in defibrillation waveforms is that of sequential pulsing (Fig. 31.11c). This method requires the addition of a third electrode to the defibrillation system to allow for pulsing over two different pathways. In theory, the sequential-pulse technique has the advantage of distributing current more uniformly over the myocardium than when a two-patch system is used and avoids the problem of field cancellation that would occur if one pulse were delivered to two anodes simultaneously.[42] The method involves the delivery of two pulses in rapid sequence. One pulse is delivered between the common cathode and one anode; microseconds later, the other pulse is delivered between the common cathode and the second anode.

Substantial reductions in defibrillation threshold can be achieved with sequential pulsing.[43] In one clinical study of 16 survivors of out-of-hospital VF, the standard two-patch single-pulse defibrillation system was compared to a three-patch sequential-pulse system during defibrillator surgery (Fig. 31.12). Sequential-pulse defibrillation reduced the mean defibrillation threshold by 37%, being more effective in those patients with higher defibrillation thresholds (Fig. 31.13).

Sequential-pulse defibrillation, although effective, has the limitation that it requires the use of a third electrode. This encumbrance is somewhat offset by the resulting lower defibrillation thresholds (DFTs). In patients whose DFT is already low, however, such a method is probably not necessary. The main application for sequential pulsing may prove to be in transvenous rather than epicardial systems, as described below in the section on nonthoracotomy lead systems.

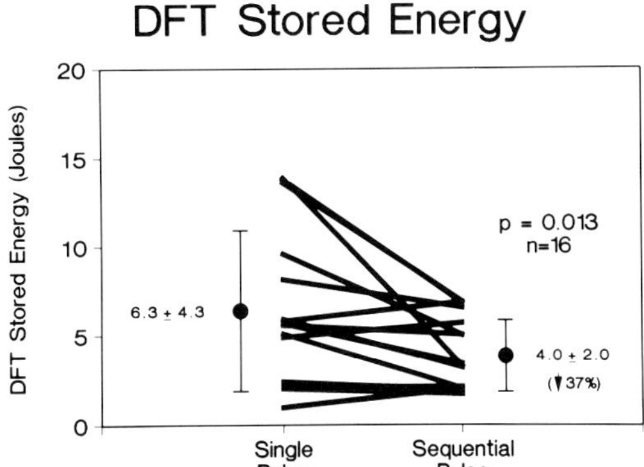

FIGURE 31.13 Defibrillation threshold (DFT) values for stored energy for single-pulse and sequential-pulse defibrillation methods. Note the decline in DFT for the sequential-pulse technique, especially in patients whose DFTs were greater than 10 J, by the single-pulse defibrillation technique. *Reprinted from Bardy et al.,*[43] *with permission.*

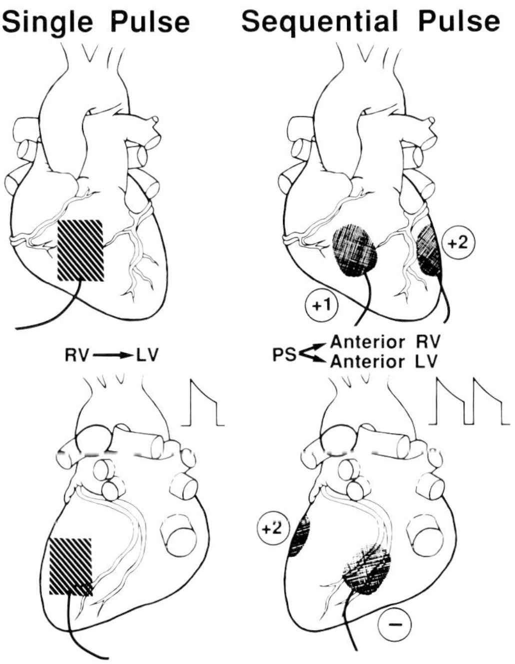

FIGURE 31.12 Comparison of patch placement and waveform pathway for standard single-pulse defibrillation versus sequential-pulse defibrillation. The standard single pulse was delivered between large patch electrodes (CPI model 0041) placed over the anterolateral right ventricle and posterolateral left ventricle. The sequential pulse technique used three electrodes (Medtronic model 6892): the posteroseptal ventricular electrode served as the common cathode (−), the anterior right ventricular electrode served as the first anode (+1), and the anterior left ventricular electrode served as the second anode (+2). *Reprinted from Bardy et al.,*[43] *with permission.*

Although most attention in waveform research has been focused on biphasic or sequential pulsing, it is important to recognize that the use of the monophasic pulse still needs to be optimized. Preliminary data suggest that simple shifts in monophasic waveform tilt are sufficient to improve defibrillation efficacy. We have observed patients in whom defibrillation threshold decreased by 10 J with a 15% change in waveform tilt from 65% to 80%. This is counter to what one might expect given that delivered energy is actually less with 65% tilt pulses of equal voltage to that of 80% tilt pulses. Nevertheless, these empiric findings suggest that much remains to be learned about waveform effect on defibrillation. Perhaps relatively minor changes in monophasic waveform shape will ultimately rival the improvements derived with biphasic- or sequential-pulse defibrillation.

LEAD SYSTEMS

The major recent goals in this area have been to improve the efficiency of defibrillation and to develop a lead system that would not require a thoracotomy for implantation. Improved efficiency means lower defibrillation thresholds, which translates into less energy expenditure for each intervention, thus prolonging generator life and/or reducing generator size. In addition, lower defibrillation thresholds provide the margin of safety needed to allow transvenous systems to be inserted with defibrillators capable of a 30-J maximum output.

Epicardial Lead Systems

The standard epicardial lead system has utilized two rectangular titanium mesh electrodes with surface areas of either 13.9 or 17.9 cm^2. The double large epicardial patch configuration has proven more efficacious than the original superior vena cava coil electrode and small apical epicardial patch.[(44)] With the large patch–large patch system most patients have had acceptable defibrillation thresholds in the 5- to 10-J range.[(38,41,43–47)] However, outliers exist and it is unclear whether increasing the surface area of the patches would further improve defibrillation efficiency. Recent changes in lead design suggest defibrillation thresholds can decrease further as patch surface area increases.[(48)] However, more lead surface area is not necessarily always better. Epicardial shunting of current between the edges of closely opposed patches of opposite polarity is a theoretical possibility that could actually increase defibrillation energy requirements. At this time, insufficient data are available to determine the optimal patch size. From experience, however, it is reasonable to expect that all patches cover approximately one-half of the ventricular epicardial surface area.

Although increasing epicardial patch area has the potential to improve defibrillation, large surface area patches around the heart can cause problems. Epicardial patch systems, for example, increase transthoracic energy requirements for defibrillation.[(49)] In addition, larger patches might also limit the feasibility of concurrent or future coronary artery bypass surgery. At this time, it is probably best to apply minimum patch size. Future epicardial leads might be improved by removing the insulative backing, which would prevent the rise in transthoracic defibrillation requirements, and by developing a variety of shapes for selection at the time of surgery that would avoid patch placement over coronary arteries or grafts.

A new structural development in epicardial lead systems is the closed-loop coil patch electrode (Fig. 31.14). The wire mesh structure of the original epicardial patch is brittle and prone to crinkling,[(50)] and its cable may be prone to stress fractures.[(51)] With the older lead design, we have observed four cable fractures in 250 defibrillator implants in patients who have had their units for more than 3 years. Because the new coil electrodes are more flexible, lead fractures and patch crinkling may be less likely. In addition, with improved pliability, the coil can better conform to the epicardial surface, particularly over bypass grafts where stiff patches could cause injury.[(26)]

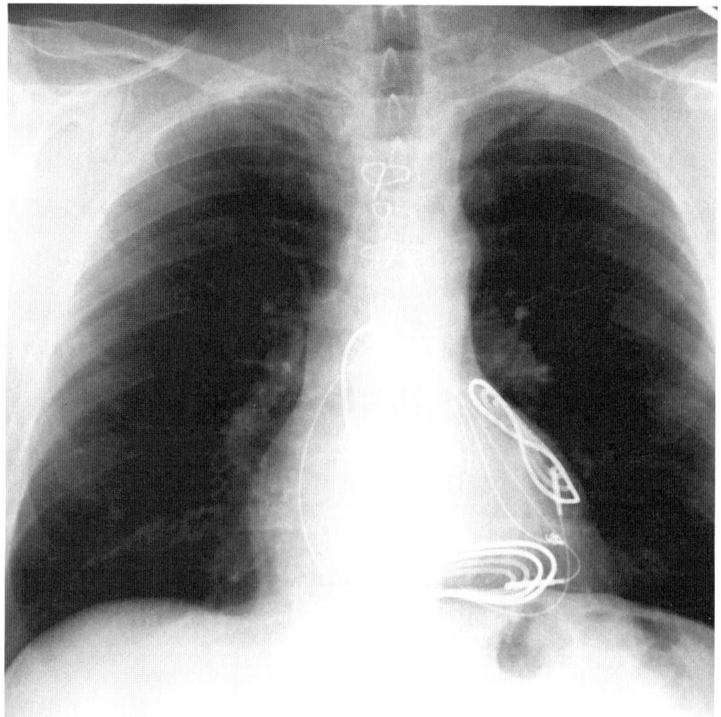

FIGURE 31.14 Example of coil design epicardial electrodes (Medtronic model 6897) in an implanted multiprogrammable antiarrhythmia device.

Nonthoracotomy Lead Systems

The concept of transvenous defibrillation has been investigated since 1954.(15,52–58) Many of these pulsing methodologies, such as pulsing between a superior vena cava (SVC) and a right ventricular electrode, have been abandoned because of defibrillation requirements above 30 J. Transvenous defibrillation, to be considered safe, should be successful at values well below a device's maximum output. The problem of obtaining adequate defibrillation thresholds poses one of the primary obstacles to successful implementation of transvenous or nonthoracotomy lead systems.

Implantation of transvenous defibrillation systems would be easier if an acceptable safety margin was known between maximal device output and defibrillation threshold. Unfortunately, these data are not available for epicardial lead systems, let alone for transvenous lead systems. Thus, we are left with only experience to guide us. Because multiple factors can alter postimplantation defibrillation threshold, including antiarrhythmic drugs, chronic electrode–tissue interactions, and progression in disease process, we have recommended that at least a 2:1 DFT safety margin be used as a reasonable guideline before proceeding with implantation of a transvenous or nonthoracotomy lead system. Therefore, it might be safe to implant a transvenous defibrillation system if the defibrillation threshold proved to be less than or equal to 15 J for a defibrillator capable of a 30-J output.

Several nonthoracotomy lead systems and pulsing methods are currently being employed in humans. Unlike epicardial defibrillation where a two-lead system is often sufficient to result in very low DFTs, in nonthoracotomy lead systems three leads are often necessary. Troup et al. reported preliminary results of implantation with a three-electrode catheter-patch nonthoracotomy system.(59) The catheter, with distal right ventricular and proximal superior vena cava coil electrodes on the same lead, is used in conjunction with a 28-cm^2 patch electrode placed subcutaneously or submuscularly in the left chest wall. Unidirectional pulsing or simultaneous bidirectional pulsing over one of four current pathways using a monophasic, fixed-tilt waveform was chosen for study. A defibrillation threshold of 15 J or less could be obtained in 25 of 36 (70%) patients. This transvenous system appeared to be suitable for many patients. Of concern, however, was a 14% incidence of lead fracture or dislodgment over 3.5 months of follow-up.

An alternative to using a SVC electrode is a coronary sinus (CS) electrode. Although the access to SVC is easy, it is far removed from the fibrillating ventricles. Pulses delivered from a right ventricular electrode to a SVC lead actually direct current away from the dominant problem tissues of the left ventricle. Pulsing between the right ventricular lead and the chest patch lead compensates somewhat for this problem and directs more of the pulse to the left ventricle. However, because of the anatomical right ventricular and chest patch location, the posterolateral left ventricle remains relatively undertreated.(60)

Pulsing from a right ventricular (RV) lead to a CS lead has the anatomic advantage of directing current to the bulk of fibrillating myocardium. Furthermore, a CS electrode provides access to the posterolateral and basal left ventricle, sectors not easily reached simply by pulsing from the right ventricle to a chest patch. There is a 20-year history of using the CS for dual-chamber permanent pacemaker insertion without any evidence of CS rupture, left ventricular dysfunction, or coronary artery injury.(2,61,62) These findings suggest that, from a mechanical perspective, the use of a CS lead is likely to cause no more difficulties than any other intracardiac or epicardial lead.

We have examined the usefulness of a coronary sinus electrode in defibrillating 20 survivors of sudden cardiac death with temporary transvenous lead systems prior to standard epicardial defibrillator surgery.(45) In these patients, pulsing between a right ventricular coil electrode and a CS electrode resulted in a defibrillation threshold that was 22% less than that found with a right ventricular chest wall patch electrode system. The mean defibrillation threshold for the RV-CS lead system was 17.5 J compared to 25.6 J for the right ventricular chest patch system and 6.0 J for the standard epicardial patch–patch system. If a 15-J cutoff were employed as a criterion for implantation of a transvenous system, 9 of 20 patients (45%) would have been candidates for the CS-based system and only 5 of 20 (25%) for the RV chest patch system.

The use of a CS electrode in isolation with a RV electrode, although an improvement, does not render the majority of defibrillator patients candidates for transvenous lead systems. However, when combined with a third electrode and alternative pulsing methods, further improvements in transvenous defibrillation efficacy have been demonstrated. We studied four different transvenous defibrillation methods using right ventricular, coronary sinus, and chest wall patch (CP) electrodes in 12 cardiac arrest survivors prior to standard epicardial defibrillator surgery (see Fig. 31.15).(63) Choosing the best of the four methods from each patient yielded a mean de-

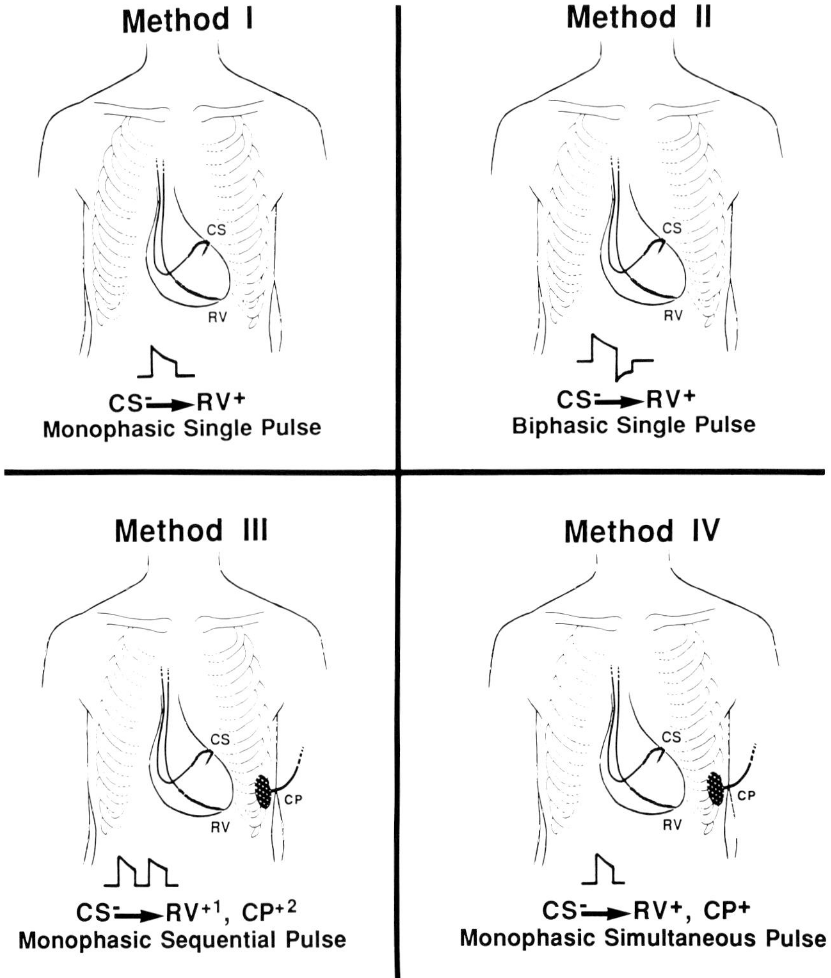

FIGURE 31.15 Flexibility in transvenous defibrillation with the use of a three-lead, multiple-pulsing-method system capable of using monophasic-, biphasic-, and sequential-pulse waveforms. *Reprinted from Bardy et al.,*[63] *with permission.*

livered energy defibrillation threshold of 11.3 J. While no method proved uniformly superior, 83% of patients had a defibrillation threshold less than 15 J with at least one of the tested configurations (Fig. 31.16).

Figure 31.17 shows an example of a patient with a permanently implanted three-lead transvenous system where flexibility in choosing defibrillation technique was important. This patient happened to defibrillate best with DFT of 7.5 J, with a single pulse delivered between the CS electrode and the right ventricular electrode. Simultaneous pulsing in this patient, for example, between the right ventricular, CS, and chest patch electrode resulted in a DFT of 10 J. Although the findings at implant showed dual-pathway pulsing to be less effective than single-pathway pulsing, a subcutaneous patch was placed. The subcutaneous patch was placed, regardless of the superiority of single-pathway pulsing to allow for flexibility in device programming should the patient's response to single-pathway pulsing change in the future. Although the patient is currently programmed to single-pulse defibrillation, he may be programmed to simultaneous- or sequential-pulse, dual-pathway defibrillation in the future.

The studies to date show that there is probably no transvenous lead system that is superior for all patients. With the versatility of multiple

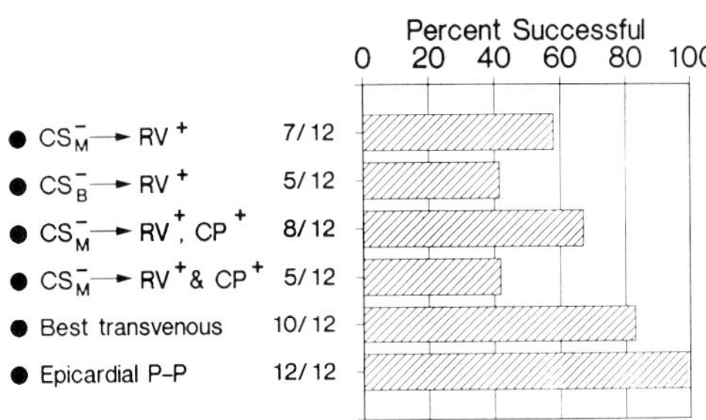

FIGURE 31.16 Percentage of patients for which defibrillation was possible at 15 J or less using the transvenous defibrillation system shown in Figure 31.15. *Reprinted from Bardy et al.,*[63] *with permission.*

lead locations and pulsing waveforms, however, a configuration with an acceptable margin of safety can probably be devised in the majority of patients. Major complications during implantation of transvenous systems have not been reported, but the long-term effectiveness is yet to be determined. Because of the potential problems of lead fracture or dislodgment, close follow-up will be necessary. Whether changes in defibrillation efficacy occur over time due to the electrode tissue interface is unknown.

SUMMARY

The advances described above have produced several multiprogrammable devices currently undergoing clinical trials. The devices offer several advantages over the traditional AICD. In general, the devices are designed with a hierarchal approach to therapy.[64] Since the rate of tachycardia is a major determinant of how hemodynamically significant the arrhythmia is likely to be, multiple rate ranges can be chosen for different sequences of therapy. Depending on the tachycardia rate and how compromising it is to the individual patient, therapy of varying aggressiveness can be chosen, from burst or ramp pacing to low-energy cardioversion, to high-energy cardioversion or defibrillation. For any single episode of tachycardia, the aggressiveness of the intervention can be programmed to increase with each attempt. For example, low-energy and then high-energy cardioversion could be tried in succession if burst pacing failed to terminate the tachycardia. Tailoring the therapy to the severity of the arrhythmia allows for better patient tolerance and longer battery life when lower-energy interventions are used.

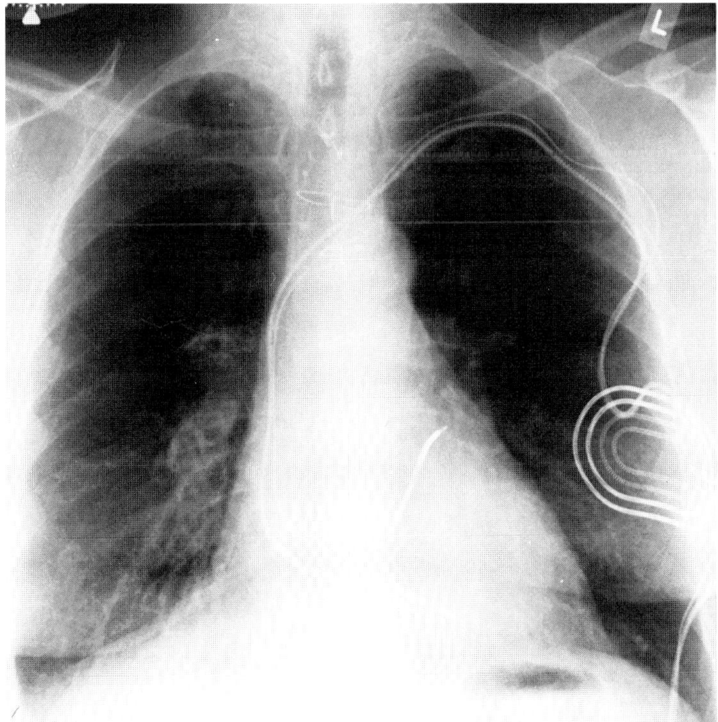

FIGURE 31.17 Permanently implanted nonthoracotomy electrode system for a multiprogrammable pacer cardioverter/defibrillator (Medtronic model 7216A). One coil electrode is positioned in the coronary sinus. Another coil electrode, with a distal pace/sense electrode, is positioned with its tip at the right ventricular apex. A large patch electrode is positioned subcutaneously on the left lateral chest wall.

Tachycardia detection based on several programmable criteria, including suddenness of onset and rate stability, may help avoid triggering interventions during SVT. In addition, allowing a programmable number of intervals needed to initiate an intervention should be helpful in avoiding unnecessary therapies in patients with short runs of nonsustained VT.

The devices also have the capability of backup VVI pacing, eliminating the need for two devices and the potential interaction between them. Eventually, VVIR and dual-chamber pacing will be part of such devices.

Versatility in choosing the type of pulsing waveform and current pathway will maximize the efficiency of the device, extending battery life and/or allowing a transvenous implantation.

Although preliminary reports on the performance of multiprogrammable devices[20,21,65] have been encouraging, these devices are complex and require careful programming and testing to assure proper function. The adequacy of tachycardia and fibrillation detection and therapy need to be assessed. Care must be taken to keep arrhythmia detection times short and choose initial therapies consistent with the hemodynamic significance of the arrhythmia. Given the risk of accelerating VT with pacing and low-energy cardioversion, as well as the possibility of hemodynamic deterioration as tachycardia persists, careful attention should be paid to the duration of successive therapies before a high-energy shock is delivered.

Several multiprogrammable antiarrhythmic devices are being evaluated in clinical trials at the present time and are likely to be approved for more widespread implantation in the near future. The implantation of transvenous lead systems will also likely become more widespread once the results of clinical trials demonstrate their long-term safety and efficacy. Investigation of the benefit of antiarrhythmic devices in patients at high risk for sudden cardiac death who have yet to experience a life-threatening arrhythmia may become feasible if transvenous systems can be implanted more safely and economically than the present devices.

Improvements in the specificity of VT and fibrillation detection and therapy based on the hemodynamic significance of the arrhythmia may be possible in future generations of devices. As advances in electronics continue, the capability for recording and retrieval of electrograms will become available and aid in patient management. Generator size should continue to decrease while longevity increases as electronics become more miniaturized and pulsing regimens become more efficient. The cost of antiarrhythmic device therapy should also fall as multiple manufacturers enter the market, generator replacements for battery depletion become less frequent, and implantation becomes simpler. The advances will allow antiarrhythmic devices to play a major role in preventing sudden cardiac death.

REFERENCES

1. Mirowski M, Reid PR, Mower MM, et al.: Termination of malignant ventricular arrhythmias with an implanted automatic defibrillator in human beings. *N Engl J Med* 1980;303:322–324.
2. Moss AJ, Rivers RJ: Termination and inhibition of recurrent tachycardias by implanted pervenous pacemakers. *Circulation* 1974;50:942–947.
3. Fisher JD, Johnston DR, Kim SG, Furman S, Mercardo AM: Implantable pacers for tachycardia termination: Stimulation techniques and long term efficacy. *PACE* 1986;9:1325–1333.
4. Swerdlow CD, Bardy GH, McAnulty J, Kron J, et al.: Determinants of induced sustained arrhythmias in survivors of out-of-hospital ventricular fibrillation. *Circulation* 1987;76:1053–1060.
5. Fisher JD, Mehra R, Furman S: Termination of ventricular tachycardia with bursts of rapid ventricular pacing. *Am J Cardiol* 1978;41:94–102.
6. Naccarelli GV, Zipes DP, Rahilly GT, Heger JJ, Prystowsky EN: Influence of tachycardia cycle length and antiarrhythmic drugs on pacing termination and acceleration of ventricular tachycardia. *Am Heart J* 1983;150:1–5.
7. Roy D, Waxman H, Buxton AE, Marchlinski FE, et al.: Termination of ventricular tachycardias: Role of tachycardia cycle length. *Am J Cardiol* 1982; 50:1346–1350.
8. Mason JW, Winkle RA: Electrode-catheter arrhythmia induction in the selection of antiarrhythmic drug therapy for recurrent ventricular tachycardia. *Circulation* 1978;58:971–985.
9. Calkins H, Brinker J, Veltri EP, Guarnieri T, Levine JH: Clinical interactions between pacemakers and automatic implantable cardioverter-defibrillators. *J Am Coll Cardiol* 1990;16:666–673.
10. Luderitz B, Gerckens U, Manz M: Automatic implantable cardioverter/defibrillator (AICD) and antitachycardia pacemaker (TACHYLOG): Combined use in ventricular tachyarrhythmias. *PACE* 1986; 9:1356–1360.
11. Josephson ME, Horowitz LN, Farshidi A, Spielman SR, et al.: Recurrent sustained ventricular tachycardia. 4. Pleomorphism. *Circulation* 1979;59:459–468.
12. Wellens HJ, Schuilenburg RM, Durer D: Electrical stimulation of the heart in patients with ventricular tachycardia. *Circulation* 1972;46:216–226.
13. Charos GS, Haffajee CI, Gold RL, Bishop RL, Berkovits B, Alpert JS: A theoretically and practically more effective method for interruption of ventricular tachycardia: Self-adapting autodecremental overdrive pacing. *Circulation* 1986;73:309–315.
14. Zipes DP, Jackman WM, Heger JJ, et al.: Clinical transvenous cardioversion of recurrent life-threatening ventricular tachyarrhythmias: Low energy synchronized cardioversion of ventricular tachycardia and termination of ventricular fibrillation in

patients using a catheter electrode. *Am Heart J* 1982;103:789–794.
15. Waspe LE, Kim SG, Matos JA, Fisher JD: Role of a catheter lead system for transvenous countershock and pacing during electrophysiologic tests: An assessment of the usefulness of catheter shocks for terminating ventricular tachyarrhythmias. *Am J Cardiol* 1983;52:477–484.
16. Ciccone JM, Saksena S, Shah Y, Pantopoulos D: A prospective randomized study of the clinical efficacy and safety of transvenous cardioversion for termination of ventricular tachycardia. *Circulation* 1985;71:571–578.
17. Saksena S, Chandran P, Shah Y, Boccadamo R, Pantopoulos D, Rothbart ST: Comparative efficacy of transvenous cardioversion and pacing in patients with sustained ventricular tachycardia: A prospective, randomized, crossover study. *Circulation* 1985;72:153–160.
18. Lindsay BD, Saksena S, Rothbart ST, Wasty N, Pantopoulos D: Prospective evaluation of a sequential pacing and high-energy bidirectional shock algorithm for transvenous cardioversion in patients with ventricular tachycardia. *Circulation* 1987; 76:601–609.
19. Zipes DP, Heger JJ, Miles WM, et al.: Early experience with an implantable cardioverter. *N Engl J Med* 1984;311:485–490.
20. Yee R, Klein GJ, Murdock C, Leitch JW, Norris C: Efficacy of antitachycardia pacing and cardioversion with the pacemaker-cardioverter-defibrillator. *J Am Coll Cardiol* 1990;15:60A.
21. Winkle RA, Mead RH, Fain ES, Cannom DS, et al.: Initial clinical experience with an implantable tiered therapy defibrillator. *J Am Coll Cardiol* 1990;15:60A.
22. Pannizzo F, Mercando AD, Fisher JD, Furman S: Automatic methods for detection of tachyarrhythmias by antitachycardia devices. *Am Coll Cardiol* 1988;11:308–316.
23. Camm AJ, Wyn DD, Wend D: Tachycardia recognition by implantable electronic devices. *PACE* 1987;10:1175–1190.
24. Manz M, Gerckens U, Ludritz B: Erroneous discharge from an implanted automatic defibrillator during supraventricular tachycardia. *Am J Cardiol* 1986;57:343–344.
25. Deborde R, Levine JH, Griffith LL, et al.: Exercise testing in patients with AICDs. *Circulation* 1985;72(suppl III):474.
26. Marchlinski FE, Flores BT, Buxton AE, Hargrove WC, et al.: The automatic implantable cardioverter-defibrillator: Efficacy, complications, and device failures. *Ann Int Med* 1986;104:481–488.
27. Fisher JD, Goldstein M, Ostrow E, Matos JA, et al.: Maximal rate of tachycardia development: Sinus tachycardia with sudden exercise vs spontaneous ventricular tachycardia. *PACE* 1983;6:221–228.
28. Schuger CD, Jackson K, Steinman RT, Lehmann MH: Atrial sensing to augment ventricular tachycardia detection by the automatic implantable cardioverter defibrillator: A utility study. *PACE* 1988; 11:1456–1463.
29. Pannizzo F, Furman S: Frequency spectra of ventricular tachycardia and sinus rhythm in human intracardiac electrograms—Application to tachycardia detection for cardiac pacemakers. *IEEE Trans Biomed Eng* 1988;35:421–425.
30. Bardy GH, Olson WH, Fishbein DL, et al.: Transvenous right ventricular impedance during spontaneous arrhythmias. *Circulation* 1985;72(suppl III):III-474.
31. Sharma AD, Bennett T, Ericson M, Yee R, Klein G, Guiraudon G: Right ventricular pressure during ventricular fibrillation. *Circulation* 1988;78:II-220.
32. Schuder JC, Stoeckle H, Gold JH, et al.: Experimental ventricular defibrillation with an automatic and completely implanted system. *Trans Am Soc Artif Organs* 1970;16:207–212.
33. Mirowski M, Mower MM, Langer A, et al.: Implantable defibrillators. Proceedings of the Purdue cardiac defibrillator conference. West Lafayette, IN. October 1–3, 1975, pp 93–96.
34. Schuder JC, Stoeckle H, Dolan AM: Transthoracic defibrillation with square wave stimuli: One-half cycle, one cycle, and multicycle waveforms. *Circ Res* 1964;15:258–264.
35. Schuder JC, Gold JH, McDaniel WC: Ultrahigh-energy hydrogen thyrztron/SCR bidirectional waveform defibrillator. *Med Bio Eng Comput* 1982;20:419–424.
36. Schuder JC, Gold JH, Stoeckle H, McDaniel WC, Cheung KN: Transthoracic ventricular defibrillation in the 100 kg calf with symmetrical 1-cycle bidirectional rectangular wave stimuli. *IEEE Trans Am Biomed Eng BME* 1983;30:415–422.
37. Schuder JC, McDaniel WC, Stoeckle H: Defibrillation of 100 kg calves with asymmetrical, bidirectional, rectangular pulses. *Cardiovasc Res* 1984;18:419–426.
38. Negovsky VA, Smerdov AA, Tabak VY, Venin IV, Bogushevich MS: Criteria of efficiency and safety of defibrillating impulse. *Resuscitation* 1980;8:53–67.
39. Tang ASL, Yabe S, Wharton M, Dolker M, Smith WM, Ideker RE: Ventricular defibrillation using biphasic waveforms: The importance of phasic duration. *J Am Coll Cardiol* 1989;13:207–214.
40. Winkle RA, Mead RH, Ruder MA, Gaudiani V, et al.: Improved low energy defibrillation efficacy in man using a biphasic truncated exponential waveform. *Am Heart J* 1989;117:122–127.
41. Bardy GH, Ivey TD, Allen MD, Johnson G, et al.: A prospective, randomized evaluation of biphasic vs monophasic waveform pulses on defibrillation efficacy in humans. *J Am Coll Cardiol* 1989;14:728–733.
42. Bourland JD, Tacker WA, Wessale JL, Kallok MJ, et al.: Sequential pulse defibrillation for implantable defibrillators. *Med Instrum* 1986;20:138–142.
43. Bardy GH, Ivey TD, Allen MD, Johnson G, Greene HL: Prospective comparison of sequential pulse and single pulse defibrillation using two different clinically available systems in man. *J Am Coll Cardiol* 1989;14:165–171.
44. Troup PJ, Chapman PD, Olinger GN, Kleinman LH: The implanted defibrillator: Relation of defibrillating lead configuration and clinical variables to defibrillation threshold. *J Am Coll Cardiol* 1985; 6:1315–1321.
45. Bardy GH, Allen MD, Mehra R, Johnson G, Feldman S, Greene HL, Ivey TD: Transvenous defibrillation in man via the coronary sinus. *Circulation* 1990;81:1252–1259.
46. Bardy GH, Ivey TD, Allen MD, Johnson G, Greene

HL: Evaluation of electrode polarity on defibrillation efficacy. *Am J Cardiol* 1989;63:433–437.
47. Bardy GH, Ivey TD, Allen MD, Johnson G: A prospective, randomized evaluation of ventricular fibrillation duration on defibrillation thresholds in humans. *J Am Coll Cardiol* 1989;13:1362–1366.
48. Dixon EG, Tang ASL, Wolf PD, Meador JT, Fine MJ, Kalfee RV, Ideker RE: Improved defibrillation thresholds with large contour epicardial electrodes and biphasic waveforms. *Circulation* 1987;76:1176–1184.
49. Lerman BB, Deale OC: Adverse effects of AICD epicardial patches on transthoracic defibrillation. *Circulation* 1988;78(suppl II):155.
50. Goodman LR, Almassi GH, Troup PJ, Gurnet JW, et al.: Complications of automatic implantable cardioverter defibrillators: Radiographic, CT, and echocardiographic evaluation. *Radiology* 1989; 170:447–452.
51. Bardy GH, Gregg MG, Johnson G, Greene HL, Ivey TD: Intraoperative identification of lead fracture during automatic implantable cardioverter/defibrillator replacement. *J Electrocardiol* 1989;3:75–80.
52. Hopps JA, Bigelow WG: Electrical treatment of cardiac arrest: A cardiac stimulator/defibrillator. *Surgery* 1954;36:833–849.
53. Jain SC, Bhatnager VM, Azami RU, Awasthey P: Elective countershock in atrial fibrillation with an intracardiac electrode/A preliminary report. *J Assoc Physicians India* 1970;18:821–824.
54. Mirowski M, Mower MM, Gott VL, Brawley RK, Denniston R: Transvenous automatic defibrillator—Preliminary clinical tests of its defibrillating subsystem. *Trans Am Soc Artif Intern Organs* 1972;18:520–525.
55. Mirowski M, Mower MM, Gott VL, Brawley RK: Feasibility and effectiveness of low-energy catheter defibrillation in man. *Circulation* 1973;47:79–85.
56. Yee R, Zipes DP, Gulamhusein S, Kallok MJ, Klein GJ: Low energy countershock using an intravascular catheter in an acute cardiac care setting. *Am J Cardiol* 1982;50:1124–1129.
57. Saksena S, Lindsay BD, Parsonnet V: Developments for future implantable cardioverters and defibrillators. *PACE* 1987;10:1342–1358.
58. Saksena S, Parsonnet V: Implantation of a cardioverter/defibrillator without thoracotomy using a triple electrode system. *JAMA* 1988;259:69–72.
59. Troup PJ, Bach SM, Barstad J, Harper N, Mayer D, Moser S, Smutka M, Theis R, Wollins J, and the Endotak Investigator Group: Non-thoracotomy catheter/chest wall patch defibrillation. *Proc Assoc Adv Med Instrm* 1989, p 92.
60. Colavita PG, Wolf P, Smith WM, Bartram FR, et al.: Determination of effects of internal countershock by direct recordings during normal sinus rhythm. *Am J Physiol* 1986;250:H736–H740.
61. Moss AJ, Rivers RJ, Kramer DH: Permanent pervenous atrial pacing from the coronary vein: Long term follow-up. *Circulation* 1974;49:222–225.
62. Moss AJ: Therapeutic uses of permanent pervenous atrial pacemakers: A review. *J Electrocardiol* 1975;8:373–380.
63. Bardy GH, Allen MD, Mehra R, Johnson G: An effective and adaptable transvenous defibrillation system using the coronary sinus in man. *J Am Coll Cardiol* 1990;16:887–895.
64. Haluska EA, Whistler SJ, Calfee RJ: A hierarchical approach to the treatment of ventricular tachycardias. *PACE* 1986;9:1320–1324.
65. Fromer M, Schlaepfer J, Goy J, Kappenberger L: Initial experience with a new tachyarrhythmia control device. *J Am Coll Cardiol* 1990;15:60A.

Chapter **32**

Surgical Therapy for Cardiac Arrhythmias: Update

T. Bruce Ferguson, Jr., MD, and James L. Cox, MD

A number of advances in the field of surgical therapy for cardiac arrhythmias have occurred over the past decade. These advances have paralleled developments in the elucidation of both supraventricular and ventricular arrhythmia mechanisms, discussed elsewhere in this book. The successful surgical therapies, in turn, have provided additional insight into a further understanding of a number of these mechanisms. The interaction between the cardiac electrophysiologic surgeon and electrophysiologist that was necessary to bring about these advances represents an almost unique arena for cooperation in the field of cardiac disease today.

GENERAL OVERVIEW

Several comprehensive reviews describing the current status of cardiac arrhythmia surgery have recently been published.[1,2] Likewise, a number of textbooks devoted to medical and surgical treatment of these rhythm disturbances contain information pertaining to the evolution and refinement of surgical techniques currently used.[3–6] The reader is referred to these sources for additional background information.

This chapter will highlight the advances in the surgical therapy of supraventricular and ventricular arrhythmias that have occurred over the past decade. Specifically, the major developments in surgical treatment of the Wolff-Parkinson-White syndrome and other accessory atrioventricular (AV) connections, AV nodal reentrant tachycardias, ectopic atrial tachycardias, atrial flutter and fibrillation, and nonischemic and ischemic ventricular tachycardias will be addressed. Initially, however, the advances in intraoperative mapping techniques, and the impact of those advances, will be discussed.

655 Avenue of the Americas, New York, NY 10010
Current Topics in Cardiology

DEVELOPMENTS IN INTRAOPERATIVE MAPPING TECHNIQUES

The initial successful attempts at clinical epicardial mapping of the heart at the time of surgery were performed in 1967 by Durrer and others.[7,8] Since that time, single-point mappings systems employing a fixed reference electrode, a roving (pencil, hand-held, finger) electrode, an activation timer, and data storage equipment (magnetic tape recorder, strip-chart recorder) have been the mainstay of intraoperative mapping systems for both supraventricular as well as ventricular tachyarrhythmia surgery. These types of systems are still probably used by the majority of institutions performing arrhythmia surgery in the United States today.

Problems associated with single-point systems have included (a) great difficulty in identifying the site of ventricular preexcitation in patients with only intermittent antegrade conduction across the accessory pathway during sinus rhythm; (b) extreme difficulty in identifying the site of ventricular preexcitation if these patients have atrial fibrillation; and (c) inability to perform presurgical mapping without causing excessive dislocation of the heart from the pericardial cavity. This latter problem has two sequelae: (a) in order to map most patients with single-point systems and not cause hemodynamic compromise, cardiopulmonary bypass is necessary, thus prolonging the total operative cardiopulmonary bypass duration, and (b) the manipulation and time required to obtain a complete map of the AV groove frequently causes temporary cessation of conduction over the accessory pathway, making mapping impossible. The only alternative in this situation is to wait until conduction over the accessory pathway returns and then to reattempt the mapping procedure.

Because of these problems, multipoint analog systems were developed by a number of institutions. In general, these were more successful for supraventricular than for ventricular applications; however, in the absence of computer support, data analysis was a prohibitively lengthy procedure in many cases.

These circumstances led to the development of a number of computer-based, multichannel mapping systems in the early part of the past decade.(9–12) The use of such systems has both impacted upon the performance of the operative procedures as well as helped to elucidate the mechanisms for a number of these arrhythmias.

The multipoint, computerized mapping system developed at Washington University in St. Louis has been described in detail elsewhere.(13) Our experience, as well as that of others, is that such systems have several beneficial effects upon the conduct of operations for both supraventricular and ventricular arrhythmias. For the Wolff-Parkinson-White syndrome, these include (a) a reduction in intraoperative mapping time by reducing both the actual time required for data acquisition as well as time for data analysis; (b) the ability to perform mapping without cardiopulmonary bypass because dislocation of the heart from the pericardial cavity is not necessary; (c) a decreased incidence of intraoperative iatrogenic block of the manifest or concealed pathway because excessive manipulation of the heart is not necessary; (d) an increased efficiency in mapping patients with multiple pathways (20% of patients in our series); and (e) the ability to map intermittently conducting accessory pathways. For other supraventricular arrhythmias, these include (a) the ability to accurately localize the site of origin of ectopic atrial tachycardias, even from a single beat of nonsustained tachycardia, and (b) the ability to map the multiple macro-reentrant circuits responsible for sustained atrial flutter. For VT, these include (a) the ability to localize more accurately the sites of origin of nonischemic ventricular tachyarrhythmias; (b) the ability to perform accurate epicardial, endocardial, and intramural maps to determine the sites of origin and the reentrant circuits involved in ischemic VT(14); (c) the ability to perform mapping without having to resort to a ventriculotomy, which results in inability to induce the clinical arrhythmia in 15 to 30% of patients(15); and (d) the ability to map and surgically ablate multiple monomorphic and/or polymorphic sustained VTs, as well as those tachycardias that are nonsustained.

Other multipoint systems, both computer and analog based, have been developed and utilized by other groups with good success.(16,17) In addition, there are several multipoint, computerized systems that are commercially available.

Do these computerized intraoperative mapping techniques allow surgeons and electrophysiologists to treat patients more effectively and efficiently? For the reasons listed above, our experience has been that this system is a definite asset in treating patients with the Wolf-Parkinson-White syndrome and with other forms of supraventricular tachycardias. The overall success rate during the first 13 years of surgery for the Wolff-Parkinson-White syndrome was only 86%, while now it approaches 100%.(13) The development of these mapping systems has undoubtedly played a role in this increase in operative success.

Likewise, these computerized systems have been demonstrated to impact upon the surgical treatment for VT. A considerable controversy has existed in the past as to whether or not intraoperative mapping actually improves the results of surgery for VT; this controversy was based on a comparison between (a) single-point, analog intraoperative mapping techniques and (b) no intraoperative mapping with performance of a "blind," nonelectrophysiologically guided resection. While electrophysiologically guided surgical procedures yielded superior results,(18) this benefit was lost in the increased operative time required for mapping and in the expense of developing an intraoperative mapping system. With better, more accurate, and affordable mapping systems becoming available, a beneficial impact similar to that seen with Wolf-Parkinson-White surgery should be able to be demonstrated for VT surgery.

SURGERY FOR THE WOLFF-PARKINSON-WHITE SYNDROME

Over the past 10 years, the indications for surgical intervention for patients with the Wolff-Parkinson-White syndrome have evolved to (a) patients who are medical refractory; (b) patients who are medication-intolerant, who have detrimental side effects from the drugs, or in whom compliance is poor; (c) patients with atrial fibrillation that conducts rapidly enough antegrade across the accessory pathway to allow induction of ventricular fibrillation from the atrium; (d) patients undergoing concomitant open-heart surgery who also have the Wolff-Parkinson-White syndrome; and (e) more recently, young, otherwise healthy patients with recurrent tachycardia in whom surgery represents the preferred alternative to a lifetime of medical therapy.

The objective of surgery for the Wolff-Parkinson-White syndrome is to divide the accessory pathway(s) responsible for the syndrome. There are two surgical approaches that have evolved and that are currently utilized. Most commonly, surgery to interrupt these pathways is performed using a standard endocardial approach originally described by Sealy et al.[19] and Iwa et al.,[20] and modified by Cox et al. beginning in 1981.[21] The alternative surgical approach is an epicardial one, described by Guiraudon et al. in 1984.[22]

The endocardial technique is designed to divide the ventricular end of the accessory pathway [4,21,23] and the epicardial technique has been directed toward division of the atrial end of the pathway[24,25] (Fig. 32.1). Although the epicardial technique has been described as a "closed-heart cryosurgical technique," it is neither closed-heart nor cryosurgery dependent in the vast majority of patients. Standard descriptions of both of these techniques can be found elsewhere.[1,2]

Important developments in the endocardial operative technique that have occurred over the past 10 years include (a) computerized intraoperative mapping, as discussed above; (b) use of 2.5× magnification to perform the dissection; (c) implementation of advances in myocardial preservation techniques that have occurred over this time interval; and most importantly, (d) conversion of the operative approach from an electrophysiologically based operation to an anatomically based procedure. The preoperative and intraoperative mapping studies tell the surgeon in which area(s) the pathway(s) are located: left free-wall, posterior septum, right free-wall, or anterior septum (Fig. 32.2). Once this area has been identified, a standard, anatomically based dissection of that area is performed in every patient to assure complete interruption of all pathways at the time of the initial operative procedure in all patients. Our intraoperative mapping studies have demonstrated that in certain patients these accessory AV connections can be in fact broad bands of conduction tissue, extending over several centimeters of the AV groove. However, pref-

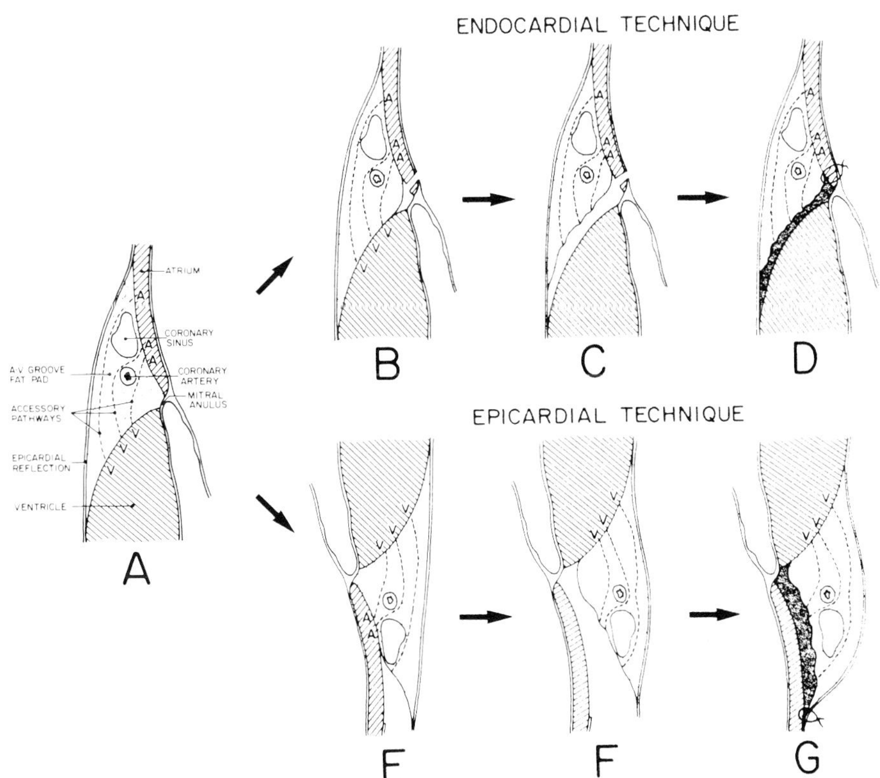

FIGURE 32.1 Diagrammatic representation of a cross-section of the posterior left heart showing the different depths at which left free-wall pathways can be located in relation to the mitral annulus and epicardial reflection (**A**). The endocardial surgical technique is depicted in **B** through **D** and the epicardial technique in **E** through **G**. See text for further discussion. *Reprinted from Cox,*[5] *with permission.*

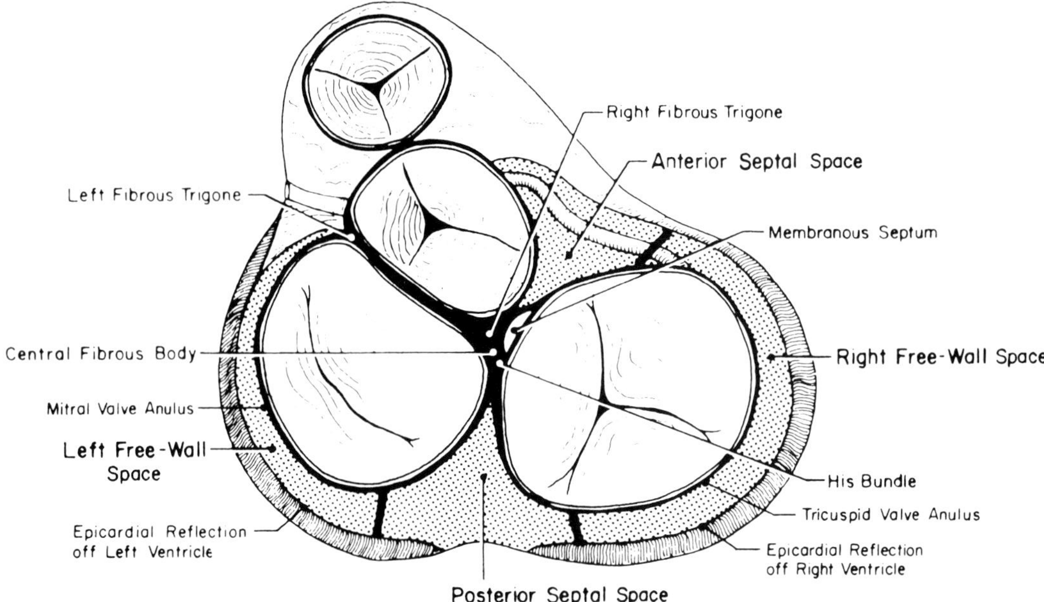

FIGURE 32.2 Diagram of the superior view of the heart with the atria cut away demonstrating the boundaries of each of the four anatomic areas where accessory pathways can occur in the Wolff-Parkinson-White syndrome. The boundaries of the left free-wall space are the mitral valve annulus and the ventricular epicardial reflection extending from the left fibrous trigone to the posterior septum. The boundaries of the posterior septal space are the tricuspid valve annulus, the mitral valve annulus, the posterior superior process of the left ventricle, and the ventricular epicardial reflection. The boundaries of the right free-wall space are the tricuspid valve annulus and the epicardial reflection extending from the posterior septum to the anterior septum. The boundaries of the anterior septal space are the tricuspid valve annulus, the membranous portion of the interatrial septum, and the ventricular epicardial reflection. All accessory atrioventricular connections must insert into the ventricle somewhere within these anatomic boundaries. *Reprinted from Cox et al.,*[21] *with permission.*

erential conduction may occur only over a small portion of this band.[26] Unless the entire band is divided at the time of surgery, preexcitation will recur postoperatively in some patients.

With the epicardial technique, its advocates have stated that the ventricular preexcitation caused by antegrade conduction across the accessory pathway disappears in virtually every case during this dissection, indicating that the pathway has been divided by the surgical dissection.[25] Despite this observation, they recommend placement of one or more cryolesions at the level of the valve annulus to destroy any accessory pathways located immediately adjacent to the valve annulus that might have survived the dissection. Some authors have expressed concern with regard to placement of cryolesions subadjacent to major coronary arteries in this technique.[27] Although the cryosurgery is unnecessary if the pathway has already been divided by the dissection (that is, the ventricular preexcitation has disappeared), this technique has nevertheless been labeled a "cryosurgical technique" by its advocates. As mentioned, the majority of patients with left free-wall, posterior septal, and anterior septal pathways require cardiopulmonary bypass but not cardioplegic arrest with this technique.

More recently, the epicardial dissection technique has been modified to perform a so-called direct approach in the majority of cases with left free-wall pathways.[25] For the majority of cases, cardiopulmonary bypass is still required. This technique requires (a) identification, ligation, and division of the obtuse margin cardiac vein, and (b) dissection of the obtuse marginal coronary artery and other ventricular branches of the circumflex coronary system over a 1- to 2-cm segment; the long-term effects of this skeletonization are not known at present. The authors have adopted this approach to decrease the incidence of injury to the coronary sinus and to decrease the amount of tension placed on the posterior left atrium; this excessive tension has produced tears in the atrial tissue at the AV junction in a small number of patients undergoing the previous type of dissection.[6] Thus, this newer "direct approach" divides the ventricular end of the accessory AV connection just as the endocardial approach does.

Finally, the excellent results obtained with surgery in adults have led pediatric electrophysiologists and surgeons to apply these same techniques in the last 5 years to an increasing number of infants and children with the Wolff-Parkinson-White syndrome. All of these patients have been refractory to medical therapy. Virtually all of these patients have been operated upon through an endocardial approach, with excellent results (see below).

SPECIAL CIRCUMSTANCES: NODOVENTRICULAR AND NODOFASCICULAR ACCESSORY CONNECTIONS AND EBSTEIN'S ANOMALY

Nodofascicular and fasciculoventricular connections, as described by Mahaim, are present between the nodal and fascicular components of the nodal bundle axis and the ventricular septum. Classically, these fibers have been depicted as originating from the atrioventricular node or penetrating (His) bundle, and then perforating the central fibrous body to insert into the ventricular myocardium (Fig. 32.3). The electrocardiographic pattern depends largely upon the site of origin of the fibers,[28,29] but rate-dependent prolongation of conduction time and a low right ventricular insertion site can usually be demonstrated on the preoperative electrophysiologic study; in addition, the electrocardiogram (ECG) demonstrates a left bundle branch configuration during preexcitation. More recently, the appropriateness of schematically picturing these accessory pathways as nodoventricular has been questioned. Tchou et al.[30] have suggested that such pathways represent atriofascicular connections with decremental conduction properties without direct anatomic connection to the AV nodal tissue, and Klein et al.[31] have suggested that "typical" nodoventricular connections may in fact be atypical AV accessory connections with decremental conduction properties and a distal right ventricular insertion site. It is worth noting that these so-called Mahaim connections are regularly found in otherwise normal hearts without arrhythmias.[32] Whatever the exact electrophysiologic substrate, from a surgical point of view Mahaim fibers may connect the His bundle to the ventricular septum by traversing the posterior septal space region[33,34] or they can traverse the anterior septal space,[35] in which case the Mahaim connection is anterior to the His bundle. In addition, right free-wall AV connections with typical "Mahaim-like" characteristics preoperatively and intraoperatively have been reported,[31] and left-sided nodoventricular connections have recently been described as well.[36]

Depending upon the spatial separation of the Mahaim fiber and the His bundle, these accessory connections can be interrupted using standard endocardial surgical dissection techniques in combination with cryosurgery. The operation performed depends upon the location of the accessory connection determined by preoperative and intraoperative mapping. When the posterior septal space is involved, a combination of discrete cryosurgery[37] and a posterior septal space dissection is required in some patients; in others, cryosurgery alone is sufficient to interrupt the pathway. In still others, a combination of cryosurgery and both anterior and posterior septal dissections is necessary. Finally, a subset of patients have what should be truly classified as "para-Hisian" connections, since despite all of these maneuvers the accessory connection is so closely juxtaposed to the His bundle that it cannot be separated, and cryosurgical ablation of the entire His bundle is the only therapeutic alternative remaining to provide control of the arrhythmia.[38]

Ebstein's anomaly is a congenital malformation in which the origins of the septal and posterior leaflets (or both) of the tricuspid valve are displaced downward into the right ventricle. There is a wide spectrum of variability, with many patients having the forme fruste of this disease.[39] In the severest forms, the proximal portion of the right ventricle is *atrialized* and may be so thin as to appear aneurysmal. The right atrium may be dilated, and an atrial septal defect of any type may be present.[40]

The position of the AV node and conduction bundles in patients with Ebstein's malformation is normal, although the right bundle branch may be compressed by thickened endocardium.[41] The incidence of arrhythmias and conduction disturbances in this patient population is high, and approximately 5% of these patients have accessory AV connections responsible for the Wolff-Parkinson-White syndrome. An additional undetermined percentage of patients with the Wolff-Parkinson-White syndrome have the forme fruste of Ebstein's malformation.

In our experience, a distinct association between patients with Ebstein's anomaly and the specific location of the accessory pathways exists. Most commonly, patients with this anomaly have posterior septal accessory AV connections.

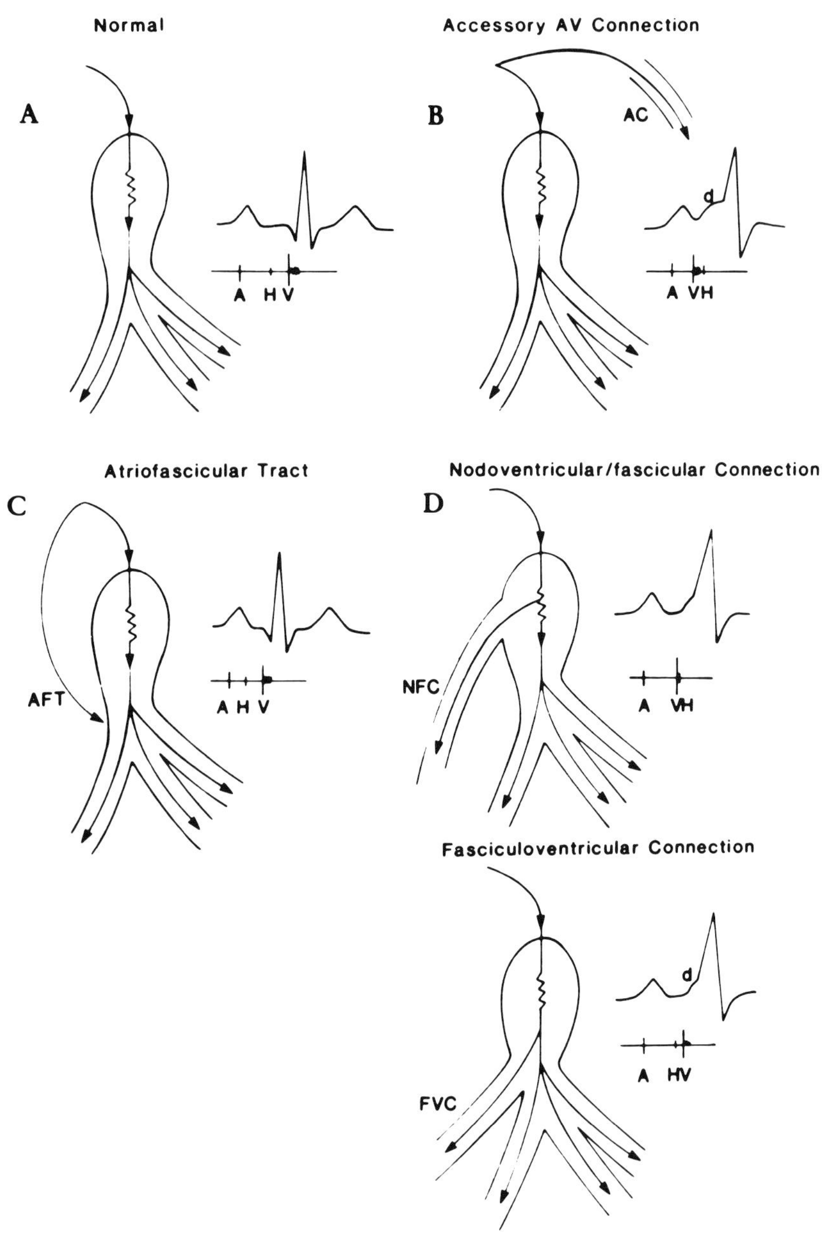

FIGURE 32.3 Configurations presumed responsible for the observed ECG patterns in preexcitation. The electrocardiogram, His-bundle electrogram, and schematic of the conduction system as occurs normally (**A**), with an accessory AV connection (**B**), an atriofascicular tract (**C**), and nodoventricular/fascicular and fasciculoventricular connections (**D**). A = atrial electrogram, H = His-bundle electrogram, V = ventricular electrogram, d = delta wave, AC = accessory AV connection, AFT = atriofascicular tract, NFC = nodoventricular/fascicular connection, FVC = fasciculoventricular connection. *Reprinted from Gornick, CC and Benson DW: Electrocardiographic aspects of the preexcitation syndromes. In Benditt DG and Benson DW:* Cardiac preexcitation syndromes. *Boston: Martinus Nijhoff Publishing, 1986, p. 45, with permission.*

However, in our series a majority of patients with Ebstein's malformation or the forme fruste of the disease have been found to have a combination of right free-wall and posterior septal accessory pathways at the time of intraoperative mapping. This correlation is so high that if a patient is found to have this combination of pathways on the preoperative electrophysiologic study, a cardiac echocardiogram is indicated to determine the status of the tricuspid valve.

Standard endocardial techniques are used to interrupt these pathways. If valve replacement or placement of an anuloplasty ring is necessary for severe Ebstein's malformation, the valve or anuloplasty ring should be placed *below* the coronary sinus by sutures placed through the true tricuspid annulus. Plication of the atrialized ventricle may or may not be necessary.[42]

SURGICAL RESULTS FOR THE WOLFF-PARKINSON-WHITE SYNDROME

Over the past 5 years, a number of reports documenting the surgical results with these two techniques have been published.[1–6] In these reports, a decreased mortality rate, an increased initial-operation efficacy, and an increased overall success rate as compared to earlier series have been achieved (Table 32.1). Cox et al.[21] and Johnson et al.[43] have reported identical initial-operation efficacy rates of 93.2%, and in the former series a total recurrence/total number of pathways rate of 0.6% was achieved. In both these series the endocardial approach was exclusively used. Iwa et al.,[44] in Japan, and Lawrie et al.[45] in this country have reported similar results using predominantly an endocardial technique. Ott et al.[46,47] and Crawford et al.[48] have achieved similar results in a series of infants and children with supraventricular tachycardias associated with accessory pathways.

Guiraudon et al. have achieved an excellent initial-operation efficacy with an epicardial dissection technique,[22,24,25] but at the expense of a significant intraoperative complication rate and reoperation rate.[22,24,49] The intraoperative complications reported with this technique include hemorrhage associated with separation of the left atrium from the left ventricle and injury to the coronary sinus in 3.8% of patients undergoing left free-wall dissections, and some authors have expressed concern over the risk of this dissection

TABLE 32.1 Surgical Results Associated with Division of Accessory Pathways

Reference:	132	21	45	44	43	49	46	48
Year Published:	1984	1985	1987	1988	1987	1989	1985	1989
Series Years:	68–83	80–85	80–86	73–86	79–85	82–88	76–85	84–88
Number of patients	267	118	57	277	118	316	79	41
Number of pathways	304	149	91	301	134	339	82	43
Pathway location (number)								
LFW	143	86	55	134	81	204	41	14
RFW	53	19	14	68	17	32	25	13
PS	81	36	18	25[a]	33	75	13	14
AS	27	8	1		3	28	3	?
Multiple	40	24	21	28	19	23	3	2
Associated diseases								
Cardiac/Oprn	74/nr	56/20	11/nr	57/nr	18/nr	22/nr	4/3	nr
Arrhyth/Oprn	nr	40/11	nr	nr	9/0[b]	nr	nr	nr
Surgical results (number)								
I	212	110	49	236	110	302	70	38
	79.4%	93.2%	85.9%	85.1%	93.2%	95.5%	88.6%	92.6%
II	7	1	0	16	6	3	3	1
III	20	0	4	6	0	10	4	2
IV	0	0	0	1	0	0	0	
Mortality (number)								
Total	11	6	2	11	1	0	1	0
Complicated[c]	6	5	2	10	0	0	0	0

[a] Combined number of septal pathways in this series.

[b] All 9 patients had AV node reentry.

[c] Complicated = additional cardiac procedure performed during operation.

LFW = left free wall, RFW = right free wall, PS = posterior septal, AS = anterior septal, arrhyth = other arrhythmia, oprn = operation, nr = not reported.

Surgical results: I = number of patients surviving first operation with successful division of the pathway(s), II = number of patients surviving first operation, with recurrence of the arrhythmia but with adequate arrhythmia control on medication alone, III = number of patients surviving first operation, with recurrence, and surviving second or third reoperation with successful division of the pathway(s), IV = number of patients surviving one or more unsuccessful operations without heart block/pacemaker therapy and poor medical control postoperatively.

from the epicardial approach. Whether the "direct" approach will decrease the incidence of these complications remains to be determined.

SUMMARY

The incidence of successful surgical correction of the Wolff-Parkinson-White syndrome now approaches 100%, with an operative mortality that ranges from 0 to 0.5%. Complication rates following these operations are likewise extremely low, and are not significantly different from other forms of closed- and open-heart surgery.[(65)] Both early and late recurrences following surgery, utilizing either the endocardial or epicardial technique, are extremely unusual. Previous problems, such as the inadvertent creation of heart block, are now of historical interest only. Whether one uses the endocardial or epicardial approach, these surgical results justify the liberalization of the previously mentioned indications for surgical intervention in this curable congenital cardiac abnormality.

AV NODE REENTRY TACHYCARDIA

Over the past 8 years, AV nodal reentrant tachycardia has become readily curable by surgical intervention. AV node reentry tachycardia is caused by a reentrant circuit that is confined to the AV node or to the perinodal tissues of the lower atrial septum. The anatomic electrophysiologic basis for this reentrant circuit is the presence of two functional conduction pathways, one slow and one fast, through the AV node, the so-called dual AV node conduction pathways.

His-Bundle Ablation

Prior to 1982, the only surgical therapy that could be offered to patients with medically refractory AV node reentrant tachycardia was surgical division of the His bundle. The objective of elective His-bundle ablation was to protect the ventricles from the AV node reentry. Because the procedure resulted in complete heart block, however, a permanent ventricular pacemaker was required postoperatively in all patients.

In 1982, Scheinman et al. described a technique for ablating the His bundle by introducing an electrical shock through a catheter placed adjacent to the His bundle.[(51)] This closed-chest procedure immediately replaced the open-heart surgical method for interrupting the His bundle for obvious reasons. However, since the catheter fulguration technique also created complete heart block, all of these patients also required permanent pacemaker systems.

Thus, both the surgical technique and the catheter ablative technique for His-bundle interruption ameliorated the unpleasant and detrimental effects of AV node reentry tachycardia; however, both procedures replaced one type of arrhythmia (tachycardia) with another (heart block). Because of this, two specific approaches to developing surgical procedures that would eliminate the substrate for the AV node reentrant tachycardia but leave the patient with intact AV conduction were developed. Following several years of animal experimentation utilizing a discrete perinodal cryosurgical technique,[(52–54)] we attained the first clinical cure of AV node reentry tachycardia utilizing this cryosurgical technique on August 13, 1982.[(55)] Ross, Johnson, and colleagues subsequently reported the cure of AV node reentry tachycardia utilizing surgical dissection of the perinodal tissues.[(56)] Guiraudon and colleagues have reported success with a similar surgical dissection technique.[(57)] Thus, at the present time, there are two well-established techniques for curing patients with AV node reentry tachycardia, perinodal cryosurgery, and perinodal surgical dissection. These techniques have been described elsewhere.[(37,58,59)]

Surgical Indications for AV Node Reentrant Tachycardia

Present indications for surgical intervention in patients with AV node reentry tachycardia are similar to those for patients with the Wolff-Parkinson-White syndrome. These include (a) medical refractoriness (the most common indication), (b) drug intolerance, (c) poor patient compliance, (d) concomitant open-heart surgery (usually the Wolff-Parkinson-White syndrome), and recently, based upon the results of surgical intervention for this arrhythmia, (d) patient preference for surgery as opposed to a lifetime of medical therapy.[(37,59)]

Of these, the indication for surgical intervention in patients with either documented AV node reentry tachycardia or simply the presence of dual AV node conduction physiology in the presence of the Wolff-Parkinson-White syndrome is controversial.[(60)] A significant percentage of patients with the Wolff-Parkinson-White syndrome also have either dual AV node physiology or documented AV node reentry tachycardia in addition to the reciprocating tachycardia associated with the Wolff-Parkinson-White syndrome. Obviously, patients who have experienced AV node reentry tachycardia before surgical correction of their Wolff-Parkinson-White syndrome can be expected to continue experiencing AV node reentry tachycardia following Wolff-Parkinson-White surgery if no concomitant attempt is made to cure the AV node reentry tachycardia. Thus, we routinely perform perinodal cryosurgery for AV node reentry tachycardia in such patients following completion of the Wolff-Parkinson-White surgical procedure. Less obvious is the desirability of performing perinodal cryosurgery for ablation of dual AV node physiology in patients undergoing surgery for the Wolff-Parkinson-White syndrome who have not previously experienced concomitant AV node reentry tachycardia. Several patients in our Wolff-Parkinson-White surgical series have been shown to have concomitant dual AV node conduction pathways in addition to the Wolff-Parkinson-White syndrome, but they had never experienced AV node reentry tachycardia clinically. Six of those patients did not undergo the perinodal cryosurgical procedure for the dual AV node pathways at the time of surgical correction of their Wolff-Parkinson-White syndrome. Three of the 6 patients subsequently developed AV node reentry tachycardia postoperatively, even though they had never experienced it preoperatively. As a result, we now routinely perform perinodal cryosurgery concomitant with Wolff-Parkinson-White surgery not only in patients who have had documented AV node reentry tachycardia previously, but also in patients with dual AV node conduction pathways who have never experienced AV node reentry tachycardia.

Both the cryosurgical technique and the surgical dissection techniques require cardiopulmonary bypass. The cryosurgical technique and Guiraudon's dissection technique are performed in the beating heart at 37°C[57,60]; Ross and Johnson's dissection technique is performed in the heart arrested with cardioplegic solution with the patient at mild hypothermia.[56,59]

TABLE 32.2 Surgical Results and Complications for Treatment of AV Node Reentry Tachycardia

Reference:	59	37	57
Year Published:	1988	1989	1990
Series Years:	83–86	82–89	85–90
Patients (number)	69	35	32
Associated diseases			
Cardiac/Oprn (number)	5/5	6/6	nr
Arrhyth/Oprn (number)	6/6	19/19	4/4
Surgical results (number)			
I	66 95.6%	35 100%	29 90.6%
II	0	0	0
III	3	0	3
Mortality			
Total (number)	0	0	0
Intraoperative complications			
Heart block	2	0	2
Pacemaker (number)			
Elective	0	0	0
Iatrogenic	2	0	0
Postoperative complications			
Late recurrence (number)	5 (7.2%)	0	3 (9.3%)

Oprn = operation, arrhyth = arrhythmia, Pacemaker-elective = insertion following His-bundle ablation because of failure to alter AV node reentry circuit and persistent tachycardia, Pacemaker-iatrogenic = inadvertent interruption of AV nodal conduction at the time of surgery, nr = not reported.

Surgical results: I = number of patients surviving first operation with successful alteration of input to the AV node and elimination of the tachycardia, II = number of patients surviving first operation with recurrence of the tachycardia postoperatively, well controlled on or off medications, III = number of patients surviving first operation, with recurrence, with successful second operative procedure and ablation of the arrhythmia.

SURGICAL RESULTS FOR AV NODE REENTRANT TACHYCARDIA

Although the surgical techniques designed to cure AV node reentry tachycardia have been employed for only a short period of time, the results have been excellent.[59,60] The results of the three largest series reported are listed in Table 32.2. There have been no operative deaths in any of the series. Following the perinodal cryosurgical procedure, smooth AV node conduction curves through the remaining single conduction pathway have been demonstrated in all patients and none of the patients have had inducible AV node reentry tachycardia postoperatively. Moreover, all patients have maintained normal conduction through the AV node–His bundle complex with no recurrent AV node reentry tachycardia. The surgical dissection technique has been associated with a low incidence of permanent complete heart block and a low incidence of recurrent AV node reentry tachycardia.

AUTOMATIC ATRIAL TACHYCARDIAS

Automatic atrial tachycardias are caused by automatic foci of arrhythmogenic tissue lying outside the region of the normal anatomic SA node. In patients with this arrhythmia, the standard ECG demonstrates a P-wave morphology that is different from that seen in sinus rhythm, suggesting the presence of an ectopic focus remote from the sinus node. Histologic examination of the atrial tissue excised at the site of origin of automatic atrial tachycardias has revealed a variety of pathological findings,[61] including monocytic infiltrates,[62] focal myocyte hypertrophy,[63] islets of fatty tissue, fibrous tissue foci,[64] proliferation of abnormal cells,[65] and small atrial wall aneurysms.[66]

A hallmark of these arrhythmias is failure of initiation or termination of the tachycardia by programmed electrical stimulation techniques on the preoperative endocardial catheter electrophysiology study. However, either premature atrial extrastimuli or direct current countershock usually will reset, rather than terminate, the tachycardia.[67] Rapid atrial pacing may result in overdrive suppression of the underlying automatic atrial tachycardia, being manifest by a pause following termination of pacing or slowing of the automatic atrial tachycardia, which then accelerates to its previous rate.

Surgical Indications for Automatic Atrial Tachycardias

Automatic atrial tachycardias frequently occur in pediatric patients in whom the tachycardia may be asymptomatic or may present with vomiting and epigastric pain. When these arrhythmias occur in children, they tend to be incessant.[68]

The indications for surgical intervention in both pediatric and adult patients with automatic atrial tachycardias include (a) medical refractoriness, (b) intolerable side effects of antiarrhythmic drugs, (c) poor patient compliance, and (d) patient preference to medical therapy. In addition, one of the most common manifestations of automatic atrial tachycardias in both adults and children is cardiomegaly and congestive heart failure, with the overall incidence being 54 to 63%.[63,69–71] It has been demonstrated experimentally that chronic atrial tachycardia may lead to significant left ventricular enlargement and dysfunction and that restoration of normal rhythm following termination of the atrial tachycardia results in restoration of normal ventricular function.[72] Therefore, additional indications for surgical intervention include (e) tachycardia-induced ventricular dysfunction, and (f) congestive heart failure.

Surgical Technique

As mentioned in the section on intraoperative mapping, computerized multipoint data acquisition is almost essential to perform ablative surgery for automatic atrial tachycardias. With these systems, frequently only one beat of tachycardia is necessary to localize the site of origin of the arrhythmia.[73] These arrhythmias are extremely vulnerable to suppression by general anesthesia.[2] The general anesthetic agents usually do not suppress the tachycardia completely, but they do frequently prevent the tachycardia from being sustained long enough intraoperatively to map and localize the site of origin with a single-point mapping system.

If the site of origin of an automatic atrial tachycardia can be localized precisely by intraoperative mapping, the arrhythmogenic focus may be either excised or cryoablated with a high degree of success.[74] Of the 125 patients with automatic atrial tachycardia reported to date, the location of the automatic focus was specified in 89 patients.[61] Sixty-one (68%) were located in the right atrial free wall, five (6%) within the atrial septum, and the remaining 23 (26%) were located within the left atrium.

Automatic foci located in the free wall of the left atrium or in either of the atrial appendages are ideal for excision or cryoablation.[74] Automatic atrial tachycardias arising near the orifices of the pulmonary veins are best treated either by pulmonary vein isolation or by isolation of the left atrial tissue from the remainder of the heart. Since the sinus node, AV node, His bundle, and bundle branches are located either in the right atrium, atrial septum, or ventricular septum, surgical isolation of the left atrium does not interfere with the normal cardiac conduction system, and no pacemaker system is necessary postoperatively.

Theoretically, if intraoperative mapping properly localized automatic foci in the free wall of the body of the right atrium, those foci can likewise be either excised or cryoablated. However, automatic right atrial tachycardias are frequently multifocal in origin and the ablation or excision of one automatic focus may be followed by the appearance of another at some later date. Thus, the recurrence rate following local excision or cryoablation of automatic right atrial tachycardias is unacceptably high and as a result, we prefer to perform a right atrial isolation procedure, even though the site of origin of the tachycardia may be well-defined by intraoperative computerized mapping.[74,75]

Surgical Results

Electrophysiologically guided operative procedures have been performed in 63 of the 125 patients available for review.[61] Fifty-six (89%) have been completely cured without the need for permanent pacemaker implantation or postoperative antiarrhythmic therapy (Table 32.3). One perioperative death occurred in a patient with

TABLE 32.3 Surgical Results and Complications Associated with Ectopic Atrial Tachycardia

Reference:	45	43	46	133
Year Published:	1987	1987	1985	1985
Series Years:	80–86	79–85	76–85	80–85
Patients (number)	8	9	18	13
Location of ectopic foci				
Right atrium (number)	8	9	13	7
Left atrium (number)	0	0	5	5
Multiple (number)	0	0	7	1
Operation(s)				
Cryoablation alone	3	0	10	7
Excision alone	2	9[a]	2	0
Excis + cryo	3	0	6	6
Reoperations (number)	1	1	5	0
Failed initial operation	excis + cryo	excis	3 cryo 2 excis + cryo	1 excis + cryo
Surgical results (number)				
I	6	8	11	12
II	0	1	2[c]	1
III	1[b]	0	3	0
IV	0	0	2[c]	0
Mortality (number)	0	0	0	0
Surgical complications				
Heart block				
Pacemaker (number)				
Elective	1	0	0	0
Sinus node dysfunction				
Pacemaker (number)				
Iatrogenic	0	0	1	0

[a] Three patients underwent sinus node excision.

[b] Atrial excision plus open AV node cryoablation/pacemaker insertion cured the patient.

[c] All four surgical failures had multiple foci of atrial ectopic tachycardia.

excis + cryo = excision plus cryoablation, Pacemaker-elective = insertion following His-bundle ablation because of failure to ablate the ectopic foci and persistent tachycardia, Pacemaker-iatrogenic = inadvertent creation of sinus node dysfunction following excision and cryoablation of a right-sided focus.

Surgical results: I = number of patients surviving first operation with successful ablation of the tachycardia, II = number of patients surviving first operation, with recurrence of the tachycardia but with control on or off medication, III = number of patients surviving first operation, with recurrence, with successful second operative procedure and ablation of the arrhythmia, IV = number of patients surviving one or more unsuccessful operations without heart block/pacemaker therapy and poor medical control postoperatively.

severe congestive heart failure and dilated cardiomyopathy who arrested during the induction of anesthesia.

ATRIAL FIBRILLATION

The development of an effective surgical procedure to cure atrial fibrillation has presented perhaps the greatest challenge to electrophysiologic surgeons. There are several reasons for this difficulty: (a) atrial fibrillation is the most complex of the supraventricular arrhythmias electrophysiologically; (b) until recently there has not been a good experimental model for this clinical arrhythmia; (c) intraoperative mapping of this arrhythmia is significantly more complex than even mapping for ventricular tachyarrhythmias; and (d) the electrophysiologic mechanism for the multiple forms of this arrhythmia are only now beginning to be understood. This section will discuss the exciting developments that have occurred in this area over the past 5 years.

The Clinical Arrhythmia of Atrial Fibrillation

Available statistics indicate that approximately 0.4% of the United States population, one million people, suffer from atrial fibrillation.[77–80] There are three detrimental side effects when one converts from normal sinus rhythm to atrial fibrillation: (a) a rapid, irregular heartbeat, (b) impaired cardiac hemodynamics due to loss of AV synchrony, and (c) an increased vulnerability to thromboembolism.

Thromboembolism occurs in approximately one-third of patients with atrial fibrillation (330,000 U.S. citizens), approximately three-fourths of all thromboembolic episodes associated with atrial fibrillation involve the brain (247,500 U.S. citizens), and approximately 60% of those cerebroembolic events result in death or permanent severe neurologic deficit (148,200 U.S. citizens).[81] Thus, an effective surgical therapy for atrial fibrillation that restores normal AV synchrony would be indicated for medical refractoriness, intolerable drug side effects, poor patient compliance, preference to medical therapy, congestive heart failure due to atrial fibrillation, and thromboembolic episodes due to atrial fibrillation.

Current Concepts Regarding the Mechanism of Atrial Fibrillation

Three theories have been proposed to explain the mechanism of atrial activation during atrial fibrillation: (a) a single automatic ectopic focus in the atrium, firing at an extremely rapid rate[82]; (b) multiple automatic foci in the atria, firing independently throughout the atria[83]; and (c) intra-atrial reentry in which multiple reentrant wavelets are present in the atria. This latter multiple wavelet theory was originally proposed by Moe[84] and has subsequently been verified by the studies of Boineau et al.[85] and Allessie et al.[86,87] These authors have demonstrated that atrial geometry, local refractory distribution, and the resultant local conduction velocity of the atrial tissues determine whether or not reentry will occur, how many wavelets will form, and whether the process will be sustained. The isolated atrial tissue studies have documented the importance of anisotropy of the tissue as a mechanism for altering conduction velocity; this probably also plays a role in sustaining atrial fibrillation.[88]

During the past 5 years, we have employed epicardial template electrodes, sutured onto the atria, to perform extensive experimental and clinical mapping of atrial fibrillation.[89] The results of these studies have documented complete macro-reentrant loops in some instances and partial loops in others. In addition, more complex patterns in which reentrant loops were not documented were consistent with the multiple wavelet hypothesis of Moe[84] and the experimental data of Allessie et al.[86,87] Although anatomic obstacles, such as the pulmonary veins, IVC, and SVC, were frequently involved in the apparent reentrant loops, complete reentrant circuits were frequently recorded in the absence of these anatomic obstacles. These reentrant loops rotated around areas of functional conduction block, the most common site being along the sulcus terminalis. Furthermore, the loops were transient over time, depending upon which area of the atrium was depolarizing at that moment.

In addition, these studies confirmed that as atrial size increases, the number of wave fronts during atrial fibrillation and the duration of atrial fibrillation increases. As the cycle lengths decrease and become more variable, various regions of the atria are completely dissociated from one another. The importance of these experimental and clinical studies lies in the fact that they have documented unequivocally that atrial flutter/fibrillation is due to intra-atrial reentry and that the concept of multifocal automaticity is not operative in the genesis or perpetuation of atrial fibrillation in man.

Finally, these studies also strongly suggested that the temporary or transient nature of the multiple reentrant circuits that were present in complex atrial fibrillation would make effective intraoperative mapping extremely difficult. Even if the circuits could be localized, by the time the data analysis had been completed the electrophysiologic nature of the fibrillation would have changed, making an electrophysiologically guided procedure for atrial fibrillation almost impossible to perform.

Surgical Options

Elective His-bundle ablation, either by open-heart surgical techniques[90] or by endocardial catheter fulguration,[51] has been the only effective means of treating atrial fibrillation nonpharmacologically up to this time. A review of the International Catheter Ablation Registry data reveals that the most common indication for endocardial catheter ablation of the His bundle, since its introduction in 1982, is atrial fibrillation. Unfortunately, His-bundle ablation does not protect the patient from *all* of the detrimental effects of atrial fibrillation; the patients remain at significant risk for thrombus formation in the fibrillating atrial cavity with the subsequent risk of thromboembolism, and the loss of atrial transport function in certain patients, particularly those with impaired ventricular function, may result in significant symptoms of congestive heart failure that are difficult to control medically.

Three other surgical procedures have been developed and attempted clinically. The first two, the left atrial isolation procedure[91] and the "corridor" procedure,[92] do not satisfy the requirements for a fully effective operative procedure for the cure of atrial fibrillation, namely (a) restoration of normal sinus rhythm mechanism; (b) restoration of atrial transport function; and (c) relief of the requirement for long-term anticoagulation with warfarin by reduction in the risk of thromboembolism.

The left atrial isolation procedure, first introduced experimentally in 1980[91] and clinically performed in 1982,[55] is capable of confining atrial fibrillation associated with mitral valve disease to the left atrium, but because the left atrium continues to fibrillate postoperatively, it has not been applied routinely to control the detrimental effects of atrial fibrillation clinically. In 1985, Guiraudon introduced the corridor procedure in which a strip of atrial septum between the sinoatrial (SA) node and AV node was isolated from the remainder of the atrial myocardium in patients with atrial fibrillation.[92] Although this procedure has now been performed in several patients, it unfortunately alleviates only the rapid, irregular ventricular response to atrial fibrillation, but it does not restore AV synchrony nor does it reduce the vulnerability of patients with atrial fibrillation to the development of thromboembolic problems; for these reasons, this operation is no better than His-bundle ablation and insertion of a pacemaker system.

The shortfalls of the above surgical procedures and the documentation that human atrial fibrillation, once established, is perhaps too complex to be amenable to map-guided surgery led to the following conclusion: The only way to prevent the atrium from fibrillating would be to direct the propagation of electrical activity from its origin in the SA node to the AV node by means of strategically placed surgical incisions across which conduction could not occur. Since maintenance of atrial transport function would be mandatory, it would be essential to have several "blind alleys" jutting off from the "major route" between the SA and AV nodes to activate all of the atrial myocardium. An electrical wave front entering these blind alleys would be incapable of turning back on itself because of the area of refractoriness that would be immediately trailing it. The wave front would be extinguished as it reached the end of the blind alley demarcated by the surgical incision, but after depolarizing the atrial musculature. In addition, it should be impossible for macro-reentrant circuits to develop anywhere in the atria, because electrical activity traveling along either the main route or the blind alleys could not return to its original site of origin.

The operative procedure that created only one entrance site (the SA node), only one exit site (the AV node), and a main route between the two with multiple blind alleys off that route was aptly named the Maze procedure.

Surgical Indications

The current surgical indications for this procedure have included patients with medically refractory atrial fibrillation, either intermittent or sustained, without other concomitant cardiac or systemic diseases.

Preoperative Evaluation

Preoperative electrophysiologic studies are of limited value in patients with atrial fibrillation, since an accurate diagnosis can be made by standard ECG. Moreover, catheter mapping is not feasible in atrial fibrillation. Nevertheless, patients who are to be subjected to surgical intervention for atrial fibrillation should undergo a preoperative electrophysiologic study to rule out concomitant electrophysiologic abnormalities. In addition, patients should undergo diagnostic studies to document the absence of coronary artery disease and/or valvular heart disease.

Intraoperative Electrophysiologic Mapping and Surgery

The Maze procedure for the surgical treatment of atrial fibrillation does not require intraoperative electrophysiologic mapping for its performance. Nevertheless, we routinely perform computerized atrial epicardial mapping in patients undergoing surgery for atrial fibrillation because occasionally the atrial fibrillation may be dependent upon a single macro-reentrant circuit for its perpetuation. In such patients, a surgical procedure less complex in which the dominant reentrant circuit is interrupted might be curative.

The Maze procedure (Fig. 32.4) is performed on cardiopulmonary bypass after standard aortic and bicaval cannulation has been completed. The SVC, IVC, and roof of the atrium between the aortic and bicaval cannulation has been completed. The SVC, IVC, and roof of the atrium between the aorta and SVC are dissected free of all surrounding connective tissue and the interatrial groove is developed as completely as possible. Cardiopulmonary bypass is instituted and the right atrial incisions are performed as diagrammed in Figure 32.4. The heart is then arrested with cold potassium cardioplegic solution, and the left atrial incisions and cryosurgery are performed under cardioplegic arrest as described in Figure 32.4. After closure of the left atrial incisions and

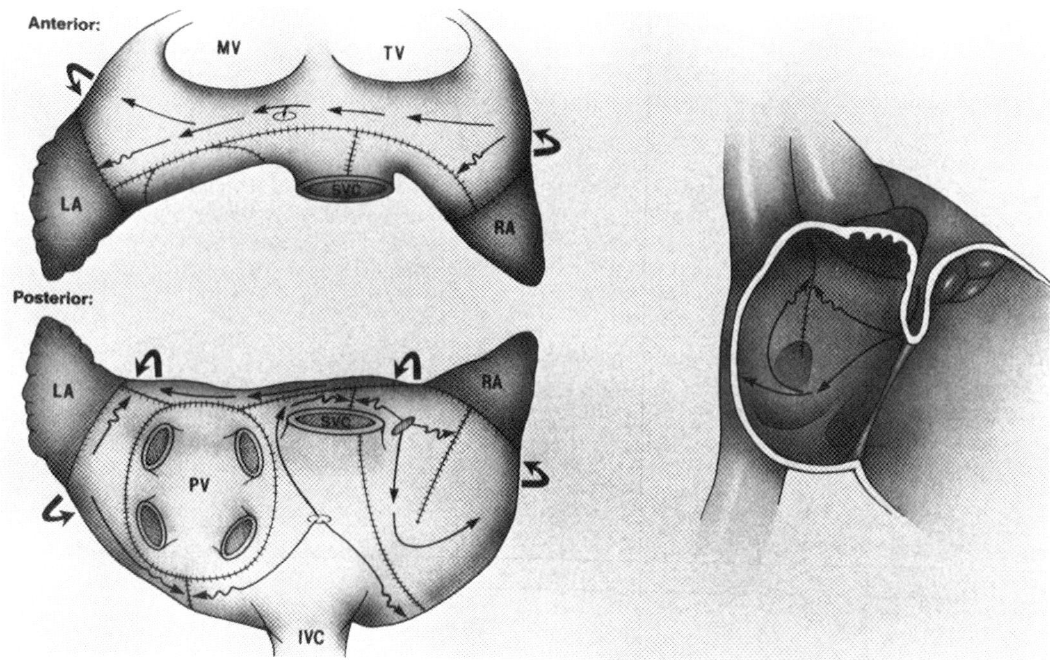

FIGURE 32.4 The Maze procedure. See text for discussion. *Reprinted from Cox,*[89] *with permission.*

the atrial septal incision, the aortic cross-clamp is released and the left side of the heart is de-aired. The right atrial incisions are closed as the heart is being rewarmed. The electrophysiologic effect of the Maze procedure is to allow a normally generated impulse to propagate from the SA node and to activate the entire atrial myocardium except for the excised atrial appendages and the pulmonary veins. At the same time, it is impossible for a large macro-reentrant circuit to exist, since the atrial impulses are precluded from turning back on themselves because of the refractory period of the tissue that has just been depolarized. Thus, regardless of the number of macro-reentrant circuits responsible for the development and perpetuation of atrial fibrillation, they cannot occur following the Maze procedure.

Surgical Results

The Maze procedure has been performed in 7 patients clinically, with no operative deaths. Complications have included drug-induced lupus erythematosus, left lower lobe pneumonia, and transient cholestasis. All patients have been cured of atrial fibrillation and none are on antiarrhythmic drugs or anticoagulants postoperatively. Five of the 7 patients are in sinus rhythm, and two of the patients who had sick-sinus syndrome preoperatively have had DDD-R pacemakers implanted postoperatively. The longest follow-up is now two and one-half years.

ISCHEMIC VENTRICULAR TACHYCARDIA

Virtually the entire development of the current therapeutic techniques for ischemic ventricular tachyarrhythmias has evolved and been tested during the past 10 to 12 years. Recently, there has been an opportunity to review this first decade of experience with these therapeutic modalities and their impact on survival in patients with these malignant arrhythmias.[93]

Anatomic-Electrophysiologic Basis

Studies in animal models, and more recently in man,[94,95] suggest that most reentrant ventricular tachycardias in ischemic heart disease are primarily myocardial in origin and that they are more complex geometrically than previously thought. The arrhythmogenic regions are located primarily in the endocardium and subendocardium, especially at the periphery of myocardial infarcts or ventricular aneurysms. In the majority of patients following acute myocardial infarction, there is subsequent progression of the acutely ischemic tissue to cell death, leaving a fibrous scar in place of the injured myocardium. The interlacing anisotropic pattern of the remaining scar and normal myocardium may harbor local areas of slow conduction, unidirectional block, uneven refractoriness, and nonuniform repolarization, the electrophysiologic substrates for the development of reentrant circuits.[96,97]

Reentrant ischemic ventricular arrhythmias result from a complex interplay between (a) a nonuniform (heterogeneous) state of repolarization, (b) slow desynchronized conduction over abnormal myocardial pathways created by fibrotic or ischemic discontinuity, and (c) ventricular ectopy. Slow desynchronized conduction is a hallmark of myocardial injury or fibrosis.[98] Conduction velocity is dependent on the synchrony of the activation process. Since fibrosis is nonuniform at the borders of infarcts, it produces complex interdigitations between normal myocardium and scar.[96,98,99] A synchronous wavefront upon approaching such an area, may become desynchronized, fragmenting into many individual wavelets. A slowing of conduction time accompanies the desynchronization.

Electrical activity is thought to be identifiable as part of the reentrant tachycardia circuit if it precedes the onset of ventricular depolarization evident on the surface ECG and is required for the initiation and perpetuation of the tachycardia. The site of origin of the tachycardia is felt to be the area exhibiting the earliest presystolic electrical activity in the latter half of diastole and represents the region of myocardium that must be identified and removed at the time of surgery to prevent the arrhythmias.[100]

Preoperative Electrophysiologic Evaluation

All patients who are potential candidates for surgical intervention for ventricular tachyarrhythmias should first undergo an endocardial catheter electrophysiology study. The objectives of the preoperative study are: (a) confirmation that the arrhythmia is ventricular rather than supraventricular in origin, (b) demonstration that the ventricular arrhythmia can be induced and terminated by programmed electrical stimulation techniques (that it is a reentrant arrhythmia), and (c) localization of the region of origin of the VT by catheter mapping when possible.

This study may demonstrate the arrhythmia to be VT of a single morphologic type, indicating that it is originating from a single region in the left or right ventricle. These *monomorphic VTs* are usually sustained for a sufficient length of time to allow endocardial catheter mapping to determine their site of origin. However, monomorphic VT may be nonsustained, thus precluding adequate mapping during the preoperative electrophysiology study.

Alternatively, the preoperative study may document the arrhythmia to be *polymorphic VT*. This term is applied not only to the VT that originates from several different regions of the left ventricle, giving rise to different morphologic types of tachycardia, but it is also applied to the tachycardia that originates from one general region of the left ventricle but is characterized electrophysiologically by excessive fragmentation, such that individual depolarization complexes may be difficult to identify. Polymorphic VT may also be either sustained or nonsustained, and it commonly deteriorates rather quickly into VF. Electrophysiological deterioration to VF may be the result of primary electrical instability or it may occur because of hemodynamic compromise associated with the onset of polymorphic VT.

The third type of ventricular tachyarrhythmia that may be identified by the preoperative electrophysiology study is *primary VF*. This arrhythmia is characterized by the absence of any type of induced VT prior to the onset of VF following programmed electrical stimulation. This distinction is important, because the surgical procedures that can be applied for monomorphic and polymorphic VT cannot be applied in the setting of primary VF.

In addition to this preoperative electrophysiologic evaluation, patients with ventricular tachyarrhythmias routinely undergo cardiac catheterization and coronary angiography prior to surgical intervention.

Surgical Indications and Contraindications

The decision regarding surgical therapy for ischemic VT is based upon a variety of preoperative clinical factors. The primary indication for surgery is refractoriness to medical therapy. The availability of computerized mapping systems that permit localization of the majority of the areas of arrhythmogenic myocardium has made the presence of nonsustained polymorphic VT no longer a contraindication to surgery.

Preoperative Variables Affecting Operative Mortality

The operative mortality for current types of VT surgery cannot be accurately predicted by any preoperative variables. Nevertheless, poor ejection fraction, absence of an aneurysm, emergency operation, preoperative heart failure class, and the functional status of the nonaneurysmal ventricular segments have all been shown to be incremental risk factors when data from individual studies were analyzed.[101–103] Of these, surgical intervention in the setting of (a) a patient with an aneurysm and class III or IV failure, (b) a patient without an aneurysm but with class III or IV failure, and (c) a patient requiring emergent intervention to control intractable ventricular arrhythmias is clearly associated with an increased operative mortality.

The only *absolute* contraindication to surgery for ischemic VT is left ventricular dysfunction so severe that the preoperative risk is judged to be prohibitive.[93] Because most patients with ischemic heart disease and VT have a left ventricular aneurysm, accurate determination of the ejection fraction in these patients is often difficult and the absolute number is not an accurate predictor of operative mortality. Poor systolic function in the nonaneurysmal portion of the ventricle increases the operative risk to the patient.

Preoperative Variables Affecting Tachycardia Recurrence

With regard to surgical failure, as determined by the recurrence of the tachycardia postoperatively, only the anatomic-electrophysiologic character of the arrhythmia itself is of importance when the various preoperative demographic, clinical, catheterization, and electrophysiologic factors are analyzed[93] (see Table 32.6, below). Of this group, VT associated with a posterior-inferior infarct or aneurysm and polymorphic VT (multiple morphologic types) are preoperative predictors of surgical failure.[102] However, several authors have indicated that intraoperative measures, such as more extensive mapping and/or the use of adjunctive procedures such as cryosurgery, may be capable of overcoming the predisposition for surgical failure in patients with posterior-inferior wall ventricular wall tachycardia.[93] Other factors suggested, but not confirmed, as preoperative predictors of a higher surgical failure rate include the absence of an aneurysm, the extent of coronary artery disease present, "disparate sites of origin" of the arrhythmia, tachycardia with a right bundle branch morphology, and the number of previous myocardial infarctions.[101]

Preoperative Amiodarone Therapy

With regard to patients considered as acceptable candidates for surgical intervention, controversy exists as to whether patients who have failed amiodarone therapy should be included in this group. Amiodarone has been shown to depress left ventricular function and this depressant effect is aggravated by ischemic cardioplegic arrest in the majority of patients on the drug.[104,105] Because therapy with amiodarone can complicate surgical intervention in patients requiring a procedure for VT and coronary revascularization, and because it fails to control the arrhythmia in a significant number of patients,[104,106] it would appear logical to make the decision regarding surgical intervention before the institution of amiodarone therapy, after the patient has been proven to be refractory to all other medications.

Surgical Options and Techniques

In 1978, Guiraudon et al. introduced the encircling endocardial ventriculotomy (EEV) as the first direct surgical technique for ablation of malignant VT.[107] Subsequent laboratory studies[108–110] and clinical experience[101,111] showed that the EEV was associated with a high incidence of postoperative left ventricular dysfunction and it has now been essentially abandoned.

In 1979, Josephson et al.[112] described the subendocardial resection procedure. This operation, based upon intraoperative mapping findings and thus an electrophysiologically guided procedure, continues to be one of the most important procedures used for control of ischemic VT. In 1982, Moran et al. reported a modification of the original subendocardial resection procedure in which all of the endocardial fibrosis associated with aneurysm or infarct was resected, the so-called extended endocardial resection procedure (EERP).[113] Other modifications of this original operation have included the addition of endocardial cryoablation as an adjunctive procedure.[111,114] In 1984, Ostermeyer et al. introduced the so-called partial EEV, with excellent surgical results.[101,115] More recently, Selle et al. have reported promising results with the Nd-YAG laser in a small series of patients.[116,117] Our approach has been to perform extensive intraoperative mapping followed by an extended endocardial resection procedure followed by extensive cryosurgical ablation. This technique has been used since 1983, with excellent results.[93] Each of these surgical techniques has been described in detail elsewhere.[101,111–117]

Intraoperative Variables Affecting Tachycardia Recurrence

Aside from performance of an EEV as originally described, analysis of the literature indicates that the only other intraoperative variable demonstrated to be associated with an increased operative mortality is the inability to perform an aneurysmectomy. Other intraoperative factors such as the duration of cardiopulmonary bypass time and aortic cross-clamp time, especially in the presence of class III or IV heart failure, may well be associated with an increased operative mortality, but these data are not conclusive.(93) Factors such as the type of operative procedure or whether the electrophysiologic portion of the operation is performed in the warm, beating heart or while under cardioplegic arrest do not affect operative mortality.

Several authors have advocated resection of the papillary muscles and replacement of the mitral valve under certain circumstances where the tachycardia involves the base of one or both of the papillary muscles.(118) Our experience, as well as that of others, is that the base of the papillary muscles can be cryoablated without causing mitral regurgitation, and thus obviating the need for valve replacement in this extremely ill group of patients.(93)

Intraoperative Factors Affecting Tachycardia Recurrence

When the direct procedures (endocardial resection, extended endocardial resection, extended endocardial resection plus cryosurgery, and partial EEV) are analyzed for intraoperative factors that affect tachycardia recurrence, it is clear that no one particular technique is superior to all others. Whether the surgical approach is a "localized" procedure guided by intraoperative mapping or a "generalized," visually guided one, the results in terms of postoperative reinducibility are roughly the same when the available data are analyzed.(93,119) Swerdlow et al. reported the largest series in which a comparison of map-guided versus visually guided surgery performed at the same institution was made; in this study, VT could be induced in 25% of patients undergoing map-guided surgery and in 64% of patients who had visually guided surgery ($p < 0.001$).(103)

Thus, it would appear that the optimal surgical approach would incorporate intraoperative mapping as a guide and that wide excision or exclusion of the suspected arrhythmogenic tissue should be performed. Krafchek has recently reported such an approach, using extensive intraoperative mapping and a "regional" procedure involving resection and cryoablation, with a surgery-alone success rate of 96%. As mentioned, this has been our approach since 1982.

Finally, the circumstances under which the electrophysiologic portion of the operative procedure is performed may influence the rate of tachycardia recurrence. Some authors have advocated performing the electrophysiologic portion of the operative procedure—intraoperative mapping and localization, resection and/or ablation, and post-procedure mapping to confirm elimination of the arrhythmogenic focus—in the normothermic, beating heart. The accompanying portion of the operation (aneurysmectomy, aneurysmectomy plus coronary bypass grafting) is performed only after the tachycardia focus has been eliminated.(93) Others have proceeded directly to arresting the heart after the intraoperative mapping and localization.(102,119) However, the effectiveness of the resection or ablative portion of the operation may not be able to be determined adequately in the operating room if this portion of the operation is performed under conditions of myocardial hypothermia and cardioplegic arrest. A false-positive result may occur in that the effect of the cold and/or cardioplegia may suppress the arrhythmogenicity of a focus that is not adequately excised during the operation.

Surgical Results

A review of the cumulative experience with these direct surgical procedures for the treatment of ventricular tachyarrhythmias during the past decade reveals that there are two problems that make VT surgery less attractive as a therapeutic option than is supraventricular tachycardia.(93) The operative mortality rate has averaged 12.4% (range, 0% to 21%), and 23.8% (range 0% to 38%) of patients still have inducible VT postoperatively (Table 32.4). Although these figures are not comparable to those that can be attained for SVT, the data must be weighed against the fact that during the decade that these procedures have been performed, the majority of patients coming to surgery for VT have failed all forms of medical therapy and most have severely depressed left ventricular function.

The assessment of long-term complications related to surgical interventions for ischemic ventricular tachycardias is drawn from the five published series in which over 50 patients have been followed for a minimum of 5 years. These include the Pennsylvania series,(102) the Stanford series,(103) the Dusseldorf series,(101) the Alabama series,(119) and the series from our institution.(93) In all five series, intraoperative mapping was performed to guide surgery whenever possible. A variety of surgical procedures were performed, and the data from several series have been recalculated to reflect the results of procedures performed in patients with ischemic VT.

The characteristics of the five series used to assess long-term results of surgery for refractory ischemic ventricular tachycardia are shown in Table 32.5. The 30-day mortality rate ranged from 5 to 23%, and the postoperative reinducibility

TABLE 32.4 Surgical Results and Complications Associated with Ischemic Ventricular Tachycardia

Reference	1	93
Procedure	Indirect[a]	Direct[b]
Number of collected patients	179	844
Operative deaths (number)	47	105
Operative mortality		
Rate	26%	12.4%
Range	22 to 41%	0 to 21%
Number of patients studied postop.	—	559
Number of patients inducible at postop. EPS	—	133
Postop. inducibility		
Rate	—	23.8%
Range		0 to 38%
Intractable arrhythmia recurrence	16%	—

[a] Surgical techniques utilized prior to 1979 including sympathectomy, coronary bypass grafting, aneurysmectomy or infarctectomy, or aortocoronary bypass grafting plus aneurysmectomy.

[b] Surgical techniques utilized after 1979 including endocardial resection, extended endocardial resection, resection plus cryoablation, cryoablation alone, partial encircling endocardial ventriculotomy, and endocardial ablation with the Nd-Yag laser.

Postop. = postoperatively; EPS = electrophysiologic study.

rate ranged from 2 to 32%. The long-term survival data from these series are shown in Table 32.6. The 30-day survival ranged from 79 to 95%, while the 5-year survival ranged from 33 to 70%. In our series, the freedom from recurrent VT and/or sudden death was 87% at 9 years of follow-up.[(93)] When the results of the Pennsylvania, Dusseldorf, and Duke-Barnes series were analyzed for the subsequent course of the operative survivors in these combined series, an overall success rate of 96% for surgery and surgery plus medical therapy for control of malignant ventricular arrhythmias was achieved, a remarkable success rate in these patients with a life-threatening problem.[(93)] Therefore, once a patient has survived the operative procedure, the prognosis for long-term control of their malignant arrhythmias is excellent.

The Automatic Internal Cardioverter-Defibrillator

Indications and Contraindications

In 1985, the Food and Drug Administration released for implantation the automatic internal cardioverter-defibrillator, or AICD. Currently, the indications for insertion of this device include (a) survivors of sudden cardiac death not associated with acute myocardial infarction, who have medically refractory ischemic or nonischemic ventricular arrhythmias and who are *not* candidates for direct surgical treatment for their arrhythmias; (b) patients with more than one cardiac arrest who have noninducible arrhythmias; (c) patients with a clinical recurrence of sustained ischemic or nonischemic VT despite adequate medical therapy, or patients with a single episode of sustained ischemic or nonischemic VT who are easily inducible despite trials of antiarrhythmic drugs, and who, in both instances, are *not* candidates for direct surgical treatment for the ar-

TABLE 32.5 Intraoperative Factors and Results of Direct Surgical Procedures for Ischemic Ventricular Tachycardia

Reference:	102	103	101	119	93
Year Published:	1984	1986	1987	1987	1989
Series Years:	77–83	nr	78–86	78–85	79–88
Number of patients	100	98	93	123	65
Intraoperative factors					
Percent map-guided	94%	79%	99%	nr	100%
Operative procedure	ERP	MP	PEEV	MP	EERP + C
Cardioplegic arrest	Yes	nr	No	Yes	No
Beating heart, normothermia	No	nr	No[a]	No	Yes
Operative mortality	9%	17%	5%	21%[b]	14%
Reinducibility rate, postoperative EPS	28%	32%[c]	19%	38%	2%

[a] Moderate hypothermic, fibrillating heart.

[b] 30-Day mortality rate; hospital mortality rate was 23%.

[c] Cryosurgery was not available in the first 46 patients in this series.

ERP = endocardial resection procedure, MP = multiple procedures; PEEV = partial encircling endocardial ventriculotomy, EERP + C = extended endocardial resection procedure plus cryosurgery, EPS = electrophysiologic study, nr = not reported.

TABLE 32.6 Survival Data (>5 years) Following Direct Surgical Procedures for Ischemic Ventricular Tachycardia

Reference: Year Published: Series Years:	102 1984 77–83	103 1986 nr	101 1987 78–86	119 1987 78–85	93 1989 79–88	Total
Number of patients	104	98	93	123	65	483
30-Day survival	91%	83%	95%	79%	86%	86%
5-Year survival	67%	53%	70%	33%	68%	56%
6-Year survival	67%	—	—	—	60%	64%
7-Year survival	—	—	—	—	60%	60%
9-Year survival	—	—	—	—	60%	60%
5-Year actuarial survival less operative mortality and noncardiac deaths	—	—	—	—	75%	—
5-Year freedom from recurrent ventricular tachycardia and/or sudden death	—	—	—	—	87%	—

nr = not reported.

rhythmias. In particular, this may include patients being evaluated for placement on amiodarone therapy due to the incidence of serious side effects associated with this drug.[105,122] Additional indications are: (d) Patients with the long QT syndrome resuscitated from sudden death; (e) patients undergoing intraoperative mapping and direct surgical treatment in whom mapping is unsuccessful and a "blinded," visually guided procedure must be performed; and, more recently, (f) patients who have survived one cardiac arrest with evidence of continuing electrical instability who, in addition, have potentially reversible myocardial ischemia but no left ventricular aneurysm; these patients would probably not be candidates for direct surgical procedures for the electrical instability, but may be candidates for coronary artery bypass grafting and AICD insertion.[93,123–125]

Contraindications to insertion include (a) patients with nonsustained episodes of VT and/or fibrillation where the frequency of the episodes cannot be controlled by medications and the device would be triggered too often, (b) patients with uncontrolled congestive heart failure, (c) patients with extreme psychological uncertainty about the device, and (d) centers not having adequate personnel and facilities for implantation and postoperative monitoring of these patients.[123]

Surgical Results

A number of complications specific to the insertion of the various components of the AICD have been reported.[6] The 30-day operative mortality in several series is 2 to 3% for patients undergoing implantation for refractory VT and no other surgical procedure, due in large part to the severity of underlying cardiac and pulmonary disease.[123] Operative mortality rates are higher in patients receiving the AICD in conjunction with endocardial resection[126] or coronary bypass grafting.[127,128] There appears to be no difference in operative mortality whether the device is inserted through a median sternotomy or a left thoracotomy.[125,129] A significant incidence of postoperative pulmonary complications has been reported, including left lower lobe atelectasis, left pleural effusion, pericardial effusion, pneumonia, and respiratory insufficiency.[129] Infection of the leads and/or the generator continues to be a serious problem; in Olinger et al.'s series where implantation of the leads and the device was staged in 47 of 106 patients, there was an infection rate of 5.6%.[127] These are life-threatening infections, requiring prompt removal of the device and all lead hardware. Importantly, a number of authors have reported late infections not related to the implantation procedure. These have occurred following other systemic infections and occasionally without antecedent evidence for an infectious etiology.[124]

Complications related to the lead systems have been reported. There is a difference of opinion regarding "prophylactic" placement of epicardial patches and sensing electrodes in patients undergoing successful direct surgery for VT as determined in the operating room.[6,93] Electrode-related complications have included lead migration of the intravenous spring-lead,[129] lead fractures, subclavian vein thrombosis, and constrictive pericarditis. Erosion of the epicardial patches into coronary arteries and/or bypass grafts has been observed by several authors, and has on occasion led to a fatal outcome.[127,129,130] This complication has militated the enthusiasm for "prophylactic" placement of patches routinely in patients considered to be at risk for malignant arrhythmias but not meeting the current implant criteria.

Complications related to the generator (inappropriate discharge, battery depletion) have been reported as well and have been recently reviewed.[6]

The Role of the AICD in the Treatment of VTs

Because the operative mortality rate associated with these direct surgical procedures for the treatment of refractory ischemic VT is higher than that for routine coronary artery bypass grafting (CABG), some authors have advocated routine implantation of an AICD and performance of coronary artery bypass surgery as the primary therapy for essentially all patients with VT.[128] Careful scrutiny of the operative and long-term follow-up results, however, argue strongly against this approach.

The first argument for routine CABG and implantation of an AICD as routine therapy for VT states that the AICD is more effective in the prevention of sudden death. The sudden death rate following AICD implantation is 1.5 to 2.0% per year or 7.5 to 10.0% over a 5-year period.[129,131] The sudden death rate in these survivors of VT surgery is less than 1% per year and less than 5% for 5 years.[93] Thus, while the sudden death rate following AICD implantation is excellent, it remains approximately twice the sudden death rate following successful VT surgery.

The second and major argument for routine implantation of the AICD for all patients with ventricular tachyarrhythmias rather than performing VT surgery is that the operative mortality rate associated with the VT surgery is higher than for CABG plus AICD insertion or AICD insertion alone.

During the first 5 to 8 years following the introduction of the direct surgical procedures, the AICD was not routinely available to most surgeons performing VT surgery. As a result, patients who were medically refractory had to be subjected to VT surgery regardless of the degree of ventricular dysfunction present. Those essentially inoperable patients who had failed all forms of medical therapy and persisted in having intractable VT accounted for the vast majority of operative deaths associated with the direct surgical procedures.[93] In "good risk" patients undergoing direct surgery for VT, the operative mortality is significantly lower than 12.3%, with a 96 to 98% chance of being cured of the tachycardia. These are far better results than can be achieved by either CABG alone (27% operative mortality, 27% recurrence rate),[55] CABG plus AICD (approximately 5% operative mortality, 2%/year sudden death rate),[126,129,131] or AICD insertion alone (2 to 3% operative mortality, 2%/year sudden death rate).[127–130]

Paradoxically, the clinical availability of the AICD is perhaps the most important single development during the past decade that should *decrease* the operative mortality rate associated with VT surgery during the next decade. The AICD represents the previously missing therapeutic option that can now be applied to patients with intractable, medically refractory VT who are inoperable because of severe left ventricular dysfunction. By selecting such patients for AICD implantation, only those patients with acceptable levels of left ventricular dysfunction should now be subjected to the direct surgical procedures for the treatment of refractory ischemic VT. Such a therapeutic approach should decrease the operative mortality rate for VT during the next decade to less than 3%, which is comparable to the operative mortality rate for AICD implantation. Since the long-term results of VT surgery are superior to the long-term results of AICD implantation in terms of the prevention of sudden death, VT surgery should be applied to any patient who is considered to be an operable candidate. Thus, the two therapeutic modalities, VT surgery and AICD implantation, should be viewed as complementary procedures for the treatment of medically refractory ischemic VT.

SUMMARY

The past decade has been an exciting one in the field of cardiac arrhythmia surgery. Well-established surgical cures for a number of both supraventricular as well as ventricular arrhythmias are now available to treat a wide spectrum of patients. Undoubtedly, additional equally significant advances will be made in the decade to come.

REFERENCES

1. Cox JL, Ferguson TB Jr: Cardiac Arrhythmia Surgery, in Wells SA (ed): *Current Problems in Surgery*. Chicago, Year Book Medical Publishers, 1989, pp 193–278.
2. Cox JL: Historic perspectives in the development of cardiac arrhythmia surgery. *Semin Thorac Cardiovasc Surg* 1989;1:3–10.
3. Ferguson TB Jr, Cox JL: The surgical treatment of cardiac arrhythmias, in Parmley WW, Chatterjee K (eds): *Cardiology*. Philadelphia, JB Lippincott, 1989, pp 1–29.
4. Ferguson TB Jr, Cox JL: Surgical therapy for patients

with supraventricular arrhythmias. *Cardiology Clinics* 1990;(in press).
5. Cox JL: Surgical management of cardiac arrhythmias, in Sabiston DC, Spencer FC (eds): *Gibbon's Surgery of the Chest*. Philadelphia, WB Saunders, 1990.
6. Ferguson TB Jr, Cox JL: Complications related to the surgical treatment of supraventricular and ventricular cardiac arrhythmias, in Waldhausen JA, Orringer MB (eds): *Complications in Cardiothoracic Surgery*. St. Louis, Mosby Year Book, 1990.
7. Durrer D, Roos JP: Epicardial excitation of the ventricles in a patient with Wolff-Parkinson-White syndrome (type B): Temporary ablation at surgery. *Circulation* 1967;35:15(abstract).
8. Boineau JP, Moore EN: Evidence for propagation of activation across an accessory atrio-ventricular connection in types A and B pre-excitation. *Circulation* 1970;41:375–397.
9. Ideker RE, Smith WM, Wallace AG, et al.: A computerized method for the rapid display of ventricular activation during the intraoperative study of arrhythmias. *Circulation* 1979;59:449–458.
10. Smith WM, Ideker RE, Kinicki RE, Harrison LA: A computer system for the intraoperative mapping of ventricular arrhythmias. *Comput Biomed Res* 1980;13:61–72.
11. deBakker JMT, Janse MJ, van Capelle FJL, Durrer D: An interactive computer system for guiding the surgical treatment of life threatening ventricular tachycardias. *IEEE Trans Biomed Eng* 1984;BME-31:362–368.
12. Witkowski FX, Corr PB: An automated simultaneous transmural cardiac mapping system. *Am J Physiol* 1984;247:H661–H668.
13. Cox JL: The evolution of intraoperative mapping techniques in cardiac arrhythmia surgery. *Semin Thorac Cardiovasc Surg* 1989;1:11–20.
14. Branyas NB, Cain ME, Cox JL, Cassidy DM: Transmural ventricular activation during consecutive cycles of sustained ventricular tachycardia. *Am J Cardiol* 1990;65:861–867.
15. Östermeyer J: Partial encircling endocardial ventriculotomy for ischemic ventricular tachycardia Presented at The first international symposium on cardiac arrhythmia surgery, Geneva, Switzerland, June, 1990.
16. Parson I, Downar E: Clinical instrumentation for the intraoperative mapping of ventricular arrhythmias. *PACE* 1984;7:683–692.
17. Mickleborough LL: Surgery for ventricular tachycardia: Intraoperative mapping techniques. *Semin Thorac Cardiovasc Surg* 1989;1:74–82.
18. Saksena S, Hussain SM, Wasty N, Gielchinsky I, Parsonnet V: Long-term efficacy of subendocardial resection in refractory ventricular tachycardia: Relationship to site of arrhythmia origin. *Ann Thorac Surg* 1986;42:685.
19. Sealy WC, Hattler BG, Blumenschein SD, Cobb FR: Surgical treatment of Wolff-Parkinson-White syndrome. *Ann Thorac Surg* 1969;8:1–11.
20. Iwa T, Kazui T, Sugii S, Wada J: Surgical treatment of Wolff-Parkinson-White syndrome. *Jpn J Thorac Surg* 1970;23:513–518.
21. Cox JL, Gallagher JJ, Cain ME: Experience with 118 consecutive patients undergoing surgery for the Wolff-Parkinson-White syndrome. *J Thorac Cardiovasc Surg* 1985;90:490.
22. Guiraudon GM, Klein GJ, Gulamhusein S, et al.: Surgical repair of Wolff-Parkinson-White syndrome: A new closed-heart technique. *Ann Thorac Surg* 1984;37:67–71.
23. Cox JL, Ferguson TB Jr. Surgery for the Wolff-Parkinson-White syndrome: The endocardial approach. *Semin Thorac Cardiovasc Surg* 1989;1:34–46.
24. Guiraudon GM, Klein GJ, Sharma AD, et al.: Closed-heart technique for Wolff-Parkinson-White syndrome: Further experience and potential limitations. *Ann Thorac Surg* 1986;42:651.
25. Guiraudon GM, Klein GJ, Sharma AD, et al.: Surgery for the Wolff-Parkinson-White syndrome: The epicardial approach. *Semin Thorac Cardiovasc Surg* 1989;1:21–33.
26. Canavan TE, Schuessler RB, Boineau JP, Corr PB, Cain ME, Cox JL: Computerized global electrophysiological mapping of the atrium in patients with the Wolff-Parkinson-White syndrome. *Ann Thorac Surg* 1989;46:223–231.
27. Kirklin JW, Barratt-Boyes BG: Tachycardias, in Kirklin JW, Barratt-Boyes BG (eds): *Cardiac Surgery*. New York, Wiley, 1986, pp 1359–1383.
28. Bardy GH, Fedor JM, German LD, Packer DL, Gallagher JJ: Surface electrocardiographic clues suggesting presence of a nodofascicular Mahaim fiber. *J Am Coll Cardiol* 1984;3:1161–1168.
29. Ellenbogen KA, Ramirez NM, Packer DL, et al.: Accessory nodoventricular (Mahaim) fibers: A clinical review. *PACE* 1986;9:868–884.
30. Tchou P, Lehmann MJ, Jazayeri M, Akhtar M: Atriofascicular connection or a nodoventricular Mahaim fiber? Electrophysiologic elucidation of the pathway and associated reentrant circuit. *Circulation* 1988;44:837–848.
31. Klein GJ, Guiraudon GM, Kerr CR, et al.: "Nodoventricular" accessory pathway: Evidence for a distinct accessory atrioventricular pathway with atrioventricular node-like properties. *J Am Coll Cardiol* 1988;11:1035–1040.
32. Davies MJ, Anderson RH, Becker AE: Morphological basis for pre-excitation, in *The Conduction System of the Heart*. London, Betterworths, 1983, pp 181–202.
33. Gallagher JJ, Selle JG, Sealy WC, et al.: Surgical interruption of nodoventricular Mahaim fibers with preservation of normal A-V conduction. *J Am Coll Cardiol* 1986;7:133A(abstract).
34. Schechtmann N, Botvinick EH, Dae M, et al.: The scintigraphic characteristics of ventricular pre-excitation through Mahaim fibers with the use of phase analysis. *J Am Coll Cardiol* 1989;13:882–891.
35. Gillette PC, Garson A, Cooley DA, McNamara DG: Prolonged and decremental antegrade conduction properties in right anterior accessory connections: Wide QRS antidromic tachycardia of left bundle branch pattern without Wolff-Parkinson-White configuration in sinus rhythm. *Am Heart J* 1982;103:66–74.
36. Abbott JA, Scheinman MM, Morady F, et al.: Coexistent Mahaim and Kent accessory connections: Diagnostic and therapeutic implications. *J Am Coll Cardiol* 1987;10:364–372.
37. Cox JL, Ferguson TB Jr. Surgery for atrioventricular node reentry tachycardia: The discrete cryosurgical technique. *Semin Thorac Cardiovasc Surg* 1989;1:47–52.

38. Klein GJ, Sealy WC, Pritchett ELC, et al.: Cryosurgical ablation of the atrioventricular node-His bundle: Long-term followup and properties of the junctional pacemaker. *Circulation* 1980;61:8–15.
39. Kirklin JW, Barratt-Boyes BG: in *Cardiac Surgery*. New York, Wiley, 1986, pp 889–909.
40. Becker AE, Becker MJ, Edwards JE: Pathologic spectrum of dysplasia of the tricuspid valve: Features in common with Ebstein's malformation. *Arch Pathol* 1971;91:167.
41. Lev M, Liberthson RR, Joseph RH, et al.: The pathologic anatomy of Ebstein's disease. *Arch Pathol* 1970;90:334.
42. Timmis HH, Hardy JD, Watson DG: The surgical management of Ebstein's anomaly. The combined use of tricuspid valve replacement, atrioventricular plication, and atrioplasty. *J Thorac Cardiovasc Surg* 1967;53:385.
43. Johnson DC, Nunn GR, Richards DA, Uther JB, Ross DL: Surgical therapy for supraventricular tachycardia, a potentially curable disorder. *J Thorac Cardiovasc Surg* 1987;93:913–918.
44. Iwa T, Mikai D, Misaki T, Mitsui T, Matsunaga Y: Surgical management of the Wolff-Parkinson-White syndrome, in Iwa T, Fontaine G (eds): *Cardiac Arrhythmias: Recent Progress in Investigation and Management*. Amsterdam, Elsevier, 1988.
45. Lawrie GM, Huang-Ta L, Wyndham CRC, DeBakey ME: Surgical treatment of supraventricular arrhythmias. *Ann Surg* 1987;205:700–711.
46. Ott DA, Garson A, Cooley DA, McNamara DG: Definitive operation for refractory cardiac tachyarrhythmias in children. *J Thorac Cardiovasc Surg* 1985;90:681–689.
47. Ott DA, Garson A, Cooley DA, Smith RT, Moak J: Cryoablative techniques in the treatment of cardiac tachyarrhythmias. *Ann Thorac Surg* 1987;43:138–143.
48. Crawford FA, Gillette PC: Pediatric electrophysiologic surgery, in *Cardiac Surgery: State of the Art Reviews*. Philadelphia, Hanley & Belfus, 1989, pp 397–310.
49. Guiraudon GM, Klein GJ, Sharma AD, Yee R, McLellan DG: Surgery for the Wolff-Parkinson-White Syndrome: The epicardial approach. Presented at clinical updates and advancements in cardiovascular diagnosis and therapy, Acapulco, Mexico, January, 1989.
50. Page PL, Pelletier LC, Kaltenbrunner W, Vitali E, Roy D, Nadeau R: Surgical treatment of the Wolff-Parkinson-White syndrome: Endocardial versus epicardial approach. *J Thorac Cardiovasc Surg* 1990; 100:83–87.
51. Scheinman MM, Morady F, Hess DS, et al.: Catheter-induced ablation of the atrioventricular junction to control refractory supraventricular arrhythmias. *JAMA* 1982;248:851.
52. Holman WL, Ikeshita M, Lease JG, et al.: Elective prolongation of atrioventricular conduction by multiple discrete cryolesions: A new technique for the treatment of paroxysmal supraventricular tachycardia. *J Thorac Cardiovasc Surg* 1982;84:554.
53. Holman WL, Ikeshita M, Lease JG, et al.: Alteration of antegrade atrioventricular conduction by cryoablation of periatrioventricular nodal tissue. *J Thorac Cardiovasc Surg* 1984;88:67–75.
54. Holman WL, Ikeshita M, Lease JG, et al.: Cryosurgical modification of retrograde atrioventricular conduction: Implications for the surgical treatment of atrioventricular node reentry tachycardia. *J Thorac Cardiovasc Surg* 1986;91:826–834.
55. Cox JL: Surgery for cardiac arrhythmias, in Harvey WP (ed): *Current Problems in Cardiology*. Chicago, Year Book Medical Publishers, 1983, pp 1–60.
56. Ross DL, Johsnon DC, Dennis AR, Cooper MJ, Richards DA, Uther JB: Curative surgery for atrioventricular junctional ("AV nodal") reentrant tachycardia. *J Am Coll Cardiol* 1985;6:1383–1392.
57. Guiraudon GM, Klein GJ, Sharma AD, et al.: Skeletonization of the atrioventricular node: Surgical alternative for AV nodal reentrant tachycardia: Experience with 32 patients. *Ann Thorac Surg* 1990;49:565–573.
58. Cox JL, Holman WL, Cain ME: Cryosurgical treatment of atrioventricular node reentry tachycardia. *Circulation* 1987;76:1329.
59. Johnson DC, Nunn GR, Meldrum-Hanna W: Surgery for atrioventricular node reentry tachycardia: The surgical dissection technique. *Semin Thorac Cardiovasc Surg* 1989;1:53–57.
60. Cox JL, Ferguson TB Jr, Lindsay BD, Cain ME: Perinodal cryosurgery for AV nodal reentrant tachycardia in 23 patients. *J Thorac Cardiovasc Surg* 1990;99:440–450.
61. Lowe JE, Hendry PJ, Packer DL, Tang AS: Surgical management of chronic ectopic atrial tachycardia. *Semin Thorac Cardiovasc Surg* 1989;1:58–66.
62. Wyndham CRC, Arnsdorf MF, Levistsky S, et al.: Successful surgical excision of focal paroxysmal atrial tachycardia. *Circulation* 1980;62:1365–1372.
63. Giorgi LV, Hartzler GO, Hamaker WR: Incessant focal atrial tachycardia: A surgically remediable cause of cardiomyopathy. *J Thorac Cardiovasc Surg* 1984;87:466–473.
64. Frank KS, Kapur S, Toomey K, et al.: Successful surgical treatment of focal atrial tachycardia: A case report and review of the literature. *Thorac Cardiovasc Surg* 1986;34:398–402.
65. Josephson ME, Spear JF, Harken AH, et al.: Surgical excision of automatic atrial tachycardia: Anatomic and electrophysiologic correlates. *Am Heart J* 1982;104:1076–1085.
66. Olsson SB, Blomstrom P, Sabel K, et al.: Incessant ectopic atrial tachycardia: Successful surgical treatment with regression of dilated cardiomyopathy picture. *Am J Cardiol* 1984;53:1465–1466.
67. Cain ME, Lindsay BD: The preoperative electrophysiologic study, in Cox JL (ed): *Cardiac Surgery: State of the Art Reviews*. Philadelphia, Hanley & Belfus, 1990, pp 1–39.
68. Gillette PC, Garson A, Hesslein PS, et al.: Successful surgical treatment of atrial, junctional and ventricular tachycardia unassociated with accessory connections in infants and children. *Am Heart J* 1984;102:984–991.
69. Borggrefe M, Breithardt G: Ectopic atrial tachycardia after transvenous catheter ablation of a posteroseptal accessory pathway. *J Am Coll Cardiol* 1986;8:441–445.
70. Garson A, Gillette PC: Electrophysiologic studies of supraventricular tachycardia in children. I. Clinical-electrophysiologic correlations. *Am Heart J* 1981;102:223–250.
71. Packer DL, Bardy GH, Worley SJ, et al.: Tachycardia-

induced cardiomyopathy: A reversible form of left ventricular dysfunction. *Am J Cardiol* 1986;57:563–570.

72. Damiano RJ, Tripp HF, Asano T, et al.: Left ventricular dysfunction and dilatation resulting from chronic supraventricular tachycardia. *J Thorac Cardiovasc Surg* 1987;94:135–143.
73. Canavan TE, Schuessler RB, Cain ME, et al.: Computerized global electrophysiological mapping of the atrium in a patient with multiple supraventricular tachyarrhythmias. *Ann Thorac Surg* 1988; 46:232–235.
74. Hendry PJ, Packer DL, Anstadt MP, Plunkett MD, Lowe JE: Surgical treatment of automatic atrial tachycardia. *Ann Thorac Surg* 1990;49:253–260.
75. Harada A, D'Agostino HJ, Schuessler RB, Boineau JP, Cox JL: Right atrial isolation: A new surgical treatment for supraventricular tachycardia. *J Thorac Cardiovasc Surg* 1988;95:643–650.
76. Harada A, D'Agostino HJ, Boineau JP, Cox JL: Right atrial isolation: A new surgical treatment for supraventricular tachycardia. II. Hemodynamic effects. *J Thorac Cardiovasc Surg* 1988;95:651–657.
77. Cameron A, Schwartz MJ, Kronmal RA, et al.: Prevalence and significance of atrial fibrillation in coronary artery disease (CASS Registry). *Am J Cardiol* 1988;61:714–717.
78. Diamantopoulos EJ, Anthopoulos L, Nanas S, et al.: Detection of arrhythmias in a representative sample of the Athens population. *Eur Heart J* 1987;8(suppl D):17–19.
79. Hirosawa K, Sekiguchi M, Kasanuki H, et al.: Natural history of atrial fibrillation. *Heart Vessels Suppl* 1987;2:14–23.
80. Onundarson PT, Thorgeirsson G, Jonmundsson E, et al.: Chronic atrial fibrillation—epidemiologic features and 14-year follow-up: A case control study. *Eur Heart J* 1987;8:521–527.
81. Fisher CM: Embolism in atrial fibrillation, in Kulbertus HE, Olsson SB, Schlepper M (eds): *Atrial Fibrillation*. Molndal, Sweden, AB Hassle, 1982, pp 192–210.
82. Rothberger CJ, Winterberg H: Uber vorhofflimmern und norhofflattern. *Pflüger's Arch* 1914;160:42–90.
83. Englemann TW: Refractare phase and compensatousche ruhe in ihrer bedeuting fur den herzrhythmus. *Pflüger's Arch due die Gesammte Physiol* 1894;59:309–349.
84. Moe GK: On the multiple wavelet hypothesis of atrial fibrillation. *Arch Int Pharmacodyn Ther* 1962;140:183–188.
85. Boineau JP, Schuessler RB, Cain ME, Corr PB, Cox JL: Activation mapping during normal atrial rhythms and atrial flutter, in Zipes DP, Jalife J (eds): *Cardiac Electrophysiology*. Philadelphia, WB Saunders, 1990.
86. Allessie MA, Bonke FIM, Schopman FJG: Circus movement in rabbit atrial muscle as a mechanism of tachycardia. III. The "leading circle" concept. A new mode of circus movement in cardiac tissue without the involvement of an anatomical obstacle. *Circ Res* 1977;41:9–18.
87. Allessie MA, Lammers WJEP, Bonke FIM, et al.: Experimental evaluation of Moe's multiple wavelet hypothesis of atrial fibrillation, in Zipes DP, Jalife J (eds): *Cardiac Electrophysiology and Arrhythmias*. Orlando, FL, Grune & Stratton, 1985, pp 265–275.
88. Boineau JP, Schuessler RB, Mooney CR, et al.: Natural and evoked atrial flutter due to circus movements in dogs. *Am J Cardiol* 1980;45:1167.
89. Cox JL, Schuessler RB, Cain ME, et al.: Surgery for atrial fibrillation. *Semin Thorac Cardiovasc Surg* 1989;1:67–73.
90. Sealy WC, Gallagher JJ, Kasell J: His bundle interruption for control of inappropriate ventricular responses to atrial arrhythmias. *Ann Thorac Surg* 1981;32:429–438.
91. Williams JM, Ungerlieder GK, Lofland GK, et al.: Left atrial isolation. New technique for the treatment of supraventricular arrhythmias. *J Thorac Cardiovasc Surg* 1980;80:373–380.
92. Fujimura O, Guiraudon GM, Yee R, Sharma AD, Klein GJ: Operative therapy of atrioventricular node reentry and results of an anatomically guided procedure. *Am J Cardiol* 1989;64:1327–1332.
93. Cox JL: Patient selection criteria and results of surgery for refractory ischemic ventricular tachycardia. *Circulation* 1989;79(suppl I):I-163–I-177.
94. Wittig JH, Boineau JP: Surgical treatment of ventricular arrhythmias using epicardial transmural and endocardial mapping. *Ann Thorac Surg* 1975; 20:117.
95. Horowitz LN, Josephson ME, Harken AH: Epicardial and endocardial activation during sustained ventricular tachycardia in man. *Circulation* 1980; 61:1227.
96. Boineau JP, Cox JL: Slow ventricular activation in acute myocardial infarction—A source of re-entrant premature contractions. *Circulation* 1973; 43:702–713.
97. Cox JL: Anatomic-electrophysiologic basis for the surgical treatment of refractory ischemic ventricular tachycardia. *Ann Surg* 1983;198:119.
98. El-Sherif N, Sherlag BJ, Lazzara R, Hopen RR: Reentrant ventricular arrhythmias in the late myocardial infarction period. I. Conduction characteristics of the infarction zone. *Circulation* 1977;55:686.
99. Boineau JP, Cox JL: Rationale for a direct surgical approach to control ventricular arrhythmias. *Am J Cardiol* 1982;49:381.
100. Marchlinski FE, Josephson ME: Surgical treatment of ventricular tachyarrhythmias, in Platia EV (ed): *Management of Cardiac Arrhythmias: The Nonpharmacologic Approach*, Philadelphia, JB Lippincott, 1987, pp 340–364.
101. Ostermeyer J, Borggrefe M, Breithardt G, et al.: Direct operations for the management of life-threatening ischemic ventricular tachycardia. *J Thorac Cardiovasc Surg* 1987;94:848–865.
102. Miller JM, Kienzle MG, Harken AH, Josephson ME: Subendocardial resection for ventricular tachycardia: Predictors of surgical success. *Circulation* 1984;70:624–631.
103. Swerdlow CD, Mason JW, Stinson EB, Oyer PE, Windel RA, Derby GC: Results of operations for ventricular tachycardia in 105 patients. *J Thorac Cardiovasc Surg* 1986;92:105–113.
104. Klein RC, Machell C, Rushforth N, et al.: Efficacy of intravenous amiodarone as short-term treatment for refractory ventricular tachycardia. *Am Heart J* 1988;115:96.
105. Landymore R, Marble A, MacKinnon G, et al.: Effects of oral amiodarone on left ventricular function in dogs: Clinical implications for patients with

life-threatening ventricular tachycardia. *Ann Thorac Surg* 1984;37:141.

106. Hockings BE, George T, Mahrous F, et al.: Effectiveness of amiodarone on ventricular arrhythmias during and after acute myocardial infarction. *Am J Cardiol* 1987;60:967.
107. Guiraudon G, Fontaine G, Frank R, et al.: Encircling endocardial ventriculotomy: A new surgical treatment of life-threatening ventricular tachycardias resistant to medical treatment following myocardial infarction. *Ann Thorac Surg* 1978;26:438.
108. Ungerlieder RM, Holman WL, Stanley TE, et al.: Encircling endocardial ventriculotomy (EEV) for refractory ischemic ventricular tachycardia. I. Electrophysiologic effects. *J Thorac Cardiovasc Surg* 1982;83:840.
109. Ungerlieder RM, Holman WL, Stanley TE, et al.: Encircling endocardial ventriculotomy (EEV) for refractory ischemic ventricular tachycardia. II. Effects on regional myocardial blood flow. *J Thorac Cardiovasc Surg* 1982;83:850.
110. Ungerleider RM, Holman WL, Calcagno D, et al.: Encircling endocardial ventriculotomy (EEV) for refractory ischemic ventricular tachycardia. I. Electrophysiologic effects. *J Thorac Cardiovasc Surg* 1982;83:857.
111. Cox JL, Gallagher JJ, Ungerlieder RM: Encircling endocardial ventriculotomy (EEV) for refractory ischemic ventricular tachycardia. IV. Clinical indications, surgical technique, mechanism of action and results. *J Thorac Cardiovasc Surg* 1982;83:865.
112. Josephson ME, Harken AH, Horowitz LN: Endocardial excision—a new surgical technique for the treatment of recurrent ventricular tachycardia. *Circulation* 1979;60:1430.
113. Moran JM, Kehoe RF, Loeb JM, et al.: Extended endocardial resection for the treatment of ventricular tachycardia and ventricular fibrillation. *Ann Thorac Surg* 1982;34:538–552.
114. Cox JL: Surgical treatment of ischemic and nonischemic ventricular tachyarrhythmias, in Cohn LH (ed): *Modern Technics in Surgery*. Mt. Kisco, NY, Futura Publishing, 1985.
115. Ostermeyer J, Breithardt G, Borggrefe M, et al.: Surgical treatment of ventricular tachycardias. Complete versus partial encircling endocardial ventriculotomy. *J Thorac Cardiovasc Surg* 1984; 87:517–525.
116. Selle JG, Svenson RH, Sealy WC, et al.: Successful clinical laser ablation of ventricular tachycardia: A promising new therapeutic method. *Ann Thorac Surg* 1986;42:380–384.
117. Svenson RH, Gallagher JJ, Selle JG, et al.: Neodymium:YAG laser photocoagulation: A successful new map-guided technique for the intraoperative ablation of ventricular tachycardia. *Circulation* 1987;76:1319–1328.
118. Moran JM, Kehoe RF, Loeb JM, et al.: The role of papillary muscle resection and mitral valve replacement in the control of refractory ventricular arrhythmias. *Circulation* 1983;68(suppl II):154.
119. McGiffin DC, Kirklin JK, Plumb VJ, Blackstone EH, Waldo AL, Kirklin JW: Relief of life-threatening ventricular tachycardia and survival after direct operations. *Circulation* 1987;76(suppl V):V93–V103.
120. Krafchek J, Lawrie GM, Roberts R, Magro SA, Wyndham CR: Surgical ablation of ventricular tachycardia: Improved results with a map-directed regional approach. *Circulation* 1986;73:1239–1247.
121. Ostermeyer J, Kirklin JK, Borggrefe M, Cox JL, Breithardt G, Bircks W: Ten years electrophysiologically guided direct operations for malignant ischemic ventricular tachycardia—Results. *Thorac Cardiovasc Surg* 1989;37:20–27.
122. Lowe JE: Surgical treatment of the Wolff-Parkinson-White syndrome and other supraventricular tachyarrhythmias. *J Cardiac Surg* 1986;1:117.
123. Cannom DS, Winkle RA: Implantation of the automatic implantable cardioverter defibrillator (AICD): Practical aspects. *PACE* 1986;9:793–809.
124. Echt DS, Armstrong K, Schmidt P, Oyer PE, Stinson EB, Winkel RA: Clinical experience, complications, and survival in 70 patients with the automatic implantable cardioverter/debrillator. *Circulation* 1985;71:289.
125. Gabry MD, Brodman R, Johnston D, et al.: Automatic implantable cardioverter-defibrillator. Patient survival, battery longevity and shock delivery analysis. *J Am Coll Cardiol* 1987;9:1349–1356.
126. Platia EV, Griffith LSC, Watkins L, et al.: Treatment of malignant ventricular arrhythmias with endocardial resection and implantation of the automatic cardioverter defibrillator. *New Engl J Med* 1986;314:213–216.
127. Olinger GN, Chapman PD, Troup PJ, Almassi GH: Stratified application of the automatic implantable cardioverter defibrillator. *J Thorac Cardiovasc Surg* 1988;96:141–149.
128. Fonger JD, Guarnieri T, Griffith LSC, et al.: Impending sudden cardiac death: Treatment with myocardial revascularization and the automatic implantable cardioverter defibrillator. *Ann Thorac Surg* 1988;46:13–19.
129. Marchlinski FE, Flores BT, Buxton AE, et al.: The automatic implantable cardioverter defibrillator: Efficacy, complications and device failures. *Ann Intern Med* 1986;104:481–488.
130. Kelly PA, Cannom DS, Garan H, et al.: The automatic implantable cardioverter defibrillator: Efficacy, complications and survival in patients with malignant ventricular arrhythmias. *J Am Coll Cardiol* 1988;11:1278–1286.
131. Mirowski M: The automatic implantable cardioverter-defibrillator: An overview. *J Am Coll Cardiol* 1985;6:461.
132. Gallagher JJ, Sealy WC, Cox JL, German LG, Kassell JH, Bardy GH, Packer DL: Results of surgery for pre-excitation caused by accessory atrioventricular pathways in 267 consecutive cases, in Josephson ME, Wellens HJJ (eds): *Tachycardias: Mechanisms, Diagnosis and Treatment*. Philadelphia, Lea & Febiger, 1984, pp 259–269.
133. Gillette PC, Wampler DG, Garson A, Zinner A, et al.: Treatment of atrial automatic tachycardia by ablation procedures. *J Am Coll Cardiol* 1985; 6:405–409.

Chapter **33**

Implantable Cardioverter: Defibrillator Devices for the Future

Sanjeev Saksena, MD, FACC

Implantable cardioverter-defibrillators (ICDs) are being clinically applied for the treatment of patients with life-threatening sustained ventricular tachyarrhythmias. Clinical experience has been obtained in patients with drug-refractory sustained ventricular tachycardia (VT)/ventricular fibrillation (VF) or in survivors of sudden cardiac arrest believed to be due to ventricular tachyarrhythmias.[1] This experience now exceeds a decade and includes most major cardiac centers in North America and Western Europe. Clinical trials of the first-generation device have demonstrated a very low incidence of sudden or arrhythmic death during follow-up in treated patients.[2] Generalized clinical application of this therapy has been recommended. Expansion to patient populations at high risk for sudden cardiac death prior to the actual symptomatic event is now being discussed. While controlled clinical trials are awaited, the makings of cardiovascular therapeutic controversy are now in progress.[3] Technologic limitations of the first-generation device are more clearly agreed upon. These limitations often result in undesirable clinical sequelae and can be related to the pulse generator or the lead system.

Nonprogrammable cardioverter-defibrillator pulse generators (for example, Ventak Model 1510 or 1520, Cardiac Pacemakers, Inc., St. Paul, MN) have definite limitations in tachycardia detection and treatment. Figure 33.1 illustrates the response of an implanted nonprogrammable cardioverter-defibrillator pulse generator to spontaneous recurrences of nonsustained and sustained VT in a patient with coronary artery disease. The top panel shows a recurrence of rapid sustained VT with a cycle length of 280 ms which is appropriately detected and treated with a high-energy shock resulting in reversion to sinus rhythm. The middle panel shows a recurrence of rapid nonsustained VT approximately 9 h later which reverts to sinus rhythm after 6 s and 16 beats. The tachyarrhythmia is, however, detected, and approximately 6 s after VT termination, a high-energy shock is delivered during sinus rhythm. This reflects the commited nature of antitachycardia therapy in such a device. The lower panel shows a subsequent recurrence of spontaneous slow VT at a rate of approximately 90 beats per minute that is neither detected nor treated by the device. This latter event is below the rate detection criteria of the device.

Other limitations of the pulse generator include the absence of pacing therapies. Figure 32.2 illustrates a chest radiogram of a patient with a first-generation endocardial defibrillation lead system that includes a transvenous catheter electrode placed in the right ventricular apex and a second defibrillation electrode in the superior vena cava/right atrium with a left thoracic patch electrode placed epicostally. This lead system is tunneled and connected to a nonprogrammable cardioverter-defibrillator. In this patient, demand ventricular pacing was required and a separate transvenous lead system was inserted from the left subclavian vein and connected to demand single-chamber pulse generator. Due to the absence of the pacing therapies in the cardioverter-defibrillator pulse generator, a second device and lead system implant for the treatment of bradyarrhythmias was necessary in this patient.

The absence of antitachycardia pacing therapy of such generators further limits their utility in patients with hemodynamically stable yet symptomatic sustained VT. Figure 33.3 shows an electrocardiogram (ECG) of a patient with recur-

655 Avenue of the Americas, New York, NY 10010
Current Topics in Cardiology

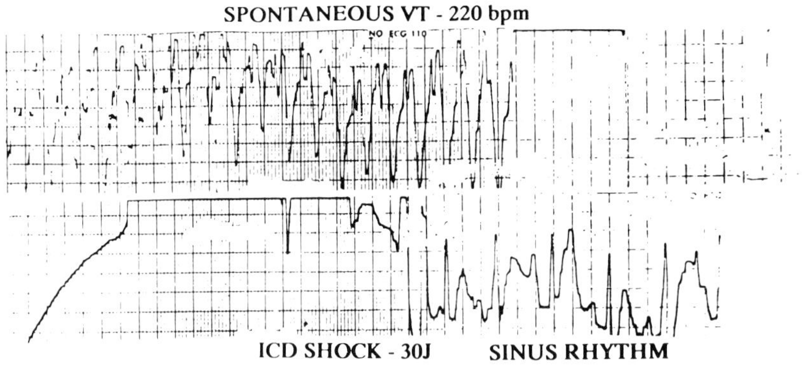

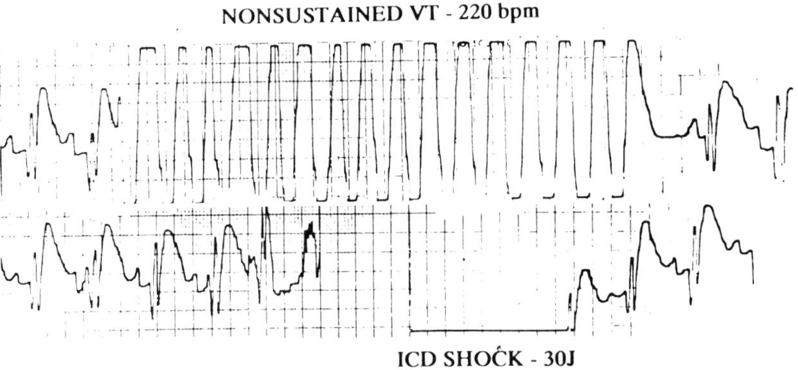

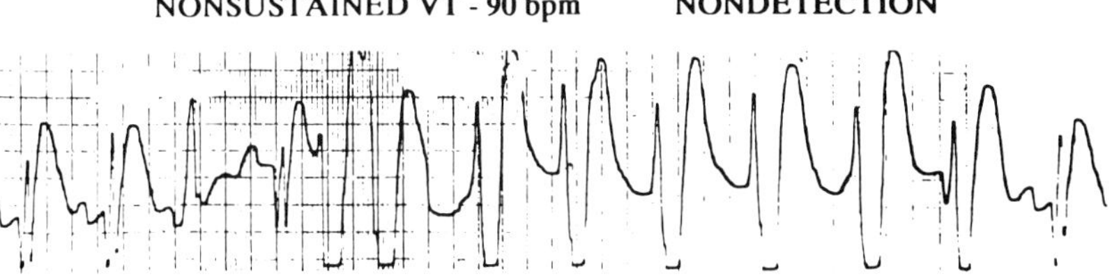

FIGURE 33.1 Function of nonprogrammable implantable cardioverter-defibrillator in response to recurrent ventricular tachyarrhythmias in a patient with a Ventak AICD (Cardiac Pacemakers, Inc., St. Paul, MN). **Top**. Spontaneous sustained rapid VT cardioverted by a single 30-J monophasic shock to sinus rhythm after automatic detection and device charging. **Middle**. Episode of rapid nonsustained VT which spontaneously terminates after approximately 5 s but nevertheless results in arrhythmia detection and device charging and shock delivery in sinus rhythm. This is due to the committed nature of device therapy. **Bottom**. Spontaneous slow sustained VT which is below the rate cutoff of the device which is not detected by the device. Nonprogrammability of rate cutoff does not permit such detection.

rent sustained VT at a rate of 140 to 150 beats per minute who experienced repeated left ventricular failure during the arrhythmia and despite maximal medical therapy had frequent recurrences of the arrhythmia. Frequent exposure to high-energy shocks was deemed undesirable. The available ICD had a nonprogrammable pulse generator and an epicardial lead system. An additional antitachycardia pacemaker (Cordis Model 284A Orthocor II, Cordis Corp., Miami, FL) was placed in the left infraclavicular region with an endocardial transvenous lead. On a combination of amiodarone and type I antiarrhythmic drug therapy, the patient continued to have frequent VT recurrences as well as severe sinus bradycardia. After pacemaker insertion, the pa-

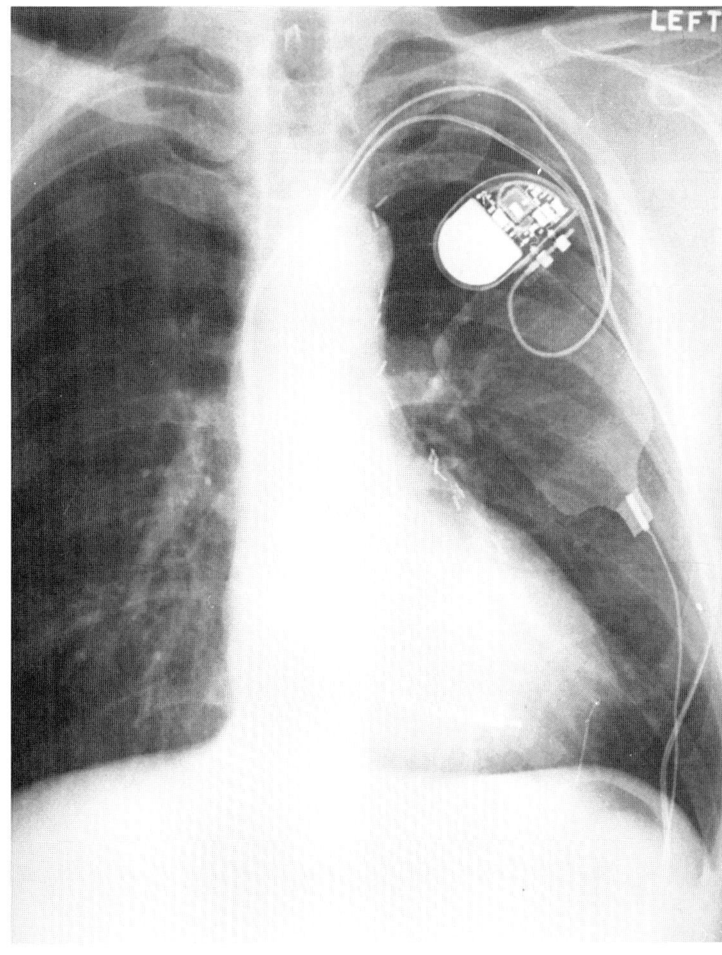

FIGURE 33.2 Insertion of demand pacemaker for bradycardia concomitantly with an implantable cardioverter-defibrillator with an endocardial lead system. Note that the tip of the defibrillation lead is positioned in the right ventricular apex while the pacing lead is fixed in the right ventricular outflow tract to minimize the effect of endocardial shocks on pacing threshold and energy passage to the pacemaker pulse generator. A subcutaneous patch is seen on the left anterior thorax.

tient used the demad ventricular pacing feature and episodes of spontaneous VT were rapidly detected and terminated with bursts of rapid ventricular pacing. VT detection was accomplished within 1.5 s and tachycardia therapy delivered in an equivalent period of time. The entire event lasted 3 s and the pacemaker reverted to demand ventricular pacing. This duration of tachycardia episode did not meet the detection requirements of the cardioverter-defibrillator pulse generator and, therefore, device charging and shock delivery was not observed. In addition, a bipolar pacing lead system was employed to avoid double counting of the pacing stimulus and ventricular

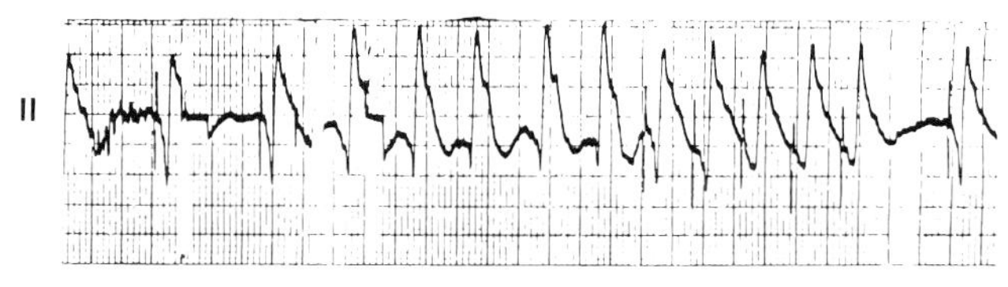

FIGURE 33.3 Combined implantation of an automatic antitachycardia pacemaker (Cordis Model 284A, Orthocor II) and AICD in a patient with frequent recurrent VT. The tachycardia rate was usually below rate selection criteria of the defibrillator but could be terminated by burst pacing, as shown. Occasionally, tachycardia acceleration was observed and terminated by the defibrillator shock. *Reprinted from Saksena and Goldschlager,[20] with permission.*

depolarization by the ICD pulse generator. This limitation of the nonprogrammable first-generation ICD pulse generator resulted in the need for implantation of an additional device and lead systems for bradyarrhythmias and antitachycardia pacing but also raises significant concerns with respect to device–device interactions.

In patients with nonsyncopal sustained ventricular tachyarrhythmias, the tachycardia rate is often slower. Such arrhythmias are often amenable to termination by low-energy shocks (Fig. 33.4). Nonprogrammable pulse generators do not provide this therapeutic option, resulting in high-energy shock delivery in a conscious patient. This results in poor patient tolerance for electrical therapy, psychological stress often resulting in fear of shock therapy, limitations in physical activity, and a reduced quality of life.

Spontaneous variability in tachyarrhythmia rate cannot at times be accommodated by the nonprogrammable tachycardia detection criteria in the first-generation cardioverter-defibrillators. Tachycardia rate may change in response to autonomic stimuli, or after alteration in the antiarrhythmic drug regimen. The latter is most often undertaken for suppression of nonsustained VT or drug side effects. Such drug–device interactions often result in the need for recurrent hospitalization for adjustment of antiarrhythmic drug regimen and, occasionally, for pulse generator revision.[4]

Other major limitations of the first-generation lead system include the need for epicardial placement by thoracotomy in this often fragile patient population. Table 33.1 outlines reported perioperative implant mortality in series from five institutions and two multicenter trials. The perioperative mortality range is from 0 to 9% and reflects a widely varying patient population as well as the frequency of concomitant cardiac surgical procedures. The earlier series, at the time when the prototype device was being developed and the procedure was being defined, reflect higher perioperative mortalities associated with a learning curve in the implant procedure and patient selection. The later series, in the latter half of the last decade, reflect lower perioperative implant mortality rates so that an average implant mortality of 2 to 3% can now be expected. In addition, implant mortality rises when concomitant cardiac surgical procedures are done. These procedures can include coronary bypass surgery, mapping-guided endocardial resection, aneurysmectomy, or valve replacement. This is reflected in our experience at two institutions where the incidence of concomitant or prior cardiac surgical procedures was 23% at one site and 4% at the other site. Multicenter experience, which reflects a continuing minority of patients (11%) who undergo defibrillator system implantation with concomitant cardiac surgery, reports perioperative mortality in the 4% to 5% range.[5]

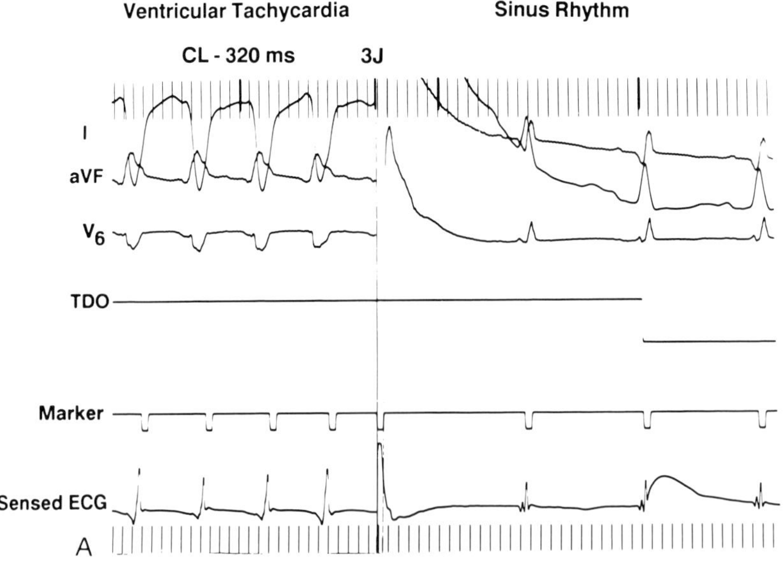

FIGURE 33.4 Cardioversion of VT by implanted pacemaker-cardioverter-defibrillator (Telectronics Model 4201). A 3-J shock terminates sustained VT in the patient. CL = cycle length, TDO = tachycardia detection output. *Reprinted from Saksena and Goldschlager,*[20] *with permission.*

TABLE 33.1 Perioperative Risk of ICD Implant

Investigator/Institution	Study Patients	Perioperative Mortality (%)	Years
Veltri, Sinai Hospital, Baltimore, MD[1]	80	9	1980 to 1987
Marchlinski, Univ of PA, Philadelphia, PA[2]	33	8	1984 to 1985
Winkle, Stanford Univ/Sequoia Hospital, CA[3]	270	1.5	1981 to 1988
Fisher, Montefiore Med Ctr, New York, NY[4]	22	0	1982 to 1986
Saksena, Beth Israel Med Ctr, Newark, NJ[5]	78	7	1983 to 1989
Saksena, Eastern Heart Inst, Passaic, NJ[6]	25	0	1989 to 1990
Guardian Model 4202 Multicenter Study (14 centers)[7]	111	5.4	1989

Sources:
1. *PACE* 1988;11:528.
2. *Ann Intern Med* 1986;104:481.
3. *J Am Coll Cardiol* 1989;13:1353.
4. *J Am Coll Cardiol* 1987;9:1349.
5. *PACE* 1990;13:528.
6. *J Am Coll Cardiol* 1991:
7. *J Am Coll Cardiol* 1990;15:55A.

These operative mortalities assume particular significance when during follow-up sudden death rate in such patients is ≤ 5% annually. Other limitations of the lead system include single-chamber epicardial sensing and pacing. Epicardial lead systems have shown inferior characteristics for demand pacing and sensing.[6] As such, they are employed either on a temporary basis for demand pacing or when adequate endocardial electrodes cannot be positioned. The prototype device utilizes epicardial sensing during the tachyarrhythmia with sensing of electrogram amplitude down to 0.5 mV with an automatic gain control. With this approach, tachyarrhythmia sensing is usually achieved within 10 s of onset in the Ventak series of devices. However, differentiation of supraventricular and ventricular tachyarrhythmias is usually not feasible and no mechanism for atrial sensing or pacing exists.

The next generation of ICD systems will incorporate many technologic advances that may resolve these clinical dilemmas. In this chapter, we shall discuss both the imminent new technology and the emerging concepts for future development. Certain special clinical situations engendered by the use of these devices also need to be considered.

NEW TECHNOLOGY

Pulse Generators

A variety of clinical trials have been recently conducted with programmable and hybrid ICD pulse generators. The first-generation programmable ICD simply permitted programming of tachycardia detection rate and, to a very limited extent, initial shock energy (for example, Ventak Model 1550 and Ventak P, Cardiac Pacemakers, Inc., St. Paul, MN). The second-generation ICD (for example, Model 4202, Telectronics Pacing Systems, Denver, CO) is a hybrid device that includes programmable bradycardia pacing, tachycardia detection rate (144 to 240 bpm), initial shock energy (3 to 30 J), sensing electrogram amplitude (0.7 to 1.8 mV), VT/VF reconfirmation, and more detailed device activation information (Fig. 33.5). These devices can be expected to be released for general clinical use in 1991. Due to limitations in sensing programmability and the well-known propensity for epicardial pacing thresholds to rise significantly over time, endocardial sensing and pacing lead systems are recommended for optimal function.[7] If epicardial leads are contemplated, a sensing and pacing safety margin of greater than or equal to 50% is advisable. Initial programmed sensitivity values should be less than or equal to 1 mV and pacing outputs equal to or greater than twice implant threshold. Experience with programmed tachycardia detection algorithms suggests that incorporation of a sensing delay or requirement of detection of 12 or more consecutive QRS complexes satisfying the rate criterion may reduce device charging in response to nonsustained arrhythmias. Initial data from our experience suggests that the VT/VF reconfirmation feature may decrease the exposure of patients to shock therapy in response to nonsustained VT due to internal diversion of the charge. Intermediate-term experience suggests that these devices have comparable efficacy and safety to those in nonprogrammable ICDs.[5] Long-term experience will be necessary for assessment of chronic performance.

Several third-generation ICDs are now in clinical trials. These devices are best character-

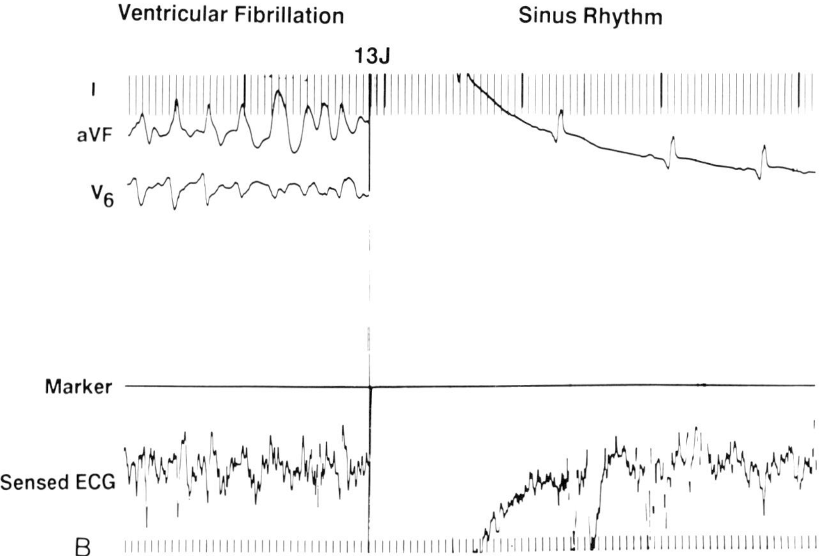

FIGURE 33.5 Termination of VF with a 13-J shock from the same implanted pacemaker-cardioverter as in Figure 33.4. Abbreviations as in Figure 33.4. *Reprinted from Saksena and Goldschlager,[20] with permission.*

TABLE 33.2 ICD Pulse Generators (1990)

	Medtronic PCD 7216/7217	Telectronics Guardian 4210	Ventritex Cadence	Intermedics Res-Q	CPI Ventak PRX	Siemens Pacesetter Thor
Tachycardia detection	Extensively programmable 2 zones	Extensively programmable >2 zones	Programmable 2 zones	Programmable 4 zones	Programmable 2 zones	Programmable 2 zones
Antitachycardia pacing	Present multiple algorithms based on ATP	Present multiple algorithms based on ATP	Present	Present multiple algorithms based on ATP	Present	Present
Single chamber bradycardia pacing	Yes	Yes	Yes	Yes	Yes	Yes
Noninvasive EPS	Yes	Yes	Yes	Yes	Yes	Yes
Programmable cardioversion/ defibrillator shocks	Yes	Yes	Yes	Yes	Yes	Yes
Sequential shocks	Yes	No	No	No	No	No
Bidirectional shocks	Yes	Yes	No	Yes	No	?
Biphasic shocks	No	No	Yes	Yes	No	No
Arrhythmic event recording	+ +	+ + +	+ +	+	+	+
Weight (gm)	281/200	270	240	220	220	200

ATP = antitachycardia pacemaker, + = limited, + + = moderate, + + + = extensive, ? = presently unknown.

ized as hybrid pacemaker-cardioverter-defibrillators. Table 33.2 summarizes the important new features of six new devices. Four of these devices (Medtronic Model 7216A, Telectronics Model 4210, Ventritex Cadence, Intermedics Res-Q) have now entered multicenter clinical trials that are planned for the remaining in 1990 to 1991. For illustration purposes, the Medtronic 7216A will serve as the model for this generation (Figs. 33.6 and 33.7). Tachycardia detection is now based on more sophisticated rate analysis utilizing tachycardia rate, duration, onset, and stability criteria to differentiate VT and supraventricular tachycardia. The tachycardia is classified on the basis of rate into VT or VF. Delivery of VT therapy requires reconfirmation of the rhythm prior to shock delivery. Detection of VF results in committed therapy. Options in the Telectronics Model 4210 require reconfirmation of VT and VF before therapy is delivered. This is accomplished rapidly over six of seven tachycardia cycles but requires accurate sensing. The previous comments regarding epicardial and endocardial sensing apply here and furthermore clinical evaluation indicates that electrogram amplitudes in VF vary greatly. Stable VF sensing must be demonstrated. This is best accomplished at implant and during chronic electrophysiologic study 4 to 6 weeks after implant. Marker channels in these devices permit accurate assessment of device sensing at the programmed parameters.

Options in VT and VF therapy differ in individual devices. The Medtronic 7216A offers independent prescriptions for VT and VF. Antitachycardia pacing is available in conjunction with monophasic shocks for VT. Pacing modes include burst and autodecremental ramp stimulation that can be programmed for number of stimuli, cycle length, interstimulus interval, and repitition sequences. The Telectronics Model 4210 incorporates pacing options available in the former PASAR and Cordis Orthocor series of devices. Extensive pacing algorithms based upon the Intermedics Intertach and Siemens Tachylog de-

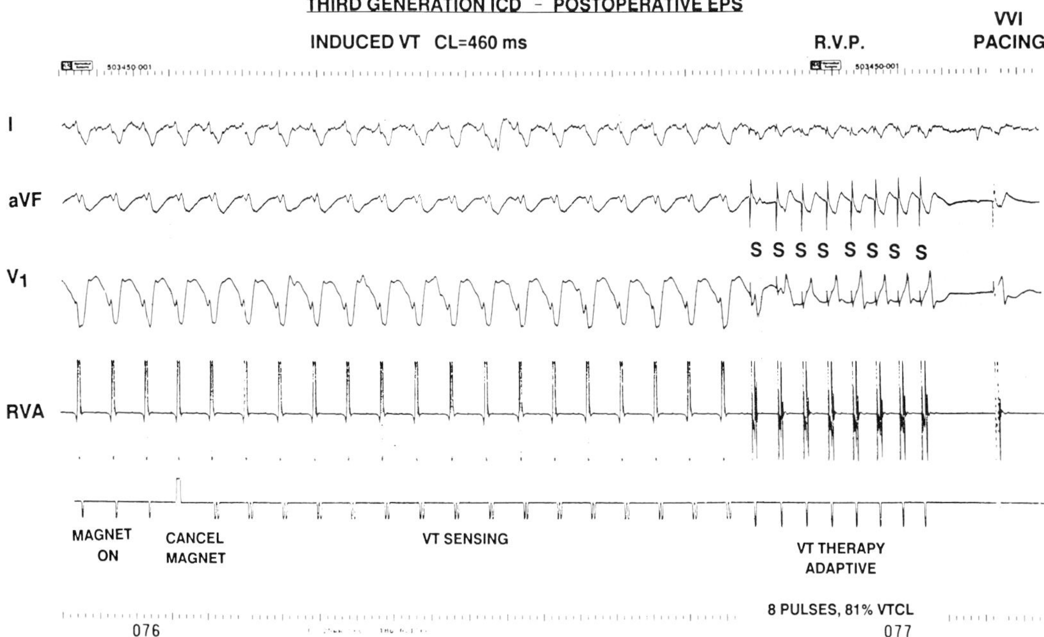

FIGURE 33.6 Induced VT at electrophysiologic study (EPS) using the programmed electrical stimulation feature in the Medtronic 7216A implantable cardioverter-defibrillator (ICD). Three surface ECG leads (I, aVF, V_1), the right ventricular apical electrogram (RVA), and the telemetric marker channel (bottom reading) are shown. The device is activated at the cancel magnet command shown on the marker channel at the bottom after VT induction. Sensing of the tachycardia in the programmed VT zone is shown by the double markers. Sixteen beats are sensed in this zone and a rate adaptive rapid ventricular pacing (RVP) burst is delivered. This was an autodecremental ramp starting at 81% of the VT cycle length for eight pulses. Following VT termination demand ventricular pacing is observed. ICD = implantable cardioverter-defibrillator, EPS = electrophysiologic study, RVP = rapid ventricular pacing, VVI = command ventricular pacing, CL = cycle length, VT = ventricular tachycardia, S = pacing stimulus, RVA = right ventricular apical electrogram.

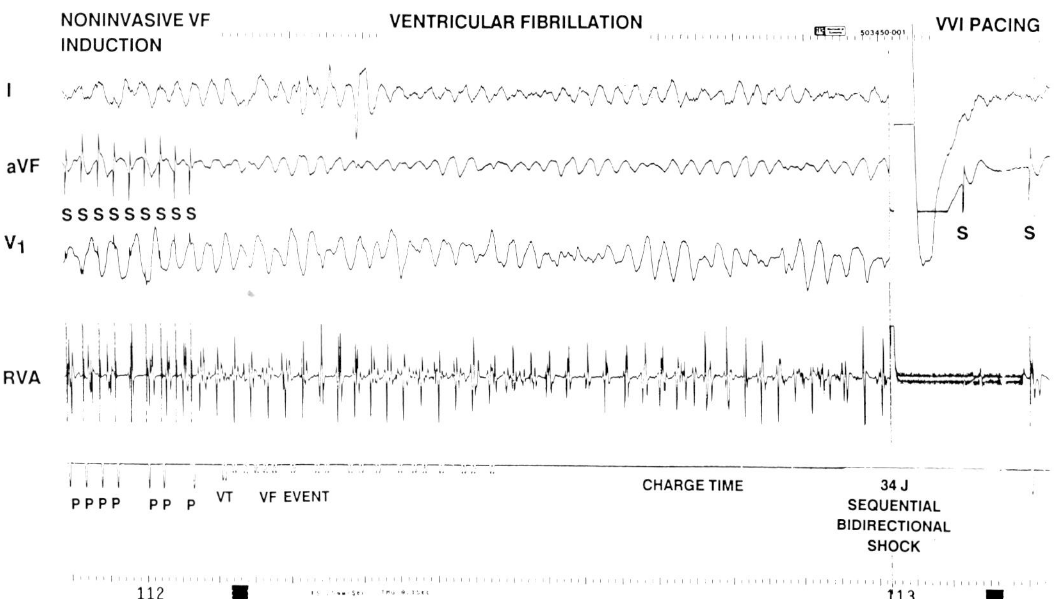

FIGURE 33.7 Induced VF at electrophysiologic study using the ramp pacing stimulation feature in the Medtronic 7216A implantable cardioverter-defibrillator. Three surface ECG leads (I, aVF, V_1) and the right ventricular apical electrogram are shown along with the marker channel at the bottom. Pacing stimuli (P and S) result in VF induction, followed by tachycardia sensing. The first tachycardia beat is in the defined VT zone while the subsequent ones are identified as VF. After 18 beats, the device charges and delivers a 34-J sequential bidirectional shock. This terminates VF and is followed by demand ventricular pacing. Abbreviations as in Figure 33.6. *Reproduced with permission from Saksena S et al: Experience with a third generation implantable cardioventer-defibrillator.* Am J Cardiol *1991 (in press).*

vices are incorporated in their ICDs. Single-chamber demand ventricular pacing is available in all devices. Shock therapy alone is used in VF by the Medtronic 7216A. Patterns of shock delivery include single unidirectional shocks for a dual electrode lead system. Simultaneous or sequential bidirectional shocks are available for triple electrode systems (Fig. 33.8A,B). Biphasic shocks are present in the Ventritex Cadence. This option may have limited value with epicardial leads but may be particularly useful with endocardial leads.[8,9] Programmable cardioversion-defibrillation shocks range in output from 0.1 to 40 J. Most devices employ trains of four to seven shocks. High-energy units with maximum outputs of 34 to 40 J are currently available with nearly all devices. Noninvasive programmed ventricular stimulation is feasible (Fig. 33.7). These can utilize extrastimulus or burst-pacing protocols which can be manually or automatically altered.

Arrhythmic event recording both before and after device activation is available in most devices. The Medtronic PCD provides this information on interrogation permitting analysis of sensing, as well as inferentially, the nature of the sensed tachycardia. Event counters for VT/VF episodes, therapies utilized for VT and VF, and outcome of individually treated VT/VF episodes are available in the Medtronic PCD. The Telectronics 4210 and the Ventritex Cadence store electrogram sequences from the last tachycardia event. Limited Holter monitoring capability is often present. Using telemetric coupling, a marker channel output is available. Device interrogation permits printing out all programmed parameters as well as data regarding arrhythmic events. Temporary modes, for example, triggered modes needed for noninvasive programmed stimulation, temporary device inactivation, or threshold determination, can be programmed as in sophisticated pacemakers.

Lead Systems

Epicardial Leads

Currently approved lead systems utilize patch defibrillation electrodes placed intra- or extrapericardially after thoracotomy (Fig. 33.8).

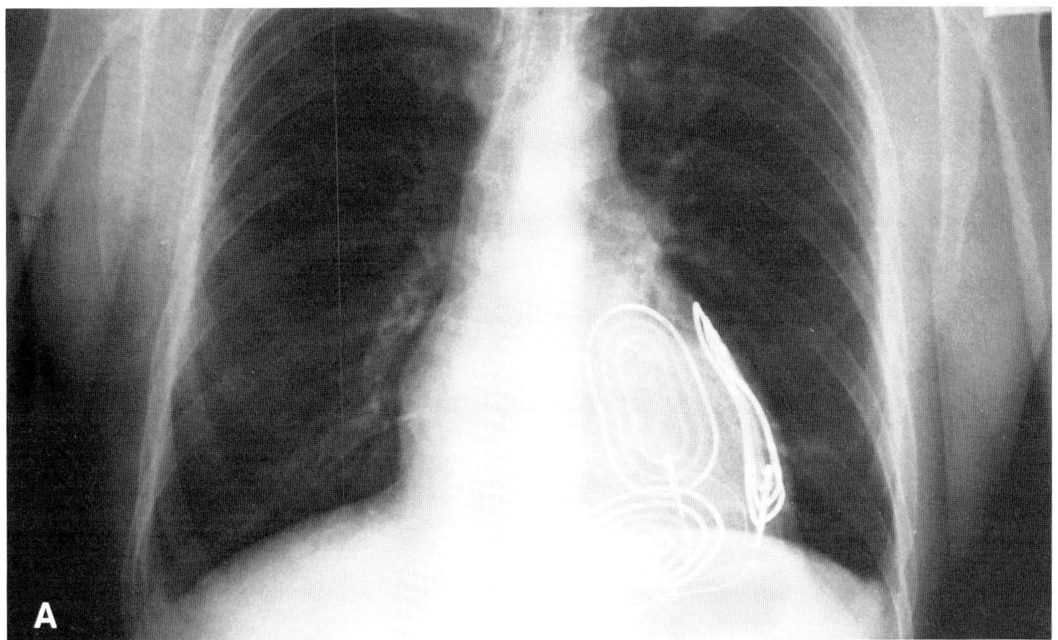

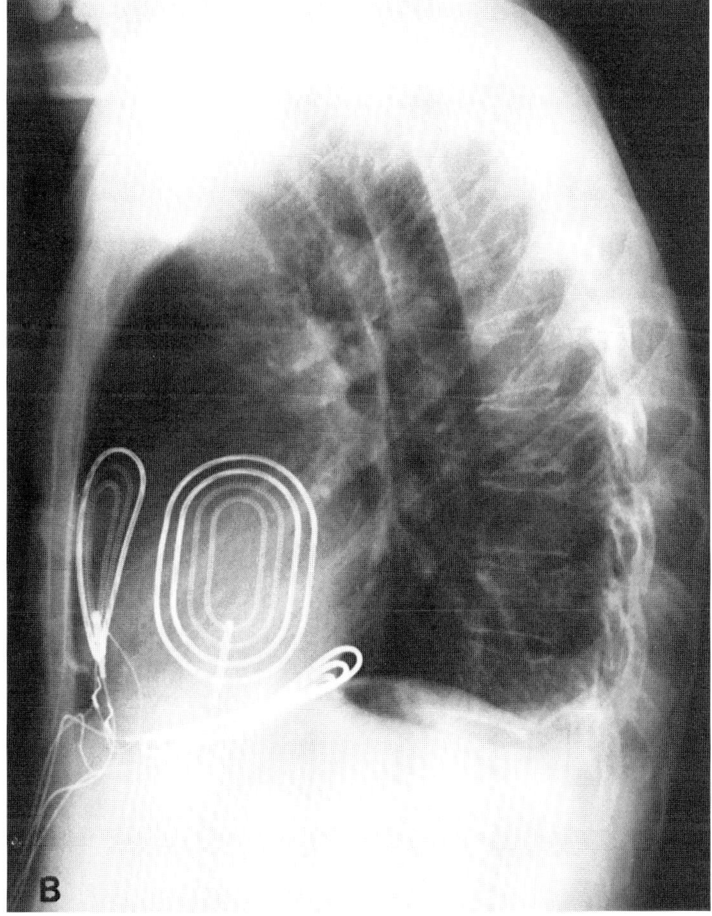

FIGURE 33.8 Chest roentgenograms. Posteroanterior view (**A**) and right lateral view (**B**) of a triple epicardial patch defibrillation electrode system (Medtronic, Inc., Minneapolis, MN) with two myocardial stab-in sensing and pacing electrodes. The left lateral patch electrode is a common cathode with the posterior and right anterior patch functioning as dual anodes. *Reproduced with permission from Saksena S et al: Experience with a third generation implantable cardioventer-defibrillator.* Am J Cardiol *1991 (in press).*

Patch electrodes utilize a wire mesh or coil. Two or three electrodes can be employed with a single generator. A dual-electrode system permits a single unidirectional shock. Triple-electrode systems permit single unidirectional or bidirectional simultaneous or sequential shocks (Fig. 33.8). Bidirectional shocks may reduce defibrillation threshold but this advantage may be inversely related to patch electrode surface area which may range from 15 to 50 cm^2 in different lead systems. Pacing and rate sensing is usually achieved by one or two myocardial electrodes of the screw-in or stab-in variety. Alternatively, an endocardial bipolar lead may be used instead. When a single epicardial electrode is used, a patch defibrillation electrode functions as the second sensing electrode. Past clinical experience has indicated that myocardial electrodes have been associated with higher chronic pacing thresholds, greater incidence of pacing exit block, and lower sensing electrogram amplitudes than endocardial leads.[(6)] Two-turn myocardial screw-in leads have the poorest track record in this regard.[(6)] Three turn screw-in leads or stab-in leads have better chronic performance in this regard.

Table 33.3 details the epicardial lead systems available for all third-generation ICD pulse generators. While a few different combinations can be used, the flexibility of using dual- or triple-electrode systems is desirable. On occasion we have found improved defibrillation thresholds with the latter approach. However, placement of three epicardial electrodes may cause problems, particularly if larger electrodes are used and cardiac size is small. This may result in arcing between adjoining electrodes. This consideration may be important in a patient with high defibrillation thresholds. Both patch electrode repositioning or addition of a third electrode may help to solve this problem. When an integrated bipolar sensing system is used as in the Medtronic 7216A cardioverter-defibrillator, the myocardial sensing electrode is usually placed in proximity (1 to 2 cm) to the indifferent patch electrode. This avoids detection of other far-field potentials, for example, atrium.

Endocardial Lead Systems

Clinical investigation of endocardial ICD lead systems is presently in progress. The prototype leads used in early studies were the Medtronic

TABLE 33.3 ICD Lead Systems (1990)

	Medtronic PCD 7216/7217	Telectronics Guardian 4210	Ventritex Cadence	Intermedics Res-Q	CPI Ventak PRX	Siemens Pacesetter Siecure
Epicardial lead system						
Defibrillator electrodes	2 or 3 Patches	2 or 3 Patches	2 Patches	?2 Patches	2 Patches	2 Patches
Rate sensing electrodes	1 Stab-in epicardial lead and patch electrode	2 Screw-in epicardial leads	2 Screw-in epicardial leads	2 Screw-in epicardial leads and patch leads	2 Screw-in epicardial leads and patch electrodes for PDF detection	2 Screw-in epicardial leads
Endocardial lead system						
Defibrillator electrodes	2 or 3 Transvenous electrodes (10 French and 8 French) with or without a thoracic patch	2 Transvenous electrodes (8 French) and thoracic patch	?	?	1 Transvenous lead (12 French) with 2 spring electrodes and thoracic patch	?
Rate sensing and pacing electrodes	Tip screw-in electrode and RV defib electrode	Tip screw-in electrode and RV defib electrode/ bipolar tip	?	?	Tip electrode and RV defib electrode	?

RV = right ventricle, defib = defibrillation, PDF = probability density function, ? = presently unknown.

6880 (Medtronic, Inc., Minneapolis, MN) and the Endotak (Cardiac Pacemakers, Inc., St. Paul, MN). These tripolar catheter electrodes consist of two defibrillation electrodes with a tip-pacing and sensing electrode (Figs. 33.9 and 33.10). In the Medtronic lead, the defibrillation electrodes had a small surface area (125 mm^2) while the Endotak lead had larger surface area electrodes (proximal = 800 mm^2, distal = 400 mm^2). These two defibrillation electrodes could be separated by variable distances (10 to 15 cm) and the distal defibrillation electrode was just proximal to the tip of the pace-sense electrode. The catheter tip was typically positioned in the right ventricular apex. The distal defibrillation electrode was thus in the right ventricle, and the proximal electrode in the right atrium and/or superior vena cava. Thus, a single unidirectional shock could be applied between these two electrodes with the right ventricle usually being the cathode and the right atrium the anode (Fig. 33.11). Addition of a third electrode permits bidirectional shocks with simultaneous or sequential monophasic shocks. It can also permit delivery of a biphasic shock in two pathways or selection of the additional pathway in a unidirectional shock. The third electrode used in most studies is a left thoracic patch electrode (Figs. 33.10 and 33.11) or a coronary sinus catheter electrode (Fig. 33.12).[10–12]

Both prototype electrodes were ineffective alone for reliable endocardial cardioversion-defibrillation of VT or VF using unidirectional monophasic shocks.[13,14] Addition of a third patch electrode improved efficacy. Reliable rapid and slow VT cardioversion using 25-J monophasic shocks was achieved with the Medtronic 6880 catheter but VF conversion remained problematic.[10,11] The larger surface area Endotak lead was more effective in VF termination in a triple-electrode configuration.[12] Chronic implant of this lead, usually in a triple-electrode configuration, was undertaken.[15,16] In our initial report, we demonstrated the efficacy of a specific triple-electrode configuration utilizing right ventricular common cathode and dual anodes (superior vena cava and left thoracic patch electrodes) for successful defibrillation using simultaneous bidirectional shocks.[16] The mean defibrillation threshold was 18 ± 5 J. This particular configuration was more successful than other dual- or triple-

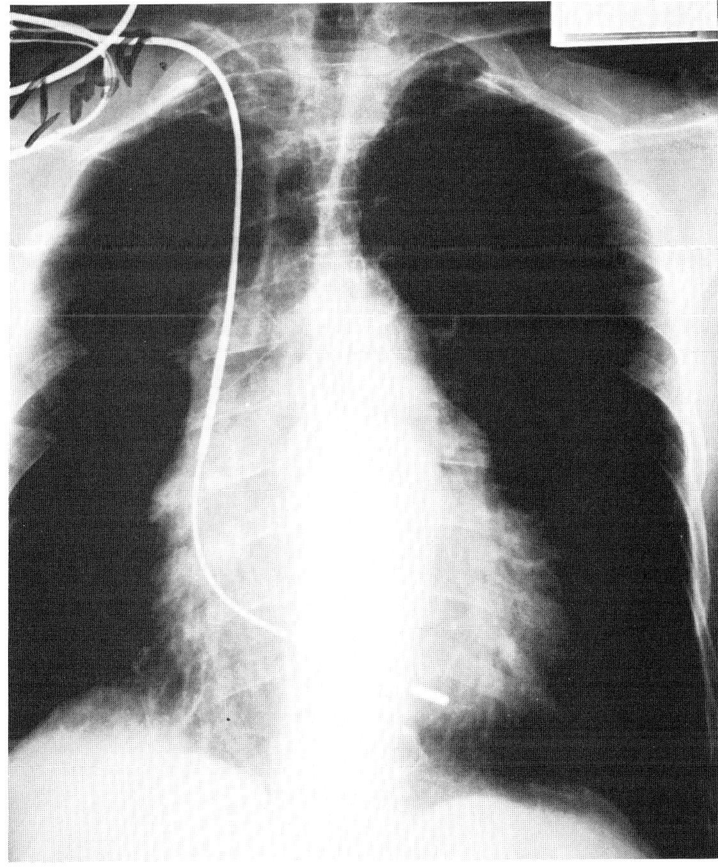

FIGURE 33.9 A tripolar Medtronic 6880 electrode catheter is introduced in the right subclavian vein, and its distal tip is positioned in the right ventricular apex. The distal two electrodes (total surface area, 125 mm^2) are used for pacing and become common during shock delivery. The proximal electrode is common with an identical surface area and is separated by a distance of 12.5 cm from the distal electrodes. It is positioned at the high right atrial–superior vena caval junction. *Reprinted from Saksena and Goldschlager,[20] with permission.*

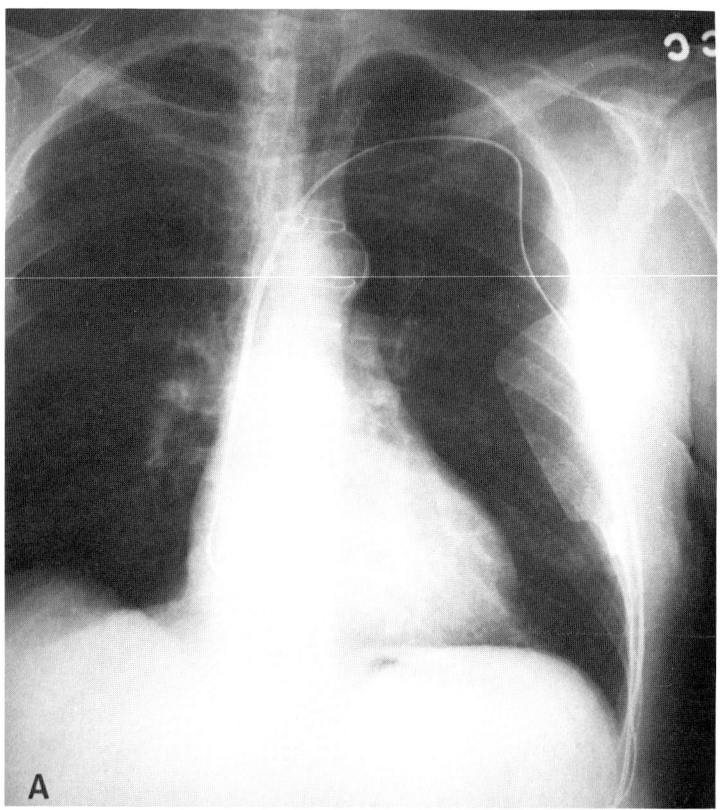

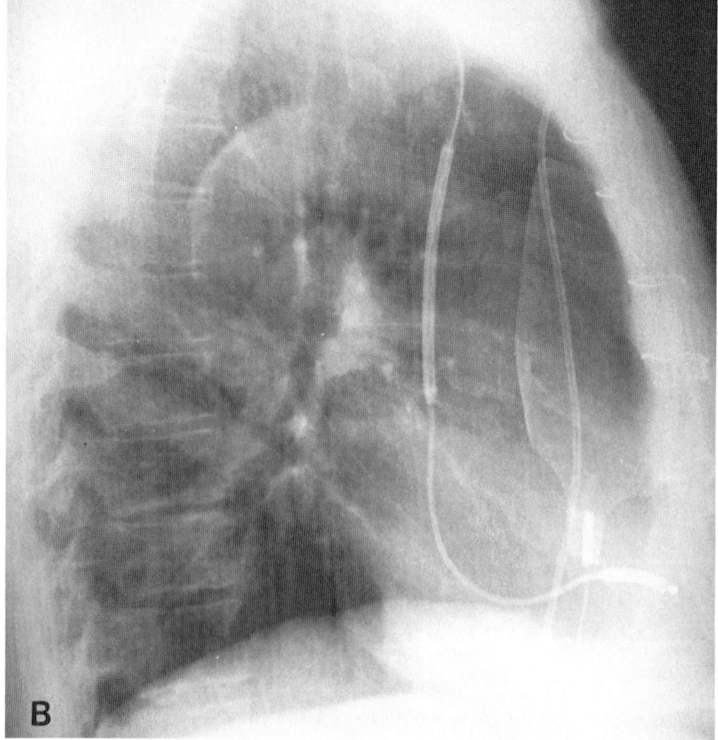

FIGURE 33.10 Chest roentgenograms. Posteroanterior view (**A**) and right lateral view (**B**) showing tripolar electrode catheter positioned in right ventricular apex and subcutaneous patch electrode in left midaxillary line at fourth intercostal space. Cables from transvenous catheter and patch electrodes have been tunneled into abdominal pocket of AICD. *Reprinted from Saksena and Goldschlager,*[20] *with permission.*

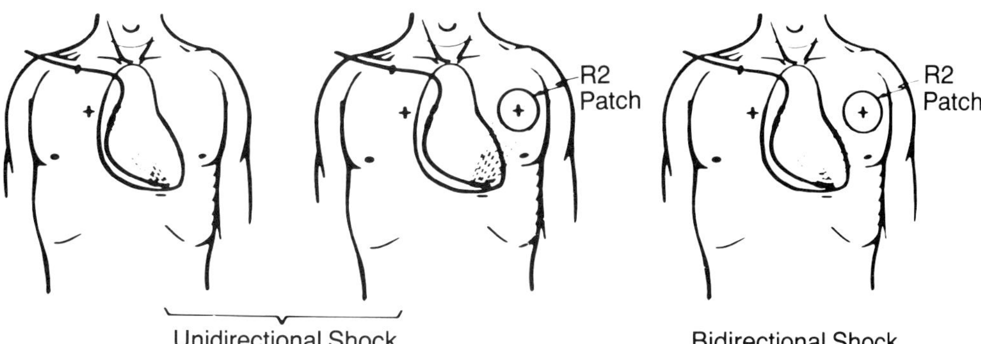

FIGURE 33.11 Different shock delivery patterns are feasible with nonepicardial electrode systems consisting of two catheter electrodes and one chest wall electrode. Unidirectional shocks employ any two electrodes, while bidirectional shocks employ a triple-electrode system. Additional configurations for bidirectional shocks (for example, using both catheter electrodes or patch electrodes as one anode or cathode) and alternative patch locations (apical, posterior, and left lateral) are also possible. *Reprinted from Saksena and Goldschlager,*[20] *with permission.*

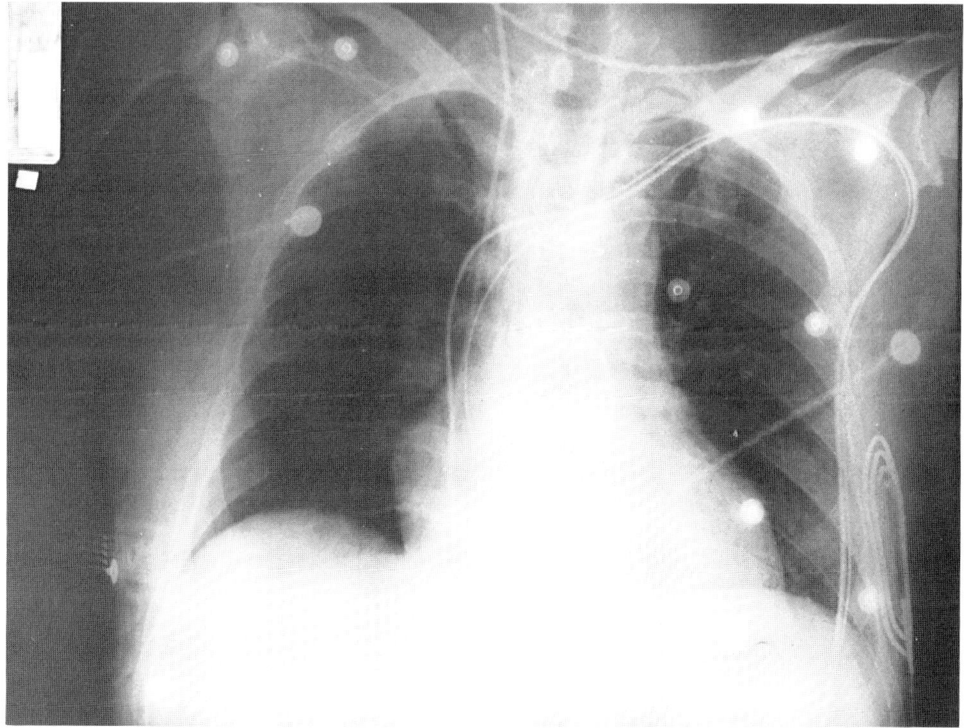

FIGURE 33.12 Chest roentgenogram (posteroanterior view) of triple nonepicardial lead system for Medtronic Model 7216A implantable cardioverter-defibrillator. Two endocardial leads (Medtronic Model 2100 and 10285, Medtronic, Inc., Minneapolis, MN) are placed with their tips in the right ventricular apex and coronary sinus, respectively. A subcutaneous patch electrode is seen on the left lateral thorax. A Swan-Ganz catheter is seen in situ. The right ventricular lead has a tip-pacing and sensing screw-in electrode. *Reprinted from Saksena et al: Experience with a third generation implantable cardioventer-defibrillator.* Am J Cardiol *1991 (in press).*

electrode configurations. Extensive evaluation of optimal patch electrode location increased the likelihood of obtaining defibrillation thresholds within the capability of ICD pulse generators.[16] This triple-electrode configuration was also more reliable than the dual-electrode systems. Acute and subacute evaluation revealed acceptable success in 69 to 90% of patients entered in the study depending on the cutoff for defibrillation thresholds, that is, 15 or 20 J, respectively.[17] However, chronic implant of the Endotak lead system was associated with instances of stress fracture of the conduction coil for both transvenous catheter and patch electrodes,[18] resulting in sudden death, inappropriate shock delivery, or unexplained elevation in defibrillation threshold during chronic electrophysiologic evaluation. Insulation breakdown between the two conductor coils in the transvenous catheter was observed probably due to the large, relatively inflexible nature of the catheter electrode, resulting in local areas of stress. The presence of a large potential difference between the two conductor coils in the same catheter body resulted in insulation breakdown.

Second-generation lead systems are designed to address these issues (Table 33.3). Smaller, flexible electrodes have been devel-

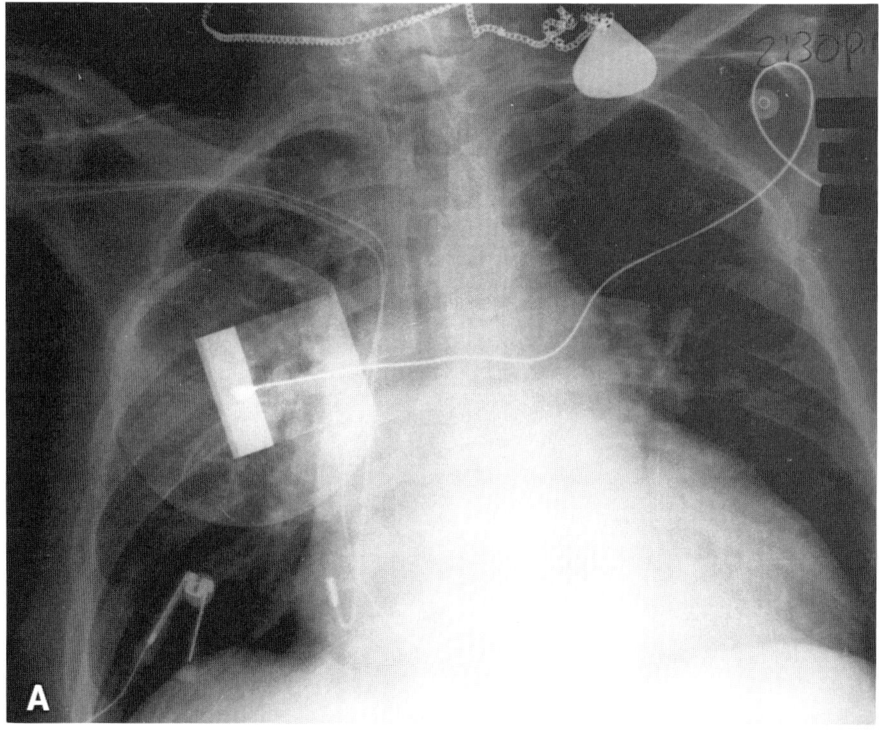

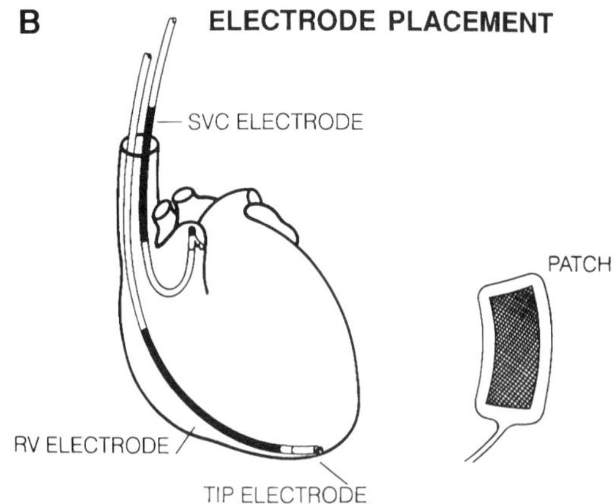

FIGURE 33.13 Chest roentgenogram (**A**) and schematic diagram (**B**) of a second-generation endocardial pacing and defibrillation lead system (Telectronics braided endocardial defibrillation lead system, Telectronics Pacing Systems, Denver, CO). This utilized two 8 French active fixation endocardial leads, each with a screw-in tip-pacing electrode and a proximal braid defibrillation electrode. One lead is fixed in the atrium while the other is fixed in the right ventricular apex, akin to conventional dual-chamber pacing leads. This can be used in conjunction with a subcutaneous patch electrode system to achieve a bidirectional shock.

oped.[9,19] Separate endocardial leads for use in different cardiac chambers are being proposed. The use of separate leads eliminates concern of insulation breakdown due to potential gradients between two conductors within a single lead body. The Telectronics Braid Endocardial Defibrillation System is a modified bipolar Accufix endocardial pacing lead with a proximal defibrillation electrode utilizing a braid material. It is a small-caliber (8 French) flexible generic pacing and defibrillation electrode. Two leads are required and can be actively fixed in atrium and ventricle (Fig. 33.13A,B). In our experimental studies, monophasic shocks provided acceptable

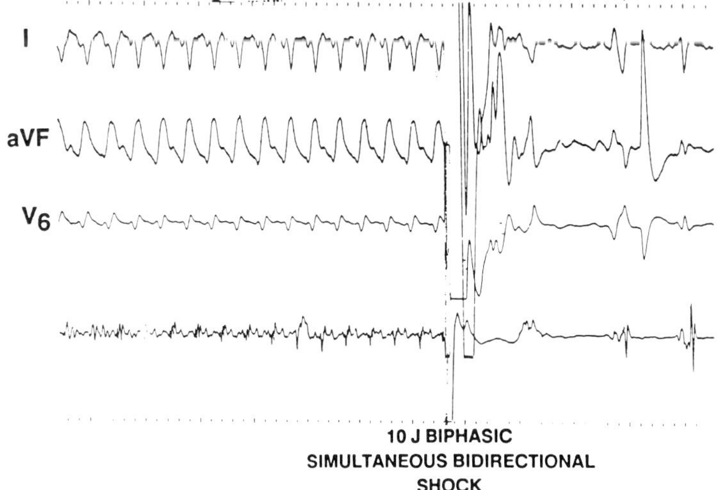

FIGURE 33.14 Epicardial cardioversion-defibrillation using monophasic and biphasic shock waveforms in a patient with recurrent VT with a triple epicardial lead system. Simultaneous bidirectional shocks were used for testing defibrillation threshold. Ventricular flutter could be induced (**top**) by alternating current and a 10-J monophasic shock was unsuccessful in tachycardia termination. VF ensued, reorganized into ventricular flutter (**bottom**) which was now terminated by a 10-J biphasic shock. CL = cycle length.

defibrillation thresholds but biphasic shocks further decreased defibrillation thresholds significantly. Stainless steel or titanium, both biocompatible materials, have been employed in prototype evaluation with comparable results. Preliminary clinical evaluation demonstrates effective endocardial cardioversion-defibrillation in 64% of patients with monophasic shocks and in more than 90% of patients with biphasic shocks at the 20-J level. Dual-chamber sensing and pacing is feasible. The Medtronic endocardial lead system uses a larger (10 French) ventricular endocardial electrode (Model 2100) and a smaller (8 French) atrial or coronary sinus electrode (Models 10285 or 6881). This can permit a totally transvenous triple-electrode system or can be used in conjunction with a thoracic patch electrode. Sequential or simultaneous monophasic or biphasic bidirectional shocks can be applied through endocardial or epicardial triple-electrode systems (Fig. 33.14). In our clinical studies, sequential and simultaneous endocardial shock patterns have proved equivalent with about 60% efficacy in VT/VF conversion during acute testing. Chronic implantation of this lead system is now being attempted (Fig. 33.12).

SPECIAL CONSIDERATIONS IN CLINICAL APPLICATION

Hybrid multiprogrammable ICDs and new lead systems raise special consideration in clinical application. Patient selection criteria for such therapy can now be widened to include patients with hemodynamically stable VT responsive to pacing termination. Patients with frequent VT recurrences may be more acceptable for ICD therapy if VT is amenable to pacing termination. Concomitant bradyarrhythmias with VT and VF can be treated with a single device. Coexistence of frequent nonsustained VT may no longer be a contraindication to ICD implant if modification of tachycardia detection criteria can be achieved noninvasively. Endocardial lead systems will allow inclusion of patients at a higher risk of thoracotomy, for example, presence of concomitant diseases such as renal or pulmonary disorders. Declining implant risk should improve both the overall patient and physician acceptance of this therapy and the total survival. Endocardial pacing and sensing will continue to prove superior to epicardial systems for both bradycardia and antitachycardia pacing and VT/VF sensing. The duration and costs associated with ICD implantation and follow-up can be expected to decline. At the same time, multiprogrammability, increased pulse generator longevity, and noninvasive testing modes will reduce hospital visits for troubleshooting, system replacement, or testing.

Implantation of an epicardial lead system for such devices causes specific concerns. Epicardial sensing during VF must be accurately assessed. Minimum VF electrogram amplitude should be measured, and programmed sensing parameters should permit reliable sensing with an anticipated up to 50% decline in electrogram amplitude. At this programmed value, oversensing during sinus rhythm and VT should be examined. Pacing thresholds should be at least one-third or less of the maximum demand pacing channel output. Evaluation of VT termination by pacing and shock therapy (Fig. 33.6) and VF termination (Fig. 33.7) by the programmed shock energy and pattern must be documented. Implant defibrillation threshold should allow for at least greater than or equal to 50% margin as compared to maximum device output. Selection of the optimal electrical therapy algorithm for VT or VF therapy needs to be examined.[20] VT with rates less than or equal to 220 beats per minute is usually amenable to pacing termination in 50 to 70% of episodes. Adaptive rate pacing using incremental bursts or autodecremental ramp methods is generally effective, although other modes are often available and can be effective. We generally employ three adaptive pacing rates (90, 80, and 70% of VT cycle length) with incremental pacing train lengths (5, 8, and 10 stimuli, respectively) and repeat the sequence at least twice if stable VT is present.[20] Faster pacing rates often encroach on ventricular refractoriness, and longer train lengths increase the incidence of VT acceleration. While low-energy shocks can also be employed, the yield of this therapy in pacing resistant VT is low. Thus,

TABLE 33.4 Implantable Cardioverter/Defibrillator

Transvenous catheter/patch lead system
Pre-implant
1. General anesthesia (continuous/intermittent)
2. Percutaneous cannulation of right subclavian vein
3. Catheter: Distal tip = RV apex; Proximal electrode = HRA/SVC
4. R_2 patch mapping
    Infraclavicular
    Left lateral
    Apical
    Left scapular region
5. VT/VF induction
6. 20-J external cardioverter/defibrillator shock
    Bidirectional
    Unidirectional
7. Defibrillation threshold determination

TABLE 33.5 Implantable Cardioverter/Defibrillator

Transvenous catheter/patch lead system
Intraoperative procedure
1. General anesthesia (continuous/?intermittent)
2. Percutaneous cannulation of left subclavian vein or cephalic vein
   Catheter tip = RV apex
   Proximal electrode = HRA/SVC
3. SQ patch insertion
   Epicostal location based on R_2 map
4. Evaluate signals
5. VT/VF induction
6. Sensing and defibrillation threshold determination
7. For DFT ≤20 J (?25 J)
   VT/VF reinduction and ICD response (sensing and arrhythmia termination)

we generally program initial shock therapy at cardioversion or defibrillation threshold. In all prescriptions of electrical therapy, we maintain the last two shocks at maximum device output. For VT rates greater than or equal to 220 beats per minute, polymorphic VT, or VF, we rely on shock therapy alone. While initial shocks may be programmed at cardioversion or defibrillation threshold, we prefer a 50% safety margin. The last two shocks are again always at maximum device output.

Endocardial lead implantation is accomplished by nonthoracotomy methods. We have proposed a screening electrophysiologic procedure for such leads using temporary endocardial and patch electrodes.[16] In our experience, this has accurately reflected implant experience. This procedure is best performed prior to the actual implant date. Table 33.4 outlines the major steps in such electrophysiologic study. Endocardial de-

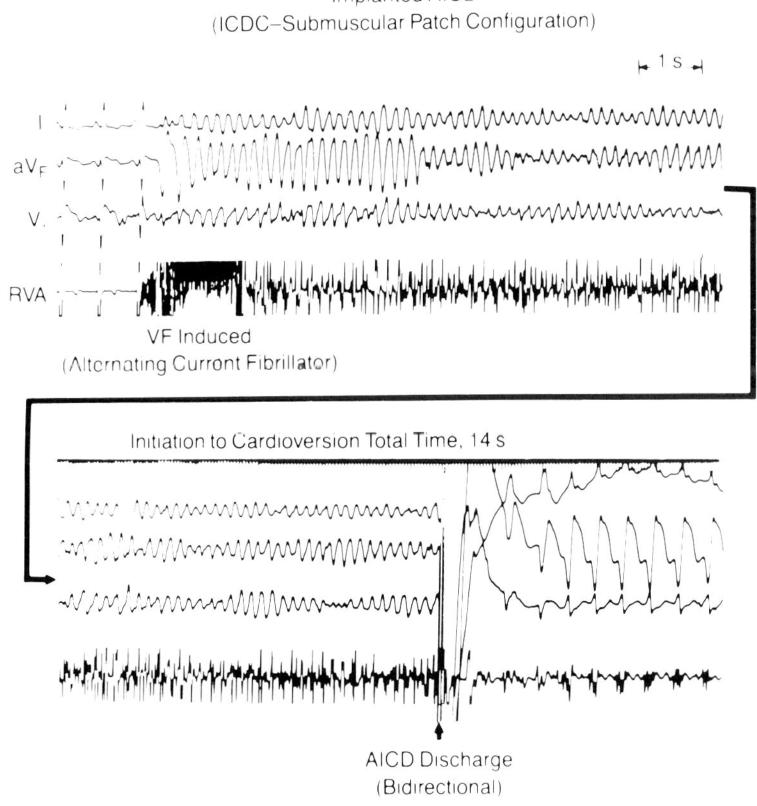

FIGURE 33.15 Postoperative induction of VF at electrophysiologic study 3 days after AICD implantation. Note that duration of sensing and charging time is 14 s and the first discharge induced supraventricular rhythm with bundle branch block pattern. Aberrant conduction subsided within a few seconds, and normal sinus rhythm was restored. RVA = right ventricular apex, ICDC = internal cardioversion/defibrillation catheter. *Reprinted from Saksena and Parsonnet,[15] with permission.*

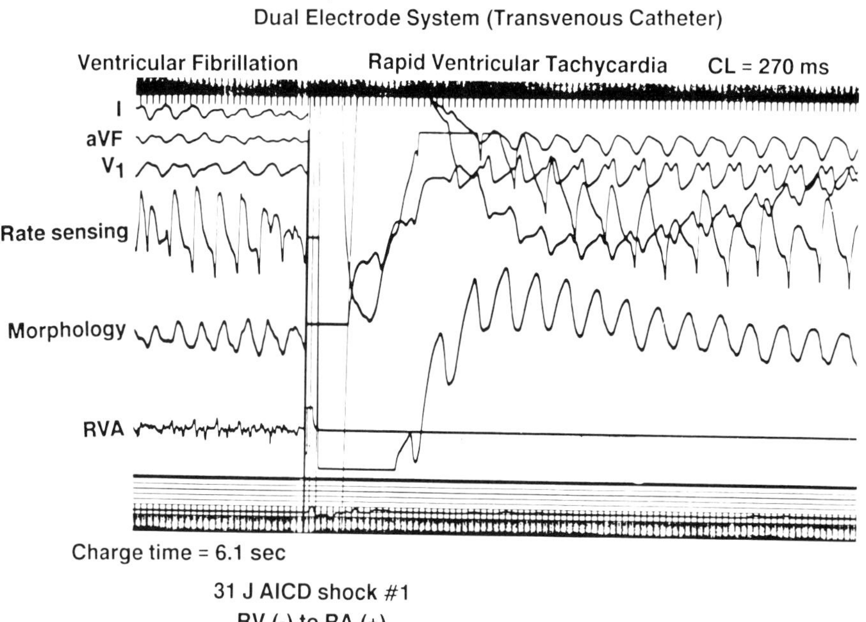

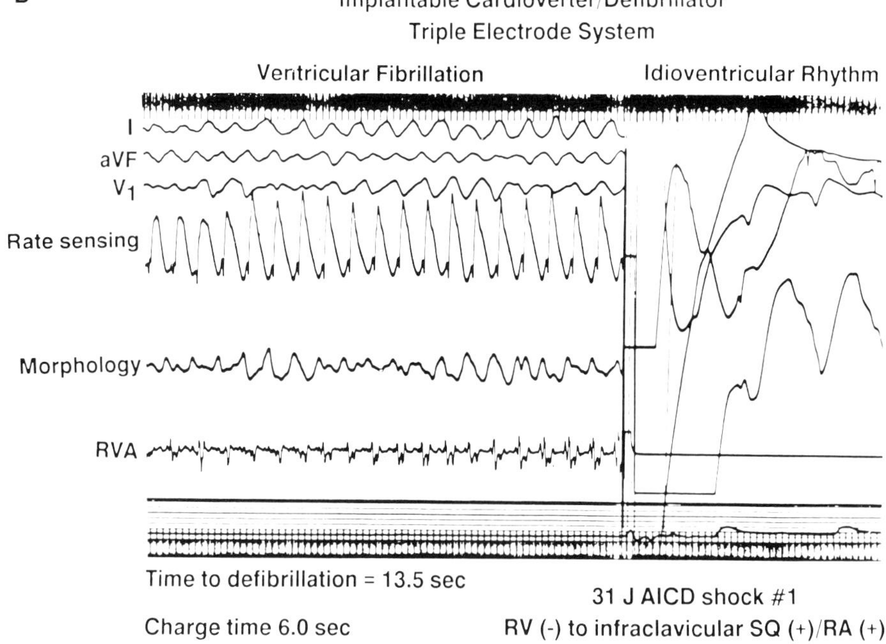

FIGURE 33.16 A. Failure to terminate VF by the first shock (31 J) of an implanted cardioverter-defibrillator using a dual-electrode system with the transvenous catheter only (right ventricular cathode and right atrial anode). Elimination of the submuscular patch and the bidirectional shock resulted in this outcome. **B.** Successful termination of VF by the first shock (31 J) of an implanted cardioverter-defibrillator with a triple nonepicardial electrode system (right ventricular cathode with right atrial and patch as dual anodes) in the same patient as in **A.** RV = right ventricle, SQ = submuscular patch, RA = right atrium, plus sign = anode, minus sign = cathode, CL = cycle length. *Reprinted from Saksena et al.,*[16] *with permission.*

TABLE 33.6 Implantable Cardioverter/Defibrillator

Transvenous catheter/patch lead system
Post-implant
1. Cardiac monitoring (24 to 48 h)
2. Stress test (48 to 72 h)
3. Predischarge EPS (48 to 72h)
4. Follow-up EPS (3 to 6 months; ?Noninvasive)

fibrillation thresholds are obtained. In the intraoperative procedure, these steps are repeated using the permanent lead system (Table 33.5). The catheter electrode is actively fixed and pacing and sensing assessed. The patch electrode is placed subcutaneously in an epicostal location. Defibrillation threshold testing and sensing evaluation is performed. Postoperative follow-up is relatively simple and usually brief (Table 33.6). Electrophysiologic testing is necessary to assess integrity of the system (Fig. 33.15).

Patients with elevated defibrillation thresholds can often be addressed by options available in these devices. Figure 33.16A shows failure of a 31-J shock to terminate VF in a patient with a dual-electrode endocardial lead system. Change in configuration to a triple-electrode system permitted reliable defibrillation (Fig. 33.16B). Similarly, biphasic shocks may increase efficacy of epicardial and endocardial defibrillation (Fig. 33.14). Optimal patch electrode location may vary in individual patients. Evaluation of the efficacy of such new ICD systems can be performed noninvasively, often as an ambulatory procedure. VT and VF should be separately induced and programmed therapies evaluated.

CONCEPTS FOR APPLICATION IN FUTURE DEVICES

A variety of new concepts are currently under evaluation to address some of the continuing limitations in electrical device therapy of VT and VF. These efforts center around certain specific issues:

- **a.** Improved detection of supraventricular and ventricular tachyarrhythmias by the device
- **b.** Tailoring electrical therapy to the clinical status of the patient during the tachycardia
- **c.** Dual-chamber pacing and sensing
- **d.** Reduction in pulse generator size for infraclavicular implantation
- **e.** Prevention of tachycardia initiation

Tachycardia Recognition

Tachycardia recognition by current implantable devices has largely focused on rate characteristics of ventricular electrograms. This approach generally provides high sensitivity for VT/VF detection but has low specificity. While the exact false-positive incidence is unknown, clinical data suggest that this could exceed 30% of all tachycardia episodes detected by ICDs. The impact of additional criteria such as onset, stability, and duration parameters remains to be assessed, but it is generally surmised that this will not fully resolve the issue. Efforts to improve tachycardia recognition have focused on two main areas: (a) more sophisticated analysis of the sensed electrograms and (b) use of other physiologic parameters.

Camm, Griffin and others[21] have suggested analysis of the single ventricular electrogram morphology by several approaches including temporal electrogram analysis, template matching, and gradient pattern detection for the electrogram potential.[21] Acute testing suggests that this is effective in many patients, but rate-related aberrancy may limit the utility of this approach. Analysis of multiple electrograms has been also suggested. The activation sequence of multiple ventricular electrograms or discordance between atrial and ventricular electrograms have been suggested as potential techniques to differentiate supraventricular tachycardia (SVT) and VT. The former is affected by aberrant ventricular conduction while the latter is of limited value in patients with one-to-one ventriculo-atrial conduction during VT.

Frequency analyses of ventricular electrograms have been performed. In 1983, we analyzed the frequency spectra of ventricular electrograms in patients with VT and noted an immediate shift to lower frequencies with VT onset reflecting slower intraventricular conduction during VT (Fig. 33.17).[20] More recently, fast Fourier transform analysis has demonstrated the ability to differentiate VT and VF from sinus rhythm.[20] Other approaches include coherence analysis of frequency spectra. This approach can differentiate disorganized rhythms such as fibrillation from regular tachycardias. While this differentiation is highly specific, the clinical problem most frequently encountered involves separation of regular SVTs and VTs.

Other physiologic parameters have been proposed. Among the most extensively evaluated have been the hemodynamic consequences of different tachyarrhythmias. Clinical observations have emphasized that sustained ventricular tachyarrhythmias are frequently associated with he-

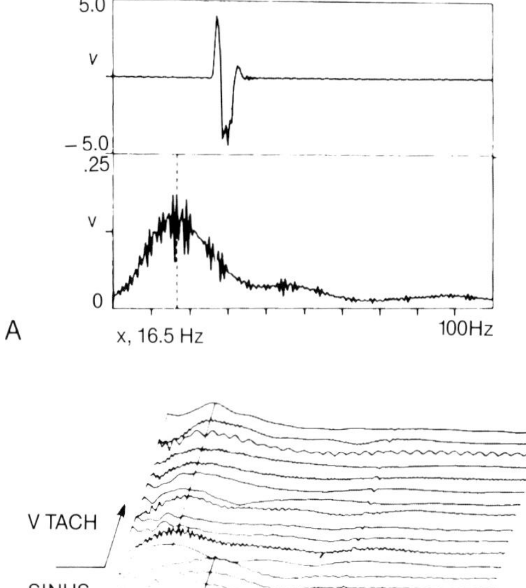

FIGURE 33.17 Spectral analysis of ventricular potentials during VT. **A.** Upper portion shows ventricular potential recorded during VT. Lower portion shows averaged spectrum of eight potentials as shown above. **B.** Spectra obtained on filtered potentials in sinus rhythm from another patient (lower portion). Spectra obtained during VT begin at the arrow. Note the regularity of peaks at 22.5 Hz obtained during sinus rhythm and the shift to 13 Hz during VT. *Reprinted from Saksena et al.,*( ) *with permission.*

modynamic collapse. Hemodynamic parameters analyzed for sensing purposes in the past include right ventricular stroke volume, right ventricular pressure, and cardiac output. In our laboratory, we observed that the onset of VT/VF was associated with an instantaneous decline in left ventricular systolic pressure, peak left ventricular dp/dt, and negative dp/dt.[22] We subsequently observed similar changes in right ventricular hemodynamic parameters.[20,23] It is being further suggested that right ventricular pulse pressure may differentiate hemodynamically stable from unstable tachycardias (Fig. 33.18A,B).[24]

Prospective evaluation of this hemodynamic approach needs to be undertaken. It follows that stable VT will be treated more aggressively with pacing therapies, whereas development of unstable VT/VF will trigger delivery of shock.

Dual-Chamber Lead Systems

From our previous discussions, we visualize development of dual-chamber pacing and defibrillation leads (Fig. 33.13). The hemodynamic advantages of dual-chamber pacing in patients with compromised left ventricular function have been documented,[25,26] and the majority of ICD recipients have compromised left ventricular function. It can be anticipated that future ICD pulse generators will have capabilities for dual-chamber pacing and sensing when interfaced with appropriate endocardial lead systems. Atrial sensing will also be valuable in detection of atrial fibrillation, flutter, or atrioventricular (AV) dissociation during VT.

Reduction in ICD Pulse Generator Size

Efforts to reduce ICD pulse generator size are in progress to permit infraclavicular implantation. This would enhance patient acceptance of this device and simplify surgical implant techniques. Development of more efficient lead systems or use of new shock waveform patterns that could lower defibrillation threshold is contemplated. Current ICD pulse generators employ two large capacitors for energy storage which comprise the bulk of the pulse generator. Biphasic pulse waveforms are being evaluated, since they can be generated with single capacitor discharges. Initial data suggest that biphasic waveforms had limited effect on epicardial defibrillation thresholds.[8] However, our experimental and clinical data suggest that they may be more valuable in reducing endocardial defibrillation thresholds.[19,21] Further evaluation of this approach should elucidate its potential usefulness.

Modification of endocardial electrode geometry and surface area has been suggested to lower defibrillation threshold. While initial data suggested that increase in endocardial electrode surface area enhances defibrillation success, this has finite limits.[27,28] Increase in cathode electrode surface area above 400 mm^2 adds little to the overall value of this approach. Experimental studies suggested that modification of cathode geometry or location may be also useful.[28] Initial clinical studies have not supported this for some dual-electrode systems.[29] Use of a large J-shaped cathode fixed in the right ventricular outflow tract had little effect on efficacy. Further investigation in this area is clearly warranted.

Tachycardia Prevention

Prevention or VT of VF recurrence by device therapy is an attractive and important concept. It has been demonstrated in experimental studies that

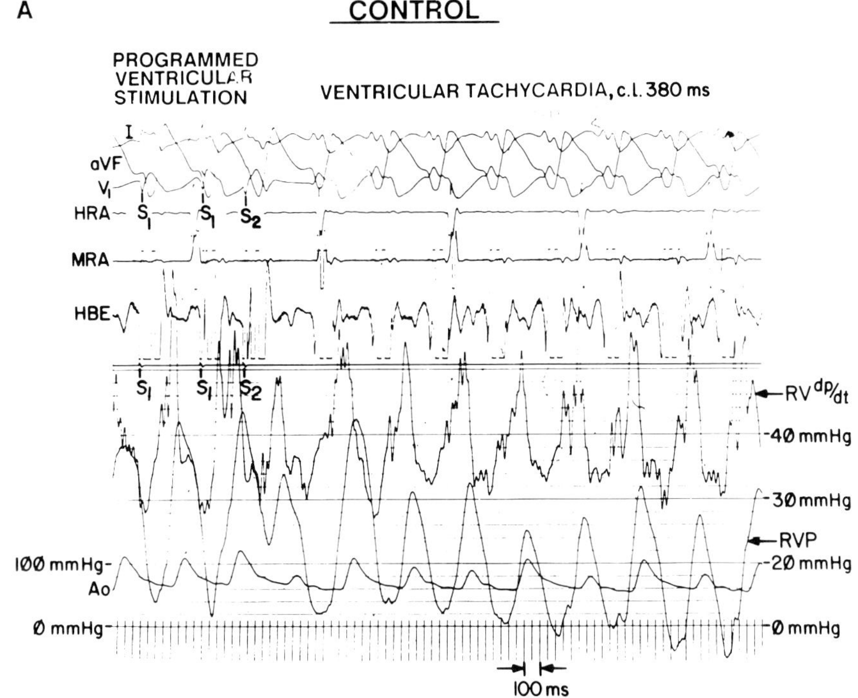

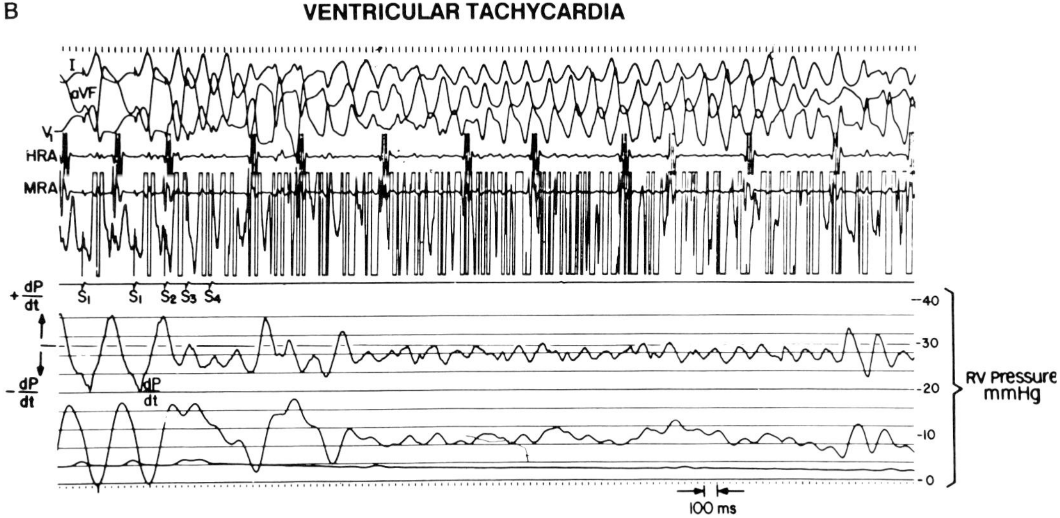

FIGURE 33.18 Right ventricular hemodynamic parameters during induced VT (**A**) and VF (**B**). The first recording shows induction of VT by a single ventricular extrastimulus (S2) at a cycle length of 380 ms. Note that the systolic aortic pressure (Ao) is slightly lower during VT, ranging from 80 to 100 mm Hg. The right ventricular (RV) systolic pressure is well preserved, ranging from 25 to 31 mm Hg, while the RV pulse pressure is also maintained. RV dp/dt and negative dp/dt are moderately reduced. In VF induced by triple ventricular extrastimuli (S2,S3,S4), systolic aortic pressure is less than 20 mm Hg, RV systolic pressure, pulse pressure, dp/dt, and negative dp/dt are essentially minimal. HRA = high right atrium, MRA = mid right atrium, HBE = His-bundle electrogram.

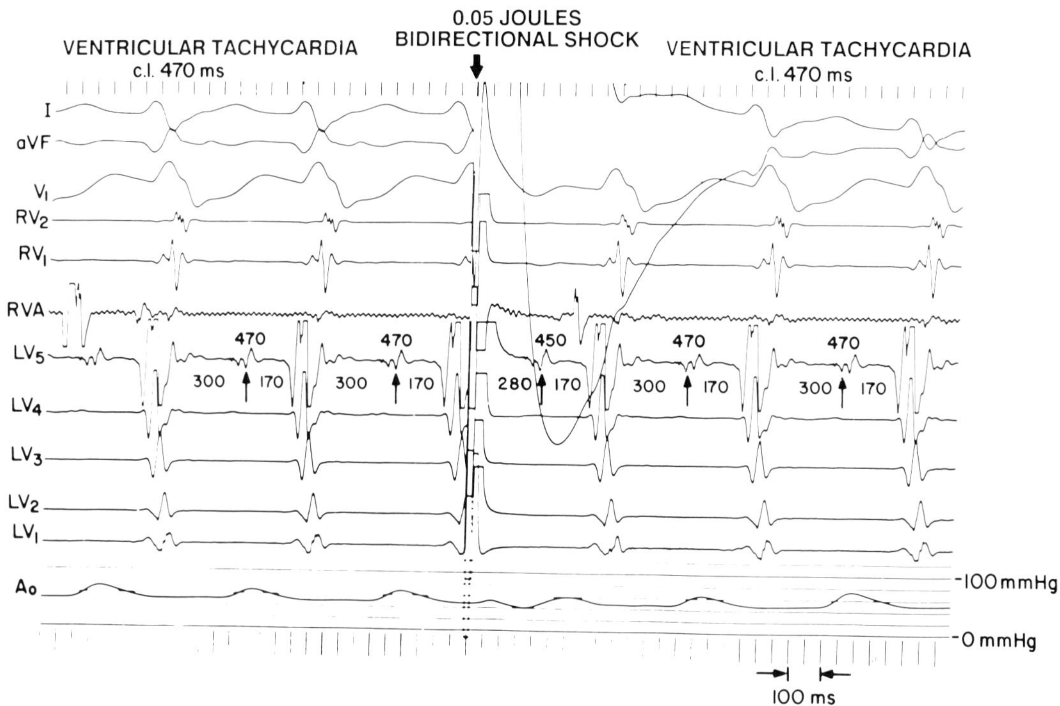

FIGURE 33.19 The VT is exposed to a 50-mJ shock. This advances the presystolic potential and VT rate by 20 ms without altering morphology, which suggests penetration into the reentrant circuit. *Reprinted from Saksena et al.,*[32] *with permission.*

subthreshold pacing stimuli or shocks can either prevent induction of VT or interrupt VT and SVT.[30,31] Such stimulation needs to be applied at a site critical to arrhythmogenesis. The electrophysiologic effects of pacing and shock stimuli applied in this fashion are still being studied. Pacing stimuli can cause local ventricular depolarization of critical diastolic conduction components of the VT circuit and make them refractory to the tachycardia wave front. Alternatively, captured paced beats may advance or reset the tachycardia and achieve a similar outcome. Endocardial shocks have been shown also to advance depolarization in diastolic elements of a tachycardia circuit, despite failing to capture a large part of the ventricular mass (Fig. 33.19). This can actually result in VT acceleration. Increasing shock energy results in conduction block and VT interruption.[32] Initial clinical application of this approach is now in progress. Should this concept be developed further and become clinically applicable, it can be visualized that the device would deliver very low-intensity stimuli at critical timing intervals to interrupt VT or prevent tachycardia generation.

A more recent suggestion is a bolus infusion of an antiarrhythmic drug such as lidocaine by a device to achieve the same goal.[33] This could be considered when a surge in ventricular ectopic activity or nonsustained VT is detected, after interruption of VT/VF by electrical therapy, or for interruption of stable VT prior to electrical therapies. This concept provides an attractive adjunct to electrical therapy and, in fact, reflects clinical physician practice in the monitored patient. It is now being widely suggested that ICD therapy be considered as a first-line approach in specific patient populations with VT and VF.[34] I would expect imminent and future developments to increase its attractiveness as a therapeutic option and compete with increasing success with other therapeutic modalities in nearly all patients with VT and VF.

REFERENCES

1. Mirowski MM, Reid PR, Watkin L, Weisfeldt ML, Mower MM: Clinical treatment of life-threatening ventricular tachyarrhythmias with the automatic implantable defibrillator. *Am Heart J* 1981;102:265–270.
2. Thomas AC, Moser SA, Smurka ML, Wilson PA: Implantable defibrillation: Eight years experience. *PACE* 1988;11:2053–2058.
3. Furman S: AICD Benefit (editorial). *PACE* 1989; 12:399–400.
4. Echt DS, Lee JT, Hammon JW: Implantation and intraoperative assessment of antitachycardia devices, in Saksena S, Goldschlager N (eds): *Textbook of Electrical Therapy for Cardiac Arrhythmias*. Philadelphia, PA, WB Saunders, 1990, pp 489–515.

5. Saksena S, Castle L, Fogoros R, Alpert B, Kron J, Pacifico A, et al.: Initial clinical experience with second generation implantable defibrillator. *J Am Coll Cardiol* 1990;(abstract).
6. Timmis GC: The electrobiology and engineering of pacemaker leads, in Saksena S, Goldschlager N (eds): *Textbook of Electrical Therapy for Cardiac Arrhythmias*. Philadelphia, PA, WB Saunders, 1990, pp 35–90.
7. Krol R, Saksena S, Tullo N, Karanam R, Gielchinsky I, Saxena A, et al.: Clinical experience with bipolar epicardial sensing for a combined implantable pacemaker-defibrillator. *PACE* 1990;13:536 (abstract).
8. Winkle RA, Mead RH, Ruder MA, et al.: Improved low energy defibrillation efficacy in man with the use of a biphasic truncated exponential waveform. *Am Heart J* 1989;117:122.
9. Saksena S, Scott SE, Accorti PR, Boveja BK, Abels D, Callaghan FJ: Efficacy and safety of monophasic and biphasic waveform shocks using a braided endocardial defibrillation lead system. *Am Heart J* 1990;120:1342–1347.
10. Saksena S, Calvo RA, Pantopoulos D, Gadhoke A, Rothbart ST: A prospective evaluation of single and dual current pathways for transvenous cardioversion in rapid ventricular tachycardia. *PACE* 1987;10:1130–1141.
11. Lindsay BD, Saksena S, Rothbart ST, Wasty N, Pantopoulos D: Prospective evaluation of a sequential pacing and high energy bidirectional shock algorithm for transvenous cardioversion in patients with ventricular tachycardia. *Circulation* 1987; 76:601–609.
12. Bardy GH, Allen MD, Mehra R, Johnson G, Feldman S, Greene HL, et al.: Transvenous defibrillation in humans via the coronary sinus. *Circulation* 1990;81:1252–1259.
13. Ciccone JM, Saksena S, Shah Y, Pantopoulos D: A prospective randomized study of the clinical efficacy and safety of transvenous cardioversion for ventricular tachycardia termination. *Circulation* 1985;71:571–578.
14. Saksena S, An H: Clinical efficacy of dual electrode systems for endocardial cardioversion of ventricular tachycardia: A prospective randomized crossover trial. *Am Heart J* 1990;119:15–23.
15. Saksena S, Parsonnet V: Implantation of a cardioverter/defibrillator without thoracotomy using a triple electrode system. *J Am Med Assoc* 1988;259:69–72.
16. Saksena S, Tullo NG, Krol RB, Mauro AM: Initial clinical experience with endocardial defibrillation using an implantable cardioverter/defibrillator with a triple-electrode system. *Arch Intern Med* 1989; 149:2333–2339.
17. Bach SM Jr, Barstad J, Harper N, et al.: Initial clinical experience: Endotak implantable transvenous defibrillator system. *J Am Coll Cardiol* 1989; 13:65A(abstract).
18. Tullo NG, Saksena S, Krol RB, Mauro AM, Kunecz D: Management of complications associated with a first-generation endocardial defibrillation lead system for implantable cardioverter-defibrillators. *Am J Cardiol* 1990;66:411–415.
19. Saksena S, Luceri R, Tullo N, Krol R, Braunstein S, Accorti P, et al.: Clinical efficacy and safety of a defibrillation lead system for ventricular tachycardia and ventricular fibrillation using monophase and biphasic shocks. *Circulation* 1990;82:111-166.
20. Saksena S: Implantable antitachycardia devices—The next generation, in Saksena S, Goldschlager N (eds): *Textbook of Electrical Therapy for Cardiac Arrhythmias*. Philadelphia, PA, WB Saunders, 1990, pp 554–573.
21. Camm AJ, Paul V, Ward DE: Recognition of tachyarrhythmias by implantable devices, in Saksena S, Goldschlager N (eds): *Textbook of Electrical Therapy for Cardiac Arrhythmias*. Philadelphia, PA, WB Saunders, 1990, pp 589–602.
22. Saksena S, Ciccone J, Craelius W, Pantopoulos D, Rothbart ST, Werres R: Studies of left ventricular function during sustained ventricular tachycardia. *J Am Coll Cardiol* 1984;4:501–508.
23. Wasty N, Pantopoulos D, Rothbart ST, Cohen L, Saksena S: Detection of sustained ventricular tachyarrhythmias using right ventricular hemodynamic parameters: A prospective study. *J Am Coll Cardiol* 1987;9:141A(abstract).
24. Sharma AD, Bennett TD, Erickson M, Klein GJ, Yee R, Guiraudon G: Right ventricular pressure during ventricular arrhythmias in humans: Implications for implantable antitachycardia devices. *J Am Coll Cardiol* 1990;15:648–655.
25. Kruse I, Arnman K, Conradson TB, Ryden L: A comparison of the acute and long-term hemodynamic effects of ventricular inhibited and atrial synchronous ventricular inhibited pacing. *Circulation* 1982;65:846–855.
26. Perrins EJ, Morley CA, Chan SL, Sutton R: Randomized controlled trial of physiological and ventricular pacing. *Br Heart J* 1983;50:112–117.
27. Saksena S, Parsonnet V, Pantopoulos D, Rothbart ST: Transvenous cardioversion and defibrillation of ventricular tachyarrhythmias using bidirectional shocks: Acute feasibility studies and chronic device implantation. *Circulation* 1987;76(II):IV-311(abstract).
28. Kavanagh KM, Tang AS, Rollins DL, Smith WM, Ideker R: Comparison of the internal defibrillation thresholds for monophasic and double single capacitor biphasic waveforms. *J Am Coll Cardiol* 1990;14:1343–1349.
29. An H, Saksena S, Mehra R, Tullo N, Krol R, Smits K: Effect of right ventricular cathode configuration on endocardial cardioversion and defibrillation with dual electrode systems and monophasic shocks. *PACE* 1990;13:511(abstract).
30. Winkle JR, Miles WM, Zipes DP, Prystowsky EN: Subthreshold conditioning stimuli prolong human ventricular refractoriness. *Am J Cardiol* 1986; 57:381.
31. Prystowsky EN, Zipes DP: Inhibition in the human heart. *Circulation* 1983;68:707–713.
32. Saksena S, Pantopoulos D, Hussain SM, Gielchinsky I: Mechanisms of ventricular tachycardia termination and acceleration during transvenous cardioversion as determined by cardiac mapping in man. *Am Heart J* 1987;113:1495–1506.
33. Camilli L: Automatic implantable pharmacologic defibrillator. *Proceedings, 9th International Congress on the New Frontiers of Arrhythmias*, Marilleva, Italy, 1990.
34. Lehmaum M, Saksena S: Implantable cardioverter-defibrillators in cardiovascular practice: Report of the NASPE policy conference. *PACE* June 1991 (in press).

Index